Post-Translational Modifications in Health and Disease

For other titles published in this series, go to
http://www.springer.com/series/6876

Cecilio J. Vidal
Editor

Post-Translational Modifications in Health and Disease

Springer

Editor
Cecilio J. Vidal
Departamento de Bioquímica y Biología Molecular-A
Campus Universitario de Espinardo
Murcia
Spain
cevidal@um.es

ISBN 978-1-4419-6381-9 e-ISBN 978-1-4419-6382-6
DOI 10.1007/978-1-4419-6382-6
Springer New York Dordrecht Heidelberg London

Library of Congress Control Number: 2010937213

Printed on acid-free paper

Springer is part of Springer Science+Business Media (www.springer.com)

Preface

When I was invited to edit this volume on posttranslational modifications (PTM) of proteins in health and disease, and after nearly 30 years of uninterrupted teaching of molecular biology of the cell and signal transduction phenomena, I knew the challenge that awaited me, considering the enormous variety of PTM that proteins can undergo, their tremendous importance for normal cell biology, and their requirement for keeping tissues in a healthy state. Nevertheless, I decided to accept the invitation as a means of learning from experts what I try to teach to my students.

Although the first step in protein diversification occurs at the transcriptional level, by mRNA splicing, the PTM of proteins at one or more sites is the way by which the number of protein variants in cells is greatly increased, so that their quantity exceeds by two to three times the number of proteins predicted by the DNA coding capacities of cells. More than 500 human protein kinases, 150 protein phosphatases and 500 proteases are engaged in protein modifications; about 5% of our genetic material encodes enzymes that intervene in PTM of proteins, and at least 1% encodes enzymes involved in the glycosylation process.

Posttranslational modifications serve many different purposes in a wide variety of cellular processes, such as protein synthesis, folding, stability, the housing of prosthetic groups, vesicular trafficking, protein targeting to particular cell stores, exocytosis and endocytosis, the biogenesis of cell organelles and basal lamina, as well as signal transduction with functional effects for enzyme regulation and metabolic control on the one hand, and for gene expression, cell division, differentiation and apoptosis on the other.

The enormous variety and versatility of the protein modifications, which may drive permanent or transient changes in conformation and physicochemical properties of the respective protein, represent a great challenge for proteomic research. There are two broad classes of covalent modifications in proteins. The first corresponds to enzyme-assisted covalent addition (or elimination) of a chemical group, frequently an electrophilic fragment of a cosubstrate, to a side chain residue in a protein. The modified side chain is usually electron-rich and acts as the nucleophile in the transfer. The second class of PTM is the covalent cleavage of peptide fragments in proteins driven by proteases or, less frequently, by autocatalytic cleavage. Phosphorylation and dephosphorylation of proteins with their gain and loss of function, glycosylation and its role in appropriate protein sorting, and secretion-associated proteolysis are the most studied topics, but further modifications, such as protein lipidation, prenylation, glypiation, acetylation, methylation, oxidation, hydroxylation, nitrosylation, sulfurylation, ubiquitinylation, sumoylation, ADP-ribosylation, the degradation of basal lamina components, prosthetic group tethering and other protein changes, are also required for essential cellular functions.

Since our current knowledge of protein PTM, including their nature and biological significance, probably only covers a small fraction of the modifications responsible for building the whole cell proteome, the goal of the present volume was to provide the reader with several updated reviews that stimulate further investigations in the proteomic field. The present volume compiles nineteen reviews focused on functional and pathological aspects of protein prenylation, the incorporation of glicosylphosphatidylinositol (GPI) moieties, oxidation, nitrosylation, glycosylation, and phosphorylation and dephosphorylation, with emphasis on their outcome for protein–protein interaction phenomena and down-stream effects. Three additional chapters are devoted to protein ubiquitination, sumoylation and endoplasmic reticulum-associated degradation (ERAD), and three more to the influence on histone modification for gene expression and DNA repair. The last two chapters are focused on proteolytic processing of intracellular and basal lamina proteins. The list of topics covered is far from complete; their selection should not be understood in the sense of merit or importance but as the result of practical limitations in the scope and assembly of this work.

The chapters of this book have been written by eminent experts to whom I have to express my most sincere gratitude. I also want to thank Dr. M. Zouhair Atassi, the publisher, Ms. Stephanie Jakob, the editor, and the Fundación Séneca de la Región de Murcia for making this work possible. I must also thank my colleagues and associates Drs. Encarnación Muñoz-Delgado, Francisco J. Campoy and María Fernanda Montenegro-Arce for their help and encouragement. I must also thank my son and daughters Federico, Ana Victoria and Belén for their patience, my little grandson Federico for his tenderness, and finally my wife María Jesús for her ability to make my life easier and happier every day.

Murcia, Spain Cecilio J. Vidal

Contents

Contributors

Daniel Abankwa
Institute for Molecular Bioscience, University of Queensland, Brisbane QLD, 4072, Australia

Kirill Alexandrov
Institute for Molecular Bioscience, University of Queensland, Brisbane QLD, 4072, Australia
k.alexandrov@uq.edu.au

Rafael Alvarez-Gonzalez
Graduate School of Biomedical Sciences, University of North Texas Health Science Center at Fort Worth, 3500 Camp Bowie Blvd., Fort Worth, TX 76107, USA
Rafael.Alvarez-Gonzalez@unthsc.edu

Emily Bernstein
Department of Oncological Sciences, Mount Sinai School of Medicine, New York, NY 10029, USA
emily.bernstein@mssm.edu

Nicholas R. Bertos
Rosalind & Morris Goodman Cancer Centre, McGill University, 1160 Ave. des Pins O, Montreal, QC, Canada, H3A 1A3
and Department of Medicine, McGill University Health Centre, Montreal, QC, Canada
nicholas.bertos@mcgill.ca

Liora Braiman-Wiksman
HealOr Ltd, P. O. Box 4115, Weitzman Science Park, Ness Ziona, 74140, Israel

Dirk Breitkreutz
DGZ, Division B050, German Cancer Research Center (DKFZ), POB 101949, D-69009, Heidelberg, Germany
dirkbtz@gmx.de

Nathalie Campagnolo
Unit of Physiological Biochemistry, Institute of Life Sciences, Catholic University of Louvain, Croix du Sud, 4-5 boîte 15, B-1348, Louvain-la-Neuve, Belgium

Kenny K. K. Chung
Department of Biochemistry, Hong Kong University of Science and Technology, Clear Water Bay, Hong Kong, People's Republic of China
bckchung@ust.hk

T.G. Cotter
Cell Development and Disease Laboratory, Biochemistry Department, Biosciences Institute, University College, Cork, Ireland
t.cotter@ucc.ie

Nicole Daum
DGZ, Division B050, German Cancer Research Center (DKFZ), P. O. Box 101949, D-69009, Heidelberg, Germany
Department of Biopharmaceutics and Pharmaceutical Technology, Saarland University, D-66123, Saarbruecken, Germany

S. Finnegan
Cell Development and Disease Laboratory, Biochemistry Department, Biosciences Institute, University College Cork, Cork, Ireland

Michel Ghislain
Université Catholique de Louvain, Institut des Sciences de la Vie, Unité de Biochimie Physiologique, Croix du Sud, 4-5, boîte 15, B-1348, Louvain-la-Neuve, Belgium
michel.ghislain@uclouvain.be

Andrew Goodall
Institute for Molecular Bioscience, University of Queensland, Brisbane, QLD, 4072, Australia

G. Groeger
Cell Development and Disease Laboratory, Biochemistry Department, Biosciences Institute, University College Cork, Cork, Ireland

Zdenko Herceg
Epigenetics Group, International Agency for Research on Cancer (IARC), 150 Cours Albert Thomas, 69008 Lyon, France
herceg@iarc.fr

Nigel M. Hooper
Proteolysis Research Group, Institute of Molecular and Cellular Biology, Faculty of Biological Sciences,
and Leeds Institute of Genetics, Health and Therapeutics, University of Leeds, Leeds, LS2 9JT, UK
n.m.hooper@leeds.ac.uk

Avnish Kapoor
Department of Oncological Sciences, Mount Sinai School of Medicine, NY, New York 10029, USA

Nobuhiko Katunuma
Institute for Health Sciences, Tokushima Bunri University, 180 Nishihamabouji, Yamashiro-cho, Tokushima 770-8514, Japan
katunuma@tokushima.bunri-u.ac.jp

Dirk J. Lefeber
Laboratory of Paediatrics and Neurology, Radboud University Nijmegen Medical Centre, Nijmegen, The Netherlands

Yihang Li
Department of Biochemistry, Hong Kong University of Science and Technology, Clear Water Bay, Hong Kong

Lisa Lukaesko
RS Dow Neurobiology Laboratories and Cancer Research Laboratories, Legacy Research, 1225 Ne 2nd Ave, Portland, OR 97232, USA

Isabelle M. Mansuy
Brain Research Institute, Medical Faculty University Zurich/Swiss Federal Institute of Technology Zurich, Winterthurerstrasse 190, CH- 8057 Zurich, Switzerland
mansuy@hifo.uzh.ch

Paul T. Martin
Departments of Pediatrics and of Physiology and Cell Biology, Center for Gene Therapy, The Research Institute at Nationwide Children's Hospital, Ohio State University College of Medicine, 700 Children's Drive, Columbus, OH 43205, USA
MartinPT@pediatrics.ohio-state.edu

Robert Meller
RS Dow Neurobiology Laboratories and Cancer Research Laboratories Legacy Research, 1225 Ne 2nd Ave, Portland, OR 97232, USA
rmeller@downeurobiology.org

Federica Montanaro
Departments of Pediatrics and of Physiology and Cell Biology, Center for Gene Therapy, The Research Institute at Nationwide Children's Hospital, Ohio State University College of Medicine, 700 Children's Drive, Columbus, OH 43205

Eva Morava
Department of Paediatrics, Radboud University Nijmegen Medical Centre, P.O. Box 9101, 6500, HB, Nijmegen, The Netherlands
E.Morava@cukz.umcn.nl

Natalia N. Nalivaeva
Institute of Evolutionary Physiology and Biochemistry, RAS, St. Petersburg, Russia

Maria Nesterova
Section on Endocrinology Genetics, Program on Developmental Endocrinology Genetics (PDEGEN), Eunice Kenney Shriver National Institute of Child Health & Human Development (NICHD), National Institute of Health (NIH), Bethesda, MD 20892, USA

Uyen T.T. Nguyen
Max-Planck-Institute for Molecular Physiology, Otto-Hahn-Str. 11, 44227 Dortmund, Germany

Morag Park
Rosalind & Morris Goodman Cancer Centre, McGill University, Montreal, QC, Canada
Department of Medicine, McGill University Health Centre, Montreal, QC, Canada
Department of Biochemistry, McGill University, Montreal, QC, Canada

C. Quiney
Cell Development and Disease Laboratory, Biochemistry Department, Biosciences Institute, University College Cork, Cork, Ireland

Kajan Ratnakumar
Department of Oncological Sciences, Mount Sinai School of Medicine, New York, NY 10029, USA

Veena Sangwan
Rosalind & Morris Goodman Cancer Centre, McGill University, Montreal, QC, Canada
Department of Medicine, McGill University Health Centre, Montreal, QC, Canada

John W. Scott
St Vincent's Institute and Department of Medicine, University of Melbourne, 41 Victoria Parade, Fitzroy 3065 VIC, Australia
jscott@svi.edu.au

C. Sehwan Park
Department of Biology of the ETH Zurich, Brain Research Institute, Medical Faculty of the University Zurich, CH-8057, Zurich, Switzerland

Constantine A. Stratakis
Section on Endocrinology Genetics, Program on Developmental Endocrinology Genetics (PDEGEN), Eunice Kenney Shriver National Institute of Child Health & Human Development (NICHD), National Institute of Health (NIH), Building 10, CRC (East Laboratories), Room 1-3330, 10 Center Dr., MSC1103, Bethesda, MD 20892, USA
stratakc@mail.nih.gov

Masae Takahashi
Institute for Health Sciences, Tokushima Bunri University, 180 Nishihamabouji Yamashiro-cho, Tokushima City, Tokushima 770-8514, Japan

David R. Taylor
Proteolysis Research Group, Institute of Molecular and Cellular Biology, Faculty of Biological Sciences, and Leeds Institute of Genetics, Health and Therapeutics, University of Leeds, Leeds, LS2 9JT, UK

Tamar Tennenbaum
HealOr Ltd, P.O. Box 4115, Weitzman Science Park Ness Ziona, 74140, Israel
Faculty of Life Sciences, Bar-Ilan University, Ramat-Gan, 52900, Israel

Tadashi Tezuka
School of Medicine, Kinki University, 377-2 Ohnohigashi, Osakasayama City, Osaka, 589-8511, Japan

Anthony J. Turner
Institute of Molecular and Cellular Biology, University of Leeds, LS2 9JT Leeds, UK
a.j.turner@leeds.ac.uk

Y. Tweedie-Cullen
Brain Research Institute, Medical Faculty of the University Zurich and Department of Biology of the ETH Zurich, CH-8057 Zurich, Switzerland

Martijn van Hagen
Department of Molecular Cell Biology, Leiden University Medical Center, P.O. Box 9600, Mailzone S-1-P, 2300RC Leiden, The Netherlands

Alfred C.O. Vertegaal
Department of Molecular Cell Biology, Leiden University Medical Center, P.O. Box 9600, Mailzone S-1-P, 2300 RC Leiden, The Netherlands
vertegaal@lumc.nl

Ron A. Wevers
Laboratory of Paediatrics and Neurology, Radboud University Nijmegen Medical Centre, Nijmegen, The Netherlands

Xiang-Jiao Yang
Rosalind & Morris Goodman Cancer Centre, McGill University, Montreal, QC, Canada
Department of Medicine, McGill University Health Centre, Montreal, QC, Canada

Chapter 1
Isoprenoid Modifications

Uyen T.T. Nguyen, Andrew Goodall, Kirill Alexandrov, and Daniel Abankwa

1.1 Introduction

Up to 2% of the mammalian proteome is post-translationally modified with isoprenes (Gelb 1997). The first prenylated polypeptides reported were secreted pheromone peptides identified in jelly fungi (Sakagami et al. 1978; Tsuchiya et al. 1978). The structure of these peptides resembled the α-factor mating pheromone from *S. cerevisiae* and contained a farnesylated cysteine methylester at the C terminus (Anderegg et al. 1988). Several parallel studies by groups working on cholesterol biosynthesis reported that a mevalonic acid derivative, other than cholesterol, could be specifically incorporated into proteins (Maltese and Sheridan 1987; Schmidt et al. 1984). Subsequently it became clear that protein prenylation is characterized by the attachment of a farnesyl or a geranylgeranyl moiety to one or two C-terminal cysteine residues via a thioether linkage. Three different protein prenyltransferases have been identified in eukaryotes: farnesyltransferase (FTase), geranylgeranyltransferase-I (GGTase-I), and Rab geranylgeranyltransferase (RabGGTase) (Casey and Seabra 1996; Maurer-Stroh et al. 2003). FTase and GGTase-I recognize their substrates via a short C-terminal recognition sequence, referred to as the CAAX box, where C is a cysteine, A is an aliphatic amino acid and X is variable but biased depending on the prenyltransferase (Table 1.1). In contrast to FTase and GGTase-I, protein substrate recognition by RabGGTase is more complex, requiring a partner protein, Rab escort protein (REP) for substrate recruitment (Fig. 1.1).

Since the initial discovery of farnesylated fungal proteins, a vast range of prenylated proteins have been discovered. The best characterized are those belonging to the Ras superfamily of small GTPases comprising Ras (Kontani et al. 2002; Shilo and Weinberg 1981), Rho and Cdc42 (Roberts et al. 2008; Yamane et al. 1991),

---

K. Alexandrov (✉)
Institute for Molecular Bioscience, University of Queensland Brisbane QLD, 4072, Australia
e-mail: k.alexandrov@uq.edu.au

C.J. Vidal (ed.), *Post-Translational Modifications in Health and Disease*, Protein Reviews 13,
DOI 10.1007/978-1-4419-6382-6_1, © Springer Science+Business Media, LLC 2011

Table 1.1 Biochemistry of the prenyltransferases

	FTase	GGTase-I	RabGGTase
Subunit composition (mammalian)	α 44 kDa β 48 kDa	α 44 kDa β 43 kDa	α 65 kDa β 37 kDa
Protein recognition motif	$-CA_1A_2X$ X = Met, Phe, Ala, Gln, Ser, Cys, Thr A_1: flexible A_2: Ile, Val preferred	$-CA_1A_2X$ X = Met, Phe, Leu, Ile, Val A_1: flexible A_2: Ile, Leu preferred	Common: –CC, –CXC, –CCX, –CCXX, –CCXXX Rare: –CXXX
Isoprenoid substrate	Farnesyl pyrophosphate (FPP) 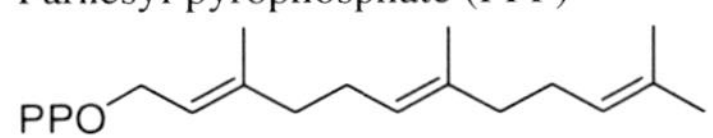 **1**	Geranylgeranyl pyrophosphate (GGPP) 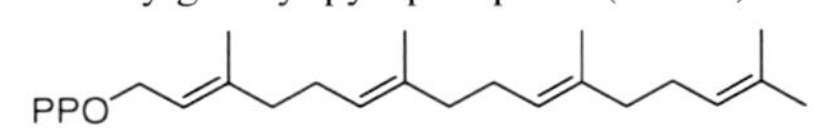 2	
Metal requirements	Zn^{2+}, Mg^{2+}	Zn^{2+}	Zn^{2+}

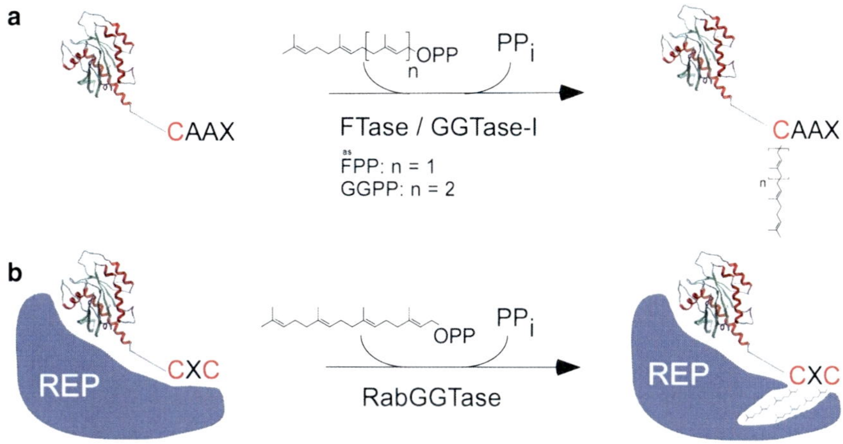

Fig. 1.1 Biochemistry of the prenyltransferases. Schematic representation of the reaction catalyzed by the two CAAX prenyltransferases FTase ($n=1$) and GGTase-I ($n=2$) (**a**) or RabGGTase in concert with REP (Rab Escort Protein) (**b**). The enzymes catalyze the formation of a thioether linkage between the prenyl group and one or two C-terminal cysteines of the protein substrate

Rac (Kinsella et al. 1991), Rap (Buss et al. 1991; Farrell et al. 1993), and Rab proteins (Farnsworth et al. 1994; Kinsella and Maltese 1991). Other prenylated proteins include the γ subunit of heterotrimeric G proteins (Lai et al. 1990; Marrari et al. 2007), centromeric proteins (Ashar et al. 2000) and several regulators of the cell cycle, apoptosis (De Smedt et al. 1996; Jefferson and Majerus 1995), cellular structure (Farnsworth et al. 1989; Kutzleb et al. 1998; Lutz et al. 1992), glycogen metabolism (Heilmeyer et al. 1992) and photoreception (Anant and Fung 1992; Inglese et al. 1992). In the following sections we will limit our discussion to the prenylation of small GTPases.

1.2 The Small GTPases of the Ras Superfamily, Their Activation and Their Biological Functions

The Ras superfamily of small GTPases is the largest group of prenylated proteins, with more than 150 members divided into four subfamilies based on sequence similarity (Takai et al. 2001; Wennerberg et al. 2005). These sub-families are designated Ras, Rho/Rac/Cdc42, Rab, and Arf/Sar1, with all members sharing common structural, biochemical and regulatory features. As Arf proteins are not prenylated they will not be discussed any further in this review. Similarly, the G protein Ran and ten other related G proteins which cannot be grouped into any of these sub-families will also not be further discussed. The interest in small GTPases and in particular Ras stems from the discovery that activating mutations in the genes encoded by rat sarcoma viruses (vH-ras, vN-Ras, and vK-Ras) can induce cell transformation and cancer (Clarke 1992; Harvey 1964). Consequently the relationship between Ras structure and function has been extensively studied and has revealed features that can be largely applied to the broad Ras superfamily.

1.2.1 The Small GTPase Activation Cycle

Two main features characterize small GTPases, a conserved GTPase domain and the specific binding of guanine nucleotides. These features enable small GTPases to function as molecular switches in signaling pathways by oscillating between an inactive GDP-bound and active GTP-bound state (Fig. 1.2) (Sprang 1997). GTP binding stimulates a conformational change which allows GTPases to interact with effector proteins and initiate downstream signaling cascades. The rate-determining step in the GTPase cycle is GDP release, having a half-life of several hours. Consequently in order to accelerate nucleotide exchange and promote an efficient response to external stimuli, additional regulatory proteins are required. The rate of GDP release is increased by specific guanine nucleotide exchange factors (GEFs) (Vetter and Wittinghofer 2001). GTP then readily binds to nucleotide-free GTPases owing to a tenfold excess of GTP over GDP (1 mM and 100 μM respectively) in cellular cytosol. The γ-phosphate group of the bound GTP induces conformational changes in the switch I and switch II loops resulting in the GTPase adopting an active conformation (Vetter and Wittinghofer 2001). The switch I and switch II

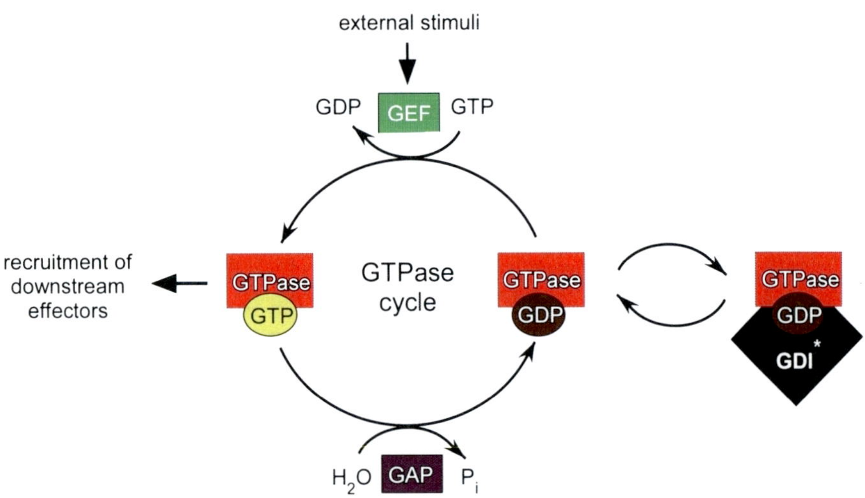

Fig. 1.2 Schematic representation of the GTPase cycle. Upon an external stimulus, the GTPase switches from an inactive GDP-bound to an active GTP-bound state which is tightly controlled by its guanine nucleotide exchange factors (GEFs). Once activated, it can recruit its effectors and enables downstream signaling processes. Inactivation is achieved through GTP hydrolysis, which is catalyzed by GTPase activating proteins (GAPs). *For Rab and Rho GTPases, an additional level of regulation is provided by GDP dissociation inhibitor (GDI), which keeps the prenylated GDP-bound protein in the cytosol

loops can then interact with downstream effector proteins, facilitating the potential regulation of a wide range of signaling pathways.

Similar to activation, the rapid "inactivation" of small GTPases also requires additional regulatory proteins. Without external regulation the longevity of the GTP-bound state is defined by the extremely slow intrinsic GTPase activity of GTPases (e.g. 0.028 min^{-1} for Ras (John et al. 1993)). This rate of GTP hydrolysis is enhanced by up to five orders of magnitude by proteins known as GTPase activating proteins (GAPs) (Bos et al. 2007; Cherfils and Chardin 1999; Scheffzek et al. 1998). In the cases of Rab and Rho, the inactive GDP-bound state is then preserved by the binding of GDP dissociation inhibitors (GDIs), providing yet another point of regulation for Rab and Rho signaling.

1.2.2 Biological Functions of Ras Family GTPases

To highlight the importance of prenylated proteins in cellular biology we will briefly summarize the cellular functions and signaling pathways regulated by the Ras, Rho/Rac/Cdc42 and Rab families of small GTPases.

In total there are 39 members of the Ras family that predominantly differ in their C-terminal hypervariable region (HVR) (Wennerberg et al. 2005). While classically known for regulating patterns of gene expression that influence cell proliferation and differentiation, the functional spectrum of Ras proteins is far broader and includes regulation of the cytoskeleton, mitogenesis, cell adhesion, protein trafficking and other cell and tissue specific events (Colicelli 2004). The best characterized Ras isoforms are the ubiquitously expressed H-Ras, N-Ras and K-Ras4B, which predominantly operate on the cytoplasmic leaflet of cell membranes. Of the wide range of signaling pathways activated by H-, N- and K-Ras, almost all are common to all three isoforms. To date the only Ras effector displaying isoforms specificity is RASSF2 which binds to K-Ras4B but not H-Ras (Devos et al. 2003). The ability of RASSF2 to bind N-Ras has not been reported. A comprehensive review of known Ras effectors can be found in Karnoub and Weinberg (2008).

The best characterized Ras signaling pathway is the Raf/MAPK pathway (Fig. 1.3). Raf proteins, of which there are three isoforms (c-Raf, A-Raf and B-Raf), are recruited to GTP-Ras in a complex with 14-3-3 (McPherson et al. 1999; Roy et al. 1998). Upon Raf binding to GTP-Ras, 14-3-3 is released leading to a complex series of phosphorylation events on Raf by various other serine/threonine and tyrosine kinases. In a basic sense this provides a docking site for MEK1/2, which upon serine phosphorylation by Raf binds ERK1/2 (MAPK) (Xiang et al. 2002). Phosphorylated ERKs form dimers that are translocated into the nucleus, where they phosphorylate Ets family transcription factors, such as Elk-1, to activate transcription. The other well characterized Ras signaling pathway is the class I PI3K system, which is initiated by GTP-Ras binding to the various p110 catalytic subunits (Rodriguez-Viciana et al. 2004). This stimulates the lipid kinase activity of class I PI3Ks providing docking sites for signaling proteins such as Akt (Chan et al. 1999).

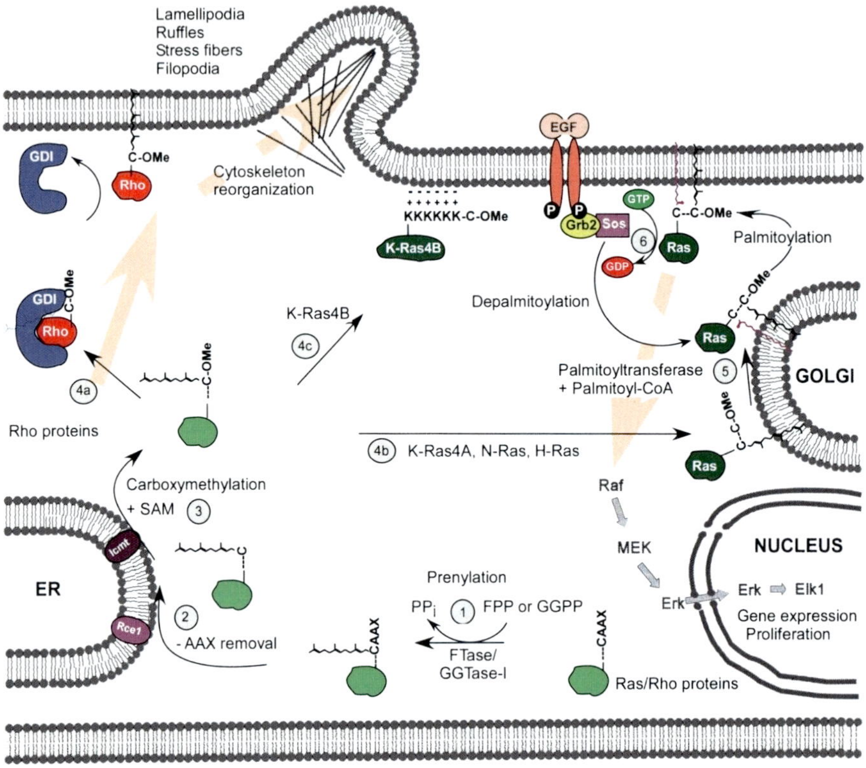

Fig. 1.3 Overview of protein prenylation, post-prenylation modifications, membrane targeting, and the function of the prenylated proteins, mediated by the CAAX prenyltransferases. A Ras or Rho protein (*light green*) is prenylated by FTase or GGTase-I with FPP or GGPP, respectively (1). –AAX removal by Rce1 (2) and carboxymethylation by Icmt (3) take place at the ER. In the case of Rho proteins (*red*), the protein is chaperoned by a Rho-specific GDI, which targets the Rho protein to the correct membrane (4a). K-Ras4A, N-Ras and H-Ras are targeted to the Golgi (4b), where they are further palmitoylated on one or two cysteines (5). Once palmitoylated the mature protein trafficks to the plasma membrane, where it recruits its GEF which activates the protein (6) and engages downstream signaling, including the MAPK signaling pathway. Through depalmitoylation-repalmitoylation, it is subjected to a dynamic acylation cycle. K-Ras4B is targeted without additional palmitoylation through its hexalysine motif to its target membrane (4c). Membrane attachment is dynamically regulated by protein kinase C and/or calmodulin. For details, see text

Like Ras, Ras homologous (Rho) proteins also play important roles within signaling networks leading to cell cycle progression, transcriptional regulation, vesicle trafficking and, most importantly actin reorganization (Fig. 1.3) (Etienne-Manneville and Hall 2002; Jaffe and Hall 2005). There are currently 23 identified members of the Rho/Rac/Cdc42 subfamily, with RhoA, Rac1 and Cdc42 being the best characterized (Colicelli 2004; Wennerberg et al. 2005). The majority of known signaling pathways activated by Rho proteins lead to a variety of cytoskeletal changes resulting in stress fibre formation (RhoA/B/C), lamellipodia extension (Rac1–3, RhoG), filopodia formation (Cdc42, TCL/RhoJ, TC10/RhoQ), cell fate determination

(Cdc42), neurite outgrowth (Cdc42, Rac1, Rac3, RhoG), endocytic trafficking (RhoB), phagocytosis (Cdc42, Rac2, RhoG) and bacterial killing via increased NADPH oxidase activity and granule release (Rac2) (Heasman and Ridley 2008). With the exception of lamellipodia and filopodia formation, the downstream signaling pathways responsible for most of these processes remain poorly understood.

At the leading edge of the cell, lamellipodia formation can occur in response to several signaling pathways. The best characterized pathway involves ARP2/3-stimulated actin polymerization downstream of Cdc42-activated WASP and/or IRSp53 and Rac-activated WAVE complex (Cory and Cullen 2007; Jaffe and Hall 2005). Evidence has also accumulated suggesting that actin polymerization leading to lamellipodia and filopodia formation also occurs via Cdc42- or Rac-mediated mDia2 activation (Cory and Cullen 2007). Both Cdc42 and Rac also activate the ser/thr kinase Pak, which in turn phosphorylates LIM kinase leading to Cofilin inhibition. The effect of this is to decrease actin depolymerization, hence promoting lamellipodia and filopodia extension (Cory and Cullen 2007). For a comprehensive list of Rho effector proteins and the cellular functions they regulate refer to the following review (Bustelo et al. 2007).

The 63 Rab proteins (Ras-related genes expressed in brain) are primarily involved in vesicular transport (Wennerberg et al. 2005). Specifics of their cellular role and effectors can be found in several excellent papers (Colicelli 2004; Fukuda et al. 2008; Grosshans et al. 2006; Novick and Zerial 1997) but as an overview they regulate vesicle transport, tethering and fusion (Zerial and McBride 2001). The distinct subcellular distribution of Rab proteins on cellular organelles has also led to suggestions that signaling by some Rab proteins contributes to organelle identity (Pfeffer 2001, 2003).

In conclusion, each branch of small GTPases has its own predominant areas of function. However, there is notable overlap in their cellular functions as well as cross-talk among GTPase-mediated signaling pathways. This suggests that significant functional synergy is likely to exist between families of small GTPases, although the specifics of this synergy remain largely unexplored. Another issue that awaits clarification is why there are so many isoforms within each GTPase branch and what the specific functions of all isoforms are. In recent years, evidence has accumulated that the specific subcellular localization of GTPases contribute to their signaling specificity (Hancock 2003; Mor and Philips 2006; Plowman and Hancock 2005; Quatela and Philips 2006). Therefore we will review what is known about the determinants of subcellular targeting of the major small GTPase subfamilies.

1.3 Subcellular Targeting of Ras, Rho and Rab Proteins

1.3.1 Ras Protein Trafficking

Ras proteins are synthesized as soluble proteins on free polysomes in the cytosol and are subsequently targeted to cellular membranes by a series of posttranslational modifications at their C-terminal hypervariable regions (Clarke 1992). The first modification involves the covalent attachment of a 15 carbon isoprenoid to the cysteine residue of the C-terminal CAAX motif by cytosolic FTase (Hancock et al. 1990; Reiss et al. 1990; Schaber et al. 1990). Alternatively some isoforms, such as e.g. K-Ras4B and the Ras-like GTPase TC21 can be geranylgeranylated by GGTase-I (Lerner et al. 1997b). Prenylation targets Ras proteins to the ER where a resident enzyme known as Ras converting enzyme 1 (Rce1) proteolytically removes the AAX sequence (Gutierrez et al. 1989; Schmidt et al. 1998). The exposed farnesylated cysteine is then carboxymethylated by isoprenylcysteine carboxymethyltransferase (Icmt), also at the ER (Clarke et al. 1988; Dai et al. 1998). While the CAAX motif alone is sufficient to target Ras proteins to the ER and Golgi, anterograde trafficking to the plasma membrane requires a second HVR signal upstream of the CAAX motif. This signal involves the palmitoylation of C181 in N-Ras and both C181 and C184 in H-Ras (Hancock et al. 1989).

While the identity of the mammalian Ras palmitoyltransferase is yet to be conclusively elucidated, a Golgi-localized complex containing GCP16 and DHHC9 has been demonstrated to palmitoylate Ras proteins in vitro (Swarthout et al. 2005). Irrespective of the identity of the Ras palmitoyltransferase, palmitoylation targets Ras proteins to the exocytic pathway and onwards to the plasma membrane (Apolloni et al. 2000; Choy et al. 1999). In contrast to H- and N-Ras, K-Ras4B is not palmitoylated but contains 11 lysine residues in its HVR. This gives the K-Ras4B HVR a net charge of +9, with the current hypothesis being that plasma membrane localization is achieved via simple diffusion through the cytosol and electrostatic interactions with acidic phospholipids such as PI(3,4,5)P3 and PI(4,5) P2 (Heo et al. 2006; Leventis and Silvius 1998; Yeung et al. 2006) (Fig. 1.4). Most other Ras isoforms appear to be predominantly targeted to the plasma membrane by similar C-terminal lipidation to H-, N- and K-Ras, although exceptions have been described (Colicelli 2004; Nomura et al. 2004).

Ras proteins have emerged as an important model system for compartmentalized signaling on organelles including the Golgi/ER (H- and N-Ras), endosomes (H-, N- and K-Ras) and mitochondria (K-Ras) (Fivaz and Meyer 2005; Rizzo et al. 2001). The pool of Ras proteins at these intracellular compartments appears to be replenished by the retrograde trafficking of Ras proteins away from the plasma membrane. For H- and N-Ras this involves the de-palmitoylation of HVR cysteine residues by an unknown mechanism. Re-palmitoylation can then re-occur creating a dynamic acylation and trafficking cycle (Goodwin et al. 2005; Rocks et al. 2005). The precise mechanism of retrograde K-Ras4B trafficking also remains to be elucidated although it has been suggested that phosphorylation of HVR serine residues disrupts electrostatic interactions with acidic phospholipids (Bivona et al. 2006b).

1.3.2 Determinants for Subcellular Membrane Targeting of Rho GTPases

Analogous to Ras proteins the subcellular targeting of Rho proteins is primarily governed by their HVR, which terminates in 16 out of 20 cases with a CAAX box (Michaelson et al. 2001). The CAAX boxes of Rho proteins can be prenylated by

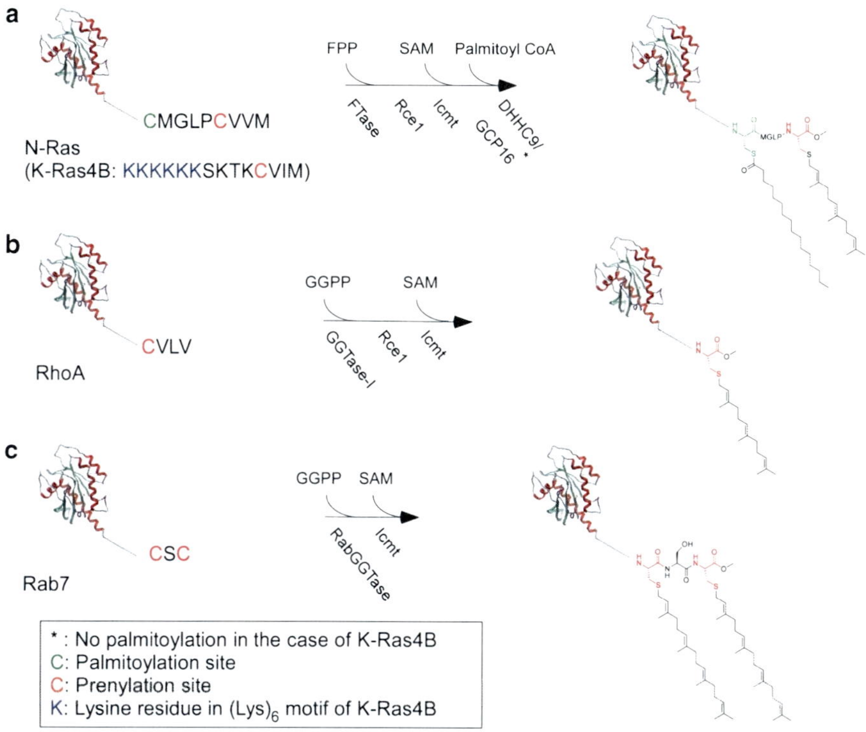

Fig. 1.4 Schematic representation of posttranslational processing of N- or K-Ras4B, RhoA and Rab7. (**a**) N-Ras is farnesylated on one cysteine (*red*) by FTase, which is followed by –AAX removal by Rce1 (Ras converting enzyme 1), carboxymethylation by Icmt (isoprenylcysteine carboxymethyltransferase) and palmitoylation on one cysteine (*green*) by DHHC9/GCP16. K-Ras4B is similarly posttranslationally processed, but does not contain any palmitoylation site. In contrast, it contains a hexalysine repeat in the membrane targeting domain (*blue*). (**b**) RhoA is geranylgeranylated by GGTase-I on one cysteine (*red*), –AAX cleaved, and carboxymethylated. (**c**) Rab 7 undergoes double geranylgeranylation on two cysteines (*red*), followed by carboxymethylation. Membrane targeting of Rho and Rab is achieved through Rho- and Rab-specific GDIs

either FTase or GGTase-I with specificity primarily conferred by the X residue (Hartman et al. 2005; Wennerberg et al. 2005). While the majority of CAAX-containing Rho proteins are specifically modified by either FTase or GGTase-I, RhoB (CKVL) and RhoH (CKIF) can be efficiently prenylated in vivo by both prenyltransferases (Baron et al. 2000; Roberts et al. 2008). At least in the case of RhoB dual prenylation is a function of its CAAX sequence. Identical to Ras proteins the majority of prenylated Rho CAAX motifs are cleaved by Rce1 and subsequently carboxymethylated by Icmt (Roberts et al. 2008) (Table 1.2). However evidence suggests that this is not the case for all Rho proteins, with the correct subcellular localization of RhoA and RhoB being not dependent on processing by Rce1 and Icmt respectively (Roberts et al. 2008).

Correct plasma membrane localization of Rho proteins also requires a second signal which comprises stretches of basic amino acids alone, or in combination with nearby palmitoylated cysteines (Dudler and Gelb 1996; Lerner et al. 1997a; Mitchell

Table 1.2 Summary of determinants for subcellular localizations of Rho-family proteins

Rho isoform	HVR Polybasic domain	P	Prenyl-moiety	Subcellular localization depends on Rcel	Icmt	Subcellular localization
Rac1	+++		GG	–	–	PM, cytosol
RhoC	+++		GG	n.d.	n.d.	PM, cytosol
Cdc42	+++		GG	n.d.	n.d.	PM, Golgi, ER, NE
Rac2	+++		GG	n.d.	n.d.	PM, Golgi, ER, NE
Rac3	+++		GG	n.d.	n.d.	PM, endomembranes
RhoG	+++		GG	n.d.	n.d.	PM, endosomes
RhoA	+++		GG	–	+	PM, cytosol
RhoH/TTF	+++		GG	+/–	+/–	Endomembranes
Rif/RhoF	+++		GG	+	+	PM
Rnd1	+++		F	+/–	+	PM
Rnd2	+++		F	+/–	+	Endosomes, cytosol
Rnd3/RhoE	+++		F	+	+	PM, Golgi, cytosol
RhoD	+++		F	+/–	+/–	PM, endomembranes
TC10/RhoQ	+++	2×	F	+/–	+/–	PM, endosomes
RhoB	+++	2×	F or GG	+	–	PM, Golgi, endosomes
TCL	+++	2×	GG	+/–	+	PM, endosomes
RhoBTB1,	Tandem-BTB domains			n.a.	n.a.	Vesicular
Chp, Wrch-1	Polybasic-Trp-polybasic-P			n.a.	n.a.	PM, endosomes

For 16 of the 20 isoforms, the C-terminal hypervariable region (HVR) can contain three types of targeting signals: a polybasic domain, palmitoylation sites and the CAAX-prenylation motif (Chenette et al. 2006). Subcellular localizations of these proteins may depend on post-prenylation processing by Rcel or Icmt (Roberts et al. 2008). The exact mechanism of targeting of non-prenylated, unusual Rho proteins, RhoBTB1-2 is unknown, while that of Chp, Wrch-1 fits into the general scheme.
P palmitoyl, *GG* geranylgeranyl, *F* farnesyl, +/– limited effect, + strong effect, – no effect, *n.d.* not determined, *n.a.* not applicable, *PM* plasma membrane, *ER* endoplasmic reticulum, *NE* nuclear envelope

et al. 1994). The mechanisms by which these secondary targeting signals confer plasma membrane localization are likely to be the same as for Ras proteins. Polybasic stretches in the HVR often serve a dual purpose in Rho proteins, with the sequence K-K/R-x-K/R constituting a canonical nuclear localization signal (NLS) (Williams 2003). The nuclear import of Rho proteins appears to be facilitated by armadillo repeat-containing proteins including karyopherin α family members (Chook and Blobel 2001) and the RhoGEF SmgGDS (Williams 2003). The Rho protein with the most characterized nuclear localization is Rac1a, although RhoC, RhoG, Cdc42 (isoform 1), TCL and Rnd1 also contain a NLS in their HVR (Williams 2003). The prominent nuclear localization of Rac1 is in contrast to RhoA, which lacks a canonical NLS and is thus largely excluded from the nucleus (Ridley 2006; Williams 2003). RhoGDI is another factor affecting the subcellular localization of Rho proteins, with increased RhoGDI interaction shifting Rho proteins into the cytoplasm (Michaelson et al. 2001). A study based on RhoA/RhoGDIα interaction demonstrated that affinity for RhoGDIs is greatly reduced by palmitoylation, thus increasing the affinity of RhoA for membranes (Michaelson et al. 2001). Additional less characterized Rho solubilizing factors may also exist, as evidenced by the unusual 73 residue N-terminal extension of RhoA which promotes the predominant cytosolic localization of the protein (Michaelson et al. 2001).

Finally, there are four non-prenylated Rho proteins, RhoBTB1, RhoBTB2, Chp/RhoV and Wrch-1/RhoU. There is evidence suggesting that RhoBTB1 and RhoBTB2 are targeted to vesicular structures via protein-protein interactions involving their C-terminal tandem broad complex (BR-C)/tramtrack (ttk)/Bric-a-brac (bab) (BTB) domain (Chang and Philips 2006). Chp and Wrch-1 are targeted to the plasma membrane and endosomes by the unusual combination of a polybasic domain in combination with palmitoylation (Berzat et al. 2005; Chenette et al. 2006). The trafficking pathway(s) utilized by Chp and Wrch-1 remains to be elucidated as it is unclear, whether the polybasic domain or palmitate groups are the primary trafficking signals.

The presence of the polybasic domain suggests that akin to K-Ras4B, Chp and Wrch-1 traffic to the plasma membrane independently of the Golgi (Apolloni et al. 2000). However based on the Golgi localization of the H- and N-Ras palmitoyltransferase complex DHHC9/GCP16, it is plausible that Chp and Wrch-1 also traffic via the Golgi (Swarthout et al. 2005). However a proteomic screen of global palmitoylation in *S. cerevisiae* demonstrated that multiple palmitoyltransferases exist with overlapping specificities and partially redundant functions (Roth et al. 2006). It is likely that a similar redundancy exists in mammalian cells and that palmitoyltransferase complexes are localized to numerous subcellular locations. This therefore raises the possibility that Chp and Wrch-1 may not need to traffic via the Golgi for palmitoylation.

1.3.3 Subcellular Membrane Targeting of Rab GTPases

Rab proteins are C-terminally modified by one or two geranylgeranyl moieties with the correct number of prenyl chains being critical for correct subcellular localization (Calero et al. 2003; Gomes et al. 2003). The mono-prenylation of normally di-prenylated Rabs leads to their non-functional targeting to the ER. This suggests that the ER is the default subcellular location of Rab proteins, akin to other mono-prenylated GTPases. Other than this, very little is known about the determinants for appropriate targeting of Rab proteins to their cognate membrane.

Initially it was proposed that similar to Ras, the C-terminal HVR directs specific targeting (Chavrier et al. 1991). However in contrast to Rho and Ras proteins it is not possible to assess the role of the HVRs of Rab proteins in isolation, as the prenylation machinery requires the intact protein as a substrate. Importantly studies tracking the localization of fluorescently tagged Rab proteins have revealed that correct subcellular targeting depends on other sequence elements outside of the

HVR (Ali et al. 2004; Pereira-Leal and Seabra 2000). This may in part explain the poor conservation of the HVR in Rab orthologues.

We have correlated the subcellular localizations of 60 Rab proteins (Ryo Misaki, personnel communication) with the biochemical characteristics of their HVR. From this correlation we could identify the determinants for plasma membrane localization of Rab proteins, which are consistent with those known from Ras and Rho proteins (Table 1.3). Strongest plasma membrane localization was observed for Rab13, Rab23, Rab34 and Rab35, followed by Rab8a/b, Rab27a/b, Rab39a/b and Rab40a/b. From the comparison of these sequences with related isoforms, it can be

Table 1.3 Rab subcellular targeting: determinants for plasma membrane targeting

Rab	C-terminal sequence	Prenylation motif	Icmt modification	Subcellular localization in COS-1
Rab34	LYLTAS**KKK**PT	CCP	–	**PM**, EE, RE
Rab36	TQES**KR**PSSLG	CC	–	Golgi
Rab40a	QSPP**K**NCT**R**NS	CKIS	+	PM, EE
Rab40b	QSPP**K**NCT**R**NS	CKIS	+	PM, EE
Rab40c	QSPPQNCS**R**SN	CKIS	+	Endosome
Rab42	**R**SPS**RK**QHSGP	CQC	+	*PM*[a]*, (perinuclear in CHO)*
Rab23	**R**TN**K**N**R**NPFSS	CSIP	+	**PM**, EE, RE
Rab22a	L**RR**QPSEP**KR**S	CC	–	EE, RE
Rab20	SH**K**PP**KR**T**R**SG	CCA	–	EE, RE
Rab27a	DQLSEE**K**E**K**GA	CGC	+	PM, EE
Rab27b	NLDGE**K**PPE**KK**	CIC	+	PM, EE
Rab38	LTST**K**VASCSG	CAKS	+	EE
Rab39a	PSEEAV**K**P**RK**E	CSC	+	PM, RE
Rab39b	SSEEVV**K**SE**RR**	CLC	+	PM, EE
Rabla	QSTPV**K**QSGGG	CC	–	Golgi
Rablb	DSTPV**K**PAGGG	CC	–	Golgi
Rab35	**K**LT**K**NS**KRKKR**	CC	–	**PM**, **nu**, EE
Rab8a	PDQQ**KR**SSFFR	CVLL	+	PM, EE, RE?
Rab8b	EN**R**S**KK**TSFF**R**	CSLL	+	PM, EE, RE?
Rab10	SGGGVTGW**K**S**K**	CC	–	RE
Rab13	L**K**TCD**KK**NTN**K**	CSLG	+	**PM**, EE, RE

Selected Rab proteins are arranged according to the phylogenetic tree in (Stenmark and Olkkonen 2001), with C-terminal sequences of human proteins as in (Colicelli 2004). Those isoforms showing plasma membrane (PM) localization are typed in *bold* and corresponding data are shaded *grey*. Some of their phylogenetic neighbours are shown for comparison. Icmt modification is predicted for CAAX and CXC motifs (Leung et al. 2006, 2007). Subcellular localizations were evaluated by expression of EGFP-tagged mouse Rab proteins in COS-1 cells, in comparison with transferrin receptor uptake (Ryo Misaki, personnel communication). Strong plasma membrane localization is indicated by *bold* typed PM. For example, Rab35 is targeted strongly to the PM and the nucleus. It has five consecutive basic residues, similar to what can be found in plasma membrane targeted Ras and Rho proteins. This sequence in combination with its –CC motif directed C-terminal di-geranylgeranylation targets it to the plasma membrane. In addition, this sequence stretch also forms a canonical nuclear localization signal. Please refer to the main text for more details, *PM* plasma membrane, *EE* early endosome, *RE* recycling endosome, *nu* nuclear, [a]predicted seen that four non-consecutive basic residues in combination with CC motif mediated di-geranylgeranylation is not sufficient for plasma membrane targeting. However Rab34, with only three consecutive basic residues in combination with CCX-mediated di-geranylgeranylation, is targeted to the plasma membrane. Similar to some Rho proteins the C-terminal polybasic region of Rab35 forms part of a canonical nuclear localization signal, resulting in Rab35 being co-distributed between the plasma membrane and nucleus.

In combination these observations suggest that the most prevalent PM-targeting signal for Rab proteins is a combination of at least two basic residues with a CCX- or a CAAX-motif. It is noteworthy, that the latter two sequences become carboxymethylated, thus increasing hydrophobicity of the C-terminus (Leung et al. 2006, 2007). Rab10 which also contains two C-terminal basic residues and a CC-motif does not become carboxymethylated. Importantly Rab10 is not targeted to the plasma membrane, highlighting the importance of prenylated cysteine carboxymethylation for the plasma membrane targeting of Rabs. Additional mechanisms for membrane targeting have also been proposed, such as GDI-displacement factors (GDFs) that dissociate Rabs from their delivering proteins once at their target membranes (Pfeffer and Aivazian 2004; Pylypenko et al. 2006). Once at target membranes the interaction of Rabs with proteins, such as GEFs and effectors, and phosphoinositides may also stabilize their attachment to specific membrane domains (Zerial and McBride 2001).

1.4 Lateral Segregation and Membrane Nanodomains

On their target membranes, prenylated proteins interact specifically with a range of proteins that mediate their activation, downstream signaling and inactivation. In the recent years it has become clear that protein-protein interactions as well as protein-lipid and lipid-lipid interactions are important determinants for these interactions (Abankwa et al. 2007; Hancock 2006; Jacobson et al. 2007). Analysis of these molecules using high resolution fluorescence techniques or electron microscopy suggest that small GTPases are non-randomly organized into proteolipid domains. Organisation into specialized nanodomains or nanoclusters is critical for the overall functioning of Ras, as their disruption severely impairs MAPK signaling (Plowman et al. 2005; Prior et al. 2003). Moreover, nanoclustering also fundamentally alters the way extracellular signals are converted into intracellular responses, by making the whole process more robust and reliable (Harding and Hancock 2008; Tian et al. 2007). The significance of nanoclusters is supported by recent data, which suggest that protein domains and not the lipids govern protein and lipid organization in the plasma membrane (Eggeling et al. 2009; Goswami et al. 2008).

The concept of membrane nanodomains and nanoclusters is a refinement of the raft-model (Hancock 2006), which was proposed about 20 years ago (Simons and van Meer 1988; van Meer and Simons 1988). It intended to explain the specific sorting of lipids and envisaged submicroscopic lipid domains containing signaling proteins and being enriched in cholesterol and sphingolipids (Simons and Ikonen 1997; Simons and Toomre 2000). While having spurred an intense and fruitful interaction between membrane biophysics and cell biology, the proposed transient nature of lipid-rafts and their inherent intangibility has led to controversy about their nature and mere existence (Hancock 2006; Jacobson et al. 2007). Here, we want to summarize evidence for functional nanodomains of prenylated small GTPases in cellular membranes. We exclude from our discussion biochemical isolates, such as detergent resistant membranes (DRMs), which, despite interesting parallels in terms of their enrichment in raft-associated biological molecules, cannot be regarded as specific isolates of any submicroscopic proteolipid structure on the plasma membrane of living cells (Lichtenberg et al. 2005).

1.4.1 Nanoclustering of Ras Proteins

Recent advances in the field were largely driven by studies of Ras proteins that laterally segregate into distinct nanodomains within the plasma membrane depending on their HVR and activation state (Fig. 1.5) (Hancock and Parton 2005; Prior et al. 2001, 2003). Both electron microscopy and Förster resonance energy transfer (FRET) analysis confirmed that the minimal membrane anchors of H- and K-ras4B are sufficient for lateral segregation and organization of Ras proteins into nanoclusters (Abankwa et al. 2008b; Abankwa and Vogel 2007; Prior et al. 2003). These nanoclusters have a radius of 6–12 nm and contain on average 6–8 Ras proteins (Plowman and Hancock 2005). Each nanocluster acts as a sensitive signal amplifier, which responds maximally even to low signal input. In this sense it is responding like a switch (nanoswitch), which releases a defined signaling quantum into the cytoplasm. The high number of Ras-GTP nanoclusters (~50,000) and their short lifetime (estimated ~0.4 s) allow a high spatial and temporal sampling rate of the incoming (analogous) signals on the plasma membrane. The outputs of the many digital nanoswitches are subsequently reassembled into a total cytoplasmic analogue signal (Harding and Hancock 2008; Tian et al. 2007).

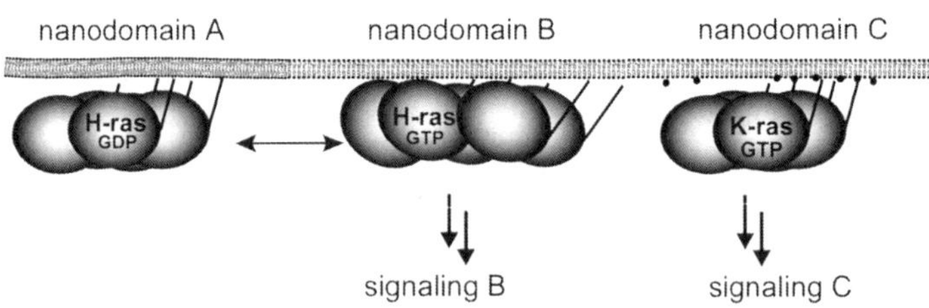

Fig. 1.5 Ras proteins laterally segregate into distinct nanoclusters. Ras isoforms H- and K-ras laterally segregated into distinct nanodomains. GTP-H-ras may furthermore be in a different nanodomain than GDP-H-ras. Both proteins are nanoclustered into submicroscopic domains of 6–12 nm of diameter, which contain on average 6–8 Ras proteins. Differences in lateral segregation can contribute to differences in signalling of the isoforms

For both H- and K-ras modulators of nanoclustering are known. The lectin Galectin-1 binds to the C-terminal farnesyl of GTP-H-ras, and modeling and mutational analysis suggest that it accommodates the farnesyl moiety in a hydrophobic pocket that is analogous to that of RhoGDI (Paz et al. 2001; Rotblat et al. 2004). Interestingly, galectin-1 also binds to non-palmitoylated H-ras and could therefore also act as a solubilizing chaperone, similar to RhoGDI. Thus, galectin-1 could shuttle H-ras from the plasma membrane to the Golgi (Belanis et al. 2008). The modulation of galectin-1 levels selectively alters H-ras-GTP nanoclustering, affects Raf-1 plasma membrane recruitment, MAPK signaling and cell transformation, correspondingly (Belanis et al. 2008; Paz et al. 2001; Prior et al. 2003). Similarly, galectin-3 binds with high selectivity to K-ras4B-GTP probably using a hydrophobic pocket that accommodates the farnesyl moiety (Shalom-Feuerstein et al. 2008). While galectin-3 augments K-ras4B-GTP levels, as well as Raf-1 and PI3K signaling, extracellular signal-regulated kinase (ERK) activity is suppressed (Elad-Sfadia et al. 2004). The increase in K-ras4B activation again appears to be mediated by the nanocluster scaffolding activity of galectin-3 (Shalom-Feuerstein et al. 2008). In addition, K-ras4B activity can be modulated by phosphorylation of S181 (Bivona et al. 2006a). Phosphorylation of this serine decreases K-ras4B nanoclustering, while blocking phosphorylation increases nanoclustering (Plowman et al. 2008). All of these data underscore the functional significance of nanoclustering.

1.4.2 Do Rho and Rab Proteins Have a Similar Submicroscopic Membrane Organization, as Ras Proteins?

It is unknown to what extent Ras nanoclustering is paradigmatic for the group of prenylated small G proteins. The fact that the minimal membrane anchors of H-, N- and K-Ras4B are sufficient for nanoclustering, may help to identify structural features in the Rho and Rab family, that would also mediate this specific membrane organization (Abankwa et al. 2007). In addition, the host membrane lipid composition may be a determinant for nanoclustering. The endocytic pathway is particularly rich in proteins that contain phosphatidylinositol-phosphate lipid binding motifs, suggesting a close collaboration of lipids and proteins in forming functional complexes (Gruenberg 2003). Based on the Ras data, it could be hypothesized that a secondary anchor such as a polybasic domain or palmitoylation and lipids from the plasma membrane are sufficient for nanoclustering. These hypothetical requirements would be satisfied for a large number of Rho proteins, for which we have a similar situation as we have for Ras isoforms in terms of membrane anchorage and subcellular targeting. Differences in subcellular targeting may also relate to the different functions of these isoforms (Hancock 2003). However, the similarity of C-termini (HVRs) of many Rho proteins with those of nanoclustering Ras, strongly suggests that at least some Rho proteins are nanoclustered on the plasma membrane. Therefore, lateral segregation into different nanoclusters may also be relevant for isoform specificity of Rho proteins (Prior and Hancock 2001). In fact, Rac1 nanoclustering is supported by a FRET-study that investigated the clustered, non-random distribution that is mediated by the HVR of several small GTPases on the plasma membrane. The minimal membrane anchor of Rac1 mediated strong self-clustering, as well as strong co-clustering with the minimal membrane anchor of K-ras4B (Abankwa and Vogel 2007).

Alternatively, the proximity of the G protein to certain membranes could suffice to induce nanoclustering (Abankwa et al. 2007). Computer simulations have shown that clustering or aggregation of proteins on the membrane can be induced by their curvature-inducing activity. This could even lead to membrane invagination and ultimately fission (Reynwar et al. 2007). In that case, nanoclustering would become a fairly widespread and common phenomenon for peripheral membrane proteins. It would also become relevant for Rab proteins, where only a few would fulfill our hypothetical nanoclustering requirements of having a second targeting signal and plasma membrane localization (Table 1.3). So far, direct evidence for Rab nanoclustering is missing. It is possible, that membrane extracting proteins, in particular GDIs for Rhos and Rabs, prevent the formation of nanoclusters or limit their lifetime. Thus, growth and stability of nanoclusters might be limited to the active GTP state of these proteins. It is important to know more about nanoclustering of prenylated small G proteins, and understand more about its possible functional implications. Conversely, identifying common structural requirements will ultimately help to unravel the mechanism of nanoclustering.

Recently, guanine nucleotide dependent membrane orientation of H-Ras was proposed to be a common feature of other Ras isoforms (Abankwa et al. 2008b). Computational simulation and FRET data have shown that the active GTP-H-ras Ras membrane orientation was stabilized by previously unappreciated membrane contacts of helix α4, while GDP-H-Ras did not show these contacts, but stabilization of a different orientation via the HVR (Gorfe et al. 2007). Therefore, the membrane contacting helix α4 of Ras could also impact on lateral segregation (Abankwa et al. 2008b). Based on the fact that helix α4 and the HVR are the most divergent sequence stretches among Ras proteins, it was proposed that isoform diversity could be defined by different membrane orientations of Ras isoforms, which would be tuned by the residues on helix α4 and in the HVR (Abankwa et al. 2008a). Moreover, membrane orientation combines with lateral segregation, suggesting a novel orientation and lateral segregation based codec for Ras isoform diversity (Abankwa et al. 2008a).

In this regard it is intriguing, that the Rho insert region is located just before the sequence stretch comprising helix α4 (Vetter and Wittinghofer 2001). Moreover, deletion or mutation of this region affects effector (Freeman et al. 1996; Nisimoto et al. 1997; Wu et al. 1997; Zong et al. 1999) or RhoGDI interactions (Wu et al. 1997), and signaling of Rho protein (Joneson and Bar-Sagi 1998). However, none of the Rho complexes with any of these interaction partners showed contacts with the insert region, maybe with the exception of some GAP-complexes (reviewed in Dvorsky and Ahmadian 2004). We propose that these data can be reconciled by a reorientation mechanism of Rho proteins that is similar to that of H-Ras. Such a mechanism would again explain Rho isoform diversity and specific functioning, by the orientation, which is tuned by variation of the residues on the balance- or switched-elements, helix α4 and the HVR. For Rho proteins, the HVR could be one such balance-element, while the Rho-insert region and helix α4 might be the other. Likewise, the only sequence stretch outside the C-terminal HVR of several Rab proteins, which is not conserved among the Rab family or subfamily members, is a sequence stretch spanning helix α4 (Merithew et al. 2001). This is again an intriguing parallel to the sequence variations in Ras and now in Rho. It will be interesting to know, whether membrane orientation is also functionally important for Rho or Rab proteins.

1.5 Structural Features of the Prenyltransferases

As mentioned above, protein prenyltransferases are responsible for the post-translational prenylation of small GTPases. All three protein prenyltransferases are α, β heterodimeric enzymes. FTase and GGTase-I share the same 44 kDa α subunit (Seabra et al. 1991), which consists of 14 α helices folded into 7 successive pairs to form a series of right-handed antiparallel coiled coils known as "helical hairpins" (Park and Beese 1997; Taylor et al. 2003).The helical hairpins form a crescent-shaped super-helix that wraps around the β subunit (Fig. 1.6a) (Lane and Beese 2006). The first 50 N-terminal amino acids of the α subunit form a proline-rich disordered domain, that is not involved in catalysis and does not influence the core enzyme structure (Andres et al. 1993a; Dunten et al. 1998).

The α subunit of RabGGTase contains both an Ig-like and a leucine-rich repeat (LRR) domain in addition to the helical domain. Despite only 22% sequence identity, the helical domain of the RabGGTase α subunit is structurally similar to that of FTase/GGTase-I (Fig. 1.6b) (Zhang et al. 2000). The function of the Ig-like and LRR domains remains unknown but as they are absent in the structures found in lower eukaryotes they are not believed to be involved in catalysis (Dursina et al. 2002; Pylypenko et al. 2003).

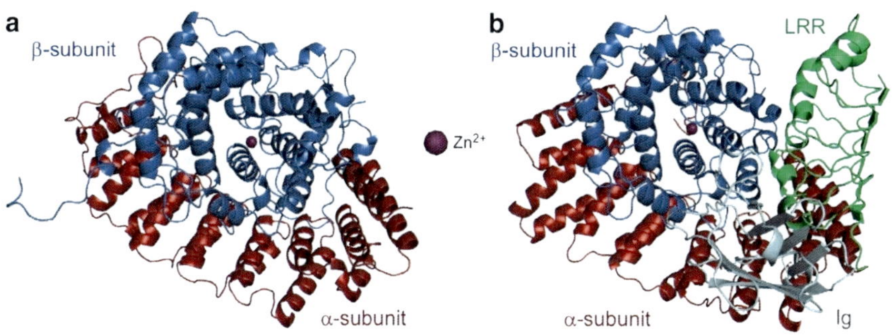

Fig. 1.6 Overall structures of FTase (**a**) and RabGGTase (**b**). *Red*: α-subunit, *blue*: β-subunit, *grey*: Ig-like domain, *green*: LRR-domain, *purple*: Zn^{2+} ion

The β subunits of all three prenyltransferases share only 25% sequence identity but are also structurally similar (Fig. 1.6), consisting of 14, 13, or 12 α helices in FTase, GGTase-I and RabGGTase, respectively. The helices fold into an α-α barrel, forming a 15 Å funnel-shaped hydrophobic cavity containing several conserved aromatic residues. This cavity represents the active site of the enzymes and contains the lipid and protein substrate binding sites. The α and β subunits form an extensive interface burying approximately 20% of the accessible surface area of each subunit (Park et al. 1997; Taylor et al. 2003).

1.5.1 Zn^{2+} Binding Site

All three prenyltransferases are zinc metalloenzymes with a single Zn^{2+} coordinated adjacent to the α, β subunit interface (Fig. 1.6) (Reiss et al. 1992; Taylor et al. 2003). Zn^{2+} is coordinated by three strictly conserved residues in the β-subunit; D297, C299 and H362 in FTase; D269, C271 and H321 in GGTase-I; and D238, C240 and H290 in RabGGTase. All three crystal structures reveal a stabilizing hydrogen bond between the Zn^{2+}-coordinating histidine residue and a conserved β-subunit aspartic acid residue (D359, D318 and D287 for FTase, GGTase-I and RabGGTase, respectively) (Fig. 1.7). High-resolution substrate and product complexes indicate that the cysteine thiol of the CAAX motif interacts with Zn^{2+} at

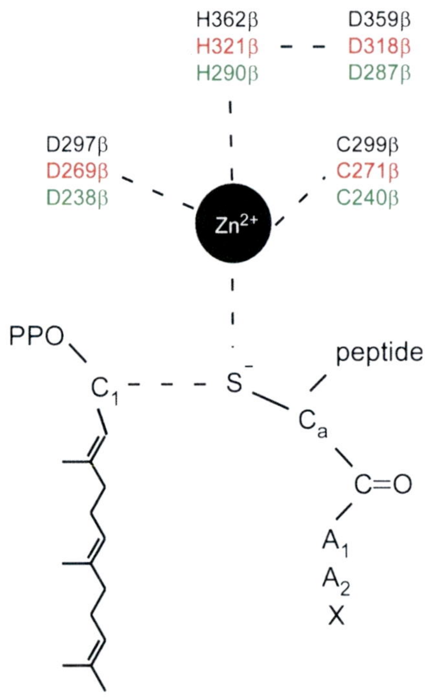

Fig. 1.7 Zn^{2+} coordination. *Black*: FTase, *red*: GGTase-I, and *blue*: RabGGTase

2.3 Å distance in the substrate-bound and 2.6 Å distance in the product-bound state, resulting in tetravalent coordination of the Zn^{2+} ion.

1.5.2 Lipid Substrate Binding

The prenyl substrates of CAAX prenyltransferases become buried adjacent to the peptide binding site in the hydrophobic cavity of the β subunit (Long et al. 1998; Taylor et al. 2003). The pyrophosphate moiety of FPP and GGPP becomes sequestered within the diphosphate binding pocket of FTase and GGTase-I. In the FTase β-subunit, residues R291 and K294 interact with the β-phosphate of FPP and Y300 forms a hydrogen bond with the α-phosphate of FPP (Pickett et al. 2003) (corresponding residues in GGTase-I are R263, K266, and Y272) (Fig. 1.8). Both the pyrophosphate and first three isoprene units of FPP and GGPP bind to FTase and GGTase-I in a similar conformation which is mediated by analogous hydrophobic interactions with conserved aromatic residues in their active sites. The additional fourth isoprene unit of GGPP is orientated 90° to the rest of the molecule (Fig. 1.9). The active site of RabGGTase is structurally similar, but the orientation of the terminal isoprene unit of GGPP differs when sequestered by RabGGTase compared to GGTase-I (Guo et al. 2008).

1.5.3 Isoprenoid Substrate Specificity

Despite prenyltransferases having nanomolar affinities for FPP and GGPP, they display vastly different catalytic efficiencies with different isoprenoids. This phenomenon is explained by the amino acid residing at the bottom of their lipid binding sites. Residues 49β (GGTase-I) and 48β (RabGGTase) are always a small

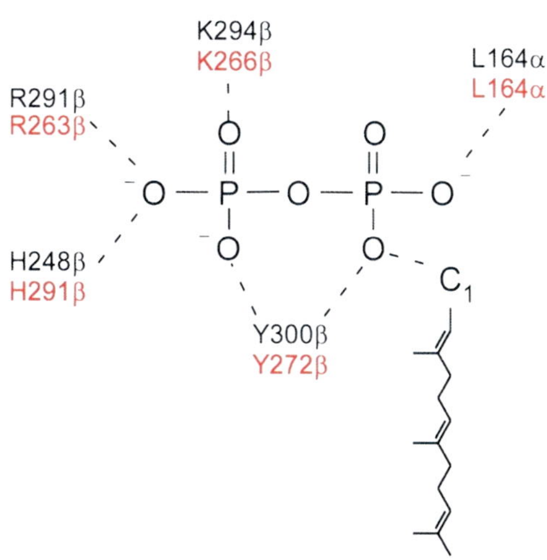

Fig. 1.8 Pyrophosphate coordination. *Black*: FTase, *red*: GGTase-I

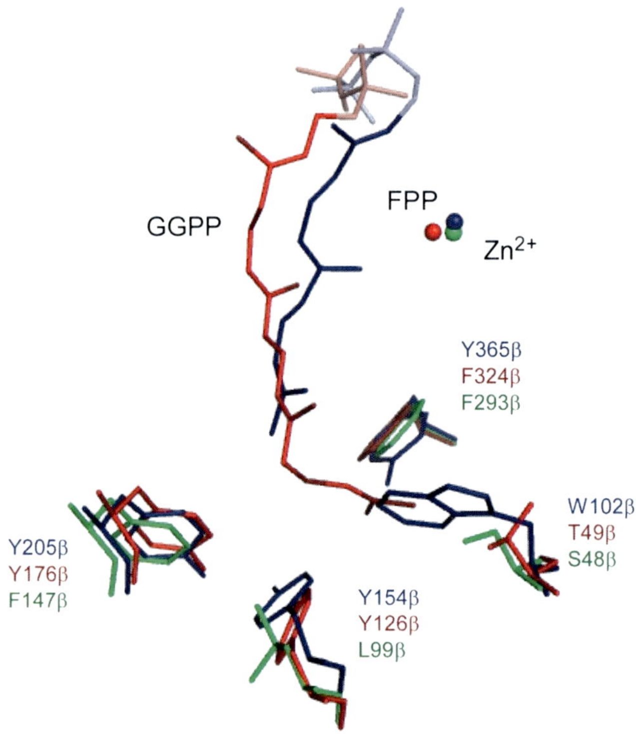

Fig. 1.9 A stick representation of the bottom lipid binding site of protein prenyltransferases. *Blue*: FTase:FPP; *red*: GGTase:GGPP; *green*: RabGGTase; *light red or blue*: pyrophosphate moiety of the respective isoprenoid. For further details, see text

amino acid like Thr or Ser, whereas the corresponding residue in FTase is Trp (W102β). The bulky side chain of W102β occupies the space used by the additional fourth isoprene unit in the GGTase-I:GGPP complex (Fig. 1.9). It has been demonstrated that this amino acid alone determines whether CAAX prenyltransferases accept FPP or GGPP as a substrate. Indeed a single W102Tβ mutation interconverts the lipid substrate preference of FTase, without changing the peptide substrate selectivity (Taylor et al. 2003). In RabGGTase the bottom of the lipid binding site is further enlarged, since Y154β (FTase) and Y126β (GGTase-I) are replaced by a smaller leucine residue (L99β) at the corresponding position (Fig. 1.9) (Leung et al. 2006, 2007).

Together these observations resulted in Beese and co-workers postulating the "molecular ruler hypothesis" as a model to explain lipid substrate discrimination by FTase and GGTase-I. This hypothesis simply postulates that the depth of the lipid binding cavity acts as a length-discriminating molecular ruler. Larger noncognate isoprenoids would fill up the lipid binding pocket such that the diphosphate and C1 of the isoprenoid are positioned incorrectly for efficient transfer onto peptide substrates (Taylor et al. 2003) (Fig. 1.10c). However by solving the structure of FTase

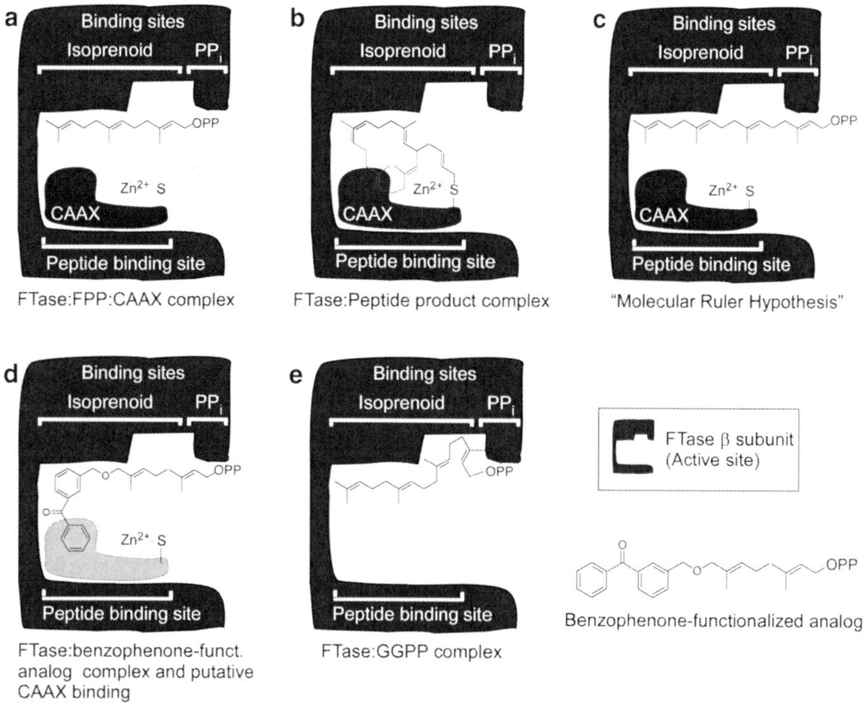

Fig. 1.10 Schematic comparison of the molecular ruler hypothesis and the second site exclusion model of lipid substrate discrimination by FTase derived from high-resolution crystal structures of FTase:substrate complexes. The β subunit of FTase (*purple*) contains the active site with the binding sites of the pyrophosphate, isoprene, and the peptide substrate. The CAAX protein/peptide is shown in *black* and comprises the nucleophilic thiol(ate), which coordinates Zn^{2+}. (**b**) Farnesylated peptide product (*black*) bound to FTase. Superimposed in *red* is the lipid conformation of the putative geranylgeranylated peptide, which clashes with peptide binding. (**c**) "Molecular Ruler" model proposed by Taylor et al. (2003). (**d**) Model of benzophenone geranyl pyrophosphate bound to FTase (Turek-Etienne et al. 2003). Superimposed is a putatively bound CAAX peptide, which leads to steric problems and prevents the formation of the catalytic ternary complex. (**e**) Model of GGPP bound to FTase. The additional units of the isoprenoid chain form a bulge. Adapted from Turek-Etienne et al., 2003

in complex with GGPP and several other lipid analogs, Distefano and co-workers revealed that the pyrophosphate and C1 atom of both GGPP and the benzophenone-functionalized GGPP are correctly orientated with respect to the FTase:FPP complex (Fig. 1.10d, e). In both these structures the conformation of the isoprene is perturbed, resulting in steric hindrance of correct CAAX peptide binding to the active site. The effect of this is to impede the critical C1 move required for complex formation between the isoprene and CAAX peptide (see Sect. 1.3.1). Based on these structures the authors proposed the "second site exclusion model" to explain the substrate specificity of FTase (Turek-Etienne et al. 2003). This model postulates that the isoprenoid substrate specificity of FTase is not mediated by misalignment of the pyrophosphate group but by either the inhibition of ternary complex formation (Fig. 1.10d) or the failure of the ternary complex to convert into product (Fig. 1.10b). The recently solved structure of FTase in complex with bulky biotin-geranyl pyrophosphate supports this model (Nguyen et al. 2009).

1.5.4 Peptide Binding and Substrate Specificity

Adjacent to the lipid substrate in the active site, the CA_1A_2X peptide forms direct van der Waals interactions with the second and third isoprene units of the lipid (Furfine et al. 1995; Pompliano et al. 1993). Extensive investigation of the CAAX requirements for FTase and GGTase-I interaction has revealed that the A_1 residue is solvent-exposed in both enzymes, allowing any amino acid to be tolerated at this position (Reid et al. 2004; Strickland et al. 1998). In contrast the A_2 residue is buried, forming a hydrophobic interaction with the "A_2 binding pocket" (Fig. 1.11). This constrains the A_2 residue to Ile, Val, Leu, Phe, Tyr, Pro, Thr or Met, although beyond this the identity of the A_2 residue has minimal effect on CAAX specificity. The terminal X residue is the primary determinant of enzyme:CAAX specificity

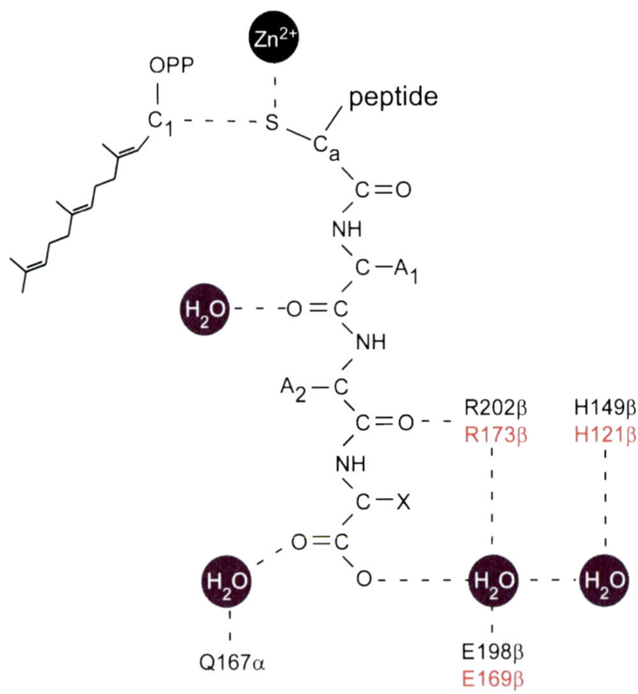

Fig. 1.11 Interaction of CAAX peptide with the active site of prenyltransferases. *Black*: FTase, *red*: GGTase-I

and binds to the "specificity pocket" at the bottom of the peptide substrate binding site. The polar FTase specificity pocket accepts Met (hydrophobic AA), Ser (small AA), Glu and Gln (polar AA) (and to a lesser extent Ala, Thr, and Cys), through a specific network of electrostatic interactions.

Interestingly despite Phe being too bulky for the specificity pocket, it is still tolerated as a viable X residue by binding within an adjacent hydrophobic cavity (Reid and Beese 2004). The crystal structure of GGTase-I reveals only one X-binding site that discriminates against polar, charged and small amino acids. Consequently the only X residues that can be accommodated are hydrophobic amino acids such as Leu or Phe and occasionally Ile or Val. The CAAX peptide is also anchored to FTase/GGTase-I through both the coordination of the cysteine thiol by Zn^{2+} and water-mediated hydrogen bonds involving the C-terminal carboxyl group (Strickland et al. 1998).

As a result of the above complex binding requirements, FTase and GGTase-I are able to self-discriminate against non-CAAX peptide substrates. In contrast RabGGTase cannot directly recognize and bind its substrate peptides but requires the assistance of REP-1/2 (Andres et al. 1993b; Alexandrov et al. 1999; Seabra et al. 2002). The need for REP-1/2 is presumably due to the diverse range of prenylation motifs RabGGTase is required to accommodate. In most cases Rab proteins do not contain a short peptide recognition sequence such as the CAAX motif but are doubly prenylated at –CC, –CXC, –CCX, –CCXX or –CCXXX. Further diversity exists with some RabGTPases mono-prenylated at a CXXX sequence and in the case of Rab8, at a CVLL C-terminal CAAX motif. Interestingly despite being a consensus GGTase-I-competent CAAX motif, GGTase-I only contributes to approximately 30% of Rab8 geranylgeranylation in vivo (Wilson et al. 1998). This may be due to Rab8 sequestration by REP-1/2.

1.5.5 Mechanism of the Prenylation Reaction

Given the importance of prenylated proteins in many signaling pathways, the mechanisms of prenylation by all three protein prenyltransferases have been exhaustively studied using multiple approaches. The prenylation reactions catalyzed by FTase and GGTase-I are very similar, employing a mainly associative mechanism with an "exploded" transition state proposed (Fig. 1.12) (Stirtan and Poulter 1997; Yokoyama et al. 1995; Zhang et al. 1994).

FPP first binds to the apo-form of FTase in the hydrophobic funnel of the active site. FPP binding is subsequently followed by CAAX peptide association. The reactive cysteine thiol is coordinated by the Zn^{2+}, which lowers the pKa of the prenylation reaction and activates the thiol for the nucleophilic attack (Huang et al. 1997; Tobin et al. 2003). Comparison of the structures of CAAX prenyltransferases in complex with isoprenoids and CAAX peptides, before and after prenylation, revealed that the conformation of the peptide backbone remains almost identical following prenylation. It is believed that during the reaction, the lipid substrate

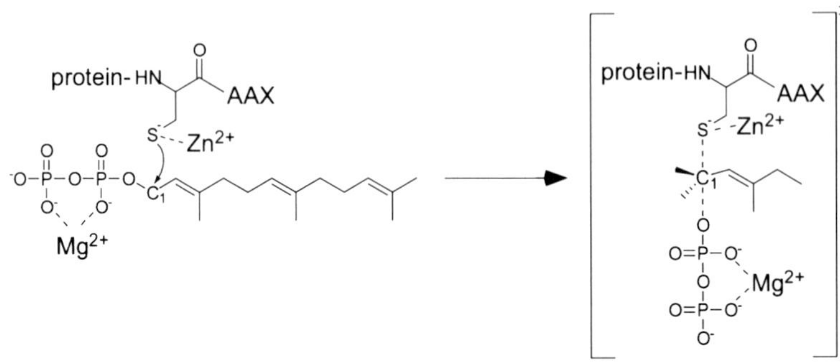

Fig. 1.12 Transition state model of the farnesylation reaction catalyzed by FTase. Adapted from Pais et al., 2007

rotates at the third isoprene unit to move the C1 atom nearer to the activated CAAX thiolate (Fig. 1.10b) (Long et al. 2002). During prenylation by GGTase-I, the GGPP rotates at the fourth isoprene unit (Taylor et al. 2003). Upon release of pyrophosphate the CAAX peptide and the farnesyl or geranylgeranyl moiety irreversibly form a stable thioether bond. Efficient catalysis by FTase is also dependent on Mg^{2+} which is coordinated by residue D352 of the β-subunit (Hartman et al. 2004; Reiss et al., 1992). GGTase-I acts independently of Mg^{2+}, with the Mg^{2+}-coordinating property of D352β is replaced by a lysine residue (L311β). The positive charge of the coordinated Mg^{2+} and L311β stabilize the developing negative charge of the FPP and GGPP diphosphate group respectively (Bowers and Fierke 2004; Hartman et al. 2004). Kinetic studies suggest that the rate-limiting step of CAAX prenylation is product release, which is stimulated by competitive binding of a new isoprenoid substrate to the enzyme. This results in the CAAX-attached lipid to redistribute into an "exit groove" on the β subunit and subsequently be released from the enzyme (Long et al. 2002; Taylor et al. 2003). The above release mechanism is supported by numerous studies but recently Troutman and co-workers postulated a more complex mechanism whereby an incoming CAAX peptide stimulated product release (Troutman et al. 2007).

The prenylation reaction catalyzed by RabGGTase substantially differs from FTase and GGTase-I, since RabGGTase requires the assistance of REP-1/2 for Rab recruitment and di-prenylation is often required. It remains unclear how the ternary complex necessary for catalysis is assembled but two possible pathways have been proposed (Shen and Seabra 1996; Thoma et al. 2001). In the classical pathway unprenylated Rab first interacts with REP-1/2 followed by RabGGTase binding. However the observation that the affinity of REP for RabGGTase increases upon RabGGTase binding either Rab or GGPP, led to the proposal of an alternative pathway in which REP binds to GGPP bound RabGGTase first followed by the association of Rab (Baron and Seabra 2008; Dursina et al. 2002).

The mechanism of RabGGTase-mediated prenylation has been studied biochemically and structurally (Guo et al. 2008; Pylypenko et al. 2003). The crystal structure of the Rab:REP complex reveals two binding interfaces between (1) the Rab binding platform (RBP) of REP and the effector loops of Rab and (2) the

C-terminal binding region (CBR) of REP and the highly conserved C-terminal CBR interacting motif (CIM) of Rab (Alory and Balch 2000; Rak et al. 2004). Biochemical studies have demonstrated that formation of the RabGGTase:REP complex is mediated through the interaction of two critical residues of REP-1 (Phe279 and Arg290) with the α-subunit of RabGGTase (Pylypenko et al. 2003). While a structure of the ternary complex has not been solved to date, modeling and biophysical studies suggest that weak associations form between the flexible C-terminus of Rab and RabGGTase. This possibly stimulates the presentation of a prenylatible Rab cysteine to the Zn^{2+}-containing active site of RabGGTase (Wu et al. 2009). REP-1 contains a five-helix lipid binding site in domain II. Upon prenyl transfer to Rab and the binding of a new isoprenoid to RabGGTase, the conjugated isoprenes move from the hydrophobic funnel in the active site of RabGGTase to the lipid binding site of REP-1. This leads to the expansion of the REP-1 lipid binding cavity and the re-orientation of Phe279 and Arg290. The reorientation of REP-1 results in a decrease in the affinity of RabGGTase for the prenylated Rab:REP complex and subsequent product release from RabGGTase.

1.6 Protein Prenyltransferases as Therapeutic Targets

As described in previous sections the GTPases of the Ras superfamily are key players in many signaling and trafficking pathways. Consequently it is not surprising that expression defects or mutations of the GTPases themselves, their effectors, regulators, or posttranslational modification machinery, can result in deregulation of signaling networks that often leads to abnormalities in humans.

Ras proteins have been linked to cancer due to the discovery of constitutively active Ras isoforms in human tumors (McBride et al. 1982; Der et al. 1982; Parada et al. 1982). This is expected, given the involvement of Ras proteins in controlling cell proliferation (Gelb et al. 2006). Indeed approximately 20% of all currently known human cancers contain an oncogenic mutation in a Ras gene but the isoforms involved and the incidence of mutations differ for various cancers (Karnoub and Weinberg 2008). Point mutations in Ras are most frequently found in genes coding for K-Ras (~85%), followed by N-Ras (~5%) and H-Ras (<1%). Commonly oncogenic Ras mutations render Ras proteins insensitive to the action of GAPs, locking Ras proteins in the GTP-bound state where they constitutively activate downstream signaling pathways (Barbacid 1987; Lowy et al. 1993).

Rho proteins have also been implicated in various types of cancer (Heasman and Ridley 2008) where they can promote both tumor progression (RhoC) and in the cases of RhoB and RhoBTB2, act as tumor suppressors (Berthold et al. 2008; Huang and Prendergast 2006; Vega and Ridley 2008). The tumor-promoting function of RhoC appears to result from its over-expression and correlates with increased metastasis (Vega and Ridley 2008; Wu et al. 2004). Evidence suggests that the tumor suppressor activity of RhoBTB2 may be due to the ability of its BTB domain to form an E3 ligase complex with Cullin3. Cullin3 regulates proteasomal degradation of the

RhoBTB/Cullin3 complex and possibly of other proteins as well. Interestingly this activity is abolished by the D299N-mutation in RhoBTB2 that has been found in some breast cancers (Berthold et al. 2008). The tumor suppressor activity of RhoB appears to involve the negative regulation of TGFβ receptor expression (Adnane et al. 2002). In addition to cancer, deregulated Rho proteins have also been associated with vascular diseases, due to their role in the reorganization of the actin cytoskeleton and other important cellular processes (Rolfe et al. 2005).

Alterations in signaling by RabGTPases have also been observed in human disease (Stein et al. 2003). For example, loss of function mutations in the Rab27A gene causes Griscelli syndrome type 2. This autosomal recessive disorder is characterized by immune impairment, increased susceptibility to infection and partial albinism, due to defects in the Rab27-mediated melanosome trafficking (Griscelli et al. 1978). Point mutation(s) in Rab regulatory molecules that cause deregulation of Rab signaling have also been associated with disease. Loss-of-function mutations in REP-1 have been associated with retinal degeneration in the X-linked disease choroideremia. Absence of REP-1 correlates with an accumulation of unprenylated Rab27 and aberrant melanosome trafficking (van den Hurk et al. 1997). Other diseases associated with genetic defects in Rab regulatory proteins include X-linked mental retardation due to genetic defects in RabGDI (D'Adamo et al. 1998) and kidney disease in tuberous sclerosis, due to mutations in the RabGAP TCS1 (van Slegtenhorst et al. 1997). Despite the fact that Rab proteins are not usually considered as oncogenic products, the emerging association of deregulated endocytosis and cancer potentially places Rabs as key mediators of tumorigenesis. Rab25 overexpression in advanced ovarian, prostate, breast and transitional cell bladder cancers has been correlated with increased anchorage-independent growth and tumor cell invasion (Cheng et al. 2004). The latter is probably due to increased delivery of integrin α5β1 to the leading edge (Caswell et al. 2007; Cheng et al. 2004). Other Rab proteins known to be over-expressed in human cancers are Rabs1, 2, 3, 4B, 5A, 5B, 7, 10, 20, 22A, 22B, 23, 24 and 31 (Chia and Tang 2009). However to date no functional link between any of these Rab proteins and tumorigenesis has been established. Rab8a/MEL has also been implicated in human cancer, as it is located in a common region of translocation breakpoints in a number of malignancies (Nimmo et al. 1991). However, again a direct link between Rab8a and tumorigenesis has not been established.

Key to the action of Ras superfamily GTPases is their organelle-specific compartmentalization, which is in turn mediated by their posttranslational modification status. Interest in the therapeutic targeting of protein prenyltransferases stems from the seminal observation that membrane detachment of Ras through inhibition of FTase can reverse the transformed phenotypes of cancer cells. Consequently a variety of small-molecule inhibitors of FTase (FTIs) have been developed, including those that compete with the CAAX motif, the lipid substrate or both (Fig. 1.13) (Basso et al. 2006; Tamanoi et al. 2001). Among FTIs three compounds, tipifarnib, lonafarnib, and BMS214662, are currently being assessed as suitable anticancer drugs in clinical trials. Although it has been demonstrated that FTIs are effective in the treatment of hematologic malignancies (Karp 2001), their therapeutic effects on solid tumors is limited (Macdonald et al. 2005). A possible explanation for this

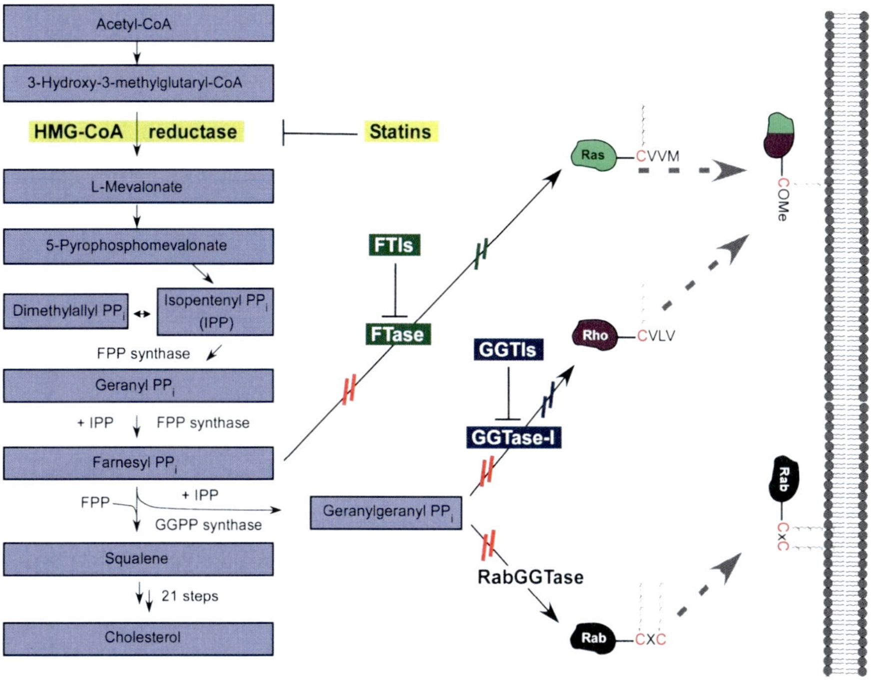

Fig. 1.13 The pathway of cholesterol and prenylated protein synthesis. Protein prenylation can be specifically inhibited by FTIs (*green*) and GGTIs (*blue*), which target the prenylation enzymes. Without isoprene modification, the protein substrates of FTase and GGTase are incapable of associating with membranes and thus cannot exert their role in signaling and trafficking pathways. Alternatively, prenylation can be modulated by statins that inhibit the conversion of HMG-CoA to L-mevalonate through competitive inhibition of the rate-limiting enzyme HMG-CoA reductase. As prenyltransferase-unspecific drugs, their action leads to a decrease in the downstream biosynthesis of cholesterol and other intermediate metabolites including FPP and GGPP (*yellow*). Consequently, the prenylation machinery of all three enzymes is shut down due to a lack of lipid substrates

limitation came from preclinical testing which revealed that the efficacy of FTIs does not entirely correlate with the presence or absence of oncogenic Ras mutations (Sepp-Lorenzino et al. 1995). This suggested that FTIs do not effectively inhibit the function of K-Ras4B which is the most commonly mutated Ras isoform in solid tumors (Karnoub and Weinberg 2008). Evidence has since accumulated that in FTI-treated cells, K-Ras4B and N-Ras can be alternatively prenylated by GGTase-I (Lerner et al. 1997b). This led to the recognition of GGTase-I as an additional therapeutic target in cancer and to the development of GGTase-I inhibitors (GGTIs). Additional recognition that GGTIs may be important for effective tumor treatment came from observations that several geranylgeranylated CAAX proteins, including RalA and RhoC, are critical for oncogenesis downstream of Ras (Chien and White 2003; Khosravi-Far et al. 1995). Subsequently GGTIs have been shown to impair tumor growth in vivo and, in contrast to FTIs, cause inhibition of Rho signaling leading to cell cycle arrest and apoptosis (Philips and Cox 2007).

Due to a lack of appropriate analytical methods, the global effects of FTIs and GGTIs on prenylation across the proteome have not been studied. Despite the fact that some FTIs and GGTIs are already used in clinical trials, the identity of the targets that mediate the therapeutic effect of those inhibitors is still disputed. For example compounds BMS1–4, initially developed by Bristol-Myers Squibb as specific FTIs, have reached the market without a complete analysis of potential cross-inhibition. Not surprisingly, Lackner et al. demonstrated that these compounds induced apoptosis by inhibiting RabGGTase rather than FTase (Lackner et al. 2005). To date the true in vivo inhibition efficiencies of these compounds towards FTase and RabGGTase remains unknown, as are other potential targets that may contribute to the clinical outcome of BMS1–4 treatment.

A second group of compounds that modulate prenylation are the cholesterol-lowering drugs known as statins (Fig. 1.13). Statins are first-line therapeutics for the prevention of coronary heart disease and atherosclerosis (Puccetti et al. 2007). By inhibiting the synthesis of mevalonate, statins prevent the formation of FPP which also serves as a precursor for GGPP and cholesterol synthesis. In addition to being clinically used to treat high cholesterol and cardiovascular disease, statins are now being used or tested for the treatment of other diseases including cancer, neurodegenerative disorders and autoimmune diseases (Denoyelle et al. 2003; Greenwood et al. 2006; Konstantinopoulos et al. 2007). It is believed that the therapeutic effects of statins for these disorders are due to their ability to suppress protein prenylation but of which targets and to what extent is also largely unknown.

1.7 Conclusions

Protein prenylation is one of the most extensively studied post-translational modifications, due to its crucial role in the regulation of diverse cellular processes that, when deregulated contribute to numerous human pathologies including cancer. Progress in understanding the mechanisms and function of protein prenylation has revealed a complex interplay of enzymatic mechanisms and subcellular distributions that connect protein prenylation to almost all essential cellular processes. The present review attempts to demonstrate the importance of understanding these processes on the systemic level, in order to explain and predict the effects of therapeutic compounds on global protein prenylation and related signaling cascades.

References

Abankwa, D., Gorfe, A.A., Hancock, J.F. (2007). Ras nanoclusters: molecular structure and assembly. *Semin. Cell Dev. Biol.* 18:599–607.

Abankwa, D., Gorfe, A.A., Hancock, J.F. (2008a). Mechanisms of Ras membrane organization and signalling: Ras on a rocker. *Cell Cycle* 7:2667–2673.

Abankwa, D., Hanzal-Bayer, M., Ariotti, N., et al. (2008b). A novel switch region regulates H-ras membrane orientation and signal output. *EMBO J.* 27:727–735.

Abankwa, D., and Vogel, H. (2007). A FRET map of membrane anchors suggests distinct microdomains of heterotrimeric G proteins. *J. Cell Sci.* 120:2953–2962.

Adnane, J., Seijo, E., Chen, Z., et al. (2002). RhoB, not RhoA, represses the transcription of the transforming growth factor beta type II receptor by a mechanism involving activator protein 1. *J. Biol. Chem.* 277:8500–8507.

Alexandrov, K., Simon, I., Yurchenko, V., et al. (1999). Characterization of the ternary complex between Rab7, REP-1 and Rab geranylgeranyl transferase. *Eur. J. Biochem.* 265:160–170.

Ali, B.R., Wasmeier, C., Lamoreux, L., et al. (2004). Multiple regions contribute to membrane targeting of Rab GTPases. *J. Cell Sci.* 117:6401–6412.

Alory, C., and Balch, W.E. (2000). Molecular basis for Rab prenylation. *J. Cell Biol.* 150:89–103.

Anant, J.S., and Fung, B.K. (1992). *In vivo* farnesylation of rat rhodopsin kinase. *Biochem. Biophys. Res. Commun.* 183:468–473.

Anderegg, R.J., Betz, R., Carr, S.A., et al. (1988). Structure of *Saccharomyces cerevisiae* mating hormone a-factor. Identification of S-farnesyl cysteine as a structural component. *J. Biol. Chem.* 263:18236–18240.

Andres, D.A., Milatovich, A., Ozcelik, T., et al. (1993a). cDNA cloning of the two subunits of human CAAX farnesyltransferase and chromosomal mapping of FNTA and FNTB loci and related sequences. *Genomics* 18:105–112.

Andres, D.A., Seabra, M.C., Brown, M.S., et al. (1993b). cDNA cloning of component A of Rab geranylgeranyl transferase and demonstration of its role as a Rab escort protein. *Cell* 73:1091–1099.

Apolloni, A., Prior, I.A., Lindsay, M., et al. (2000). H-ras but not K-ras traffics to the plasma membrane through the exocytic pathway. *Mol. Cell Biol.* 20:2475–2487.

Ashar, H.R., James, L., Gray, K., et al. (2000). Farnesyl transferase inhibitors block the farnesylation of CENP-E and CENP-F and alter the association of CENP-E with the microtubules. *J. Biol. Chem.* 275:30451–30457.

Barbacid, M. (1987). ras genes. *Annu. Rev. Biochem.* 56:779–827.

Baron, R., Fourcade, E., Lajoie-Mazenc, I., et al. (2000). RhoB prenylation is driven by the three carboxyl-terminal amino acids of the protein: evidenced in vivo by an anti-farnesyl cysteine antibody. *Proc. Natl. Acad. Sci. U.S.A.* 97:11626–11631.

Baron, R.A., and Seabra, M.C. (2008). Rab geranylgeranylation occurs preferentially via the preformed REP-RGGT complex and is regulated by geranylgeranyl pyrophosphate. *Biochem. J.* 415:67–75.

Basso, A.D., Kirschmeier, P., Bishop, W.R. (2006). Lipid posttranslational modifications. Farnesyl transferase inhibitors. *J. Lipid Res.* 47:15–31.

Belanis, L., Plowman, S.J., Rotblat, B., et al. (2008). Galectin-1 is a novel structural component and a major regulator of h-ras nanoclusters. *Mol. Biol. Cell* 19:1404–1414.

Berthold, J., Schenkova, K., Rivero, F. (2008). Rho GTPases of the RhoBTB subfamily and tumorigenesis. *Acta Pharmacol. Sin.* 29:285–295.

Berzat, A.C., Brady, D.C., Fiordalisi, J.J., et al. (2005). Using inhibitors of prenylation to block localization and transforming activity. *Meth. Enzymol.* 407:575–597.

Bivona, T.G., Quatela, S., Philips, M.R. (2006a). Analysis of Ras activation in living cells with GFP-RBD. *Meth. Enzymol.* 407:128–143.

Bivona, T.G., Quatela, S.E., Bodemann, B.O., et al. (2006b). PKC regulates a farnesyl-electrostatic switch on K-Ras that promotes its association with Bcl-XL on mitochondria and induces apoptosis. *Mol. Cell* 21:481–493.

Bos, J.L., Rehmann, H., Wittinghofer, A. (2007). GEFs and GAPs: critical elements in the control of small G proteins. *Cell* 129:865–877.

Bowers, K.E., and Fierke, C.A. (2004). Positively charged side chains in protein farnesyltransferase enhance catalysis by stabilizing the formation of the diphosphate leaving group. *Biochemistry* 43:5256–5265.

Buss, J.E., Quilliam, L.A., Kato, K., et al. (1991). The COOH-terminal domain of the Rap1A (Krev-1) protein is isoprenylated and supports transformation by an H-Ras:Rap1A chimeric protein. *Mol. Cell Biol.* 11:1523–1530.

Bustelo, X.R., Sauzeau, V., Berenjeno, I.M. (2007). GTP-binding proteins of the Rho/Rac family: regulation, effectors and functions in vivo. *Bioessays* 29:356–70.

Calero, M., Chen, C.Z., Zhu, W., et al. (2003). Dual prenylation is required for Rab protein localization and function. *Mol. Biol. Cell* 14:1852–1867.

Casey, P.J., and Seabra, M.C. (1996). Protein prenyltransferases. *J. Biol. Chem.* 271:5289–5292.

Caswell, P.T., Spence, H.J., Parsons, M. (2007). Rab25 associates with alpha5beta1 integrin to promote invasive migration in 3D microenvironments. *Dev. Cell* 13:496–510.

Chan, T.O., Rittenhouse, S.E., Tsichlis, P.N. (1999). AKT/PKB and other D3 phosphoinositide-regulated kinases: kinase activation by phosphoinositide-dependent phosphorylation. *Annu. Rev. Biochem.* 68:965–1014.

Chang, E.C., and Philips, M.R. (2006). Spatial segregation of Ras signaling: new evidence from fission yeast. *Cell Cycle* 5:1936–1939.

Chavrier, P., Gorvel, J.P., Stelzer, E., et al. (1991). Hypervariable C-terminal domain of rab proteins acts as a targeting signal. *Nature* 353:769–772.

Chenette, E.J., Mitin, N.Y., Der, C.J. (2006). Multiple sequence elements facilitate Chp Rho GTPase subcellular location, membrane association, and transforming activity. *Mol. Biol. Cell* 17:3108–3121.

Cheng, K.W., Lahad, J.P., Kuo, W.L., et al. (2004). The RAB25 small GTPase determines aggressiveness of ovarian and breast cancers. *Nat. Med.* 10:1251–1256.

Cherfils, J., and Chardin, P. (1999). GEFs: structural basis for their activation of small GTP-binding proteins. *Trends Biochem. Sci.* 24:306–311.

Chia, W.J., and Tang, B.L. (2009). Emerging roles for Rab family GTPases in human cancer. *Biochim. Biophys. Acta.* 1795:110–116.

Chien, Y., and White, M.A. (2003). RAL GTPases are linchpin modulators of human tumour-cell proliferation and survival. *EMBO Rep.* 4:800–806.

Chook, Y.M., and Blobel, G. (2001). Karyopherins and nuclear import. *Curr. Opin. Struct. Biol.* 11:703–715.

Choy, E., Chiu, V.K., Silletti, J., et al. (1999). Endomembrane trafficking of ras: the CAAX motif targets proteins to the ER and Golgi. *Cell* 98:69–80.

Clarke, S. (1992). Protein isoprenylation and methylation at carboxyl-terminal cysteine residues. *Annu. Rev. Biochem.* 61:355–386.

Clarke, S., Vogel, J.P., Deschenes, R.J., et al. (1988). Posttranslational modification of the Ha-ras oncogene protein: evidence for a third class of protein carboxyl methyltransferases. *Proc. Natl. Acad. Sci. U.S.A.* 85:4643–4647.

Colicelli, J. (2004). Human RAS superfamily proteins and related GTPases. Sci. STKE. 250:RE13.

Cory, G.O., and Cullen, P.J. (2007). Membrane curvature: the power of bananas, zeppelins and boomerangs. *Curr. Biol.* 17:R455–R457.

D'Adamo, P., Menegon, A., Lo, N.C., et al. (1998). Mutations in GDI1 are responsible for X-linked non-specific mental retardation. *Nat. Genet.* 19:134–139 [see erratum in *Nat. Genet.* 1998, Jul;19(3):303].

Dai, Q., Choy, E., Chiu, V., et al. (1998). Mammalian prenylcysteine carboxyl methyltransferase is in the endoplasmic reticulum. *J. Biol. Chem.* 273:15030–15034.

De Smedt, F., Boom, A., Pesesse, X., et al. (1996). Post-translational modification of human brain type I inositol-1,4,5-trisphosphate 5-phosphatase by farnesylation. *J. Biol. Chem.* 271:10419–10424.

Denoyelle, C., Hong, L., Vannier, J.P., et al. (2003). New insights into the actions of bisphosphonate zoledronic acid in breast cancer cells by dual RhoA-dependent and -independent effects. *Br. J. Cancer.* 88:1631–1640.

Der, C.J., Krontiris, T.G., Cooper, G.M. (1982). Transforming genes of human bladder and lung carcinoma cell lines are homologous to the ras genes of Harvey and Kirsten sarcoma viruses. *Proc. Natl. Acad. Sci. U.S.A.* 79:3637–3640.

Devos, S.A., Van Den Bossche, N., De Vos, M., et al. (2003). Adverse skin reactions to anti-TNF-alpha monoclonal antibody therapy. *Dermatology* 206:388–390.

Dudler, T., and Gelb, M.H. (1996). Palmitoylation of Ha-Ras facilitates membrane binding, activation of downstream effectors, and meiotic maturation in Xenopus oocytes. *J. Biol. Chem.* 271:11541–11547.

Dunten, P., Kammlott, U., Crowther, R., et al. (1998). Protein farnesyltransferase: structure and implications for substrate binding. *Biochemistry* 37:7907–7912.

Dursina, B., Thoma, N.H., Sidorovitch, V., et al. (2002). Interaction of yeast rab geranylgeranyl transferase with its protein and lipid substrates. *Biochemistry* 41:6805–6816.

Dvorsky, R., and Ahmadian, M.R. (2004). Always look on the bright site of Rho: structural implications for a conserved intermolecular interface. *EMBO Rep.* 5:1130–1136.

Eggeling, C., Ringemann, C., Medda, R., et al. (2009). Direct observation of the nanoscale dynamics of membrane lipids in a living cell. *Nature* 457:1159–1162.

Elad-Sfadia, G., Haklai, R., Balan, E., et al. (2004). Galectin-3 augments K-Ras activation and triggers a Ras signal that attenuates ERK but not phosphoinositide 3-kinase activity. *J. Biol. Chem.* 279:34922–34930.

Etienne-Manneville, S., and Hall, A. (2002). Rho GTPases in cell biology. *Nature* 420:629–635.

Farnsworth, C.C., Seabra, M.C., Ericsson, L.H., et al. (1994). Rab geranylgeranyl transferase catalyzes the geranylgeranylation of adjacent cysteines in the small GTPases Rab1A, Rab3A, and Rab5A. *Proc. Natl. Acad. Sci. U.S.A.* 91:11963–11967.

Farnsworth, C.C., Wolda, S.L., Gelb, M.H, et al. (1989). Human lamin B contains a farnesylated cysteine residue. *J. Biol. Chem.* 264:20422–20429.

Farrell, F.X., Yamamoto, K., Lapetina, E.G. (1993). Prenyl group identification of rap2 proteins: a ras superfamily member other than ras that is farnesylated. *Biochem. J.* 289(Pt 2):349–355.

Fivaz, M., and Meyer, T. (2005). Reversible intracellular translocation of KRas but not HRas in hippocampal neurons regulated by Ca2+/calmodulin. *J. Cell Biol.* 170:429–441.

Freeman, J.L., Abo, A., Lambeth, J.D. (1996). Rac "insert region" is a novel effector region that is implicated in the activation of NADPH oxidase, but not PAK65. *J. Biol. Chem.* 271:19794–19801.

Fukuda, M., Kanno, E., Ishibashi, K., et al. (2008). Large scale screening for novel rab effectors reveals unexpected broad Rab binding specificity. *Mol. Cell. Proteomics* 7:1031–1042.

Furfine, E.S., Leban, J.J., Landavazo, A., et al. (1995). Protein farnesyltransferase: kinetics of farnesyl pyrophosphate binding and product release. *Biochemistry* 34:6857–6862.

Gelb, M.H. 1997. Protein prenylation, etcetera: signal transduction in two dimensions. *Science* 275:1750–1751.

Gelb, M.H., Brunsveld, L., Hrycyna, C.A., et al. (2006). Therapeutic intervention based on protein prenylation and associated modifications. *Nat. Chem. Biol.* 2:518–528.

Gomes, A.Q., Ali, B.R., Ramalho, J.S., et al. (2003). Membrane targeting of Rab GTPases is influenced by the prenylation motif. *Mol. Biol. Cell* 14:1882–1899.

Goodwin, J.S., Drake, K.R., Rogers, C., et al. (2005). Depalmitoylated Ras traffics to and from the Golgi complex via a nonvesicular pathway. *J. Cell Biol.* 170:261–272.

Gorfe, A.A., Bayer, M.-H, Abankwa, D., et al. (2007). Structure and dynamics of the full-length lipid-modified H-Ras protein in a 1,2-dimyristoylglycero-3-phosphocholine bilayer. *J. Med. Chem.* 50:674–684.

Goswami, D., Gowrishankar, K., Bilgrami, S., et al. (2008). Nanoclusters of GPI-anchored proteins are formed by cortical actin-driven activity. *Cell* 135:1085–1097.

Greenwood, J., Steinman, L., Zamvil, S.S. (2006). Statin therapy and autoimmune disease: from protein prenylation to immunomodulation. *Nat. Rev. Immunol.* 6:358–370.

Griscelli, C., Durandy, A., Guy-Grand, D., et al. (1978). A syndrome associating partial albinism and immunodeficiency. *Am. J. Med.* 65:691–702.

Grosshans, B.L., Ortiz, D., Novick, P. (2006). Rabs and their effectors: achieving specificity in membrane traffic. *Proc. Natl. Acad. Sci. U.S.A.* 103:11821–11827.

Gruenberg, J. (2003). Lipids in endocytic membrane transport and sorting. *Curr. Opin. Cell Biol.* 15:382–388.

Guo, Z., Wu, Y.W., Das, D., et al. (2008). Structures of RabGGTase-substrate/product complexes provide insights into the evolution of protein prenylation. *EMBO J.* 27:2444–2456.

Gutierrez, L., Magee, A.I., Marshall, C.J., et al. (1989). Post-translational processing of p21ras is two-step and involves carboxyl-methylation and carboxy-terminal proteolysis. *EMBO J.* 8:1093–1098.

Hancock, J.F. (2003). Ras proteins: different signals from different locations. *Nat. Rev. Mol. Cell Biol.* 4:373–384.

Hancock, J.F. (2006). Lipid rafts: contentious only from simplistic standpoints. *Nat. Rev. Mol. Cell Biol.* 7:456–462.

Hancock, J.F., Magee, A.I., Childs, J.E., et al. (1989). All ras proteins are polyisoprenylated but only some are palmitoylated. *Cell* 57:1167–1177.

Hancock, J.F., and Parton, R.G. (2005). Ras plasma membrane signalling platforms. *Biochem. J.* 389:1–11.

Hancock, J.F., Paterson, H., Marshall, C.J. (1990). A polybasic domain or palmitoylation is required in addition to the CAAX motif to localize p21ras to the plasma membrane. *Cell* 63:133–139.

Harding, A., and Hancock, J.F. (2008). Ras nanoclusters: combining digital and analog signaling. *Cell Cycle* 7:127–134.

Hartman, H.L., Bowers, K.E., Fierke, C.A. (2004). Lysine beta311 of protein geranylgeranyltransferase type I partially replaces magnesium. *J. Biol. Chem.* 279:30546–30553.

Hartman, H.L., Hicks, K.A., Fierke, C.A. (2005). Peptide specificity of protein prenyltransferases is determined mainly by reactivity rather than binding affinity. *Biochemistry* 44:15314–15324.

Harvey, J.J. (1964). An unidentified virus which causes the rapid production of tumours in mice. *Nature* 204:1104–1105.

Heasman, S.J., and Ridley, A.J. (2008). Mammalian Rho GTPases: new insights into their functions from in vivo studies. *Nat. Rev. Mol. Cell. Biol.* 9:690–701.

Heilmeyer, L.M. Jr., Serwe, M., Weber, C., et al. (1992). Farnesylcysteine, a constituent of the alpha and beta subunits of rabbit skeletal muscle phosphorylase kinase: localization by conversion to S-ethylcysteine and by tandem mass spectrometry. *Proc. Natl. Acad. Sci. U.S.A.* 89:9554–9558.

Heo, W.D., Inoue, T., Park, W.S., et al. (2006). PI(3,4,5)P3 and PI(4,5)P2 lipids target proteins with polybasic clusters to the plasma membrane. *Science* 314:1458–1461.

Huang, C.C., Casey, P.J., Fierke, C.A. (1997). Evidence for a catalytic role of zinc in protein farnesyltransferase. Spectroscopy of Co2+-farnesyltransferase indicates metal coordination of the substrate thiolate. *J. Biol. Chem.* 272:20–23.

Huang, M., and Prendergast, G.C. (2006). RhoB in cancer suppression. *Histol. Histopathol.* 21:213–218.

Inglese, J., Koch, W.J., Caron, M.G., et al. (1992). Isoprenylation in regulation of signal transduction by G-protein-coupled receptor kinases. *Nature* 359:147–150.

Jacobson, K., Mouritsen, O.G., Anderson, R.G. (2007). Lipid rafts: at a crossroad between cell biology and physics. *Nat. Cell Biol.* 9:7–14.

Jaffe, A.B., and Hall, A. (2005). Rho GTPases: biochemistry and biology. *Annu. Rev. Cell. Dev. Biol.* 21:247–269.

Jefferson, A.B., Majerus, P.W. (1995). Properties of type II inositol polyphosphate 5-phosphatase. *J. Biol. Chem.* 270:9370–9377.

John, J., Rensland, H., Schlichting, I., et al. (1993). Kinetic and structural analysis of the Mg(2+)-binding site of the guanine nucleotide-binding protein p21H-ras. *J. Biol. Chem.* 268:923–929.

Joneson, T., and Bar-Sagi, D. (1998). A Rac1 effector site controlling mitogenesis through superoxide production. *J. Biol. Chem.* 273:17991–17994.

Karnoub, A.E., and Weinberg, R.A. (2008). Ras oncogenes: split personalities. *Nat. Rev. Mol. Cell Biol.* 9:517–531.

Karp, J.E. (2001). Farnesyl protein transferase inhibitors as targeted therapies for hematologic malignancies. *Semin. Hematol.* 38:16–23.

Khosravi-Far, R., Solski, P.A., Clark, G.J., et al. (1995). Activation of Rac1, RhoA, and mitogen-activated protein kinases is required for Ras transformation. *Mol. Cell Biol.* 15:6443–6453.

Kinsella, B.T., Erdman, R.A., Maltese, W.A. (1991). Carboxyl-terminal isoprenylation of ras-related GTP-binding proteins encoded by rac1, rac2, and ralA. *J. Biol. Chem.* 266:9786–9794.

Kinsella, B.T., and Maltese, W.A. (1991). rab GTP-binding proteins implicated in vesicular transport are isoprenylated *in vitro* at cysteines within a novel carboxyl- terminal motif. *J. Biol. Chem.* 266:8540–8544.

Konstantinopoulos, P.A., Karamouzis, M.V., Papavassiliou, A.G. (2007). Post-translational modifications and regulation of the RAS superfamily of GTPases as anticancer targets. *Nat. Rev. Drug Discov.* 6:541–555.

Kontani, K., Tada, M., Ogawa, T., et al. (2002). Di-Ras, a distinct subgroup of ras family GTPases with unique biochemical properties. *J. Biol. Chem.* 277:41070–41078.

Kutzleb, C., Sanders, G., Yamamoto, R., et al. (1998). Paralemmin, a prenyl-palmitoyl-anchored phosphoprotein abundant in neurons and implicated in plasma membrane dynamics and cell process formation. *J. Cell Biol.* 143:795–813.

Lackner, M.R., Kindt, R.M., Carroll, P.M., et al. (2005). Chemical genetics identifies Rab geranylgeranyl transferase as an apoptotic target of farnesyl transferase inhibitors. *Cancer Cell* 7:325–336.

Lai, R.K., Perez-Sala, D., Canada, F.J., et al. (1990). The gamma subunit of transducin is farnesylated. *Proc. Natl. Acad. Sci. U.S.A.* 87:7673–7677.

Lane, K.T., and Beese, L.S. (2006). Thematic review series: lipid posttranslational modifications. Structural biology of protein farnesyltransferase and geranylgeranyltransferase type I. *J. Lipid Res.* 47:681–699.

Lerner, E.C., Hamilton, A.D., Sebti, S.M. (1997a). Inhibition of Ras prenylation: a signaling target for novel anti-cancer drug design. *Anticancer Drug Des.* 12:229–238.

Lerner, E.C., Zhang, T.T., Knowles, D.B., et al. (1997b). Inhibition of the prenylation of K-Ras, but not H- or N-Ras, is highly resistant to CAAX peptidomimetics and requires both a farnesyltransferase and a geranylgeranyltransferase I inhibitor in human tumor cell lines. *Oncogene* 15:1283–1288.

Leung, K.F., Baron, R., Ali, B.R., et al. (2007). Rab GTPases containing a CAAX motif are processed post-geranylgeranylation by proteolysis and methylation. *J. Biol. Chem.* 282:1487–1497.

Leung, K.F., Baron, R., Seabra, M.C. (2006). Thematic review series: lipid posttranslational modifications. Geranylgeranylation of Rab GTPases. *J. Lipid Res.* 47:467–475.

Leventis, R., and Silvius, J.R. (1998). Lipid-binding characteristics of the polybasic carboxy-terminal sequence of K-ras4B. *Biochemistry* 37:7640–7648.

Lichtenberg, D., Goni, F.M., Heerklotz, H. (2005). Detergent-resistant membranes should not be identified with membrane rafts. *Trends Biochem. Sci.* 30:430–436.

Long, S.B., Casey, P.J., Beese, L.S. (1998). Cocrystal structure of protein farnesyltransferase complexed with a farnesyl diphosphate substrate. *Biochemistry* 37:9612–9618.

Long, S.B., Casey, P.J., Beese, L.S. (2002). Reaction path of protein farnesyltransferase at atomic resolution. *Nature* 419:645–650.

Lowy, D.R., Johnson, M.R., DeClue, J.E., et al. (1993). Cell transformation by ras and regulation of its protein product. *Ciba Found. Symp.* 176:67–80; discussion 80–84.

Lutz, R.J., McLain, T.M., Sinensky, M. (1992). Feedback inhibition of polyisoprenyl pyrophosphate synthesis from mevalonate *in vitro*. Implications for protein prenylation. *J. Biol. Chem.* 267:7983–7986.

Macdonald, J.S., McCoy, S., Whitehead, R.P., et al. (2005). A phase II study of farnesyl transferase inhibitor R115777 in pancreatic cancer: a Southwest oncology group (SWOG 9924) study. *Invest. New Drugs* 23:485–487.

Maltese, W.A., and Sheridan, K.M. (1987). Isoprenylated proteins in cultured cells: subcellular distribution and changes related to altered morphology and growth arrest induced by mevalonate deprivation. *J. Cell Physiol.* 133:471–481.

Marrari, Y., Crouthamel, M., Irannejad, R., et al. (2007). Assembly and trafficking of heterotrimeric G proteins. *Biochemistry* 46:7665–7677.

Maurer-Stroh, S., Washietl, S., Eisenhaber, F. (2003). Protein prenyltransferases. *Genome Biol.* 4:212.

McBride, O.W., Swan, D.C., Santos, E., et al. (1982). Localization of the normal allele of T24 human bladder carcinoma oncogene to chromosome 11. *Nature* 300:773–774.

McPherson, R.A., Harding, A., Roy, S., et al. (1999). Interactions of c-Raf-1 with phosphatidylserine and 14-3-3. *Oncogene* 18:3862–3869.

Merithew, E., Hatherly, S., Dumas, J.J., et al. (2001). Structural plasticity of an invariant hydrophobic triad in the switch regions of Rab GTPases is a determinant of effector recognition. *J. Biol. Chem.* 276:13982–13988.

Michaelson, D., Silletti, J., Murphy, G., et al. (2001). Differential localization of Rho GTPases in live cells: regulation by hypervariable regions and RhoGDI binding. *J. Cell Biol.* 152:111–126.

Mitchell, D.A., Farh, L., Marshall, T.K., et al. (1994). A polybasic domain allows nonprenylated Ras proteins to function in Saccharomyces cerevisiae. *J. Biol. Chem.* 269:21540–21546.

Mor, A., and Philips, M.R. (2006). Compartmentalized Ras/MAPK signaling. *Annu. Rev. Immunol.* 24:771–800.

Nguyen, U.T., Guo, Z., Delon, C. et al. (2009). Analysis of the eukaryotic prenylome by isoprenoid affinity tagging. *Nat. Chem. Biol.* 5:227–235.

Nimmo, E.R., Sanders, P.G., Padua, R.A. et al. (1991). The MEL gene: a new member of the RAB/YPT class of RAS-related genes. *Oncogene* 6:1347–1351.

Nisimoto, Y., Freeman, J.L., Motalebi, S.A., et al. (1997). Rac binding to p67(phox). Structural basis for interactions of the Rac1 effector region and insert region with components of the respiratory burst oxidase. *J. Biol. Chem.* 272:18834–18841.

Nomura, K., Kanemura, H., Satoh, T., et al. (2004). Identification of a novel domain of Ras and Rap1 that directs their differential subcellular localizations. *J. Biol. Chem.* 279:22664–226673.

Novick, P., and Zerial, M. (1997). The diversity of Rab proteins in vesicle transport. *Curr. Opin. Struct. Cell Biol.* 9:496–504.

Parada, L.F., Tabin, C.J., Shih, C. (1982). Human EJ bladder carcinoma oncogene is homologue of Harvey sarcoma virus ras gene. *Nature* 297:474–478.

Park, H.W., and Beese, L.S. (1997). Protein farnesyltransferase. *Curr. Opin. Struct. Biol.* 7:873–880.

Park, H.W., Boduluri, S.R., Moomaw, J.F., et al. (1997). Crystal structure of protein farnesyltransferase at 2.25 angstrom resolution. *Science* 275:1800–1804.

Paz, A., Haklai, R., Elad-Sfadia, G., et al. (2001). Galectin-1 binds oncogenic H-Ras to mediate Ras membrane anchorage and cell transformation. *Oncogene* 20:7486–7493.

Pereira-Leal, J.B., and Seabra, M.C. (2000). The mammalian Rab family of small GTPases: definition of family and subfamily sequence motifs suggests a mechanism for functional specificity in the Ras superfamily. *J. Mol. Biol.* 301:1077–1087.

Pfeffer, S. (2003). Membrane domains in the secretory and endocytic pathways. Cell 112:507–517.

Pfeffer, S., and Aivazian, D. (2004). Targeting Rab GTPases to distinct membrane compartments. *Nat. Rev. Mol. Cell Biol.* 5:886–896.

Pfeffer, S.R. (2001). Rab GTPases: specifying and deciphering organelle identity and function. *Trends Cell Biol.* 11:487–491.

Philips, M.R., and Cox, A.D. (2007). Geranylgeranyltransferase I as a target for anti-cancer drugs. *J. Clin. Invest.* 117:1223–1225.

Pickett, J.S., Bowers, K.E., Fierke, C.A. (2003). Mutagenesis studies of protein farnesyltransferase implicate aspartate beta 352 as a magnesium ligand. *J. Biol. Chem.* 278:51243–51250.

Plowman, S.J., Ariotti, N., Goodall, A., et al. (2008). Electrostatic interactions positively regulate K-Ras nanocluster formation and function. *Mol. Cell Biol.* 28:4377–4385.

Plowman, S.J., and Hancock, J.F. (2005). Ras signaling from plasma membrane and endomembrane microdomains. *Biochim. Biophys. Acta.* 1746:274–283.

Plowman, S.J., Muncke, C., Parton, R.G. (2005). H-ras, K-ras, and inner plasma membrane raft proteins operate in nanoclusters with differential dependence on the actin cytoskeleton. *Proc. Natl. Acad. Sci. U.S.A.* 102:15500–15505.

Pompliano, D.L., Schaber, M.D., Mosser, S.D., et al. (1993). Isoprenoid diphosphate utilization by recombinant human farnesyl-protein transferase. Interactive binding between substrates and a preferred kinetic pathway. *Biochemistry* 32:8341–8347.

Prior, I.A., and Hancock, J.F. (2001). Compartmentalization of Ras proteins. *J. Cell Sci.* 114:1603–1608.

Prior, I.A., Harding, A., Yan, J., et al. (2001). GTP-dependent segregation of H-ras from lipid rafts is required for biological activity. *Nat. Cell Biol.* 3:368–375.

Prior, I.A., Muncke, C., Parton, R.G., et al. (2003). Direct visualization of Ras proteins in spatially distinct cell surface microdomains. *J. Cell Biol.* 160:165–170.

Puccetti, L., Acampa, M., Auteri, A. (2007). Pharmacogenetics of statins therapy. *Recent Patents Cardiovasc. Drug Discov.* 2:228–236.

Pylypenko, O., Rak, A., Durek, T., et al. (2006). Structure of doubly prenylated Ypt1:GDI complex and the mechanism of GDI-mediated Rab recycling. *EMBO J.* 25:13–23.

Pylypenko, O., Rak, A., Reents, R., et al. (2003). Structure of rab escort protein-1 in complex with rab geranylgeranyltransferase. *Mol. Cell* 11:483–494.

Quatela, S.E., and Philips, M.R. (2006). Ras signaling on the Golgi. *Curr. Opin. Cell Biol.* 18:162–167.

Rak, A., Pylypenko, O., Niculae, A., et al. (2004). Structure of the Rab7:REP-1 complex: insights into the mechanism of Rab prenylation and choroideremia disease. *Cell* 117:749–760.

Reid, T.S., and Beese, L.S. (2004). Crystal structures of the anticancer clinical candidates R1 15777 (Tipifarnib) and BMS-214662 complexed with protein farnesyltransferase suggest a mechanism of FTI selectivity. *Biochemistry* 43:6877–6884.

Reid, T.S., Long, S.B., Beese, L.S. (2004). Crystallographic analysis reveals that anticancer clinical candidate L-778,123 inhibits protein famesyltransferase and geranylgeranyltransferase-I by different binding modes. *Biochemistry* 43:9000–9008.

Reiss, Y., Brown, M.S., Goldstein, J.L. (1992). Divalent cation and prenyl pyrophosphate specificities of the protein farnesyltransferase from rat brain, a zinc metalloenzyme. *J. Biol. Chem.* 267:6403–6408.

Reiss, Y., Goldstein, J.L., Seabra, M.C., et al. (1990). Inhibition of purified p21ras farnesyl protein transferase by Cys-AAX tetrapeptides. *Cell* 62:81–88.

Reynwar, B.J., Illya, G., Harmandaris, V.A., et al. (2007). Aggregation and vesiculation of membrane proteins by curvature-mediated interactions. *Nature* 447:461–464.

Ridley, A.J. (2006). Rho GTPases and actin dynamics in membrane protrusions and vesicle trafficking. *Trends Cell Biol.* 16:522–529.

Rizzo, M.A., Kraft, C.A., Watkins, S.C., et al. (2001). Agonist-dependent traffic of raft-associated Ras and Raf-1 is required for activation of the mitogen-activated protein kinase cascade. *J. Biol. Chem.* 276:34928–34933.

Roberts, P.J., Mitin, N., Keller, P.J., et al. (2008). Rho Family GTPase modification and dependence on CAAX motif-signaled posttranslational modification. *J. Biol. Chem.* 283:25150–25163.

Rocks, O., Peyker, A., Kahms, M., et al. (2005). An acylation cycle regulates localization and activity of palmitoylated Ras isoforms. *Science* 307:1746–1752.

Rodriguez-Viciana, P., Sabatier, C., McCormick, F. (2004). Signaling specificity by Ras family GTPases is determined by the full spectrum of effectors they regulate. *Mol. Cell Biol.* 24:4943–4954.

Rolfe, B.E., Worth, N.F., World, C.J., et al. (2005). Rho and vascular disease. *Atherosclerosis* 183:1–16.

Rotblat, B., Niv, H., Andre, S., et al. (2004). Galectin-1(L11A) predicted from a computed galectin-1 farnesyl-binding pocket selectively inhibits Ras-GTP. *Cancer Res.* 64:3112–3118.

Roth, A.F., Wan, J., Bailey, A.O., et al. (2006). Global analysis of protein palmitoylation in yeast. *Cell* 125:1003–1013.

Roy, S., McPherson, R.A., Reiss, A., et al. (1998). 14-3-3 facilitates Ras-dependent Raf-1 activation in vitro and in vivo. *Mol. Cell Biol.* 18:3947–3955.

Sakagami, Y., Isogai, A., Suzuki, A., et al. (1978). Amino acid sequence of tremerogen a-10, a peptidal hormone inducing conjugation tube formation in tremella mesenterica. *Fr. Agric. Biol. Chem.* 42:1301–1302.

Schaber, M.D., O'Hara, M.B., Garsky, V.M., et al. (1990). Polyisoprenylation of Ras in vitro by a farnesyl-protein transferase. *J. Biol. Chem.* 265:14701–14704.

Scheffzek, K., Ahmadian, M.R., Wiesmuller, L., et al. (1998). Structural analysis of the GAP-related domain from neurofibromin and its implications. *EMBO J.* 17:4313–4327.

Schmidt, R.A., Schneider, C.J., Glomset, J.A. (1984). Evidence for post-translational incorporation of a product of mevalonic acid into Swiss 3T3 cell proteins. *J. Biol. Chem.* 259:10175–10180.

Schmidt, W.K., Tam, A., Fujimura-Kamada, K., et al. (1998). Endoplasmic reticulum membrane localization of Rce1p and Ste24p, yeast proteases involved in carboxyl-terminal CAAX protein processing and amino-terminal a-factor cleavage. *Proc. Natl. Acad. Sci. U.S.A.* 95:11175–11180.

Seabra, M.C., Mules, E.H., Hume, A.N. (2002). Rab GTPases, intracellular traffic and disease. *Trends Mol. Med.* 8:23–30.

Seabra, M.C., Reiss, Y., Casey, P.J., et al. (1991). Protein farnesyltransferase and geranylgeranyltransferase share a common alpha subunit. *Cell* 65:429–434.

Sepp-Lorenzino, L., Ma, Z., Rands, E., et al. (1995). A peptidomimetic inhibitor of farnesyl protein transferase blocks the anchorage-dependent and -independent growth of human tumor cell lines. *Cancer Res.* 55:5302–5309.

Shalom-Feuerstein, R., Plowman, S.J., Rotblat, B., et al. (2008). K-ras nanoclustering is subverted by overexpression of the scaffold protein galectin-3. *Cancer Res.* 68:6608–6616.

Shen, F., and Seabra, M.C. (1996). Mechanism of digeranylgeranylation of Rab proteins. Formation of a complex between monogeranylgeranyl-Rab and Rab escort protein. *J. Biol. Chem.* 271:3692–3698.

Shilo, B.Z., and Weinberg, R.A. (1981). Unique transforming gene in carcinogen-transformed mouse cells. *Nature* 289:607–609.

Simons, K., and Ikonen, E. (1997). Functional rafts in cell membranes. *Nature* 387:569–572.

Simons, K., and Toomre, D. (2000). Lipid rafts and signal transduction. *Nat. Rev. Mol. Cell Biol.* 1:31–39.

Simons, K., and van Meer, G. (1988). Lipid sorting in epithelial cells. *Biochemistry* 27:6197–6202.

Sprang, S.R. (1997). G proteins, effectors and GAPs: structure and mechanism. *Curr. Opin. Struct. Biol.* 7:849–856.

Stein, M.P., Dong, J., Wandinger-Ness, A. (2003). Rab proteins and endocytic trafficking: potential targets for therapeutic intervention. *Adv. Drug Deliv. Rev.* 55:1421–1437.

Stenmark, H., and Olkkonen, V.M. (2001). The Rab GTPase family. *Genome Biol.* 2:REVIEWS3007.

Stirtan, W.G., and Poulter, C.D. (1997). Yeast protein geranylgeranyltransferase type-I: steady-state kinetics and substrate binding. *Biochemistry* 36:4552–4557.

Strickland, C.L., Windsor, W.T., Syto, R., et al. (1998). Crystal structure of farnesyl protein transferase complexed with a CaaX peptide and farnesyl diphosphate analogue. *Biochemistry* 37:16601–16611.

Swarthout, J.T., Lobo, S., Farh, L., et al. (2005). DHHC9 and GCP16 constitute a human protein fatty acyltransferase with specificity for H- and N-Ras. *J. Biol. Chem.* 280:31141–31148.

Takai, Y., Sasaki, T., Matozaki, T. (2001). Small GTP-binding proteins. *Physiol. Rev.* 81:153–208.

Tamanoi, F., Gau, C.L., Jiang, C.H., et al. (2001). Protein farnesylation in mammalian cells: effects of farnesyltransferase inhibitors on cancer cells. *Cell. Mol. Life Sci.* 58:1636–1649.

Taylor, J.S., Reid, T.S., Terry, K.L., et al. (2003). Structure of mammalian protein geranylgeranyltransferase type-I. *EMBO J.* 22:5963–5974.

Thoma, N.H., Iakovenko, A., Kalinin, A., et al. (2001). Allosteric regulation of substrate binding and product release in geranylgeranyltransferase type II. *Biochemistry* 40:268–274.

Tian, T., Harding, A., Inder, K., et al. (2007). Plasma membrane nanoswitches generate high-fidelity Ras signal transduction. *Nat. Cell Biol.* 9:905–914.

Tobin, D.A., Pickett, J.S., Hartman, H.L., et al. (2003). Structural characterization of the zinc site in protein farnesyltransferase. *J. Am. Chem. Soc.* 125:9962–9969.

Troutman, J.M., Andres, D.A., Spielmann, H.P. (2007). Protein farnesyl transferase target selectivity is dependent upon peptide stimulated product release. *Biochemistry* 46:11299–11309.

Tsuchiya, E., Fukui, S., Kamiya, Y., et al. (1978). Requirements of chemical structure of hormonal activity of lipopeptidyl factors inducing sexual differentiation in vegetative cells of heterobasidiomycetous yeasts. *Biochem. Biophys. Res. Commun.* 85:459–463.

Turek-Etienne, T.C., Strickland, C.L., Distefano, M.D. (2003). Biochemical and structural studies with prenyl diphosphate analogues provide insights into isoprenoid recognition by protein farnesyl transferase. *Biochemistry* 42:3716–3724.

van den Hurk, J.A., Hendriks, W., van de Pol, D.J., et al. (1997). Mouse choroideremia gene mutation causes photoreceptor cell degeneration and is not transmitted through the female germline. *Hum. Mol. Genet.* 6:851–858.

van Meer, G., and Simons, K. (1988). Lipid polarity and sorting in epithelial cells. *J. Cell Biochem.* 36:51–58.

van Slegtenhorst, M., de Hoogt, R., Hermans, C., et al. (1997). Identification of the tuberous sclerosis gene TSC1 on chromosome 9q34. *Science* 277:805–808.

Vega, F.M., and Ridley, A.J. (2008). Rho GTPases in cancer cell biology. *FEBS Lett.* 582:2093–2101.

Vetter, I.R., and Wittinghofer, A. (2001). The guanine nucleotide-binding switch in three dimensions. *Science* 294:1299–1304.

Wennerberg, K., Rossman, K.L., Der, C.J. (2005). The Ras superfamily at a glance. *J. Cell Sci.* 118:843–846.

Williams, C.L. (2003). The polybasic region of Ras and Rho family small GTPases: a regulator of protein interactions and membrane association and a site of nuclear localization signal sequences. *Cell. Signal.* 15:1071–1080.

Wilson, A.L., Erdman, R.A., Castellano, F., *et al.* (1998). Prenylation of Rab8 GTPase by type I and type II geranylgeranyl transferases. *Biochem. J.* 333:497–504.

Wu, M., Wu, Z.F., Kumar-Sinha, C., *et al.* (2004). RhoC induces differential expression of genes involved in invasion and metastasis in MCF10A breast cells. *Breast Cancer Res. Treat.* 84:3–12.

Wu, W.J., Leonard, D.A., A-Cerione, R., *et al.* (1997). Interaction between Cdc42Hs and RhoGDI is mediated through the Rho insert region. *J. Biol. Chem.* 272:26153–26158.

Wu, Y.W., Goody, R.S., Abagyan, R., *et al.* (2009). Structure of the disordered C-terminus of Rab7 GTPase induced by binding to the Rab geranylgeranyl transferase catalytic complex reveals the mechanism of Rab prenylation. *J. Biol. Chem.*, 284:13185–13192.

Xiang, X., Zang, M., Waelde, C.A., *et al.* (2002). Phosphorylation of 338SSYY341 regulates specific interaction between Raf-1 and MEK1. *J. Biol. Chem.* 277:44996–45003.

Yamane, H.K., Farnsworth, C.C., Xie, H.Y., *et al.* (1991). Membrane-binding domain of the small G protein G25K contains an S-(all-trans-geranylgeranyl)cysteine methyl ester at its carboxyl terminus. *Proc. Natl. Acad. Sci. U.S.A.* 88:286–290.

Yeung, T., Terebiznik, M., Yu, L., *et al.* (2006). Receptor activation alters inner surface potential during phagocytosis. *Science* 313:347–351.

Yokoyama, K., McGeady, P., Gelb, M.H. (1995). Mammalian protein geranylgeranyltransferase-I: substrate specificity, kinetic mechanism, metal requirements, and affinity labeling. *Biochemistry* 34:1344–1354.

Zerial, M., and McBride, H. (2001). Rab proteins as membrane organizers. *Nat. Rev. Mol. Cell Biol.* 2:107–117.

Zhang, F.L., Moomaw, J.F., Casey, P.J. (1994). Properties and kinetic mechanism of recombinant mammalian protein geranylgeranyltransferase type I. *J. Biol. Chem.* 269:23465–23470.

Zhang, Y.W., Li, X.Y., Koyama, T. (2000). Chain length determination of prenyltransferases: both heteromeric subunits of medium-chain (E)-prenyl diphosphate synthase are involved in the product chain length determination. *Biochemistry* 39:12717–12722.

Zong, H., Raman, N., Mickelson-Young, L.A., *et al.* (1999). Loop 6 of RhoA confers specificity for effector binding, stress fiber formation, and cellular transformation. *J. Biol. Chem.* 274:4551–4560.

Chapter 2
GPI-Anchored Proteins in Health and Disease

David R. Taylor and Nigel M. Hooper

2.1 Introduction

Many proteins are associated with cell membranes via stretches of transmembrane spanning hydrophobic amino acids; however, other forms of membrane anchorage exist. Of these, one of the best characterised examples is a glycolipid structure which is covalently attached to the C-terminus of many proteins, known as a glycosyl-phosphatidylinositol (GPI) anchor. This moiety allows the attached protein to anchor to the outer leaflet of the cell membrane. GPI-anchored proteins are produced by most eukaryotic cells, ranging in complexity from protozoa through to vertebrates, and perform a diverse set of functions including roles in signal transduction, cell adhesion and antigen presentation (McConville and Ferguson 1993; Paulick and Bertozzi 2008). The importance of GPI-anchored proteins is evidenced by the embryonic lethality of GPI-deficient mice (Nozaki et al. 1999).

In addition to their normal physiological functions, GPI-anchored proteins are also associated with a range of diseases. For example, variant surface glycoprotein (VSG), a GPI-anchored protein from *Trypanosoma brucei*, is involved in the pathobiology of the parasite by forming a protective coating around it (Ferguson 1999). Moreover, reduced expression of GPI-anchored proteins in hematopoietic stem cells is associated with the human disease paroxysmal nocturnal hemoglobinuria (Brodsky 2006). Perhaps the most notorious of the GPI-anchored proteins is the prion protein (PrP), which is the causative agent of the prion diseases including Creutzfeldt-Jakob disease in humans and bovine spongiform encephalopathy in cattle (Prusiner 1998).

N.M. Hooper (✉)
Protedysis Research Group, Institute of Molecular and Cellular Biology,
Faculty of Biological Sciences, and Leeds Institute of Genetics, Health and Therapeutics,
University of Leeds, Leeds LS2 9JT, UK
e-mail: n.m.hooper@leeds.ac.uk

C.J. Vidal (ed.), *Post-Translational Modifications in Health and Disease*, Protein Reviews 13,
DOI 10.1007/978-1-4419-6382-6_2,

2.2 Historical Origins and Structure of the GPI Anchor

The discovery of the GPI anchor can be traced back over 30 years ago to 1976 with the discovery that a newly purified enzyme from bacteria, phosphatidylinositol-specific phospholipase C (PI-PLC), was capable of releasing alkaline phosphatase from tissue samples (reviewed in Paulick and Bertozzi 2008). Subsequently, it was observed that other proteins could be released in such a manner leading to the hypothesis that these proteins were covalently attached to the cell membrane via a phosphatidylinositol moiety. In separate studies, structural information gleaned from the C-terminus of a mammalian protein, Thy-1, as well as from *T. brucei* VSG revealed the attachment of ethanolamine, various sugars and certain lipid species (Ferguson et al. 1988; Homans et al. 1988). The original PI-PLC data was rationalised with respect to the latter structural data to create a general structure for a GPI anchor (Fig. 2.1).

Detailed analysis of the structure of GPI anchors from a variety of organisms has revealed a common core motif. Through a phosphoethanolamine bridge, the α-carboxyl group at the C-terminus of the mature protein is linked to a highly conserved core glycan: -6mannose(α1-2)mannose(α1-6)mannose(α1-4)glucosamine(α1-6)*myo*-inositol. In turn, a lipid moiety is linked by a phosphodiester bridge to the inositol ring. The lipid moiety attached to inositol ranges from ceramide (slime mold proteins) to diacylglycerol (protozoa) to 1-alkyl-2-acylglycerol (most mammalian proteins; McConville and Ferguson 1993). The core glycan can be extensively modified by the addition of both ethanolamine and sugar side chains e.g. *N*-acetylglucosamine, mannose and galactose. The types of GPI anchor modification can be protein, tissue

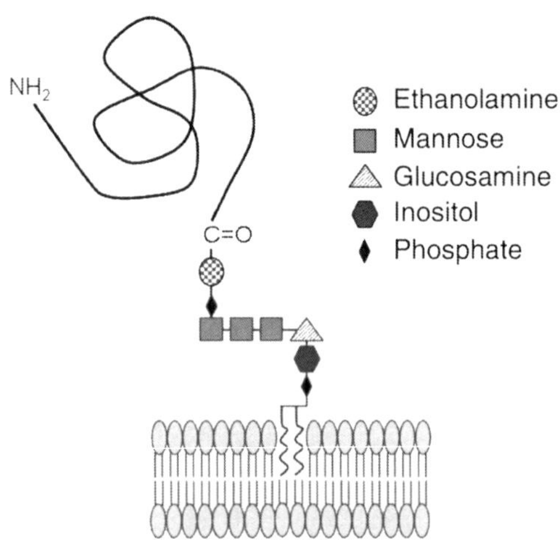

Fig. 2.1 Schematic representation of a GPI anchor. The polypeptide chain is covalently linked at the C-terminus to the core GPI anchor structure ethanolamine-phosphate-(mannose)$_3$-glucosamine-phosphatidyinositol. The lipid tail is attached to the inositol ring and in turn incorporates into the outer leaflet of the lipid bilayer. In this case the lipid is alkyl-acyl-glycerol as found on mammalian proteins

or species specific. A second lipid group, such as palmitic acid, can also be attached to the inositol ring and this modification leaves the GPI anchor resistant to cleavage by PI-PLC (McConville and Ferguson 1993). Another common modification is the remodelling of the fatty acids within the GPI anchor, which must take place to allow the protein to associate with lipid rafts (Maeda et al. 2007).

2.3 Cellular Synthesis of GPI Anchors

Synthesis of the GPI anchor shares similarities to the pathway leading to N-linked protein glycosylation. Both pathways utilise preassembled complexes that are subsequently added to the targeted protein. Furthermore, both pathways are initiated on the cytoplasmic face of the endoplasmic reticulum (ER) and require lipid flippases to translocate a glycolipid intermediate to the luminal face of the ER membrane where the processes are continued (Pomorski and Menon 2006). During synthesis both pathways share certain activated sugars (Dol-P-Man and UDP-GlcNAc) although the synthesis of the GPI anchor differs in its requirement for phosphatidylethanolamine. The core of the GPI anchor is built on the cytoplasmic face of the ER by the sequential addition of sugar moieties and phosphoethanolamine to a phosphatidylinositol molecule (Ferguson 1999; McConville and Menon 2000). Most of the enzymes carrying out the steps involved are well characterised and reviewed elsewhere (McConville and Menon 2000). Initially, UDP-GlcNAc transfers *N*-acetylglucosamine to phosphatidylinositol, and the *N*-acetylglucosamine subsequently undergoes de-*N*-acetylation. Prior to the addition of further sugar molecules inositol acylation must take place in yeast and mammalian cells. Subsequently, the Glc-PI is mannosylated by the addition of three mannose residues that are donated by dolichol-phosphomannose. The GPI anchor precursor is completed by the addition of a phosphoethanolamine molecule derived from phosphatidylethanolamine, and the whole GPI core is flipped across the ER membrane to the lumen where the transamidase enzyme facilitates addition to a protein containing the necessary GPI anchor addition signal.

All GPI-anchored proteins studied to date are resident on either the outer leaflet of the plasma membrane or contained within the lumen of intracellular organelles or vesicles. Therefore, to achieve this orientation GPI-anchored proteins possess two signal sequences. The first is a cleavable N-terminal signal sequence of hydrophobic amino acids that directs the cotranslational entry of the protein into the lumen of the ER through the Sec61 translocon complex (Rapoport 2007). The second sequence directs the addition of the GPI anchor and lies at the very C-terminus of the nascent chain. The sequence comprises 15–25 amino acid residues with a stretch of hydrophobic residues at the C-terminal end. The C-terminal hydrophobic sequence is preceded by a consensus sequence (ω, $\omega+1$ and $\omega+2$) that directs GPI anchor addition. Studies have shown that the ω amino acid residue may only possess a small side chain (Ala, Asn, Asp, Cys, Gly or Ser), with $\omega+1$ being any residue except Pro or Trp and $\omega+2$ being Gly or Ala (or occasionally Ser or

Thr) (Udenfriend and Kodukula 1995). The preassembled GPI anchor is added on the C-terminal side of the ω residue by the transamidase enzyme after the cleavage of the polypeptide chain at this location. The transamidase itself is a complex of several proteins. The protein Gpi8p contains the enzyme's active site, with other components including Gaa1p, Gpi16p (PIG-T) and Gpi17p (PIG-S) (Fraering et al. 2001). Modifications to the GPI anchor of the newly formed GPI-anchored protein occur as the protein is trafficked through the ER and Golgi.

2.4 Identification of a GPI Anchor

The presence of a GPI anchor on a protein can be investigated experimentally by a number of means. As mentioned, these include the common approach of assessing the protein's susceptibility to release from the cell membrane using bacterial PI-PLC. Other strategies include antibody detection of the cross-reacting determinant, inositol-1,2-cyclic monophosphate, exposed after cleavage by PLC (Broomfield and Hooper 1993). Alternatively, metabolic labelling using radiolabelled GPI anchor constituents or detergent insolubility of the protein can be used to assess whether it possesses a GPI anchor (Hooper 2001). Latterly, the potential for GPI anchor addition to a protein is often assessed using bioinformatics approaches such as the Web-based 'big Π predictor' algorithm (Eisenhaber and Eisenhaber 2007; Eisenhaber et al. 2004).

2.5 Properties Conveyed on a Protein by a GPI Anchor

In addition to its primary function in membrane anchorage of its attached protein, the GPI anchor can convey additional properties to the protein which may regulate its function.

2.5.1 Phospholipase Cleavage

In addition to their release by bacterial PI-PLCs, endogenous phospholipases exist that can cleave the GPI anchor in vivo and release the attached protein, perhaps in response to a stimulus (Fig. 2.2a). The secreted protein may possess the same or slightly different functions to the membrane anchored protein. Several secreted mammalian GPI-anchored proteins have been identified which possess the cross-reacting determinant suggesting cleavage by a PLC, though the identity of this enzyme remains unknown (Movahedi and Hooper 1997). To date, three other enzymes have been suggested to cleave GPI anchors. A GPI-specific phospholipase D, which is present in blood plasma, has been shown to cleave GPI-anchored proteins

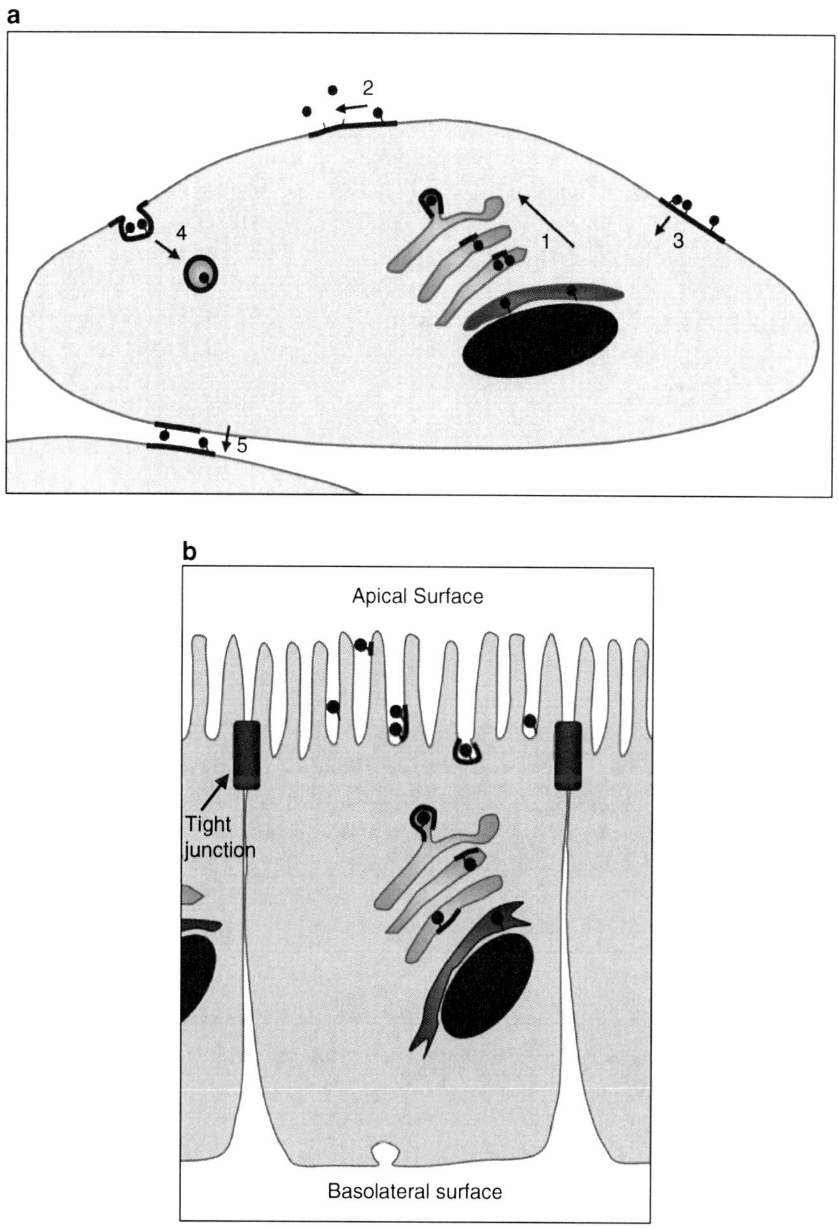

Fig. 2.2 Functions conveyed on a protein by the GPI anchor. (**a**) The GPI anchor is rapidly added to proteins possessing the necessary signal peptides within the ER. The GPI anchor conveys the following properties on its attached protein. (1) Association of the protein with lipid rafts in the Golgi on route to the cell surface. (2) Regulated release of the attached protein by enzymatic cleavage within the GPI anchor. (3) Transduction of signals across cell membranes. (4) Targeting of the attached protein for endocytosis. (5) Intercellular transfer of the protein from one cell to another in the process of cell painting. (**b**) In polarised cells the plasma membrane is spatially separated into apical and basolateral domains by tight junctions. The GPI-anchor is proposed to act as an apical targeting determinant owing to the association of GPI-anchored proteins with lipid rafts in the secretory pathway

in vitro (Low and Huang 1991). Additionally, phospholipase D was able to cleave GPI anchor synthesis intermediates and therefore may provide a mechanism by which the levels of such intermediates can be regulated (Mann et al. 2004). A protein first identified in a screen of Wnt signalling inhibitors, Notum, has also been shown to be capable of cleaving the GPI anchor of several proteins including members of the glypican family which regulate Wnt signalling (Traister et al. 2008). In contrast to phospholipase D, Notum was able to cleave its substrates extracellularly (Traister et al. 2008). Angiotensin-converting enzyme (ACE) has also recently been reported to be capable of cleaving GPI anchors (Kondoh et al. 2005), although the validity of this report remains controversial (Kondoh et al. 2005; Leisle et al. 2005).

2.5.2 *Lipid Raft Association*

There is substantial evidence for the compartmental organisation of the plasma membrane into specialised microdomains termed lipid rafts (Simons and Vaz 2004). Lipid rafts are enriched in sphingomyelin, glycosphingolipids and cholesterol and are believed to serve as locations for diverse cellular functions such as signalling and endocytosis (Simons and Ikonen 1997; Simons and Toomre 2000). Owing to favourable hydrophobic interactions between the saturated acyl chains of the GPI anchor and lipid raft-resident membrane lipids, GPI-anchored proteins are enriched within lipid rafts (Fig. 2.2a). The long saturated chains of sphingolipids self-associate within rafts to form a more tightly packed gel-like phase than the unsaturated kinked side chains of the surrounding phospholipid bilayer (Simons and Vaz 2004). Cholesterol preferentially integrates into the sphingolipid-rich phase hence lipid raft integrity is sensitive to cholesterol depletion (Rajendran and Simons 2005).

The close packing of raft lipids renders them partially resistant to solubilisation with certain non-ionic detergents. Indeed, Triton X-100 insolubility at 4°C has become the most widely used assessment of lipid raft association for a protein (Munro 2003). GPI-anchored proteins in particular are enriched in the isolated detergent resistant membranes (DRMs) owing to favourable interactions between the saturated acyl chains of the GPI anchor and the surrounding lipids (Mayor and Riezman 2004). However, whether DRMs equate to physiologically-relevant lipid rafts is a subject of much debate in the field (Lichtenberg et al. 2005).

2.5.3 *Signal Transduction*

Antibody cross-linking of GPI-anchored proteins has long been known to evoke signalling responses, including rises in intracellular Ca^{2+} or tyrosine phosphorylation (Fig. 2.2a) (Kasahara and Sanai 2000). The importance of the GPI anchor is evidenced by an absence of signalling when the same proteins are anchored by a transmembrane domain (Jones and Varela-Nieto 1998). The targeting of GPI-anchored proteins to lipid rafts appears to be critical for their signalling functions, and this is not surprising given the enrichment of signalling molecules in such domains (Allen et al. 2007; Hugel et al. 2004). The mechanism by which the GPI anchor transduces signals across the membrane is not fully understood, although direct association with signalling transmembrane proteins in lipid rafts is one possibility (Simons and Toomre 2000).

In recent studies it appears that the functional specificity of the signalling mediated by a GPI-anchored protein may reside within the GPI anchor signal sequence. When the GPI anchor signal sequence of differentiation-promoting neural cell adhesion molecule (NCAM) was exchanged for the signal sequence from carcinoembryonic antigen (CEA), a mature protein with NCAM external domains but CEA-like differentiation-blocking activity was produced (Screaton et al. 2000). The altered signalling capabilities of the chimeras appeared to be due to GPI anchor mediated association with specific subpopulations of membrane microdomains (Nicholson and Stanners 2006). Surprisingly, the function of NCAM could be changed by the insertion of the CEA GPI signal sequence residues GLSAG 6–10 amino acids downstream of the GPI attachment site in NCAM (Nicholson and Stanners 2007).

2.5.4 GPI Anchor Dependent Endocytic Targeting

Like all proteins, once at the cell surface GPI-anchored proteins are targeted for down-regulation and degradation through endocytosis. The best characterised endocytic mechanism is clathrin-dependent endocytosis. Clathrin-dependent endocytosis involves the recognition of endocytic motifs in the cytoplasmic tail of transmembrane proteins by endocytic adaptor proteins which act as a bridge between the transmembrane protein and clathrin. Indeed, there is evidence that despite lacking a cytoplasmic tail, GPI-anchored proteins can be targeted for clathrin-dependent endocytosis. Clathrin-dependent endocytosis of PrP and the GPI-anchored urokinase plasminogen activator receptor (uPAR) is facilitated by interaction with the transmembrane low density lipoprotein receptor-related protein-1 (Conese et al. 1995; Parkyn et al. 2008; Taylor and Hooper 2007). However, in the absence of lateral association with transmembrane proteins, other mechanisms of endocytosis for GPI-anchored proteins exist, which are dependent on the GPI anchor itself (Fig. 2.2a).

Initial studies suggested that GPI anchored proteins were internalised through caveolae, a subpopulation of lipid rafts coated with caveolin (Anderson 1998). Nonetheless, it is now established that GPI-anchored proteins are not normally enriched in caveolae (Mayor et al. 1994). Many endocytic pathways, including clathrin- and caveolin-dependent endocytosis require scission of the endocytic vesicle from the cell membrane by a GTPase, dynamin. However, it is now known that many GPI-anchored proteins are capable of undergoing endocytosis in a dynamin-independent manner (Ricci et al. 2000; Sabharanjak et al. 2002). It is proposed that GPI-anchored proteins are internalised through a pathway known as the GPI-anchored-protein-enriched early endosomal compartments (GEEC) pathway (Mayor and Riezman 2004). Uptake of GPI-anchored proteins via this pathway is dependent on the GPI anchor and is susceptible to inhibition of the Rho-family GTPase cdc42 (Sabharanjak et al. 2002). It appears that recruitment of GPI-anchored proteins into the GEEC pathway is sensitive to cholesterol depletion and requires the recruitment of the actin polymerisation machinery by cdc42 (Chadda et al. 2007). The protein GRAF1 appears to be a specific non-cargo marker for the GEEC pathway and appears to coordinate small G protein signalling and membrane remodelling to enable internalisation of GPI-anchored proteins into the pathway (Lundmark et al. 2008). Recently, it has been shown that GPI-anchored proteins are enriched in another novel clathrin and dynamin-independent pathway mediated by the raft protein, flotillin-1 (Glebov et al. 2006). Whether the GPI anchor directly interacts with flotillin-1 remains to be determined.

2.5.5 *Intercellular Transfer*

A number of GPI-anchored proteins have been shown to transfer between one cell and another in a process termed 'cell painting' (Fig. 2.2a) (Anderson et al. 1996; Kooyman et al. 1995; Liu et al. 2002). The lack of a cytoplasmic tail is a key to this event and it is well established that GPI-anchored proteins can spontaneously insert into lipid bilayers (Milhiet et al. 2002). The physiological significance of this process is poorly understood but may provide a means by which cells that cannot synthesise their own proteins under normal circumstances (e.g. mature erythrocytes) could express GPI-anchored proteins.

2.5.6 *Apical Targeting Signal*

Within polarised cells there are marked differences in both the lipid and protein composition between the apical and basolateral cell surfaces which are separated by tight junctions (Rodriguez-Boulan et al. 2005). The establishment of polarity allows for the compartmentalisation of specialised functions and the GPI anchor itself has extensively been suggested to impart apical targeting of the attached protein (Fig. 2.2b) (Folsch 2008; Schuck and Simons 2006). In 1992, Brown and Rose demonstrated that GPI-anchored proteins are targeted to the apical surface owing to their association with detergent insoluble lipid rafts in the Golgi (Brown and Rose 1992). A large body of work supports this observation (Schuck and Simons 2006).

Nonetheless there are several caveats to the notion that a GPI anchor confers apical targeting. The majority of GPI-anchored proteins are targeted to the basolateral membrane of Fisher rat thyroid cells (Zurzolo et al. 1993). Moreover, both N-glycans and protein oligomerisation have been suggested to impart apical targeting on GPI-anchored proteins (Benting et al. 1999; Paladino et al. 2007; Pang et al. 2004). Recently, it has been shown that sorting of GPI-anchored proteins to the apical or basolateral membrane of polarised cells is dependent on the GPI anchor attachment sequence itself (Paladino et al. 2008). In this study, the GPI anchor addition sequences for either the folate receptor or PrP were fused to the C-terminus of green fluorescent protein (GFP) and targeting of the resulting fusion proteins assessed (Paladino et al. 2008). The GFP-folate receptor fusion oligomerised readily and was targeted to the apical surface, however, as is observed with PrP^{C}, the GFP-PrP fusion was targeted to the basolateral surface (Paladino et al. 2008; Sarnataro et al. 2002). It was speculated that differences in the targeting may be a consequence of differences in the structure of the GPI anchor and/or in the surrounding lipid environment.

2.6 GPI-Anchored Proteins and Disease

A body of evidence has demonstrated that GPI-anchored proteins are involved in a variety of disease states. In the following sections some examples are provided with particular emphasis on the role of the GPI anchor itself.

2.6.1 Paroxysmal Nocturnal Hemoglobinuria

Paroxysmal nocturnal hemoglobinuria (PNH) is a rare clonal haematopoietic stem cell disorder which can arise either de novo or develop from acquired aplastic anaemia, a bone marrow failure resulting after certain drug treatments (Nagarajan et al. 1995; Young and Maciejewski 1997). PNH leads to complement-system mediated intravascular hemolysis, thrombosis and bone marrow failure (Moyo et al. 2004). The underlying cause of PNH is a somatic mutation in stem cells that means they and all their resultant progeny completely lack GPI-anchored proteins at their cell surface (Brodsky 2006). This defect is the consequence of mutation of the *PIG-A* gene, required for the final step in GPI anchor synthesis (Miyata et al. 1994). A large number of genes are involved in the biosynthesis of the GPI anchor, yet the deficiency in GPI-anchored proteins is due solely to mutations in *PIG-A* (Bessler et al. 1994; Takeda et al. 1993). To date over 100 mutations have been identified that span the whole of *PIG-A*, with most mutations being small insertions or deletions which result in frameshifts or premature termination of transcription (Nafa et al. 1995).

Regarding PNH pathology, erythrocytes originating from PNH stem cells are particularly vulnerable to lysis mediated by the complement system. This is a consequence of a lack of two GPI-anchored proteins, membrane inhibitor of reactive lysis (CD59) and decay accelerating factor (CD55), which are both involved in complement system regulation. As a result of increased erythrocyte lysis, large amounts of free hemoglobin are released into the plasma which subsequently mediates scavenging of nitric oxide (Rother et al. 2005). The increased consumption of nitric oxide results in the clinical symptoms associated with PNH such as abdominal pain, fatigue, esophageal spasm and possibly thrombosis.

One unanswered question in the PNH field remains the issue of clonal dominance. Why are PNH stem cells spared from immunological attack (e.g. as seen in aplastic anaemia)? It is speculated that PNH cells are spared from the immunological attack owing to the absence of certain GPI-anchored proteins (Young and Maciejewski 1997). In a recent study it was shown that there is an ongoing immune attack against haematopoietic stem cells in both PNH and aplastic anaemia, however, PNH stem cells were resistant to attack (Savage et al. 2009). Furthermore, *PIG-A* mutations appear to confer a cellular resistance to stresses that would otherwise lead to apoptosis (Savage et al. 2009).

2.6.2 *Prion Diseases*

The prion diseases are characterised by the post-translational misfolding of the normal cellular form of the prion protein (PrP^{C}) into the infectious isoform (PrP^{Sc}). According to the protein-only hypothesis, interaction between PrP^{Sc} and endogenous PrP^{C} is sufficient to result in the template-driven formation of more PrP^{Sc} (Prusiner 1998). A wealth of literature suggests that lipid rafts may act as a site for the conformational conversion of PrP^{C} to PrP^{Sc} (Campana et al. 2005). Given that PrP's GPI anchor is one of the strongest determinants of its lipid raft targeting, it is not surprising that a body of literature has attempted to assess the role of the GPI anchor in prion disease (Taylor and Hooper 2006).

Like other GPI-anchored proteins, PrP^{C} can be transferred efficiently between cells (Liu et al. 2002). This process requires cell contact, an intact GPI anchor and activation of protein kinase C (Liu et al. 2002). Given that PrP^{Sc} has to be inserted into a contiguous lipid raft membrane with PrP^{C} for misfolding to occur, GPI anchor dependent cell transfer of PrP^{Sc} as hypothesised by Liu et al. from infected to uninfected cells could provide a mechanism for prion spread (Baron et al. 2002; Liu et al. 2002).

Recently, the GPI anchor of PrP has been suggested to play a role in the toxicity associated with prion disease (Chesebro et al. 2005). In this study transgenic mice expressing an anchorless, secreted form of PrP^{C} were shown to accumulate PrP^{Sc} in plaques yet the mice never developed clinical disease although the PrP^{Sc} produced was capable of infecting other transgenic mice expressing GPI-anchored PrP (Chesebro et al. 2005). Therefore, the authors of this study were able to discriminate between prion conversion/infectivity and prion toxicity. One interpretation of these data is that the GPI anchor itself is important in transducing neurotoxic signals that result in the onset of clinical symptoms (Aguzzi 2005). Removal of the GPI anchor of PrP^{Sc} by the aspartic endoprotease cathepsin D was shown in in vivo and cell-based assays to have little effect on either PrP^{Sc} propagation or infectivity supporting the conclusions of Chesebro et al. (Chesebro et al. 2005; Lewis et al. 2006).

2.6.3 Malaria

Malaria is an infectious disease responsible for the deaths of two million people a year, with more than 80% of cases occurring in Africa (Nebl et al. 2005). There are four species of malarial parasites known to infect humans, with the bulk of cases attributed to *Plasmodium falciparum* infection. Disease symptoms include acute respiratory disease, renal failure, pulmonary oedema and seizures (White and Ho 1992). Site specific localisation of parasites coupled with both local and global inflammatory response to cytokine production is thought to underlie disease (Stevenson and Riley 2004). A number of years ago it was suggested that GPI anchors play a key role in *P. falciparum* infection as purified GPI-anchored proteins from this parasite were shown to be potent activators of immune cells capable of producing pro-inflammatory cytokines such as TNF-α (Schofield and Hackett 1993; Vijaykumar et al. 2001). The effect was due to the GPI anchor and not the attached protein as the agent was susceptible to reagents that disrupt the integrity of GPI anchors, but not to proteinase digestion (Schofield and Hackett 1993). Further studies have shown that the GPI anchor of *P. falciparum* alone is sufficient to cause malarial symptoms in various host tissues and cell types (Schofield and Hackett 1993; Schofield et al. 1996; Tachado and Schofield 1994). It is interesting to note that the GPI anchors from *T. brucei*, *Trypanosoma cruzi* and *Toxoplasma gondii* all possess similar properties to the GPI anchor from *P. falciparum*, suggesting that there are shared pathogenic mechanisms between trypanosomiasis, Chagas' disease, toxoplasmosis and malaria.

2.6.4 Pore Forming Toxins and the GPI Anchor

The bacterial toxin aerolysin produced by the human pathogen *Aeromonas hydrophilia* has its cytolytic effect by forming oligomers that insert into the cell membrane and form a channel (van der Goot et al. 1994). A number of GPI-anchored proteins are known to bind the toxin, such as Thy-1, and the determinant for binding was shown to be the GPI anchor itself (Diep et al. 1998). Aerolysin could not bind to all GPI-anchored proteins implying that certain GPI anchor modifications impart binding specificity (Diep et al. 1998). One such motif may be the β-*N*-acetylglucosamine sidechain present on the GPI anchor of placental-like alkaline phosphatase (Fukushima et al. 2003).

2.7 Conclusions and Future Perspectives

The GPI anchor is a complex post-translational modification that is implicated in the regulation of a host of physiological processes by conveying properties to the protein to which it is attached. However, there remain large gaps in our knowledge of GPI anchors, both in health and disease. As the phosphoinositol, glucosamine and mannose moieties within the core of the GPI anchor can be extensively modified by the addition of phosphoethanolamine and sugars, diverse functional properties on the GPI anchor would be expected to be imparted, beyond a simple role in membrane anchorage. However, due precisely to this structural heterogeneity, the roles of different GPI anchor modifications have been difficult to assess (Paulick and Bertozzi 2008).

Attempts to manipulate GPI anchor modifications in a cellular context have failed owing both to incomplete knowledge of the biosynthetic enzymes as well as the fact that their disruption often results in complete loss of GPI anchorage (Bastisch et al. 2000; Kawagoe et al. 1996; Kinoshita and Inoue 2000). Recently, several groups have adopted an alternative approach to understanding GPI anchor function to advance the field by attempting its chemical synthesis (Paulick and Bertozzi 2008). Currently, the results of such experiments are unsatisfactory owing to the complicated nature of the synthesis pathways and the fact that the synthetic product is often not able to be coupled to a protein. Nonetheless, advances are being made; a study showed that a short synthetic peptide of 12 amino acids could be added to a synthetically produced GPI anchor (Shao et al. 2004).

Moreover, the behaviour of GFP fused to GPI anchor analogues by chemical ligation was recently studied in a cell model and provided evidence that individual components of the GPI anchor impart certain properties on the attached protein (Paulick et al. 2007). The lateral mobility of the GFP conjugates in the membrane was shown to be drastically reduced by deletions within the glycan core, whilst endocytosis was unaltered (Paulick et al. 2007). However, although such studies are promising synthetic production of a native GPI-anchored protein has not been reported. Significantly though, semi synthesis of a GPI-anchored prion protein was recently reported (Becker et al. 2008). In this study, a synthetic cysteine-tagged GPI anchor was fused with recombinant PrP containing a C-terminal thioester (Becker et al. 2008). The quest to produce synthetic GPI anchors may have potential in disease therapeutics. This is evidenced by the fact that synthetic GPI anchors have been suggested as a candidate anti-toxic vaccine for malaria (Kamena et al. 2008; Schofield et al. 2002).

Acknowledgements The authors wish to acknowledge the funding of the Wellcome Trust and the Medical Research Council of Great Britain.

References

Aguzzi, A. (2005). Cell biology. Prion toxicity: all sail and no anchor. *Science* 308:1420–1421.

Allen, J.A., Halverson-Tamboli, R.A., *et al.* (2007). Lipid raft microdomains and neurotransmitter signalling. *Nat. Rev. Neurosci.* 8:128–140.

Anderson, R.G. (1998). The caveolae membrane system. *Annu. Rev. Biochem.* 67:199–225.

Anderson, S.M., Yu, G., Giattina, M., *et al.* (1996). Intercellular transfer of a glycosylphosphatidylinositol (GPI)-linked protein: release and uptake of CD4-GPI from recombinant adeno-associated virus-transduced HeLa cells. *Proc. Natl. Acad. Sci. U.S.A.* 93:5894–5898.

Baron, G.S., Wehrly, K., Dorward, D.W., *et al.* (2002). Conversion of raft associated prion protein to the protease-resistant state requires insertion of PrP-res (PrP(Sc)) into contiguous membranes. *EMBO J.* 21:1031–1040.

Bastisch, I., Tiede, A., Deckert, M., *et al.* (2000). Glycosylphosphatidylinositol (GPI)-deficient Jurkat T cells as a model to study functions of GPI-anchored proteins. *Clin. Exp. Immunol.* 122:49–54.

Becker, C.F., Liu, X., Olschewski, D., *et al.* (2008). Semisynthesis of a glycosylphosphatidylinositol-anchored prion protein. *Angew. Chem. Int. Ed. Engl.* 47:8215–8219.

Benting, J.H., Rietveld, A.G., Simons, K. (1999). N-Glycans mediate the apical sorting of a GPI-anchored, raft-associated protein in Madin-Darby canine kidney cells. *J. Cell Biol.* 146:313–320.

Bessler, M., Mason, P.J., Hillmen, P., *et al.* (1994). Paroxysmal nocturnal haemoglobinuria (PNH) is caused by somatic mutations in the PIG-A gene. *EMBO J.* 13:110–117.

Brodsky, R.A. (2006). New insights into paroxysmal nocturnal hemoglobinuria. *Hematology Am. Soc. Hematol. Educ. Program* 24–28:516.

Broomfield, S.J., and Hooper, N.M. (1993). Characterization of an antibody to the cross-reacting determinant of the glycosyl-phosphatidylinositol anchor of human membrane dipeptidase. *Biochim. Biophys. Acta* 1145:212–218.

Brown, D.A., and Rose, J.K. (1992). Sorting of GPI-anchored proteins to glycolipid-enriched membrane subdomains during transport to the apical cell surface. *Cell* 68:533–544.

Campana, V., Sarnataro, D., Zurzolo, C. (2005). The highways and byways of prion protein trafficking. *Trends Cell Biol.* 15:102–111.

Chadda, R., Howes, M.T., Plowman, S.J., *et al.* (2007). Cholesterol-sensitive Cdc42 activation regulates actin polymerization for endocytosis via the GEEC pathway. *Traffic* 8:702–717.

Chesebro, B., Trifilo, M., Race, R., *et al.* (2005). Anchorless prion protein results in infectious amyloid disease without clinical scrapie. *Science* 308:1435–1439.

Conese, M., Nykjaer, A., Petersen, C.M., *et al.*, (1995). alpha-2 Macroglobulin receptor/Ldl receptor-related protein (Lrp)-dependent internalization of the urokinase receptor. *J. Cell Biol.* 131:1609–1622.

Diep, D.B., Nelson, K.L., Raja, S.M., *et al.* (1998). Glycosylphosphatidylinositol anchors of membrane glycoproteins are binding determinants for the channel-forming toxin aerolysin. *J. Biol. Chem.* 273:2355–2360.

Eisenhaber, B., and Eisenhaber, F. (2007). Posttranslational modifications and subcellular localization signals: indicators of sequence regions without inherent 3D structure? *Curr. Protein Pept. Sci.* 8:197–203.

Eisenhaber, B., Schneider, G., Wildpaner, M., *et al.* (2004). A sensitive predictor for potential GPI lipid modification sites in fungal protein sequences and its application to genome-wide studies for *Aspergillus nidulans, Candida albicans, Neurospora crassa, Saccharomyces cerevisiae and Schizosaccharomyces pombe. J. Mol. Biol.* 337:243–253.

Ferguson, M.A. (1999). The structure, biosynthesis and functions of glycosylphosphatidylinositol anchors, and the contributions of trypanosome research. *J. Cell Sci.* 112:2799–2809.

Ferguson, M.A., Homans, S.W., Dwek, R.A. *et al.* (1988). Glycosyl-phosphatidylinositol moiety that anchors *Trypanosoma brucei* variant surface glycoprotein to the membrane. *Science* 239:753–759.

Folsch, H. (2008). Regulation of membrane trafficking in polarized epithelial cells. *Curr. Opin. Cell Biol.* 20:208–213.

Fraering, P., Imhof, I., Meyer, U., *et al.* (2001). The GPI transamidase complex of *Saccharomyces cerevisiae* contains Gaa1p, Gpi8p, and Gpi16p. *Mol. Biol. Cell* 12:3295–3306.

Fukushima, K., Ikehara, Y., Kanai, M., *et al.* (2003). A beta-N-acetylglucosaminyl phosphate diester residue is attached to the glycosylphosphatidylinositol anchor of human placental

alkaline phosphatase: a target of the channel-forming toxin aerolysin. *J. Biol. Chem.* 278:36296–36303.
Glebov, O.O., Bright, N.A., Nichols, B.J. (2006). Flotillin-1 defines a clathrin-independent endocytic pathway in mammalian cells. *Nat. Cell Biol.* 8:46–54.
Homans, S.W., Ferguson, M.A., Dwek, R.A., *et al.* (1988). Complete structure of the glycosyl phosphatidylinositol membrane anchor of rat brain Thy-1 glycoprotein. *Nature* 333: 269–272.
Hooper, N.M. (2001). Determination of glycosyl-phosphatidylinositol membrane protein anchorage. *Proteomics* 1:748–755.
Hugel, B., Martinez, M.C., Kunzelmann, C., *et al.* (2004). Modulation of signal transduction through the cellular prion protein is linked to its incorporation in lipid rafts. *Cell. Mol. Life Sci.* 61:2998–3007.
Jones, D.R., and Varela-Nieto, I. (1998). The role of glycosyl-phosphatidylinositol in signal transduction. *Int. J. Biochem. Cell Biol.* 30:313–326.
Kamena, F., Tamborrini, M., Liu, X., *et al.* (2008). Synthetic GPI array to study antitoxic malaria response. *Nat. Chem. Biol.* 4:238–240.
Kasahara, K., and Sanai, Y. (2000). Functional roles of glycosphingolipids in signal transduction via lipid rafts. *Glycoconj. J.* 17:153–162.
Kawagoe, K., Kitamura, D., Okabe, M., *et al.* (1996). Glycosylphosphatidylinositol-anchor-deficient mice: implications for clonal dominance of mutant cells in paroxysmal nocturnal hemoglobinuria. *Blood* 87:3600–3606.
Kinoshita, T., and Inoue, N. (2000). Dissecting and manipulating the pathway for glycosylphosphatidylinositol-anchor biosynthesis. *Curr. Opin. Chem. Biol.* 4:632–638.
Kondoh, G., Tojo, H., Nakatani, Y., *et al.* (2005). Angiotensin-converting enzyme is a GPI-anchored protein releasing factor crucial for fertilization. *Nat. Med.* 11:160–166.
Kooyman, D.L., Byrne, G.W., McClellan, S., *et al.* (1995). In vivo transfer of GPI-linked complement restriction factors from erythrocytes to the endothelium. *Science* 269:89–92.
Leisle, L., Parkin, E.T., Turner, A.J. *et al.* (2005). Angiotensin-converting enzyme as a GPIase: a critical reevaluation. *Nat. Med.* 11:1139–1140.
Lewis, P.A., Properzi, F., Prodromidou, K., *et al.* (2006). Removal of the glycosylphosphatidylinositol anchor from PrP(Sc) by cathepsin D does not reduce prion infectivity. *Biochem. J.* 395:443–448.
Lichtenberg, D., Goni, F.M., Heerklotz, H. (2005). Detergent-resistant membranes should not be identified with membrane rafts. *Trends Biochem. Sci.* 30:430–436.
Liu, T., Li, R., Pan, T., *et al.* (2002). Intercellular transfer of the cellular prion protein. *J. Biol. Chem.* 277:47671–47678.
Low, M.G., and Huang, K.S. (1991). Factors affecting the ability of glycosylphosphatidylinositol-specific phospholipase D to degrade the membrane anchors of cell surface proteins. *Biochem. J.* 279:483–493.
Lundmark, R., Doherty, G.J., Howes, M.T., *et al.* (2008). The GTPase-activating protein GRAF1 regulates the CLIC/GEEC endocytic pathway. *Curr. Biol.* 18:1802–1808.
Maeda, Y., Tashima, Y., Houjou, T., *et al.* (2007). Fatty acid remodeling of GPI-anchored proteins is required for their raft association. *Mol. Biol. Cell* 18:1497–1506.
Mann, K.J., Hepworth, M.R., Raikwar, N.S., *et al.* (2004). Effect of glycosylphosphatidylinositol (GPI)-phospholipase D overexpression on GPI metabolism. *Biochem. J.* 378:641–648.
Mayor, S., and Riezman, H. (2004). Sorting GPI-anchored proteins. *Nat. Rev. Mol. Cell. Biol.* 5:110–120.
Mayor, S., Rothberg, K.G., Maxfield, F.R. (1994). Sequestration of GPI-anchored proteins in caveolae triggered by cross-linking. *Science* 264:1948–1951.
McConville, M.J., and Ferguson, M.A. (1993). The structure, biosynthesis and function of glycosylated phosphatidylinositols in the parasitic protozoa and higher eukaryotes. *Biochem. J.* 294:305–324.
McConville, M.J., and Menon, A.K. (2000). Recent developments in the cell biology and biochemistry of glycosylphosphatidylinositol lipids. *Mol. Membr. Biol.* 17:1–16.

Milhiet, P.E., Giocondi, M.C., Baghdadi, O., *et al.* (2002). Spontaneous insertion and partitioning of alkaline phosphatase into model lipid rafts. *EMBO Rep.* 3:485–490.

Miyata, T., Yamada, N., Iida, Y., *et al.* (1994). Abnormalities of PIG-A transcripts in granulocytes from patients with paroxysmal nocturnal hemoglobinuria. *N. Engl. J. Med.* 330:249–255.

Movahedi, S., and Hooper, N.M. (1997). Insulin stimulates the release of the glycosyl phosphatidylinositol-anchored membrane dipeptidase from 3T3-L1 adipocytes through the action of a phospholipase C. *Biochem. J.* 326:531–537.

Moyo, V.M., Mukhina, G.L., Garrett, E.S. *et al.* (2004). Natural history of paroxysmal nocturnal haemoglobinuria using modern diagnostic assays. *Br. J. Haematol.* 126:133–138.

Munro, S. (2003). Lipid rafts: elusive or illusive? *Cell* 115:377–388.

Nafa, K., Mason, P.J., Hillmen, P., *et al.* (1995). Mutations in the PIG-A gene causing paroxysmal nocturnal hemoglobinuria are mainly of the frameshift type. *Blood* 86:4650–4655.

Nagarajan, S., Brodsky, R.A., Young, N.S. *et al.* (1995). Genetic defects underlying paroxysmal nocturnal hemoglobinuria that arises out of aplastic anemia. *Blood* 86:4656–4661.

Nebl, T., De Veer, M.J., Schofield, L. (2005). Stimulation of innate immune responses by malarial glycosylphosphatidylinositol via pattern recognition receptors. *Parasitology* 130 Suppl:S45–62.

Nicholson, T.B., and Stanners, C.P. (2006). Specific inhibition of GPI-anchored protein function by homing and self-association of specific GPI anchors. *J. Cell Biol.* 175:647–659.

Nicholson, T.B., and Stanners, C.P. (2007). Identification of a novel functional specificity signal within the GPI anchor signal sequence of carcinoembryonic antigen. *J. Cell Biol.* 177:211–218.

Nozaki, M., Ohishi, K., Yamada, N., *et al.* (1999). Developmental abnormalities of glycosylphosphatidylinositol-anchor-deficient embryos revealed by Cre/loxP system. *Lab. Invest.* 79:293–299.

Paladino, S., Sarnataro, D., Tivodar, S., *et al.* (2007). Oligomerization is a specific requirement for apical sorting of glycosyl-phosphatidylinositol-anchored proteins but not for non-raft-associated apical proteins. *Traffic* 8:251–258.

Paladino, S., Lebreton, S., Tivodar, S., *et al.* (2008). Different GPI-attachment signals affect the oligomerisation of GPI-anchored proteins and their apical sorting. *J. Cell Sci.* 121:4001–4007.

Pang, S., Urquhart, P., and Hooper, N.M. (2004). N-glycans, not the GPI anchor, mediate the apical targeting of a naturally glycosylated, GPI-anchored protein in polarised epithelial cells. *J. Cell Sci.* 117:5079–5086.

Parkyn, C.J., Vermeulen, E.G., Mootoosamy, R.C., *et al.* (2008). LRP1 controls biosynthetic and endocytic trafficking of neuronal prion protein. *J. Cell Sci.* 121:773–783.

Paulick, M.G., and Bertozzi, C.R. (2008). The glycosylphosphatidylinositol anchor: a complex membrane-anchoring structure for proteins. *Biochemistry* 47:6991–7000.

Paulick, M.G., Forstner, M.B., Groves, J.T. *et al.* (2007). A chemical approach to unraveling the biological function of the glycosylphosphatidylinositol anchor. *Proc. Natl. Acad. Sci. U.S.A.* 104:20332–20337.

Pomorski, T., and Menon, A.K. (2006). Lipid flippases and their biological functions. *Cell. Mol. Life Sci.* 63:2908–2921.

Prusiner, S.B. (1998). Prions. *Proc. Natl. Acad. Sci. U.S.A.* 95:13363–13383.

Rajendran, L., and Simons, K. (2005). Lipid rafts and membrane dynamics. *J. Cell Sci.* 118:1099–1102.

Rapoport, T.A. (2007). Protein translocation across the eukaryotic endoplasmic reticulum and bacterial plasma membranes. *Nature* 450:663–669.

Ricci, V., Galmiche, A., Doye, A., *et al.* (2000). High cell sensitivity to *Helicobacter pylori* VacA toxin depends on a GPI-anchored protein and is not blocked by inhibition of the clathrin-mediated pathway of endocytosis. *Mol. Biol. Cell* 11:3897–3909.

Rodriguez-Boulan, E., Kreitzer, G., Musch, A. (2005). Organization of vesicular trafficking in epithelia. *Nat. Rev. Mol. Cell. Biol.* 6:233–247.

Rother, R.P., Bell, L., Hillmen, P. *et al.* (2005). The clinical sequelae of intravascular hemolysis and extracellular plasma hemoglobin: a novel mechanism of human disease. *JAMA* 293:1653–1662.

Sabharanjak, S., Sharma, P., Parton, R.G. *et al.* (2002). GPI-anchored proteins are delivered to recycling endosomes via a distinct cdc42-regulated, clathrin-independent pinocytic pathway. *Dev. Cell* 2:411–423.

Sarnataro, D., Paladino, S., Campana, V., *et al.* (2002). PrPC is sorted to the basolateral membrane of epithelial cells independently of its association with rafts. *Traffic* 3:810–821.

Savage, W.J., Barber, J.P., Mukhina, G.L., *et al.* (2009). Glycosylphosphatidylinositol-anchored protein deficiency confers resistance to apoptosis in PNH. *Exp. Hematol.* 37:42–51.

Schofield, L., and Hackett, F. (1993). Signal transduction in host cells by a glycosylphosphatidylinositol toxin of malaria parasites. *J. Exp. Med.* 177:145–153.

Schofield, L., Novakovic, S., Gerold, P., *et al.* (1996). Glycosylphosphatidylinositol toxin of Plasmodium up-regulates intercellular adhesion molecule-1, vascular cell adhesion molecule-1, and E-selectin expression in vascular endothelial cells and increases leukocyte and parasite cytoadherence via tyrosine kinase-dependent signal transduction. *J. Immunol.* 156:1886–1896.

Schofield, L., Hewitt, M.C., Evans, K. *et al.* (2002). Synthetic GPI as a candidate anti-toxic vaccine in a model of malaria. *Nature* 418:785–789.

Schuck, S., and Simons, K. (2006). Controversy fuels trafficking of GPI-anchored proteins. *J. Cell Biol.* 172:963–965.

Screaton, R.A., DeMarte, L., Draber, P., *et al.* (2000). The specificity for the differentiation blocking activity of carcinoembryonic antigen resides in its glycophosphatidyl-inositol anchor. *J. Cell Biol.* 150:613–626.

Shao, N., Xue, J., Guo, Z. (2004). Chemical synthesis of a skeleton structure of sperm CD52-A GPI-anchored glycopeptide. *Angew. Chem. Int. Ed. Engl.* 43:1569–1573.

Simons, K., and Ikonen, E. (1997). Functional rafts in cell membranes. *Nature* 387:569–572.

Simons, K., and Toomre, D. (2000). Lipid rafts and signal transduction. *Nat. Rev. Mol. Cell. Biol.* 1:31–39.

Simons, K., and Vaz, W.L. (2004). Model systems, lipid rafts, and cell membranes. *Annu. Rev. Biophys. Biomol. Struct.* 33:269–295.

Stevenson, M.M. and Riley, E.M. (2004). Innate immunity to malaria. *Nat. Rev. Immunol.* 4:169–180.

Tachado, S.D., and Schofield, L. (1994). Glycosylphosphatidylinositol toxin of *Trypanosoma brucei* regulates IL-1 alpha and TNF-alpha expression in macrophages by protein tyrosine kinase mediated signal transduction. *Biochem. Biophys. Res. Commun.* 205:984–991.

Takeda, J., Miyata, T., Kawagoe, K., *et al.*, (1993). Deficiency of the GPI anchor caused by a somatic mutation of the PIG-A gene in paroxysmal nocturnal hemoglobinuria. *Cell* 73:703–711.

Taylor, D.R., and Hooper, N.M. (2006). The prion protein and lipid rafts. *Mol. Membr. Biol.* 23:89–99.

Taylor, D.R., and Hooper, N.M. (2007). The low-density lipoprotein receptor-related protein 1 (LRP1) mediates the endocytosis of the cellular prion protein. *Biochem. J.* 402:17–23.

Traister, A., Shi, W. and Filmus, J. (2008). Mammalian Notum induces the release of glypicans and other GPI-anchored proteins from the cell surface. *Biochem. J.* 410:503–511.

Udenfriend, S., and Kodukula, K. (1995). How glycosylphosphatidylinositol-anchored membrane proteins are made. *Annu. Rev. Biochem.* 64:563–591.

van der Goot, F.G., Pattus, F., Parker, M., *et al.* (1994). The cytolytic toxin aerolysin: from the soluble form to the transmembrane channel. *Toxicology* 87:19–28.

Vijaykumar, M., Naik, R.S. and Gowda, D.C. (2001). *Plasmodium falciparum* glycosylphosphatidylinositol-induced TNF-alpha secretion by macrophages is mediated without membrane insertion or endocytosis. *J. Biol. Chem.* 276:6909–6912.

White, N.J., and Ho, M. (1992). The pathophysiology of malaria. *Adv. Parasitol.* 31:83–173.
Young, N.S. and Maciejewski, J. (1997). The pathophysiology of acquired aplastic anemia. *N. Engl. J. Med.* 336:1365–1372.
Zurzolo, C., Lisanti, M.P., Caras, I.W., *et al.* (1993). Glycosylphosphatidylinositol-anchored proteins are preferentially targeted to the basolateral surface in Fischer rat thyroid epithelial cells. *J. Cell. Biol.* 121:1031–1039.

Chapter 3
Protein Oxidation

C. Quiney, S. Finnegan, G. Groeger, and T.G. Cotter

Abbreviations

AD	Alzheimer's disease
CF	Cystic fibrosis
COX	Cyclooxygenase
DUOX	Dual oxidase
$HO^{\bullet}$	Hydroxyl radical
H_2O_2	Hydrogen peroxide
LOX	Lipoxygenase
MPO	Myeloperoxidase
$NO^{\bullet}$	Nitric oxide
NOS	Nitric oxide synthase
Nox	NADPH Oxidase
$O_2^{\bullet-}$	Superoxide
PD	Parkinson's disease
RNS	Reactive nitrogen species
ROS	Reactive oxygen species
SOD	Superoxide dismutase
Trx2	Thioredoxin2
XDH	Xanthine dehydrogenase
XO	Xanthine oxidase
XOR	Xanthine oxidoreductase

T.G. Cotter (✉)
Cell Development and Disease Laboratory, Biochemistry Department,
Biosciences Institute, University College, Cork, Ireland
e-mail: t.cotter@ucc.ie

C.J. Vidal (ed.), *Post-Translational Modifications in Health and Disease*, Protein Reviews 13,
DOI 10.1007/978-1-4419-6382-6_3, © Springer Science+Business Media, LLC 2011

3.1 Introduction

Protein oxidation occurs when reactive oxygen species (ROS) donate electrons to proteins. ROS are a group of molecules derived from oxygen, with the two main members being superoxide ($O_2^{\bullet -}$) and hydrogen peroxide (H_2O_2). They are generated under many conditions and can be either deleterious or beneficial for cells. If they are produced at high concentrations, cells undergo oxidative stress, which can result in damage to proteins, lipids and DNA and ultimately cause cell death. Consequently, cells have many different systems capable of neutralising ROS, e.g. superoxide dismutase (SOD), catalase and glutathione (Fig. 3.1). Therefore, in order for meaningful signaling to occur through protein oxidation, any ROS produced needs to interact with specific proteins and not result in general, non-specific protein oxidation.

This review is divided into three main sections. Mitochondria and NADPH oxidase (Nox) proteins are the two primary sources of ROS generation and will be given

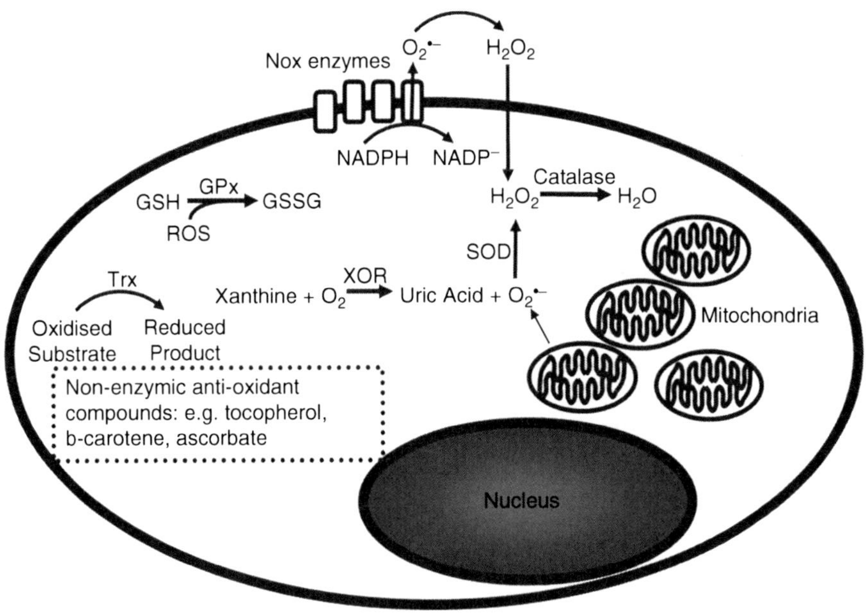

Fig. 3.1 The predominant sources and sinks of ROS in a typical cell. Mitochondria and Nox enzymes generate superoxide ($O_2^{\bullet -}$), which can be dismutated to hydrogen peroxide (H_2O_2) through the action of superoxide dismutase (SOD) and then converted to water by catalase. Xanthine oxidoreductase (XOR) can also generate $O_2^{\bullet -}$. The glutathione system, consisting of reduced glutathione (GSH), oxidised glutathione (GSSG), and several enzymes such as glutathione peroxidase (GPx), can convert ROS to less reactive molecules. Thioredoxin (Trx) performs a similar function but using a different mechanism. There are also several non-enzymic anti-oxidant compounds in cells that also reduce ROS, such as tocopherol and ascorbate

particular attention (Fig. 3.1). A second focus is the different ways in which proteins can be oxidized, either reversibly which generally allows a signaling pathway to operate, or irreversibly which tends to result in proteasomal degradation. Finally the role of protein oxidation in health and disease is examined.

3.2 Sources of ROS

3.2.1 Mitochondria

Mitochondria are the power-house of cells as they provide most of the energy required by eukaryotic cells. In their production of ATP, a process known as oxidative phosphorylation, they consume large quantities of oxygen and form ROS as a by-product. Oxidative phosphorylation uses the electron transport chain of mitochondria, which consists of several enzyme complexes in the inner mitochondrial membrane, namely: complex I (NADH-CoQ reductase), complex II (succinate-CoQ reductase), complex III (reduced CoQ-cytochrome c reductase), complex IV (cytochrome c oxidase) and complex V (ATP synthase) (Fig. 3.2). Together these generate a proton gradient, which allows the movement of key molecules into and out of the mitochondrial matrix (Kakkar and Singh 2007). Complexes I + III have proven to be primary sources for $O_2^{\bullet-}$ (Turrens and Boveris 1980; Turrens et al. 1985; Barja and Herrero 1998). Therefore, the main sources of ROS in mitochondria are few, but the electron transport chain is in constant use, and so there exists a great potential for ROS production and hence protein oxidation.

ROS produced in mitochondria tend either to be metabolized within these organelles or result in oxidisation of other mitochondrial molecules. The mitochondrial matrix contains manganese SOD (MnSOD) which facilitates the dismutation of $O_2^{\bullet-}$ to H_2O_2 (Kakkar and Singh 2007), whilst the inter-membrane space contains a different isozyme, CuZnSOD (Okado-Matsumoto and Fridovich 2001). The inter-membrane space also contains high levels of cytochrome c, which can be oxidized by $O_2^{\bullet-}$ regenerating O_2 (Butler et al. 1975), and has a low pH, facilitating the spontaneous dismutation of $O_2^{\bullet-}$ (Guidot et al. 1995). Mitochondria contain two other main anti-oxidant systems, those of glutathione and thioredoxin 2 (Trx2) (Go and Jones 2008). Glutathione exists mainly in its reduced form under normal conditions, but can be oxidized when it interacts with ROS. The mitochondrial specific Trx2 acts as all other thioredoxins by reducing oxidized proteins through cysteine thiol-disulfide exchange and has proven to be essential in cellular metabolism (Chen et al. 2002; Tanaka et al. 2002). Through all of these systems, mitochondria generally maintain their redox status, contributing minimally to the overall redox status of cells (St-Pierre et al. 2002).

When any of the systems previously mentioned malfunction, it may lead to the overproduction of ROS by mitochondria which shifts the cellular redox status towards oxidative stress. In fact, mitochondria play a key role in the intrinsic pathway of apoptosis (Fleury et al. 2002). In this pathway, ROS tend to accumulate in

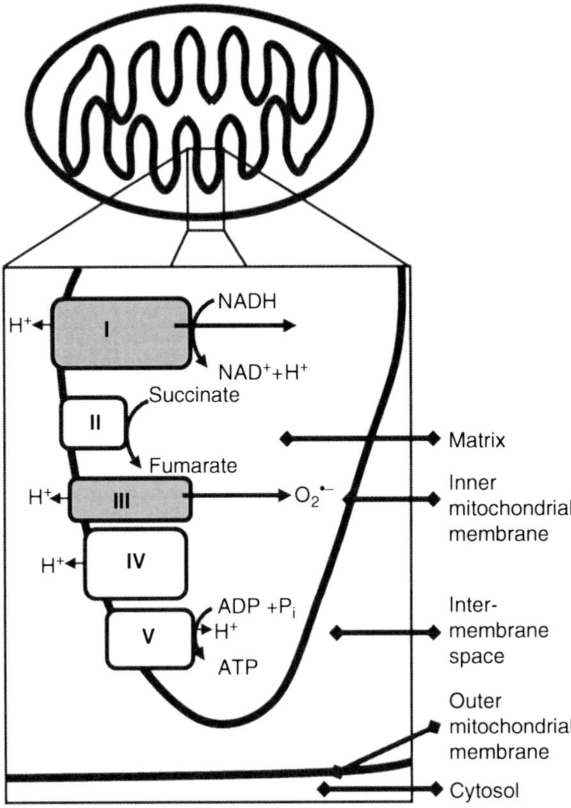

Fig. 3.2 The movement of electrons in mitochondria and the locations of superoxide ($O_2^{\bullet-}$). Roman numerals indicate complex number. Complexes I, III and IV transport protons out of the mitochondrial matrix, but complex V transports them back into the matrix to enable it to produce ATP. Complexes I and III (*shaded in grey*) are the primary sources of $O_2^{\bullet-}$ production

mitochondria and once the mitochondrial membrane potential depolarizes, there is a release of ROS and cytochrome c from these organelles and cells undergo apoptosis. This overproduction of ROS by mitochondria leading to increased cell death has been proposed as one of the factors leading to various neurodenegerative diseases, but the link has yet to be conclusively proven (Mancuso et al. 2006). So even though mitochondria generally do not contribute to the overall redox status of cells, their production of ROS can play a vital role to cellular health in certain conditions.

3.2.2 Nox Proteins

As mitochondria generally tend to have a balanced redox status, other sources of ROS need to be considered when discussing protein oxidation. The best known non-mitochondrial source of ROS is the Nox family of proteins. These proteins have 6 transmembrane domains, which form a channel to allow the successive transfer of electrons from NADPH to FAD to heme and then to O_2 to make $O_2^{\bullet-}$ (Fig. 3.3) (Sumimoto 2008). In total, there are seven members of the Nox family, Nox1-5 and Dual Oxidases (Duox) 1 and 2. Nox1-4 are similar in sequence (Suh et al. 1999; Shiose et al. 2001), whereas Nox5 is significantly different due to its four EF domains which confer a dependency on calcium (Banfi et al. 2001, 2004a, b). Duox1 and 2 are similar to Nox5, as they also have EF domains (Dupuy et al. 1999), but they stand apart from the others because of having a peroxidase-like domain (Edens et al. 2001), which means they produce H_2O_2 as their final product (Geiszt et al. 2003). All Nox proteins generate some form of ROS and so can be involved in protein oxidation. Indeed, ROS produced by Nox proteins have been shown to play a role in cell survival through protein oxidation (Vaquero et al. 2004; Mackey et al. 2008).

Nox family members have different tissue expression patterns, which determine where they produce ROS and so which proteins they can oxidize. Nox2, originally called gp91phox, is the prototype member of this family. It is highly expressed in phagocytes where it produces ROS to destroy engulfed pathogens (Iyer et al. 1961; Royerpokora et al. 1986). Nox1 has high expression in the colon (Cheng et al. 2001). Nox3 is expressed predominantly in the inner ear (Banfi et al. 2004a, b) and in some foetal tissues (Cheng et al. 2001). Nox4 has high expression in the kidney (Geiszt et al. 2000), but is found at low levels in many tissues (Cheng et al. 2001). Nox5 is highly expressed in testes, spleen and lymph nodes (Banfi et al. 2001). Duox1 and 2 are mainly found in the thyroid and in airway epithelia (Donko et al. 2005). Due to the widespread distribution of Nox proteins, they may produce ROS in a range of conditions and hence affect protein oxidation in a variety of circumstances.

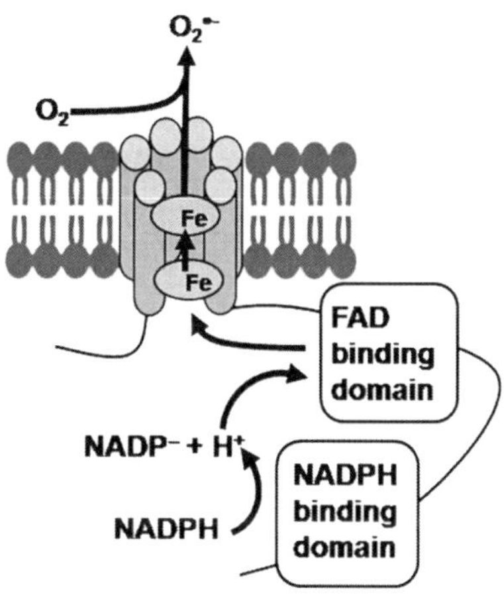

Fig. 3.3 A schematic representation of Nox1-4 illustrating the transfer of electrons (*thick arrows*) from NADPH to FAD to the first iron containing heme group (Fe) to the second heme and finally to oxygen to form superoxide ($O_2^{\bullet-}$)

To understand exactly how Nox proteins produce ROS, which may lead to protein oxidation, it is important to understand their regulation (Lambeth et al. 2007). Nox2 usually resides in the cytoplasmic membrane with p22phox. In the resting cell, the other signaling partners, p47phox, p67phox and p40phox, form a large complex in the cytoplasm (Wientjes et al. 1993). During cellular stimulation, the auto-inhibition of p47phox is removed through phosphorylation (Inanami et al. 1998). This phosphorylation is correlated with recruitment of Rac to this complex and together these four proteins translocate to Nox2 in the membrane resulting in the production of $O_2^{\bullet-}$. The regulation of the other Nox family members has been less comprehensively investigated, although that of Nox1 appears similar with NoxO1 being homologous to p47phox, and NoxA1 homologous to p67phox. Nox5, Duox1 and 2 do not require any of the known regulatory subunits (Kawahara et al. 2005) and their activity is dependent on calcium (Banfi et al. 2001; Banfi et al. 2004a, b). There is still much debate over the regulation of Nox3 and Nox4 (Bedard and Krause 2007). To date many varied events such as insulin signaling (Mahadev et al. 2004) and adherence to fibronectin (Edderkaoui et al. 2005) have been shown to activate various Noxes upstream of the events detailed and lead to ROS formation. Once the regulation of these proteins is understood fully, better control of overall cellular redox status and hence protein oxidation will be possible.

3.2.3 *Other Non-Mitochondrial Sources*

There are sources of ROS other than mitochondria and Nox proteins, which can lead to protein oxidation, but much less is known about them. Xanthine oxidoreductase (XOR) is an enzyme that can be found in two forms, xanthine oxidase (XO) and xanthine dehydrogenase (XDH) (Chung et al. 1997). XO is capable of producing $O_2^{\bullet-}$ and uric acid from xanthine and O_2 XDH, however, produces NADH and uric acid using xanthine and NAD. The switch between these two forms is complex (Nishino et al. 2008). XOR is found in many mammalian tissues, predominantly in epithelia and in the liver (Chung et al. 1997). Its precise physiological function remains unknown, but it is thought to contribute to liver metabolism through the degradation of xanthine to uric acid and is important in some cardiovascular disorders (Boueiz et al. 2008).

Myeloperoxidase (MPO) is an ROS generating enzyme mainly expressed in macrophages (Malle et al. 2007). It combines H_2O_2 with chloride (or other halides) to form hypochlorous acid, which is very toxic and acts as an anti-microbial agent. Despite its limited presence amongst cells in the body, MPO can affect many different organs, due to the ability of macrophages to travel and secrete MPO (Miyasaki et al. 1991). Hence, it has been listed as a causative agent in several diseases e.g. renal injury, cancer and multiple sclerosis (Klebanoff 2005).

Other enzymes contribute to ROS production indirectly, through generating different types of reactive species that can eventually interact with O_2 to form ROS. The three main families of this type are lipoxygenases (LOX), cyclooxygenases (COX) and nitric oxide synthases (NOS). The LOX family uses O_2 and arachidonic acid to form leukotriene, lipoxins and their derivatives (Kuhn and Thiele 1999). During this process unstable intermediaries and by-products can form and these act as ROS (Kim et al. 2008). Some of these products are also known to directly activate Nox proteins and so generate more ROS (Sadok et al. 2008). COX enzymes also use arachidonic acid, but to create prostaglandins again via reactive intermediates (Suleyman et al. 2007). NOS usually generate nitric oxide ($NO^{\bullet}$) which can interact with $O_2^{\bullet-}$ to form the highly reactive molecule peroxynitrite that is implicated in many cardiovascular diseases (Puddu et al. 2008). Also, if NOS become uncoupled, they no longer transfer electrons to arginine, but to O_2 to form $O_2^{\bullet-}$ directly (Vasquez-Vivar et al. 1998). All of the enzymes that contribute indirectly to ROS formation may be the ultimate cause of protein oxidation and they should therefore be given due consideration when examining this post-translational modification.

3.3 Targets of ROS

ROS can react with amino acid and oxidize both the backbone and their side chain leading to a loss of function of proteins and deactivation of enzymes. High concentration of H_2O_2 may lead to irreversible damage, followed by cell death; however it is not always the case and H_2O_2 is capable of reversible inhibition of many proteins (i.e. phosphatases) along the main survival pathways.

3.3.1 Amino Acid Oxidation

Almost all the amino acids residues in the proteins can be oxidized by ROS. However the sulfur containing amino acids (cysteine and methionine) and the aromatic amino acid (tyrosine and tryptophan) are the most sensitive to oxidation. Oxidation produces hydroxyl and carbonyl groups on proteins and can also induce more significant changes such as intra or inter molecular crosslinking. Secondary oxidations, whose consequences are not necessarily less important, include the lipid oxidation leading to the release of lipid peroxidation products such as the malondialdehyde, particularly reactive with amino acids like cysteine or lysine. In turn, they will form what is called the additions of Michael (for review see Stadtman 2001).

The aromatic amino acid residues of protein are prime target for oxidation by various forms of ROS. As shown in Fig. 3.4 phenylalanine residues are oxidized to ortho- and meta-tyrosine derivatives. Tyrosine residues are converted to the 3, 4-dihydroxy derivative (DOPA) and also to bi-tyrosine cross-linked derivatives. Tryptophan residues are converted to either 2-, 4-, 5-, 6-, or 7-hydroxyl derivatives, and also to N-formylkynurenine. Histidine residues are oxidized to 2-oxohistidine (Stadtman and Levine 2003).

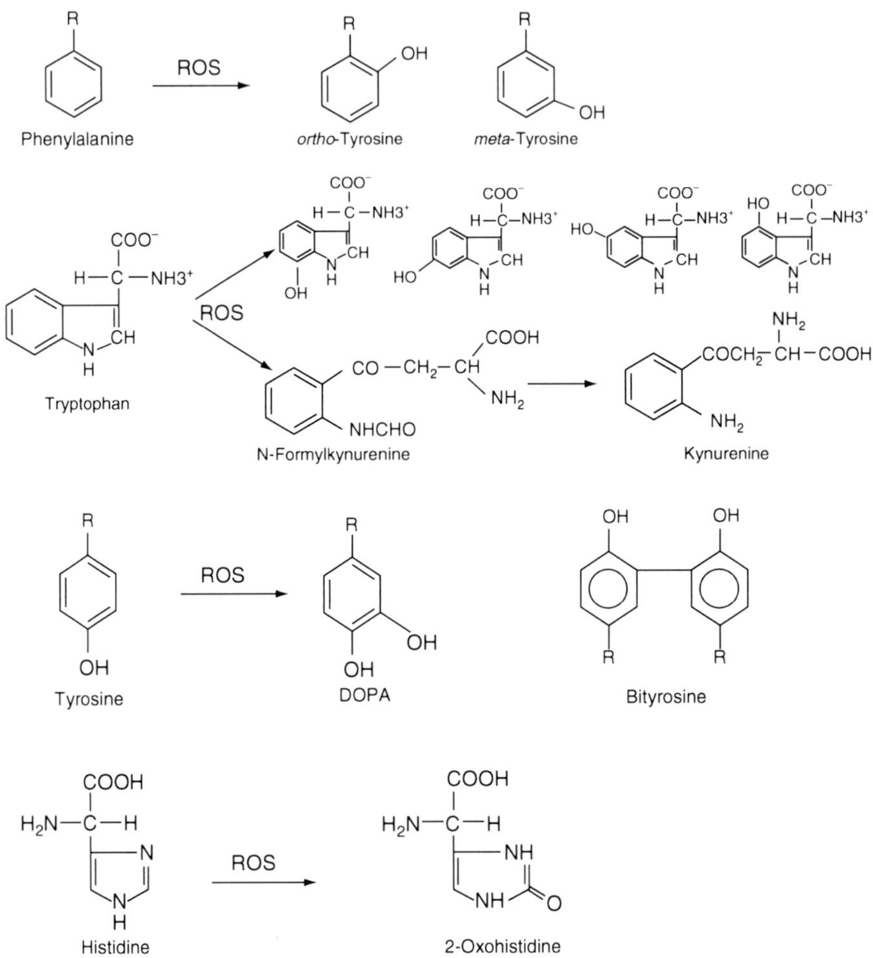

Fig. 3.4 Oxidation of aromatic amino acid residues

The sulfur containing amino acid residues are particularly susceptible to oxidation by various ROS. Methionine is a sulfur containing amino acid residue, and though its role in proteins is not well defined it is likely that methionine can function as a key component in the regulation of cellular metabolism. Methionine is oxidized to methionine sulfoxide by many different ROS and reactive nitrogen species (RNS). The methionine sulfoxide reductases have the potential to reduce the residue back to methionine. Such reversible modifications have been recognized to provide the mechanistic basis for most cellular regulation (phosphorylation/dephosphorylation being another way of regulation which is well studied). Methionine/methionine sulfoxide interconversion can thus function to regulate biological activity of proteins through modification of the catalytic efficiency or through modification of the surface hydrophobicity of the protein (Levine et al. 1996).

3.3.2 *Cysteine Oxidation*

Cysteine oxidation is of particular interest because it is present in the active site of several enzymes involved in a cell survival response to stress. For a long time ROS were considered harmful for the cell, but it is increasingly clear ROS (in particular H_2O_2) plays a significant role as a signaling molecule in different survival pathways.

3.3.2.1 Phosphatases

The regulation of cell function through redox-sensitive cysteine residues has been most convincingly demonstrated in protein tyrosine phosphatases (PTP). The main survival pathways, being the mitogen-activated protein kinase (MAPK) or the Pi3kinase/Akt pathway, transduce their signal for cell survival mainly through phosphorylation of target molecules. The MAPK pathways operate in a cascade fashion with MAPKKK phosphorylating and activating MAPKK which then activates MAPK; the PI3K/Akt pathway operates by regulating the phosphorylation of Akt and GSK3β. As a consequence of the survival pathway transducing its survival signal via phosphorylation of key proteins, phosphatases are potent negative regulators. Multiple steps along the PI3-kinase/Akt survival pathway are negatively regulated by protein phosphatases (PP).

The PP superfamily can be categorized into three smaller subfamilies based on substrate specificity; (a) the classical protein phosphatase which is a tyrosine-specific phosphatase (PTP); (b) the dual-specificity phosphatases (DSP) which can dephosphorylate phospho-tyrosine, phospho-serine and phospho-threonine containing susbstrates; and (c) serine/threonine (Ser/Thr)-specific phosphatases which are further divided into two major classes. Type I phosphatase includes PP1; Type II phosphatase include spontaneously active phosphatase (PP2A), calcium dependent (PP2B) or magnesium dependent (PP2C) classes of phosphatases (for more information regarding classification and actions of protein phosphatases see Chap. 10 of this book).

All PTPs contain an essential cysteine residue in the signature motif C[X]5R that exists as a thiolate anion at neutral pH (Denu and Dixon 1998). This thiolate anion contributes to the formation of a thiol-phosphate intermediate in the catalytic mechanism of PTPs. Oxidation of the active-site cysteine of PTPs to a sulfenic derivative leads to enzymatic inactivation, however this modification can be reversed by incubation with thiol compounds. In some cases, the sulphydryl group is open to further irreversible oxidation if no cysteine derivatives or thiols are close enough to facilitate the formation of a disulphide bridge. The addition of another oxygen molecule or two additional oxygen molecules results in the formation of sulphinic and sulphonic acid, respectively. These oxidative modifications are irreversible and the phosphatase will be unable to become active again even in a reducing environment. It is highly likely that all phosphatases are sensitive to oxidative inhibition to some degree, as they all require a reduced cysteine for catalysis.

Reversible oxidation was first demonstrated for PTP-1B during EGF (Lee et al. 1998) and insulin (Mahadev et al. 2001) signaling. The same regulation was then demonstrated for low molecular weigh-PTP (LMW-PTP) during PDGF stimulation (Chiarugi et al. 2001). Both PTP-1B and LMW-PTP rescued their phosphatase activity thanks to a re-reduction 30 min after receptor activation (Barrett et al. 1999; Caselli et al. 1998). Reversible oxidation of the Scr-homology-2 domain (SHP-2) PTP has also been reported (Meng et al. 2002).

The dual specificity phosphatases have also been reported to be oxidized, but in contrast with the classical PTP member like PTP-1B that forms a sulphenyl amide linkage between the active-site cysteine and an adjacent main-chain nitrogen (Salmeen and Barford 2005), oxidation of PTEN in vitro with H_2O_2 leads to the formation of a disulphide bond between the active site cysteine (Cys-124) and another cysteine residue (Cys-71) which is close by in the three dimensional structure of PTEN; this disulphide bond prevents further irreversible oxidation of the cysteine residues and PTEN can retain its phosphatase ability. PTEN has been demonstrated with many studies to be the main phosphatase negatively regulating the Pi3-kinase/Akt pathway. PTEN can dephosphorylate the lipid PIP3 to PIP2 preventing the recruitment of PH containing proteins to the plasma membrane. This results in a decrease in the survival signal transduced by Akt. PTENs importance in regulating this pathway is highlighted by its classification as a tumour suppressor molecule. Savitsky and Finkel (2002) have provided evidence for the degradation of cdc25 phosphatase (another DSP) which is the result of H_2O_2-induced disulfide bond formation between the active-site cysteine and another invariant cysteine residue.

The serine/threonine phosphatases [the main members being Protein Phosphatase 1 (PP1), Protein Phosphatase 2A (PP2A), Protein Phosphatase 2B (PP2B) and Protein Phosphatase 2C (PP2C)] dephosphorylate serine and threonine which are the main phosphorylation events in the transduction of the PI3-kinase/Akt survival pathway. Various studies have demonstrated links between these phosphatases and the Pi3-kinase/Akt pathway. For example the calcium activated PP2B has been shown to be a direct phosphatase of Akt, Gsk3β, and Bad (Millward et al. 1999; Klumpp et al. 2003). PP1α has been shown to dephosphorylate Akt and Bad, while PP2A has been demonstrated to be a key Akt phosphatase. PP2A can dephosphorylate Akt on both threonine 308 and serine 473 blocking the Pi3-kinase/Akt pathway. Immunoprecipitation studies have shown that PP2A can co-localize with Akt. Recently a novel phosphatase belonging to the PP2C family of phosphatases has been shown to dephosphorylate Akt on Ser473 only. This novel phosphatase (PHLPP) contains a PH domain that localizes it in the vicinity of activated Akt. Expression of PHLPP in cells derived from a small cell lung cancer that have constitutively active Akt (phosphorylated on Ser473 and Threo409), led to a 50% decrease in phosphoserine-473 Akt levels. This resulted in a correlating decrease in the phosphorylation levels of an Akt substrate, and in increase in the basal levels of apoptosis in this cell line (Gao et al. 2005; Vandermoere et al. 2005). These experiments highlight the importance of the phosphorylated serine residue in Akt signaling and demonstrate a clear link between the increased activity of PHLPP with an induction of apoptosis via the PI3-kinase/Akt pathway.

Inactivation of PP2A has been shown in cells treated with Tumor Necrosis Factor-alpha (TNF-α) or interleukin-1 (Guy et al. 1995) which can both induce H_2O_2 production. As described for the PTPs and DSP, the Ser/Thr phosphatases are thought to be sensitive to redox modifications. It has been demonstrated in vitro that hydrogen peroxide can reversibly block some of the main Ser/Thr phosphatases like PP2A, and PP1α (Rao and Clayton 2002; O'Loghlen et al. 2003). PP1 and PP2A contain redox-sensitive Cys residues (Fetrow et al. 1999; Guy et al. 1995). Structure based analysis has identified a potential disulfide oxidoreductase active site, Cys-X-X-Cys, in members of the PP1 subfamily (Fetrow et al. 1999). It is still not clear whether the oxidation of this pair of cysteine residues can result in PP1 inactivation in vivo.

3.3.2.2 Caspases

Under certain conditions ROS may directly affect the activity of cell death effector proteins. Several effectors of apoptosis are redox sensitive and their functions can be directly modulated by intracellular ROS, those effectors are caspases, Bcl-2 and cytochrome c.

A hallmark of apoptosis is the activation of caspases which requires sequential proteolysis of the initiator caspases and effector caspases. Stimuli that induce apoptosis can trigger caspase activation either through the extrinsic (death receptor-mediated activation) or the intrinsic (mitochondria mediated activation) pathways. Activation of the caspase cascade ultimately leads to the cleavage of different target proteins such as poly(ADP-ribose) polymerase (PARP) and α-fodrin leading to cell death. ROS can directly affect functions of caspases; the reduced state of the cysteine in the active site is necessary for the catalytic activity of caspases, thus depending on the degree of intracellular oxidative stress caspases can be activated or inhibited. Using different concentration of exogenous H_2O_2 Hampton et al. demonstrated that a low dose of H_2O_2 can activate caspases and induce apoptosis, but high dose of H_2O_2 inhibits caspases and cells undergo necrosis (Hampton and Orrenius 1997).

3.3.2.3 Transcription Factors

Another way for the cell to regulate cell survival through protein oxidation is through activation of transcription factors. Different transcription factors are known to be redox sensitive; these include p53, HIF, AP-1 and NF-κB.

The heterodimeric protein NF-κB is a ubiquitous redox-regulated transcription factor that remains in the cytoplasm as an inactive complex with its inhibitory counterpart IκBα. Exposure to oxidative stimuli leads to phosphorylation and subsequent proteasomal degradation of IκBα, thereby releasing free NF-κB dimers for translocation to the nucleus. Experimental evidence suggests that ROS seem to have paradoxical effects on NF-κB regulation. ROS can either activate or inhibit

NF-κB activity depending on the ROS levels, type of stimuli and cell types. Moderate level of ROS generally leads to NF-κB activation. On the contrary high level of ROS could inactivate NF-κB leading to cell death. In the nucleus, direct oxidation of the redox-sensitive Cys62 of the p50 subunit inhibits its availability to bind DNA (Toledano and Leonard 1991). This oxidation is reversible and DNA binding can be restored. Besides direct structural modifications, DNA binding activity of NF-κB can be modulated by chromatin remodelling (Rahman et al. 2004). Thus, the enzyme histone deacetylase (HDAC), which catalyses the removal of an acetyl group from histone can be inactivated by oxidative stress allowing histone acetylation, chromatin uncoiling and increased accessibility for NF-κB (Rahman et al. 2002). In the cytosol, NF-κB is sequestrated as a complex formed with its inhibitor IκB, its activation is then regulated by phosphorylation of NF-κB itself or phosphorylation of its inhibitor. Under certain condition H_2O_2 is able to directly activate NF-κB activity through phosphorylation of IκB-kinase (IκB) (Kamata et al. 2002) or indirectly through activation of Akt and/or MEKK1 which then phosphorylates and activates IκB. The transactivation of NF-κB induced by Akt or MEKK1 has been shown really important in NF-κB anti-apoptotic effects (Nawata et al. 2003; Vandermoere et al. 2005). Active IκB phosphorylates IκB and liberates active NF-κB from the complex to translocate to the nucleus. Phosphorylated IκB is then degraded by the proteasome. Since the proteasome system is also redox-sensitive, ROS can regulate NF-κB activity by affecting IκB stability.

3.3.3 Carbonylation/Nitrosylation of Proteins

As described above some ROS-induced protein modifications are benign events and can even promote cell survival. However, irreversible modifications can also result in unfolding or alteration of protein structure leading to inactivation of various proteins.

Carbonylation is an irreversible, non enzymatic modification of proteins. The chemistry of the reaction that gives rise to carbonyl groups is well described in reviews by Stadtman and coworkers. Lysine, arginine, proline and threonine residues of proteins are particularly sensitive to metal-catalysed oxidation leading in each case to the formation of carbonyl derivatives (Dalle-Donne et al. 2006). Briefly there are four different oxidative pathways by which carbonyl groups are introduced into proteins: (a) direct oxidation of Lys, Arg, Pro and Thr residues' side chain especially via metal-catalyzed oxidation; (b) oxidative cleavage of the protein backbone by the α-amidation pathway or by oxidation of glutamyl residues; (c) adduction of reactive aldehyde derived from the metal-catalyzed oxidation of polyunsaturated fatty acids (Lys, His and Cys are the residues reacting preferentially with the lipoxidation products) (Refsgaard et al. 2000); (d) reaction with reactive carbonyl derivatives generated as a consequence of the reaction of reducing sugars or their oxidation products with lysine residue of proteins.

The chemical modification of protein by reactive carbonyl compounds derived from lipid peroxidation reactions results in the formation of advanced lipoxidation end-products (ALEs) while the one derived from sugar oxidation results in advanced glycation-products (AGEs). Whereas moderately carbonylated proteins are degraded by the proteasome, heavily carbonylated proteins tend to form aggregates that are resistant to degradation and accumulate as damaged or unfolded proteins; those aggregates can even lead to an inactivation of the proteasome. The presence of carbonyl groups in proteins is used as a marker of ROS-mediated protein oxidation. Considering the presence of carbonyl groups it has been established that protein oxidation is associated with aging, oxidative stress and a number of diseases; the identification of specific carbonylated proteins should then provide new diagnostic tools for human diseases (Dalle-Donne et al. 2003).

Nitric oxide ($NO^{\bullet}$) is generated from arginine by the action of nitric oxide synthase. It is an important signaling molecule playing a major role in physiological processes such as smooth muscle relaxation and neurotransmission. However, depending on the redox state of the cell $NO^{\bullet}$ can also induce oxidative stress even though it can not be categorized as a classical ROS it is part of the reactive nitrogen species (RNS) (for a review see Moncada and Erusalimsky 2002). $NO^{\bullet}$ reacts with superoxide to give peroxynitrite ($ONOO^-$) which under physiological conditions can react with CO_2 to form nitrosoperoxocarboxylate (NPC or $ONOOCO_2^-$); both those RNS are particularly harmful for the cell.

Cysteine and methionine residues are particularly sensitive to oxidation by peroxynitrite. Peroxynitrite is responsible for the oxidation of methionine residues to methionine sulfoxide and the nitration of protein sulfhydryl groups to form S-nitrosothiol derivatives (Leseney et al. 1999). The covalent attachment of a nitrogen monoxide group to the thiol side chain of cysteine is called an S-nitrosylation and has emerged as an important post-transcriptional modification of signaling proteins. Both tyrosine and tryptophane residues are selective targets for nitrosoperoxocarboxylate (NPC)-dependent nitration. The nitration of tyrosine residue is particularly important because nitration prevents tyrosine residue to undergo cyclic interconversion between phosphorylated and unphosphorylated forms (Hunter 1995); it is an irreversible process thus locking the targeted enzyme into an inactive configuration. Accordingly transduction signals involving a phospho-tyrosine (such as interferon-α signaling) could be inactivated by nitration of the same tyrosine.

3.4 Protein Oxidation and Disease

Reactive oxygen species such as H_2O_2, $O_2^{\bullet-}$ and hydroxyl radicals ($HO^{\bullet}$) are all biologically relevant oxygen radicals which are routinely produced as normal by-products of many different metabolic processes, and small amounts of ROS are needed by cells to act as signaling molecules (Halliwell and Gutteridge 1999; Hensley et al. 2000). However ROS have the potential to induce significant biological damage. If the sensitive redox status of a cell is perturbed in some way this can result in an imbalance in the delicate intracellular oxidant/antioxidant status and can lead to accumulation of ROS and ultimately to oxidative stress-induced injury and cell death. It is the increased concentration and the body's inability to effectively deal with these ROS that can often lead to the oxidation of various biochemicals, such as lipids, nucleic acids, sugars and proteins.

It is still remains relatively unclear whether oxidative stress is actually the cause of or the consequence of many human pathologies; nevertheless it is now widely accepted that aberrant oxidative stress is intricately linked to the progression of specific diseases such as Alzheimer's and Parkinson's. ROS can directly affect proteins by oxidation of both the backbone and amino acid side chains, and can also interact with other biomolecules such as sugars and lipids generating products that can then go on to also interact with proteins. Oxidative modification of proteins can have many detrimental biological affects and as such this post-translational modification often has a role to play in the pathogenesis several diseases.

As protein activity and function are both tightly regulated by protein conformation, oxidative damage of proteins can in some cases lead to abnormal formation of aggregated cross-linked protein which may be resistant to proteinase degradation and thus lead to accumulation of such aggregates (Butterfield and Stadtman 1997). In some cases oxidation can convert proteins into forms that are more susceptible to degradation by proteases (Stadtman 1990). This interference in normal protein turnover and loss of protein and enzyme catalytic function can ultimately impede the normal functioning of a cell and lead to pathological alterations.

Increased levels of oxidised proteins have indeed been associated with a number of diseases including Alzheimer's disease (AD) (Smith et al. 1991; Markesbery and Lovell 2007), muscular dystrophy (Murphy and Kehrer 1989), cataractogenesis (Davies and Truscott 2001), rheumatoid arthritis (Hitchon and El-Gabalawy 2004), respiratory distress syndrome (Lamb et al. 1999) and progeria (Hutchinson-Gilford progeria syndrome) (Oliver et al. 1987). Although not directly implicated there is also substantial evidence to suggest that oxidatively modified proteins are implicated in the pathogenesis of cystic fibrosis (CF) (Salh et al. 1989; Starosta et al. 2006), Parkinson's disease (PD) (Danielson and Andersen 2008), diabetes (Cakatay 2005), atherosclerosis (Brennan and Hazen 2005), essential hypertension (Puddu et al. 2008), diabetes (Kubisch et al. 1994) and ulcerative colitis (Babbs 1992).

AD is the most prevalent dementia in the elderly population. It affects approximately 16 million people worldwide and beyond the age of 65 the incidence doubles every 5 years (Goedert and Spillantini 2006). AD brains show evidence of ROS-mediated injury (Praticò and Sung 2004) and there is an increasing body of evidence indicating protein oxidation as one of the main contributors towards Alzheimer's associated neurofibrillary degeneration and concurrent cognitive deterioration (Butterfield and Stadtman 1997; Aksenov et al. 2001).

The irreversible formation and accumulation of protein carbonyls is very often a strong indicator of severe oxidative protein damage and protein carbonylation is thus the most commonly used marker for protein oxidation (Dalle-Donne et al. 2003). The production of protein carbonyls can arise via the direct oxidation of amino-acid side chains and through oxidative cleavage of proteins. Carbonyl groups can also be added to proteins via reactions of unsaturated aldehydes which are derived from lipid peroxidation, and can also be introduced by addition of reactive carbonyl derivatives produced by the reaction of reducing sugars or their oxidation products (Dalle-Donne et al. 2003).

Carbonylation of proteins often leads to loss of protein function and accumulation of proteolysis-resistant aggregates of carbonylated proteins in tissues has been noted for a large number of neurodegenerative diseases (Dalle-Donne et al. 2006). When ROS attack protein side-chains it can cause specific chemical alterations such as the formation of hydroxyl and carbonyl groups as already discussed. Such changes in turn can lead to loss of protein function, and the identification of specific functionally impaired carbonylated proteins is just one hallmark of the AD brain (Stadtman 1990). This oxidative modification of proteins is pertinent to the pathogenesis of AD and it is these protein carbonyls that have been detected in significantly increased amounts in the AD brain (Markesbery and Carney 1999; Starosta et al. 2006).

Parkinson disease (PD) is the second most common neurodegenerative disorder (Farrer 2006). PD is a severe and debilitating neurodegenerative syndrome caused by various factors including genetic susceptibility, the ageing process and various environmental factors (Danielson and Andersen 2008). There is no cure as yet and the disease affects around 1–2% of the population over 50 years old (Thomas and Beal 2007). The exact molecular pathways involved in the pathological progression of this disease are still obscure. There is substantial evidence however that increased oxidative stress and decreased levels of antioxidants (Pearce et al. 1997) appear to be common underlying factors involved in the loss of dopaminergic neurons from the substantia nigra (SN), the region of the brain most prominently affected by the disease (Farrer 2006). As is the case with Alzheimer's increased levels of oxidized protein carbonyls have also been detected in the brain of PD sufferers (Alam et al. 1997) and in a more recent study it was demonstrated that antioxidants efficiently prevented protein carbonylate formation (Esteves et al. 2009) suggesting this as a potential treatment.

There are also a number of specific proteins that have been found to be post-translationally modified via oxidation in the brains of PD patients, for example the protein DJ-1 is thought to function as an antioxidant (Taira et al. 2004) and mass spectrometric data have shown increased amounts of methionine oxidized DJ-1 in the brains of patients with sporadic PD thus suggesting a role of methionine oxidation of this particular protein in disease pathogenesis (Choi et al. 2006). Although it is unclear as to whether oxidative stress and subsequent oxidative modification of proteins play a primary role in the initiation of PD or whether it is simply a secondary effect due to the disease progression, the fact that several oxidised proteins have been described in PD brains would suggest that these modifications seem to be highly relevant to and contributors towards PD (Danielson and Andersen 2008).

Not only does protein oxidation seem to play a significant role in the aforementioned neurodegenerative disorders but also in the more common condition of cataractogenesis. Cataract development in the ageing population affects more than 20 million people worldwide and is the leading cause of blindness in developing countries (West 2007). Cataracts form when the lens of the eye becomes opaque and it is believed that oxidative stress is an initiating factor in the development of age-related cataracts (Spector 1995). The protein redox status seems to be crucial to maintain proper function of the lens and lens transparency, and, when this delicate balance is upset it can lead to development of cataracts. In the lens, one particular group of proteins, crystalins, constitute around 90% of the total protein content and are susceptible to age-related oxidative changes (Boscia et al. 2000), including disulfide bond formation, inter- and intramolecular cross-linking, and methionine oxidation, all of which result in production and accumulation of high molecular weight aggregates (Siew et al. 1981). Generally any protein that has been abnormally modified will be degraded via proteasomal degradation (Jung and Grune 2008). However, in the case of cataractogenesis these oxidatively modified proteins are not recognized by the host, are not degraded and thus accumulate in the lens over time causing the formation of cataracts (Spector 1995; Williams 2006).

Cystic fibrosis (CF) is a condition in which protein oxidation is believed to contribute towards disease progression. CF is a chronic, progressive, genetic disease (Schidlow 2000) involving persistent inflammation and continuous periods of infection leading to progressive damage of the lungs and pulmonary fibrosis which is eventually responsible for over 90% of the mortality linked with this disease (Davis 1993). There is now evidence to support the hypothesis that free radical protein damage is involved in the pathogenesis of the disease. In a study by Starosta et al. (2006) the level of oxidative stress was measured in CF patients by assessing protein oxidation from the content of protein carbonyls in their bronchoalveolar fluid. It was found that CF patients had significantly higher levels of protein carbonyls in their bronchoalveolar fluid when compared to healthy test subjects, these data thus supported the long held hypothesis that an over abundance of ROS may be one of the major contributing factors towards the gradual destructive pulmonary damage seen in CF patients (Starosta et al. 2006).

It is widely accepted that protein oxidation has a major role to play in the normal process of aging (Chakravarti and Chakravarti 2007) so it is interesting to note that the premature aging disease progeria is also characterized by increased levels of protein oxidation in the form of protein carbonyls in sufferers of the disease (Oliver et al. 1987). Progeria is a rare autosomal-dominant disorder characterized by the appearance of premature signs of aging (Hennekam 2006; Kieran et al. 2007). In a study by Oliver et al. (1987) the levels of oxidized proteins in cultured human fibroblasts from healthy individuals and from progeria sufferers were determined by measuring carbonyl content. It was discovered that the carbonyl content of the progeria samples was significantly higher than that of the age-matched controls; in fact the levels were comparable to those found in cultured fibroblasts from 80-year-old subjects (Oliver et al. 1987).

Generally the intracellular level of an oxidized protein is determined by the balance between the rate of protein oxidation and the rate at which the oxidized protein is degraded. The two processes, protein oxidation and degradation of oxidized proteins, are multifactorial activities controlled by a number of different events including the production of ROS, the intracellular redox potential and the availability and concentration of antioxidants: It is the fine balance between these myriad of factors that determines the steady state level of oxidized protein in cells

(Stadtman and Berlett 1997). As previously stated there is now sufficient evidence to demonstrate that protein oxidation is most likely a key contributor to the dysfunction associated with the aforementioned pathologies although it is important to bear in mind that a definitive causal relationship between protein oxidation and the etiology or progression of the aforementioned diseases has not been thoroughly established. Nevertheless in the cases of neurodegenerative diseases such as AD there is a definite positive correlation between increased amounts of oxidized proteins in the brain of sufferers and increased progression of the disease (Forster et al. 1996; Carney et al. 1991).

3.5 Concluding Remarks

As previously discussed there are many sources of ROS in a biological system, the two primary sources being mitochondria and the Nox proteins. Protein oxidation seems to play a vital role in many different biological functions; the reversible oxidation of proteins which usually takes place under normal physiological conditions allows various signaling pathways to operate, however, during conditions of oxidative stress the irreversible oxidation of proteins is thought to play a significant role in the etiology and or progression of certain diseases. The challenges that lie ahead are to understand the mechanisms that drive disease-associated oxidative stress, to elucidate diagnostic biomarkers for oxidative damage during disease and develop effective therapeutic options such as antioxidant treatments.

Acknowledgments This work was supported by Science Foundation Ireland, Children Leukemia Research Project and the Irish Cancer Society.

References

Aksenov, M.Y., Aksenova, M.V., Butterfield, D.A., et al. (2001). Protein oxidation in the brain in Alzheimer's disease. *Neuroscience* 103:373–383.

Alam, Z.I., Daniel, S.E., Lees, A.J., et al. (1997). Generalised increase in protein carbonyls in the brain in Parkinson's but not incidental Lewy body disease. *J. Neurochem.* 69:1326–1329.

Babbs, C.F. (1992). Oxygen radicals in ulcerative colitis. *Free Radical Biol. Med.* 13:169-181.

Banfi, B., Molnar, G., Maturana, A., et al. (2001). A Ca2+-activated NADPH oxidase in testis, spleen, and lymph nodes. *J. Biol. Chem.* 276:37594–37601.

Banfi, B., Malgrange, B., Knisz, J., et al. (2004). NOX3, a superoxide-generating NADPH oxidase of the inner ear. *J. Biol. Chem.* 279:46065–46072.

Banfi, B., Tirone, F., Durussel, I., et al. (2004). Mechanism of Ca2+ activation of the NADPH oxidase 5 (NOX5). *J. Biol. Chem.* 279:18583–18591.

Barja, G., and Herrero, A. (1998). Localization at complex I and mechanism of the higher free radical production of brain nonsynaptic mitochondria in the short-lived rat than in the longevous pigeon. *J. Bioenerg. Biomembr.* 30:235–243.

Barrett, W.C., DeGnore, J.P., Konig, S., et al. (1999). Regulation of PTP1B via glutathionylation of the active site cysteine 215. *Biochemistry* 38:6699–6705.

Bedard, K., and Krause, K.H. (2007). The NOX family of ROS-generating NADPH oxidases: Physiology and pathophysiology. *Physiol. Rev.* 87:245–313.

Boscia, F., Grattagliano, I., Vendemiale G., et al. (2000). Protein oxidation and lens opacity in humans. *Invest. Ophthalmol. Vis. Sci.* 41:2461–2465.

Boueiz, A., Damarla, M., and Hassoun, P.M. (2008). Xanthine oxidoreductase in respiratory and cardiovascular disorders. *Am. J. Physiol. Lung Cell. Mol. Physiol.* 294:L830–L840.

Brennan, M.L., and Hazen, S.L. (2005). Amino acid and protein oxidation in cardiovascular disease. *Amino Acids* 25:365–374.

Butler, J., Jayson, G.G., and Swallow, A.J. (1975). Reaction between superoxide anion radical and cytochrome C. *Biochim. Biophys. Acta* 408:215–222.

Butterfield, D.A., and Stadtman, E.R. (1997). Protein oxidation processes in aging brain. *Adv. Cell Aging Gerontol.* 2:161–191.

Cakatay, U. (2005). Protein oxidation parameters in type 2 diabetic patients with good and poor glycaemic control. *Diabetes Metab.* 31:551–557.

Carney, J. M., Starke-Reed, P. E., Oliver, C. N., et al. (1991) Reversal of age-related increase in brain protein oxidation, decrease in enzyme activity loss and loss of temporal and spatial memory by chronic administration of the spin-trapping compound N-tert-butyl-alpha-phenylnitrone. *Proc. Natl. Acad. Sci. U.S.A.* 88:3633–3636.

Caselli, A., Marzocchini, R., Camici, G., et al. (1998). The inactivation mechanism of low molecular weight phosphotyrosine-protein phosphatase by H_2O_2. *J. Biol. Chem.* 273:32554–32560.

Chakravarti, B., and Chakravarti, D.N. (2007). Oxidative modification of proteins: age-related changes. *Gerontology* 53:128–139.

Chen, Y., Cai, J.Y., Murphy, T.J., et al. (2002). Overexpressed human mitochondrial thioredoxin confers resistance to oxidant-induced apoptosis in human osteosarcoma cells. *J. Biol. Chem.* 277:33242–33248.

Cheng, G.J., Cao, Z.H., Xu, X.X., et al. (2001). Homologs of gp91phox: cloning and tissue expression of Nox3, Nox4, and Nox5. *Gene* 269:131–140.

Chiarugi, P., Fiaschi, T., Taddei, M.L., et al. (2001). Two vicinal cysteines confer a peculiar redox regulation to low molecular weight protein tyrosine phosphatase in response to platelet-derived growth factor receptor stimulation. *J. Biol. Chem.* 276:33478–33487.

Choi, J., Sullards, M.C., Olzmann, J.A., et al. (2006). Oxidative damage of DJ-1 is linked to sporadic Parkinson and Alzheimer diseases. *J. Biol. Chem.* 281:10816–10824.

Chung, H.Y., Baek, B.S., Song, S.H., et al. (1997). Xanthine dehydrogenase, xanthine oxidase and oxidative stress. *Age* 20:127–140.

Dalle-Donne, I., Giustarini, D., Colombo, R., et al. (2003). Protein carbonylation in human diseases. *Trends Mol. Med.* 9:169–176.

Dalle-Donne, I., Aldini, G., Carini, M., et al. (2006). Protein carbonylation, cellular dysfunction, and disease progression. *J. Cell. Mol. Med.* 10:389–406.

Danielson, S.R., and Andersen, J.K. (2008). Oxidative and nitrative protein modifications in Parkinson's disease. *Free Radical Biol. Med.* 44:1787–1794.

Davies, M.J., and Truscott, R.J. (2001). Photo-oxidation of proteins and its role in cataractogenesis. *J. Photochem. Photobiol. B* 63:114–125.

Davis, P.B.O. Pathophysiology of the lung disease in cystic fibrosis. In: Davis P.B., ed. Cystic fibrosis. New York: Marcel Dekker, 1993, pp. 193–218.

Denu, J.M., and Dixon, J.E. (1998). Protein tyrosine phosphatases: mechanisms of catalysis and regulation. *Curr. Opin. Chem. Biol.* 2:633–641.

Donko, A., Peterfi, Z., Sum, A., et al. (2005). Dual oxidases. *Philos. Trans. R. Soc. Lond. B. Biol. Sci.* 360:2301–2308.

Dupuy, C., Ohayon, R., Valent, A., et al. (1999). Purification of a novel flavoprotein involved in the thyroid NADPH oxidase. Cloning of the porcine and human cDNAs. *J. Biol. Chem.* 274:37265–37269.

Edderkaoui, M., Hong, P., Vaquero, E.C., et al. (2005). Extracellular matrix stimulates reactive oxygen species production and increases pancreatic cancer cell survival through 5-lipoxygenase and NADPH oxidase. *Am. J. Physiol. Gastrointest. Liver Physiol.* 289:G1137–G1147.

Edens, W.A., Sharling, L., Cheng, G.J., et al. (2001). Tyrosine cross-linking of extracellular matrix is catalyzed by Duox, a multidomain oxidase/peroxidase with homology to the phagocyte oxidase subunit gp91 phox. *J. Cell Biol.* 154:879–891.

Esteves, A.R., Arduíno, D.M., Swerdlow, R.H., et al. (2009). Oxidative stress involvement in α-synuclein oligomerization in Parkinson disease cybrids. *Antioxid. Redox Signal* 11:439–448.

Farrer, M.J. (2006). Genetics of Parkinson disease: paradigm shifts and future prospects. *Nat. Rev. Genet.* 7:306–318.

Fetrow, J.S., Siew, N., Skolnick, J. (1999). Structure-based functional motif identifies a potential disulfide oxidoreductase active site in the serine/threonine protein phosphatase-1 subfamily. *FASEB J.* 13:1866–1874.

Fleury, C., Mignotte, B., and Vayssiere, J.L. (2002). Mitochondrial reactive oxygen species in cell death signaling. *Biochimie* 84:131–141.

Forster, M. J., Dubey, A., Dawson, K. M., et al. (1996). Age-related losses of cognitive function and motor skills in mice are associated with oxidative protein damage in the brain. *Proc. Natl. Acad. Sci. U.S.A.* 93:4765–4769.

Gao, T., Furnari, F., and Newton, A.C. (2005). PHLPP: a phosphatase that directly dephosphorylates Akt, promotes apoptosis, and suppresses tumor growth. *Mol. Cell* 18:13–24.

Geiszt, M., Kopp, J.B., Varnai, P., et al. (2000). Identification of renox, an NAD(P)H oxidase in kidney. *Proc. Natl. Acad. Sci. U.S.A.* 97:8010–8014.

Geiszt, M., Witta, J., Baffi, J., et al. (2003). Dual oxidases represent novel hydrogen peroxide sources supporting mucosal surface host defense. *FASEB J.* 17:1502–1504.

Go, Y.M., and Jones, D.P. (2008). Redox compartmentalization in eukaryotic cells. *Biochim. Biophys. Acta* 1780:1271–1290.

Goedert, M., and Spillantini, M.G. (2006). A century of Alzheimer's disease. *Science* 314:777–781.

Guidot, D.M., Repine, J.E., Kitlowski, A.D., et al. (1995). Mitochondrial respiration scavenges extramitochondrial superoxide anion via a nonenzymatic mechanism. *J. Clin. Invest.* 96:1131–1136.

Guy, G.R., Philp, R., and Tan, Y.H. (1995). Activation of protein kinases and the inactivation of protein phosphatase 2A in tumour necrosis factor and interleukin-1 signal-transduction pathways. *Eur. J. Biochem.* 229:503–511.

Halliwell, B., and Gutteridge, J. (1999). Free radicals in biology and medicine, 3rd ed. Oxford: Oxford University Press.

Hampton, M.B., and Orrenius, S. (1997). Dual regulation of caspase activity by hydrogen peroxide: implications for apoptosis. *FEBS Lett.* 414:552–556.

Hennekam, R.C. (2006). Hutchinson-Gilford progeria syndrome: review of the phenotype. *Am. J. Med. Genet. A*, 140:2603–2624.

Hensley, K., Robinson, K.A., Gabbita, S.P., et al. (2000). Reactive oxygen species, cell signaling, and cell injury. *Free Radical Biol. Med.* 28:1456–1462.

Hitchon, C.A., and El-Gabalawy, H.S. (2004).Oxidation in rheumatoid arthritis. *Arthritis Res. Ther.* 6:265–278.

Hunter, T. (1995) Protein kinases and phosphatases: The yin and yang of protein phosphorylation and signalling. *Cell* 80:225–236.

Inanami, O., Johnson, J.L., McAdara, J.K., et al. (1998). Activation of the leukocyte NADPH oxidase by phorbol ester requires the phosphorylation of p47(PHOX) on serine 303 or 304. *J. Biol. Chem.* 273:9539–9543.

Iyer, G.Y., Islam, M.F., and Quastel, J.H. (1961). Biochemical aspects of phagocytosis. *Nature* 192:535–542.

Jung T., and Grune T.T. (2008). The proteasome and its role in the degeneration of oxidised proteins. *IUBMB Life* 60:743–752.

Kakkar, P., and Singh, B.K. (2007). Mitochondria: a hub of redox activities and cellular distress control. *Mol. Cell. Biochem.* 305:235–253.

Kamata, H., Manabe, T., Oka, S., et al. (2002). Hydrogen peroxide activates IkappaB kinases through phosphorylation of serine residues in the activation loops. *FEBS Lett.* 519:231–237.

Kawahara, T., Ritsick, D., Cheng, G.J., et al. (2005). Point mutations in the proline-rich region of p22(phox) are dominant inhibitors of Nox1- and Nox2-dependent reactive oxygen generation. *J. Biol. Chem.* 280:31859–31869.

Kieran, M.W., Gordon, L., and Kleinman, M. (2007). New approaches to Progeria. *Pediatrics* 120:834–841.

Kim, C., Kim, J.Y., and Kim, J.H. (2008). Cytosolic phospholipase A(2), lipoxygenase metabolites, and reactive oxygen species. *BMB Rep.* 41:555–559.

Klebanoff, S.J. (2005). Myeloperoxidase: friend and foe. *J. Leukoc. Biol.* 77:598–625.

Klumpp, S., Selke, D., and Krieglstein, J. (2003). Protein phosphatase type 2C dephosphorylates BAD. *Neurochem. Int.* 42:555–560.

Kubisch, H. M., Wang, J., Luche, R., et al. (1994). Transgenetic copper/zinc superoxide dismutase modulates susceptibility to type 1 diabetes. *Proc. Natl. Acad. Sci. U.S.A.* 91:9956–9959.

Kuhn, H., and Thiele, B.J. (1999). The diversity of the lipoxygenase family: Many sequence data but little information on biological significance. *FEBS Lett.* 449:7–11.

Lamb, N.J., Gutteridge, J.M., Baker, C., et al. (1999). Oxidative damage to proteins of bronchoalveolar lavage fluid in patients with acute respiratory distress syndrome: Evidence for neutrophil-mediated hydroxylation, nitration, and chlorination. *Critical Care Med.* 27:1738–1744.

Lambeth, J.D., Kawahara, T., and Diebold, B. (2007). Regulation of Nox and Duox enzymatic activity and expression. *Free Radical Biol. Med.* 43:319–331.

Lee, S.R., Kwon, K.S., Kim, S.R., et al. (1998). Reversible inactivation of protein-tyrosine phosphatase 1B in A431 cells stimulated with epidermal growth factor. *J. Biol. Chem.* 273:15366–15372.

Leseney, A.M., Deme, D., Legue, O., et al. (1999). Biochemical characterization of a Ca2+/NAD(P)H-dependent H_2O_2 generator in human thyroid tissue. *Biochimie* 81:373–380.

Levine, R.L., Mosoni, L., Berlett, B.S., et al. (1996). Methionine residues as endogenous antioxidants in proteins. *Proc. Natl. Acad. Sci. U.S.A.* 93:15036–15040.

Mackey, A.M., Sanvicens, N., Groeger, G., et al. (2008). Redox survival signalling in retina-derived 661W cells. *Cell Death Differ.* 15:1291–1303.

Mahadev, K., Zilbering, A., Zhu, L., et al. (2001). Insulin-stimulated hydrogen peroxide reversibly inhibits protein-tyrosine phosphatase 1B in vivo and enhances the early insulin action cascade. *J. Biol. Chem.* 276:21938–21942.

Mahadev, K., Motoshima, H., Wu, X.D., et al. (2004). The NAD(P)H oxidase homolog Nox4 modulates insulin-stimulated generation of H_2O_2 and plays an integral role in insulin signal transduction. *Mol. Cell. Biol.* 24:1844–1854.

Malle, E., Furtmuller, P.G., Sattler, W., et al. (2007). Myeloperoxidase: a target for new drug development? *Br. J. Pharmacol.* 152:838–854.

Mancuso, M., Coppede, F., Migliore, L., et al. (2006). Mitochondrial dysfunction, oxidative stress and neurodegeneration. *J. Alzheimer's Dis.* 10:59–73.

Markesbery, W.R., and Carney, J.M. (1999). Oxidative alterations in Alzheimer's disease. *Brain Pathol.* 9:133–146.

Markesbery, W.R., and Lovell, M.A. (2007). Damage to lipids, proteins, DNA, and RNA in mild cognitive impairment. *Arch. Neurol.* 64:954–956.

Meng, T.C., Fukada, T., and Tonks, N.K. (2002). Reversible oxidation and inactivation of protein tyrosine phosphatases in vivo. *Mol. Cell* 9:387–399.

Millward, T.A., Zolnierowicz, S., and Hemmings, B.A. (1999). Regulation of protein kinase cascades by protein phosphatase 2A. *Trends Biochem. Sci.* 24:186–191.

Miyasaki, K.T., Song, J.P., and Murthy, A.R.K. (1991). Secretion of myeloperoxidase isoforms by human neutrophils. *Anal. Biochem.* 193:38–44.

Moncada, S., and Erusalimsky, J.D. (2002). Does nitric oxide modulate mitochondrial energy generation and apoptosis? *Nat. Rev. Mol. Cell Biol.* 3:214–220.

Murphy, M.E., and Kehrer, J.P. (1989). Oxidative stress and muscular dystrophy. *Chem. Biol. Interact.* 69:101–173.

Nawata, R., Yujiri, T., Nakamura, Y., et al. (2003). MEK kinase 1 mediates the antiapoptotic effect of the Bcr-Abl oncogene through NF-kappaB activation. *Oncogene* 22:7774–7780.

Nishino, T., Okamoto, K., Eger, B.T., et al. (2008). Mammalian xanthine oxidoreductase. Mechanism of transition from xanthine dehydrogenase to xanthine oxidase. *FEBS J.* 275:3278–3289.

Okado-Matsumoto, A., and Fridovich, I. (2001). Subcellular distribution of superoxide dismutases (SOD) in rat liver -Cu,Zn-SOD- in mitochondria. *J. Biol. Chem.* 276:38388–38393.

Oliver, C. N., Ahn, B.W., Moerman, E. J., Goldstein, S., and Stadtman, E. R. (1987). Age-related changes in oxidized proteins. *J. Biol. Chem.* 262:5488–5491.

O'Loghlen, A., Perez-Morgado, M.I., Salinas, M. et al. (2003). Reversible inhibition of the protein phosphatase 1 by hydrogen peroxide. Potential regulation of eIF2 alpha phosphorylation in differentiated PC12 cells. *Arch. Biochem. Biophys.* 417:194–202.

Pearce, R.K., Owen, A., Daniel, S., et al. (1997). Alterations in the distribution of glutathione in the substantia nigra in Parkinson's disease. *J. Neural Transm.* 104:661–677.

Praticò, D., and Sung, S. (2004). Lipid peroxidation and oxidative imbalance: early functional event in Alzheimer's disease. *J. Alzheimers Dis.* 6:171–175.

Puddu, P., Puddu, G.M., Cravero, E., et al. (2008). The molecular sources of reactive oxygen species in hypertension. *Blood Press.*17:70–77.

Rahman, I., Gilmour, P.S., Jimenez, L.A. et al. (2002). Oxidative stress and TNF-alpha induce histone acetylation and NF-kappaB/AP-1 activation in alveolar epithelial cells: potential mechanism in gene transcription in lung inflammation. *Mol. Cell. Biochem.* 234–235:239–248.

Rahman, I., Marwick, J., and Kirkham, P. (2004). Redox modulation of chromatin remodeling: impact on histone acetylation and deacetylation, NF-kappaB and pro-inflammatory gene expression. *Biochem. Pharmacol.* 68:1255–1267.

Rao, R.K., and Clayton, L.W. (2002). Regulation of protein phosphatase 2A by hydrogen peroxide and glutathionylation. *Biochem. Biophys. Res. Commun.* 293:610–616.

Refsgaard, H.H., Tsai, L., Stadtman, E.R. (2000). Modifications of proteins by polyunsaturated fatty acid peroxidation products. *Proc. Natl. Acad. Sci. U.S.A.* 97:611–616.

Royerpokora, B., Kunkel, L.M., Monaco, A.P., et al. (1986). Cloning the gene for an inherited human disorder -Chronic Granulomatous-Disease- on the basis of its chromosomal location. *Nature* 322:32–38.

Sadok, A., Bourgarel-Rey, V., Gattacceca, F., et al. (2008). Nox1-dependent superoxide production controls colon adenocarcinoma cell migration. *Biochim. Biophys. Acta* 1783:23–33.

Salh, B., Webb, K., Guyan, P.M., et al. (1989). Aberrant free radical activity in cystic fibrosis. *Clin. Chim. Acta* 181:65–74.

Salmeen, A., and Barford, D. (2005). Functions and mechanisms of redox regulation of cysteine-based phosphatases. *Antioxid. Redox Signal.* 7:560–577.

Savitsky, P.A., and Finkel, T. (2002). Redox regulation of Cdc25C. *J. Biol. Chem.* 277:20535–20540.

Schidlow, D.V. (2000). Newer therapies for cystic fibrosis. *Paediatr. Respir. Rev.* 1:107–113.

Shiose, A., Kuroda, J., Tsuruya, K., et al. (2001). A novel superoxide-producing NAD(P)H oxidase in kidney. *J. Biol. Chem.* 276:1417–1423.

Siew, E.L., Opalecky, D., and Bettelheim, F.A. (1981). Light scattering of normal human lens. II. Age dependence of the light scattering parameters. *Exp. Eye Res.* 33:603–614.

Smith, C. D., Carney, J. M., Starke-Reed, P. E., et al. (1991). Excess brain protein oxidation and enzyme dysfunction in normal and Alzheimer's disease. *Proc. Natl. Acad. Sci. U.S.A.* 88:10540–10543.

Spector, A. (1995). Oxidative stress-induced cataract: mechanism of action. *FASEB J.* 9:1173–1182.

Stadtman, E.R. (1990). Metal ion-catalyzed oxidation of proteins: biochemical mechanisms and biological consequences. *Free Radical Biol. Med.* 9:315–325.

Stadtman, E.R. (2001). Protein oxidation in aging and age-related diseases. *Ann. New York Acad. Sci.* 928:22–38.

Stadtman, E.R., and Berlett, B.S. (1997). Reactive oxygen-mediated protein oxidation in aging and disease. *Chem. Res. Toxicol.* 10:485–494.

Stadtman, E.R., and Levine, R.L. (2003). Free radical-mediated oxidation of free amino acids and amino acid residues in proteins. *Amino Acids* 25:207–218.

Starosta, V., Rietschel, E., Paul, K., et al. (2006). Oxidative changes of bronchoalveolar proteins in cystic fibrosis. *Chest* 129:431–437.

St-Pierre, J., Buckingham, J.A., Roebuck, S.J., et al. (2002). Topology of superoxide production from different sites in the mitochondrial electron transport chain. *J. Biol. Chem.* 277:44784–44790.

Suh, Y.A., Arnold, R.S., Lassegue, B., et al. (1999). Cell transformation by the superoxide-generating oxidase Mox1. *Nature* 401:79–82.

Suleyman, H., Demircan, B., and Karagoz, Y. (2007). Anti-inflammatory and side effects of cyclooxygenase inhibitors. *Pharmacol. Rep.* 59:247–258.

Sumimoto, H. (2008). Structure, regulation and evolution of Nox-family NADPH oxidases that produce reactive oxygen species. *FEBS J.* 275:3249–3277.

Taira, T., Saito, Y., Niki, T., et al. (2004). DJ-1 has a role in antioxidative stress to prevent cell death. *EMBO Rep.* 5:213–218.

Tanaka, T., Hosoi, F., Yamaguchi-Iwai, Y., et al. (2002). Thioredoxin-2 (TRX-2) is an essential gene regulating mitochondria-dependent apoptosis. *EMBO J.* 21:1695–1703.

Thomas, B., and Beal, M.F. (2007). Parkinson's disease. *Human. Mol. Genet.* 16:183–194.

Toledano, M.B., and Leonard, W.J. (1991). Modulation of transcription factor NF-kappa B binding activity by oxidation-reduction in vitro. *Proc. Natl. Acad. Sci. U.S.A.* 88:4328–4332.

Turrens, J.F., and Boveris, A. (1980). Generation of superoxide anion by the NADH dehydrogenase of bovine heart-mitochondria. *Biochem. J.* 191:421–427.

Turrens, J.F., Alexandre, A., and Lehninger, A.L. (1985). Ubisemiquinone is the electron-donor for superoxide formation by Complex III of heart-mitochondria. *Arch. Biochem. Biophys.* 237:408–414.

Vandermoere, F., El Yazidi-Belkoura, I., Adriaenssens, E., et al. (2005). The antiapoptotic effect of fibroblast growth factor-2 is mediated through nuclear factor-kappaB activation induced via interaction between Akt and IkappaB kinase-beta in breast cancer cells. *Oncogene* 24:5482–5491.

Vaquero, E.C., Edderkaoui, M., Pandol, S.J., et al. (2004). Reactive oxygen species produced by NAD(P)H oxidase inhibit apoptosis in pancreatic cancer cells. *J. Biol. Chem.* 279:34643–34654.

Vasquez-Vivar, J., Kalyanaraman, B., Martasek, P., et al. (1998). Superoxide generation by endothelial nitric oxide synthase: the influence of cofactors. *Proc. Natl. Acad. Sci. U.S.A.* 95:9220–9225.

West, S. (2007). Epidemiology of cataract: accomplishments over 25 years and future directions. *Ophthalmic Epidemiol.* 14:173–178.

Wientjes, F.B., Hsuan, J.J., Totty, N.F., et al. (1993). P40phox, a third cytosolic component of the activation complex of the NADP oxidase to contain Src homology 3 domains. *Biochem. J.* 296:557–561.

Williams, D.L. (2006). Oxidation, antioxidants and cataract formation: a literature review 1: *Vet. Ophthalmol.* 9:292–298.

Chapter 4
Involvement of S-Nitrosylation in Neurodegeneration

Yihang Li and Kenny K.K. Chung

4.1 Introduction

Nitric oxide (NO) is a diffusible signaling molecule that regulates various physiological functions in the biological system. For example, nitric oxide signals vasodilation through the activation of soluble guanylate cyclase (sGC) by binding to the enzyme's heme group directly. Activation of sGC increases the production of cyclic guanosine monophosphate (cGMP) which signals the relaxation of smooth muscle in the blood vessels for vasodilation. Another signal mechanism mediated by NO is through the attachment of NO covalently to the cysteine residues in proteins to modulate their activities directly. This modification, designated as S-nitrosylation, is now recognized as an important post-translational modification that is comparable to phosphorylation in regulating physiological responses such as vasodilation, neurotransmission, vesicle trafficking and apoptosis (Fig. 4.1). However dysregulation of S-nitrosylation is emerging as a contributor in the development of a number of diseases such as neurodegeneration.

Neurodegenerative disorders such as Alzheimer's disease (AD), Parkinson's disease (PD), Huntington's disease (HD) and amyotrophic lateral sclerosis (ALS) are characterized by progressive degeneration of specific groups of neurons. The pathogenic mechanisms of these disorders are not completely understood, but NO induced nitrosative stress has long been regarded as an important contributing factor. It is believed that elevated nitrosative stress through free radicals can damage protein, lipid and DNA which can ultimately lead to neurodegeneration. However, recent studies suggest that NO induced S-nitrosylation of neuroprotective proteins can also contribute to the process of neurodegeneration.

K.K.K. Chung (✉)
Department of Biochemistry, Hong Kong University of Science and Technology, Clear Water Bay, Kowloon, Hong Kong, People's Republic of China
e-mail: bckchung@ust.hk

C.J. Vidal (ed.), *Post-Translational Modifications in Health and Disease*, Protein Reviews 13,
DOI 10.1007/978-1-4419-6382-6_4, © Springer Science+Business Media, LLC 2011

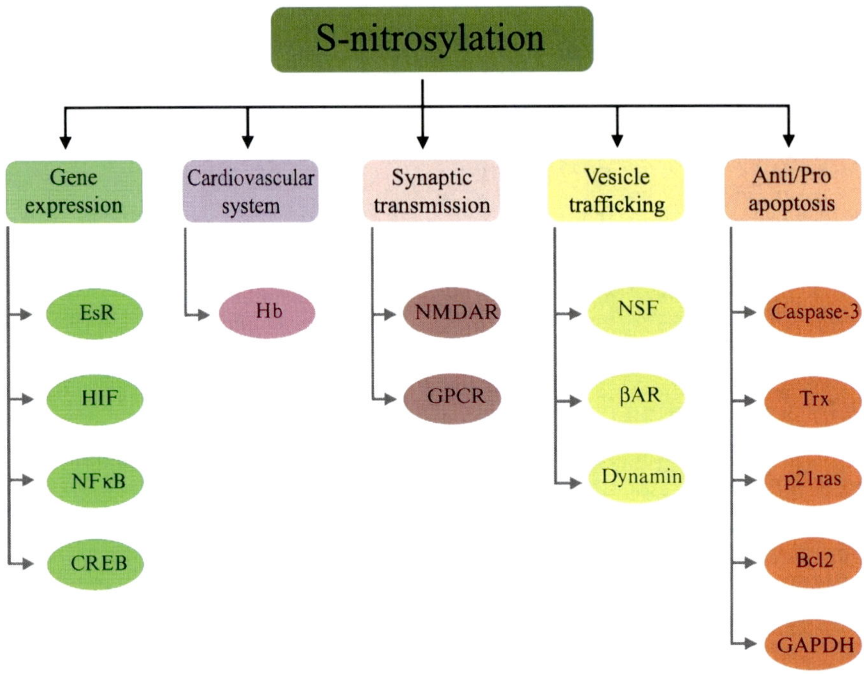

Fig. 4.1 Physiological functions of S-nitrosylation. S-nitrosylation is involved in the regulation of several important cellular systems including gene expression, cardiovascular system, synaptic transmission, vesicle trafficking and regulation of apoptosis. *EsR* estrogen receptor, *HIF* hypoxia inducible factor, *NFκB* nuclear factor kappa B, *CREB* cAMP responsive element binding, *Hb* hemoglobin, *NMDAR* N-methyl D-aspartate receptor, *βAR* beta-adrenergic receptor, *GPCR* G-protein-coupled receptor, *NSF* *N*-ethylmaleimide-sensitive factor, *Trx* thioredoxin, *GAPDH* glyceraldehyde-3-phosphate dehydrogenase. See text for the details

4.2 Biochemistry of S-Nitrosylation

S-nitrosylation occurs when a NO molecule is covalently attached to the thiol group of cysteine residues in proteins. In the biological system, proteins can be S-nitrosylated through a number of mechanisms which include transnitrosylation, reaction with electrophilic nitrogen oxide species, or through metalloprotein catalyzed reactions. Transnitrosylation, also called transnitrosation, is a process by which a NO group is exchanged for a hydrogen group between cysteine residues. This reaction is largely depending on the pH, in which the reaction rate slows dramatically when pH decreases (Arnelle and Stamler 1995). The generation of nitrosothiols directly from NO under physiological condition requires an electron acceptor to act as an electrophilic nitrosating intermediate before the formation of nitrosothiols. In the biological system, oxygen is a common electron acceptor that reacts with NO to generate the N_2O_3 intermediate which then reacts with cysteine resides in proteins forming the nitrosothiols (Gaston et al. 2003). Apart from N_2O_3, a metal nitrosyl can also act as an intermediate in the process of nitrosothiol formation. For example, Cu^+ or Fe^{2+} in metalloproteins can serve as electron acceptors to facilitate the generation of nitrosothiols (Gaston et al. 2003). In summary, S-nitrosylation can be carried out by transnitrosylation or through the oxidation of a thiol group by NO via intermediates such as N_2O_3 and metal nitrosyls.

4.3 Specificity of S-Nitrosylation In Vivo

Up to date, more than 100 proteins have been demonstrated to be S-nitrosylated. The site of S-nitrosylation is usually depending on a consensus acid-base motif in the target protein (Stamler et al. 1997a). For example, in the beta-subunit of hemoglobin, the S-nitrosylated cysteine (Cys 93) is flanked by an acidic and a basic amino acid (His92-Cys93-Asp94) (Stamler et al. 1997a, b; Chung et al. 2005a). Interestingly, later studies found that this acid-base motif is not necessarily based on the primary amino acid sequence, but it can be formed by the secondary or tertiary protein structure (Hess et al. 2001; Chung 2007). For example, the enzymatic activity of methionine adenosyltransferase (MAT) is inhibited by S-nitrosylation at Cys121 (Perez-Mato et al. 1999). However, Cys121 in MAT is not flanked by acidic and basic amino acids in the primary sequence. However, replacement of the acidic (Asp355) or basic (Arg357 and Arg363) amino acids located in the vicinity of Cys121 due to protein folding reduces S-nitrosylation of MAT (Perez-Mato et al. 1999), indicating the importance of the three-dimensional structure in the formation of the acid-base motif for S-nitrosylation.

4.4 Detection of Protein S-Nitrosylation

Because of the labile nature of S-nitrosylation and the lack of sensitive method in detecting this modification, it is still difficult to identify and study how S-nitrosylation can regulate the function of proteins in different cellular pathways. A number of methods have been developed to detect S-nitrosylation in vitro and in vivo (Table 4.1) (Chung et al. 2005b; Torta et al. 2008). The most sensitive method is the photolysis or the chemical-induced release of NO from nitrosothiols, which is coupled with chemiluminescence to detect the amount of NO released from S-nitrosylated proteins (Chung et al. 2005b). This approach is sensitive but it requires expensive instrumental setup and it also lacks the ability to distinguish which protein is S-nitrosylated. A less expensive approach is to use a colorimetric or fluorometric method instead of chemiluminescence to detect the NO released from nitrosothiols, but the sensitivity of this approach is much lower (Chung et al. 2005b). A recently developed biotin switch assay provides an inexpensive biochemical method to detect protein S-nitrosylation in a reasonably sensitive approach to characterize protein S-nitrosylation (Chung et al. 2005b). In this assay, the

Table 4.1 Principles, advantages and limitations of methods in studying S-nitrosylation

Methods	Principles	Advantages	Limitations
Chemiluminescence, colorimetric and fluorometric methods	Photolysis or chemical induced release of NO from nitrosothiols	Sensitive and quantitative	Non-specific and indirect measurement
Anti-nitrosothiol antibodies	Antibodies against S-nitrosylated proteins	Convenient for Western blot or immuno-histochemistry	Non-specific reactivity to cysteine residues in proteins
Biotin switch assay	S-nitrosylated cysteine is converted to biotinylated cysteine	Convenient to be detected by biochemical methods	Low sensitivity and complicated procedure
Mass spectrometry	Detection of S-nitrosylation based on changes in mass	Identification of the site of modification	Purified proteins required

S-nitrosylated cysteine residue is switch to a biotin group, which can then be detected by biochemical methods. Although the biotin switch is still not as sensitive as the chemiluminescence approach, it has greatly assisted the study and identification of proteins that are S-nitrosylated in different cellular pathways. Lastly, antibodies have also been developed to detect S-nitrosylated proteins by Western blot or immunohistochemistry. However, these antibodies usually suffer from non-specific reactivity to cysteine residues in proteins (Chung et al. 2005b).

4.5 Physiological Function of S-Nitrosylation

4.5.1 S-Nitrosylation and Gene Expression

S-nitrosylation has been shown to affect gene transcription by modulating the activity of a number of transcriptional factors such as hypoxia inducible factor-1 (HIF-1), estrogen receptor (EsR), nuclear factor kappa B (NFκB) and cAMP responsive element binding (CREB) (Garban et al. 2005; Li et al. 2007; Marshall et al. 2004; Reynaert et al. 2004; Riccio et al. 2006; Nott et al. 2008). For instance, HIF-1 is a heterodimeric transcriptional factor that is composed of HIF-1α and HIF-1β with HIF-1α functions as the regulatory subunit (Yee-Koh et al. 2008). Under normoxic condition, HIF-1α is hydroxylated at conserved proline residues (Pro402, Pro564) at the oxygen-dependent degradation (ODD) domain by prolyl hydroxylases (PHDs). This modified HIF-1α will be targeted by the von Hippel-Lindau protein (pVHL), an E3 ligase in the ubiquitin proteasomal system (UPS), for degradation (Yee-Koh et al. 2008). However, a recent study also showed that HIF-1α can be S-nitrosylated at Cys533 within the ODD domain and this prevents its degradation by the UPS which is independent of the PHD pathway (Li et al. 2007). This stabilization of HIF-1α by S-nitrosylation has been suggested to be the mechanism of how tumor cells can survival after radiation treatment (Li et al. 2007).

EsR is an important transcriptional factor that mediates cellular changes in response to the hormone estrogen. It is known that EsR is redox-sensitive, potentially because of the cysteine residues within the two highly conserved zinc finger domains. In fact, it was reported that the cysteine residues within EsR can be S-nitrosylated and S-nitrosylation of EsR inhibits the binding of EsR to specific estrogen-responsive elements (EREs) (Garban et al. 2005). This provides another example of how direct S-nitrosylation of transcriptional factor can affect gene transcription at the cellular level.

In contrast to HIF-1 and EsR, S-nitrosylation modulates the activity of NFκB indirectly. Under basal condition, NFκB is sequestered to the cytoplasm through its interaction with inhibitory κB (IκB) (Marshall et al. 2004; Reynaert et al. 2004). When the NFκB is activated, IκB kinase (IKK) phosphorylates IκB which leads to its degradation by the UPS. Without binding to IκB, NFκB translocates to the nucleus and initiates gene transcription (Marshall et al. 2004; Reynaert et al. 2004). S-nitrosylation of IKK decreases its phosphorylation activity towards IkB, thus preventing NFκB's translocation into the nucleus for gene transcriptional activation (Marshall et al. 2004; Reynaert et al. 2004).

CREB is a well-known transcriptional factor that regulates important cellular processes in neurons such as synaptic plasticity, axonal growth, and survival (Lonze and Ginty 2002). For example, brain-derived neurotrophic factor (BDNF) is a growth factor that activates CREB-mediated gene transcription and promotes neuronal survival. It was initially reported that the activation of CREB by BDNF was dependent on the S-nitrosylation of components in the CREB DNA-binding complex (Riccio et al. 2006). A more recent study showed that BDNF-induced S-nitrosylation of histone deacetylase 2 (HDAC2) at Cys262 and Cys274 increases acetylation of histones and promotes CREB dependent gene transcription which promotes dendritic growth and neuronal survival (Nott et al. 2008).

4.5.2 S-Nitrosylation and Cardiovascular System

NO was first identified as the vasodilation signaling molecule produced by eNOS in the endothelium to induce smooth muscle relaxation by the activation of sGC. However, recent studies suggest that NO plays a more important role in modulating the cardiovascular system. Studies suggest that red blood cells carry NO at the Cys93 of the beta subunit of hemoglobin (Luchsinger et al. 2003). Under low O_2 tension, the release of O_2 from the hemoglobin in tissues with high metabolic rate will trigger a conformational change that also promotes the release of NO (Pawloski et al. 2001; McMahon et al. 2002). The released NO can induce local vasodilation so as to facilitate increased blood flow to active tissues (Pawloski et al. 2001;

McMahon et al. 2002). In agreement with the hypothesis that nitrosothiols are important in blood flow regulation, mice that are defective in S-nitrosogluthione (GSNO) metabolism exhibit an increased level of nitrosothiols in the blood and a higher blood pressure under anesthesia (Liu et al. 2004). Interestingly, in another report, it was found that infusion of deoxygenated blood or nitrosothiols to the nucleus tractus solitarius (NTS) can induce hyperventilatory response as observed in animals exposed to hypoxic condition (Lipton et al. 2001). These results suggest that although the role of NO in cardiovascular system is not completely clear, NO through S-nitrosylation is an important component in the regulation of the cardiovascular system.

4.5.3 S-Nitrosylation and Neurotransmission

In the central nervous system, N-methyl D-aspartate receptor (NMDAR) is a well-characterized target for S-nitrosylation. S-nitrosylation of NMDAR subunits has been identified at Cys744 and Cys798 in NR1, and Cys87, Cys320, and Cys399 in NR2 (Nakamura and Lipton 2008). Because nNOS is linked to the NMDAR through PSD-95, and activation of NMDAR activates nNOS, the S-nitrosylation of NMDAR due to increased NO production serves as a negative feedback mechanism for NMDAR activation (Nakamura and Lipton 2008). Downstream of NMDAR, several components are activated through S-nitrosylation. For instance, H-ras and Dexras1 are S-nitrosylated and activated upon NMDAR activation to facilitate downstream signaling (Fang et al. 2000). Activation of ras family proteins by S-nitrosylation has been implicated in tumorigenesis (Lim et al. 2008), whereas S-nitrosylation of Dexras1 has been linked to increased iron uptake in neurons (Cheah et al. 2006). Although the long-term consequence of how the activation of these molecules by S-nitrosylation can modulate the synaptic plasticity is not completely clear, it is generally agreed that S-nitrosylation is an important component in the normal synaptic function.

In the peripheral system, NO has been shown to modulate the beta-adrenergic receptor (βAR) signaling through S-nitrosylation (Whalen et al. 2007). βAR is a G-protein-coupled receptor (GPCR) that is involved in the sympathetic nervous system to regulate autonomic responses such as heart rate and bronchial relaxation. The activity of βAR is in part regulated by consistent internalization from or reinsertion to the cytoplasmic membrane. The internalization of βAR is modulated by (G protein coupled receptors (GPCR) kinases (GRKs) in response to agonist stimulation. For example, GKR2 phosphorylates βAR after agonist stimulation and this will recruit β-arrestin to promote βAR internalization. Previous studies have implicated that NO can affect the GPCR signaling and a recent report showed that this is mediated through S-nitrosylation of GRK2 (Whalen et al. 2007). GRK2 is S-nitrosylated at Cys340 and this modification inhibits its kinase activity, which reduces the agonist induced desensitization and internalization of βAR (Whalen et al. 2007).

4.5.4 S-Nitrosylation and Vesicle Trafficking

Vesicle trafficking in cells is a complex process that involves a number of components. For instance, in exocytosis, the fusion of vesicles to the cytoplasmic membrane is mediated by the formation of a SNARE (soluble N-ethylmaleimide-sensitive-factor attachment protein receptor) complex. Once the process of exocytosis is finished, the SNARE complex is recycled by the *N*-ethylmaleimide-sensitive factor (NSF). NSF is an ATPase that couples the hydrolysis of ATP to the conformational changes that is required for the recycling of SNARE complex. NO is known to affect exocytosis and it was recently found that NSF can be S-nitrosylated at Cys91 and Cys264 and this has been shown to affect both exocytosis and receptor trafficking (Matsushita et al. 2003; Huang et al. 2005). For example, during inflammation, Weibel-Palade bodies are released by exocytosis to mediate vascular thrombosis and inflammation (Matsushita et al. 2003). NO is known to inhibit vascular inflammation, but the exact mechanism was not known. It is now observed that S-nitrosylation of NSF inhibits the release of Weibel-Palade bodies (Matsushita et al. 2003). Interestingly S-nitrosylation of NSF does not affect its ATPase activity, but instead inhibits NSF's normal function in the recycling of SNARE complex (Matsushita et al. 2003).

S-nitrosylation of NSF has also been found to affect trafficking of AMPA (α-amino-3-hydroxy-5-methyl-4-isoxazolepropionic acid) receptor (Huang et al. 2005). It was reported that S-nitrosylation of Cys91 of NSF enhances the interaction between NSF and GluR2 subunit of AMPA receptor (AMPAR) and this facilitates the surface expression of AMPAR (Huang et al. 2005). Since the neurotransmission of NMDAR is associated with the activation of nNOS, the S-nitrosylation of NSF also provides a mechanism of how activation of NMDAR can promote the surface expression of AMPAR (Huang et al. 2005).

Apart from exocytosis and AMPAR trafficking, NO has been found to S-nitrosylate dynamin at the Cys607 and this increases the GTPase activity of dynamin and also its ability to self-assembly (Wang et al. 2006). This explains how NO can enhance endocytosis through S-nitrosylation of dynamin in the regulation of βAR2 and endothelial growth factor receptor (EGFR) (Wang et al. 2006).

4.5.5 S-Nitrosylation and Apoptosis

NO is a well-known regulator for cell survival through a number of mechanisms. For instance, under extensive oxidative stress NO can react with superoxide in mediating damaging effects on protein, lipid and DNA in the cells. For S-nitrosylation, studies have shown that both survival and death inducing factors in the cell can be modulated by S-nitrosylation. This discrepancy may be explained by the sensitivity of different proteins to nitrosative stress. For instance, moderate level of NO may inhibit apoptosis, whereas elevated level of nitrosative stress can induce cell cytotoxicity.

One of the examples that NO enhances cell survival is the S-nitrosylation of caspase-3. Caspase-3 is one of the final executioners in apoptosis. Like the other members of caspase family, caspase-3 exists as an inactive zymogen under normal cellular condition. Apoptotic signal first triggers the activation of initiator caspases, such as caspase-8, -9 and -10, which in turn activate the effector caspases such as caspase-3, -6 and -7. The activated effector caspases will then cleave other cellular targets, which finally will lead to apoptosis (Thornberry and Lazebnik 1998). Under normal cellular conditions, caspase-3 is S-nitrosylated at the catalytic cysteine residues (Mannick et al. 1999). However, upon stimulation of the cell death inducing factors, such as Fas ligand, caspase-3 is denitrosylated and activated caspase-3 will initiate apoptosis (Mannick et al. 1999, 2001). S-nitrosylation and denitrosylation of cellular or mitochondrial caspase-3 has been shown to be regulated by thioredoxins (Trxs) (Mitchell and Marletta 2005; Benhar et al. 2008). In fact S-nitrosylation of Trx-1 at different cysteine residues have been shown to be important for apoptosis in different models of cell death (Haendeler et al. 2002; Mitchell and Marletta 2005; Mitchell et al. 2007; Benhar et al. 2008).

Other proteins that have been found to enhance cell survival after S-nitrosylation include p21ras and Bcl-2. The activation of the ras protein family is known to promote cell survival. However, deregulation of ras activity is also a major component in the process of tumorigenesis. For instance, activation of p21ras leads to activation of downstream signaling in promoting cell survival. S-nitrosylation of p21ras has been associated with pro-survival signal mechanism and this has been linked to the process of tumorigenesis (Raines et al. 2007; Lim et al. 2008). S-nitrosylation of Bcl-2 is another pro-survival modification induced by NO. Under normal cellular condition, Bcl-2 inhibits cell death by binding to proteins such as BAX and BAK in the Bcl-2 protein family (Zhai et al. 2008). However, the protein level of Bcl-2 is controlled by the UPS through proteasomal degradation of Bcl-2 (Zhai et al. 2008). S-nitrosylation inhibits proteasomal degradation of Bcl-2 and thus protects cells against apoptosis induced by BAX and BAK (Azad et al. 2006).

Glyceraldehyde-3-phosphate dehydrogenase (GAPDH) is a well-studied cytosolic enzyme that converts glyceradehyde-3-phosphate to glycerate 1,3-bisphosphate in the glycolytic pathway. Apart from the normal metabolic function of GAPDH, recent studies suggest that GAPDH plays an active role in apoptosis. It has been observed that GAPDH is translocated to the nucleus after apoptotic stimuli and the mechanism has been elucidated recently by a number of studies (Hara et al. 2005; Sen et al. 2008). It was initially reported that GAPDH can be S-nitrosylated at Cys150 and this modification enhances its binding to Siah1, an E3 ligase in the UPS (Hara et al. 2005). The translocation of S-nitrosylated GAPDH to the nucleus is mediated by the binding with Siah1 which possesses a nuclear localization signal (Hara et al. 2005). Once in the nucleus, GAPDH stabilizes Siah1 to degrade nuclear proteins that inhibit apoptosis (Hara et al. 2005). GAPDH in the nucleus is also acetylated at Lys160 by p300/CBP and this acetylation promotes the interaction between GAPDH and p300/CBP (Sen et al. 2008). The binding of GAPDH activates the acetyltransferase activity of p300/CBP which facilitates the p53 dependent apoptotic pathways (Fig. 4.2) (Sen et al. 2008). Taken together, these studies show a new role of GAPDH in mediating cell death.

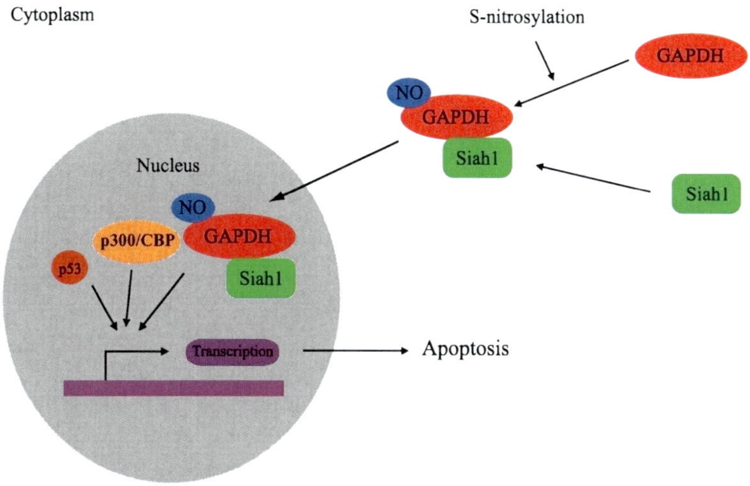

Fig. 4.2 S-nitrosylated GAPDH is translocated to the nucleus during apoptosis. S-nitrosylation of GAPDH enhances its interaction with Siah1, which translocates GAPDH into the nucleus. Nuclear GAPDH interacts with and activates p300/CBP to facilitate the p53 dependent apoptosis

4.6 S-Nitrosylation and Neurodegeneration

NO is produced actively by nNOS in neurons as a signaling molecule especially in the excitatory synapses. Various studies have shown that excessive production of NO is a major source of oxidative stress that contributes to neurodegeneration. This is particularly important as neuronal cell death is usually associated with inflammatory response which can induce the expression of iNOS and further increase the concentration of NO in the affected areas. Excessive amount of NO can react with other free radicals such as superoxide anion to form the more reactive nitrogen species such as peroxynitrite. These free radicals can induce protein nitration, lipid peroxidation, and DNA damage, which ultimately lead to neuronal degeneration (Ischiropoulos and Beckman 2003). This pathogenic mechanism has been demonstrated in a number of neurodegenerative disorders. However, more recent studies suggest that S-nitrosylation of proteins can also contribute to a number of neurodegenerative disorders. For instance, recent studies have found that S-nitrosylation can contribute to the development of ALS, AD and PD.

ALS is the most common motor neuronal disorder marked by degeneration of motor neurons in the spinal cord, brain stem and cortex (Rothstein 2009). Although most of ALS cases are sporadic, mutations of superoxide dismutase 1 (SOD1), have been found in some patients with familial history of ALS (Rothstein 2009). How mutations in SOD1 can cause ALS is not fully understood, but studies found that SOD1 mutants possess increased denitrosylase activity when compared to wild-type (WT) SOD1 (Schonhoff et al. 2006). In cell expressing SOD1 mutants, S-nitrosothiol (SNO) levels are decreased in the mitochondria (Schonhoff et al. 2006). Similarly, decreased levels of SNO are observed in the spinal cords of mutant SOD1 transgenic mice (Schonhoff et al. 2006). These results suggest that imbalance of SNO levels caused by mutation in SOD1 is one of the mechanisms that lead to motor neuron degeneration in ALS.

The role of S-nitrosylation is less defined in AD, but a recent report sheds light on how NO can affect mitochondrial fission and induce neuronal injury through the S-nitrosylation of dynamin-related protein 1 (Drp1) in the pathogenesis of AD (Cho et al. 2009). AD is the most common neurodegenerative disorder that affects increasing number of elderly as the lifespan of general population is improved. A number of mechanisms have been proposed for the development of AD and the β-amyloid (Aβ) induced neurodegeneration is widely believed to be a major contributor (Cho et al. 2009). How Aβ can induce neuronal injury has been studied intensively and a number of hypotheses have been suggested which include excitotoxicity, oxidative stress, mitochondrial dysfunction, and apoptosis. In particular, oxidative stress in connection with mitochondrial dysfunction has been proposed to be an important mechanism in causing neurodegeneration in AD. A recent study provides a pathogenic mechanism of how Aβ can induce nitrosative stress that subsequently leads to abnormal mitochondrial fission/fusion that affects neuronal survival (Cho et al. 2009). Increased nitrosative stress is known to be induced by Aβ and this has been linked to mitochondrial dysfunction. The linkage of nitrosative stress and mitochondrial dysfunction is recently found to be caused by S-nitrosylation of Drp1 (Cho et al. 2009). Drp1 is an important molecule that regulates mitochondria fission and dysfunction of Drp1 has been implicated in neurodegenerative disorders (Cho et al. 2009). Aβ induces production of NO and this leads to the S-nitrosylation of Drp1 (Cho et al. 2009). S-nitrosylation of Drp-1 at Cys644 enhances dimerization and increases GTPase activity of Drp-1 (Cho et al. 2009). S-nitrosylation of Drp-1 induces mitochondrial fragmentation in relation to neurodegeneration in AD (Cho et al. 2009). Taken together, these results provide a new mechanism of how nitrosative stress through S-nitrosylation of Drp1 can contribute to the pathogenesis of AD.

In PD, S-nitrosylation has been recently emerging as an important contributor in the pathogenesis of the disease. PD is the second most common neurodegenerative disorders marked by movement impairment. The disease is caused by a selective degeneration of dopaminergic neurons in the substantia nigra (SNc) with the presence of intraneuronal protein aggregates designated as Lewy bodies (LB) (Savitt et al. 2006). The mechanism of neuronal cell death in the SNc is not completely clear, but studies suggest that oxidative stress, mitochondria dysfunction, protein aggregation, and dysfunction of the UPS are the major contributors for neurodegeneration in PD (Savitt et al. 2006; Tsang and Chung 2009). Recent identification of mutations in genes that cause the rare familial form of PD (FPD) has improved our understanding of PD mechanism. For instance, mutations in α-synuclein (α-syn) and LRRK2 cause autosomal dominant form of FPD, whereas mutations in parkin, DJ1 and PINK1 cause recessive form of FPD (Thomas and Beal 2007). The functional studies of these genes agree with the established pathogenic mechanism of PD. For instance, α-syn is a protein that is prone for aggregation and soon after its initial identification as the FPD linked gene product, α-syn was found to be the major component of LB (Savitt et al. 2006). On the other hand, parkin was found to be an E3 ligase in the UPS, DJ1 was shown to be an oxidative chaperone, and PINK1 was identified as a mitochondrial kinase that is important for the normal function of mitochondria (Savitt et al. 2006; Tsang and Chung 2009). All these findings support the importance of protein aggregation, oxidative stress, mitochondrial or UPS dysfunction in the development of PD.

Nitrosative stress is known to be an important contributor for PD because nitrated protein aggregates are consistently reported to be a prominent feature of brain tissues in PD patients (Giasson et al. 2000). This notion is further supported by studies showing that nitrated α-syn is prone to aggregation, and nitrated α-syn is commonly found in the LB (Giasson et al. 2000; Tsang and Chung 2009). These findings suggested that nitrosative stress can induce the aggregation of α-syn, which can subsequently lead to the formation of LB. However, more recent studies found that NO can also contribute to PD through the S-nitrosylation of different components in neuroprotective pathways.

We initially reported that parkin can be S-nitrosylated and this modification compromises parkin's neuroprotective function (Chung et al. 2004). Parkin is an E3 ligase in the UPS and mutations in parkin were first identified in a group of Japanese patients that developed early-onset autosomal recessive form of FPD (Savitt et al. 2006). Various studies have demonstrated parkin's protective ability against different kinds of cellular insults (Feany and Pallanck 2003). For instance, as an E3 ligase, parkin ubiquitinates its substrates and directs them for UPS-dependent degradation, thus prevents the accumulation of misfolded or aggregated toxic protein species in neurons (Kahle and Haass 2004). We and another group first reported that parkin can be S-nitrosylated both in vitro and in vivo (Chung et al. 2004; Yao et al. 2004). S-nitrosylation of parkin inhibits its E3 ligase activity and compromises its protective function (Chung et al. 2004; Yao et al. 2004). More importantly, we observed increased levels of parkin S-nitrosylation in the brain tissues of PD animal models and PD patients (Chung et al. 2004). Another FPD linked gene that was found to be S-nitrosylated is DJ-1. A study showed that DJ-1 can be S-nitrosylated at Cys46 and Cys53 (Ito et al. 2006). Further study suggested that Cys46 is an important residue for the dimerization of DJ-1, but how DJ-1 S-nitrosylation can affect its dimerization is not clear (Ito et al. 2006). Since studies have shown that DJ-1 dimerization is crucial for its normal function as a chaperone, DJ-1 S-nitrosylation might affect its protective function in dopaminergic neurons. These results suggest that nitrosative stress through the modification of FAD gene products by S-nitrosylation is an important contributor for the PD pathogenesis.

Apart from parkin and DJ-1, a number of proteins with neuroprotective functions have been demonstrated to be the targets of S-nitrosylation. For instance, protein-disulphide isomerase (PDI), an endoplasmic reticulum (ER) protein that catalyzes disulfide bond formation and assists proper protein folding and maturation, was found to be S-nitrosylated in high levels of nitrosative stress (Uehara et al. 2006). S-nitrosylation of PDI suppresses both the chaperone and isomerase activities of the enzyme (Uehara et al. 2006). Under normal condition, PDI inhibits the formation of Lewy body-like aggregates in a cellular model of PD (Chung et al. 2001; Uehara et al. 2006).

S-nitrosylation of PDI attenuates this anti-aggregate formation function of PDI (Uehara et al. 2006). In SH-SY5Y neuroblastoma, S-nitrosylation of PDI abrogates its neuroprotective function against ER stress induced by unfolded proteins or proteasomal inhibition (Uehara et al. 2006). In neurons exposed to excitotoxicity induced by NMDA, S-nitrosylated PDI was increased and accompanied by accumulation of unfolded and polyubiquitinated proteins and the subsequent neuronal cell death (Uehara et al. 2006). In the post-mortem brain tissues of AD and PD patients, increased levels of S-nitrosylated PDI were observed. Taken together, this study established a link between nitrosative stress and protein misfolding in neurodegenerative disorders through the impairment of PDI.

The S-nitrosylation of parkin, DJ-1 and PDI aggravates cellular insults to neurons because these proteins safeguard cells against protein misfolding or aggregation. In addition, studies have also found that NO can also affect other neuroprotective pathways such as anti-oxidative stress protein or protein that has anti-apoptotic function. For instance, peroxiredoxin 2 (Prx2), an intracellular peroxidase, was found to be S-nitrosylated and this affects its normal function as an anti-oxidative stress enzyme (Fang et al. 2007). Prx2 is the most abundant peroxidase in neurons that metabolizes peroxides and protects neurons against oxidative injuries (Fang et al. 2007). Prx2 was found to be S-nitrosylated at Cys51 and Cys172, and these cysteines are critical residues within the catalytic domain of Prx2 (Fang et al. 2007). The peroxidase activity of Prx2 is attenuated after S-nitrosylation and this affects its neuroprotective function against oxidative stress in neurons (Fang et al. 2007). The levels of S-nitrosylated Prx2 are increased in brain tissues of PD patients which suggest that neurons under such condition are more vulnerable to oxidative stress (Fang et al. 2007).

Our recent study has also identified X-linked inhibitor of apoptosis (XIAP) as another protein whose S-nitrosylation is correlated with the neurodegenerative process in PD (Tsang et al. 2009). XIAP belongs to a highly conserved protein family which promotes cell survival through their baculoviral IAP repeat (BIR) domains (Vaux and Silke 2005; Srinivasula and Ashwell 2008). Similar to parkin, XIAP also possesses a RING domain at the C-terminal which enables XIAP to function as an E3 ligase in the UPS to target a number of substrates including XIAP itself (Vaux and Silke 2005; Srinivasula and Ashwell 2008). Studies have shown that the anti-apoptotic activity mediated by XIAP is through the highly conserved BIR domains of XIAP in antagonizing the pro-apoptotic activities of caspases (Vaux and Silke 2005; Srinivasula and Ashwell 2008). Caspases are the executioners of apoptosis that mediate the process of program cell death. Not surprisingly, overexpression of XIAP is commonly observed in tumors, which is one of the major contributors for the process of tumorigenesis. In our study, we found that XIAP can be S-nitrosylated both in vitro and in vivo. In contrast to parkin, XIAP is not S-nitrosylated at the RING domain but is modified at the BIR domains by NO (Tsang et al. 2009). S-nitrosylation of XIAP impairs its ability to inhibit caspase-3 activity and in turn down-regulates its anti-apoptotic functions. Different from parkin, the E3 ligase activity of XIAP is not affected by S-nitrosylation (Tsang et al. 2009). In the striatum of animal model of PD, elevated levels of S-nitrosylated XIAP were observed when compared to the control. In addition, increased XIAP

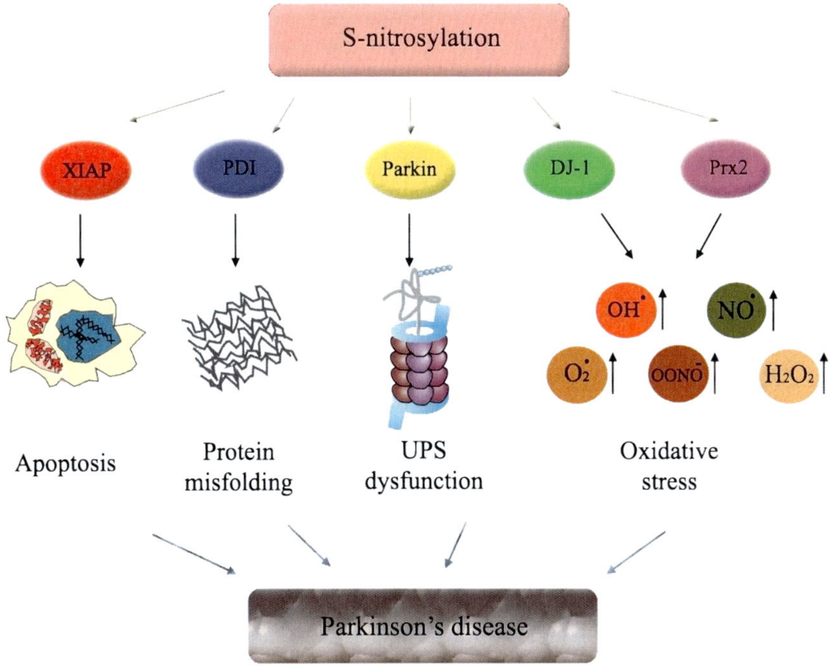

Fig. 4.3 Involvement of S-nitrosylation in the pathogenesis of Parkinson's disease (PD). S-nitrosylation of several proteins including parkin, PDI, DJ-1 and Prx2 and XIAP can lead to UPS dysfunction, protein misfoldong, increased oxidative stress and apoptosis which can contribute to the pathogenesis of PD

S-nitrosylation was also found in the brain tissues of PD patients (Tsang et al. 2009). These results suggested that the pro-survival function of XIAP in neurons is compromised by S-nitrosylation in the development of PD.

In summary, nitrosative stress can contribute to PD via a number of mechanisms. For instance, protein nitration or peroxynitrite-induced damages in lipids and DNA can promote neuronal cell death. In addition, S-nitrosylation of proteins that possess neuroprotective function such as parkin, DJ-1, Prx2 and XIAP can compromise the survival of neurons under various cytotoxic insults (Fig. 4.3), which make neurons more vulnerable to degeneration.

4.7 Conclusion

S-nitrosylation is a post-translation modification that regulates a number of cellular and physiological functions such as gene expression, cell survival, neurotransmission, vesicle trafficking and cardiovascular function. However imbalance of nitrosative stress can also compromise a number of neuroprotective proteins through

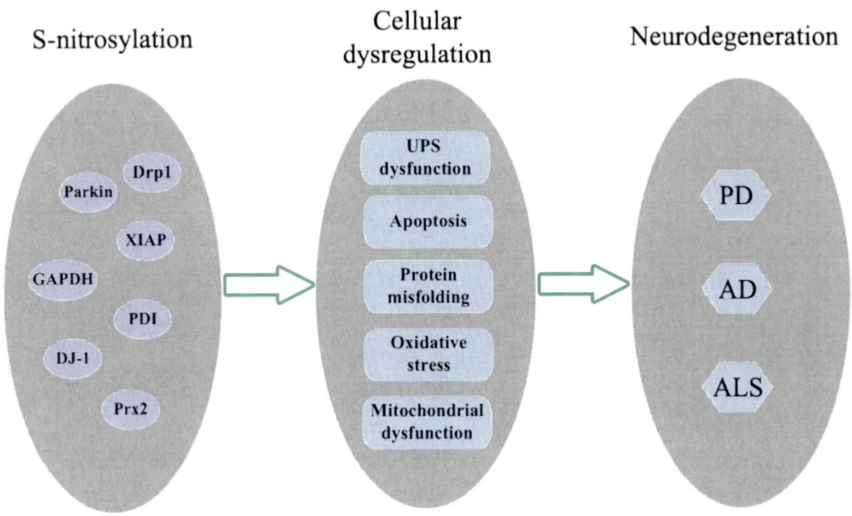

Fig. 4.4 Involvement of S-nitrosylation in neurodegeneration. Dysregulation of S-nitrosylation is involved in the causation of several neurodegenerative diseases including ALS, AD and PD through interconnecting pathways involving UPS dysfunction, protein misfolding, oxidative stress, mitochondrial dysfunction and apoptosis

S-nitrosylation and contribute to the neurodegenerative process in ALS, AD and PD (Fig. 4.4). By more thorough understanding in how S-nitrosylation can affect pathways that contribute to these disorders, new therapeutic targets can be developed for the treatment of these diseases in the future.

Acknowledgments KKKC is supported by grants and the Area of Excellence Scheme established under the University Grants Committee of the Hong Kong Special Administrative Region (HKUST6435/06M, HIA05/06.SC04, AoE/B-15/01)

References

Arnelle, D.R., and Stamler, J.S. (1995). NO+, NO, and NO- donation by S-nitrosothiols: implications for regulation of physiological functions by S-nitrosylation and acceleration of disulfide formation. *Arch. Biochem. Biophys.* 318:279–285.

Azad, N., Vallyathan, V., Wang, L., et al. (2006). S-nitrosylation of Bcl-2 inhibits its ubiquitin-proteasomal degradation. A novel antiapoptotic mechanism that suppresses apoptosis. *J. Biol. Chem.* 281:34124–34134.

Benhar, M., Forrester, M.T., Hess, D.T., et al. (2008). Regulated protein denitrosylation by cytosolic and mitochondrial thioredoxins. *Science* 320:1050–1054.

Cheah, J.H., Kim, S.F., Hester, L.D., et al. (2006). NMDA receptor-nitric oxide transmission mediates neuronal iron homeostasis via the GTPase Dexras1. *Neuron* 51:431–440.

Cho, D.H., Nakamura, T., Fang, J., et al. (2009) S-nitrosylation of Drp1 mediates beta-amyloid-related mitochondrial fission and neuronal injury. *Science* 324:102–105.

Chung, K.K. (2007). Say NO to neurodegeneration: role of S-nitrosylation in neurodegenerative disorders. *Neurosignals* 15:307–313.

Chung, K.K., Zhang, Y., Lim, K.L., et al. (2001). Parkin ubiquitinates the alpha-synuclein-interacting protein, synphilin-1: implications for Lewy-body formation in Parkinson disease. *Nat. Med.* 7:1144–1150.
Chung, K.K., Thomas, B., Li, X., et al. (2004). S-nitrosylation of parkin regulates ubiquitination and compromises parkin's protective function. *Science* 304:1328–1331.
Chung, K.K., Dawson, T.M., Dawson, V.L. (2005a). Nitric oxide, S-nitrosylation and neurodegeneration. *Cell. Mol. Biol. (Noisy-le-grand)* 51:247–254.
Chung, K.K., Dawson, V.L., Dawson, T.M. (2005b). S-nitrosylation in Parkinson's disease and related neurodegenerative disorders. *Meth. Enzymol.* 396:139–150.
Fang, M., Jaffrey, S.R., Sawa, A., et al. (2000). Dexras1: a G protein specifically coupled to neuronal nitric oxide synthase via CAPON. *Neuron* 28:183–193.
Fang, J., Nakamura, T., Cho, D.H., et al. (2007). S-nitrosylation of peroxiredoxin 2 promotes oxidative stress-induced neuronal cell death in Parkinson's disease. *Proc. Natl. Acad. Sci. U.S.A.* 104:18742–18747.
Feany, M.B., and Pallanck, L.J. (2003). Parkin: a multipurpose neuroprotective agent? *Neuron* 38:13–16.
Garban, H.J., Marquez-Garban, D.C., et al. (2005). Rapid nitric oxide-mediated S-nitrosylation of estrogen receptor: Regulation of estrogen-dependent gene transcription. *Proc. Natl. Acad. Sci. U.S.A.* 102:2632–2636
Gaston, B.M., Carver, J., Doctor, A. et al. (2003). S-nitrosylation signaling in cell biology. *Mol. Interv.* 3:253–263.
Giasson B. I., Duda J. E., Murray I. V., et al. (2000). Oxidative damage linked to neurodegeneration by selective alpha-synuclein nitration in synucleinopathy lesions. *Science* 290:985–989.
Haendeler, J., Hoffmann, J., Tischler, V., et al. (2002). Redox regulatory and anti-apoptotic functions of thioredoxin depend on S-nitrosylation at cysteine 69. *Nat. Cell Biol.* 4:743–749.
Hara, M.R., Agrawal, N., Kim, S.F., et al. (2005). S-nitrosylated GAPDH initiates apoptotic cell death by nuclear translocation following Siah1 binding. *Nat. Cell Biol.* 7:665–674.
Hess, D.T., Matsumoto, A., Nudelman, R., et al. (2001). S-nitrosylation: spectrum and specificity. *Nat. Cell Biol.* 3:E46–49.
Huang, Y., Man, H.Y., Sekine-Aizawa, Y., et al. (2005). S-nitrosylation of N-ethylmaleimide sensitive factor mediates surface expression of AMPA receptors. *Neuron* 46:533–540.
Ischiropoulos, H., and Beckman, J.S. (2003). Oxidative stress and nitration in neurodegeneration: cause, effect, or association? *J. Clin. Invest.* 111:163–169.
Ito, G., Ariga, H., Nakagawa, Y., et al. (2006). Roles of distinct cysteine residues in S-nitrosylation and dimerization of DJ-1. *Biochem. Biophys. Res. Commun.* 339:667–672.
Kahle, P.J., and Haass, C. (2004). How does parkin ligate ubiquitin to Parkinson's disease? *EMBO Rep.* 5:681–685.
Li, F., Sonveaux, P., Rabbani, Z.N., et al. (2007). Regulation of HIF-1alpha stability through S-nitrosylation. *Mol. Cell* 26:63–74.
Lim, K.H., Ancrile, B.B., Kashatus, D.F., et al. (2008). Tumour maintenance is mediated by eNOS. *Nature* 452:646–649.
Lipton, A.J., Johnson, M.A., Macdonald, T., et al. (2001). S-nitrosothiols signal the ventilatory response to hypoxia. *Nature* 413:171–174.
Liu, L., Yan, Y., Zeng, M., et al. (2004). Essential roles of S-nitrosothiols in vascular homeostasis and endotoxic shock. *Cell* 116:617–628.
Lonze, B.E., and Ginty, D.D. (2002). Function and regulation of CREB family transcription factors in the nervous system. *Neuron* 35:605–623.
Luchsinger B.P., Rich, E.N., Gow, A.J., et al. (2003). Routes to S-nitroso-hemoglobin formation with heme redox and preferential reactivity in the beta subunits. *Proc. Natl. Acad. Sci. U.S.A.* 100:461–466
Mannick, J.B., Hausladen, A., Liu, L., et al. (1999). Fas-induced caspase denitrosylation. *Science* 284:651–654.
Mannick, J.B., Schonhoff, C., Papeta, N., et al. (2001). S-Nitrosylation of mitochondrial caspases. *J. Cell Biol.* 154:1111–1116.

Marshall, H.E., Hess, D.T., Stamler, J.S. (2004). S-nitrosylation: physiological regulation of NF-kappaB. *Proc. Natl. Acad. Sci. U.S.A.* 101:8841–8842.

Matsushita, K., Morrell, C.N., Cambien, B., et al. (2003). Nitric oxide regulates exocytosis by S-nitrosylation of N-ethylmaleimide-sensitive factor. *Cell* 115:139–150.

McMahon, T.J., Moon, R.E., Luschinger, B.P., et al. (2002). Nitric oxide in the human respiratory cycle. *Nat. Med.* 8:711–717.

Mitchell, D.A., and Marletta, M.A. (2005). Thioredoxin catalyzes the S-nitrosation of the caspase-3 active site cysteine. *Nat. Chem. Biol.* 1:154–158.

Mitchell, D.A., Morton, S.U., Fernhoff, N.B., et al. (2007). Thioredoxin is required for S-nitrosation of procaspase-3 and the inhibition of apoptosis in Jurkat cells. *Proc. Natl. Acad. Sci. U.S.A.* 104:11609–11614

Nakamura, T., and Lipton, S.A. (2008). Emerging roles of S-nitrosylation in protein misfolding and neurodegenerative diseases. *Antiox. Redox Sign.* 10:87–101.

Nott, A., Watson, P.M., Robinson, J.D., et al. (2008). S-Nitrosylation of histone deacetylase 2 induces chromatin remodelling in neurons. *Nature* 455:411–415.

Pawloski, J.R., Hess, D.T., Stamler, J.S. (2001). Export by red blood cells of nitric oxide bioactivity. *Nature* 409:622–626.

Perez-Mato, I., Castro, C., Ruiz, F.A., et al. (1999). Methionine adenosyltransferase S-nitrosylation is regulated by the basic and acidic amino acids surrounding the target thiol. *J. Biol. Chem.*274:17075–17079.

Raines, K.W., Bonini, M.G., Campbell, S.L. (2007). Nitric oxide cell signaling: S-nitrosation of Ras superfamily GTPases. *Cardiovasc. Res.* 75:229–239.

Reynaert, N.L., Ckless, K., Korn, S.H., et al. (2004). Nitric oxide represses inhibitory kappaB kinase through S-nitrosylation. *Proc. Natl. Acad. Sci. U.S.A.* 101:8945–8950

Riccio, A., Alvania, R.S., Lonze, B.E., et al. (2006). A nitric oxide signaling pathway controls CREB-mediated gene expression in neurons. *Mol. Cell* 21:283–294.

Rothstein, J.D. (2009). Current hypotheses for the underlying biology of amyotrophic lateral sclerosis. *Ann. Neurol.* 65 Suppl 1:S3–9.

Savitt, J.M., Dawson, V.L., Dawson, T.M. (2006). Diagnosis and treatment of Parkinson disease: molecules to medicine. *J. Clin. Invest.* 116:1744–1754.

Schonhoff, C.M., Matsuoka, M., Tummala, H., et al. (2006). S-nitrosothiol depletion in amyotrophic lateral sclerosis. *Proc. Natl. Acad. Sci. U.S.A.* 103:2404–2409.

Sen, N., Hara, M.R., Kornberg, M.D., et al. (2008). Nitric oxide-induced nuclear GAPDH activates p300/CBP and mediates apoptosis. *Nat. Cell Biol.* 10:866–873.

Srinivasula, S.M., and Ashwell, J.D. (2008). IAPs: what's in a name? *Mol. Cell* 30:123–135.

Stamler, J.S., Toone, E.J., Lipton, S.A., et al. (1997a). (S)NO signals: translocation, regulation, and a consensus motif. *Neuron* 18:691–696.

Stamler, J.S., Jia, L., Eu, J.P., et al. (1997b). Blood flow regulation by S-nitrosohemoglobin in the physiological oxygen gradient. *Science* 276:2034–2037.

Thomas, B., and Beal, M. F. (2007). Parkinson's disease. *Hum. Mol. Genet.* 16(Spec No. 2):-R183–194.

Thornberry, N.A., and Lazebnik, Y. (1998). Caspases: enemies within. *Science* 281:1312–1316.

Torta, F., Usuelli, V., Malgaroli, A. et al. (2008). Proteomic analysis of protein S-nitrosylation. *Proteomics* 8:4484–4494.

Tsang, A.H., and Chung, K.K. (2009). Oxidative and nitrosative stress in Parkinson's disease. *Biochim. Biophys. Acta* 1792:643–650.

Tsang, A.H., Lee, Y.I., Ko, H.S., et al. (2009). S-nitrosylation of XIAP compromises neuronal survival in Parkinson's disease. *Proc. Natl. Acad. Sci. U.S.A.* 106:4900–4905

Uehara, T., Nakamura, T., Yao, D., et al. (2006). S-nitrosylated protein-disulphide isomerase links protein misfolding to neurodegeneration. *Nature* 441:513–517.

Vaux, D.L., and Silke, J. (2005). IAPs, RINGs and ubiquitylation. *Nat. Rev. Mol. Cell Biol.* 6:287–297.

Wang, G., Moniri, N.H., Ozawa, K., et al. (2006). Nitric oxide regulates endocytosis by S-nitrosylation of dynamin. *Proc. Natl. Acad. Sci. U.S.A.* 103:1295–1300

Whalen, E.J., Foster, M.W., Matsumoto, A., et al. (2007). Regulation of beta-adrenergic receptor signaling by S-nitrosylation of G-protein-coupled receptor kinase 2. *Cell* 129:511–522.

Yao, D., Gu, Z., Nakamura, T., et al. (2004). Nitrosative stress linked to sporadic Parkinson's disease: S-nitrosylation of parkin regulates its E3 ubiquitin ligase activity. *Proc. Natl. Acad. Sci. U.S.A.* 101:10810–10814

Yee-Koh, M., Spivak-Kroizman, T.R., Powis, G. (2008). HIF-1 regulation: not so easy come, easy go. *Trends Biochem. Sci.* 33:526–534.

Zhai, D., Jin, C., Huang, Z., et al. (2008). Differential regulation of Bax and Bak by anti-apoptotic Bcl-2 family proteins Bcl-B and Mcl-1. *J. Biol. Chem.* 283:9580–9586.

Chapter 5
Protein Glysosylation and Congenital Disorders of Glycosylation

Eva Morava, Dirk J. Lefeber, and Ron A. Wevers

Abbreviations

CDG	Congenital disorders of glycosylation
COG	Conserved oligomeric Golgi
UDP	Uridine 5′-diphosphate
ER	Endoplasmatic reticulum
IEP	Isoelectric point
TIEF	Transferrin isoelectric focusing
IEF	Isoelectric focusing
LLO	Lipid linked oligosaccharides
apoC-III	Apolipoprotein C-III
ARCL	Autosomal recessive cutis laxa
	(For Abbreviations of enzymes see Table 5.1)

5.1 Introduction

Glycosylation is a post-translational modification defining the final structure and function of many proteins in the body (Freeze and Aebi 2005). Glycosylated proteins play a role in biochemical, endocrine, immunological and transport processes, define cell-cell interactions, and determine developmental pathways during embryogenesis. Cells respond to environmental stimuli by altering surface glycan structures. Normal cells influence their development and malignant cells regulate their growth by remodeling their glycans (Tao et al. 2008). More than 50% of our proteins are glycosylated and 1% of our genetic material is involved in the glycosylation-process (Lowe and Marth 2003). The cytoplasm, the endoplasmatic reticulum (ER) and the Golgi-apparatus play a role in the biosynthesis of glycans

E. Morava (✉)
Department of Paediatrics, Radboud University Nijmegen Medical Centre, P.O. Box 9101, 6500 HB Nijmegen, The Netherlands
e-mail: E.Morava@cukz.umcn.nl

C.J. Vidal (ed.), *Post-Translational Modifications in Health and Disease*, Protein Reviews 13,
DOI 10.1007/978-1-4419-6382-6_5,

(Wopereis et al. 2006). Based on the attachment site of the glycan on the protein two common types of glycosylation can be discriminated: N-linked and O-linked glycosylation. The current review focuses on these two, clinically highly relevant sub-types. Congenital Disorders of Glycosylation (CDG) form a group of inborn errors of metabolism due to defects in the biosynthesis of glycans. The first disease in this group was described by Jaeken (Jaeken et al. 1987). Glycoproteins play an essential role in many different processes in the organism. Therefore, it is not surprising that inborn errors of glycosylation result in a wide variety of multisystem diseases and congenital malformation syndromes (De Lonlay and Seta 2008). Due to the clinical variability in CDG finding discriminative symptoms and a specific clinical pattern become essential to find the diagnosis (Grünewald et al. 2000).

Since 1987, the molecular basis of 29 different CDG subtypes has been delineated (Jaeken and Matthijs 2007; Table 5.1). A recent classification of CDGs divides the defects into four major biochemical categories: three involving proteins (disorders of N-linked glycosylation, O-linked glycosylation and combined N- and O-glycosylation) and one involving lipid-glycosylation (Jaeken et al. 2008). CDG syndrome is a rare, so called "Orphan disease", usually presenting as a multisystem disease.

The Golgi apparatus houses parts of the N-glycan biosynthesis pathway and the full mucin type O-glycan biosynthetic pathway (Wopereis et al. 2003, 2006). Some Golgi processes and features are relevant for both biosynthetic pathways. A genetic defect in such a shared step of the two pathways will result in a *combined* N- and O-glycan biosynthesis defect (Wu et al. 2004; Wopereis et al. 2005a; Morava et al. 2007; Zeevaert et al. 2008). Recently the clinical/biochemical phenotype has been unraveled in several unsolved CDG patients leading to the finding of new disorders (Freeze 2007; Morava et al. 2006, 2008a; Van Maldergem et al. 2008; Kornak et al. 2008; Zeevaert et al. 2008).

5.2 The General Biosynthesis of Protein-Linked Glycans

The biosynthesis of protein-linked glycans can be subdivided into four stages (Fig. 5.1). In the first stage, the biosynthesis of nucleotide sugars occurs in the cytoplasm of the cell. Monosaccharides used for the biosynthesis of nucleotide sugars derive from dietary sources and salvage pathways. Series of phosphorylation, epimerization, and acetylation reactions convert the monosaccharides into various high-energy nucleotide sugar donors. Specific nucleotides are used for each monosaccharide required in glycan biosynthesis: Uridine 5′-diphosphate (UDP) is linked to the monosaccharides galactose (UDP-Gal), glucose (UDP-Glc), *N*-acetylgalactosamine (UDP-GalNAc), *N*-acetylglucosamine (UDP-GlcNAc) and xylose (UDP-Xyl); guanidine 5′diphosphate (GDP) is linked to mannose (GDP-Man) and fucose (GDP-Fuc) and cytidine 5′-monophosphate (CMP) is linked to neuraminic acid (CMP-NeuAc). As observed in CDG-Ia and CDG-Ib, aberrant glycosylation can result from an insufficient availability of a specific nucleotide sugar.

The second stage represents the transport mechanisms that are responsible for the translocation of the activated monosaccharides into the lumen of the endoplasmic

Table 5.1 Overview of CDG subtypes

Gene, Protein	EC	Former nomenclature (CDG type)	Nr on Fig. 5.1
PMM2, Phosphomannomutase II	5.4.2.8	Ia	1
MPI, Phosphomannose isomerase	5.3.1.8	Ib	2
h*ALG6*, *G*lucosyltransferase I	2.4.1.–	Ic	3
h*ALG3*, Mannosyltransferase VI	2.4.1.130	Id	4
DPM1, Dol-P-Man synthase subunit 1	2.4.1.83	Ie	5a
LEC35, Dol-P-Man utilization protein	–	If	6
h*ALG12*, Mannosyltransferase VIII	2.4.1.130	Ig	7
h*ALG8*, Glucosyltransferase II	2.4.1.–	Ih	8
h*ALG2*, Mannosyltransferase II	2.4.1.132	Ii	9
DPAGT1, GlcNActransferase I	2.7.8.15	Ij	10
h*ALG1*, Mannosyltransferase I	2.4.1.142	Ik	11
h*ALG9*, Mannosyltransferase VII	2.4.1.130	Il	12
DOLK, Dolichol kinase	2.7.1.108	Im	13
RFT1, RFT1 protein	–	In	14a
DPM3, Dol-P-Man synthase subunit 3	2.4.1.83	Io	5c
h*ALG11*, GDP-Man: $Man_3GlcNAc_2$-PP-dolichol-α 1,2-mannosyltransferase	2.4.1.0	Ip	14b
MGAT2, GlcNActransferase II	2.4.1.143	IIa	15
N33/TUSC, Oligosaccharide transferase subunit N33/TUSC	–	–	16
GCS1, Glucosidase I	3.2.1.106	IIb	17
SLC35C1, GDP-fucose transporter	–	IIc	18
β4GALT1, Galactosyltransferase	2.4.1.38	IId	19
COG7, COG 7 subunit	–	IIe	20
SLC35A1, CMP-NeuAc transporter	–	IIf	21
COG1, COG 1 subunit	–	IIg	22
COG8, COG8 subunit	–	IIh	23
ATP6V0A2, vesicular $H^{(+)}$-ATPase subunit a2	3.6.3.14	–	27
COG4, COG 4 subunit	–	III	24
COG5, COG 5 subunit	–	–	25
COG6, COG 6 subunit	–	–	26

reticulum (ER) and the Golgi. In the ER, the nucleotide sugars bind to the cytosolic side of the membrane-bound lipid dolichol-phosphate (Dol-P). Subsequently, the nucleotide moiety is cleaved off. A "flippase" then mediates the turnover of the Dol-P-monosaccharide complex from the cytoplasmic leaflet to the lumenal leaflet of the ER. Abnormal glycosylation can result from a reduced Dol-P-monosaccharide synthesis or transport as is seen in patients with DPM1 deficiency (former CDG-Ie) and LEC35 defect (former CDG-If) (Kim et al. 2000; Schenk et al. 2001). In the Golgi, nucleotide sugars enter the lumen via specific nucleotide sugar transporters (NSTs). These NSTs are antiporters in which the nucleotide sugar entry into the ER/Golgi is coupled to the equimolar exit of the corresponding nucleoside monophosphate from the ER/Golgi lumen (Hirschberg et al. 1998). Abnormal glycosylation can result from reduced NST function as observed in deficiency of the GDP-Fuc transporter

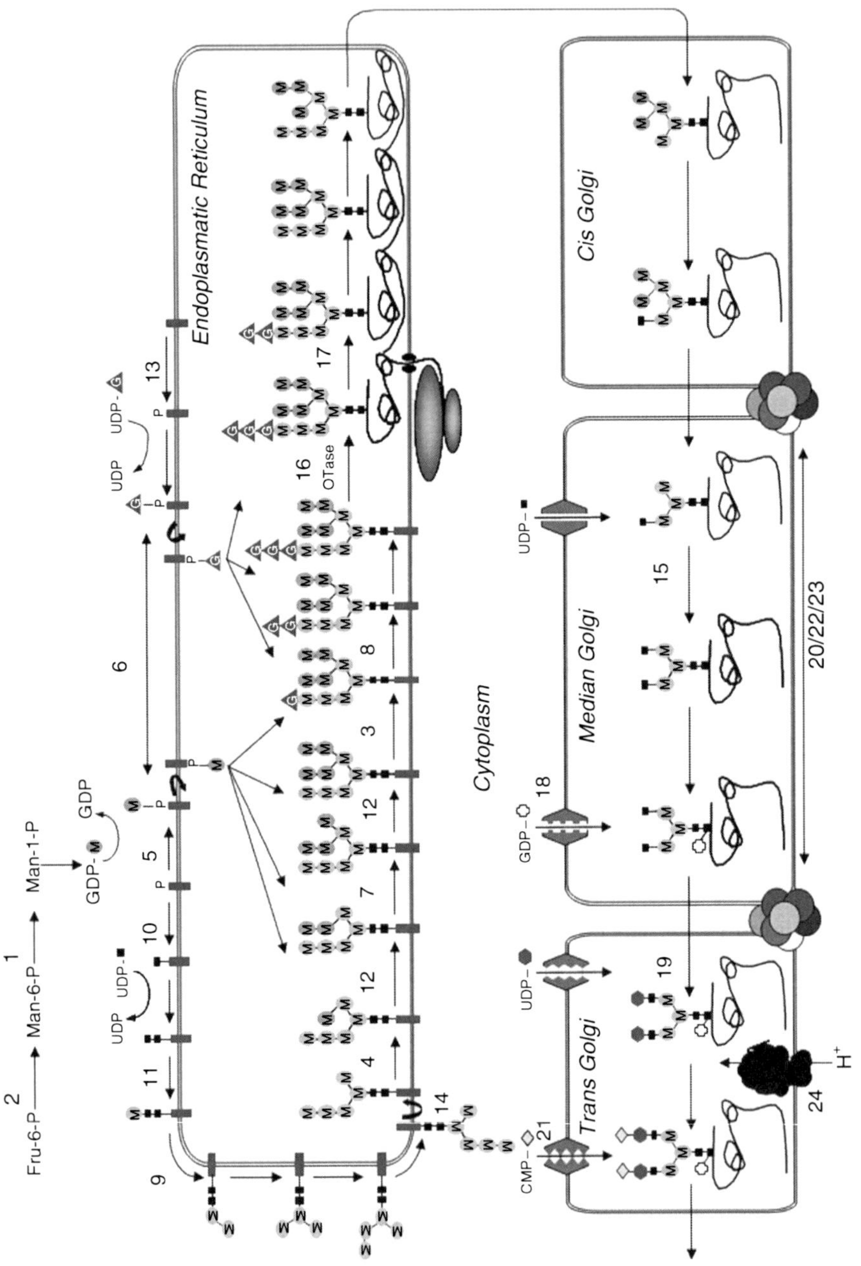

Fig. 5.1 Overview of the glycan synthesis pathway

(former CDG-IIc) (Lubke et al. 2001), and the CMP-NeuAc transporter (former CDG-IIf) (Martinez-Duncker et al. 2005).

In the third stage, specific transferases attach the glycan in the ER and the Golgi. N- and O-glycans are synthesized by the sequential action of a number of specific membrane-bound glycosyltransferases in a highly controlled fashion. The pathways of glycosylation are determined by the distinct substrate specificities of glycosyltransferases. Most CDG subtypes and O-glycosylation disorders described to date are caused by deficiencies in glycosyltransferases.

The last stage represents the Golgi traffic and Golgi integrity mechanisms. The Golgi apparatus exists of several cisternae organized in the form of a stack. Glycoproteins and also glycosyltransferases, for example, are transported to their destination via the cisternae and series of Golgi vesicles. Defects in the subunits of the Conserved Oligomeric Golgi (COG) complex were shown to cause abnormal glycosylation. Dysfunction of the COG complex leads to separation of glycosyltransferases from anterograde cargo molecules passing along secretory pathway, thus affecting normal protein glycosylation. The first defect in a subunit of the COG complex was in the subunit 7 of this complex (former CDG-IIe) (Wu et al. 2004).

5.3 The Biosynthesis of N-Glycans

The N-glycosylation pathway comprises the assembly and processing of glycans and extends over three subcellular compartments: the cytoplasm, the ER and the various Golgi compartments. The N-glycosylation biosynthesis routing has been excellently reviewed by Marquardt and Denecke (2003). In short, the assembly of N-glycans is initiated in the cytosol. Two specific cytosolic UDP-GlcNAc transferases deliver the first two GlcNAc monosaccharides to Dol-P. Biosynthesis is continued on the cytosolic side with the attachment of five consecutive Man residues donated by the nucleotide sugar GDP-Man. The lipid bound heptasaccharide is then flipped to the ER lumen by the flippase. In contrast to cytosolic glycan biosynthesis, luminal glycan biosynthesis uses Dol-P-bound saccharides as precursor. The $Man_5GlcNAc_2$-Dol-P is subsequently extended with four additional Man and three Glc residues donated by Dol-P-Man and Dol-P-Glc, respectively. This results in a $Glc_3Man_9GlcNAc_2$-Dol-P structure. This lipid-linked oligosaccharide (LLO) is the substrate for the enzyme oligosaccharyltransferase (OTase). Defects in the assembly of this mature LLO-form are referred to as CDG type I.

The processing of N-glycans starts with the removal of the $Glc_3Man_9GlcNAc_2$ from Dol-P and its transfer to a nascent protein chain containing the consensus sequence Asn-X-Ser/Thr (X denotes any amino acid). This transfer is catalyzed by the OTase complex. Directly after the transfer of the N-glycan chain to the protein, the terminal glucose residues are removed by two membrane-bound glucosidases followed by the cleavage of the terminal-linked Man residues by several mannosidases. Several lectins recognizing terminal bound Man residues promote the transport of synthesized polypeptides to the Golgi apparatus. In the different Golgi compartments the glycoprotein can be further processed. Defects in the processing of N-glycans are referred to as CDG type II.

In general, three protein-linked N-glycan types can be distinguished (Fig. 5.2). The first type is the "high mannose structure", which is an N-glycan containing between five and nine mannose residues (Fig. 5.2a). The glycans attached to some proteins remain in this state when the glycoprotein moves through the various Golgi compartments to the cell surface. The second type is the "hybrid structure", which is an N-glycan with one antenna of the complex type and one antenna of the high mannose type (Fig. 5.2b). Hybrid structures are formed when Golgi mannosidases do not act after the addition of a single GlcNAc residue by GlcNAc-transferase I. The last N-glycan type is the "complex structure", which is an N-glycan with antennae that are fully processed (Fig. 5.2c). The complex type N-glycan is very heterogeneous. Variations are seen in the terminal residues and in the branching of N-glycans; besides the bi-antennary N-glycan, also tri-antennary and tetra-antennary N-glycans can be distinguished. Subsequently, N-glycans can be modified by addition of a bisecting GlcNAc residue, an N-glycan core fucose residue, fucosylation, or a polylactosamine epitope, which is a series of repeated Gal-GlcNAc residues.

In summary, N-glycans share a common core structure and their diversity results from the differential action of glycosidases and glycosyltransferases in the Golgi apparatus (Drickamer and Taylor, 2002). It is interesting that glycan structures are

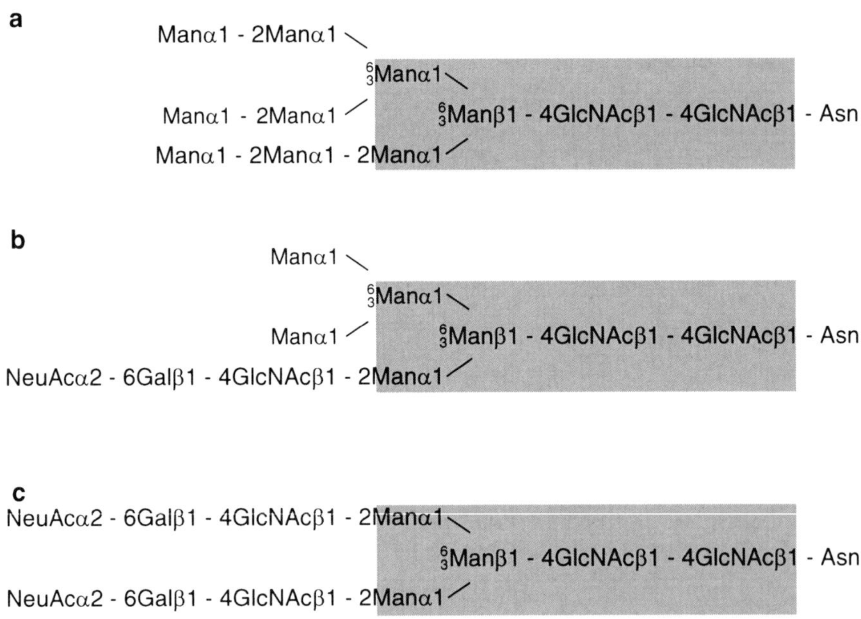

Fig. 5.2 The three N-glycan types. (**a**) "high mannose type" (**b**) "hybrid type" (**c**) "complex type". The common N-glycan core structure is *shaded*. Because the two branching mannose residues in the core are in different linkages to the innermost mannose residue, these two branches are known as the 1–3 antennae and the 1–6 antennae of an oligosaccharide

determined in this way and that there is no template available for their biosynthesis. While DNA forms the template for the sequence of amino acids in a protein, there is no such equivalent for the design of glycans. The sequence of monosaccharides on a protein is determined by a number of factors, such as subcellular compartimentalisation, glycosyltransferase specificity and expression, and the availability of monosaccharides.

5.4 The Biosynthesis of O-Glycans

The biosynthesis of O-glycans is initiated after the folding and oligomerization of proteins and can set off in the late ER or in one of the Golgi compartments (Peters et al. 1989; Röttger et al. 1998; Spiro, 2002; Vertel et al. 1993). Therefore, O-glycosylation mainly takes place in coil, turn and linker regions of proteins. In short, seven different types of O-glycan attachments are known in man, classified on basis of the first sugar (GalNAc, Xyl, GlcNAc, Gal, Man, Glc, Fuc) bound to the serine, threonine or hydroxylysine residue of the protein. In man, the mucin-type O-glycan (with GalNAc as the first sugar) and the glycosaminoglycans (GAGs; with Xyl as the first sugar) are by far the most common structures. The total amount of monosaccharide residues in mucin-type O-glycans may vary between one and about 10, whereas GAG chains are about 100 or more monosaccharides long. Mucin-type O-glycans and GAGs occur in various different forms. Mucin-type O-glycans, for example, can be further subdivided into eight core structures depending on the second sugar(s) and/or sugar binding and further elongation of these core structures give rise to at least 50 different mucin-type O-glycan structures. The structures of the other five O-glycan types (starting with GlcNAc, Gal, Man, Glc, and Fuc) seem to show less variability and mostly occur in one conformation (van den Steen et al. 1998; 2000).

The O-glycosylation process differs from the N-glycosylation process in various aspects. N-linked glycans share a common protein-glycan linkage, have a common core structure and only few N-glycan structures exist. This reflects a common biosynthetic pathway that only diverges in its late stages. In contrast, O-glycans have different protein-glycan linkages in which GalNAc, Xyl, Fuc, Man, Glc, GlcNAc and Gal can be attached to serine, threonine or hydroxylysine resulting in many different glycan types.

A specific type of O-glycosylation involves the attachment of a mannose residue to Ser/Thr amino acids. The best characterised protein substrate is alpha-dystroglycan, although a few other O-mannosylated proteins are known. This glycan type is initiated in the endoplasmic reticulum by the combined action of the tissue-specific protein-O-mannosyltransferases POMT1/POMT2 using dolichol-P-mannose as a monosaccharide donor. The glycan is further elongated by the GlcNAc-transferase (*POMGnT1* gene) in the Golgi, which specifically acts on O-linked mannose. The enzymes involved in further extending the glycan with galactose and sialic acid residues are currently unknown.

5.5 Diagnostic Procedures

5.5.1 *Isoelectric Focusing of Transferrin*

Plasma transferrin isoelectric focusing (TIEF) is generally applied in the diagnosis of defects in the biosynthesis of N-glycans. Before TIEF is performed the protein has to be saturated with iron. Transferrin is a plasma protein that contains two N-glycan structures (positioned at Asn 413 and 611) of the "complex type" with two or three antennae (Fig. 5.3a, b). Sialic acid, or neuraminic acid has a negative charge and is always positioned at the terminus of the N-glycan structure. The charge of the sialic acid residues contributes to the isoelectric point (IEP) of the protein. In human plasma the tetrasialotransferrin fraction is the most abundant form, corresponding with the transferrin isoform that contains two fully sialylated biantenarry N-glycans. Also the trisialo-, pentasialo- and hexasialotransferrin isoforms can be observed in human plasma (Fig. 5.3c, lane 1). Children younger than 1 month can have a different transferrin isoform ratio that may mimick a CDG TIEF profile. A defect in the biosynthesis of N-glycans is picked up by abnormal quantities of the different transferrin isoforms.

Transferrin IEF can also be used to distinguish between CDG type I and CDG type II defects (Fig. 5.3c). Defects in the assembly of N-glycans result in unglycosylated proteins, thus in proteins that lack complete N-glycan structures. This is reflected in the characteristic CDG type I profile, with increased amounts of asialo- and disialotransferrin and decreased amounts of tetrasialo and pentasialotransferrin (Fig. 5.3c, lane 2). Defects in the processing of N-glycans often lead to hyposialylation of proteins, thus in proteins lacking their terminal sialic acid residues. All profiles that are not a type I profile are classified as a CDG type II profile (Fig. 5.3c, lanes 3 and 4). N-glycosylation biosynthesis defects can be confirmed by performing IEF of thyroxine-binding globulin, another N-glycosylated plasma protein.

There are some pitfalls of the transferrin IEF test. Not all CDG type II defects can be picked up with transferrin IEF. Patients with *GLS1*, *SLC35C1* and *SLC35A1* defects (former CDG-IIb,CDG-IIc and CDG-IIf) have a normal transferrin IEF profile. Polymorphisms in the protein part of transferrin can complicate the interpretation of transferrin IEF profiles. Polymorphisms can result in a different IEP of the protein and thus can have an influence on the TIEF profile. The double bands in Fig. 5.3c, lanes 3 and 4 represent a frequently occurring polymorphism. Some of the rare polymorphisms result in a TIEF profile that easily may be misinterpreted as pathological and indicative for CDG. To exclude these false-positive results transferrin should be incubated with neuraminidase. This enzyme catalyzes the hydrolysis of NeuAc residues from glycans. In a neuraminidase treated sample all transferrin isoforms migrate to the asialo position on TIEF, whereas a neuraminidase treated sample with a hetero-allelic polymorphism in transferrin gives two bands on TIEF. The paper of Weidinger et al. gives an overview of known polymorphisms and their influence on the electrophoretic behaviour of the protein (Weidinger et al. 1984).

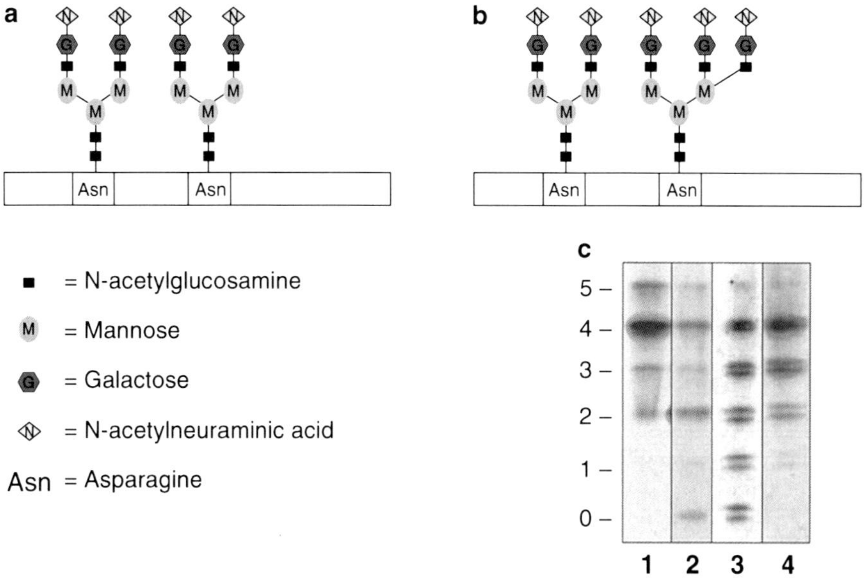

Fig. 5.3 N-glycosylation of transferrin. (**a**) "tetrasialo transferrin" (**b**) "pentasialo transferrin" (**c**) isoelectric focusing of human plasma transferrin. *Lane 1*: normal TIEF profile, *lane 2*: CDG type I TIEF profile, *lane 3*: CDG type II profile, *lane 4*: CDG type II profile

Some patients secondarily have an abnormal biosynthesis of N-glycans. In samples of patients with galactosemia, fructosemia, alcohol abuse and severe liver disease TIEF can result in a CDG type 1 profile. TIEF performed on samples from Haemolytic Uraemic Syndrome patients result in a CDG type II profile.

5.5.2 *Enzymatic Measurements*

Phosphomannomutase (PMM2) and phosphomannose isomerase (PMI) are the enzymes that relate to the former CDG types Ia and Ib. Their enzymatic activity can be measured in fibroblasts or leukocytes (Van Schaftingen and Jaeken 1995; Körner et al. 1998a). Leukocytes seem to be more reliable than fibroblasts for PMM measurements, as high residual activity has been observed in the fibroblasts of some patients, whereas PMM activity was always abnormal in leukocytes of these patients (Grünewald et al. 2001).

5.5.3 *Lipid-Linked Oligosaccharide Analysis*

LLO analysis can be applied on patients with CDG type I who have normal PMM and PMI activities. The Dol-P bound oligosaccharides, or LLO's, are released from the patients' fibroblasts by chloroform/methanol/water extractions. Subsequently, the glycans are released from Dol-P by mild acid hydrolysis and analyzed with HPLC. The short LLO's are released by a chloroform/methanol extraction and analyzed by thin layer chromatography. The assembly of N-glycans is highly conserved in eukaryotic cells and therefore yeast mutants have been very useful in the identification of new N-glycosylation disorders in human. By comparing LLO structures of patients with those of yeast mutant strains, most CDG type I defects have been elucidated (Burda et al. 1998; Körner et al. 1998b, 1999; Frank et al. 2004; Morava et al. 2008b).

5.5.4 Protein Glycan Structure Analysis

Protein glycan structure analysis can be applied on plasma samples from patients with a CDG type II or a defect in the biosynthesis of O-glycans. The determination of glycoprotein glycans requires their enzymatic or chemical release. Enzymatic release of N-glycans is accomplished by the endoglycosidases peptide N-glycosidase F and A, that catalyze the release of intact reducing N-glycans from the protein backbone. Two endo-α-*N*-acetylgalactosaminidases have been reported to facilitate O-glycan removal from proteins, although their specificity is narrowed to unsubstituted core 1 O-glycans (Carlson, 1966; Ishii-Karakasa et al. 1997). A general endoglycosidase for release of all O-glycans, if any, remains to be discovered. Therefore, release of O-glycans is accomplished by chemical methods like reductive β-elimination, hydrazinolysis and nonreductive β-elimination (Carlson 1966; Royle et al. 2002). Both N- and O-glycans are analyzed by methods like HPLC or mass spectrometry (Körner et al. 1998b; Mills et al. 2003; Guillard et al. 2009).

5.5.5 Isoelectric Focusing of Apolipoprotein C-III

Plasma apolipoprotein C-III (apoC-III) IEF can be applied in the diagnosis of defects in the biosynthesis of core 1 O-glycans. ApoC-III is a plasma protein that contains 1 core 1 O-glycan (positioned at Thr 94). The core 1 O-glycan can be elongated with one or two NeuAc residues (see also Fig. 5.3a–c). Three isoforms of the protein can be distinguished; apoC-III_0, apoC-III_1 and apoC-III_2. The number in the isoform correlates with the number of NeuAc residues that is attached to the protein. As NeuAc has a negative charge it contributes to the IEP of the protein so that the three isoforms can be separated from each other with IEF. In human plasma the apoC-III_1 and apoC-III_2 fractions are about equally distributed, whereas apoC-III_0 is the minor fraction. In controls, the ratio of the different apoC-III isoforms is constant, although it changes with age. ApoC-III glycans in very young children carry more NeuAc residues than in adults. A defect in the biosynthesis of core 1 O-glycans can be picked up by an abnormal ratio of the different apoC-III isoforms.

Just like in TIEF, polymorphisms in the protein part can alter the protein's IEP and thus have an influence on the apoC-III IEF profile. To exclude these false-positive results the protein can be treated with neuraminidase.

5.6 Classification of CDG Subtypes

The number of CDG subtypes has increased tremendously in the past 10 years. Including the 23 known "classic" CDG defects more than 40 different types have been discovered, including protein and lipid-glycosylation defects. At least 200 genes seem to be involved in glycosylation, and the number of novel subtypes is growing continuously. This changing and growing patient cohort makes the classification difficult. Classification of the different CDG types in the past has been applied based on the biochemical analytical pattern (CDG type I pattern on the TIEF suggesting a primary defect in the cytoplasm or endoplasmatic reticulum), and CDG type II pattern, suggesting a primary defect in the Golgi apparatus), followed by a letter according to the time of discovery (chronological labeling with letters of the alphabet) (Jaeken et al. 2008). This classification has some disadvantages and restrictions. The subtyping does not include the genetic background or the name of the underlying enzyme defect. Based on the type of pattern (CDG type I or II) it is not easy to include additional novel entities, especially correctly name the group with combined N- and O-linked glycosylation defects. This led to a new nomenclature including four major disease groups: Defects of protein N-glycosylation, defects of protein O-glycosylation, defects of glycosphyngolipid and glycosylphosphatidylinositol anchor glycosylation and defects of multiple glycosylation and other pathways (Jaeken et al. 2008).

5.6.1 N-glycosylation Defects: Clinical Features

Most patients demonstrate an early onset, multisystem disease. In general, there are two major clinical groups; children, presenting with a severe, neonatal form with organ failure and children with developmental delay and a more restricted, so-called neurological form. There is almost no human organ, which has not been described with functional defects in the most common form of CDG syndrome caused by *PMM2* deficiency (former CDG I type Ia; Table 5.2). In the severe neonatal form patients are frequently diagnosed with hypothyroidism, growth delay, failures to thrive, edema forming including pericardial fluid collection, ascites and eventual pleural fluid collection. Muscle hypotonia is variable, and mostly becomes apparent at a couple of months of age. The children have feeding problems, frequent diarrhea and vomiting. Cardiac involvement has been also described, mostly as hyperthrophic cardiomyopathy. Some of the patients develop severe liver dysfunction with hypoalbuminaemia and decreased synthesis of coagulation factors.

Table 5.2 Clinical features in N-linked glycosylation disorders

Disorder (underlying gene defect)	MIM	Clinical features
PMM2	212065	Mental retardation, hypotonia, strabismus, lipodystrophy, coagulopathy, hypothyreoidism, cerebellar hypoplasia, stroke-like episodes, seizures, ascites, pericardial fluid, skeletal anomalies, cataract, retinitis pigmentosa, inverted nipples
MPI	602579	Coagulopathy, hepatic fibrosis, hepatomegaly, protein-losing enteropathy, hypoglycaemia, no mental retardation
h*ALG6*	603147	Epilepsy, moderate mental retardation, hypotonia, strabismus, epilepsy
h*ALG3*	601110	Severe psychomotor delay, microcephaly, optic atrophy, blindness, iris colobomas, epilepsy (hypsarrhythmia)
DPM1	603503	Severe mental retardation, muscle hypotonia, epilepsy, dysmorphism, coagulopathy, brain migration defects, high CK levels
MPDU1	608799	Short stature, icthyosis, psychomotor retardation, visual loss, retinitis pigmentosa
h*ALG12*	607143	Muscle hypotonia, facial dysmorphism, psychomotor retardation, microcephaly, recurrent infections, hypoglobulinemia, skeletal dysplasia
h*ALG8*	608104	Hepatic dysfunction, protein-losing enteropathy, renal failure, hypoalbuminaemia, oedema, ascites
h*ALG2*	607906	Mental retardation, hypomyelination, intractable seizures, iris colobomas, hepatomegaly, coagulopathy
DPAGT1	608093	Severe mental retardation, hypotonia, seizures, microcephaly, strabismus
h*ALG1*	608540	Severe psychomotor retardation, muscle hypotonia, microcephaly, seizures, coagulopathy, nephrotic syndrome, early death
h*ALG9*	608776	Mental retardation, severe microcephaly, hypotonia, seizures, hepatomegaly
DOLK	610768	Cardiomyopathy, ichthyosis, seizures, hypoglycemia
RFT1	612015	Developmental delay, hypotonia, seizures, hepatomegaly, coagulopathy
DPM3	612937	Mild mental retardation, muscledystrophy, cardiomyopathy, high CK levels
h*ALG11*	–	Muscular hypotonia, seizures, developmental retardation and lethality

Coagulopathy is common in CDG due to abnormal glycosylation of several factors, like factor IX and XI. An increased APT/APTT is very common. Still thrombotic events frequently occur due to abnormal glycosylation of protein C and S. Besides the common problems with bleeding diathesis and a higher incidence of vascular thrombi patients have an extra risk for coagulation problems due to an associated thrombocytopenia and loosing coagulation factors due to protein loosing enteropathy. Further possible symptoms include renal tubulopathy, with severe proteinuria and various endocrine abnormalities due to abnormal central hormonal regulation.

Dysmorphic features are also characteristic in PMM2 deficiency. The pathognomonic feature in PMM2 deficiency is abnormal fat distribution, characteristically around the genital region and the buttocks, and the presence of inverted nipples.

Table 5.3 Disorders of O-linked glycosylation

Disorders of O-linked glycosylation	MIM	Molecular defect	Clinical features
Ehlers–Danlos syndrome	130070	*B4GALT7*	Progeroid Ehlers–Danlos syndrome
Heriditary multiple exostosis	133700	*EXT1/EXT2*	Multiple exostoses syndrome
Chondrodysplasias	222600	*DTDST/SLC26A2*	Diastrophic dysplasia
	600972		Achondrogenesis Ib
	256050		Atelosteogenesis II
Spondylo-epimetaphyseal dysplasia	603005	*ATPSK2*	Abnormal skeletal development, dwarfism
Macular corneal dystrophy types I and II	217800	*CHST6*	Corneal erosions, photophobia
Familial tumoral calcinosis	211900	*GALNT3*	Calcium deposits in skin and tissue
O-mannosylation disorders	236670 253800 606612 608840	*POMT1, POMT2, POMGnT1, FCMD, FKRP, LARGE*	Muscular dystrophies, variable severity, variable presence of associated brain and eye malformations

Some of the other, rare types of N-glycosylation defects might have a very different clinical presentation, and some of the unique features might lead to a clinical diagnosis (e.g. normal mental development, protein loosing enteropathy and coagulation problems in PMI defect, severe epilepsy without systemic involvement in *glucosyltransferase I deficiency* (former CDG Ic), skeletal dysplasia in *mannosyltransferase VIII deficiency* (*former CDG Ig*), etc (Table 5.3).

Mental development is delayed in most patients, although patients with normal intellectual abilities have been observed. Speech delay is characteristic in both the "multisystem-type" and the "neurologic-type" presentation. Patients, with the more restricted neurological form have frequently visual impairment (eye malformations and central defects) (Morava et al. 2008b) and hearing problems (sensorineural deafness), ataxia due to cerebellar vermis hypoplasia and occasional peripheral neuropathy. Epilepsy is common in most CDG types. The neurological features might present additional to the organ involvement in the multisystem type as well. Except for symptomatic therapy there is no treatment for most of the N-glycosylation defects. The only exception so far is PMI deficiency (former CDG Ib), where patients benefit from high dose of oral mannose supplementation.

5.6.2 O-glycosylation Defects: Clinical Features

Patients with disorders of O-linked glycosylation defects demonstrate a highly variable clinical picture according to the type and the expression pattern of the glycosylation defect. The most common forms leading to human disease are congenital defects of O-mannosylation, and disorders of the O-xylose synthetic pathway. The clinical presentation is very different to that of the N-glycosylation defects in O-mannosylation disorders leading to muscle dystrophy, and the involvement of the skeletal system, cartilage and extracellular matrix in *O*-xylose-based defects (Okajima et al. 1999; Wuyts and van Hul 2000; Yoshida et al. 2001; Topaz et al. 2004; Table 5.3).

5.6.3 O-mannosyl-glycosylation Defects

Genetic defects in the biosynthesis of the O-mannosyl glycan are collectively called the alpha-dystroglycanopathies. Mutations in POMT1 and POMT2 lead to Walker-Warburg Syndrome, whereas POMGnT1 mutations can be found in Muscle-Eye-Brain disease. Defects in three additional gene products are known to result in a dystroglycanopathy: fukutin, fukutin-related protein (FKRP) and LARGE. The role of these proteins in the biosynthesis of the O-mannosyl glycan remains obscure. There is no strict correlation of the disease phenotype with the genetic defect, since mutations in fukutin, FKRP and LARGE have been found in Walker-Warburg patients as well. Mutations in these six genes approximately account for 50% of the patients (Mercuri et al. 2009). Apart from direct sequencing of the known dystroglycanopathy genes, immunohistochemical staining of muscle biopsies is possible using IIH6 or VIA4 antibodies specific for glycosylated alpha-dystroglycan. Enzymatic assays have been described for protein-O-mannosyltransferase in lymfoblasts and POMGnT1 in muscle homogenates and lymfoblasts (Endo and Manya 2006).

Children with O-mannosyl glycan biosynthesis defects characteristically have congenital muscular dystrophy (CMD) characterized by a decreased alpha-dystroglycan staining in the muscle tissue (Okajima et al. 1999). Some of the known alpha-dystroglycanopathies, like Fukuyama muscular dystrophy, muscle-eye-brain disease or Walker-Warburg syndrome, are multiple malformation syndromes associated with brain migration defects and congenital eye anomalies. Defective biosynthesis of O-mannosyl glycans leads to a hypoglycosylation specifically of the extracellular matrix protein alfa-dystroglycan, causing abnormal binding of the sarcolemma to the extracellular matrix (ECM). This leads to central nervous system developmental defects through disrupted cell interactions with the ECM and abnormal receptor signaling in the embryonic period (Michele et al. 2002). Pachygyria and brain migration defects have been reported in children with O-glycan biosynthesis defects, especially in O-mannosyl glycan biosynthesis defects, whereas cerebellar hypoplasia is common in the congenital disorders of N-linked glycosylation (CDG). Previous studies revealed that dystroglycan, a protein with many O-mannosyl-glycans, targets different proteins to their functional sites in the central nervous system through interactions with extracellular matrix proteins (Côté et al. 1999; Beltran-Valero de Bernabe et al. 2002; Martin 2003). An inborn error of glycosylation leads to defective receptor signaling and developmental defects, like pachygyria, already in the early embryonic period (Moore et al. 2002; Wu et al.

2004) One of the most intriguing clinical picture, bridging the different types of glycosylation is DMP3-CDG. This disorder shows clinical and biochemical symptoms of the classic N-linked CDG defects and also those of the dystroglycanopathies (Table 5.2; Lefeber et al). Additional to the neural migration defects, in both Walker-Warburg syndrome and Muscle-Eye-Brain disease, congenital myopia is a characteristic finding. In Walker-Warburg syndrome microphthalmia and cataract occur as well. The underlying etiology of the development of the myopia in O-mannosyl glycan disorders is not fully understood (Moore et al. 2002; Wu et al. 2004; Morava et al. 2006). Alpha-dystroglycan is a muscle membrane protein, carrying both mucin type and mannosyl type O-glycans. Abnormal biosynthesis of O-mannosyl glycans altering posttranslational modification disrupts the binding activity to its ligands, known to cause a congenital muscular dystrophy.

5.6.4 Combined Defects of N-Linked and O-Linked Glycosylation

A number of factors involved in glycan biosynthesis are common to the N- and O-glycosylation process, such as nucleotide sugars and their transporters. Also, Golgi trafficking and structural integrity are crucial for both types of glycosylation. Combined glycosylation defects have been observed in several patients, including a group with mutations in one of the subunits of the Conserved Oligomeric Golgi (COG) complex. This octameric complex is essential for the structure of the Golgi complex and is thought to have a role in trafficking between ER and Golgi and amongst Golgi cisternae (Axelsson et al. 2001; Smith and Lupashin 2008). Based on transferrin iso-electric focusing patterns as well as apoC-III iso-electric focusing and SDS-PAGE profiles of serum from CDG-II patients, Wopereis et al. (2005b) defined six biochemical subgroups among patients with combined N- and O-glycan biosynthesis defects.

Surprisingly, many patients in the subgroup showing disialotransferrin IEF and apoC-III_1 profiles, displayed a unique multiple malformation phenotype associated with cutis laxa. The observed phenotype was identical with a known genetic syndrome; autosomal recessive cutis laxa type II (ARCL type II) (Van Maldergem et al. 2008). These patients turned out to have a defect in the a2 subunit of the V-ATPase (Van Maldergem et al. 1989; Kornak et al. 2008). Subsequently, all patients with cutis laxa due to *ATP6V0A2* mutations known to date have been shown to have an abnormal transferrin isofocusing profile (Morava et al. 2005, 2008a; Kornak et al. 2008). Alongside the N-glycan biosynthesis defect, patients also have a core one mucin-type O-glycan biosynthesis defect. This is shown by the typical apoC-III_1 profile, characterized by decreased apoC-III2 and increased apoC-III_1 isoforms (Fig. 5.4d, lane 3). It has been observed that patients may have a normal transferrin isofocusing profile in the first months of life, but develop the typical transferrin abnormality later on. In these patients, the apoC-III isofocusing was already abnormal in the first months of life.

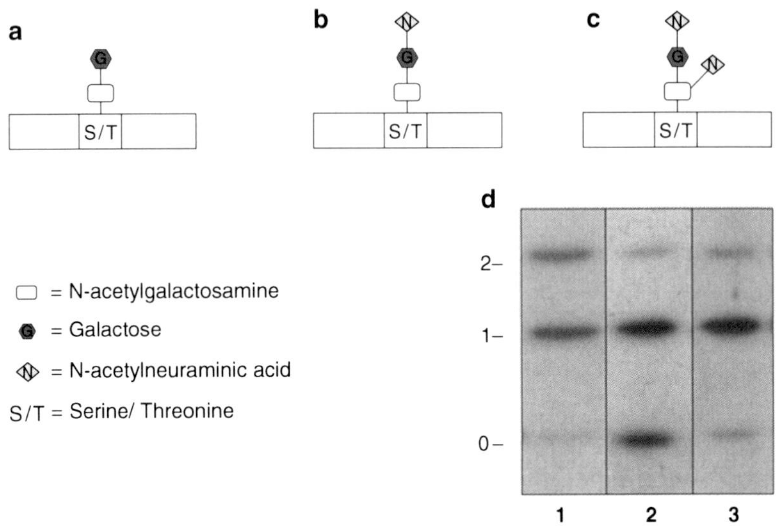

Fig. 5.4 O-glycosylation of apolipoprotein C-III. (**a**) "ApoC-III$_0$" (**b**) "ApoC-III$_1$" (**c**) "ApoC-III$_2$" (**d**) isoelectric focusing of human plasma apoC-III. *Lane 1*: normal apoC-III isoform profile, *lane 2*: abnormal apoC-III$_0$ profile, *lane 3*: abnormal apoC-III$_1$ profile (patient with *ATP6V0A2* defect)

5.6.5 Combined Defects of N-Linked and O-Linked Glycosylation: Clinical Features

Several novel disorders have been described with a combined N- and O-glycosylation disorder. After the discovery of the first COG complex-related defect in COG7, novel mutations have been observed in the subunits COG1, COG4, COG5, COG6 and GOG8 as well (Freeze 2007; Zeevaert et al. 2008). Interestingly the clinical features in these children are somewhat less pathognomonic. COG4 patients have microcephaly, hypotonia, seizures, liver disease, and coagulation defects. COG5 is very unspecific, presenting with mild psychomotor retardation with delayed motor and language development. The recently discovered COG6 defect leads to vitamin K deficiency, vomiting, intractable focal seizures, intracraninal bleeding and fatal outcome. The symptoms include developmental and growth delay, mental retardation, hypotonia, ataxia, frequently similar to the common phenotype in N-linked glycosylation disorders. The most distinctive features are present in the so far most common COG7 deficiency. Here, we discuss only the clinical picture in two of the combined N- and O-linked glycosylation defects.

5.6.6 Clinical Features in COG Mutations, the Common COG7 Mutation

Several patients have been diagnosed with a combined N- and O-linked glycosylation defect (former CDG IIe) due to a defect in COG7; one of the eight subunits of the

Conserved Oligomeric Golgi complex. The patients presented with growth retardation, progressive, severe microcephaly, hypotonia, adducted thumbs, feeding problems by gastrointestinal pseudoobstruction, failure to thrive, cardiac anomalies, wrinkled skin and episodes of extreme hyperthermia. Western blot analysis showed a severe reduction in the COG five and seven subunits of the Conserved Oligomeric Golgi complex. A homozygous, intronic splice site mutation (c.169+4A>C) of the *COG7* gene was identified in all patients. The initially reported two children had skeletal anomalies as well, which appeared to be a unique finding and has not been observed in other patients with the same mutation. The presence of the common, founder mutation makes direct DNA analysis possible in children presenting with the pathognomonic phenotype, even in the absence of protein glycosylation analysis.

5.6.7 Clinical Features in Patients with ATP6V0A2 Mutations (ARCL type II)

Recently, missense and nonsense mutations were discovered in the *ATP6V0A2* gene in several families diagnosed with autosomal recessive cutis laxa type II and wrinkly skin syndrome, suggesting that both are phenotypic variants of the same disorder (Van Maldergem et al. 2008; Kornak et al. 2008). ARCL II and WSS patients with *ATP6V0A2* mutations have generalized skin abnormalities, skeletal, neuromuscular and central nervous system involvement in variable degree with or without mental retardation. Their perinatal history is mostly normal, and generalized cutis laxa is present already at birth. Skin biopsy in the most severe cases demonstrates an abnormal, broken, shortened and fuzzy elastic fiber structure with a significantly decreased amount of elastin. The skin anomalies became less obvious, and sometimes disappear with age. Cardiac anomalies are rare. Many of the patients have congenital hip dislocations and increased joint laxity. The systemic involvement is very mild, including elevated liver enzyme activities and slightly abnormal coagulation in a very few patients. Common ophthalmologic abnormalities are strabismus, myopia or amblyopia. Congenital or progressive microcephaly is the most common associated feature in patients. Congenital brain malformations are extremely rare. Although developmental delay can be observed in the majority of the children the motor development improves in most cases with age. One might suspect that the motor delay was partially due to muscle hypotonia, and in some degree to the hyperelastic joints. Most of the children demonstrate a normal mental development and only a minority shows intellectual disability (Morava et al. 2005, 2008a). So far, all patients with *ATP6V0A2* mutations show a combined defect in the biosynthesis of N- and O-linked glycans. This can easily be picked up with plasma transferrin isofocusing.

No specific clinical features have been found distinguishing the group of *ATP6V0A2*-patients from other patients with autosomal recessive cutis laxa (ARCL II). Apparently, the frequency of central nervous system malformations is somewhat higher in the group with CDG but congenital brain malformations have been described in the ARCLII group without glycosylation defect as well (personal communications, Van Maldergem).

5.7 Epilogue

The number of different CDG types has duplicated in the last 10 years. Nevertheless, there is a lot to learn about glycosylation, demonstrated by the fact that the molecular basis of Golgi-related glycosylation defects in most of the patients is still unknown. With international collaboration, we have established a database for a unique, large cohort of CDG patients (EUROGLYCAN*ET* network). Solving the biochemical/genetic puzzle in these cases using a comprehensive, interdisciplinary, translational approach forms the objective of our future plans. Discovering glycosylation-related genes and syndromes has significance far beyond unraveling individual defects. In view of the significance of the biochemical process "glycosylation", finding novel defects will help us understand several basic pathways in the processing, trafficking and maturation of proteins and the importance of glycosylation in organ development, function and regulation. Parallel with better understanding of cellular pathways and networks orphan disease research might also lead to the development of orphan drugs.

Acknowledgement This work was supported by a grant from the European Commission (FP6 – Euroglycanet – contract no. 512131).

References

Axelsson, M.A., Karlsson, N.G., Steel, D.M., et al. (2001). Neutralization of pH in the Golgi apparatus causes redistribution of glycosyltransferases and changes in the O-glycosylation of mucins. Glycobiology 11:633–644.

Beltran-Valero de Bernabe, D., Currier, S., Steinbrecher, A., et al. (2002). Mutations in the O-mannosyltransferase gene POMT1 give rise to the severe neuronal migration disorder Walker-Warburg syndrome. *Am. J. Hum. Genet.* 71:1033–1043.

Burda, P., Borsig, L., deRijkvanAndel, J., et al. (1998). A novel carbohydrate-deficient glycoprotein syndrome characterized by a deficiency in glucosylation of the dolichol-linked oligosaccharide. *J. Clin. Invest.* 102:647–652.

Carlson, D.M. (1966). Oligosaccharides isolated from pig submaxillary mucin. J. Biol. Chem. 241:2984–2986.

Côté, P.D., Moukhles, H., Lindenbaum, M., et al. (1999). Chimaeric mice deficient in dystroglycans develop muscular dystrophy and have disrupted myoneural synapses. *Nat. Genet.* 23:338–342.

De Lonlay, P., and Seta, N. (2008). The clinical spectrum of phosphomannose isomerase deficiency, with an evaluation of mannose treatment for CDG-Ib. *Biochim. Biophys. Acta*, 1792:841–843, doi:10.1016/j.bbadis.2008.11.012.

Drickamer, K., and Taylor, M.E. (2002). Glycan arrays for functional glycomics. *Genome Biol.* 3(12):1034.1–1034.4.

Endo,T., and Manya, H. (2006). O-mannosylation in mammalian cells. *Methods Mol. Biol.* 347:43–56.

Frank, C.G., Grubenmann, C.E., Eyaid, W. et al. (2004). Identification and functional analysis of a defect in the human ALG9 gene: definition of congenital disorder of glycosylation type IL. Am. J. Hum. Genet. 75:146–150.

Freeze, H. (2007). Congenital disorders of glycosylation: CDG-I, CDG-II, and beyond. *Curr. Mol. Med.* 7:389–396.

Freeze, H.H., and Aebi, M. (2005). Altered glycan structures: the molecular basis of congenital disorders of glycosylation. *Curr. Opin. Struct. Biol.* 15:490–498.

Grünewald, S., Imbach, T., Huijben, K., et al. (2000). Clinical and biochemical characteristics of congenital disorder of glycosylation type Ic, the first recognized endoplasmic reticulum defect in N-glycan synthesis. *Ann. Neurol.* 47:776–781.

Grünewald, S., Schollen, E., Van Schaftingen, E., et al. (2001). High residual activity of PMM2 in patients' fibroblasts: possible pitfall in the diagnosis of CDG-Ia (phosphomannomutase deficiency). *Am. J. Hum. Genet.* 68:347–354.

Guillard, M., Gloerich, J., Wessels, H.J., Morava, E., Wevers, R.A., Lefeber, D.J. (2009). Automated measurement of permethylated serum N-glycans by MALDI-linear ion trap mass spectrometry. *Carbohydr Res.* 17;344(12):1550–1557.

Hirschberg, C.B., Robbins, P.W., Abeijon, C. (1998). Transporters of nucleotide sugars, ATP, and nucleotide sulfate in the endoplasmic reticulum and Golgi apparatus. *Annu. Rev. Biochem.* 67:49–69.

Ishii-Karakasa, I., Iwase, H., Hotta, K. (1997). Structural determination of the O-linked sialyl oligosaccharides liberated from fetuin with endo-alpha-N-acetylgalactosaminidase-S by HPLC analysis and 600-MHz 1H-NMR spectroscopy. *Eur. J. Biochem.* 247:709–715.

Jaeken, J., and Matthijs, G. (2007). Congenital disorders of glycosylation: a rapidly expanding disease family. *Annu. Rev. Genomics Hum. Genet.* 8:261–278.

Jaeken, J., Eggermont, E., Stibler, H. (1987). An apparent homozygous X-linked disorder with carbohydrate-deficient serum glycoproteins. Lancet 2:1398.

Jaeken, J., Hennet, T., Freeze, H.H., et al. (2008). On the nomenclature of congenital disorders of glycosylation (CDG). *J. Inherit. Metab. Dis.* 31:669–672.

Kim, S., Westphal, V., Srikrishna, G., et al. (2000). Dolichol phosphate mannose synthase (DPM1) mutations define congenital disorder of glycosylation Ie (CDG-Ie). *J. Clin. Invest.* 105:191–198.

Kornak, U., Reynders, E., Dimopoulou, A., et al. (2008). Impaired glycosylation and cutis laxa caused by mutations in the vesicular H^+-ATPase subunit ATP6V0A2. *Nat. Genet.* 40:32–34.

Körner, C., Lehle, L., von Figura, K. (1998a). Abnormal synthesis of mannose 1-phosphate derived carbohydrates in carbohydrate-deficient glycoprotein syndrome type I fibroblasts with phosphomannomutase deficiency. *Glycobiology* 8:165–171.

Körner, C., Knauer, R., Holzbach, U., et al. (1998b). Carbohydrate-deficient glycoprotein syndrome type V: deficiency of dolichyl-P-Glc:Man9GlcNAc2-PP-dolichyl glucosyltransferase. *Proc. Natl. Acad. Sci. U.S.A.* 95:13200–13205.

Körner, C., Knauer, R., Stephani, U., et al. (1999). Carbohydrate deficient glycoprotein syndrome type IV: deficiency of dolichyl-P-Man:Man(5)GlcNAc(2)-PP-dolichyl mannosyltransferase. *EMBO J.* 18:6816–6822.

Lefeber, D.J., Schönberger, J., Morava, E., Guillard, M., Huyben, K.M., Verrijp, K., Grafakou, O., Evangeliou, A., Preijers, F.W., Manta, P., Yildiz, J., Grünewald, S., Spilioti, M., van den Elzen, C., Klein, D., Hess, D., Ashida, H., Hofsteenge, J., Maeda, Y., van den Heuvel, L., Lammens, M., Lehle, L., Wevers, R.A. (2009). Deficiency of Dol-P-Man synthase subunit DPM3 bridges the congenital disorders of glycosylation with the dystroglycanopathies. *Am J Hum Genet.* 85(1):76–86.

Lowe, J.B., Marth, J.D. (2003). A genetic approach to Mammalian glycan function. Annu. Rev. Biochem. 72:643–691.

Lubke, T., Marquardt, T., Etzioni, A., et al. (2001). Complementation cloning identifies CDG-IIc, a new type of congenital disorders of glycosylation, as a GDP-fucose transporter deficiency. *Nat. Genet.* 28:73–76.

Marquardt, T., and Denecke, J. (2003). Congenital disorders of glycosylation: review of their molecular bases, clinical presentations and specific therapies. *Eur. J. Pediatr.* 162:359–379

Martin, P.T. (2003). Dystroglycan glycosylation and its role in matrix binding in skeletal muscle. *Glycobiology* 13:55R–66R.

Martinez-Duncker, I., Dupre, T., Piller, V., et al. (2005). Genetic complementation reveals a novel human congenital disorder of glycosylation of type II, due to inactivation of the Golgi CMP-sialic acid transporter. *Blood* 105:2671–2676.

Mercuri, E., Messina, S., Bruno, C., et al. (2009). Congenital muscular dystrophies with defective glycosylation of dystroglycan. A population study. *Neurology* doi:10.1212/01.wnl.00003456518.68110.60.

Michele, D.E., Barresi, R., Kanagawa, M., et al. (2002). Post-translational disruption of dystroglycan-ligand interactions in congenital muscular dystrophies. *Nature* 418:417–422

Mills, P.B., Mills, K., Mian, N., et al. (2003) Mass spectrometric analysis of glycans in elucidating the pathogenesis of CDG type IIx. *J. Inherit. Metab. Dis.* 26:119–134.

Moore S.A., Saito, F., Chen, J., et al. (2002). Deletion of brain dystroglycan recapitulates aspects of congenital muscular dystrophy. *Nature* 418:422–425.

Morava, E., Wopereis, S., Coucke, P., et al. (2005). Defective protein glycosylation in patients with cutis laxa syndrome. Eur. J. Hum. Genet. 13:414-421.

Morava, E., Willemsen, M.A., Wopereis, S., et al. (2006). High myopia and congenital myopathy with partial pachygyria in cutis laxa syndrome. *Eur. J. Ophthalmol.* 16:190–194.

Morava, E., Zeevaert, R., Korsch, E., et al. (2007). A common mutation in the COG7 gene with a consistent phenotype including microcephaly, adducted thumbs, growth retardation, VSD and episodes of hyperthermia. *Eur. J. Hum. Genet.* 15:638–645.

Morava, E., Lefeber, D.J., Urban, Z., et al. (2008a). Defining the phenotype in an autosomal recessive cutis laxa syndrome with a combined congenital defect of glycosylation. *Eur. J. Hum. Genet.* 16:28–35.

Morava, E., Wosik, H., Karteszi, J., et al. (2008b). Congenital disorder of glycosylation type Ix: review of clinical spectrum and diagnostic steps. J. Inherit. Metab. Dis. 31:450–456.

Okajima, T., Fukumoto, S., Furukawa, K., et al. (1999). Molecular basis for the progeroid variant of Ehlers-Danlos syndrome. *J. Biol. Chem.* 274:2841–2844.

Peters, B.P., Krzesicki, R.F., Perini, F., et al. (1989). O-glycosylation of the alpha-subunit does not limit the assembly of chorionic gonadotropin alpha beta dimer in human malignant and nonmalignant trophoblast cells. *Endocrinology* 124:1602–1612.

Röttger, S., White, J., Wandall, H.H., et al. (1998). Localization of three human polypeptide GalNAc-transferases in HeLa cells suggests initiation of O-linked glycosylation throughout the Golgi apparatus. *J. Cell Sci.* 111:45–60.

Royle, L., Mattu, T.S., Hart, E., et al. (2002). An analytical and structural database provides a strategy for sequencing O-glycans from microgram quantities of glycoproteins. *Anal. Biochem.* 304:70–90.

Schenk, B., Imbach, T., Frank, C.G., et al. (2001). MPDU1 mutations underlie a novel human congenital disorder of glycosylation, designated type If. *J. Clin. Invest.* 108:1687–1695.

Smith, R.D., and Lupashin, V.V. (2008). Role of the conserved oligomeric Golgi (COG) complex in protein glycosylation. *Carbohydr. Res.* 343:2024–2031.

Spiro, R.G. (2002) Protein glycosylation: nature, distribution, enzymatic formation, and disease implications of glycopeptide bonds. *Glycobiology* 12:43R–56R.

Tao, S.C., Li, Y., Zhou, J., et al. (2008). Lectin microarrays identify cell-specific and functionally significant cell surface glycan markers. *Glycobiology* 18:761–769.

Topaz, O., Shurman, D.L., Bergman, R., et al. (2004). Mutations in GALNT3, encoding a protein involved in O-linked glycosylation, cause familial tumoral calcinosis. *Nat. Genet.* 36:579–581.

Van den Steen, P., Rudd, P.M., Dwek, R.A., et al. (1998). Concepts and principles of O-linked glycosylation. *Crit. Rev. Biochem. Mol. Biol.* 33:151–208.

Van den Steen, P., Rudd, P., Wormald, M., et al. (2000). O-linked glycosylation in focus. *Trends Glycosci. Glycotechnol.* 63:35–49.

Van Maldergem, L., Og r, G., Yüksel, M. (1989). Facial anomalies in congenital cutis laxa with retarded growth and skeletal dysplasia. *Am. J. Med. Genet.* 32:265.

Van Maldergem, L., Yuksel-Apak, M., Kayserili, H., et al. (2008). Cobblestone-like brain dysgenesis and altered glycosylation in congenital cutis laxa, Debre type. *Neurology* 71:1602–1608.

Van Schaftingen, E., and Jaeken, J. (1995). Phosphomannomutase deficiency is a cause of carbohydrate-deficient glycoprotein syndrome type I. *FEBS Lett.* 377:318–320.

Vertel, B.M., Walters, L.M., Flay, N., et al. (1993). Xylosylation is an endoplasmic reticulum to Golgi event. *J. Biol. Chem.* 268:11105–11112.

Weidinger, S., Cleve, H., Schwarzfischer, F., et al. (1984). Transferrin subtypes and variants in Germany; further evidence for a Tf null allele. *Hum. Genet.* 66:356–360.

Wopereis, S., Grünewald, S., Morava, E., et al. (2003). Apolipoprotein C-III isofocusing in the diagnosis of genetic defects in O-glycan biosynthesis. *Clin. Chem.* 49:1839–1845.

Wopereis, S., Morava, E., Grünewald, S., et al. (2005a). A combined defect in the biosynthesis of N- and O-glycans in patients with cutis laxa and neurological involvement: the biochemical characteristics. Biochim. Biophys. Acta 1741:156–164.

Wopereis, S., Morava, E., Grünewald, S., et al. (2005b). Patients with unsolved congenital disorders of glycosylation type II can be subdivided in six distinct biochemical groups. *Glycobiology* 15:1312–1319.

Wopereis, S., Lefeber, D.J., Morava, E., et al. (2006). Mechanisms in protein O-glycan biosynthesis and clinical and molecular aspects of protein O-glycan biosynthesis defects: a review. Clin. Chem. 52:574–600.

Wu, X., Steet, R.A., Bohorov, O., et al. (2004). Mutation of the COG complex subunit gene COG7 causes a lethal congenital disorder. *Nat. Med.* 10:518–523.

Wuyts, W., and Van Hul, W. (2000). Molecular basis of multiple exostoses: mutations in the EXT1 and EXT2 genes. *Hum. Mutat.* 15:220–227.

Yoshida, A., Kobayashi, K., Manya, H., et al. (2001). Muscular dystrophy and neuronal migration disorder caused by mutations in a glycosyltransferase, POMGnT1. *Dev. Cell* 1:717–724.

Zeevaert, R., Foulquier, F., Jaeken, J., et al. (2008). Deficiencies in subunits of the Conserved Oligomeric Golgi (COG) complex define a novel group of Congenital Disorders of Glycosylation. *Mol. Genet. Metab.* 93:15–21.

Chapter 6
Defective Glycosylation of Dystroglycan in Muscular Dystrophy and Cancer

Federica Montanaro and Paul T. Martin

6.1 Introduction

Almost all proteins present on the extracellular surface of cells or in the extracellular matrix are glycoproteins. As such, they are modified either with N-linked glycans (on asparagines), O-linked glycans (on serines or threonines), glycosaminoglycans (on serines), or with glycosylphosphatidylinositol (GPI, on variable amino acids). These glycoproteins, in combination with glycolipids present in the plasma membrane, contribute to the expression of a glycocalyx, a highly concentrated halo of glycan that surrounds the extracellular surface of all cells. It is in this intensely carbohydrate-rich environment that all ligand-receptor signaling is initiated, all infectious processes begin, and in which all cellular movements and adhesive changes take place. As such, it should not be surprising that the carbohydrate environment of proteins provides a rich and essential context in which to understand their function. There are many examples of essential roles for glycosylation in health and disease (Freeze 2006). Examples where glycosylation is directly involved include the lysosomal storage disorders, which often arise from the inability of mutant glycosidases to break down glycans on glycolipids or glycoproteins (including proteoglycans), congenital disorders glycosylation, many of which arise from failure of enzymes in N-linked glycan biosynthesis but that also include defects in enzymes regulating glycosoaminoglycan synthesis, Golgi localization of glycosyltransferases, and O-linked biosynthesis. In addition, there are myriad examples where protein glycosylation plays a secondary role in dictating disease behavior, particularly in inflammatory diseases, blood disorders, and cancer. Tumor cell extravasation from the blood to the lymph, for example, is greatly

P.T. Martin (✉)
Departments of Pediatrics and of Physiology and Cell Biology, Center for Gene Therapy, The Research Institute at Nationwide Children's Hospital, Ohio State University College of Medicine, 700 Children's Drive, Columbus, OH 43205, USA
e-mail: MartinPT@pediatrics.ohio-state.edu

C.J. Vidal (ed.), *Post-Translational Modifications in Health and Disease*, Protein Reviews 13, DOI 10.1007/978-1-4419-6382-6_6,

affected by glycan repertoire present on cancer cells and by lectins on endothelial cells that bind to them. One very important membrane glycoprotein that plays both direct roles and indirect roles in human disease is dystroglycan (Michele and Campbell 2003; Higginson and Winder 2005; Sgambato and Brancaccio 2005; Barresi and Campbell 2006; Martin 2006). It has become clear from an amazing convergence of studies utilizing human and mouse genetics as well as biochemistry and cell biology that dystroglycan is an essential protein in mammals whose glycosylation is necessary for its function and whose aberrant glycosylation can cause disease.

6.2 The Dystroglycan Glycoprotein

In mammals, the dystroglycan gene (*Dag1*) is encoded by only two exons, both of which contain protein-coding sequence (Ibraghimov-Beskrovnaya et al. 1993). As such, splicing is not a regulatory mechanism important to dystroglycan biology (though splicing is more complex in some organisms (for example *Drosophila melanogaster* (Schneider and Baumgartner 2008))). Dystroglycan contains a signal peptide and becomes modified by N- and O-linked glycosylation as it makes its way through the endoplasmic reticulum (ER) and Golgi apparatus to the plasma membrane (Ervasti and Campbell 1991; Ibraghimov-Beskrovnaya et al. 1992; Ervasti and Campbell 1993; Holt et al. 2000). It is also proteolytically cleaved into two polypeptide chains, termed α and β dystroglycan (Ervasti and Campbell 1991). α dystroglycan is a membrane-associated extracellular protein that binds tightly, but non-covalently, to β dystroglycan, a transmembrane protein. α dystroglycan contains a mucin-like domain with as many as 55 serines or threonines (Ibraghimov-Beskrovnaya et al. 1992) and becomes heavily glycosylated with O-linked glycans in the ER and Golgi (Martin 2003a). The extent of this type of glycosylation on α dystroglycan varies between different tissues and also within the same tissue during development. α dystroglycan is converted from an unglycosylated polypeptide of 72 kDa to a glycoprotein of 120 kDa in brain, 140 kDa in cardiac muscle, and 156 kDa in skeletal muscle (Ervasti and Campbell 1993; Gee et al. 1993; Ervasti et al. 1997; Barresi and Campbell 2006). The extent of glycosylation becomes higher as skeletal muscle development proceeds (Leschziner et al. 2000) and also varies with stage of pregnancy in the placenta (Santhanakrishnan et al. 2008). In most tissues, however, it ultimately becomes a glycoprotein that is, on average, half carbohydrate by molecular weight. α Dystroglycan can also be cleaved by furin in the trans-Golgi or at the plasma membrane, thereby eliminating the N-terminal third of the protein sequence (Singh et al. 2004). The extent of glycosylation on β dystroglycan, by contrast, is far less, with the protein likely only modified on several N-linked sites. The enzyme that cleaves the protein into α and β chains has not been identified. Mutation of serine 654, the first amino acid of β dystroglycan, to alanine, however, inhibits cleavage of all endogenous α/β dystroglycan protein chains and causes muscular dystrophy when overexpressed in skeletal muscle

(Jayasinha et al. 2003). Some human tissues, in particular pediatric bone, have a 160 kDa species that can be recognized by antibodies to β dystroglycan, and this is consistent with the presence of uncleaved α/β dystroglycan in this tissue (Martin et al. 2007).

Glycan sequencing of the O-linked chains on α dystroglycan, both by Smalheiser and Dell, from sheep brain (Smalheiser et al. 1998) and by Endo and colleagues, from bovine peripheral nerve (Chiba et al. 1997) and rabbit skeletal muscle (Sasaki et al. 1998), revealed a mixture of relatively common core 1 glycans (Galβ1,3GalNAcα-O-Ser/Thr or T antigen) and an unusual O-linked mannose tetrasaccharide (NeuAcα2,3Galβ1,4GlcNAcβ1,2Manα-O-Ser/Thr). Smalheiser and Dell also identified O-mannosyl-linked Lewis X (Galβ1,4[Fucα1,3] GlcNAcβ1,2Manα-O) in brain (Smalheiser et al. 1998). These studies show that α dystroglycan contains an unusual sialylated O-linked mannose structure that has not been commonly found in mammals. Such structures, however, may not be uncommon. Feizi and colleagues, for example, showed that O-linked mannose may represent as much as the third of all O-linked glycan on proteins in rabbit brain (Chai et al. 1999), an amount far in excess of the preponderance of dystroglycan protein. In addition, O-linked mannose is a relatively common cell surface modification in lower eukaryotes such as yeast (Willer et al. 2003). While these elegant glycan sequencing papers leave no doubt that dystroglycan contains both O-linked mannose and O-linked GalNAc structures, less convincing antibody or lectin blotting suggests other structures may also be present, including the HNK-1 epitope (S04-GlcAβ1,3Galβ1,4-) (Smalheiser and Kim 1995), the Sda/CT carbohydrate epitope (Neu5Ac (or 5Gc)α2,3[GalNAcβ1,4]Galβ1,4GlcNAcβ-) (Xia et al. 2002), and Tn Antigen (GalNAc-α-O) (Ervasti et al. 1997). Both HNK-1 and CT carbohydrates have the potential to be present on O-linked mannose structures, while the Tn antigen may reflect incompletely galactosylated core 1 structures. Some of these structures may be present on small amounts of the total dystroglycan protein, and as such may not have been identified in the original sequencing studies. For example, the CT carbohydrate is present in skeletal muscle only at the neuromuscular synapse (Martin et al. 1999), which comprises about 0.1% of the total muscle membrane protein.

The original studies of Ervasti and Campbell showed that removal of the N-linked chains from α dystroglycan removed only 10 kDa of glycan from the 156 kDa polypeptide (Ervasti and Campbell 1993), an amount that reflects approximately 2–3 N-linked chains, which is consistent with the number of predicted N-linked sites (Ibraghimov-Beskrovnaya et al. 1992). These N-linked glycans, however, were not required for laminin binding to α dystroglycan. By contrast, acid digestion of all glycans (in this case both N- and O-linked) led to the loss of all laminin binding, suggesting an essential role for the O-linked chains. This seminal study has now been borne out by human genetics studies showing that genes required for O-linked mannose biosynthesis on α dystroglycan are required for the binding of laminin and other ECM proteins (Michele et al. 2002). Thus, the O-linked mannosylation of α dystroglycan is essential to its function as a receptor for the extracellular matrix.

6.3 The Dystrophin-Associated Glycoprotein Complex

Dystroglycan associates with a large number of extracellular and intracellular proteins as an essential member of transmembrane protein complexes (Fig. 6.1). The composition of these complexes differs depending on the cell type and its subcellular

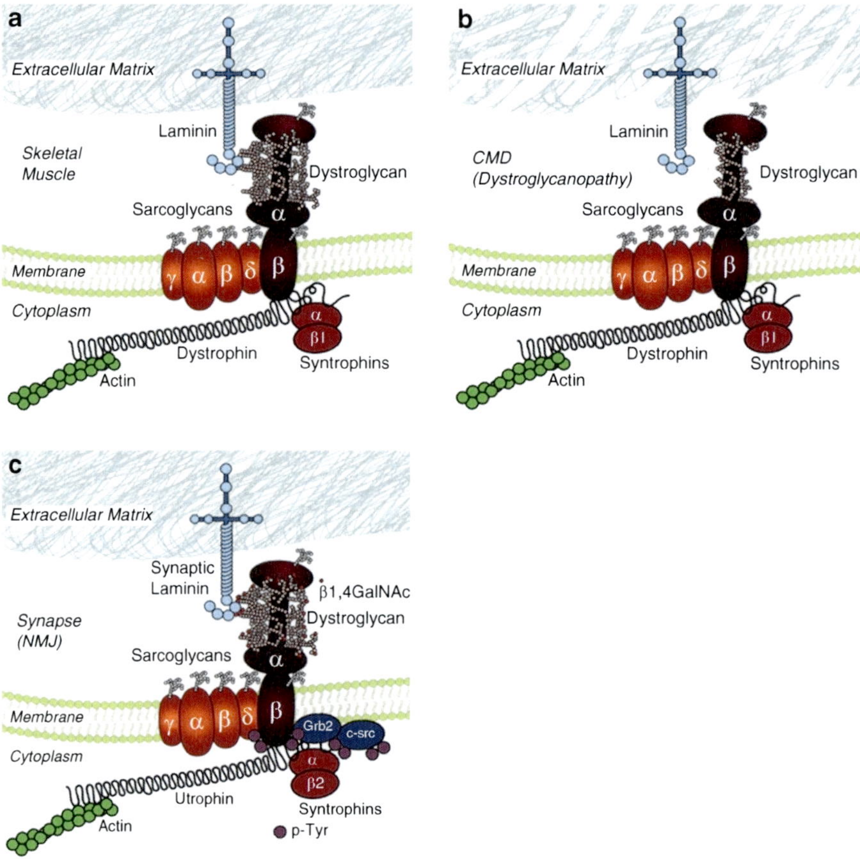

Fig. 6.1 The dystrophin-associated glycoprotein (DAG) complex in normal skeletal muscle, in skeletal muscle with dystroglycanopathy, and at the neuromuscular junction. (**a**) Normal skeletal muscle. Dystroglycan at the sarcolemmal membrane links laminin in the extracellular matrix, sarcoglycans in the membrane, and dystrophin and ultimately F-actin in the cytoplasm. α and β1 syntrophin, as well as other molecules (dystrobrevins, plectin1, ankyrin, etc., not shown), also bind dystrophin; (**b**) Patients with dystroglycanopathy (congenital muscular dystrophy, CMD) have reduced glycosylation of the O-linked structures on α dystroglycan. This can have the secondary consequence of lowering laminin expression in the extracellular matrix; (**c**) α dystroglycan at the neuromuscular junction (NMJ) is differentially glycosylated by Galgt2 with terminal β1, 4GalNAc structures. Synaptic forms of laminin bind to α dystroglycan. β dystroglycan is phosphorylated on tyrosine and interacts with synaptic linker and signaling molecules. Utrophin, a synaptic orthologue of dystrophin, is present, as is β2 syntrophin, as a uniquely synaptic form. Other synaptic proteins (e.g. α1 dystrobrevin) are not shown

localization (Martin 2003a, b). In the sarcolemmal membrane of skeletal muscle, dystroglycan is a central component of the dystrophin-associated glycoprotein (DAG) complex (Fig. 6.1a). Here, dystroglycan binds to the principal extrasyanptic muscle laminin (laminin-2 or α2,β1,γ1 laminin), and this binding requires the O-mannosyl-linked glycans present in its mucin-like domain(Ervasti and Campbell 1993; Michele et al. 2002). α/β dystroglycan interacts within the membrane with sarcoglycans, which are a four protein complex of α–δ sarcoglycan in skeletal muscle, and via the cytoplasmic domain of β dystroglycan with dystrophin, which ultimately links the complex to filamentous actin and other structural and signaling components (Fig. 6.2). The DAG complex was originally purified and

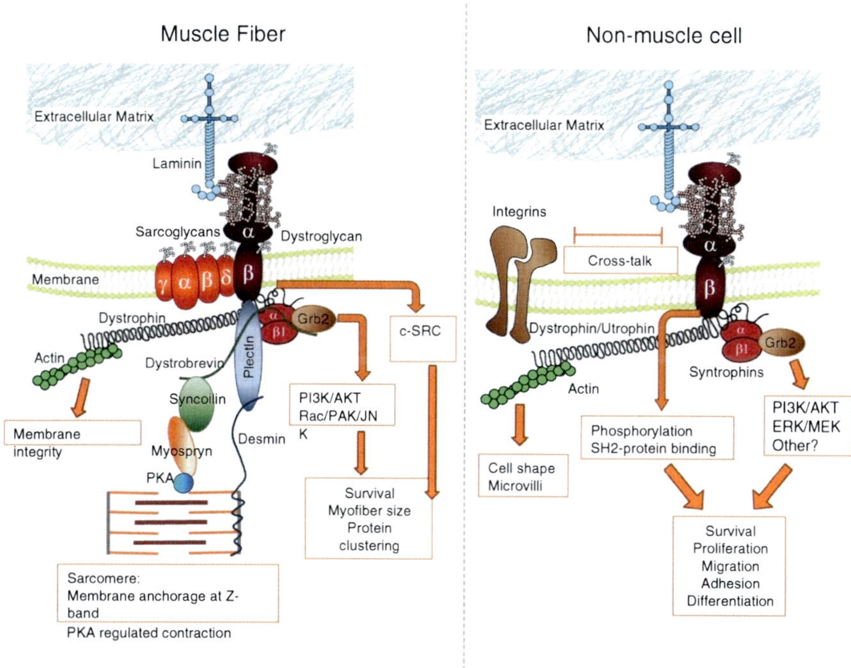

Fig. 6.2 Functions of dystroglycan in muscle and non-muscle cells. In skeletal muscle fibers, dystroglycan is part of a large complex that links laminin to the actin cytoskeleton via dystrophin. This dystrophin-associated glycoprotein (DAG) complex plays a structural role protecting the membrane from mechanical damage during repeated cycles of contraction. In addition, the DAG proteins are linked to the underlying sarcomeres at the Z-lines via desmin and regulate the phosphorylation of several sarcomeric proteins by anchoring protein kinase A (PKA). Dystroglycan also mediates activation of intracellular signaling pathways. Phosphorylation of β dystroglycan by c-src modulates its interactions with dystrophin while laminin-binding to α dystroglycan regulates activation of the PI3K/AKT and Rac/PAK/JNK pathways via the syntrophins and Grb2. In non-muscle cells and in muscle precursor cells, the dystroglycan complex antagonizes the actions of integrins. Binding of laminin to α dystroglycan regulates cell shape via re-organization of the actin cytoskeleton, and promotes differentiation and cell survival over proliferation and migration. Dystroglycan interacts with the PI3K/AKT and ERK/MEK signaling pathways, and β dystroglycan can be tyrosine phosphorylated leading to binding of SH2 domain proteins. Hypoglycosylation of α dystroglycan in cancers disrupts laminin binding and leads to loss of cell polarity, proliferation and migration

characterized by James Ervasti, Kevin Campbell, and colleagues (Ervasti et al. 1990; Ervasti and Campbell 1991).

Mutations affecting expression of almost all of DAGs cause forms of muscular dystrophy, strongly demonstrating a functional role for these protein associations. Loss of dystrophin causes Duchenne muscular dystrophy (Hoffman et al. 1987; Koenig et al. 1987), a severe X-linked myopathy, while partial loss of dystrophin causes Becker muscular dystrophy, which typically has a milder clinical progression than DMD (Love et al. 1989; Blake et al. 2002). Similarly, loss of any of the four muscle sarcoglycans (α–δ) causes Limb-girdle muscular dystrophy (LGMD2D, 2E, 2C, and 2F, respectively) (Vainzof et al. 1996; Angelini 2004; Rezniczek et al. 2007), and loss of laminin α2 causes Congenital muscular dystrophy 1A (Mendell et al. 2006). Complete loss of dystroglycan is lethal in mice from an early embryonic stage (Williamson et al. 1997), and this may explain why human *DAG1* mutations have not been identified as causing muscular dystrophy. However, loss of proteins that glycosylate α dystroglycan cause forms of congenital muscular dystrophy or limb-girdle muscular dystrophy by disrupting laminin binding (Fig. 6.1b) and will be discussed in the next section (Martin 2006; Mendell et al. 2006).

The common feature of these muscular dystrophies is a progressive degeneration of muscle fibers, often linked to increased fragility of the cell membrane in the face of repeated mechanical stress imposed by muscle contraction. All give rise to chronic cycles of muscle degeneration and regeneration, with additional immune components, that in their most severe forms lead to wasting of the skeletal muscle tissue and replacement of muscle with fat or extracellular matrix. Such muscle wasting leads to progressive muscle weakness which in the most severe forms results in wheelchair dependence and ultimately early mortality due to complications arising from respiratory and/or cardiac failure. Thus, a wealth of both biochemical and genetic data points to the importance of the DAG complex as a stabilizer of the muscle membrane.

At the neuromuscular junction, dystroglycan is differentially glycosylated with synaptic carbohydrates by Galgt2, a UDP-GalNAc:Neu5Acα2,3Galβ1,4GlcNAc-β1,4 N-acetylgalactosaminyltransferase (Nguyen et al. 2002; Xia et al. 2002). Galgt2 (Xia et al. 2002), like the CT carbohydrate it creates (Martin et al. 1999), is normally confined in skeletal muscle to the area including the neuromuscular synapse, thereby defining a uniquely synaptic glycoform of dystroglycan protein (Fig. 6.1c). At the neuromuscular synapse, there are unique isoforms of laminin (laminin 9 (α4,β2,γ1) and laminin-11 (α5,β2,γ1)) (Patton et al. 1997), now called laminin 421 and 521 respectively, that may interact with the unique glycoform of dystroglycan, and also utrophin, a synaptic orthologue of dystrophin (Ohlendieck et al. 1991). Additionally, dystroglycan may be modified by tyrosine kinases such as c-src (Sadasivam et al. 2005) via interactions with synaptic adaptor proteins, such as rapsyn (Apel et al. 1995) or Grb2 (Yang et al. 1995). Further evidence of a unique synaptic complex is the fact that transgenic overexpression of Galgt2, which creates the CT carbohydrate on α dystroglycan along the entire muscle membrane, leads to the ectopic expression of synaptic laminins and utrophin (Nguyen et al. 2002; Xia et al. 2002). The ectopic expression of this synaptic glycan further prevents dystrophin-deficient muscles from developing muscular dystrophy, which may be in part due to the extrasynaptic expression of novel synaptic DAGs (Nguyen et al. 2002). That dystroglycan is essential for proper synaptic structure is further supported by studies where it is deleted in the skeletal muscles of mice and by studies where genes that control its O-linked mannosylation are altered (Cote et al. 1999; Saito et al. 2007). Similarly complex novel protein associations may be found in other subcellular regions of skeletal myofibers where dystroglycan is present, such as the costamere or the myotendinous junction.

The list of proteins found to interact, either directly or indirectly, with the DAG complex continues to expand. This is perhaps not surprising given the number of DAGs involved in the complex. Many interactions are mediated by dystrophin, which in addition to F-actin also binds syntrophins (Peters et al. 1997), dystrobrevins (Butler et al. 1992; Peters et al. 1998), plectin 1 (Rezniczek et al. 2007), ankyrin (Ayalon et al. 2008), and cytokeratin 19 (Ursitti et al. 2004; Stone et al. 2005). Plectin1 and ankyrin also can associate directly with the cytoplasmic face of β dystroglycan (Rezniczek et al. 2007; Ayalon et al. 2008), as can a host of signaling or adaptor proteins, including Grb2 (Yang et al. 1995; Russo et al. 2000), dynamin (Zhan et al. 2005), c-src (Sotgia et al. 2001), rapsyn (Apel et al. 1995; Cartaud et al. 1998), ezrin (Batchelor et al. 2007), and MAP kinase (Spence et al. 2004b). Additional signaling and channel proteins, including neuronal nitric oxide synthase (Brenman et al. 1995), voltage-gated Na+ channels (Gee et al. 1998), aquaporin-4 (Adams et al. 2001), TRPC1 calcium channel (Vandebrouck et al. 2007), and syncoilin (a desmin-binding protein) (Newey et al. 2001; Poon et al. 2002), may enter the complex via interactions with dystrobrevins or syntrophins. Plectin 1 and ankyrin also bind to cytoskeletal proteins including microtubules, microtubule-binding proteins, and intermediate filaments. Plectin1 also binds F-actin and loss of plectin1 causes muscular dystrophy (with epidermolysis bullosa). Filamin C, which binds sarcoglycans and F-actin, could serve similar structural roles (Thompson et al. 2000). These proteins provide additional scaffolds that may work even in the absence of dystrophin to link dystroglycan to the cytoskeleton.

Dystroglycan shows equally complex associations with the extracellular matrix proteins that reside in the basal lamina that surrounds each myofiber (Martin 2003a). Outside the cell, α dystroglycan binds to β dystroglycan, via a protein-protein interface requiring the C-terminal third of α dystroglycan, and with extracellular matrix proteins, including multiple forms of laminin, agrin, and perlecan (Henry and Campbell 1999). α Dystroglycan also interacts with specialized ECM proteins, such as pikachurin (at ribbon synpases in the retina (Sato et al. 2008)), biglycan (Bowe et al. 2000), transmembrane proteins including neurexins (in neurons) (Sugita et al. 2001), infectious agents including Lassa Fever virus, Lymphocytic choriomeningitis virus (Cao et al. 1998; Kunz et al. 2001) (but see (Imperiali et al. 2008)), and *Micobacterium leprae* (Rambukkana et al. 1998), and also Golgi proteins involved in its own glycosylation (LARGE (Kanagawa et al. 2004)). Most ECM proteins known to bind α dystroglycan contain laminin G domain motifs (Timpl et al. 2000), and these binding interactions require glycosylation of α dystroglycan with O-linked mannose structures (Michele et al. 2002). Whether these glycans mediate direct binding, however, remains a matter of debate.

6.4 The Dystroglycanopathies

Dystroglycanopathies are neuromuscular disorders defined by altered glycosylation of α dystroglycan (Martin and Freeze 2003; Jimenez-Mallebrera et al. 2005; Martin 2006; Mendell et al. 2006; Moore et al. 2008; Muntoni et al. 2008). There has been an increasing focus given to using dystroglycan glycosylation as a diagnostic, and therefore the number of patients identified continues to expand. The more severe end of the dystroglycanopathies encompasses congenital muscular dystrophies, diseases that are present at or before birth. These diseases include Walker Warburg syndrome (WWS), muscle eye brain disease (MEB), Fukuyama congenital muscular dystrophy (FCMD), and congenital muscular dystrophy 1C and 1D (MDC1C and MDC1D). These diseases arise from mutations in genes affecting dystroglycan glycosylation, including *POMT1* (Beltran-Valero de Bernabe et al. 2002), *POMT2* (van Reeuwijk et al. 2005), *POMGnT1* (Yoshida et al. 2001), *FKTN* (Kobayashi et al. 1998), *FKRP* (Brockington et al. 2001b; Topaloglu et al. 2003; Beltran-Valero de Bernabe et al. 2004), and *LARGE* (Longman et al. 2003; van Reeuwijk et al. 2007). WWS, MEB, and FCMD patients display a variety of neurological findings in the brain, including type 2-like lissencephaly (or "cobblestone cortex"), cerebellar cysts, pontocerebellar hypoplasia, hydrocephalus, reduced or absent corpus collosum, and white matter changes (hypomyelination), and also changes in the eye, including congenital glaucoma, retinal dysplasia or detachment, microopathalmia, myopia, atrophy of the optic nerve, buphthalmos, and anterior chamber defects. All also include muscle pathology consistent with severe muscular dystrophy as well as variably present cardiomyoapthy and also defects in neuromuscular synapses (Jimenez-Mallebrera et al. 2005; Taniguchi et al. 2006).

While the range of clinical presentation can be quite variable, children with WWS rarely live beyond a year of age and show severe hypotonia (Jimenez-Mallebrera et al. 2005). Children with other CMD variants can live into their teens, typically show significant muscle weakness, and can have loss of ambulation and mental retardation. These diseases differ from other muscular dystrophies involving loss of other members of the DAG complex (e.g. dystrophin and sarcoglycans) in that they show multiple brain phenotypes in addition to muscular dystrophy, arguing that dystroglycan may have unique functions in the brain that cannot be compensated for by other DAG members. In patients and in animal models where lissencephaly is present, the observed defect in neuronal migration is not cell autonomous but rather arises from defects in the formation of the glial limitans-pial membrane surface, where dystroglycan is essential for proper ECM expression and integrity (Holzfeind et al. 2002; Moore et al. 2002). This has the effect of causing a fraction of the cortical neurons to continue to migrate through holes in the pial surface, ultimately leaving the brain.

At the other end of the clinical spectrum for the dystroglycanopathies are the Limb-girdle muscular dystrophies LGMD2I, LGMD2K, LGMD2L, LGMD2M, and LGMD2N (arising from mutations in *FKRP* (Brockington et al. 2001a), *POMT1* (Balci et al. 2005; D'Amico et al. 2006), *FKTN* (Godfrey et al. 2006),

POMGnT1 (Clement et al. 2008), and *POMT2* (Biancheri et al. 2007), respectively). These diseases typically do not show brain or eye changes found in the congenital muscular dystrophies and have milder muscle and heart disease as well. There is a founder effect mutation present in the human population of *FKRP* carriers (826C>A) that leads to an increased incidence of LGMD2I relative to other LGMD dystroglycanopathies (Louhichi et al. 2004; Frosk et al. 2005). Patients with LGMD dystroglycanopathies can often ambulate as children, but muscle weakness is often progressive and can be associated with cardiomyopathy (Straub and Bushby 2006). These diseases typically show no neurological findings (Jimenez-Mallebrera et al. 2005; Muntoni et al. 2008). Because the molecular weight of skeletal muscle α dystroglycan in these LGMDs appears only to be reduced to the molecular weight normally seen in brain, brain dystroglycan function may be unaffected in these diseases.

There are at least six genes identified where loss of function mutations give rise to an aberrantly underglycosylated form of α dystroglycan with defective ECM binding properties and also cause forms of congenital or limb-girdle muscular dystrophy (Moore and Hewitt 2009; Muntoni et al. 2008). These include three genes known to be essential for the biosynthesis of O-linked mannose chains on α dystroglycan: Protein O-mannosyltransferase 1 and 2 (POMT1 and POMT2) are a dimeric protein complex required for O-linked mannose biosynthesis on α dystroglycan in mammals (Akasaka-Manya et al. 2004; Manya et al. 2004; Akasaka-Manya et al. 2006) and in flies (Ichimiya et al. 2004). POMT1 and POMT2 act to specifically glycosylate O-linked peptides from the mucin region of α dystroglycan (Manya et al. 2007). Protein O-mannosyl-β1,2-N-acetylglucosaminyltransferase (POMGnT1) synthesizes the second sugar of the O-mannosyl tetrasaccharide structure on α dystroglycan (Yoshida et al. 2001; Manya et al. 2003). The second three genes known to cause disease are fukutin (FKTN), fukutin-related protein (FKRP), and LARGE. Here again, loss of function causes underglycosylation of α dystroglycan and disease (Grewal et al. 2001; Hayashi et al. 2001; Brown et al. 2004), but the exact function of these genes is not known. FKRP, FKTN, and LARGE proteins are localized to the Golgi in muscle tissue and all three contain motifs (DxD) found in glycosyltransferases that are involved in binding sugar nucleotide substrates (Esapa et al. 2002; Brockington et al. 2005; Grewal et al. 2005; Torelli et al. 2005). Thus, they are localized and have structural motifs consistent with their either being glycosyltransferases or in mediating glycosyltransferase activity. These genes share no sequence similarity to sialyltransferases or β-galactosyltransferases that synthesize the outer glycans of the O-mannosyl tetrasaccharide on α dystroglycan.

Recent studies have shown that fukutin-deficient mice have reduced POMGnT1 activity and that fukutin can interact with POMGnT1 (Xiong et al. 2006), suggesting that fukutin may be a mediator of POMGnT1 activity or subcellular localization. Indeed, mutations in FKRP and FKTN can cause protein mislocalization from the Golgi to the ER, and FKTN mutations additionally cause POMGnT1 mislocalization (Esapa et al. 2002; Xiong et al. 2006). Further evidence of a protein targeting role is the finding that FKRP binds dystroglycan and can co-localize with it at the sarcolemmal membrane in skeletal muscle (Beedle et al. 2007). LARGE is perhaps the most intriguing of these genes with unknown function. Campbell and colleagues have shown that overexpression of *LARGE* in cells deficient in *POMGnT1* or *FKTN* can stimulate glycosylation of α dystroglycan, converting it from its underglycosylated pathologic form to its native state. In doing so, LARGE overexpression restores laminin binding (Barresi et al. 2004). LARGE2, described by Hewitt and colleagues (Grewal et al. 2005), has a similar function, but is not normally expressed at appreciable levels in skeletal muscle. It is clear from studies using CHO cell glycosylation mutants and also from digestion of α dystroglycan with glycosidases that the type of glycosylation LARGE stimulates is not one common to other mammalian glycoproteins (Combs and Ervasti 2005; Patnaik and Stanley 2005).

Importantly, muscles from most patients with dystroglycanopathies show normal expression of α and β dystroglycan protein at the sarcolemmal membrane, only having reduced α dystroglycan glycosylation. This is typically demonstrated by showing altered migration of α dystroglycan on SDS-PAGE gels and by absent or reduced binding of carbohydrate-dependent monoclonal antibodies, such as IIH6, to α dystroglycan (using both immunostaining and immunoblotting). A defect in α dystroglycan glycosylation was first described by Hayashi et al. in FCMD patients (Hayashi et al. 2001). There are a number of patients with dystroglycanopathies where all six of the genes known to cause disease are not mutated. Thus, additional genes in this class of disorders have yet to be discovered. Indeed, there may also be patients where α dystroglycan expression is reduced without altering protein glycosylation. Such a molecular change has been recently described for a novel dystroglycanopathy in the Sphinx and Devon Rex breeds of cat (Martin et al. 2008).

Because loss of dystroglycan leads to early embryonic lethality in mice (Williamson et al. 1997), an increasing number of tissue-specific knockouts, coupled with genetic chimeras, have been created to allow viability into adulthood. These include loss of *DAG1* in skeletal muscle, astrocytes, Schwann cells, and the embryonic nervous system (Cote et al. 1999; Cohn et al. 2002; Moore et al. 2002; Saito et al. 2003; Satz et al. 2008). These valuable mouse models have shown that most of the phenotypes found in patients can be mimicked by loss of the dystroglycan protein. Work by Campbell and colleagues has conclusively demonstrated that dystroglycan can be the primary mediator of glycosylation defect because loss of dystroglycan protein in the affected tissues phenocopies loss of genes affecting its glycosylation (Cohn et al. 2002; Moore et al. 2002; Saito et al. 2003; Satz et al. 2008). Jarad and Miner (2009) used an unusual Pax3-Cre transgenic mouse that has a rostral-caudal gradient of Cre transgene expression to knock out dystroglycan specifically in caudal skeletal muscles, which allows the mice to live up to a year of age. Such animals should be especially useful in exploring the role of *DAG1* in skeletal muscle. In particular, α dystroglycan can, in some circumstances, stimulate the polymerization of proteins, in particular laminins, into the extracellular matrix (Yurchenco et al. 2004; Nishimune et al. 2008). Since laminin binding is dependent upon glycosylation, this polymerization process may be affected in dystroglycanopathy patients, where laminin α2 protein is often reduced in the muscle basal lamina (Brockington et al. 2001a, b; Brown et al. 2004). Thus, altered α dystroglycan glycosylation may work in a bidirectional manner, leading to both loss of ECM ligand binding and ECM polymerization. Indeed, lethality in *DAG1*, POMT1, and FKTN null mice appears to be due to loss of laminin expression in the basal lamina (in particular Reichert's membrane) and resultant fracturing of surrounding basement membranes (Williamson et al. 1997; Willer et al. 2004; Kurahashi et al. 2005).

As is the case with all of the muscular dystrophies, the relative severity of dystroglycanopathies can vary between affected tissues, as can the glycosylation changes. For example, Nishino and colleagues have described patients with *FKTN* mutations that have minimal muscle weakness but with severe cardiomyopathy, much as can be seen in some Becker MD cases (Murakami et al. 2006). Likewise, Muntoni and collaborators have found several patients with *FKTN* and *FKRP* mutations where dystroglycan glycosylation is greatly reduced but that have only clinically mild forms of LGMD (Jimenez-Mallebrera et al. 2008). While such data would seem to argue that glycosylation might not be an absolute predictor of disease, the carbohydrate reagents used for such studies are often fraught with quality control issues, as the specific glycans they bind to are unknown. Toda and colleagues have made a knock-in model of the most common FCMD gene deletion in Japan, which is retrotransponsal insertion in the 3′ untranslated region of the *FKTN* gene (Kanagawa et al. 2009). This in effect knocks down FKTN mRNA expression in the affected tissues. Mice with this insertion, unlike FCMD patients with the same genetic abnormality, show only a 50% decrease in glycosylated α dystroglycan, and show no disease (Kanagawa et al. 2009). This appears to be an increasingly common theme in mouse models. In mice, quality control in protein synthesis and/or folding may be more efficient than in humans, thus leading to a relatively reduced loss of function for missense mutations or insertions. Perhaps the best recent example of this is that mice with a knock-in of the most common LGMD2D mutation in α sarcoglycan (R77C) show normal expression of the mutant protein on the sarcolemmal membrane, and thus no disease (Kobuke et al. 2008), while humans with the same mutation have reduced α sarcoglycan expression (and have LGMD2D) (Vainzof et al. 1996). It may be that fukutin protein is similarly stabilized in mice with FCMD-like insertions relative to humans, thus making the reduction in its protein levels less impact-full than would be the case in human muscle.

The secondary pathological or molecular consequence of dystroglycanopathy mutations is similarly becoming increasingly variable. For example, Reilich et al. (2006) have identified LGMD2I patients that have inclusion bodies in their muscles, a finding not present in most muscular dystrophies, including CMDs. While such changes could be secondary, for example to muscle inflammation (which can be present in LGMD2I (Darin et al. 2007)), this may also reflect the fact that the pathological consequences of FKRP mutations have not yet be fully elucidated. Similarly, Topolalglu and colleagues have identified a patient with a POMGnT1 mutation where the dominant presenting sign is severe autistic features (Haliloglu et al. 2004), a finding very different from most CMD patients. At the molecular level, the notion that α dystroglycan is normally expressed at the membrane is questioned by the study of McNalley and colleagues where some patients with FKRP mutations have reduced expression of both α and β dystroglycan on the sarcolemmal membrane (MacLeod et al. 2007). Again, this is contrary to the original published reports showing that glycosylation did not affect membrane expression of α and β dystroglycan protein (Michele et al. 2002).

Similar findings, including reduced expression also of sarcoglycans and sarcospans, have been found in FCMD muscles (Wakayama et al. 2008). Such studies add potentially new insights to the current molecular dogmas of how these diseases are caused. In a similar vein, Hewitt and colleagues has shown that IIH6, the antibody that recognizes the laminin-binding carbohydrate epitope of α dystroglycan, shows a very restricted pattern that does not coincide with LARGE but rather with LARGE2 in zebrafish (Moore et al. 2008). As such, LARGE2 may be important for stimulating the glycosylation-dependent laminin binding epitopes on α dystroglycan, consistent with its activity in cultured cells (Brockington et al. 2005; Fujimura et al. 2005; Grewal et al. 2005). Knock-down of LARGE2 in eliminates IIH6 staining, demonstrating a clear requirement of LARGE2 for dystroglycan glycosylation in fish (Moore et al. 2008). Because LARGE2, like LARGE, can stimulate α dystroglycan glycosylation when overexpressed, it may be an important new target for therapeutic intervention in the CMDs. Additionally, fish may be used to identify genetic suppressors of LARGE that inhibit its activity.

6.5 Dystroglycan and Cancer

The development and progression of many types of cancers involves changes in cell-extracellular matrix receptors that normally function in the maintenance of normal tissue cytoarchitecture and adhesion. In the context of cancer, the most studied extracellular matrix receptors are integrins, α and β heterodimeric protein receptors involved in regulating cell adhesion, migration, survival and proliferation. A growing body of evidence, however, now also implicates dystroglycan in tumor cell biology. In fact, disruption of the dystroglycan complex appears to be a widespread phenomenon in cancers of varied cellular origins.

In vitro and in vivo studies have implicated dystroglycan in epithelial cell growth inhibition, cell polarity, tissue-specific gene expression, differentiation, basement membrane formation, and survival. These functions are all affected during tumor progression and their disruption parallels the loss of dystroglycan in a growing list of epithelial tumors and tumor cell lines, including breast cancer (Cross et al. 2008) and oral squamous cell carcinoma (Jing et al. 2004). Loss of dystroglycan expression or altered dystroglycan glycosylation has also been demonstrated in non-epithelial cancers, including pediatric rhabdomyosarcoma and neuroblastoma (Martin et al. 2007) and also adult glioma (Calogero et al. 2006). Exceptions to this list include pediatric liver cancers, osteosarcomas, yolk sack tumors, and Hodgkin's lymphomas (Martin et al. 2007). Thus disruption of dystroglycan function appears in many, but not all, cancer types.

Three major mechanisms have been proposed in cancer cells for disruption of dystroglycan function: proteolytic cleavage of the dystroglycan complex, altered glycosylation of α dystroglycan, and overall reduction or loss of dystroglycan expression (Losasso et al. 2000; Yamada et al. 2001; Singh et al. 2004; Martin et al. 2007; Cross et al. 2008). The first two mechanisms converge on the fact that they disrupt the ability of α dystroglycan to tether the cell to the surrounding basement membrane, either by removing carbohydrates essential for its interactions with laminin or by inducing shedding of α dystroglycan from the membrane. The latter can involve cleavage of β dystroglycan, leading to loss of the extracellular domain of β dystroglycan, which also binds α dystroglycan, from the membrane (Yamada et al. 2001; Singh et al. 2004).

Cleavage of the extracellular domain of β dystroglycan coincides with the appearance of a 31 kDa protein fragment recognized by monoclonal antibodies to the extreme C-terminus of the β dystroglycan protein (in its cytoplasmic domain) (Yamada et al. 2001; Singh et al. 2004). This protein fragment has been detected in primary oral squamous cell carcinomas (Jing et al. 2004) and cervical cancers (Sgambato et al. 2006) as well as in mammary epithelial and prostate tumor cell lines (Losasso et al. 2000; Sgambato et al. 2007b). Of note, this 31 kDa β dystroglycan protein is not unique to tumors and has been detected in normal tissues (Yamada et al. 2001; Sgambato et al. 2006). The function of this 31 kDa β dystroglycan protein is unknown (Losasso et al. 2000; Yamada et al. 2001), but it appears to be more readily targeted for degradation than intact β dystroglycan since its levels are increased following treatment with proteasome inhibitors (Singh et al. 2004). The identity of the protease mediating this cleavage of β dystroglycan remains controversial; Some studies have implicated matrix metalloproteases (MMPs) (Yamada et al. 2001; Singh et al. 2004) while others reported no (or partial) effect of MMP inhibitors (Jing et al. 2004). Adding to the confusion, MMP inhibitors can restore association of α dystroglycan with the cell membrane in cells where only full-length 43 kDa β dystroglycan is present (Singh et al. 2004). This result indicates the existence of a second proteolytic mechanism for disruption of the dystroglycan complex by MMPs that does not involve direct cleavage of either α or β dystroglycan.

Many studies have reported a specific loss of α dystroglycan in primary tumors and cancer cell lines however these often have used the IIH6 or VIA4-1 monoclonal antibodies that recognize only natively glycosylated α dystroglycan. Thus the apparent loss of α dystroglycan, as evidenced by loss of IIH6 or VIA4-1 binding, accompanied with preservation of β dystroglycan, could reflect hypoglycosylation of α dystroglycan or loss of α dystroglycan protein expression. Primary tumors where α dystroglycan appears to be affected include oral squamous cell carcinoma (Jing et al. 2004), cervical cancer (Sgambato et al. 2006), renal carcinoma (Sgambato et al. 2007a), glioma (Calogero et al. 2006), prostate cancer (Sgambato et al. 2007b), and pediatric rhabdomyosarcoma and neuroblastoma (Martin et al. 2007).

Some studies using antibodies to the core peptide of α dystroglycan have shown that this hypoglycosylated form of α dystroglycan remains associated with the cell membrane but is unable to bind laminin (Beltran-Valero de Bernabe et al. 2009), and similar studies have also utilized antibodies to β dystroglycan, which is co-translated with α dystroglycan, to show maintained dystroglycan expression (Martin et al. 2007). Recent studies in tumor cell lines have attributed the hypoglycosylation of α dystroglycan to a selective loss of LARGE expression. Altered LARGE function is attributed to epigenetic modifications rather than mutations in the LARGE gene (Beltran-Valero de Bernabe et al. 2009). Finally, expression of β dystroglycan has been found to be reduced or absent in a relatively large set of primary cancers. These include prostate cancer (15 out of 15 tumors (Henry et al. 2001)), breast cancer (337/343 tumors (Cross et al. 2008); 6/6 tumors (Henry et al. 2001)), esophageal adenocarcinomas (10/10 tumors (Cross et al. 2008)), colorectal adenocarcinomas (102/105 tumors (Cross et al. 2008)), and ureteric transitional cell carcinomas (55/55 tumors (Cross et al. 2008)). Expression of glycosylated α dystroglycan and core protein was not assayed in most of these tumors. It is interesting that the same β dystroglycan antibody was found to strongly stain most pediatric cancers (Martin et al. 2007), suggesting adult and pediatric cancers may be fundamentally different with regard to dystroglycan expression.

Cancer progression involves loss of the differentiated phenotype as well as uncontrolled cellular proliferation and, in more advanced stages, acquisition of a metastatic potential associated with altered cellular interactions with the extracellular matrix. In normal mammary epithelial cells, the interaction of dystroglycan with laminin has been implicated in cellular differentiation, including the establishment of cell polarity and expression of β-casein, a gene expressed during the process of maturation (Weir et al. 2006). Down-regulation of dystroglycan expression by siRNA in non-tumorigenic mouse mammary epithelial cells leads to inhibition of lactogenic differentiation and apoptotic cell death. Given the known roles of dystroglycan in epithelial cell differentiation and maturation, it is important to determine whether the loss of dystroglycan is a secondary consequence of cellular transformation or a significant contributor to the malignant phenotype. Forced expression of dystroglycan in prostate and mammary epithelial tumor cell lines leads to inhibition of cell cycle progression, loss of anchorage-independent growth, and impairment in tumor formation in vivo (Sgambato et al. 2004, 2007b; Calogero et al. 2006). Interestingly, up-regulation of β dystroglycan, without restoration of glycosylated α dystroglycan, was sufficient to inhibit cell proliferation and significantly reduce tumorigenicity in these cancer cell lines. However, expression of markers of cell polarity and differentiation did depend on expression of glycosylated α dystroglycan capable of interacting with laminin (Muschler et al. 2002).

Taken together these observations suggest that dystroglycan performs a dual function in epithelial cells; the interaction of α dystroglycan with laminin is important for cellular polarization and maturation while signaling through β dystroglycan regulates cellular proliferation. These experiments are particularly relevant in light of the suggested correlation for some primary tumors between loss of dystroglycan expression and tumor grade or stage (Sgambato et al. 2003, 2006, 2007a; Martin et al. 2007). For example, a study of oral squamous cell carcinomas found that while primary tumors consistently lost expression of α dystroglycan, tumors showing metastasis lost both α and β dystroglycan expression (Jing et al. 2004). Thus, in these tumors, metastasis was accompanied by a complete loss of dystroglycan protein. The possible correlation between dystroglycan expression and tumor grade or stage is tantalizing, however, the number of tumors used in many of these studies is often too small to support any clinical relevance. In addition, such results may be complicated by residual normal tissue within the samples and by the absence of generally available antibodies to the α dystroglycan core protein suitable for immunohistochemistry. Nonetheless, these studies support an important role for dystroglycan as a tumor suppressor and warrant further exploration as to dystroglycan's diagnostic and prognostic significance.

6.6 Dystroglycan and Signaling

Dystroglycan was first described as a core member of the dystrophin-associated glycoprotein (DAG) complex in skeletal muscle, connecting the cytoskeletal actin-binding protein dystrophin, or its synaptic orthologue utrophin, through the membrane to the extracellular matrix. The proteins liking to dystroglycan in the sarcolemmal membrane involve laminin and other ECM proteins in the basal lamina surrounding the myofiber, sarcoglycans within the membrane, and cytoplasmic structural proteins at the Z-band including dystrophin, desmin, plectin, myospryn, syncoilin and filamin C, among others (Fig. 6.2). The linkage to the sarcomere is more than just structural since myospryn also anchors protein kinase A to the sarcomere (Reynolds et al. 2008), which modulates contraction via phosphorylation of sarcomeric proteins including myosin and titin. In addition, the DAG complex is linked to proteins that control calcium dynamics, and calcium homeostasis is dysregulated in skeletal and cardiac muscles lacking dystrophin (Constantin et al. 2006; Williams et al. 2006; Williams and Allen 2007).

In non-muscle cells, where contraction is not an issue, dystroglycan has also been shown to perform a structural role in determining cell shape, such as polarity of mammary epithelial cells (Weir et al. 2006), and in the formation of cellular extensions such as microvilli (Spence et al. 2004a) (Fig. 6.2). In these processes dystroglycan also regulates the actin cytoskeleton via its interactions with laminin. It is therefore not surprising that restoration of cellular polarity in mammary tumor cell lines depends on glycosylation of α dystroglycan allowing binding to laminin. In this respect it is interesting that utrophin, the intracellular ligand of dystroglycan in epithelial cells, is mutated in some breast cancers (21/61 cancers screened) (Li et al. 2007) and its expression is lost in breast and prostate cancers in a manner similar to α dystroglycan (Henry et al. 2001).

The intracellular interactions of β dystroglycan are modulated by two processes (Fig. 6.2): engagement of α dystroglycan by extracellular ligands and phosphorylation of the cytoplasmic region of β dystroglycan on tyrosines, particularly tyrosine 892. In muscle cells, binding of laminin to α dystroglycan has been shown to induce activation of the PI3K/AKT and the Rac1/PAK1/JNK signaling pathways (Langenbach and Rando 2002; Oak et al. 2003). Experimental evidence includes biochemical studies of protein interactions via pull-downs (Oak et al. 2003; Xiong et al. 2009) as well as altered regulation of these signaling pathways in muscles of *mdx* mice (Xiong et al. 2009), which lack dystrophin (Hoffman et al. 1987) and therefore have a destabilized DAG complex (Matsumura et al. 1992). The actions of dystroglycan on these signaling pathways appear to mediate muscle fiber survival and protection from atrophy (Chockalingam et al. 2002; Langenbach and Rando 2002). In non-muscle cells, dystroglycan can also act as a scaffold for the ERK/MAP kinase signaling cascade (Fig. 6.2). Here, dystroglycan can negatively modulate the ERK/MAP kinase activity by physically segregating the cellular compartments, thereby preventing their interaction (Ferletta et al. 2003; Spence et al. 2004b). Furthermore, dystroglycan can inhibit the ERK/MEK pathway in pancreatic cells to promote differentiation (Jiang et al. 2001). As the PI3K/AKT and MEK/ERK signaling cascades mediate cancer cell proliferation and survival (McCubrey et al. 2007), it is tempting to speculate that dystroglycan may have anti-proliferative and tumor suppressor activities that could be manipulated to alter tumor behavior or growth.

Phosphorylation of β dystroglycan has been reported to disengage dystroglycan from dystrophin and utrophin (James et al. 2000; Ilsley et al. 2001), rendering its cytoplasmic region available for interaction with SH2 domain containing proteins involved in signal transduction (Sotgia et al. 2001). Proteins that may bind to the cytoplasmic domain of β dystroglycan include Src, Fyn, Csk, Nck and Shc. The biological significance of these interactions is currently not known. One biological process linked to β dystroglycan phosphorylation involves the assembly of podosomes (Thompson et al. 2008). Podosomes are transient adhesion structures that mediate directional cellular migration and tissue invasion via protrusions called invadopodia. Podosomes are often formed on migrating tumor cells and their assembly is regulated by Rho family GTPases and tyrosine kinases including src. Overexpression of dystroglycan leads to inhibition of podosome formation by sequestering proteins essential for podosome assembly. Overexpression of a mutant dystroglycan lacking a src phosphorylation site is unable to interfere with podosome formation, indicating that this process relies on β dystroglycan phosphorylation.

The current knowledge on the connections of dystroglycan to major signaling pathways as well as its interplay with integrins favor the notion of different biological functions of dystroglycan depending on the cellular context (Fig. 6.2). In particular, dystroglycan appears to cooperate with integrins in muscle fibers where activation of the ERK/MEK and PI3K/AKT signaling pathways contributes to survival and prevent atrophy. In other cell types, dystroglycan appears to antagonize integrin's functions by inhibiting cell proliferation, promoting differentiation, and interfering with the formation of cell-matrix adhesion structures important for cell migration and tissue invasion.

6.7 Conclusions

The glycosylation of dystroglycan is complex and contains an unusual sialylated O-linked mannose structure not commonly found in mammals. This unique pattern of glycosylation involves a series of genes, *POMT1*, *POMT2*, *POMGnT1*, *FKTN*,

FKRP, and *LARGE*, some of which encode specialized glycosyltransferases. Mutations affecting the activity of any one of the proteins encoded by these genes leads to a group of diseases commonly referred to as dystroglycanopathies. Clinical features of these disorders include progressive and usually severe muscular dystrophy accompanied in some cases by neurological involvement and/or cardiomyopathy. While it is not clearly established whether dystroglycan is the only protein glycosylated by these specialized glycosyltransferases, tissue-specific dystroglycan knockout mice phenocopy many of the clinical pathological aspects associated with human disease. Thus, the disruption of dystroglycan function is likely a key determinant of phenotype. α dystroglycan is a receptor for a subset of extracellular matrix proteins and most of these interactions depend on or are modulated by glycosylation. The effects of hypoglycosylation of α dystroglycan in disease states such as dystroglycanopathies and cancer is providing insights on how glycosylation modulates key aspects of cellular function. Dystroglycan has also been implicated in extracellular matrix assembly, protection of the plasma membrane from mechanical stress, and intracellular signaling via a variety of pathways regulating cell survival, proliferation, differentiation and migration. Resolving the composition and structure of the different carbohydrate moieties on α dystroglycan in normal and disease states is essential to gain a clearer and broader picture of how the structural and intracellular signaling functions of the dystroglycan complex are regulated.

References

Adams, M.E., Mueller, H.A., Froehner, S.C. (2001). In vivo requirement of the alpha-syntrophin PDZ domain for the sarcolemmal localization of nNOS and aquaporin-4. *J. Cell Biol.* 155:113–122.

Akasaka-Manya, K., Manya, H., Endo, T. (2004). Mutations of the POMT1 gene found in patients with Walker-Warburg syndrome lead to a defect of protein O-mannosylation. *Biochem. Biophys. Res. Commun.* 325:75–79.

Akasaka-Manya, K., Manya, H., Nakajima, A., et al. (2006). Physical and functional association of human protein O-mannosyltransferases 1 and 2. *J. Biol. Chem.* 281:19339–19345.

Angelini, C. (2004). Limb-girdle muscular dystrophies: heterogeneity of clinical phenotypes and pathogenetic mechanisms. *Acta Myol.* 23:130–136.

Apel, E.D., Roberds, S.L., Campbell, K.P., et al. (1995). Rapsyn may function as a link between the acetylcholine receptor and the agrin-binding dystrophin-associated glycoprotein complex. *Neuron* 15:115–126.

Ayalon, G., Davis, J.Q., Scotland, P.B., et al. (2008). An ankyrin-based mechanism for functional organization of dystrophin and dystroglycan. *Cell* 135:1189–1200.

Balci, B., Uyanik, G., Dincer, P., et al. (2005). An autosomal recessive limb girdle muscular dystrophy (LGMD2) with mild mental retardation is allelic to Walker-Warburg syndrome (WWS) caused by a mutation in the POMT1 gene. *Neuromuscular Disord.* 15:271–275.

Barresi, R., and Campbell, K.P. (2006). Dystroglycan: from biosynthesis to pathogenesis of human disease. *J. Cell Sci.* 119:199–207.

Barresi, R., Michele, D.E., Kanagawa, M. et al. (2004). LARGE can functionally bypass alpha-dystroglycan glycosylation defects in distinct congenital muscular dystrophies. *Nat. Med.* 10:696–703.

Batchelor, C.L., Higginson, J.R., Chen, Y.J., et al. (2007). Recruitment of Dbl by ezrin and dystroglycan drives membrane proximal Cdc42 activation and filopodia formation. *Cell Cycle* 6:353–363.

Beedle, A.M., Nienaber, P.M., Campbell, K.P. (2007). Fukutin-related protein associates with the sarcolemmal dystrophin-glycoprotein complex. *J. Biol. Chem.* 282:16713–16717.

Beltran-Valero de Bernabe, D., Currier, S., Steinbrecher, A., et al. (2002). Mutations in the O-mannosyltransferase gene POMT1 give rise to the severe neuronal migration disorder Walker-Warburg syndrome. *Am. J. Hum. Genet.* 71:1033–1043.

Beltran-Valero de Bernabe, D., Voit, T., Longman, C., et al. (2004). Mutations in the FKRP gene can cause muscle-eye-brain disease and Walker-Warburg syndrome. *J. Med. Genet.* 41: e61.

Beltran-Valero de Bernabe, D., Inamori, K.I., Yoshida-Moriguchi, T., et al. (2009). Loss of alpha-dystroglycan laminin binding in epithelium-derived cancers is caused by silencing of large. *J. Biol. Chem.* 284(17):11279–11284.

Biancheri, R., Falace, A., Tessa, A., et al. (2007). POMT2 gene mutation in limb-girdle muscular dystrophy with inflammatory changes. *Biochem. Biophys. Res. Commun.* 363:1033–1037.

Blake, D.J., Weir, A., Newey, S.E., et al. (2002). Function and genetics of dystrophin and dystrophin-related proteins in muscle. *Physiol. Rev.* 82:291–329.

Bowe, M.A., Mendis, D.B., Fallon, J.R. (2000). The small leucine-rich repeat proteoglycan biglycan binds to alpha-dystroglycan and is upregulated in dystrophic muscle. *J. Cell Biol.* 148:801–810.

Brenman, J.E., Chao, D.S., Xia, H., et al. (1995). Nitric oxide synthase complexed with dystrophin and absent from skeletal muscle sarcolemma in Duchenne muscular dystrophy. *Cell* 82:743–752.

Brockington, M., Blake, D.J., Prandini, P., et al. (2001a). Mutations in the fukutin-related protein gene (FKRP) cause a form of congenital muscular dystrophy with secondary laminin alpha2 deficiency and abnormal glycosylation of alpha-dystroglycan. *Am. J. Hum. Genet.* 69:1198–1209.

Brockington, M., Yuva, Y., Prandini, P., et al. (2001b). Mutations in the fukutin-related protein gene (FKRP) identify limb girdle muscular dystrophy 2I as a milder allelic variant of congenital muscular dystrophy MDC1C. *Hum. Mol. Genet.* 14:2851–2859.

Brockington, M., Torelli, S., Prandini, P., et al. (2005). Localization and functional analysis of the LARGE family of glycosyltransferases: significance for muscular dystrophy. *Hum. Mol. Genet.* 14:657–665.

Brown, S.C., Torelli, S., Brockington, M., et al. (2004). Abnormalities in alpha-dystroglycan expression in MDC1C and LGMD2I muscular dystrophies. *Am. J. Pathol.* 164:727–737.

Butler, M.H., Douville, K., Murnane, A.A., et al. (1992). Association of the Mr 58,000 postsynaptic protein of electric tissue with Torpedo dystrophin and the Mr 87,000 postsynaptic protein. *J. Biol. Chem.* 267:6213–6218.

Calogero, A., Pavoni, E., Gramaglia, T., et al. (2006). Altered expression of alpha-dystroglycan subunit in human gliomas. *Cancer Biol. Ther.* 5:441–448.

Cao, W., Henry, M.D., Borrow, P., et al. (1998). Identification of alpha-dystroglycan as a receptor for lymphocytic choriomeningitis virus and Lassa fever virus. *Science* 282:2079–2081.

Cartaud, A., Coutant, S., Petrucci, T.C., et al. (1998). Evidence for *in situ* and *in vitro* association between beta-dystroglycan and the subsynaptic 43K rapsyn protein. Consequence for acetylcholine receptor clustering at the synapse. *J. Biol. Chem.* 273:11321–11326.

Chai, W., Yuen, C.T., Kogelberg, H., et al. (1999). High prevalence of 2-mono- and 2,6-di-substituted manol-terminating sequences among O-glycans released from brain glycopeptides by reductive alkaline hydrolysis. *Eur. J. Biochem.* 263:879–888.

Chiba, A., Matsumura, K., Yamada, H., et al. (1997). Structures of sialylated O-linked oligosaccharides of bovine peripheral nerve alpha-dystroglycan. The role of a novel O-mannosyl-type oligosaccharide in the binding of alpha-dystroglycan with laminin. *J. Biol. Chem.* 272:2156–2162.

Chockalingam, P.S., Cholera, R., Oak, S.A., et al. (2002). Dystrophin-glycoprotein complex and Ras and Rho GTPase signaling are altered in muscle atrophy. *Am. J. Physiol. Cell Physiol.* 283:C500–C511.

Clement, E.M., Godfrey, C., Tan, J., et al. (2008). Mild POMGnT1 mutations underlie a novel limb-girdle muscular dystrophy variant. *Arch. Neurol.* 65:137–141.

Cohn, R.D., Henry, M.D., Michele, D.E., et al. (2002). Disruption of DAG1 in differentiated skeletal muscle reveals a role for dystroglycan in muscle regeneration. *Cell* 110:639–648.

Combs, A.C., and Ervasti, J.M. (2005). Enhanced laminin binding by alpha-dystroglycan after enzymatic deglycosylation. *Biochem. J.* 390:303–309.

Constantin, B., Sebille, S., Cognard, C. (2006). New insights in the regulation of calcium transfers by muscle dystrophin-based cytoskeleton: implications in DMD. *J. Muscle Res. Cell Motil.* 27:375–386.

Cote, P.D., Moukhles, H., Lindenbaum, M., et al. (1999). Chimaeric mice deficient in dystroglycans develop muscular dystrophy and have disrupted myoneural synapses. *Nat. Genet.* 23:338–342.

Cross, S.S., Lippitt, J., Mitchell, A., et al. (2008). Expression of beta-dystroglycan is reduced or absent in many human carcinomas. *Histopathology* 53:561–566.

D'Amico, A., Tessa, A., Bruno, C., et al. (2006). Expanding the clinical spectrum of POMT1 phenotype. *Neurology* 66:1564–1567; Discussion 1461.

Darin, N., Kroksmark, A.K., Ahlander, A.C., et al. (2007). Inflammation and response to steroid treatment in limb-girdle muscular dystrophy 2I. *Eur. J. Paediatr. Neurol.* 11:353–357.

Ervasti, J.M., and Campbell, K.P. (1991). Membrane organization of the dystrophin-glycoprotein complex. *Cell* 66:1121–1131.

Ervasti, J.M., and Campbell, K.P. (1993). A role for the dystrophin-glycoprotein complex as a transmembrane linker between laminin and actin. *J. Cell Biol.* 122:809–823.

Ervasti, J.M., Ohlendieck, K., Kahl, S.D., et al. (1990). Deficiency of a glycoprotein component of the dystrophin complex in dystrophic muscle. *Nature* 345:315–319.

Ervasti, J.M., Burwell, A.L., Geissler, A.L. (1997). Tissue-specific heterogeneity in alpha-dystroglycan sialoglycosylation. Skeletal muscle alpha-dystroglycan is a latent receptor for Vicia villosa agglutinin b4 masked by sialic acid modification. *J. Biol. Chem.* 272:22315–22321.

Esapa, C.T., Benson, M.A., Schroder, J.E., et al. (2002). Functional requirements for fukutin-related protein in the Golgi apparatus. *Hum. Mol. Genet.* 11:3319–3331.

Ferletta, M., Kikkawa, Y., Yu, H., et al. (2003). Opposing roles of integrin alpha6Abeta1 and dystroglycan in laminin-mediated extracellular signal-regulated kinase activation. *Mol. Biol. Cell* 14:2088–2103.

Freeze, H.H. (2006). Genetic defects in the human glycome. *Nat. Rev. Genet.* 7:537–551.

Frosk, P., Greenberg, C.R., Tennese, A.A., et al. (2005). The most common mutation in FKRP causing limb girdle muscular dystrophy type 2I (LGMD2I) may have occurred only once and is present in Hutterites and other populations. *Hum. Mutat.* 25:38–44.

Fujimura, K., Sawaki, H., Sakai, T., et al. (2005). LARGE2 facilitates the maturation of alpha-dystroglycan more effectively than LARGE. *Biochem. Biophys. Res. Commun.* 329:1162–1171.

Gee, S.H., Blacher, R.W., Douville, P.J., et al. (1993). Laminin-binding protein 120 from brain is closely related to the dystrophin-associated glycoprotein, dystroglycan, and binds with high affinity to the major heparin binding domain of laminin. *J. Biol. Chem.* 268:14972–14980.

Gee, S.H., Madhavan, R., Levinson, S.R., et al. (1998). Interaction of muscle and brain sodium channels with multiple members of the syntrophin family of dystrophin-associated proteins. *J. Neurosci.* 18:128–137.

Godfrey, C., Escolar, D., Brockington, M., et al. (2006). Fukutin gene mutations in steroid-responsive limb girdle muscular dystrophy. *Ann. Neurol.* 60:603–610.

Grewal, P.K., Holzfeind, P.J., Bittner, R.E., et al. (2001). Mutant glycosyltransferase and altered glycosylation of alpha-dystroglycan in the myodystrophy mouse. *Nat. Genet.* 28:151–154.

Grewal, P.K., McLaughlan, J.M., Moore, C.J., et al. (2005). Characterization of the LARGE family of putative glycosyltransferases associated with dystroglycanopathies. *Glycobiology* 15:912–923.

Haliloglu, G., Gross, C., Senbil, N., et al. (2004). Clinical spectrum of muscle-eye-brain disease: from the typical presentation to severe autistic features. *Acta Myol.* 23:137–139.

Hayashi, Y.K., Ogawa, M., Tagawa, K., et al. (2001). Selective deficiency of alpha-dystroglycan in Fukuyama-type congenital muscular dystrophy. *Neurology* 57:115–121.

Henry, M.D., and Campbell, K.P. (1999). Dystroglycan inside and out. *Curr. Opin. Cell Biol.* 11:602–607.

Henry, M.D., Cohen, M.B., Campbell, K.P. (2001). Reduced expression of dystroglycan in breast and prostate cancer. *Hum. Pathol.* 32:791–795.

Higginson, J.R., and Winder, S.J. (2005). Dystroglycan: a multifunctional adaptor protein. *Biochem. Soc. Trans.* 33:1254–1255.

Hoffman, E.P., Brown, R.H., Jr., Kunkel, L.M. (1987). Dystrophin: the protein product of the Duchenne muscular dystrophy locus. *Cell* 51:919–928.

Holt, K.H., Crosbie, R.H., Venzke, D.P., et al. (2000). Biosynthesis of dystroglycan: processing of a precursor propeptide. *FEBS Lett.* 468:79–83.

Holzfeind, P.J., Grewal, P.K., Reitsamer, H.A., et al. (2002). Skeletal, cardiac and tongue muscle pathology, defective retinal transmission, and neuronal migration defects in the Large (myd) mouse defines a natural model for glycosylation-deficient muscle – eye – brain disorders. *Hum. Mol. Genet.* 11:2673–2687.

Ibraghimov-Beskrovnaya, O., Ervasti, J.M., Leveille, C.J., et al. (1992). Primary structure of dystrophin-associated glycoproteins linking dystrophin to the extracellular matrix. *Nature* 355:696–702.

Ibraghimov-Beskrovnaya, O., Milatovich, A., Ozcelik, T., et al. (1993). Human dystroglycan: skeletal muscle cDNA, genomic structure, origin of tissue specific isoforms and chromosomal localization. *Hum. Mol. Genet.* 2:1651–1657.

Ichimiya, T., Manya, H., Ohmae, Y., et al. (2004). The twisted abdomen phenotype of Drosophila POMT1 and POMT2 mutants coincides with their heterophilic protein O-mannosyltransferase activity. *J. Biol. Chem.* 279:42638–42647.

Ilsley, J.L., Sudol, M., Winder, S.J. (2001). The interaction of dystrophin with beta-dystroglycan is regulated by tyrosine phosphorylation. *Cell Signal* 13:625–632.

Imperiali, M., Sporri, R., Hewitt, J., et al. (2008). Post-translational modification of (alpha)-dystroglycan is not critical for lymphocytic choriomeningitis virus receptor function in vivo. *J. Gen. Virol.* 89:2713–2722.

James, M., Nuttall, A., Ilsley, J.L., et al. (2000). Adhesion-dependent tyrosine phosphorylation of (beta)-dystroglycan regulates its interaction with utrophin. *J. Cell Sci.* 113 (Pt 10): 1717–1726.

Jarad, G., and Miner, J.H. (2009). The Pax3-Cre transgene exhibits a rostrocaudal gradient of expression in the skeletal muscle lineage. *Genesis* 47:1–6.

Jayasinha, V., Nguyen, H.H., Xia, B., et al. (2003). Inhibition of dystroglycan cleavage causes muscular dystrophy in transgenic mice. *Neuromuscular Disord.* 13:365–375.

Jiang, F.X., Georges-Labouesse, E., Harrison, L.C. (2001). Regulation of laminin 1-induced pancreatic beta-cell differentiation by alpha6 integrin and alpha-dystroglycan. *Mol. Med.* 7:107–114.

Jimenez-Mallebrera, C., Brown, S.C., Sewry, C.A., et al. (2005). Congenital muscular dystrophy: molecular and cellular aspects. *Cell. Mol. Life Sci.* 62:809–823.

Jimenez-Mallebrera, C., Torelli, S., Feng, L., et al. (2008). A comparative study of alpha-dystroglycan glycosylation in dystroglycanopathies suggests that the hypoglycosylation of alpha-dystroglycan does not consistently correlate with clinical severity. *Brain Pathol.* 19(4):596–611.

Jing, J., Lien, C.F., Sharma, S., et al. (2004). Aberrant expression, processing and degradation of dystroglycan in squamous cell carcinomas. *Eur. J. Cancer* 40:2143–2151.

Kanagawa, M., Saito, F., Kunz, S., et al. (2004). Molecular recognition by LARGE is essential for expression of functional dystroglycan. *Cell* 117:953–964.

Kanagawa, M., Nishimoto, A., Chiyonobu, T., et al. (2009). Residual laminin-binding activity and enhanced dystroglycan glycosylation by LARGE in novel model mice to dystroglycanopathy. *Hum. Mol. Genet.* 18:621–631.

Kobayashi, K., Nakahori, Y., Miyake, M., et al. (1998). An ancient retrotransposal insertion causes Fukuyama-type congenital muscular dystrophy. *Nature* 394:388–392.

Kobuke, K., Piccolo, F., Garringer, K.W., et al. (2008). A common disease-associated missense mutation in alpha-sarcoglycan fails to cause muscular dystrophy in mice. *Hum. Mol. Genet.* **17**:1201–1213.

Koenig, M., Hoffman, E.P., Bertelson, C.J., et al. (1987). Complete cloning of the Duchenne muscular dystrophy (DMD) cDNA and preliminary genomic organization of the DMD gene in normal and affected individuals. *Cell* **50**:509–517.

Kunz, S., Sevilla, N., McGavern, D.B., et al. (2001). Molecular analysis of the interaction of LCMV with its cellular receptor [alpha]-dystroglycan. *J. Cell Biol.* **155**:301–310.

Kurahashi, H., Taniguchi, M., Meno, C., et al. (2005). Basement membrane fragility underlies embryonic lethality in fukutin-null mice. *Neurobiol. Dis.* **19**:208–217.

Langenbach, K.J., and Rando, T.A. (2002). Inhibition of dystroglycan binding to laminin disrupts the PI3K/AKT pathway and survival signaling in muscle cells. *Muscle Nerve* **26**:644–653.

Leschziner, A., Moukhles, H., Lindenbaum, M., et al. (2000). Neural regulation of alpha-dystroglycan biosynthesis and glycosylation in skeletal muscle. *J. Neurochem.* **74**:70–80.

Li, Y., Huang, J., Zhao, Y.L., et al. (2007). UTRN on chromosome 6q24 is mutated in multiple tumors. *Oncogene* **26**:6220–6228.

Longman, C., Brockington, M., Torelli, S., et al. (2003). Mutations in the human LARGE gene cause MDC1D, a novel form of congenital muscular dystrophy with severe mental retardation and abnormal glycosylation of alpha-dystroglycan. *Hum. Mol. Genet.* **12**:2853–2861.

Losasso, C., Di Tommaso, F., Sgambato, A., et al. (2000). Anomalous dystroglycan in carcinoma cell lines. *FEBS Lett.* **484**:194–198.

Louhichi, N., Triki, C., Quijano-Roy, S., et al. (2004). New FKRP mutations causing congenital muscular dystrophy associated with mental retardation and central nervous system abnormalities. Identification of a founder mutation in Tunisian families. *Neurogenetics* 5:27–34.

Love, D.R., Forrest, S.M., Smith, T.J., et al. (1989). Molecular analysis of Duchenne and Becker muscular dystrophies. *Br. Med. Bull.* 45:659–680.

MacLeod, H., Pytel, P., Wollmann, R., et al. (2007). A novel FKRP mutation in congenital muscular dystrophy disrupts the dystrophin glycoprotein complex. *Neuromuscular Disord.* **17**:285–289.

Manya, H., Sakai, K., Kobayashi, K., et al. (2003). Loss-of-function of an N-acetylglucos aminyltransferase, POMGnT1, in muscle-eye-brain disease. *Biochem. Biophys. Res. Commun.* **306**:93–97.

Manya, H., Chiba, A., Yoshida, A., et al. (2004). Demonstration of mammalian protein O-mannosyltransferase activity: coexpression of POMT1 and POMT2 required for enzymatic activity. *Proc. Natl. Acad. Sci. U.S.A.* **101**:500–505.

Manya, H., Suzuki, T., Akasaka-Manya, K., et al. (2007). Regulation of mammalian protein O-mannosylation: preferential amino acid sequence for O-mannose modification. *J. Biol. Chem.* **282**:20200–20206.

Martin, P.T. (2003a). Dystroglycan glycosylation and its role in matrix binding in skeletal muscle. *Glycobiology* **13**:55R–66R.

Martin, P.T. (2003b). Glycobiology of the neuromuscular junction. *J. Neurocytol.* **32**:915–929.

Martin, P.T. (2006). Mechanisms of disease: congenital muscular dystrophies-glycosylation takes center stage. *Nat. Clin. Pract. Neurol.* **2**:222–230.

Martin, P.T., and Freeze, H.H. (2003). Glycobiology of neuromuscular disorders. *Glycobiology* **13**:67R–75R.

Martin, P.T., Scott, L.J., Porter, B.E., et al. (1999). Distinct structures and functions of related pre- and postsynaptic carbohydrates at the mammalian neuromuscular junction. *Mol. Cell. Neurosci.* **13**:105–118.

Martin, L.T., Glass, M., Dosunmu, E., et al. (2007). Altered expression of natively glycosylated alpha dystroglycan in pediatric solid tumors. *Hum. Pathol.* **38**:1657–1668.

Martin, P.T., Shelton, G.D., Dickinson, P.J., et al. (2008). Muscular dystrophy associated with alpha-dystroglycan deficiency in Sphynx and Devon Rex cats. *Neuromuscular Disord.* **18**:942–952.

Matsumura, K., Ervasti, J.M., Ohlendieck, K., et al. (1992). Association of dystrophin-related protein with dystrophin-associated proteins in mdx mouse muscle. *Nature* 360:588–591.

McCubrey, J.A., Steelman, L.S., Chappell, W.H., et al. (2007). Roles of the Raf/MEK/ERK pathway in cell growth, malignant transformation and drug resistance. *Biochim. Biophys. Acta* 1773:1263–1284.

Mendell, J.R., Boue, D.R., Martin, P.T. (2006). The congenital muscular dystrophies: recent advances and molecular insights. *Pediatr. Dev. Pathol.* 9:427–443.

Michele, D.E., and Campbell, K.P. (2003). Dystrophin-glycoprotein complex: post-translational processing and dystroglycan function. *J. Biol. Chem.* 278:15457–15460.

Michele, D.E., Barresi, R., Kanagawa, M., et al. (2002). Post-translational disruption of dystroglycan-ligand interactions in congenital muscular dystrophies. *Nature* 418:417–422.

Moore, C.J., and Hewitt, J.E. (2009). Dystroglycan glycosylation and muscular dystrophy. *Glycoconj. J.* 26:349–357.

Moore, S.A., Saito, F., Chen, J., et al. (2002). Deletion of brain dystroglycan recapitulates aspects of congenital muscular dystrophy. *Nature* 418:422–425.

Moore, C.J., Goh, H.T., and Hewitt, J.E. (2008). Genes required for functional glycosylation of dystroglycan are conserved in zebrafish. *Genomics* 92:159–167.

Muntoni, F., Torelli, S., Brockington, M. (2008). Muscular dystrophies due to glycosylation defects. *Neurotherapeutics* 5:627–632.

Murakami, T., Hayashi, Y.K., Noguchi, S., et al. (2006). Fukutin gene mutations cause dilated cardiomyopathy with minimal muscle weakness. *Ann. Neurol.* 60:597–602.

Muschler, J., Levy, D., Boudreau, R., Henry, M., et al. (2002). A role for dystroglycan in epithelial polarization: loss of function in breast tumor cells. *Cancer Res.* 62:7102–7109.

Newey, S.E., Howman, E.V., Ponting, C.P., et al. (2001). Syncoilin, a novel member of the intermediate filament superfamily that interacts with alpha-dystrobrevin in skeletal muscle. *J. Biol. Chem.* 276:6645–6655.

Nguyen, H.H., Jayasinha, V., Xia, B., et al. (2002). Overexpression of the cytotoxic T cell GalNAc transferase in skeletal muscle inhibits muscular dystrophy in mdx mice. *Proc. Natl. Acad. Sci. U.S.A.* 99:5616–5621.

Nishimune, H., Valdez, G., Jarad, G., et al. (2008). Laminins promote postsynaptic maturation by an autocrine mechanism at the neuromuscular junction. *J. Cell Biol.* 182:1201–1215.

Oak, S.A., Zhou, Y.W., Jarrett, H.W. (2003). Skeletal muscle signaling pathway through the dystrophin glycoprotein complex and Rac1. *J. Biol. Chem.* 278:39287–39295.

Ohlendieck, K., Ervasti, J.M., Matsumura, K., et al. (1991). Dystrophin-related protein is localized to neuromuscular junctions of adult skeletal muscle. *Neuron* 7:499–508.

Patnaik, S.K., and Stanley, P. (2005). Mouse large can modify complex N- and mucin O-glycans on alpha-dystroglycan to induce laminin binding. *J. Biol. Chem.* 280:20851–20859.

Patton, B.L., Miner, J.H., Chiu, A.Y., et al. (1997). Distribution and function of laminins in the neuromuscular system of developing, adult, and mutant mice. *J. Cell Biol.* 139:1507–1521.

Peters, M.F., Adams, M.E., Froehner, S.C. (1997). Differential association of syntrophin pairs with the dystrophin complex. *J. Cell Biol.* 138:81–93.

Peters, M.F., Sadoulet-Puccio, H.M., Grady, M.R., et al. (1998). Differential membrane localization and intermolecular associations of alpha-dystrobrevin isoforms in skeletal muscle. *J. Cell Biol.* 142:1269–1278.

Poon, E., Howman, E.V., Newey, S.E., et al. (2002). Association of syncoilin and desmin: linking intermediate filament proteins to the dystrophin-associated protein complex. *J. Biol. Chem.* 277:433–3439.

Rambukkana, A., Yamada, H., Zanazzi, G., et al. (1998). Role of alpha-dystroglycan as a Schwann cell receptor for *Mycobacterium leprae*. *Science* 282:2076–2079.

Reilich, P., Petersen, J.A., Vielhaber, S., et al. (2006). LGMD 2I due to the common mutation 826C>A in the FKRP gene presenting as myopathy with vacuoles and paired-helical filaments. *Acta Myol.* 25:73–76.

Reynolds, J.G., McCalmon, S.A., Donaghey, J.A., et al. (2008). Deregulated protein kinase A signaling and myospryn expression in muscular dystrophy. *J. Biol. Chem.* 283:8070–8074.

Rezniczek, G.A., Konieczny, P., Nikolic, B., et al. (2007). Plectin 1f scaffolding at the sarcolemma of dystrophic (mdx) muscle fibers through multiple interactions with beta-dystroglycan. *J. Cell Biol.* 176:965–977.

Russo, K., Di Stasio, E., Macchia, G., et al. (2000). Characterization of the beta-dystroglycan-growth factor receptor 2 (Grb2) interaction. *Biochem. Biophys. Res. Commun.* 274:93–98.

Sadasivam, G., Willmann, R., Lin, S., et al. (2005). Src-family kinases stabilize the neuromuscular synapse in vivo via protein interactions, phosphorylation, and cytoskeletal linkage of acetylcholine receptors. *J. Neurosci.* 25:10479–10493.

Saito, F., Moore, S.A., Barresi, R., et al. (2003). Unique role of dystroglycan in peripheral nerve myelination, nodal structure, and sodium channel stabilization. *Neuron* 38:747–758.

Saito, F., Masaki, T., Saito, Y., et al. (2007). Defective peripheral nerve myelination and neuromuscular junction formation in fukutin-deficient chimeric mice. *J. Neurochem.* 101:1712–1722.

Santhanakrishnan, M., Ray, K., Oppenheimer, K., et al. (2008). Dynamic regulation of alpha-dystroglycan in mouse placenta. *Placenta* 29:932–936.

Sasaki, T., Yamada, H., Matsumura, K., et al. (1998). Detection of O-mannosyl glycans in rabbit skeletal muscle alpha-dystroglycan. *Biochim. Biophys. Acta* 1425:599–606.

Sato, S., Omori, Y., Katoh, K., et al. (2008). Pikachurin, a dystroglycan ligand, is essential for photoreceptor ribbon synapse formation. *Nat. Neurosci.* 11:923–931.

Satz, J.S., Barresi, R., Durbeej, M., et al. (2008). Brain and eye malformations resembling Walker-Warburg syndrome are recapitulated in mice by dystroglycan deletion in the epiblast. *J. Neurosci.* 28:10567–10575.

Schneider, M., and Baumgartner, S. (2008). Differential expression of Dystroglycan-spliceforms with and without the mucin-like domain during Drosophila embryogenesis. *Fly (Austin)* 2, 29–35.

Sgambato, A., and Brancaccio, A. (2005). The dystroglycan complex: from biology to cancer. *J. Cell Physiol.* 205:163–169.

Sgambato, A., Migaldi, M., Montanari, M., et al. (2003). Dystroglycan expression is frequently reduced in human breast and colon cancers and is associated with tumor progression. *Am. J. Pathol.* 162:849–860.

Sgambato, A., Camerini, A., Faraglia, B., et al. (2004). Increased expression of dystroglycan inhibits the growth and tumorigenicity of human mammary epithelial cells. *Cancer Biol. Ther.* 3: 967–975.

Sgambato, A., Tarquini, E., Resci, F., et al. (2006). Aberrant expression of alpha-dystroglycan in cervical and vulvar cancer. *Gynecol. Oncol.* 103:397–404.

Sgambato, A., Camerini, A., Amoroso, D., et al. (2007a). Expression of dystroglycan correlates with tumor grade and predicts survival in renal cell carcinoma. *Cancer Biol. Ther.* 6:1840–1846.

Sgambato, A., De Paola, B., Migaldi, M., et al. (2007b). Dystroglycan expression is reduced during prostate tumorigenesis and is regulated by androgens in prostate cancer cells. *J. Cell Physiol.* 213:528–539.

Singh, J., Itahana, Y., Knight-Krajewski, S., et al. (2004). Proteolytic enzymes and altered glycosylation modulate dystroglycan function in carcinoma cells. *Cancer Res.* 64:6152–6159.

Smalheiser, N.R. and Kim, E. (1995). Purification of cranin, a laminin binding membrane protein. Identity with dystroglycan and reassessment of its carbohydrate moieties. *J. Biol. Chem.* 270:15425–15433.

Smalheiser, N.R., Haslam, S.M., Sutton-Smith, M., et al. (1998). Structural analysis of sequences O-linked to mannose reveals a novel Lewis X structure in cranin (dystroglycan) purified from sheep brain. *J. Biol. Chem.* 273:23698–23703.

Sotgia, F., Lee, H., Bedford, M.T., et al. (2001). Tyrosine phosphorylation of beta-dystroglycan at its WW domain binding motif, PPxY, recruits SH2 domain containing proteins. *Biochemistry* 40:14585–14592.

Spence, H.J., Chen, Y.J., Batchelor, C.L., et al. (2004a). Ezrin-dependent regulation of the actin cytoskeleton by beta-dystroglycan. *Hum. Mol. Genet.* 13:1657–1668.

Spence, H.J., Dhillon, A.S., James, M., et al. (2004b). Dystroglycan, a scaffold for the ERK-MAP kinase cascade. *EMBO Rep.* 5:484–489.

Stone, M.R., O'Neill, A., Catino, D., et al. (2005). Specific interaction of the actin-binding domain of dystrophin with intermediate filaments containing keratin 19. *Mol. Biol. Cell* 16:4280–4293.

Straub, V., and Bushby, K. (2006). The childhood limb-girdle muscular dystrophies. *Semin Pediatr. Neurol.* 13:104–114.

Sugita, S., Saito, F., Tang, J., et al. (2001). A stoichiometric complex of neurexins and dystroglycan in brain. *J. Cell Biol.* 154:435–445.

Taniguchi, M., Kurahashi, H., Noguchi, S., et al. (2006). Aberrant neuromuscular junctions and delayed terminal muscle fiber maturation in alpha-dystroglycanopathies. *Hum. Mol. Genet.* 15:1279–1289.

Thompson, T.G., Chan, Y.M., Hack, A.A., et al. (2000). Filamin 2 (FLN2): A muscle-specific sarcoglycan interacting protein. *J. Cell Biol.* 148:115–126.

Thompson, O., Kleino, I., Crimaldi, L., et al. (2008). Dystroglycan, Tks5 and Src mediated assembly of podosomes in myoblasts. *PLoS ONE* 3: e3638.

Timpl, R., Tisi, D., Talts, J.F., et al. (2000). Structure and function of laminin LG modules. *Matrix Biol.* 19:309–317.

Topaloglu, H., Brockington, M., Yuva, Y., et al. (2003). FKRP gene mutations cause congenital muscular dystrophy, mental retardation, and cerebellar cysts. *Neurology* 60:988–992.

Torelli, S., Brown, S.C., Brockington, M., et al. (2005). Sub-cellular localisation of fukutin related protein in different cell lines and in the muscle of patients with MDC1C and LGMD2I. *Neuromuscular Disord.* 15:836–843.

Ursitti, J.A., Lee, P.C., Resneck, W.G., et al. (2004). Cloning and characterization of cytokeratins 8 and 19 in adult rat striated muscle. Interaction with the dystrophin glycoprotein complex. *J. Biol. Chem.* 279:41830–41838.

Vainzof, M., Passos-Bueno, M.R., Canovas, M., et al. (1996). The sarcoglycan complex in the six autosomal recessive limb-girdle muscular dystrophies. *Hum. Mol. Genet.* 5:1963–1969.

van Reeuwijk, J., Janssen, M., van den Elzen, C., et al. (2005). POMT2 mutations cause alpha-dystroglycan hypoglycosylation and Walker-Warburg syndrome. *J. Med. Genet.* 42:907–912.

van Reeuwijk, J., Grewal, P.K., Salih, M.A., et al. (2007). Intragenic deletion in the LARGE gene causes Walker-Warburg syndrome. *Hum. Genet.* 121:685–690.

Vandebrouck, A., Sabourin, J., Rivet, J., et al. (2007). Regulation of capacitative calcium entries by alpha1-syntrophin: association of TRPC1 with dystrophin complex and the PDZ domain of alpha1-syntrophin. *FASEB J.* 21:608–617.

Wakayama, Y., Inoue, M., Kojima, H., et al. (2008). Reduced expression of sarcospan in muscles of Fukuyama congenital muscular dystrophy. *Histol. Histopathol.* 23:1425–1438.

Weir, M.L., Oppizzi, M.L., Henry, M.D., et al. (2006). Dystroglycan loss disrupts polarity and beta-casein induction in mammary epithelial cells by perturbing laminin anchoring. *J. Cell Sci.* 119:4047–4058.

Willer, T., Valero, M.C., Tanner, W., et al. (2003). O-mannosyl glycans: from yeast to novel associations with human disease. *Curr. Opin. Struct. Biol.* 13:621–630.

Willer, T., Prados, B., Falcon-Perez, J.M., et al. (2004). Targeted disruption of the Walker-Warburg syndrome gene Pomt1 in mouse results in embryonic lethality. *Proc. Natl. Acad. Sci. U.S.A.* 101:14126–14131.

Williams, I.A., and Allen, D.G. (2007). Intracellular calcium handling in ventricular myocytes from mdx mice. *Am J Physiol Heart Circ Physiol* 292: H846–H855.

Williams, J.C., Armesilla, A.L., Mohamed, T.M., et al. (2006). The sarcolemmal calcium pump, alpha-1 syntrophin, and neuronal nitric-oxide synthase are parts of a macromolecular protein complex. *J. Biol. Chem.* 281:23341–23348.

Williamson, R.A., Henry, M.D., Daniels, K.J., et al. (1997). Dystroglycan is essential for early embryonic development: disruption of Reichert's membrane in Dag1-null mice. *Hum. Mol. Genet.* 6:831–841.

Xia, B., Hoyte, K., Kammesheidt, A., et al. (2002). Overexpression of the CT GalNAc transferase in skeletal muscle alters myofiber growth, neuromuscular structure, and laminin expression. *Dev. Biol.* 242:58–73.

Xiong, H., Kobayashi, K., Tachikawa, M., et al. (2006). Molecular interaction between fukutin and POMGnT1 in the glycosylation pathway of alpha-dystroglycan. *Biochem. Biophys. Res. Commun.* 350:935–941.

Xiong, Y., Zhou, Y., Jarrett, H.W. (2009). Dystrophin glycoprotein complex-associated Gbetagamma subunits activate phosphatidylinositol-3-kinase/Akt signaling in skeletal muscle in a laminin-dependent manner. *J. Cell Physiol.* 219:402–414.

Yamada, H., Saito, F., Fukuta-Ohi, H., et al. (2001). Processing of beta-dystroglycan by matrix metalloproteinase disrupts the link between the extracellular matrix and cell membrane via the dystroglycan complex. *Hum. Mol. Genet.* 10:1563–1569.

Yang, B., Jung, D., Motto, D., et al. (1995). SH3 domain-mediated interaction of dystroglycan and Grb2. *J. Biol. Chem.* 270:11711–11714.

Yoshida, A., Kobayashi, K., Manya, H., et al. (2001). Muscular dystrophy and neuronal migration disorder caused by mutations in a glycosyltransferase, POMGnT1. *Dev. Cell* 1:717–724.

Yurchenco, P.D., Cheng, Y.S., Campbell, K., et al. (2004). Loss of basement membrane, receptor and cytoskeletal lattices in a laminin-deficient muscular dystrophy. *J. Cell Sci.* 117: 735–742.

Zhan, Y., Tremblay, M.R., Melian, N., et al. (2005). Evidence that dystroglycan is associated with dynamin and regulates endocytosis. *J. Biol. Chem.* 280:18015–18024.

Chapter 7
Protein Kinase A: The Enzyme and Cyclic AMP Signaling

Maria Nesterova and Constantine A. Stratakis

7.1 Introduction

Reversible phosphorylation is the most common type of posttranslational modifications of proteins and key to the regulation of most aspects of cell function. It is known today that about one third of proteins encoded by the mammalian genome contain covalently bound phosphate. Abnormalities in phosphorylation can cause (or be caused by) disease, but this was not appreciated at first: when the first protein kinase was found, it was not realized that phosphorylation as a mechasnism was such a fundamental way of biological regulation. First, any extracellular signal can be the first messenger. Then, a second messenger, a substance mediating the effects of binding of a first messenger to a receptor, turns on a cascade of events that through phosphorylation transmit signals that started outside of the cell membrane all the way to the cell nucleus. In recent years, there has been remarkable progress in the identification and characterization of the molecular mechanisms by which these factors regulate cell expression (Conkright et al. 2003; Mayr and Montminy 2001; Zhang et al. 2005); however, relatively little is known about upstream and parallel events that increase the complexity of these signaling events.

Discovered by Sutherland in 1957 as the first molecule to act as a "second messenger," cAMP is present in all cells and tissues, from bacteria to humans (Robison 1973). cAMP-mediated pathways remain the best characterized second-messenger systems leading to activation of protein kinase A (PKA or cAMP-dependent kinase) (Fig. 7.1). Today it is generally accepted that, in keeping with Greengard's hypothesis, most cAMP effects are mediated through PKA and the phosphorylation of its protein substrates (Kuo and Greengard 1969). In this chapter,

C.A. Stratakis (✉)
Section on Endocrinology Genetics, Program on Developmental Endocrinology Genetics (PDEGEN), Eunice Kenney Shriver National Institute of Child Health & Human Development (NICHD), National Institute of Health (NIH), Building 10, CRC (East Laboratories), Room 1-3330, 10 Center Dr., MSC1103, Bethesda, MD 20892, USA
e-mail: stratakc@mail.nih.gov

C.J. Vidal (ed.), *Post-Translational Modifications in Health and Disease*, Protein Reviews 13, DOI 10.1007/978-1-4419-6382-6_7,

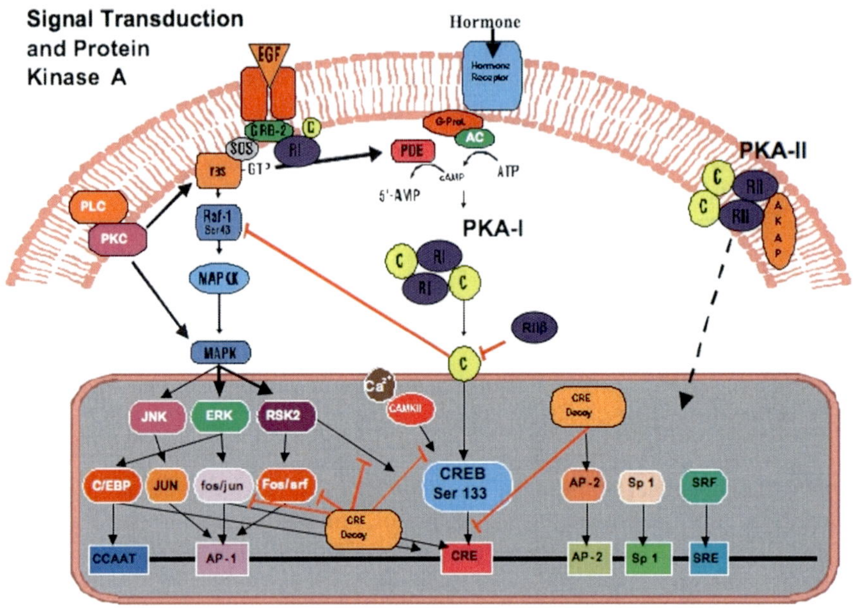

Fig. 7.1 Cyclic AMP signaling starts after a ligand stimulates a G-protein coupled receptor (GPCR); the final steps are completed by phosphorylation of a number of substrate proteins by its serine-threonine protein kinase (PKA) mediator. The best characterized among these substrates is CREB which binds to the CREs in many PKA-regulated genes. There is crosstalk between the cAMP signaling pathway and other cellular pathways. Here we present potential interactions of PKA and the EGF receptor, MAP-kinase and PKC-signaling pathways

we will focus on aspects of the complex regulation and factors that affect specificity of PKA signaling; these features make PKA a suitable therapeutic target in various diseases.

7.2 PKA's Multiple Isoforms

PKA holoenzyme is an enzymatically inactive complex of two catalytic (C) subunits, each bound to one of the two regulatory (R) subunits that together form a homodimer (Fig. 7.2). cAMP binds in a cooperative manner to two sites on each of the R subunits. Upon binding of cAMP, the enzyme dissociates into an R subunit dimer with cAMP bound and two active C subunits. Free catalytic subunits phosphorylate serine and threonine residues on specific substrate proteins (Beebe and Corbin 1986). How one enzyme with broad substrate specificity like PKA can mediate – so precisely – such a variety of diverse physiological events? First, PKA exists in two isoforms, termed type I and type II (PKA-I and PKA-II, respectively) (Fig. 7.2). These isoforms were found based on elution pattern during DEAE-cellulose chromatography (Corbin et al. 1975). By molecular cloning, various types of R and

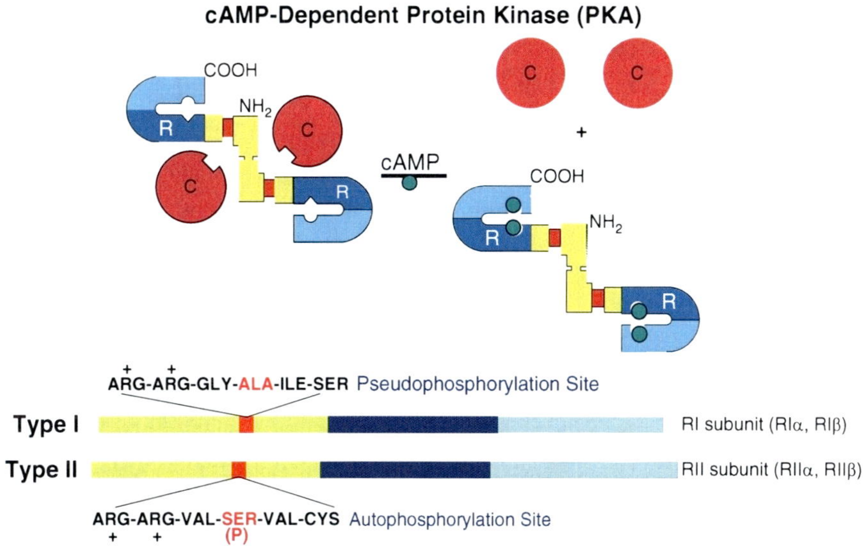

Fig. 7.2 The PKA holoenzyme is an enzymatically inactive complex of two catalytic (C) subunits bound to a dimer of regulatory (R) subunits. cAMP binds in a cooperative manner to two sites on each R subunit. Upon binding of cAMP, the enzyme dissociates into an R subunit dimer with cAMP bound and two active C subunits. Free C subunits phosphorylate substrate proteins. PKA exists in two different isozymes, designated as type I and type II. These two isoforms are distinct in their biochemical properties and RI and RII differ significantly in the N-terminus at a proteolytically-sensitive hinge region that binds the catalytic domain of the C subunit in the holoenzyme complex. In this segment, RII contains the sequence that can undergo autophosphorylation at Ser, an essential step for dissociation of the PKA-II holoenzyme

C subunits were identified: two RI subunits, RIα (Lee et al. 1983; Sandberg et al. 1987) and RIβ (Clegg et al. 1988; Solberg et al. 1991) and two RII subunits, RIIα (Øyen et al. 1989; Scott et al. 1987) and RIIβ (Jahnsen et al. 1986; Levy et al. 1988), all coded by different genes, *PRKAR1A*, *PRKAR1B*, *PRKAR2A*, and *PRKAR2B*, respectively. Alternative splice variants of the RIα subunit have also been demonstrated (Solberg et al. 1997), increasing the complexity of this system. In addition, three distinct C subunits were identified and designated as Cα (Uhler et al. 1986a), Cβ (Uhler et al. 1986b), and Cγ (Beebe et al. 1990), also coded by different genes (*PRKARCA*, *PRKACB*, and *PRKACG*, respectively). Multiple catalytically active and inactive splice variants from Cα and Cβ have also been reported (Lange-Carter and Malkinson 1991; McKnight et al. 1996) and a third, distinct, X-chromosome-located, cAMP-dependent protein kinase catalytic subunit, PrKX (coded by the *PRKX* gene), was more recently identified in both the human (Klink et al. 1995) and mouse genome (Blaschke et al. 2000). With the exception of the Cα2 splice variant, all C subunits posses the catalytic core motif common to all protein kinases (Hanks et al. 1988; Taylor et al. 1992). Both R and C subunits are major determinants of PKA specificity (Cheley and Bayley 1991; Beebe et al. 1992; Gamm et al. 1996).

Cγ is unique when compared to the other catalytic subunits: the Cγ (*PRKACG*) gene probably is created from a reverse-transcribed Cα mRNA produced by a retrotransposition event that is timed approximately 65 million years ago and during primate evolution (Reinton et al. 1998). The *PRKACG* gene is highly homologous to the *PRKACA* one: they share more than 85% of their open reading frame and 3′ untranslated sequences, while the two proteins are approximately 83% identical (Beebe et al. 1990; Zhang et al. 2004). However, Cγ differs from Cα in both substrate and pseudosubstrate specificity: in contrast to Cα, Cγ phosphorylates histone proteins better than it phosphorylates the "classic" PKA substrate kemptide; Cγ has also lower affinity for all R subunit isoforms in vitro than Cα (Beebe et al. 1992).

PKA regulates transcription through the phosphorylation of several different transactivating and/or transcription factors. The best characterized such mediator of cAMP action is the cAMP-responsive element (CRE) binding protein (CREB): CREB binds to CRE sequences in PKA-targeted genes (Gonzalez and Montminy 1989). Interestingly, catalytic subunits differ in their effects on CREB activity: transient transfection studies have shown that Cα and Cβ, but not Cγ, can activate reporter genes that contain CREs in their regulatory sequences (Huggenvik et al. 1991; Mellon et al. 1989). On the other hand, both Cα and Cγ activated a CRE-containing fragment of the α-glycoprotein hormone promoter, although activation by Cα was greater than that by Cγ (Morris et al. 2002). Cγ, but not Cα, activated the herpes simplex virus thymidine kinase promoter in Sp1-dependent manner (Morris et al. 2002). Cα1 exhibited a lower K_m for certain peptide substrates and a lower IC_{50} for inhibition by protein kinase inhibitor (PKI) and RIIα (Gamm et al. 1996). These results suggested that cAMP signaling through the expression of a variable catalytic subunit can regulate separate sets of genes and thus serve distinct functions.

High affinity interactions were demonstrated for PrKX with RIα and heat-stable PKI; in contrast, PrKX did not associate with RIIβ under physiological conditions (Zimmermann et al. 1999). Kemptide and autophosphorylation activities of PrKX were strongly inhibited by the RIα subunit and by PKI in vitro, but only weakly by the RIIα subunit. Thus, PrKX is in essence a type I cAMP-dependent protein kinase that is activated at lower cAMP concentrations than the PKA holoenzyme with the more common and abundantly expressed Cα subunit. These examples suggest unique features associated with the various C subunits, which implies different functions of these proteins in vivo.

The PKA-I and PKA-II holoenzymes have also distinct biochemical properties (Beebe and Corbin 1986; Cheng et al. 2001; Dostman et al. 1990). It has been shown that association and dissociation rates of PKA isozymes are affected differently by factors such as salt and Mg-ATP concentrations (Herberg and Taylor 1993). RI and RII differ significantly in the N-terminus at a proteolytically sensitive hinge region that occupies the catalytic domain of the C subunit in the holoenzyme complex (Taylor et al. 1988). In this segment, RII contains the sequence that undergoes autophosphorylation at *Ser* which is essential for the dissociation of PKA-II (Rosen and Erlichman 1975; Titani et al. 1984) (Fig. 7.2). PKA-I holoenzymes are more easily dissociated by cAMP in vitro than PKA-II holoenzymes

(Zawadzki and Taylor 2004). Recent experiments demonstrated that the substrate may also play a role in defining the type of holoenzyme activation (PKA-I versus PKA-II) (Vigil et al. 2004): for example, low concentrations of cAMP activate mainly PKA-I, whereas the concentration of cAMP does not affect the cAMP-sensitivity of PKA-II (Viste et al. 2005).

7.3 Subcellular Distribution of PKA Isozymes

The compartmentalization of PKA isoforms within a cell introduces another level of specificity for PKA, and for cAMP signaling in general. The RI isoforms were found originally primarily in cytoplasm in soluble condition. On the contrary, major pools of RII were found in particulate compartments and in association with certain subcellular structures; the molecules mediating this localization were found to be the A-kinase anchor proteins (AKAPs) (Colledge and Scott 1999). More recently, a group of dual specificity AKAPs has been identified on the basis of their ability to bind both isoforms, PKA-I and PKA-II (Jarnaess et al. 2008). Dual specificity AKAPs contain an additional PKA-binding determinant called the RI Specifier Region (RISR). A variety of protein interaction assays and immuno-precipitation and -localization experiments indicated that RISR augments RI binding in vitro and inside cells. In T-cells lipid rafts, where cAMP acts as a negative regulator of T cell immune responses through phosphorylation of C-terminal Src kinase, a novel AKAP termed *Ezrin* that targets PKA-I (Ruppelt et al. 2007) was identified. Thus, different subcellular distribution has been demonstrated for the different PKA isozymes: in LS-174T colon cancer cells, plasma membranes, Golgi, and mitochondria contained mainly RIIα whereas in lysosomes and microsomes both RIα and RIIα were found in nearly equal amounts (Kondrashin et al. 1998). RIIα and RIIβ subunits were found associated with pericentriolar matrix of the centrosome during interphase; also, the catalytic subunit (Cα) binds to microtubules and/or the mitotic spindle (Carlson et al. 2001; Imazumi-Scherrer et al. 2001). RIIα is tightly bound to centrosomal structures during interphase through interaction with AKAP450, but dissociates and redistributes from centrosomes at mitosis (Carlson et al. 2001). We also recently showed that RIα binds tightly to the late endosome (Mavrakis et al. 2006) where it interacts with mTOR and may regulate autophagy (Mavrakis et al. 2007), although it is unclear if this binding is mediated through an AKAP.

Special attention deserves nuclear cAMP signaling: the prevailing theory is that in the basal state, PKA resides in the cytoplasm as an inactive holoenzyme and induction of cAMP frees the catalytic subunit, which passively diffuses into the nucleus and induces cellular gene expression (Neary and Cho-Chung 2001; Riabowol et al. 1988; Solberg et al. 1994). Recently, it was shown that the two types of PKA in NG108-15 cells (a neuroblastoma-glioma hybrid cell line) differentially mediate the forskolin- and ethanol-induced CREB phosphorylation and CRE-mediated gene transcription (Constantinescu et al. 2004): the authors found that activated PKA-II is translocated into the nucleus (where it phosphorylates CREB), whereas activated PKA-I remains mainly cytoplasmic and induces the activation of a coactivator CBP to eventually lead to CRE-mediated gene transcription. This is different in cells harboring a mutant or deficient RIα in which nuclear cAMP-signaling increases (Groussin et al. 2002; Nesterova et al. 2008). Other molecules, such as the v-Ki-Ras oncogene may also alter nuclear cAMP signaling by affecting the location and/or expression of RIIβ (Feliciello et al. 1996).

7.4 Mouse Models: An Aid in Understanding PKA Functions

Gene knockout (KO) experiments in mice led to the understanding of functions of PKA that in certain cases were unexpected (Table 7.1). A KO of the RIβ gene does not affect total PKA activity, at least in brain, where it was studied. In the RIα-, RIIα-, and RIIβ-subunit-deficient mice, total and cAMP-stimulated PKA activities are affected to a different degree: these mice have reduced cAMP-stimulated and (protein kinase A-specific inhibitor-PKI) PKI-inhibited PKA activity, but only the RIα-deficient mice display significantly increased baseline (non cAMP-stimulated but PKI-inhibited) PKA activity (Amieux et al. 1997; Burton et al. 1999; Cummings et al. 1996). A small increase of this baseline activity is also seen in the neonate muscle of RIIα-KO, as well as the fat tissue of RIIβ-KO mice.

Interestingly, the RIIβ-deficient mice display a lean phenotype, resistance to diet-induced obesity, increased lipolysis, reduced insulin and cholesterol levels, and

Table 7.1 Phenotypes of animal models of PKA regulatory subunit deficiency

PKA subunit	Phenotype	Total PKA activity
Prkar1a$^{-/-}$	Early embryonic lethality due to heart and other developmental defects	Increased
Prkar1b$^{-/-}$	1. Defective hippocampal depotentiation, long-term-depression and other neuronal functional abnormalities 2. Reduced inflammation and response to pain	Unchanged
Prkar2a$^{-/-}$	Normal; long-term follow-up	Decreased
Prkar2b$^{-/-}$	1. Lean phenotype, resistance to diet-induced obesity, increased lipolysis, reduced plasma insulin and cholesterol 2. Diminished motor learning 3. Loss of haloperidol-induced catalepsy 4. Increased alcohol consumption and decreased sedation	Decreased
Prkaca$^{-/-}$	1. Background-dependent lethality (C57BL/6) 2. Eliminates most of PKA activity except for that mediated by Cβ1; very low R1α (except brain, kidney and spleen) 3. Surviving animals 65% of wt; infertile, decreased IGF-1; survived up to 3 months	Decreased (very low)

diminished motor learning and neuronal gene expression (Brandon et al. 1998; Cummings et al. 1996; Planas et al. 1999; Schreyer et al. 2001). Brown adipose tissue of RIIβ KO mice exhibited a compensatory increase in RIα, which almost entirely replaced lost RIIβ, leading to primarily PKA-I *only* in these cells. This model was the first in vivo one that showed a switch to a different type of PKA when one main subunit is inactivated. Indeed, the PKA-I holoenzyme from mutant adipose tissue binds cAMP more efficient and is more easily activated than the wild-type enzyme (Cummings et al. 1996), explaining at least in part the lean phenotype.

The RIIα deficient mice were mostly normal, while the RIα deficient mice displayed the most severe phenotype with gross developmental defects in mesodermal morphogenesis and early embryonic lethality due to incomplete development of the primitive heart tube (Amieux et al. 1997; Burton et al. 1997; Burton et al. 1999; Kirschner et al. 2005). Mice with either loss of one allele for RIα (Kirschner et al. 2005) or a transgenically down-regulated RIα allele by as much as 60–70% (Griffin et al. 2004a, b) developed tumors in cAMP-rensponsive tissues. The later observation and the absence of severe phenotype in the remaining PKA regulatory subunit KOs (RIIα-, RIIβ and RIβ-deficient mice) point to the essential role of the RIα regulatory subunit in maintaining the PKA catalytic subunit under cAMP control during cellular growth and proliferation. More recent experiments with tissue-specific KO of the RIα subunit point to the Schwan cell-, neuroendocrine-, pituitary-, heart-, and bone-specific roles of RIα (Jones et al. 2008; Pavel et al. 2008; Yin et al. 2008a, b), as well as to its possible role in mesenchymal-to-epithelial transition (Nadella et al. 2008).

7.5 Differential Expression of PKA Isozymes in Normal and Malignant Cells

The four types of regulatory subunits have different expression patterns in mammals (Hofman et al. 1977). Every tissue is characterized by a specific PKA-I/PKA-II ratio that is further more specific to a particular physiological condition. This balance of PKA isozymes apparently is essential for the regulation of diverse processes such as ontogeny and differentiation, growth and proliferation, and possibly cell death and even the transcriptional machinery.

PKA-I mediates the inhibitory effect of cAMP on cell replication in human T lymphocytes (Skalhegg et al. 1992). The importance of type I PKA-mediated effects in inhibition of T cell replication has led to its consideration as a therapeutic target in combined variable immunodeficiency (CVI) and acquired immune deficiency syndrome (AIDS) (Bryn et al. 2008; Tasken et al. 2000). PKA-I in T cells may also serve as potential therapeutic target in systemic lupus erythromatosis (Khan et al. 2001; Laxminarayana and Kammer 2000).

Unbalanced expression of PKA isozymes has been detected in a variety of types of human cancer tissues and cell lines (Fossberg et al. 1978; Handschin et al. 1983;

Miller 2002). Recently it was shown that the differential expression of PKA isozymes affected the growth of prostate carcinoma cells (Neary et al. 2004). DNA microarray analysis demonstrated that RIIβ overexpression led to up-regulation of a cluster of differentiation genes and the down-regulation of transformation and proliferation gene signatures. Overexpression of RIα and Cα (which upregulated PKA-I) elicited expression signatures opposite of those found in RIIβ-expressing cells. Accordingly, modulation of the regulatory isoforms (RI versus RII) of PKA influenced the ability of PKA to regulate cancer cell growth and to induce tumor reversion.

Recent studies on ovarian cancer cells revealed that regulatory subunits of PKA define an axis of cellular proliferation/differentiation state (Cheadle et al. 2008). RIα-transfected cells exhibit hyper-proliferative growth and RIIβ-transfected cells revert to a relatively quiescent state. Profiling by microarray demonstrated equally profound changes in gene expression between RIα, RIIβ, and parental OVCAR cells. Genes specifically up-regulated in RIα cells were highly enriched for pathways involved in cell growth while genes up-regulated in RIIβ cells were enriched for pathways involved in differentiation. A large group of genes (~3,600) was regulated along an axis of proliferation/differentiation between RIα, parental, and RIIβ cells. In summary, over-expression of PKA regulatory subunits in an ovarian cancer cell line dramatically influenced the cell gene signature. The proliferation phenotype is strongly correlated with recently identified clinical biomarkers predictive of poor prognosis in ovarian cancer suggesting a possible pivotal role for PKA regulation in disease progression. It should be noted that high levels of RIα have been seen in many malignant tumors and cancer cell lines (Cho-Chung 1990). In contrast, endocrine and other tumors from RIα-KO mice, as discussed above, and from patients with Carney Complex (see below) exhibit down-regulation of RIα or even complete absence of RIα (Kirschner et al. 2000; Lania et al. 2004). In addition, mouse embryonic fibroblasts (MEFs) with complete absence of RIα become immortalized and demonstrate activation of cyclin D1 (Nadella and Kirschner 2005).

At first glance, it appears hard to reconcile these data. But, as we discussed above, the PKA-I/-II balance is of paramount importance in many cell functions, and changes towards the one or the other PKA isozyme could be consistent with induction or inhibition of growth or proliferation depending on the tissue, the developmental phase and/or the cell macro- and micro-environment (Nadella et al. 2008), and possibly even the stage of cell division. Most of the earlier data, and the gene-signature experiments described above were obtained in immortalized cancer cell lines or in tissues from highly aggressive malignant tumors, where the PKA system may have already been bypassed by the activation of other signaling pathways, whereas the mouse and human data are from tissues with RIα deficiency. Thus, although we still do not understand exactly how RIα works in its states of excess of deficiency, it is clear that small changes in the PKA-I/PKA-II balance have large effects of growth and proliferation and corresponding dramatic changes in gene signatures that are very cell-specific.

7.6 Mutations in *PRKAR1A* Gene: Source of New PKA Variants

Carney complex (CNC) was the first human disease to be linked to mutations of the PKA enzyme, and in particular inactivating mutations of the RIα subunit (coded by the *PRKAR1A* gene) (Groussin et al. 2002; Kirschner et al. 2000; Stratakis 2000; Stratakis et al. 2001). CNC describes the association "of spotty skin pigmentation, myxomas, and endocrine overactivity." Primary pigmented nodular adrenocortical disease (PPNAD), a micronodular form of bilateral adrenal hyperplasia that causes a unique, inherited form of Cushing syndrome, is also the most common endocrine manifestation of CNC. CNC and PPNAD are genetically heterogeneous, but *PRKAR1A* is the most frequently responsible gene, at least for those families that map to 17q22-24 (the chromosome region that harbors the *PRKAR1A* gene). The 578delTG mutation in exon 5 (which corresponds to position 163 of the protein) is the most frequent *PRKAR1A* mutation and leads to a predicted truncated RIα protein (Groussin et al. 2002; Kirschner et al. 2000). This mutation was also found in sporadic cases, where it appears to have occurred de novo because it was absent in the parental peripheral blood DNA.

Like the 578delTG mutation, most of the reported *PPKARIA* mutations result in premature stop codons, and the predicted mutant protein products cannot be found in affected cells secondary to nonsense-mediated mRNA decay (NMD) of the mutant sequence (Kirschner et al. 2000; Stratakis et al. 2001). Loss of RIα protein leads to increased cAMP-stimulated total kinase activity, and an increase in CREB phosphorylation in CNC tissues (Groussin et al. 2002; Stratakis 2000).

We recently described seven *PRKAR1A* mutations whose mRNAs do not seem to undergo NMD and instead result in an expressed mutant RIα protein (Greene et al. 2008). These "expressed" mutations (p.Ser9Asn, p.Glu60_Lys116del [D-exon 3], p.Arg74Cys, p.Arg146Ser, p.Asp183Tyr, p.Ala213Asp, and p.Gly289Trp) were spread over the entire functional RIα domains (Fig. 7.3), and all of them exhibited increased PKA activity in vitro; this was attributed to decreased binding to cAMP and/or the catalytic subunit. These data corroborated previous findings: altered *PRKAR1A* function, and not only haploinsufficiency, is enough to elevate PKA activity which is apparently associated with tumor formation in tissues affected by CNC. In fact, in at least one mutation, the presence of a mutant PRKAR1A protein appeared to lead to a more serious clinical phenotype.

Another interesting RIα defect, is the first *PRKAR1A* mutation to be described that did not undergo NMD and led to an expressed protein. The IVS6+1G>T *PRKAR1A* mutation leads to a protein that lacks exon 6 sequences (Meoli et al. 2008) (Fig. 7.4). We assessed PKA activity, subunit expression, phosphorylation of target molecules, and the properties of both wild type RIα and mutant RIα under various conditions. By confocal microscopy, RIα tagged with green fluorescent protein was studied alone and on its interactions with Cerulean-tagged catalytic subunit. Introduction of the mutant RIα led to aberrant cellular morphology and higher PKA activity but no increase in type II PKA subunits. There was diffuse,

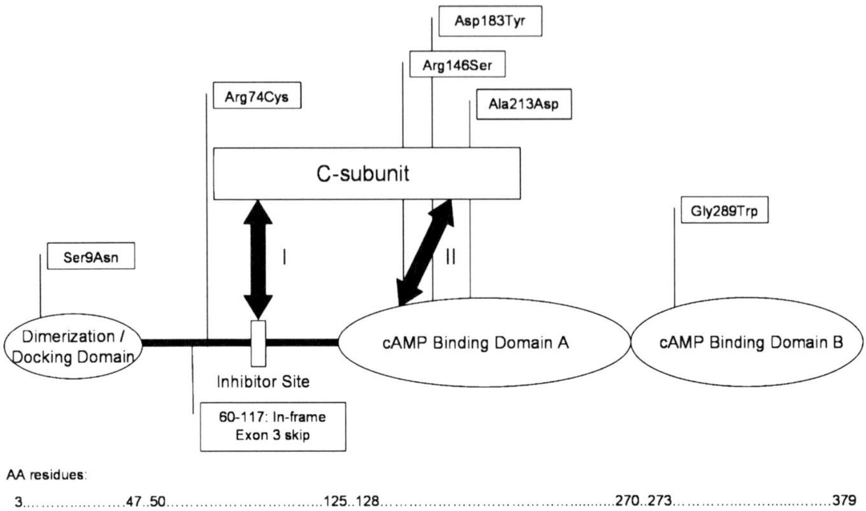

Fig. 7.3 Functional domains of *PRKAR1A* and expressed mutations: *PRKAR1A* is composed of three structural domains: a dimerization/docking domain at the N-terminal, two tandem cAMP-binding domains at the C-terminal, and a linker region in between. The catalytic subunit binds to the R-subunit at two locations: the primary site (I) at the inhibitor sequence in the linker region and the secondary site (II) within cAMP binding domain A. The mutations that are expressed at the protein level are spread throughout the entire functional domain of *PRKAR1A* and include six missense substitutions and one on-frame deletion that eliminates exon 3 (Greene et al. 2008)

cytoplasmic localization of R1α protein in wild type RIα and mutant RIα-transfected cells but the former also exhibited discrete aggregates of RIα that bound catalytic subunit; these were absent in mutant RIα-transfected cells and did not bind catalytic subunit at baseline or in response to cyclic AMP. Other changes induced by the mutant RIα included decreased nuclear catalytic subunit levels; the authors concluded that mutated RIα led to increased PKA activity through decreased binding of the catalytic subunit and did not involve changes in other PKA subunits, suggesting that at least for this pathogenic mutation a switch to type II PKA activity was not necessary for increased kinase activity or tumorigenesis.

Large germline *PRKAR1A* deletions were also found in two patients with CNC, both sporadic cases (Horvath et al. 2008). In the first patient, the deletion is expected to lead to decreased *PRKAR1A* mRNA levels but no other effects on the protein; the molecular phenotype is predicted to be *PRKAR1A* haploinsufficiency, consistent with the majority of *PRKAR1A* mutations causing CNC. In the second patient, the deletion led to an in-frame elimination of exon 3 and the expression of a shorter protein, lacking the primary site for interaction with the catalytic subunit. In vitro transfection studies of the mutant *PRKAR1A* showed impaired ability to bind cyclic AMP and PKA activation. The patient bearing this mutation had a more-severe-than-average CNC phenotype that included the relatively rare psammomatous melanotic schwannoma. Thus, large *PRKAR1A* deletions may be responsible for

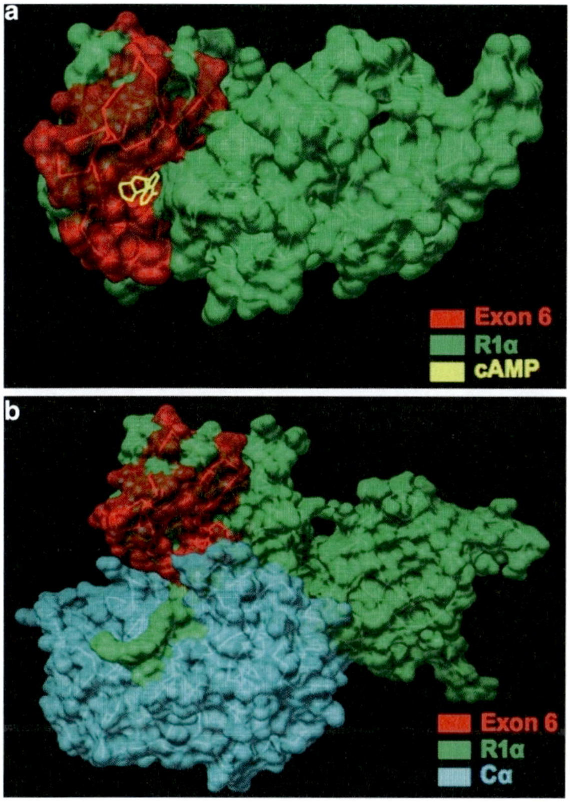

Fig. 7.4 The published (see Groussin et al. 2002) *PRKAR1A* deletion (R1Δ6) highlighted within an available model of RIα when bound to cAMP and in complex with Cα. R1Δ6 lacks the most conserved region of RIα, one that is essential for both cAMP binding and its interaction with Cα (Meoli et al. 2008)

CNC in patients that do not have *PRKAR1A* gene defects identifiable by sequencing. Preliminary data indicate that these patients may have a different phenotype especially if their defect results in an expressed, abnormal version of the *PRKAR1A* protein (Horvath et al. 2008).

The study of the mechanism of *PRKAR1A* involvement in CNC revealed that interactions of PKA with other pathways such as mitogen-activated protein kinase (MAPK), protein kinase B, protein kinase C were different from those of normal, control cells (Robinson-White et al. 2003, 2006a, b; Robinson-White and Stratakis 2002). For example, lymphocytes from CNC patients with known *PRKAR1A* mutations showed altered PKA activity and increased MAPK extracellular signal-regulated kinase phosphorylation (Robinson-White et al. 2003). Cell metabolism and proliferation studies suggested that altered PKA activity is associated with reversal of PKA-mediated inhibition of the MAPK pathway resulting in increased cell proliferation (Robinson-White et al. 2003).

Recently, we reported the association of RIα with mTOR kinase within autophagosomes (Mavrakis et al. 2006, 2007). The number of autophagosomes was significantly reduced in mouse embryonic fibroblasts that were devoid of any RIα protein. In biochemical assays, RIα was found to interact with mTOR and affect significantly its kinase activity: down-regulation of RIα in HEK293 cells increased phosphorylation of mTOR and mTOR kinase activity. Accordingly, PPNAD tissues with *PRKAR1A* mutations showed increased mTOR phosphorylation and activity. These data suggested that indeed RIα and mTOR are involved in a common pathway regulating mammalian autophagy (Mavrakis et al. 2007).

Taken together our data indicate that RIα may also have PKA-independent effects and that mutations of *PRKAR1A* gene may affect not only the balance between PKA isozymes, but may also modulate the activity of other signaling pathways.

7.7 PKA: Target for Anticancer Drugs

7.7.1 Site-Selective cAMP Analogs

In earlier studies it was shown that growth and morphology of a number of cultured cell lines can be influenced by cAMP or its derivatives (Cho-Chung 1979; Pastan, et al. 1975; Sheppard 1974). Analogs of cAMP that can activate only one specific PKA isozyme and change the balance between isoforms were synthesized and were tested in these and other experiments. cAMP binds to the regulatory subunits of PKA, at two different binding sites, termed site A (site I; or rapid site) and site B (Site II; or slow site), arranged in tandem (Døskeland 1978; Øgreid and Døskeland 1980; Øgreid et al. 1983). These binding sites have been identified by differences in their rate of cyclic nucleotide dissociation and their specificity for cyclic nucleotide binding. Unlike parental cAMP, site-selective cAMP analogs demonstrate selective binding for either one of the two known cAMP binding sites in the R subunit, resulting in preferential binding and activation of either protein kinase isozymes (Robinson-Steiner and Corbin 1983; Øgreid et al. 1985).

Numerous cAMP analogs were synthesized and screened for their therapeutic potential (Schwede et al. 2000). Of these site-selective cAMP analogs, very potent in growth inhibition was 8-Cl-cAMP (Cho-Chung et al. 1991; Tortora et al. 1988). Other cAMP analogs, such as 8-pCPTcAMP, 8-thiomethyl-cAMP, 8-thioisopropyl-cAMP, *N*6-benzyl-cAMP, *N*6-carbamoylphenyl-cAMP, or the doubly modified analog *N*6-phenyl-8-pCPT-cAMP, exhibited similar potencies in the micromolar range, but further studies were conducted mainly with 8-Cl-cAMP, most readily available by synthesis.

8-Cl-cAMP binds with high affinity to both sites A and B of RI. In contrast, it binds with high affinity to site B of RII, but with low affinity to site A, which may keep this isozyme in its nonactivated holoenzyme form (Ally et al. 1988).

8-Cl-cAMP was found to modulate RI and RII levels, leading to the restoration of a more natural RI/RII balance in cancer cells (Noguchi et al. 1998; Rohlff et al. 1993).

Site-selective cAMP analogs demonstrate a major regulatory effect on growth in a broad spectrum of human cancer cell lines, including breast, colon, lung and gastric carcinomas, fibrosarcomas, gliomas, and leukemias, as well as athymic mice with human cancer xenografts of various cell types, including breast, colon, and lung carcinomas (Ally et al. 1988; Cho-Chung 1989; Handschin et al. 1983). The effect of the analogs on growth inhibition appeared to be selective toward transformed cancer cells as opposed to nontransformed cells (Cho-Chung 1989).

Several studies reported apoptotic cell death related to 8-Cl-cAMP application to cancer cells (Hoffmann et al. 1996; Krett et al. 1997; Vintermyr et al. 1995). Recently it was found, that activation of PKA by 8-Cl-cAMP selectively causes rapid apoptosis in v-*abl* transformed fibroblasts by inhibiting the Raf-1 kinase (Weissinger et al. 2004). The authors investigated whether 8-Cl-cAMP is useful for the treatment of chronic myelogenous leukemia (CML), which is marked by the expression of the $p210^{bcr/abl}$ oncogene. They studied the effects of 8-Cl-cAMP on primary leukemia cells and bone marrow cells (BMCs) from eight patients with CML. Ex vivo treatment of BMCs (obtained in chronic CML) with 8-Cl-cAMP for 24–48 h led to the selective purging of Philadelphia Chromosome (Ph I chromosome), without toxic side effects on BMCs of healthy donors as measured by colony-forming unit (CFU) assays. These data indicated that 8-Cl-cAMP could be useful to purge malignant cells from the bone marrow of patients with CML and certain other forms of leukemia (Weissinger et al. 2004).

8-Cl-cAMP has also been tested in combination with paclitaxel, one of the first-line drugs against recurrent ovarian carcinomas (Einzig et al. 1992), in human cancer cells, including ovary, head, colon, lung carcinomas, and melanomas (Di Isernia et al. 1996; McDaid and Johnston 1999; Tortora et al. 1997a, b, c), and in vivo against human colon cancer xenografts in mice (Tortora et al. 1997a, b, c). The effect of both drugs was either additive (Di Isernia et al. 1996) or, in most cases, synergistic (McDaid and Johnston 1999; Tortora et al. 1997a, b, c). The synergism was highest in protocols with paclitaxel application prior to 8-Cl-cAMP. Cell-cycle analysis of several cell lines demonstrated accumulation of cells in the G2/M phase and a highly increased level of apoptotic cells. Beside paclitaxel, studies with docetaxel, cisplatin, carboplatin, and different retinoic acid derivatives have been conducted to test the effectiveness of 8-Cl-cAMP in combination with current standard chemotherapeutics (Tortora et al. 1997a, b, c; Srivastava et al. 1999). 8-Cl-cAMP appears to inhibit cancer cell growth through both an antiproliferation mechanism and a proapototic mechanism (Kim et al. 2000). Further studies are required to refine the mechanism of action of 8-Cl-cAMP action. Although it is still not completely clear whether 8-Cl-cAMP acts trough a PKA-activating mechanism or only as a pro-drug for its effective metabolite, 8-Cl-adenosine (Schwede et al. 2000), these studies showed that 8-Cl-cAMP could be useful in some cancers (Tortora and Ciardiello 2002).

7.7.2 Therapeutic Oligonucleotides

7.7.2.1 Antisense RIα

In some cancer cells, RIα is a positive effector of cancer cell growth (Cho-Chung 1990). Down-regulation of RIα by 21-mer antisense oligonucleotide directed to codons 1–7 of human RIα led to growth arrest and differentiation in leukemia cells and inhibition of growth in human cancer cells of epithelial origin, including breast, colon, and gastric carcinoma and neuroblastoma cells (Tortora et al. 1991; Yokozaki et al. 1993). Sequence-specific decreases of RIα inhibited in vivo tumor growth (Nestereova and Cho-Chung 1995) and subcutaneous injections into nude mice bearing LS-174T human colon carcinoma cells with RIα antisense phosphorothiate oligonucleotide (directed to 8–13 codons of human RIα) resulted in an almost complete suppression of tumor growth for 7 days without an apparent sign of systemic toxicity.

The "second generation" of structurally modified oligonucleotides has improved their effectiveness through a wide variety of sugar modifications. The most important modifications involve the 2′ position: 2′O-methyl, 2′O-methoxy-ethyl, 2′O-alkyl or other groups. These analogs generally have increased affinity for RNA, and are more resistant to nucleases. Nevertheless, these oligonucleotides do not support RNase H activity, and for this reason their antisense effect is limited to a physical block of translation. Other representatives of "second generation" oligonucleotides have modified phosphate linkages or ribosyl moieties, as well as oligonucleotides with an altered backbone (Agrawal et al. 1997). One of the most interesting examples of such second generation ODN is the RNA-DNA mixed backbone ODN of RIα antisense, HYB165 (GEM 231) (Chen et al. 2000). The polyanionic nature of the antisense RIα PS-ODN is minimized, and the immunostimulatory effect (GCGT motif) is blocked in RNA-DNA mixed-backbone RIα antisense ODN (Agrawal et al. 1997). Such second-generation ODNs have been shown to improve antisense activity, be more resistant to nucleases, form more stable duplexes with RNA and retain the capability to induce RNase H (Metelev et al. 1994; Monia et al. 1993). Studies conducted both in vitro and in animal models have demonstrated that, following treatment with GEM231, down-regulation of PKA-I is balanced by a rapid compensatory increase of PKA-IIβ isoform and is associated with early inhibition of expression of growth factors and their receptors (TGFα, EGFR and erbB-2), oncogenes (myc and ras), and angiogenic factors (VEGF and basic fibroblast growth factor, bFGF), induction of apoptosis and, finally, led to growth arrest (Ciardiello et al. 1993; Cho et al. 2001).

Several studies have shown that antisense RIα is able to work with a variety of anticancer drugs of different classes, following intraperitoneal as well as oral administration of the antisense. In particular, synergistic antitumor activity associated with increased apoptosis can be obtained with taxanes, topoisomerase I and II inhibitors, and platinum derivatives, both in vitro and in nude mice bearing a wide variety of human cancer types (Tortora et al. 1997a, b, c, 2000, 2001; Wang et al.

2002). The biochemical and molecular basis of the cooperative effect that was observed probably include the sensitization of cells to certain anticancer agents following down-regulation of PKA-I, the pharmacokinetic interactions of antisense moleculaes with certain drugs, and finally, the involvement of PKA-I in signaling pathways affected by cytotoxic drugs. Phase I studies with antisense RIα have now been done (Chen et al. 2000).

7.7.2.2 CRE-Transcription Factor Decoy

The cAMP response element (CRE) consensus sequence is intimately involved in the transcription of a wide range of genes (Mayr and Montminy 2001). All of the cAMP responsive gene promoter regions have the same eight-base enhancer CRE sequence, which is the imperfect palindrome 5′-TGACGTCA-3′ (Conkright et al. 2003). The main protein that binds to the CREs is the 43 kDa, basic leucine zipper DNA-binding motif-containing, CRE-binding protein (CREB), (Shaywitz and Greenberg 1999). CREB coordinates a multitude of genes that regulate numerous cellular processes, including cell growth and differentiation (Droogmans et al. 1992; Moriuchi et al. 2003; Russell et al. 2003). The ubiquitous nature of the CRE consensus site makes it a good target for chemotherapy. Synthetic double-stranded phosphorothioate oligonucleotoides with high affinity for a target transcription factor can be introduced into cells as decoy cis-elements to bind to these factors and alter gene expression. Because the CRE cis-element is palindromic, a synthetic single-stranded oligonucleotide composed of the CRE sequence self-hybrydizes to form a duplex/hairpin. It has been shown that a palindromic trioctamer of this sequence can interfere with CREB binding, and specifically inhibits PKA-subunit expression, interfering with the CRE-PKA pathway (Park et al. 1999, 2001; Cho et al. 2002). This oligonucleotide restrained tumor cell proliferation, without affecting the growth of noncancerous cells. Furthermore, CRE-decoy oligonucleotides inhibited tumor growth in nude mice without obvious toxicity (Park et al. 1999) and in a panel of three colorectal cancer cell lines in combination with etoposide and 5-fluorouracil inhibited cell proliferation (Liu et al. 2004). These studies showed that combining CRE-decoy oligo with chemotherapy resulted in an enhancement of antiproliferative effects without increasing toxicity.

7.7.3 Strategy for Targeting Protein-Protein Interactions

Signaling pathways represent complex systems of scaffolding, anchoring and adaptor proteins that mediate assembly of multi-protein complexes containing receptors, second messenger-generating enzymes, kinases, phosphatases, and substrates (Beene and Scott 2007). Modulation of proteins included in such complexes can lead to changes in the activity of a whole system. For example, in excitatory synapses glutamate receptors are linked to signaling proteins, the actin cytoskeleton, and synaptic adhesion molecules on dendritic spines through a network of scaffolding proteins that may play important roles regulating synaptic structure and receptor functions in synaptic plasticity underlying learning and memory (Dell'Acqua et al. 2006). Translocation of the AKAP79/150–PKA complex may shift the balance of PKA and phosphatase activity at the postsynapse in favor of phosphatases. This loss of PKA could then promote protein dephosphorylation during induction of long-term potentiation including maintaining glutamate receptor dephosphorylation, endocytosis, and preventing recycling. These findings challenge the concept that AKAPs are static anchors that bind signaling proteins near fixed targets; they suggest that AKAPs can function in more dynamic manners to regulate local signaling events.

A similar example is provided by a recent study of AKAPs in heart failure (Lygren and Tasken 2008): it is known that the adrenaline-mediated β-adrenoreceptor–cAMP–PKA signaling regulates heart rate and contractility. Although changes in contractility are associated with cardiovascular disease, surprisingly few drugs are available that modulate the cardiac myocyte cAMP system. β-blocking agents reduce cAMP levels only by 50%. It is important to develop such compounds that interfere with the pathway at other levels and make therapy of heart failure more potent, targeted and reduce side effects. The A-kinase anchoring protein (AKAP18δ) is important for organizing the molecular machinery that mediates adrenergic control of calcium re-absorption into sarcoplasmic reticulum and it is essential for relaxation and heart filling; this is the rate-limiting step for making the heart beat faster in response to adrenaline or noradrenaline. The authors concluded that targeting AKAP18δ may be possible in manipulating calcium re-absorption and protecting the heart from adrenergic pacing.

There are growing indications that many more proteins, besides AKAPs and mTOR (Mavrakis et al. 2006), can interact with both PKA R and C subunits: the second subunit of Replication Factor C (RFC) complex, RFC40 (RFC2) (Gupte et al. 2005); cytochrome c oxidase subunit Vb, inhibiting cytochrome oxidase activity (Yang et al. 1998); PAP7, a peripheral-type benzodiazepine receptor that is involved in cholesterol transport and so facilitates human chorionic gonadotropin-stimulated steroidogenesis (Liu et al. 2006); and the ribosomal S6 kinase (RSK1) (Chaturverdi et al. 2006).

Finally, a group of investigators proposed that functional interactions of EGFR and PKA-I (Fig. 7.1) may provide the rational basis to develop a therapeutic strategy based on the combination of their selective inhibitors (Ciardiello and Tortora 1998; Tortora et al. 1997a, b, c). The combination of MabC225 (chimeric monoclonal antibody against EGFR – erbitux) with either the site-selective cAMP analog 8-Cl-cAMP or, later, with GEM231, has been the first demonstration of the feasibility of such approaches (Tortora and Ciardiello 2002). We can expect in the future, an increasing number of drug targeting strategies aimed at protein-protein interactions in disease.

7.8 Conclusion

The cAMP-mediated signaling pathway is one of the most important regulatory systems of all cells. Its specificity is highly and very effectively regulated by a multitude of ways, from tissue-specific expression of partner molecules to compartmentalization and cell-stage specific interactions. The dual and often contradictory effects of cAMP on growth, proliferation and differentiation are mediated by cAMP levels, changes in the amount or activity of PKA isozymes, and the expression of the partnering molecules and interacting pathways. PKA has emerged as a major therapeutic target. Therapeutic use of specific PKA modulators has to take into account the complexity of the system; but the encouraging news is that these studies have started and some molecules have already completed phase I clinical trials. However, designing novel kinase-specific modulators is required and the use of new technologies, hopefully, will accelerate this process.

Acknowledgments This work was supported by NICHD, NIH intramural project Z01-HD-000642-04 to Dr. C.A. Stratakis.

References

Agrawal, S., Jiang, Z., Zhao Q, et al. (1997). Mixed-backbone oligonucleotides as second generation antisense oligonucleotides: *in vitro* and *in vivo* studies. *Proc. Natl. Acad. Sci. U.S.A.* 94:2620–2625.

Ally, S., Tortora, G., Clair, T., et al. (1988). Selective modulation of protein kinase isozymes by the site-selective analog 8-chloroadenosine 3′,5′-cyclic monophosphate provides a biological means for control of human colon cancer cell growth. *Proc. Natl. Acad. Sci. U.S.A.* 85:6319–6322.

Amieux, P.S., Cummings, D.E., Motamed, K., et al. (1997). Compensatory regulation of RI alpha protein levels in protein kinase A mutant mice. *J. Biol. Chem.* 272:3993–3998.

Beebe, S.J., and Corbin, J.D. (1986). Cyclic nucleotide-dependent protein kinases. In: Boyer, P.D., Krebs, E.G. (eds) The Enzymes: Control by Phosphorylation, Part A, vol 17. Academic Press, New York pp. 43–111.

Beebe, S.J., Øyen, O., Sandberg, M., et al. (1990). Molecular cloning of tissue-specific protein kinase (Cγ) from human testis representing a third isoform for the catalytic subunit of cAMP-dependent protein kinase. *Mol. Endocrinol.* 4:465–475.

Beebe, S.J., Salomonsky, T., Jahnsen, T., et al. (1992). The C gamma subunit is a unique isozyme of the cAMP-dependent protein kinase. *J. Biol. Chem.* 267:25505–25512.

Beene, D.L., and Scott, J.D. (2007). A-kinase anchoring proteins take shape. *Curr. Opin. Cell Biol.* 19:192–198.

Blaschke, R.J., Monaghan, P., Bock, D., et al. (2000). A novel murine PKA-related protein kinase involved in neuronal differentiation. *Genomics* 64:187–194.

Brandon, E.P., Logue, S.F., Adams, M.R., et al. (1998). Defective motor behavior and neural gene expression in RIIbeta-protein kinase A mutant mice. *J. Neurosci.* 18:3639–3649.

Bryn, T., Mahic, M., Aandahl, E.M., et al. (2008). Inhibition of protein kinase A improves effector function of monocytes from HIV-infected patients. *AIDS Res. Hum. Retrov.* 24:1013–1015.

Burton, K.A., Johnson, B.D., Hausken, Z.E., et al. (1997). Type II regulatory subunits are not required for the anchoring-dependent modulation of $Ca2^{+}$ channel activity by cAMP-dependent protein kinase. *Proc. Natl. Acad. Sci. U.S.A.* 94:11067–11072.

Burton, K.A., Treash-Osio, B., Muller, C.H., et al. (1999). Deletion of type II alpha regulatory subunit delocalizes protein kinase A in mouse sperm without affecting motility or fertilization. *J. Biol. Chem.* 274:24131–24136.

Carlson, C.R., Witczak, O., Vossebein, L., et al. (2001). CDK1-mediated phosphorylation of the RIIα regulatory subunit of PKA works as a molecular switch that promotes dissociation of RIIα from centrosomes at mitosis. *J. Cell Sci.* 114:3243–3254.

Chaturverdi, D., Poppleton, H.M., Stringfield, T., et al. (2006). Subcellular localization and biological actions of activated RSK1 are determined by its interactions with subunits of cyclic AMP-dependent protein kinase. *Mol. Cell. Biol.* 26:4586–4600.

Cheadle, C., Nesterova, M., Watkins, T., et al. (2008). Regulatory subunits of PKA define an axis of cellular proliferation/differentiation in ovarian cancer cells. *BMC Med. Genomics* 1:1–14.

Cheley, S., and Bayley, H. (1991). Kinetics and regulation of two catalytic subunits of cAMP-dependent protein kinase from *Aplysia californica*. *Biochemistry* 30:10246–10255.

Chen, H.X., Marshall, J.L., Ness, E., et al. (2000). A safety and pharmacokinetic study of a mixed-backbone oligonucleotide (GEM 231) targeting the type I protein kinase A by 2-hour infusions in patients with refractory solid tumors. *Clin. Cancer Res.* 6:1259–1266.

Cheng, X., Phelps, C., Taylor, S. (2001). Differential binding of cAMP-dependent protein kinase regulatory subunit isoforms Iα and IIβ to the catalytic subunit. *J. Biol. Chem.* 276:4102–4108

Cho, Y.S., Kim, M.K., Cheadle, C., et al. (2001). Antisense DNAs as multisite genomic modulators identified by DNA microarray. *Proc. Natl. Acad. Sci. U.S.A.* 98:9819–9823.

Cho, Y.S., Kim, M.K., Cheadle, C., et al. (2002). A genomic-scale view of the cAMP response element-enhancer decoy: a tumor target-based genetic tool. *Proc. Natl. Acad. Sci. U.S.A.* 99:15626–15631.

Cho-Chung, Y.S. (1979). On the interaction of cyclic AMP-binding protein and estrogen receptor in growth control. *Life Sci.* 24:1231–1240.

Cho-Chung, Y.S. (1989). Site-selective 8-chloro-cyclic adenosine 3′, 5′-monophosphate as a biologic modulator of cancer: restoration of normal control mechanisms. *J. Natl. Cancer Inst.* 81:982–987.

Cho-Chung, Y.S. (1990). Role of cyclic AMP receptor proteins in growth, differentiation, and suppression of malignancy new approaches to therapy. *Cancer Res.* 50:7093–7100.

Cho-Chung, Y.S., Clair, T., Tortora, G., et al. (1991). Role of site-selective cAMP analogs in the control and reversal of malignancy. *Pharmacol. Ther.* 50:1–33.

Ciardiello, F. and Tortora, G. (1998). Interactions between the epidermal growth factor receptor and type I protein kinase A: biological significance and therapeutic implications. *Clin. Cancer Res.* 4:821–828.

Ciardiello, F., Pepe, S., Bianco. C., et al. (1993). Down-regulation of RIα subunit of cAMP-dependent protein kinase induces growth inhibition of human mammary epithelial cells transformed by c-Ha-ras and c-erbB-2 proto-oncogenes. *Int. J. Cancer* 53:438–443.

Clegg, C.H., Cadd, G.G., McKnight, G.S. (1988). Genetic characterization of a brain-specific form of the type I regulatory subunit of cAMP-dependent protein kinase. *Proc. Natl. Acad. Sci. U.S.A.* 85:3703–3707.

Colledge, M., and Scott, J.D. (1999). Akaps: from structure to function. *Trends Cell Biol.* 9:216–221.

Conkright, M.D., Guzmán, E., Flechner, L., et al. (2003). Genome-wide analysis of CREB target genes reveals a core promoter requirement for cAMP responsiveness. *Mol. Cell* 11:1101–1108.

Constantinescu, A., Wu, M., Asher, O., et al. (2004). cAMP-dependent protein kinase type I regulates ethanol-induced cAMP response element-mediated gene expression via activation of CREB-binding protein and inhibition of MAPK. *J. Biol. Chem.* 279:43321–43329.

Corbin, .J.D, Keely, S.L., Park, C.R. (1975). The distribution and dissociation of cyclic adenosine 3′, 5′-monophosphate-dependent protein kinases in adipose, cardiac and other tissues. *J. Biol. Chem.* 250:218–225.

Cummings, D.E., Brandon, E.P., Planas, J.V., et al. (1996). Genetically lean mice result from targeted disruption of the RII beta subunit of protein kinase A. *Nature* 382:622–626.

Dell'Acqua, M.L., Smith, K.E., Gorski, J., et al. (2006). Regulation of neuronal PKA signaling through AKAP targeting dynamics. *Eur. J. Cell Biol.* 85:627–633.

Di Isernia, G., Ciardiello, F., Sandomenico, C., et al. (1996). 8-Chloro-cAMP enhances the growth-inhibitory effect of cytotoxic drugs in human colon cancer cells. *Int. J. Oncol.* 9:1233–1237.

Døskeland, S.O. (1978). Evidence that rabbit muscle protein kinase has two kinetically distinct binding sites for adenosine 3′, 5′-cyclic monophosphate. *Biochem. Biophys. Res. Commun.* 83:542–549.

Dostman, W.R., Taylor, S.S., Genieser, H.G., et al. (1990). Probing the cyclic nucleotide binding sites of cAMP-dependent protein kinase I and II with analogs of adenosine 3′, 5′-cyclicphosphorothioates. *J. Biol. Chem.* 265:10484–10491.

Droogmans, L., Cludts, I., Cleuter, Y., et al. (1992). Nucleotide sequence of the bovine interleukine-6 gene promoter. *DNA Seq.* 3:115–117.

Einzig, A.I., Wiernik, P.H., Sasloff, J. (1992). Phase II study and long term follow-up of patients treated with taxol for advanced ovarian adenocarcinoma. *J. Clin. Oncol.* 10:1748–1753.

Feliciello, A., Giuliano, P., Porcellini, A., et al. (1996). The v-Ki-Ras oncogene alters cAMP nuclear signaling by regulating the location and the expression of cAMP-dependent protein kinase IIbeta. *J. Biol. Chem.* 271:25350–25359.

Fossberg, T.M., Doskeland, S.O., Ueland, P.M. (1978). Protein kinases in human renal cell carcinoma and renal cortex. A comparison of isozyme distribution and of responsiveness to adenosine 3′, 5′-cyclic monophosphate. *Arch. Biochem. Biophys.* 189:372–381.

Gamm, D.M., Baude, E.J., Uhler, M.D. (1996). The major catalytic subunit isoforms of cAMP-dependent protein kinase have distinct biochemical properties *in vitro* and *in vivo*. *J. Biol. Chem.* 271:15736–15742.

Gonzalez, G.A., and Montminy, M.R. (1989). Cyclic AMP stimulates somatostatin gene transcription by phosphorylation of CREB at serine 133. *Cell* 59:675–680.

Greene, E.L., Horvath, A.D., Nesterova, M., et al. (2008). In vitro functional studies of naturally occurring pathogenic PRKAR1A mutations that are not subject to nonsense mRNA decay. *Hum. Mut.* 29:633–639.

Griffin, K.J., Kirschner, L.S., Matyakhina, L., et al. (2004a). Down-regulation of regulatory subunit type 1A of protein kinase A leads to endocrine and other tumors. *Cancer Res.* 64:8811–8815.

Griffin, K.J., Kirschner, L.S., Matyakhina, L., et al. (2004b). A transgenic mouse bearing an antisense construct of regulatory subunit type 1A of protein kinase A develops endocrine and other tumours: comparison with Carney complex and other PRKAR1A induced lesions. *J. Med. Genet.* 41:923–931.

Groussin, L., Kirschner, L.S., Vincent-Dejean, C., et al. (2002). Molecular analysis of the cyclic AMP-dependent protein kinase A (PKA) regulatory subunit 1A (PRKAR1A) gene in patients with Carney complex and primary pigmented nodular adrenocortical disease (PPNAD) reveals novel mutations and clues for pathophysiology: augmented PKA signaling is associated with adrenal tumorigenesis in PPNAD. *Am. J. Hum. Genet.* 71:1433–1442.

Gupte, R.S., Weng, Y., Liu, L., et al. (2005). The second subunit of the replication factor C complex (RFC40) and the regulatory subunit (RIalpha) of protein kinase A form a protein complex promoting cell survival. *Cell Cycle* 4:323–329.

Handschin, J.C., Handloser, K., Takahashi, A., et al. (1983). Cyclic adenosine 3′, 5′-monophosphate receptor proteins in dysplastic and neoplastic human breast tissue cytosol and their inverse relationship with estrogen receptors. *Cancer Res.* 43:2947–2954.

Hanks, S.K., Qinn, A.M., Hunter, T. (1988). The protein kinase family: conserved features and deduced phylogeny of the catalytic domains. *Science* 241:42–52.

Herberg, F.W. and Taylor, S.S. (1993). Physiological inhibitors of the catalytic subunit of cAMP-dependent protein kinase: effect of MgATP on protein-protein interactions. *Biochemistry* 32:14015–14022.

Hoffmann, C., Raffel, S., Ruchaud, S., et al. (1996). Chloro-substituted cAMP analogues and their adenosine metabolites induce apoptosis of the human promyelocytic leukaemia cell line NB4: molecular basis for cell type selectivity. *Cell. Pharmacol.* 3:417–427.

Hofman, F., Bechtel, P.J., Krebs, E.G. (1977). Concentrations of cyclic AMP-dependent protein kinase subunits in various tissues. *J. Biol. Chem.* 252:1441–1447.

Horvath, A., Bossis, I., Giatzakis, C., et al. (2008). Large deletions of the PRKAR1A gene in Carney complex. *Clin. Cancer Res.* 14:388–395.

Huggenvik, J.I., Collard, M.W., Stofko, R.E., et al. (1991). Regulation of the human enkephalin promoter by two isoforms of the catalytic subunit of cyclic adenosine 3′, 5′-monophosphate-dependent protein kinase. *Mol. Endocrinol.* 5:921–930.

Imazumi-Scherrer, T., Faust, D.M., Barradeau, S. et al. (2001). Type I protein kinase A is localized to interphase microtubules and strongly associated with the mitotic spindle. *Exp. Cell Res.* 264:250–265.

Jahnsen, T., Hedin, L., Kidd, V.J., et al. (1986). Molecular cloning, cDNA structure and regulation of the regulatory subunit of type II cAMP-dependent protein kinase from rat ovarian granulose cells. *J. Biol. Chem.* 261:12352–12361.

Jarnaess, E., Ruppelt, A., Stokka, A.J., et al. (2008). Dual specificity A-kinase anchoring proteins (AKAPs) contain an additional binding region that enhances targeting of protein kinase A type I. *J. Biol. Chem.* 283:33708–33718.

Jones, G.N., Tep, C., Towns, W.H. 2nd, et al. (2008). Tissue-specific ablation of Prkar1a causes schwannomas by suppressing neurofibromatosis protein production. *Neoplasia* 10:1213–1221.

Khan, I.U., Laxminarayana, D., Kammer, G.M. (2001). Protein kinase A RI beta subunit deficiency in lupus T lymphocytes: bypassing a block in RI beta translation reconstitutes protein kinase A activity and augments IL-2 production. *J. Immunol.* 166:600–7605.

Kim, S.N., Kim, S.G., Park, J.H., et al. (2000). Dual anticancer activity of 8-Cl-cAMP: inhibition of cell proliferation and induction of apoptotic cell death (2000). *Biochem. Biophys. Res. Commun.* 73:404–410.

Kirschner, L.S., Carney, J.A., Pack, S.D. et al. (2000). Mutations of the gene encoding the protein kinase A type I-alpha regulatory subunit in patients with the Carney complex. *Nat. Genet.* 26:89–92.

Kirschner, L.S., Kusewitt, D.F., Matyakhina, L., et al. (2005). A mouse model for the Carney complex tumor syndrome develops neoplasia in cyclic AMP-responsive tissues. *Cancer Res.* 65:4506–4514.

Klink, A., Schiebel, K., Winkelmann, M., et al. (1995). The human protein kinase gene PKX1 on Xp22.3 displays Xp/Yp homology and is a site of chromosomal instability. *Hum. Mol. Genet.* 4:869–878.

Kondrashin, A.A., Nesterova, M.V., Cho-Chung, Y.S. (1998). Subcellular distribution of the R-subunits of cAMP-dependent protein kinase in LS-174T human colon carcinoma cells. *Biochem. Mol. Biol. Intern.* 45:237–244.

Krett, N.L., Zell, J.L., Halgren, R.G., et al. (1997). Cyclic adenosine-3′, 5′-monophosphate-mediated cytotoxicity in steroid sensitive and resistant myeloma. *Clin. Cancer Res.* 3:1781–1787.

Kuo, G.F., and Greengard, P. (1969). Cyclic nucleotide-dependent protein kinases. IV. Widespread occurrence of adenosine 3′, 5′-monophosphate-dependent protein kinase in various tissues and phyla of the animal kigdom. *Proc. Natl. Acad. Sci. U.S.A.* 64:1349–1355.

Lange-Carter, C.A., and Malkinson, A.M. (1991). Alterations in the cAMP signal transduction pathway in mouse lung tumoriginesis. *Exp. Lung Res.* 17:341–357.

Lania, A.G., Montovani, G., Ferrero, S., et al. (2004). Proliferation of transformed somatotroph cells related tp low or absent expression of protein kinase a regulatory subunit 1A protein. *Cancer Res.* 64:9193–9198.

Laxminarayana, D., and Kammer G.M. (2000). mRNA mutations of type I protein kinase A regulatory subunit alpha in T lymphocytes of a subject with systemic lupus erythematosus. *Int. Immunol.* 12:1521–1529.

Lee, D.C., Carmichael, D.F., Krebs, E.G., et al. (1983). Isolation of a cDNA clone for the type I regulatory subunit of bovine cAMP-dependent protein kinase. *Proc. Natl. Acad. Sci. U.S.A.* 80:3608–3612.

Levy, F.O., Øyen, O., Sandberg, M., et al. (1988). Molecular cloning, complementary deoxyribonucleic acid structure and predicted full-length amino acid sequence of the hormone-inducible regulatory subunit of 3′, 5′-cyclic adenosine monophosphate-dependent protein kinase from human testis. *Mol. Endocrinol.* 2:1364–1373.

Liu, W.M., Scott, K.A., Shahin, S., et al. (2004). The *in vitro* effects of CRE-decoy oligonucleotides in combination with conventional chemotherapy in colorectal cancer cell lines. *Eur. J. Biochem.* 271:2773–2781.

Liu, J., Rone, M.B., Papadopoulos, V. (2006). Protein-protein interactions mediate mitochondrial cholesterol transport and steroid biosynthesis. *J. Biol. Chem.* 281:38879–38893.

Lygren, B., and Tasken, K. (2008). The potential use of AKAP18δ as a drug target in heart failure patients. *Expert Opin. Biol. Ther.* 8:1099–1108.

Mavrakis, M., Lippincott-Schwartz, J., Stratakis, C.A., et al. (2006). Depletion of type IA regulatory subunit (RIα) of protein kinase A (PKA) in mammalian cells and tissues activates mTOR and causes autophagic deficiency. *Hum. Mol. Genet.* 15:2962–2971.

Mavrakis, M., Lippincott-Schwartz, J., Stratakis, C.A., et al. (2007). mTOR kinase and the regulatory subunit of protein kinase A (PRKAR1A) spatially and functionally interact during autophagosome maturation. *Autophagy* 3:151–153.

Mayr, B., and Montminy, M. (2001). Transcriptional regulation by the phosphorylation-dependent factor CREB. *Nat. Rev. Mol. Cell. Biol.* 2:599–609.

McDaid, H.M., and Johnston, P.G. (1999). Synergistic interaction between paclitaxel and 8-chloro-adenosine 3′, 5′-monophosphate in human ovarian carcinoma cell lines. *Clin. Cancer Res.* 5:215–220.

McKnight, G.S., Idzerda, R.L., Kandel, E.R., et al. (1996). Targeted disruption of the protein kinase A system in mice. In: Hansson, V., Levy, F.O., Tasken, K. (eds) Signal transduction in testicular cells. Basic and Clinical Aspects, vol 1. Springer-Verlag, Berlin, pp. 95–122.

Mellon, P.L., Clegg, C.H., Correll, L.A. et al. (1989). Regulation of transcription by cyclic AMP-dependent protein kinase. *Proc. Natl. Acad. Sci. U.S.A.* 86:4887–4891.

Meoli, E., Bossis, I., Cazabat, L. et al. (2008). Protein kinase A. Effects of an expressed *PRKAR1A* mutation associated with aggressive tumors. *Cancer Res.* 68:3133–3141.

Metelev, V., Liszlewicz, J., Agrawal, S. (1994). Study of antisense oligonucleotide phosphorothioates containing segments of oligodeoxynucleotides and 2′ *O*-methyloligoribonucleotides. *Bioorg. Med. Chem. Lett.* 4:2929–2934.

Miller, W.R. (2002). Regulatory subunits of PKA and breast cancer. *Ann. NY Acad. Sci.* 968:37–48.

Monia, B.P., Lesmik, E.A., Gonzalez, C., et al. (1993). Evaluation of 2′-modified oligonucleotides containing 2′-deoxygaps as antisense inhibitors of gene expression. *J. Biol. Chem.* 268:14514–14522.

Moriuchi, A., Ido, A., Nagata, Y. et al. (2003). A CRE and the region occupied by a protein induced by growth factor contribute to up-regulation of cyclin D1 expression in hepatocytes. *Biochem. Biophys. Res. Commun.* 300:415–421.

Morris, R.C., Morris, G.Z., Zhang, W., et al. (2002). Differential transcriptional regulation by the alpha- and gamma-catalytic subunit isoforms of cAMP-dependent protein kinase. *Arch. Biochem. Biophys.* 403:219–228.

Nadella, K.S., and Kirschner, L.S. 2005 Disruption of protein kinase a regulation causes immortalization and dysregulation of D-type cyclins. *Cancer Res.* 65:10307–10315.

Nadella, K.S., Jones, G.N., Trimboli, A. et al. (2008). Targeted deletion of Prkar1a reveals a role for protein kinase A in mesenchymal-to-epithelial transition. *Cancer Res.* 68:2671–2677.

Neary, C.L., and Cho-Chung, Y.S. (2001). Nuclear translocation of the catalytic subunit of protein kinase A induced by an antisense oligonucleotide directed against the RI alpha regulatory subunit. *Oncogene* 20:8019–8024.

Neary, C.L., Nesterova, M., Cho, Y.S., et al. (2004). Protein kinase A isozyme switching: eliciting differential cAMP signaling and tumor reversion. *Oncogene* 23:8847–8856.

Nestereova, M. and Cho-Chung, Y.S. (1995). A single-injection protein kinase A-directed antisense treatment to inhibit tumour growth. *Nature Med.* 1:528–533.

Nesterova, M., Bossis, I., Wen, F., et al. (2008). An immortalized human cell line bearing a PRKAR1A-inactivating mutation: effects of overexpression of the wild-type allele and other protein kinase A subunits. *J. Clin. Endocrinol. Metab.* 93:565–571.

Noguchi, K., Murata, T., Cho-Chung, Y.S. (1998). 8-Chloradenosine 3′, 5′-monophosphate (8-Cl cAMP). Selectively eliminates protein kinase A type I to induce growth inhibition in c-ras.-transformed fi broblasts. *Eur. J. Cancer* 34:1260–1267.

Øgreid, D., and Døskeland, S.O. (1980). Protein kinase II has two distinct binding sites for cyclic AMP, only one which is detectable by the conventional membrane filtration method. *FEBS Lett.* 121:340–344.

Øgreid, D., Døskeland, S.O., Miller, J.P. (1983). Evidence that cyclic nucleotides activating rabbit muscle protein kinase I interact with both types of cAMP binding sites associated with the enzyme. *J. Biol. Chem.* 258:1041–1049.

Øgreid, D., Ekanger, R., Suva, R.H., et al. (1985). Activation of protein kinase izosymes by cyclic nucleotide analogs used singly or in combination. *Eur. J. Biochem.* 150:219–227.

Øyen, O., Myklebust, F., Scott, J.D., et al. (1989). Human testis cDNA for the regulatory subunit RIIα of cAMP-dependent protein kinase encodes an alternate amino-terminal region. *FEBS Lett.* 246:57–64.

Park, Y.G., Nesterova, M., Agrawal, S., et al. (1999). Dual blockade of cyclic AMP response element-(CRE) and AP-1-directed transcription by CRE-transcription gfactor decoy oligonucleotide. Gene-specific inhibition of tumor growth. *J. Biol. Chem.* 274:1573–1580.

Park, Y.G., Park, S., Lim, S.O., et al. (2001). Reduction in cyclin D1/Sdk4/retinoblastoma protein signaling by CRE-decoy oligonucleotide. *Biochem. Biophys. Res. Commun.* 281:1213–1219.

Pastan, I., Johnson, G.S., Anderson, W.B. (1975). Role of cyclic nucleotides in growth control. *Annu. Rev. Biochem.* 44:491–522.

Pavel, E., Nadella, K., Towns, W.H. 2nd, Kirschner, L.S. (2008). Mutation of Prkar1a causes osteoblast neoplasia driven by dysregulation of protein kinase A. *Mol. Endocrinol.* 22:430–440.

Planas, J.V., Cummings, D.E., Idzerda, R.L., et al. (1999). Mutation of the RII beta subunit of protein kinase A differentially affects lipolysis but not gene induction in white adipose tissue. *J. Biol. Chem.* 274:36281–36287.

Reinton, N., Haugen, T.B., Ørstavik, S., et al. (1998). The gene encoding the C gamma catalytic subunit of cAMP-dependent protein kinase is a transcribed retroposon. *Genomics* 49:290–297.

Riabowol, K.T., Fink, J.S., Oilman, M.Z., et al. (1988). The catalytic subunit of cAMP-dependent protein kinase induces expression of genes containing cAMP-responsive enhancer elements. *Nature* 336:83–86.

Robinson-Steiner, A.M., and Corbin, J. (1983). Probable involvement of both intrachain cAMP binding sites in activation of protein kinase. *J. Biol. Chem.* 259:1032–1040.

Robinson-White, A., and Stratakis, C.A. (2002). Protein kinase A signaling: "cross-talk" with other pathways in endocrine cells. *Ann. N Y Acad. Sci.* 968:256–270.

Robinson-White, A., Hundley, T.R., Shiferaw, M., et al. (2003). Protein kinase-A activity in PRKAR1A-mutant cells, and regulation of mitogen-activated protein kinases ERK1/2. *Hum. Mol. Genet.* 12:1475–1484.

Robinson-White, A.J., Leitner, W.W., Aleem, E., et al. (2006a). PRKAR1A inactivation leads to increased proliferation and decreased apoptosis in human B lymphocytes. *Cancer Res.* 66:10603–10612.

Robinson-White, A., Meoli, E., Stergiopoulos, S., et al. (2006b). PRKAR1A Mutations and protein kinase A interactions with other signaling pathways in the adrenal cortex. *J. Clin. Endocrinol. Metab.* 91 2380–2388.

Robison, G.A. (1973). The biological role of cyclic AMP: an updated review. In: Kahn, RH., and Lands, W.E.M. (eds) Prostaglandins and Cyclic AMP: Biological Actions and Clinical Applications, Academic Press, New York, pp. 229–247.

Rohlff, C., Clair, T., Cho-Chung, Y.S. (1993). 8-Cl-cAMP induces truncation and down-regulation of the RI α subunit and up-regulation of the RII β subunit of cAMP-dependent protein kinase leading to type II holoenzyme-dependent growth inhibition and differentiation of HL-60 leukemia cells. *J. Biol. Chem.* 268:5774–5782.

Rosen, O.M., and Erlichman, J. (1975). Reversible autophosphorylation of a cyclic 3′, 5′-AMP-dependent protein kinase from bovine cardiac muscle. *J. Biol. Chem.* 250:7788–7794.

Ruppelt, A., Mosenden, R., Gronholm, M. et al. (2007). Inhibition of T cell activation by cyclic adenosine 5 -monophosphate requires lipid raft targeting of protein kinase A type I by the A-kinase anchoring protein Ezrin1. *J. Immunol.* 179:5159–5168.

Russell, D.L., Doyle, K.M., Gonzales-Robayna, I. et al. (2003). Egr-1 induction in rat granulose cells by follicle-stimulating hormone and luteinizing hormone: combinatorial regulation by transcription factors cyclic adenosine 3′, 5′-monophosphate regulatory element binding protein, serum response factor, sp1, and early growth response factor-1. *Mol. Endocrinol.* 17:520–533.

Sandberg, M., Tasken, K., Øyen, O., et al. (1987). Molecular cloning, cADNA structure and deduced amino acid sequence for a type I regulatory subunit of cAMP-dependent protein kinase from human testis. *Biochem. Biophys. Res. Commun.* 149:939–945.

Schreyer, S.A., Cummings, D.E., McKnight, G.S., et al. (2001). Mutation of the RII beta subunit of protein kinase A prevents diet-induced insulin resistance and dyslipidemia in mice. *Diabetes* 50:2555–2562.

Schwede, F., Maronde, E., Genieser, H., et al. (2000). Cyclic nucleotide analogs as biochemical tools and prospective drugs. *Pharmacol. Ther.* 87:199–226.

Scott, J.D., Glaccum, M.B., Zoller, M.J. et al. (1987). The molecular cloning of a type II regulatory subunit of the cAMP-dependent protein kinase from rat skeletal muscle and mouse brain. *Proc. Natl. Acad. Sci. U.S.A.* 84:5192–5196.

Shaywitz, A.J., and Greenberg, M.E. (1999). CREB: a stimulus-induced transcription factor activated by a diverse array of extracellular signals. *Annu. Rev. Biochem.* 68:821–861.

Sheppard, J.R. (1974). The role of cyclic AMP in the control of cell division. In: Braun, W., Lichtenstein, L.M., Parker, C.W. (eds.). Cyclic AMP, Cell Growth, and the Immune Response. Springer, Berlin, pp 290–301.

Skalhegg, B.S., Landmark, B.F., Doskeland, S.O., et al. (1992). Cyclic AMP-dependent protein kinase type I mediates the inhibitory effect of 3′, 5′-cyclic adenosine monophosphate on cell replication in human T lymphocytes. *J. Biol. Chem.* 267:15707–15714.

Solberg, R., Tasken, K., Keiserud, A., et al. (1991). Molecular cloning, cDNA structure and tissue-specific expression of the human regulatory subunit RIβ of cAMP-dependent protein kinases. *Biochem. Biophys. Res. Commun.* 176:166–172.

Solberg, R., Tasken, K., Wen, W., et al. (1994). Human regulatory subunit RIβ of cAMP-dependent protein kinases: expression, holoenzyme formation and microinjection into living cells. *Exp. Cell Res.* 214:595–605.

Solberg, R., Sandberg, M., Natarajan, V., et al. (1997). The human gene for the regulatory subunit RIα of cAMP-dependent protein-kinase. Two distinct promoters provide differential regulation of alternately spliced mRNAs. *Endocrinology* 138:169–181.

Srivastava, R.K., Srivastava, A.R., Cho-Chung, Y.S., et al. (1999). Synergistic effects of retinoic acid and 8-Cl-cAMP on apoptosis require caspase-3 activation in human ovarian cancer cells. *Oncogene* 18:1755–1763.

Stratakis, C.A. (2000). Genetics of Carney complex and related familial lentiginoses, and other multiple tumor syndromes. *Front Biosci* 5:353–366.

Stratakis, C.A., Kirshner, L.S., Carney, J.A. (2001). Clinical and molecular features of the Carney complex: diagnostic criteria and recommendations for patient evaluation. *J. Clin. Endocrinol. Metab.* 86:4041–4046.

Tasken, K., Hansson, V., Aukrust, P., et al. (2000). PKAI as a potential target for therapeutic intervention. *Drug News Perspect.* 13:12–18.

Taylor, S.S., Bubis, J., Toner-Webb, J., et al. (1988). cAMP-dependent protein kinase: prototype for a family of enzymes. *FASEB J.* 2:2677–2685.

Taylor, S.S., Knighton, D.R., Zheng, J., et al. (1992). Structural framework for the protein kinase family. *Annu. Rev.Cell Biol.* 8:429–462.

Titani, K., Sasagawa, T., Ericsson, L.H., et al. (1984). Amino acid sequence of the regulatory subunit of bovine type I cAMP-dependent protein kinase. *Biochemistry* 23:4193–4199.

Tortora, G., and Ciardiello, F. (2002). Protein kinase A as target for novel integrated strategies of cancer therapy. In: Stratakis, C.A., Cho-Chung, Y.S. (eds) Protein Kinase A in human disease. *Ann. NY Acad. Sci.* 968:139–147.

Tortora, G., Tagliaferri, P., Clair, T., et al. (1988). Site-selective cAMP analogs at micromolar concentrations induce growth arrest and differentiation of acute promyelocytic, chronic myelocytic and acute lymphocytic human leukaemia cell lines. *Blood* 71:230–233.

Tortora, G., Yokozaki, H., Pepe, S., et al. (1991). Differentiation of HL-60 leukemia cells by type I regulatory subunit antisense oligodeoxynucleotide of cAMP-dependent protein kinase *Proc. Natl. Acad. Sci. U.S.A.* 88:2011–2015.

Tortora, G., Caputo, R., Damiano, V. et al. (1997a). Synergistic inhibition of human cancer cell growth by cytotoxic drugs and mixed backbone antisense oligonucleotide targeting protein kinase A. *Proc. Natl. Acad. Sci. U.S.A.* 94:12586–12591.

Tortora, G., Damiano, V., Bianco, C., et al. (1997b). The RIalpha subunit of protein kinase A (PKA) binds to Grb2 and allows PKA interaction with the activated EGF-Receptor. *Oncogene* 14:923–928.

Tortora, G., Di Isernia, G., Sandomenico, C., et al. (1997c). Synergistic inhibition of growth and induction of apoptosis by 8-chloro-cAMP and paclitaxel or cisplatin in human cancer cells. *Cancer Res.* 57:5107–5111.

Tortora, G., Bianco, C., Damiano, V. et al. (2000). Oral antisense that targets protein kinase A cooperates with taxol and inhibits tumor growth, angiogenesis, and growth factor production. *Clin. Cancer Res.* 6:2506–2512.

Tortora, G., Caputo, R., Damiano, V. et al. (2001). Combined blockade of protein kinase A and Bcl-2 by antisense strategy induces apoptosis and inhibits tumor growth and angiogenesis. *Clin. Cancer Res.* 7:2537–2544.

Uhler, M.D., Carmichael, D.M., Lee, D.C., et al. (1986a). Isolation of cDNA clones coding for the catalytic for the catalytic subunit of mouse cAMP-dependent protein kinase. *Proc. Natl. Acad. Sci. U.S.A.* 83:1300–1304.

Uhler, M.D., Chrivia, J.C., McKnight, G.S. (1986b). Evidence for a second isoform of the catalytic subunit of cAMP-dependent protein kinase. *J. Biol. Chem.* 261:15360–15363.

Vigil, D., Blumenthal, D.K., Brown, S. et al. (2004). Differential effects of substrate on type I and type II PKA holoenzyme dissociation. *Biochemistry* 43:5629–5636.

Vintermyr, O.K., Bøe, R., Brustugun, O.T., et al. (1995). Cyclic adenosine monophosphate (cAMP) analogs 8-Cl-cAMP and 8-NH2-cAMP induce cell death independently of cAMP kinase-mediated inhibition of the G1/S transition in mammary carcinoma cells (MCF-7). *Endocrinology* 136:2513–2520.

Viste, K., Kopperud, R.K., Christensen, A.E., et al. (2005). Substrate enhances the sensitivity of type I protein kinase A to cAMP. *J. Biol. Chem.* 280:13279–13284.

Wang, H., Hang, J., Shi, Z., et al. (2002). Antisense oligonucleotide targeted to RI alpha subunit of cAMP-dependent protein kinase (GEM231) enhances therapeutic effectivness of cancer chemotherapeutic agent irinotecan in nude mice bearing human cancer xenografs: in vivo synergistic activity, pharmacokinetics and host toxicity. *Int. J. Oncol.* 21:73–80.

Weissinger, E.M., Oettrich, K., Evans, C., et al. (2004). Activation of protein kinase A (PKA) by 8-Cl-cAMP as a novel approach for antileukaemic therapy. *Br. J. Cancer* 91:186–192.

Yang, W.L., Iacono, L., Tang, W.M., et al. (1998). A novel mechanism of cAMP signaling through the interaction of the regulatory subunit of protein kinase A with cytochrome c oxidase subunit Vb. *Biochemistry* 37:14175–14180.

Yin, Z., Jones, G.N., Towns, W.H. 2nd, et al. (2008a). Heart-specific ablation of Prkar1a causes failure of heart development and myxomagenesis. *Circulation.* 117:414–1422.

Yin, Z., Williams-Simons, L., Parlow, A.F. et al. (2008b). Pituitary-specific knockout of the Carney complex gene Prkar1a leads to pituitary tumorigenesis. *Mol. Endocrinol.* 22:380–387.

Yokozaki, H., Budillon, A., Tortora, G. et al. (1993). An antisense oligodeoxynucleotide that depletes RIα subunit of cyclic AMP-dependent protein kinase induces growth inhibition in human cancer cells. *Cancer Res.* 53:868–872.

Zawadzki, K.M. and Taylor, S.S. (2004). cAMP-dependent protein kinase regulatory subunit type II active site mutations define an isoform-specific network for allosteric signaling by cAMP. *J. Biol. Chem.* 20:7029–7036

Zhang, W., Morris, G.Z., Beebe, J. (2004). Characterization of the cAMP-dependent protein kinase catalytic subunit Cγ expressed and purified from sf9 cells. *Protein Expr. Purif.* 35:156–169.

Zhang, X., Odom, D.T., Koo, S.H., et al. (2005). Genome-wide analysis of cAMP-response element binding protein occupancy, phosphorylation, and target gene activation in human tissues. *Proc. Natl. Acad. Sci. U.S.A.* 102:4459–4464.

Zimmermann, B., Chiorini, J.A., Ma, Y., et al. (1999). PrKX is a novel catalytic subunit of the cAMP-dependent protein kinase regulated by the regulatory subunit Type I. *J. Biol. Chem.* 274:5370–5378.

Chapter 8
The Protein Kinase C Family: Key Regulators Bridging Signaling Pathways in Skin and Tumor Epithelia

Dirk Breitkreutz, Liora Braiman-Wiksman, Nicole Daum, and Tamar Tennenbaum

8.1 Discovery, Structure and Properties of the Protein Kinase C Family

The first protein kinase C (PKC) was described by Nishizuka and coworkers (Takai et al. 1979), who initially discovered that this serine/threonine (Ser/Thr) kinase is activated by diacylglycerol (DAG), a cleavage product of plasma membrane phosphatidylinositol generated by phospholipase C (PLC) (Inoue et al. 1977). Subsequently PKC was identified as intracellular receptor of tumor promoting phorbol esters (Castagna et al. 1982). Altogether ten isoforms/isozymes are known, PKCβI/βII representing splice variants and PKCι/λ human/mouse orthologues (Nishizuka 1988, 1995), while PKCμ/PKD1 belongs to the related protein kinase D (PKD) family (also phosphoinositide-dependent kinase/PDK; Dekker et al. 1995; Parker and Murray-Rust 2004; Rozengurt et al. 2005; Wang 2006).

All PKC isoforms share a regulatory moiety at the N-terminal and a kinase core at the C-terminal half connected by a hinge region (V3), both consisting of conserved (C1–C4) and variable regions (V1–V5) which also determine isoforms-specific binding properties (Fig. 8.1) (Nishizuka 1988; Mellor and Parker 1998; Newton 2003). The regulatory moiety starts with a pseudosubstrate sequence which blocks kinase activity by binding to the substrate recognition site (House and Kemp 1987). Three PKC subgroups are distinguishable according to their domain substructure and cofactor requirements. All three, the classical cPKC (α, βI, βII and γ), novel

D. Breitkreutz (✉)
DGZ, Division B050, German Cancer Research Center (DKFZ), POB 101949, D-69009 Heidelberg, Germany
e-mail: dirkbtz@gmx.de

C.J. Vidal (ed.), *Post-Translational Modifications in Health and Disease*, Protein Reviews 13, DOI 10.1007/978-1-4419-6382-6_8, © Springer Science+Business Media, LLC 2011

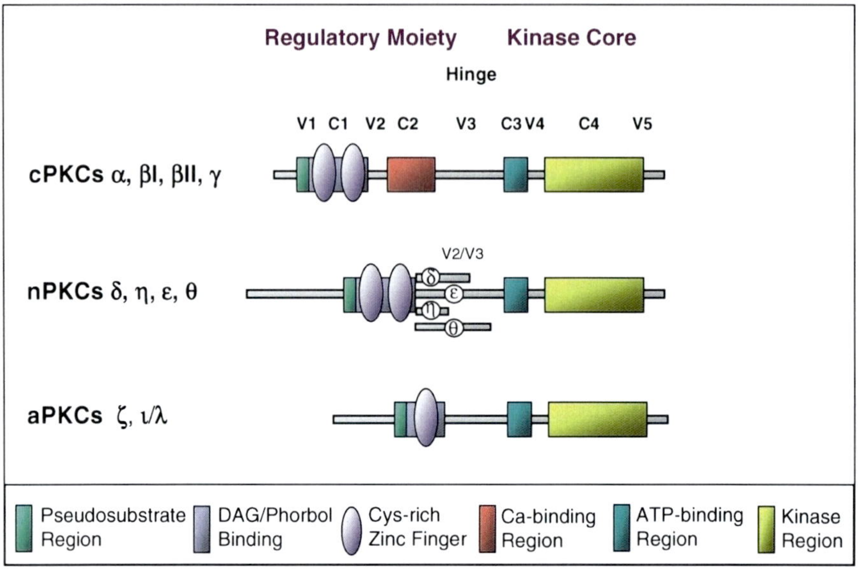

Fig. 8.1 PKC family isoforms

nPKC (δ, ε, η, θ), and atypical aPKC (ζ and ι/λ) isoforms depend for activation on specific lipid binding to phosphatidyl-serine, a constituent of the phospholipid bilayer (Svetek et al. 1995). The cPKCs require in addition calcium (Ca^{2+}) and DAG or phorbol esters, while nPKCs are Ca^{2+} independent (Kazanietz et al. 1993), and aPKCs need only phosphatidyl-serine for activation (Chauhan et al. 1990). In nPKCs the C2 (Ca^{2+} binding) domain of cPKCs is replaced by a C2-like domain placed in front of C1 (tandem repeat C1A, C1B; Griner and Kazanietz 2007). These distinct modular arrays provide a basis for broad-range, though specific interactions with a large variety of activators, inhibitors, membrane structures (Johnson et al. 2000; Giorgione et al. 2006), and other cell components (further references: Breitkreutz et al. 2007; Griner and Kazanietz 2007; Dempsey et al. 2000; Jaken and Parker 2000; Newton 1997, 2003).

8.2 Regulation of PKC Activity and Substrate Specificity

The specific activities of PKC isoforms rely on phosphorylation of distinct target substrates, but crucial for substrate binding is the relief of the autoinhibitory pseudosubstrate domain. Using chimeras as well as by mutational analysis the role of the pseudosubstrate sequence was confirmed for the selectivity of PKCα, ε, and η

(Dekker et al. 1995). However, most PKC isoforms phosphorylate similar target sequences and substrate selectivity is apparently quantitative in nature, Km values determining PKC affinity to specific substrates. This implies further levels of control, including relative quantities of individual PKC isoforms in specific cellular settings, distinct tissue distribution, intracellular compartmentalization mediated by adapter or scaffolding proteins, and the phosphorylation state of PKCs and those adapters. Thus, the formation of distinct signaling complexes defines substrate selectivity as well as functional specificity (Newton 2003; Parekh et al. 2000; Parker and Murray-Rust 2004).

8.2.1 PKC Phosphorylation

Representing prominent Ser/Thr kinases, PKCs themselves require Thr phosphorylation at the activation loop by PKD1, followed by Ser/Thr autophosphorylation of the turn and hydrophobic motif for a signaling-competent "maturation" state (Dutil et al. 1994; Sumandea et al. 2008). This "priming" allows PKC activation at the plasma membrane by DAG, being reversible due to the short DAG half-life (contrasting sustained activation by phorbol esters), and/or phospholipids (detailed reviews: Griner and Kazanietz 2007; Newton 2003; Parekh et al. 2000). However, several publications have documented PKCδ to be unique in this respect, showing that ser/thr phosphorylation of PKCδ, unlike that of PKCα and β, is not essential for its permissive activation. Responsible for this is the glutamic acid residue 500, mutation of which is causing an 80% loss of kinase activity (Steinberg 2004). All this indicates an alternative post-translational modification for PKCδ activation. In contrast, tyrosine (Tyr) phosphorylation has been primarily linked to inhibition/turnover particularly of PKCδ (Denning et al. 1998). Thus, constitutive phosphorylation of certain tyrosines by src kinases permanently inactivated PKCδ in v-$_{Ha}$ras transfected keratinocytes (Denning et al. 1993; Joseloff et al. 2002). But Tyr phosphorylation (e.g. by sarc) can also activate PKCδ (Tapia et al. 2003) as observed for PKCα, β, ε, and ζ (Joseloff et al. 2002; Tan et al. 2006). PKCδ activation was also seen in response to manifold agents like H_2O_2 (Konishi et al. 1997; Efimova et al. 2004), ionizing or UV irradiation (Fukunaga et al. 2001), further inducers of apoptosis (Blass et al. 2002; Brodie and Blumberg 2003), insulin, insulin-like growth factor-1 (IGF-1) or other factors (Braiman et al. 1999a, b; Czifra et al. 2006; Gschwendt et al. 1994; Jackson and Foster 2004; Li et al. 1994a, b). Apparently, phosphorylation of distinct tyrosine residues induces conformational changes depending on the particular PKC isoform (Braiman et al. 2001a, b; Joseloff et al. 2002; Le Good and Brindley 2004). This directs PKCs towards specific substrates, thereby addressing diverse signaling routes (Gschwendt 1999; Jaken and Parker 2000), or even changing activation mechanisms in case of PKCδ (Steinberg 2004).

8.2.2 Intracellular PKC Distribution and Molecular Associations

Complexed with phospholipids, PKC activation at the plasma membrane leads to Ser/Thr phosphorylation and endocytosis of various transmembrane proteins and receptors, like internalization of fibroblast growth factor receptor (FGFR) by PKCα (Asakai et al. 1995) and of insulin receptor (IR) by PKCδ (Braiman et al. 2001b; Shen et al. 2001). Furthermore, PKCα and PKCδ phosphorylate insulin receptor substrate-1 (IRS-1) in muscle cells (Rosenzweig et al. 2002), which may mediate a negative feedback terminating insulin signaling, shown for PKCβII in adipocytes (Liberman et al. 2008) or PKCζ in hepatoma cells (Liu et al. 2001). In return insulin activates aPKCs via IRS-1-dependent phospatidylinositol 3-kinase/PI3K and/or IRS-2/PI3K in classical insulin-sensitive tissues regulating glucose transport (Farese et al. 2005; Sajan et al. 2006). Addressing extracellular matrix (ECM) receptors, PKCδ triggers integrin α6β4 internalization (Alt et al. 2004), thus destabilizing firm adhesion of epidermal basal cells to basement membrane (BM). Promoting cell spreading or migration, complexes of PKCε were found with β1-integrins (Berrier et al. 2000), being further associated with the cytoskeletal component vimentin (Ivaska et al. 2005). Likewise, PKCα can interact with β1-integrins (Ng et al. 1999; Zhang et al. 2001) and syndecans representing another ECM receptor family (Jaken and Parker 2000; Keum et al. 2004). Contributing to epithelial polarity, PKCα associates with both the Par-3/ASIP (atypical PKC isotype-specific interacting protein) and Par-6 cell polarity proteins, in common with PKCζ and PKCι/λ (Helfrich et al. 2007; Izumi et al. 1998; Suzuki et al. 2001, 2002, 2004). Concerning cell compartments, PKCα is collocalized with keratin filaments, tight junctions, caveolae (like PKCδ), and desmosomes (Mineo et al. 1998; Ohmori et al. 1998), PKCγ with Golgi bodies, and PKCη the perinuclear rough endoplasmic reticulum (Cardell et al. 1998; Chida et al. 1994; Lehel et al. 1995a, b; Rybin et al. 2008). The translocation of PKCα, β, δ, ε, and ζ to mitochondria, the Golgi, nuclear or perinuclear regions regulates mitosis, apoptosis, and cell survival (Buchner 2000; Denning et al. 2002; DeVries et al. 2002; Dries et al. 2007; Li et al. 1999; Sitailo et al. 2006; Wang et al. 2004).

8.2.3 Specific Modifier and Adapter Proteins

Representative examples for typical PKC substrates are STICKs (substrates that interact with C kinase) like the actin-capping protein adducin (Kiley et al. 1999a, b). Phosphorylation of these phospholipid-binding proteins by PKC modifies their activity, reducing STICK binding to calmodulin and actin (Jaken and Parker 2000). Interestingly, a variety of actin-binding and modulating proteins such as vinculin, talin, and filamin belong to the group of STICK proteins (Tigges et al. 2003; Ziegler et al. 2002). Calponin, another actin-binding protein connects PKCα, βII, and ε to the extracellular signal-regulated kinase 1 (ERK-1) pathway, facilitating PKC activation (Leinweber et al. 2000). Critical for specific PKC translocation from cytosol to particulate compartments are the receptors for activated/inactive kinases (RACKs/RICKs; Parker and Murray-Rust 2004; Schechtman and Mochly-Rosen 2001). By binding to specific sequences in distinct PKC isoforms, RACKs direct their activation state and define their subcellular distribution. Further substrates or adapter proteins like PICK1 (protein interacting with C-kinase-1; Masukawa et al. 2006) and ZIP (zeta interacting protein; Puls et al. 1997) were found to affect PKC localization in their activated/resting state (Jaken and Parker 2000; Schechtman and Mochly-Rosen 2001). Many of these adapters interact with different proteins or complexes like the aPKC interacting, but very versatile p62/sequestosome 1 (Hirai and Chida 2003; Sanchez et al. 1998) involved in NFκB activation (Moscat and Diaz-Meco 2000; Moscat et al. 2003). Some detailed overviews cover structural aspects, biochemical properties, and substrate specificity of PKC isoforms (Griner and Kazanietz 2007; Le Good et al. 1998; Newton 2003; Parekh et al. 2000).

8.2.4 PKC Desensitization and Turnover

There are several mechanisms to downregulate PKC activity and for maintaining balanced levels of signaling-competent isozymes (Chen et al. 2007; Leontieva and Black 2004). When chronically activated (e.g. by TPA) PKCs become prone for dephosphorylation at their autophosphorylation sites by membrane-bound phosphatases (Hansra et al. 1999), guiding inactive forms to the endosomal/ubiquitination pathway (Prevostel et al. 2000). A very specific desensitization of cPKCs and nPKCs has been shown by the *PH* domain *l*eucin-rich repeat *p*rotein *p*hosphatase1 and -2 (PHLPP1/-2; Gao et al. 2008), rendering those PKCs susceptible to degradation. Alternatively, perhaps the major route, active PKCs associate with caveoli (within lipid rafts) and navigate themselves to the lysosomal compartment in the perinuclear region. Furthermore, RINCK (*RIN*G-finger protein binding *C k*inase) can control the bulk level of all PKC isoforms in a cell/tissue type-dependent context. By binding to the highly conserved C1a domain, RINCK mediates PKC ubiquitination and proteolysis (Chen et al. 2007). Conversely, the heatshock proteins Hsp70 (Gao and Newton 2002; Mashukova et al. 2009) and Hsp90 (Gould et al. 2009) stabilize dephosphorylated forms, allowing reactivation, Hsp90 also providing protection during the PKC maturation process.

8.3 PKC Tissue Distribution

Generally, PKCs are ubiquitous, some (PKCα, δ, and ζ) exist in all tissues, while others are tissue-specific. PKCγ for example is confined to brain and neuronal tissue (Cardell et al. 1998), the shortened PKCζ-transcript only to brain (Hernandez et al. 2003). PKCι was initially identified in testis and insulin secreting cells (Selbie et al. 1993), and PKCθ in skeletal muscle and T cells (Berry and Nishizuka 1990; Osada et al. 1992), but both were later detected also in epidermal cells (Papp et al. 2004). Frequently PKC functions are directed in a tissue-specific manner. For instance, PKCδ controls both proliferation and apoptosis in various epithelial models (Denning et al. 1998; Lu et al. 1997), whereas in the "classical" insulin responsive tissues, muscle, liver, and adipocytes PKCδ regulates glucose transport and metabolism (Braiman et al. 1999a, 2001b; Rosenzweig et al. 2002). In keratinocytes PKCδ stimulates the Na^+/K^+ pump (Shen et al. 2001), and in C6 glioma cells the Na^+/H^+ exchanger (Chen and Wu 1995).

8.4 PKC Isoforms in Skin, Focusing on Epidermis

In skin, the stratified epidermis covering the dermis provides protection against environmental, chemical, and physical impacts. This requires continuous epidermal renewal (basal cells) and sequential differentiation stages leading to terminal differentiation/keratinization (spinous, granular, cornified layers). Together with tight junctions underneath and lipids from lamellar bodies, this constitutes the protective water barrier (Helfrich et al. 2007; Langbein et al. 2002). The basal cells at the basement membrane (BM; dermo-epidermal junction) cease proliferation when leaving this compartment (Alt et al. 2001, 2004; Fuchs and Raghavan 2002). Loss of BM-contact is linked to initial (spinous) differentiation processes, indicated by keratin 1 (K1)/K10 expression replacing basal K5/K14. In upper spinous layers involucrin and epidermal transglutaminase K (TG-K) arise, followed by filaggrin, loricrin, repetin in the granular layer (Eckert et al. 2005; Huber et al. 2005). Finally, formation of the rigid cornified envelopes, correlating with autolysis of cell organelles, gives rise to keratinized surface squames (Fuchs and Raghavan 2002). Investigations in murine or human keratinocyte cultures indicated that extracellular Ca^{2+} and high cell density induce differentiation, mimicking epidermal keratinization (Breitkreutz et al. 1984, 1993; Hennings et al. 1980; Lee et al. 1998; Ryle et al. 1989; Yuspa et al. 1989). Therefore, the Ca^{2+} sensitive PKCα was considered as key regulator, initiating differentiation (Denning et al. 1995a). Furthermore, the Ca^{2+} independent isoforms PKCδ, ε, ζ, η, and θ are differentially expressed in keratinocyte cultures (Denning et al. 1995a; Papp et al. 2004; Stanwell et al. 1996), largely resembling their patterns in vivo (Wang et al. 1993, 1999). The specific roles of these isoforms direct signaling pathways controlling skin physiology, aberrations thereof being hallmarks in carcinogenesis.

8.4.1 PKCα

PKCα was the first classical PKC detected in keratinocytes in vitro and in vivo (Dlugosz et al. 1992), while PKCβI and βII are mainly confined to epidermal

Langerhans cells (marginal in keratinocytes; Fisher et al. 1993). Despite its assigned role in Ca^{2+} induced differentiation, most PKCα is seen in suprabasal layers (Denning et al. 1995a, Denning 2004), primarily associated with keratin filaments and desmosomal cell junctions (Szalay et al. 2001), being functionally involved in Ca^{2+}/transforming growth factor β (TGFβ)-induced growth inhibition (mediated by S100C/A11; Sakaguchi et al. 2004) or irreversible cell cycle withdrawal (Tibudan et al. 2002). Apparently keratins are phosphorylated by PKCs, which deserves further investigation (Mashukova et al. 2009; Paramio et al. 2001). Since exposure to the classical PKC activator TPA/PMA (12-*O*-tetradecanoylphorbol-13-acetate) specifically suppressed spinous markers, PKCα was thought triggering the shift to granular differentiation (Lee et al. 1998). Indeed, blocking PKCα activity or synthesis (antisense-RNA) abolished granular and revived spinous markers like K1/K10 or, by implementing dominant negative PKCα, involucrin (Deucher et al. 2002). Contrarily, the influence of PKCα on cellular traffic and membrane recruitment of β1-integrin during migration (Ng et al. 1999) presumably promotes wound reepithelialization. Still another ECM-receptor, syndecan-4 binds to the catalytic domain of PKCα, thereby also potentiating kinase activity (Jaken and Parker 2000; Keum et al. 2004; Lim et al. 2003). Correspondingly, the lipid-bound and oligomerized vinculin in the internal cell-matrix adhesion complexes is phosphorylated by PKCα (Ziegler et al. 2002). Altogether, PKCα is a potent inducer of granular and suppressor of spinous differentiation, whereas its interactions with ECM receptors presumably reflect responses to impacts by wounding or tumorigenesis. While consequences on the crosstalk with epidermis are still elusive, PKCα specifically increases in mouse skin fibroblasts with age (Bossi et al. 2008) which may have pro-inflammatory and tumor-promoting effects.

8.4.2 PKCδ

The novel nPKC PKCδ responds to different stimuli by very distinct translocation patterns with diverse biological effects. In epidermis PKCδ is acting in basal cells at the onset of differentiation (Denning et al. 1995a, 2004; Lee et al. 1998), accompanied by translocation to cell membranes and G1 growth arrest. Severe epidermal damage by higher UV doses for example, which leads to apoptosis (Denning et al. 1998), has been associated with generation of the constitutively active catalytic fragment PKMδ/PKCδ-cat (Denning et al. 2002; see also "tumorigenesis"), but also with distinct tyrosine phosphorylation as upon H_2O_2 treatment (Fukunaga et al. 2001). Generally, activation of PKCδ is mediated by p38 MAPK or ERK signaling (Efimova et al. 2004; Gonzales-Guerrico et al. 2005) and involves redistribution to various cellular compartments such as mitochondria, the Golgi complex, perinuclear and nuclear sites (Cross et al. 2000; Denning et al. 2002; Farshori et al. 2003; Kajimoto et al. 2004; Li et al. 1999). This seems to be mediated or guided by associations with the cell cytoskeleton (actin, tubulin, intermediate filaments like vimentin; Szalay et al. 2001). Tyr phosphorylation of PKCδ occurs very rapidly in response to apoptotic agents, as mediated by the Tyr kinase c-abl, and plays a role in targeting of PKCδ to the endoplasmic reticulum/ER and the nucleus (Humphries et al. 2008; Qi and Mochly-Rosen 2008; Reyland 2007). This clearly contrasts the anti-apoptotic function of PKCβII which associates with v-abl instead, getting translocated to the nucleus.

The rapid breakdown of the nuclear lamina in cells undergoing apoptosis is caused by proteolysis of lamin B after its phosphorylation by PKCδ (Cross et al. 2000). Enhancing the response, caspase3 cleavage of PKCδ leads to the release of the catalytic domain which is constitutively active and gets strictly localized to the nucleus (Reyland 2007). PKCδ-cat strongly promotes apoptosis by reduction of Mcl-1 (anti-apoptotic~), activation of Bax (pro-apoptotic Bcl-2 family member), and cytochrome c release (Sitailo et al. 2004, 2006). Finally, most important for maintaining skin integrity is the upregulation of p53 in response to stress (Abbas et al. 2004). In epidermis activated PKCδ also targets hemidesmosomes, the stable adhesions to BM, phosphorylating α6 (Alt et al. 2001; Gimond et al. 1995) or both integrin α6β4 chains (Alt et al. 2004). Getting dissociated, α6 and β4 are internalized and deactivated, which causes hemidesmosome disintegration thus allowing cells to enter suprabasal compartments (differentiation) or to migrate along the ECM (as in wounds). On the contrary, blockade of PKCδ completely inhibited the migration of keratinocytes on type I collagen (Li et al. 2002). Also the PDGF-BB (platelet derived growth factor)-mediated migration of fibroblasts requires activation of PKCδ which is translocated to the leading edge. There the kinase seems to activate signal transducer and activator of transcription 3 (STAT3) which is transducing migratory signals (Fan et al. 2006). PKCδ participates in various signalling pathways such as ras-raf-MEK-ERK (Farshori et al. 2003; Fedorov et al. 2002; Jackson and Foster 2004; Steinberg 2004). At the level of transcriptional control regulation by PKC involves Sp-1 and particularly AP-1 sites, representing response elements for TPA regulated genes (Rutberg et al. 1996). Thus, PKCδ increases the levels of c-fos and junB while lowering c-jun and fra-1 (Rutberg et al. 1997), representing a switch for initiating differentiation. In keratinocytes, PKCδ is deactivated through Tyr phosphorylation (by src) in response to ras transformation or growth factors like TGFα and EGF (Denning et al. 1993, 2000; Joseloff et al. 2002). Conversely, PKCδ phosphotyrosines are dephosphorylated upon insulin-induced activation. However, as outlined above, PKCδ phosphorylation at distinct tyrosine residues has opposing effects on activity or activating mechanisms and physiological outcome (Steinberg 2004). Collectively, PKCδ is crucial for the balance between growth, differentiation and cell death, generally antagonizing PKCα (Denning 2004; Hornia et al. 1999) as well as PKCε (Griner and Kazanietz 2007). This applies also for PKCδ interactions with PKD1 in epidermis (Rozengurt et al. 2005) or with IRS-1 (Rosenzweig et al. 2002). Underlining further the contrary PKCα and PKCδ functions, in murine fibroblasts only PKCδ reduces reactive oxygen species (ROS) after UVA radiation (penetrating into skin), getting translocated to nuclei and down-regulated (Bossi et al. 2008).

8.4.3 PKCη

The second nPKC identified in squamous epithelia (Kuroki et al. 2000; Osada et al. 1993) is PKCη, which is primarily present in the uppermost granular layer of epidermis and marginally in hair follicle inner root sheaths (Koizumi et al. 1993). Being concentrated in perinuclear regions and rough endoplasmic reticulum, PKCη gets associated with the keratin cytoskeleton upon activation (Chida et al. 1994), correlating with an increase of granular differentiation markers (Kuroki et al. 2000; Takahashi et al. 1998). Thus, overexpression in human keratinocytes induced involucrin and transglutaminase-1 (TG-K), crucial for cornified envelope formation (Ohba et al. 1998; Ueda et al. 1996). Correspondingly, PKCη is activated by cholesterol sulfate, a second messenger accumulating in upper layers (Ikuta et al. 1994; Denning et al. 1995b), leading to cell cycle arrest in G1 (review: Kuroki et al. 2000). Possible mechanisms are that PKCη (1) binds to the cyclin E/cdk2 complex, affecting its nuclear translocation, or (2) phosphorylates p21 which inhibits cdk2 kinase activity (Kashiwagi et al. 2000; Shtutman et al. 2003). Both differentiation and growth arrest could be further linked to activation of fyn (src kinase family) and down modulation of epidermal growth factor receptor (EGFR) signaling (Cabodi et al. 2000). Thus, PKCη drives several mechanisms to prevent cell cycle progression and to advance terminal differentiation.

8.4.4 PKCε

PKCε, another nPKC, is expressed in basal keratinocytes where increased levels and membrane translocation correlate to proliferative activity. Coinciding with proposed mutual transactivation (Rozengurt et al. 2005), PKD1 is collocated in the basal layer (Dodd et al. 2005). Several pathways critical for cell growth could be linked to PKCε function like induction of autocrine growth factors, activation of ras associated raf-1 and MAPK signaling (Hamilton et al. 2001), regulation of transcriptional activators such as c-fos and c-jun, and NFκB (Flescher and Rotem 2002; Razin et al. 1994). Thus, in PKCε transfected Rat6 fibroblasts constitutively activated raf-1 was demonstrated (Cacace et al. 1998) and similarly PKCε-induced growth of NIH3T3 cells was raf-1 dependent, implying a direct interaction (Cai et al. 1997). However, in transiently transfected A293 cells PKCε activated raf-1 indirectly through secretion of factors like TGFβ, being associated with oncogenic activity (Ueffing et al. 1997). According to a recent report active complexes of PKCε with 14-2-3 (phospho-peptide binding protein family) are essential for completion of cytokinesis in cell division (Saurin et al. 2008). In addition, PKCε regulates integrin mediated cell adhesion and spreading (Berrier et al. 2000) which involves vimentin phosphorylation (Ivaska et al. 2005) enhancing cytoskeletal plasticity.

8.4.5 *PKCζ and PKCι/λ*

The atypical PKCs PKCζ and PKCι/λ share important roles in determining cell and tissue polarity (Horikoshi et al. 2009; Suzuki and Ohno 2006; Yamanaka et al. 2006). While PKCζ may drive polar assembly and distribution of premature junctional complexes (Suzuki et al. 2002), PKCι/λ directly associates with tight junction proteins (Helfrich et al. 2007), both outlined above ("molecular interactions"). In contrast, the shortened form PKCζII (regulatory domain transcribed from a distinct gene) suppresses epithelial polarization (Parkinson et al. 2004), presumably counteracting PKCζ by competitive binding to effector targets and thus facilitating tissue assembly and remodeling. Generally, PKCζ can mediate growth stimulation or inhibition, depending on the cell type, and among other functions glucose uptake or IR-signaling (Braiman et al. 2001a; Liu et al. 2001). In K10 overexpressing epidermal cells PKCζ became activated with onset of differentiation, mediating cell cycle arrest (Nishikawa et al. 1997). This was confirmed in transgenic mice overexpressing K10, implying functional links of PKCζ with the keratin cytoskeleton (Paramio et al. 2001). PKCζ activation was associated with reduction of cyclin D1, sequestration of Akt/PKB, and impaired phosphoinositide 3-kinase (PI3K) signaling. Overall, both aPKCs are indispensable for epithelial architecture and barrier function, while PKCζ promotes in addition early differentiation steps. In a very recent paper tight association of PKCι with keratin and Hsp70 has been documented which stabilizes steady state kinase activity (Mashukova et al. 2009). In fibroblasts both aPKCs, more pronounced though PKCι/λ, cause loss of actin stress fibers by interacting with Cdc42 or ras (Coghlan et al. 2000b) increasing cell motility.

8.5 Ambivalent Role of Specific PKC Isoforms in Skin Tumorigenesis

The contribution of PKC signaling in the etiology of skin cancer became apparent in the classical two-stage chemical carcinogenesis model in mice. Herein a low dose of the carcinogen 7,12-dimethylbenz(a)anthracene (DMBA) is topically applied to skin (initiation step), followed by repeated applications of a tumor promoter (promotion phase), usually the phorbol ester TPA/PMA (Hecker 1985; comprehensive review: Griner and Kazanietz 2007). This leads mostly to benign (papillomas) and eventually to malignant skin tumors (squamous cell carcinomas/SCCs). Searching for molecular targets revealed that phorbol esters are broad-spectrum PKC activators (Hansen et al. 1990). Short TPA exposure induces activation and translocation of PKC isoforms from cytosol to membrane or cytoskeletal compartments, but also nuclear translocation as demonstrated for PKCδ in human keratinocyte (HaCaT cells) and fibroblast cultures, similarly to density-mediated growth reduction (Dietrich et al. 2001; Heit et al. 2001). Repeated TPA treatment, comparable to ionizing or UV radiation, can induce cleavage and release of the respective catalytic PKM domains (like PKCδ-cat), resulting in constitutive activity (Denning et al. 1998; Inoue et al. 1977). Processing enzymes for PKM generation include calpains/calcium activated proteases (Al and Cohen 1993; Cressman et al. 1995; Kishimoto et al. 1989), caspases (Akkaraju and Basu 2000; Datta et al. 1997; Denning et al. 2002), and interleukin-1β converting enzyme (ICE) (Emoto et al. 1995). But further prolonging TPA exposure can also downregulate PKC activity (Hansen et al. 1990) through complete PKM degradation by calpains (Cressman et al. 1995), other internal proteases like caspases (Denning et al. 1998) or via the ubiquitin/proteasome pathway (Chen et al. 2007; Griner and Kazanietz 2007; Lu et al. 1998). Finally, calpains can also release regulatory domains, inhibiting PKC by competitive substrate binding (Parissenti et al. 1998).

Underlining the PKC contribution, the endogenous PKC activator DAG, generated from membrane lipids and competing with TPA binding (Sharkey et al. 1984), promotes tumor formation in mouse skin (Mills et al. 1993; Verma 1988). But unlike TPA, DAG is turned over very rapidly and physiological concentrations are extremely low. One hallmark of early events in classical promotion experiments was the induction of ornithin decarboxylase (ODC), a key enzyme in polyamine synthesis. An ODC increase could be also mediated by various PKCs (Jansen et al. 2001a) in murine keratinocytes in vivo and in vitro, and human skin organ cultures (Fischer et al. 1993; Verma et al. 1986). However, mouse strain-specific TPA responses have been found with variable ODC induction (Coghlan et al. 2000a; Mills and Smart 1989) and in several studies ODC levels did neither correspond to tumor yield nor overall PKC activity (Imamoto et al. 1993; Jansen et al. 2001a). Most strikingly, PKCδ induces ODC despite its tumor suppressor function (Wheeler et al. 2002), implying isoform-specific supportive functions of ODC.

Addressing *individual PKC isoforms*, acute TPA exposure increases particulate *PKCα* activity in epidermis, delaying differentiation. So, PKCα has been found constitutively activated in cutaneous SCCs, corresponding to advanced stages of human malignant melanomas (Lahn et al. 2004) and tumors of the airway-gastro-intestinal tract (Lahn and Sundell 2004). However, conflicting experimental and clinical data imply that this greatly depends on the particular tumor type or staging (listing in: Griner and Kazanietz 2007). Thus, PKCα activates the breast cancer associated Ets1 protein (e.g. enhancing matrix-metalloprotease-1 (MMP-1) and MMP-9) in several human cancer cell lines (Vetter et al. 2005), while it suppressed for example colon cancer in established transgenic mouse models (Oster and Leitges 2006). Delayed differentiation, particularly inhibition of spinous markers, observed also in experimental cutaneous SCCs (Tennenbaum et al. 1993; Tomakidi et al. 2003), correlated with elevated PKCα activity in tumor cells in vitro (Dlugosz et al. 1992; Yang et al. 2003). Furthermore, complexes of PKCα, tetraspanins (TM4SF transmembrane proteins), and β1-integrins (mainly α3β1 and α6β1) were detected after TPA stimulation (Zhang et al. 2001). Actually, PKCα driven tread-milling of β1-integrin for migration (Ng et al. 1999) may propel invading tumor cells. Similar to β1-integrins, a dramatic redistribution of BM-associated integrin α6β4 into suprabasal areas occurred in mouse and human tumor models

(Tennenbaum et al. 1993; Tomakidi et al. 1999; references therein). In many epithelial tumor cells and melanomas PKCα induces (in vivo/in vitro) MMP-1, -2, and -9 as well as urokinase plasminogen activator (uPA), all responsible for extracellular matrix degradation and thus promoting tumor invasion (Lahn et al. 2004; Lahn and Sundell 2004 and references; Kermorgant et al. 2001). Similarly, a concerted action of ErbB2 (HER2/neu) was demonstrated in human breast cancer cells, activating both PKCα and src and thus uPA receptor (uPAR), responsible for invasion (Tan et al. 2006).

In addition to PKCα, an increase of the classical *PKCβ* can affect malignancy but apparently restricted mainly to internal epithelia such as colon. Comparing *PKCβI* and *PKCβII*, only the latter was significantly increasing early on in tumor development, whereas the differentiation associated PKCβI was drastically down-regulated (Murray et al. 1999, 2009; Gokmen-Polar et al. 2001). Regarding therapeutic approaches, for their anti-apoptotic/pro-survival function (favouring maintained tumor growth) PKCα and PKCβII are considered as promising drug targets, though particularly in case of PKCα with the caveat of tissue/tumor diversity. Thus, PKCα is growth inhibitory in colon cancers (or cell lines) linked to β-catenin/adenomatous polyposis coli (APC) mutations (Oster and Leitges 2006) by downregulating expression and translation of cyclin D1 (Pysz et al. 2009). Apparently this plays a crucial role for differentiation of intestinal cyst cells (Pysz et al. 2009). Furthermore activation of PKCα can promote dephosphorylation and inactivation of pro-survival kinase Akt/PKB (Gonzales-Guerrico et al. 2005).

Conversely, *PKCδ* is diminished in murine papillomas and human SCCs (D'Costa et al. 2006), like other nPKCs and aPKCs are reduced in certain stages of skin tumor progression, by TPA activating PKCδ and apoptosis via the p38 MAPK pathway (Gonzales-Guerrico et al. 2005). So, expanding α6β4 and β1-integrin patterns in cutaneous tumors actually reflected gain (PKCα) and loss of PKC function (PKCδ). Accordingly, PKCδ suppresses (maybe PKCα-induced) expression of uPA and MMPs as reported for human and murine mammary tumor cells (Grossoni et al. 2007; Jackson et al. 2005). Besides its modification in pathways like ras-raf1-MEK-ERK (Farshori et al. 2003), frequently activated in early tumorigenesis (Steinberg 2004), PKCδ also interacts with signaling elements linked to cancer progression including src, ras, STAT-1 and -3 (Gschwendt et al. 1994; Joseloff et al. 2002; Novotny-Diermayr et al. 2002; Uddin et al. 2002), which may erase its pro-apoptotic activity. PKCδ can also regulate the transcription of death genes through activation or inactivation of STAT1, p53, and p73 (Abbas et al. 2004).

Likewise, *PKCη* exerts predominantly suppressive effects and so, PKCη disruption enhanced murine tumorigenesis (Chida et al. 2003) or accompanied malignant progression in several human tumors (Masso-Welch et al. 2001). But as a further example for tissue-specific functional diversity, in the breast cancer MCF7 cells PKCη promoted cell cycle progression, increasing cyclin E levels either by activating its promoter or inhibiting its degradation (Fima et al. 2001).

Also *PKCζ* is drastically reduced in many squamous tumors (Verma et al. 2006). Thus, down-regulation of *PKCζ* promotes skin tumorigenesis possibly by lowering constraints on Akt/PKB signaling driven by insulin like growth factor-1 (IGF-1) which gets activated during tumor promotion and progression (Segrelles et al. 2006; Wilker et al. 2005). Similarly, the PKCζ/Par-4 complex down-regulates Akt/PKB activation in a murine lung tumor model (Joshi et al. 2008). Changing cell targets, PKCζ suppresses IL-6 production in fibroblasts, an important mediator in tumor stroma interactions (Galvez et al. 2009). Alternatively, PKCζ can be antagonized by its short form PKCζII, most likely contributing to epithelial tumor plasticity or invasion, and consequently elevated PKCζII was found in several autochthonous tumors (Parkinson et al. 2004). But also pro-neoplastic effects of PKCζ via MAP kinase have been documented such as growth stimulation in head and neck SCCs (Cohen et al. 2006), enhancement of uPA and MMP-9 by PKCζ over-expression in murine mammary cells (Urtreger et al. 2005), and a crucial role in EGF-induced chemotaxis in human breast cancer cells (Sun et al. 2005).

PKCε usually correlates positively with proliferation and has been implicated as a powerful oncogene. Thus, it is constitutively active in small cell lung cancer cell lines and overexpression of PKCε in Rat6 embryo fibroblasts or colon epithelial cells increased their growth and invasive capacity (Cacace et al. 1998; Ueffing et al. 1997). On the molecular level PKCε directly interacts with Akt in several cell lines including recurrent prostate cancer CWR-R1 cells, breast cancer MCF-7 cells and rat glomerular mesangial cells. In many cases PKCε is actively involved in development of malignant tumors, for example enhancing UV-induced SCCs (Wheeler et al. 2005). Finally, disturbed PKCε/14-3-3 complexes (Saurin et al. 2008), interfering with complete cytokinesis, result in accumulation of tetraploid cells prone to chromosomal aberrations and malignancy (Fujiwara et al. 2005).

In *basal cell carcinomas* (BCCs) typical PKCs are downregulated like PKCα and PKCδ. In these undifferentiated tumors, evolving from hair follicles or interfollicular epidermis, the sonic hedgehog (sHh) pathway is aberrantly activated involving the transcription factors GLI-1 and GLI-2 (Beachy et al. 2004; Kasper et al. 2006; Riobo and Manning 2007). So far PKCα was shown to suppress GLI-1 activity (keeping it balanced in hair follicles), thus inhibiting BCC growth (Neill et al. 2003). The adverse effect of PKCδ stimulating GLI-1 function may explain controversial PKC influences on some other Hh-dependent tumors (Riobo et al. 2006; also references therein).

To summarize the observations in skin, activation of some TPA-responsive PKCs, especially PKCα and presumably PKCε, and deactivation of others, like PKCδ and PKCη, is apparently crucial for epidermal (squamous epithelial) tumor induction and progression.

8.6 Functional PKC Specificity in Transgenic Mouse Models

The contribution of individual PKC isoforms to skin physiology and pathology, including tumorigenesis and malignant progression, has been investigated by overexpression or suppression in transgenic mice (Jansen et al. 2001a; Leitges et al. 2001; Oster and Leitges 2006; Reddig et al. 1999; Wang and Smart 1999) and mostly utilizing two-stage carcinogenesis protocols (Griner and Kazanietz 2007). While TPA induced skin hyperplasia in mice overexpressing any PKC isoform, each isoform specifically governed the resulting tumor phenotype (summarized in Table 8.1). Overexpression of *PKCα* in transgenic mice (driven by a K5-promoter, enforcing premature expression in basal cells) caused primarily severe inflammatory responses. Increased epidermal thickening and edema correlated to neutrophil infiltration, multiple micro-abscesses, and a marked increase of typical mediators like cyclooxygenase-2 (COX-2) or macrophage inflammatory protein (MIP) (Wang and Smart 1999). TPA treatment induced epidermal hyperplasia, intra-epidermal inflammation, but also massive apoptosis (Cataisson et al. 2003; Jansen et al. 2001a), so neither tumor promotion nor progression was enhanced (Wang and Smart 1999). Of note in this context, in human keratinocytes a synergism of PKCα with TGFβ (an epithelial growth inhibitor) has been found, both targeting the Ca^{2+}-binding protein S100C/A11, alternatively to the TGFβ/smad-pathway (Sakaguchi et al. 2004). Furthermore, persisting hyperplasia and inflammatory cell infiltration certainly represent risk factors for human cancers which generally develop more slowly. As expected TPA-induced hyperplasia was abolished in PKCα-deficient mice, however, an increase of benign tumors (papillomas) was observed but no progression towards malignancies (Hara et al. 2005). Alternatively, these tumors may arise from hair follicles due to unbalanced Hh-signaling (lack of PKCα control on GLI, see above).

In accord to previous data, *PKCδ* overexpressing mice are extremely resistant to chemically induced skin tumorigenesis for the strong effect of PKCδ in pro-apoptotic signaling. Both incidence of papillomas and progression towards malignancy were reduced (Reddig et al. 1999), though this refers to chemically but not UV-induced carcinogenesis (Aziz et al. 2006). Consequently, in PKCδ knockout mice apoptosis

Table 8.1 Role of PKC isoforms in epidermis, skin pathology and cancer

PKC isoform	Regulatory roles in skin physiology	Activation state in transformed keratinocytes	Skin pathology and cancer in transgenic and *knockout*-mice
PKCα	Induction of granular, inhibition of spinous differentiation	Elevated expression and activity	Increased inflammatory response in epidermis (*tumor promotion**)
PKCδ	Migration, proliferation, apoptosis	Reduced expression and activity	Tumor suppression (*tumor promotion and inhibited apoptosis**)
PKCη	Cornified envelopes, growth arrest and cell cycle regulation	Reduced expression	N.D. (*tumor promotion**)
PKCε	Cell adhesion and spreading	Elevated expression	Reduced papilloma and increased SCC formation malignant progression
PKCζ	Early stage of skin differentiation	Reduced expression	N.D.

Effects in transformed cells, transgenic, and knockout* mice (**in parenthesis*)

was suppressed which may enhance tumorigenesis (Humphries et al. 2006). In these mice the response to apoptotic stimuli was inhibited such as caspase activation, DNA fragmentation or loss of mitochondrial membrane potential, confirming the anti-apoptotic PKCδ function (Reyland 2007).

While over-expression of *PKCη* is apparently difficult to accomplish in vivo (no published data), deficient mice show prolonged epidermal hyperplasia upon TPA treatment and an increased susceptibility to tumor formation. So PKCη seems to strengthen resistance against tumors (Chida et al. 2003) like PKCδ.

In contrast, *PKCε* overexpressing mice revealed epidermal hyperproliferation and skin ulceration at 4 months of age. In two-stage skin carcinogenesis, papilloma formation was lowered but progression to eventually metastasizing SCCs increased (Jansen et al. 2001b; Reddig et al. 2000). Carcinomas also developed independently of papillomas (Li et al. 2005) and these mice were sensitized to UV radiation-induced SCC formation (Aziz et al. 2007). This clearly illustrates the complexity of PKCε function in skin physiology and tumor development. Alternative carcinogenesis models have been created by cross-breeding various PKC-mice (e.g. PKCα over-expression) with other transgenic/cancer prone strains, but so far this is addressing only visceral and not cutaneous tumors (Oster and Leitges 2006).

So what can be learned from these mouse models? In case of over- and under-expression of PKCs the balance is changed dramatically between these and related kinases, competing also for adapter or signaling molecules which themselves can be TPA-targets (Gonzales-Guerrico et al. 2005; Griner and Kazanietz 2007). This can switch pathways in adverse directions (e.g. anti- versus pro-neoplastic) differently from internal physiologic regulation. Anyhow, what became clear is the strong impact of PKCs on inflammatory skin reactions, certainly deserving particular attention also in the context of tumor progression.

8.7 Conclusion and Future Prospects

Skin physiology and cancer development are largely affected by isoform-specific PKC signaling but the combined effects of various PKCs can have diverse phenotypic outcome. Besides absolute and relative levels of different PKC isoforms in a given cellular and tissue context, this will be also controlled by changing microenvironmental influences. For better understanding of epithelial/tumor-stroma interactions, three-dimensional skin-organotypic co-cultures should be generated employing genetically altered epithelial and stromal cells. This supplements well transgenic approaches particularly in cases of embryonic/perinatal lethality (Nischt et al. 2007; Sadagurski et al. 2006) and may allow to analyze PKC signaling under controlled skin- or tumor-like conditions (Gutschalk et al. 2006). Of particular interest would be the influence of PKCs on matrix proteases like MMPs in this setting with defined cell populations (Baumann et al. 2000). Also isoform-specific inhibitors (or activators; Way et al. 2000) will be applicable in this model, ranging from selective kinase inhibitors like enzostaurin (PKCβ; Liu et al. 2004; Fields et al. 2009) or specific blocking peptides (V1-2ε for PKCε; Gray et al. 1997) to small interfering (siRNA) (Cameron et al. 2008; Irie et al. 2002) or antisense oligonucleotides/AOR (aprinocarsen, PKCα; Lahn et al. 2004; Lahn and Sundell 2004; Michie and Nakagawa 2005; Shimao et al. 1999). Additional transgenic mice with conditional over-expression or suppression of PKCs (tissue- or differentiation stage-specific) shall further elucidate the contribution of individual PKCs for the dynamics of skin carcinogenesis. Still another option, mainly for elucidating tumor/stroma-interactions may offer SCID (severe combined immune deficiency) mice, "humanized" by installing elements of the human immune system. Challenging with human tumor cells should provoke more authentic inflammatory cell infiltrates, enhancing the tumor stroma reaction and matrix turnover (Mueller 2006; Tomakidi et al. 1999; Willhauck et al. 2007). The synopsis of all these approaches should further disclose the interplay between PKC-signaling and other regulatory networks. Of particular interest will be the crosstalk with other DAG mediated pathways (Griner and Kazanietz 2007; Newton 2003; Wang 2006) as well as interactions with further substrates and adapter proteins (Moscat et al. 2006). Some effects, formerly assigned to PKCs, are actually mediated by those other DAG receptors, all sharing with PKCs the C1 domain responsible for DAG binding. Such DAG effector molecules include PKDs (protein kinase D1, 2, 3) (Le Good et al. 1998; Rozengurt et al. 2005; Wang et al. 2003), DGKs (DAG kinases β and γ) regulating mitogenic pathways (Crotty et al. 2006), and the modulators of small G proteins (exchange factors) RasGRPs and chimaerins (Griner and Kazanietz 2007; Wang et al. 2006). While RasGRP isozymes (1–4) are activating small GTPases and promote ras-GTP loading (Ebinu et al. 1998), chimaerins (α1, 2 and β1, 2) accelerate rac-GTP hydrolysis which deactivates rac (Wang et al. 2006). Consequently, increased RasGRP1 gave rise to spontaneous skin tumors in untreated mice (Oki-Idouchi and Lorenzo 2007) and promoted invasive carcinomas in the DMBA/TPA model (Luke et al. 2007), whereas β2-chimaerin inhibited growth of breast cancer cells (Yang et al. 2005). Presumably, many of the involved adapter proteins are modified by PKCs as demonstrated for PKDs (Rozengurt et al. 2005), thus bridging individual pathways in the entire frame of signaling networks (Griner and Kazanietz 2007; Kanzaki et al. 2004). What should be kept in mind for future therapeutic approaches is that PKCs and related kinases have apparently a Janus-head function, acting largely as interpreters of external or internal signaling cues in a cell- and tissue-specific manner.

References

Abbas, T., White, D., Hui, L., et al. (2004). Inhibition of human p53 basal transcription by down-regulation of protein kinase Cδ. *J. Biol. Chem.* 279:9970–9977.

Akkaraju, G.R., and Basu, A. (2000). Overexpression of protein kinase C-eta attenuates caspase activation and tumor necrosis factor-alpha-induced cell death. *Biochem. Biophys. Res. Commun.* 279:103–107.

Al, Z., and Cohen, C.M. (1993). Phorbol 12-myristate 13-acetate-stimulated phosphorylation of erythrocyte membrane skeletal proteins is blocked by calpain inhibitors: possible role of protein kinase M. *Biochem. J.* 296:675–683.

Alt, A., Gartsbein, M., Ohba, M., et al. (2004). Differential regulation of α6β4 integrin by PKC isoforms in murine keratinocytes. *Biochem. Biophys. Res. Commun.* 314:17–23.

Alt, A., Ohba, M., Li, L., et al. (2001). Protein kinase Cδ-mediated phosphorylation of α6β4 is associated with reduced integrin localization to the hemidesmosome and decreased keratinocyte attachment. *Cancer Res.* 61:4591–4598.

Asakai, R., Akita, Y., Tamura, K., et al. (1995). Protein kinase C-dependent down-regulation of basic fibroblast growth factor (FGF-2) receptor by phorbol ester and epidermal growth factor in porcine granulosa cells. *Endocrinology* 136:3470–3479.

Aziz, M.H., Manoharan, H.T., and Verma, A.K. (2007). Protein kinase Cε, which sensitizes skin to sun's UV radiation-induced cutaneous damage and development of squamous cell carcinomas, associates with Stat3. *Cancer Res.* 67:1385–1394.

Aziz, M.H., Wheeler, D.L., Bhamb, B., et al. (2006). Protein kinase Cδ overexpressing mice are resistant to chemically but not to UV radiation-induced development of squamous cell carcinomas: a possible link to cytokines and cyclooxygenase-2. *Cancer Res.* 66:713–722.

Baumann, P., Zigrino, P., Mauch, C., et al. (2000). Membrane-type 1 matrix metallo-proteinase-mediated progelatinase-A activation in non-tumorigenic and tumorigenic human keratinocytes. *Br. J. Cancer* 83:1387–1393.

Beachy, P.A., Karhadkar, S.S., and Berman, D.M. (2004). Tissue repair and stem cell renewal in carcinogenesis. *Nature* 432:324–331.

Berrier, A.L., Mastrangelo, A.M., Downward, J., et al. (2000). Activated R-ras, Rac1, PI 3-kinase and PKC-epsilon can each restore cell spreading inhibited by isolated integrin beta1 cytoplasmic domains. *J. Cell Biol.* 151:1549–1560.

Berry, N., and Nishizuka, Y. (1990). Protein kinase C and T cell activation. *Eur. J. Biochem.* 189:205–214.

Blass, M., Kronfeld, I., Kazimirski, G., et al. (2002). Tyrosine phosphorylation of protein kinase Cδ is essential for its apoptotic effect in response to etoposide. *Mol. Cell. Biol.* 22:182–195.

Bossi, O., Gartsbein, M., Leitges, M., et al. (2008). UV irradiation increases ROS production via PKCδ signaling in primary murine fibroblasts. *J. Cell Biochem.* 105:194–207.

Braiman, L., Alt, A., Kuroki, T., et al. (1999a). Protein kinase Cdelta mediates insulin-induced glucose transport in primary cultures of rat skeletal muscle. *Mol. Endocrinol.* 13:2002–2012.

Braiman, L., Alt, A., Kuroki, T., et al. (2001a). Activation of protein kinase Czeta induces serine phosphorylation of vamp2 in the glut4 compartment and increases glucose transport in skeletal muscle. *Mol. Cell. Biol.* 21:7852–7861.

Braiman, L., Alt, A., Kuroki, T., et al. (2001b). Insulin induces specific interaction between insulin receptor and protein kinase C delta in primary cultured skeletal muscle. *Mol. Endocrinol.* 15:565–574.

Braiman, L., Sheffi-Friedman, L., Bak, A., et al. (1999b). Tyrosine phosphorylation of specific protein kinase C isoenzymes participates in insulin stimulation of glucose transport in primary cultures of rat skeletal muscle. *Diabetes* 48:1922–1929.

Breitkreutz, D., Bohnert, A., Herzmann, E., et al. (1984). Differentiation specific functions in cultured and transplanted mouse keratinocytes: environmental influences on ultrastructure and keratin expression. *Differentiation* 26:154–169.

Breitkreutz, D., Braiman-Wiksman, L., Daum, N., et al. (2007). Protein kinase C family: on the crossroads of cell signaling in skin and tumor epithelium. *J. Cancer Res. Clin. Oncol.* 133:793–808.

Breitkreutz, D., Stark, H.J., Plein, P., et al. (1993). Differential modulation of epidermal keratinization in immortalized (HaCaT) and tumorigenic human skin keratinocytes (HaCaT-ras) by retinoic acid and extracellular Ca2+. *Differentiation* 54:201–217.

Brodie, C., and Blumberg, P.M. (2003). Regulation of cell apoptosis by protein kinase c δ. *Apoptosis* 8:19–27.

Buchner, K. (2000). The role of protein kinase C in the regulation of cell growth and in signalling to the cell nucleus. *J. Cancer Res. Clin. Oncol.* 126:1–11.

Cabodi, S., Calautti, E., Talora, C., et al. (2000). A PKC-eta/Fyn-dependent pathway leading to keratinocyte growth arrest and differentiation. *Mol. Cell* 6:1121–1129.

Cacace, A.M., Ueffing, M., Han, E.K.H., et al. (1998). Overexpression of PKCε in R6 fibroblasts causes increased production of active TGFβ. *J. Cell Physiol.* 175:314–322.

Cai, H., Smola, U., Wixler, V., et al. (1997). Role of diacylglycerol-regulated protein kinase C isotypes in growth factor activation of the raf-1 protein kinase. *Mol. Cell. Biol.* 17:732–741.

Cameron, A.J., Procyk, K.J., Leitges, M., et al. (2008). PKC alpha protein but not kinase activity is critical for glioma proliferation and survival. *Int. J. Cancer* 123:769–779.

Cardell, M., Landsend, A.S., Eidet, J., et al. (1998). High resolution immunogold analysis reveals distinct subcellular compartmentalization of protein kinase C gamma and delta in rat Purkinje cells. *Neuroscience* 82:709–725.

Castagna, M., Takai, Y., Kaibuchi, K., et al. (1982). Direct activation of calcium-activated, phospholipid-dependent protein kinase by tumor-promoting phorbol esters. *J. Biol. Chem.* 257:7847–7851.

Cataisson, C., Joseloff, E., Murillas, R., et al. (2003). Activation of cutaneous protein kinase Calpha induces keratinocyte apoptosis and intraepidermal inflammation by independent signaling pathways. *J. Immunol.* 171:2703–2713.

Chauhan, V.P.S., Chauha, A., Deshmukh, D.S., et al. (1990). Lipid activators of protein kinase C. *Life Sci.* 47:981–986.

Chen, C.C., and Wu, M.L. (1995). Protein kinase C isoform delta is involved in the stimulation of the Na(+)-H+ exchanger in C6 glioma cells. *Mol. Pharmacol.* 48:995–1003.

Chen, D., Gould, C., Garza, R., et al. (2007). Amplitude control of protein kinase C by RINCK, a novel E3 ubiqutin ligase. *J. Biol. Chem.* 282:33776–33787.

Chida, K., Hara, T., Hirai, T., et al. (2003). Disruption of protein kinase Ceta results in impairment of wound healing and enhancement of tumor formation in mouse skin carcinogenesis. *Cancer Res.* 63:2404–2408.

Chida, K., Sagara, H., Suzuki, Y., et al. (1994). The η isoform of protein kinase C is localized on rough endoplasmic reticulum. *Mol. Cell. Biol.* 14:3782–3790.

Coghlan, L.G., Gimenez-Conti, I., Kleiner, H.E., et al. (2000a). Development and initial characterization of several new inbred strains of SENCAR mice for studies of multistage skin carcinogenesis. *Carcinogenesis* 21:641–646.

Coghlan, M.P., Chou, M.M., and Carpenter, C.L. (2000b). Atypical protein kinases Cλ and -ζ associate with the GTP-binding protein Cdc42 and mediate stress fiber loss. *Mol. Cell. Biol.* 20:2280–2289.

Cohen, E.E.W., Lingen, M.W., Zhu, B., et al. (2006). Protein kinase Cζ mediates epidermal growth factor-induced growth of head and neck tumor cells by regulating mitogen-activated protein kinase. *Cancer Res.* 66:6296–6303.

Cressman, C.M., Mohan, P.S., Nixon, R.A., et al. (1995). Proteolysis of protein kinase C: mM and μM calcium-requiring calpains have different abilities to generate, and degrade the free catalytic subunit, protein kinase M. *FEBS Lett.* 367:223–227.

Cross, T., Griffiths, G., Deacon, E., et al. (2000). PKC-delta is an apoptotic lamin kinase. *Oncogene* 19:2331–2337.

Crotty, T., Cai, J., Sakane, F., et al. (2006). Diacylglycerol kinase δ regulates protein kinase C and epidermal growth factor receptor signaling. *Proc. Natl. Acad. Sci. U. S. A.* 103:5485–15490.

Czifra, G., Toth, I.B., Marincsak, R., et al. (2006). Insulin-like growth factor-I-coupled mitogenic signaling in primary cultured human skeletal muscle cells and in C2C12 myoblasts. A central role of protein kinase Cδ. *Cell Signal* 18:1461–1472.

D'Costa, A.M., Robinson, J.K., Maududi, T., et al. (2006). The proapoptotic tumor suppressor protein kinase C-delta is lost in human squamous cell carcinomas. *Oncogene* 25:378–386.

Datta, R., Kojima, H., Yoshida, K., et al. (1997). Caspase-3-mediated cleavage of protein kinase C theta in induction of apoptosis. *J. Biol. Chem.* 272:20317–20320.

Dekker, L.V., Palmer, R.H., Parker, P.J. (1995). The protein kinase C and protein kinase C related gene families. *Curr. Opin. Struct. Biol.* 5:396–402.

Dempsey, E.C., Newton, A.C., Mochly-Rosen, D., et al. (2000). Protein kinase C isozymes and the regulation of diverse responses. *Am. J. Physiol. Lung Cell Mol. Physiol.* 279:L429–L438.

Denning, M.F. (2004). Epidermal keratinocytes: regulation of multiple cell types by multiple protein kinase isoforms. *Int. J. Biochem. Cell Biol.* 36:1141–1146.

Denning, M.F., Dlugosz, A.A., Cheng, C., et al. (2000). Cross-talk between epidermal growth factor receptor and protein kinase C during calcium-induced differentiation of keratinocytes. *Exp. Dermatol.* 9:192–199.

Denning, M.F., Dlugosz, A.A., Howett, M.K., et al. (1993). Expression of an oncogenic ras^{Ha} gene in murine keratinoctyes induces tyrosine phosphorylation and reduced activity of protein kinase C δ. *J. Biol. Chem.* 268:26079–26081.

Denning, M.F., Dlugosz, A.A., Williams, E.K., et al. (1995a). Specific protein kinase C isozymes mediate the induction of keratinocyte differentiation markers by calcium. *Cell Growth Differ.* 6:149–157.

Denning, M.F., Kazanietz, M.G., Blumberg, P.M., et al. (1995b). Cholesterol sulfate activates multiple protein kinase C isoenzymes and induces granular cell differentiation in cultured murine keratinocytes. *Cell Growth Differ.* 6:1619–1626.

Denning, M.F., Wang, Y., Nickoloff, B.J., et al. (1998). Protein kinase C δ is activated by caspase-dependent proteolysis during ultraviolet radiation-induced apoptosis of human keratinoctyes. *J. Biol. Chem.* 273:29995–30002.

Denning, M.F., Wang, Y., Tibudan, S., et al. (2002). Caspase activation and disruption of mitochondrial membrane potential during UV radiation-induced apoptosis of human keratinocytes requires activation of protein kinase C. *Cell Death Differ.* 9:40–52.

Deucher, A., Efimova, T., Eckert, R.L. (2002). Calcium-dependent involucrin expression is inversely regulated by protein kinase C (PKC) alpha and PKC delta. *J. Biol. Chem.* 277:17032–17040.

DeVries, T.A., Neville, M.C., Reyland, M.E. (2002). Nuclear import of PKCdelta is required for apoptosis: identification of a novel nuclear import sequence. *EMBO J.* 21:6050–6060.

Dietrich, C., Gumpert, N., Heit, I., et al. (2001). Rottlerin induces a transformed phenotype in human keratinocytes. *Biochem. Biophys. Res. Commun.* 282:575–579.

Dlugosz, A.A., Mischak, H., Mushinski, J.F., et al. (1992). Transcripts encoding protein kinase C-alpha, -delta, -epsilon, -zeta, and -eta are expressed in basal and differentiating mouse keratinocytes in vitro and exhibit quantitative changes in neoplastic cells. *Mol. Carcinog.* 5:286–292.

Dodd, M.E., Ristich, V.L., Ray, S., et al. (2005). Regulation of protein kinase D during differentiation and proliferation of primary mouse keratinocytes. *J. Invest. Dermatol.* 125:294–306.

Dries, D.R., Gallegos, L.L., Newton, A.C. (2007). A single residue in the C1 domain sensitizes novel protein kinase C isoforms to cellular diacylglycerol production. *J. Biol. Chem.* 282:826–830.

Dutil, E.M., Keranen, L.M., DePaoli-Roach, A.A., et al. (1994). *In vivo* regulation of protein kinase C by trans-phosphorylation followed by autophosphorylation. *J. Biol. Chem.* 269:29359–29362.

Ebinu, J.O., Bottorf, D.A., Chan, E.Y.W., et al. (1998). RasGRP, a ras guanyl nucleotide-releasing protein with calcium - and diacylglycerol-binding motifs. *Science* 280:1082–1086.

Eckert, R.L., Sturniolo, M.T., Broome, A.M., et al. (2005). Transglutaminase function in epidermis. *J. Invest. Dermatol.* 124:481–492.

Efimova, T., Broome, A.M., Eckert, R.L. (2004). Protein kinase Cdelta regulates keratinocyte death and survival by regulating activity and subcellular localization of a p38δ-extracellular signal-regulated kinase 1/2 complex. *Mol. Cell. Biol.* 24:8167–8183.

Emoto, Y., Manome, Y., Meinhardt, G., et al. (1995). Proteolytic activation of protein kinase Cδ by an ICE-like protease in apoptotic cells. *EMBO J.* 14:6148–6156.

Fan, J., Guan, S., Cheng, C.F., et al. (2006). PKCδ clustering at the leading edge and mediating growth factor-enhanced, but not ECM-initiated, dermal fibroblast migration. *J. Invest. Dermatol.* 126:1233–1243.

Farese, R.V., Sajan, M.P., Standaert, M.L. (2005). Insulin-sensitive protein kinases (atypical protein kinase C and protein kinase B/Akt): actions and defects in obesity and type II diabetes. *Exp. Biol. Med.* 230:593–605.

Farshori, P.Q., Shah, B.H., Arora, K.K., et al. (2003). Activation and nuclear translocation of PKCdelta, Pyk2 and ERK1/2 by gonadotropin releasing hormone in HEK293 cells. *J. Steroid Biochem. Mol. Biol.* 85:337–347.

Fedorov, Y.V., Jones, N.C., Olwin, B.B. (2002). Atypical protein kinase Cs are the Ras effectors that mediate repression of myogenic satellite cell differentiation. *Mol. Cell. Biol.* 22:1140–1149.

Fields, A.P., Calcagno, S.R., Krishna, M., et al. (2009). Protein kinase Cβ is an effective target for chemoprevention of colon cancer. *Cancer Res.*69:1643–1650.

Fima, E., Shtutman, M., Libros, P., et al. (2001). PKCeta enhances cell cycle progression, the expression of G1 cyclins and p21 in MCF-7 cells. *Oncogene* 20:6794–6804.

Fischer, S.M., Lee, M.L., Maldve, R.E., et al. (1993). Association of protein kinase C activation with induction of ornithine decarboxylase in murine but not human keratinocyte cultures. *Mol. Carcinog.* 7:228–237.

Fisher, G.J., Tavakkol, A., Leach, K., et al. (1993). Differential expression of protein kinase C isoenzymes in normal and psoriatic adult human skin: reduced expression of protein kinase C-beta II in psoriasis. *J. Invest. Dermatol.* 101:553–559.

Flescher, E., and Rotem, R. (2002). Protein kinase Cε mediates the induction of P-glycoprotein in LNCaP prostate carcinoma cells. *Cell Signal* 14:37–43.

Fujiwara, T., Bandi, M., Nitta, M., et al. (2005). Cytokinesis failure generating tetraploids promotes tumorigenesis in p53-null cells. *Nature* 437:1043–1047.

Fuchs, E., and Raghavan, S. (2002). Getting under the skin of epidermal morphogenesis. *Nat. Rev. Genet.* 3:199–209.

Fukunaga, M., Oka, M., Ichihashi, M., et al. (2001). UV-induced tyrosine phosphorylation of PKC delta and promotion of apoptosis in the HaCaT cell line. *Biochem. Biophys. Res. Commun.* 289:573–579.

Galvez, A.S., Duran, A., Linares, J.F., et al. (2009). Protein kinase Cζ represses the interleukine-6 promoter and impairs tumorigenesis in vivo. *Mol. Cell. Biol.* 29:104–115.

Gao, T., Brognard, J., Newton, A.C. (2008). The phosphatase PHLPP controls the cellular levels of protein kinase C. *J. Biol. Chem.* 283:6300–6311.

Gao, T., and Newton, A.C. (2002). The turn motif is a phosphorylation switch that regulates the binding of Hsp70 to protein kinase C. *J. Biol. Chem.* 277:31585–31592.

Gimond, C., de Melker, A., Aumailley, M., et al. (1995). The cytoplasmic domain of α6A integrin subunit is an in vitro substrate for protein kinase C. *Exp. Cell Res.* 216:232–235

Giorgione, J.R., Lin, J.H., McCammon, J.A., et al. (2006). Increased membrane affinity of the C1 domain of protein kinase Cδ compensates for the lack of involvement of its C2 domain in membrane recruitment. *J. Biol. Chem.* 281:1660–1669.

Gokmen-Polar, Y., Murray, N.R., Velasco, M.A., et al. (2001). Elevated protein kinase C βII is an early promotive event in colon carcinogenesis. *Cancer Res.* 61:1375–1381.

Gonzales-Guerrico, A.M., Meshki, J., Xiao, L., et al. (2005). Molecular mechanisms of protein kinase C-induced apoptosis in prostate cancer cells. *J. Biochem. Mol. Biol.* 38:639–645.

Gould, C.M., Kannan, N., Taylor, S.S., et al. (2009). The chaperons Hsp90 and Cd37 mediate the maturation and stabilization of protein kinase C through a conserved PXXP motif in the C-terminal tail. *J. Biol. Chem.* 284:4921–4935.

Gray, M.O., Karliner, J.S., Mochly-Rosen, D. (1997). A selective ε-protein kinase C antagonist inhibits protection of cardiac myocytes from hypoxia-induced cell death. *J. Biol. Chem.* 272:30945–30951.

Griner, E.M., and Kazanietz, M.G. (2007). Protein kinase C and other diacylglycerol effectors in cancer. *Nat. Rev. Cancer* 7:281–293.

Grossoni, V.C., Falbo, K.B., Mauro, L.V., et al. (2007). Protein kinase Cδ inhibits the production of proteolytic enzymes in murine mammary cells. *Clin. Exp. Metastas.* 24:513–520.

Gschwendt, M. (1999). Protein kinase C delta. *Eur. J. Biochem.* 259:555–564.

Gschwendt, M., Kielbassa, K., Kittstein, W., et al. (1994). Tyrosine phosphorylation and stimulation of protein kinase Cd from porcine spleen by src *in vitro*. Dependence on the activated state of protein kinase Cδ. *FEBS Lett.* 347:85–89.

Gutschalk, C.M., Herold-Mende, C.C., Fusenig, N.E., et al. (2006). Granulocyte colony-stimulating factor and granulocyte-macrophage colony-stimulating factor promote malignant growth from head and neck squamous cell carcinomas in vivo. *Cancer Res.* 66:8026–8036.

Hamilton, M., Liao, J., Cathcart, M.K., et al. (2001). Constitutive association of c-N-Ras with c-Raf-1 and protein kinase C epsilon in latent signaling modules. *J. Biol. Chem.* 276:29079–29090.

Hansen, L.A., Monteiro-Riviere, N.A., Smart, R.C. (1990). Differential down-regulation of epidermal protein kinase C by 12-O-tetradecanoylphorbol-13-acetate and diacylglycerol: association with epidermal hyperplasia and tumor promotion. *Cancer Res.* 50:5740–5745.

Hansra, G., Garcia-Paramio, P., Prevostel, C., et al. (1999). Multisite dephosphorylation and desensitation of conventional protein kinase C isotypes. *Biochem. J.* 342:337–344.

Hara, T., Saito, Y., Hirai, T., et al. (2005). Deficiency of protein kinase Cα in mice results in impairment of epidermal hyperplasia and enhancement of tumor formation in two-stage skin carcinogenesis. *Cancer Res.* 65:7356–7362.

Hecker, E. (1985). Cell membrane associated protein kinase C as receptor of diterpene ester co-carcinogens of the tumor promoter type and the phenotypic expression of tumors. *Arzneimittelforschung* 35:1890–1903.

Heit, I., Wieser, R.J., Herget, T., Faust, D., et al. (2001). Involvement of protein kinase Cδ in contact-dependent inhibition of growth in human and murine fibroblasts. *Oncogene* 20:5143–5154.

Helfrich, I., Schmitz, A., Zigrino, P., et al. (2007). Role of aPKC isoforms and their binding partners Par3 and Par6 in epidermal barrier formation. *J. Invest. Dermatol.* 127:782–791.

Hennings, H., Michael, D., Cheng, C., et al. (1980). Calcium regulation of growth and differentiation of mouse epidermal cells in culture. *Cell* 19:245–254.

Hernandez, A.I., Blace, N., Crary, J.F., et al. (2003). Protein kinase Mζ synthesis from a brain mRNA encoding an independent protein kinase Cζ catalytic domain. *J. Biol. Chem.* 278:40305–40316.

Hirai, T., and Chida, K. (2003). Protein kinase Cζ (PKCζ): activation mechanisms and cellular functions. *J. Biochem.* 133:1–7.

Horikoshi, Y., Suzuki, A., Yamanaka, T., et al. (2009). Interaction between PAR-3 and the aPKC-PAR-6 complex is indispensable for apical domain development of epithelial cells. *J. Cell Sci.* 122:1595–1606.

Hornia, A., Lu, Z., Sukezane, T., et al. (1999). Antagonistic effects of protein kinase C alpha and delta on both transformation and phospholipase D activity mediated by the epidermal growth factor receptor. *Mol. Cell. Biol.* 19:7672–7680.

House, C., and Kemp, B.E. (1987). Protein kinase C contains a pseudosubstrate prototype in its regulatory domain. *Science* 238:1726–1728.

Huber, M., Siegenthaler, G., Mirancea, N., et al. (2005). Isolation and characterization of human repetin, a member of the fused gene family of the epidermal differentiation complex. *J. Invest. Dermatol.* 124:998–1007.

Humphries, M.J., Limesand, K.H., Schneider, J.C., et al. (2006). Suppression of apoptosis in the protein kinase Cδ null mouse in vivo. *J. Biol. Chem.* 281:9728–9737.

Humphries, M.J., Ohm, A.M., Schaack, J., et al. (2008). Tyrosine phosphorylation regulates nuclear translocation of PKCδ. *Oncogene* 27:3045–3953.

Ikuta, T., Chida, K., Tajima, O., et al. (1994). Cholesterol sulfate, a novel activator for the eta isoform of protein kinase C. *Cell Growth Differ.* 5:943–947.

Imamoto, A., Wang, X.J., Fujiki, H., et al. (1993). Comparison of 12-O-tetradecanoylphor-bol-13-acetate and teleocidin for induction of epidermal hyperplasia, activation of epidermal PKC isozymes and skin tumor promotion in SENCAR and C57BL/6 mice. *Carcinogenesis* 14:719–724.

Inoue, M., Kishimoto, A., Takai, Y., et al. (1977). Studies on a cyclic nucleotide-independent protein kinase and its proenzyme in mammalian tissues. *J. Biol. Chem.* 252:7610–7616.

Irie, N., Sakai, N., Ueyama, T., et al. (2002). Subtype- and species-specific knockdown of PKC using short interfering RNA. *Biochem. Biophys. Res. Commun.* 298:738–743.

Ivaska, J., Vuoriluoto, K., Huovinen, T., et al. (2005). PKCε-mediated phosphorylation of vimentin controls integrin recycling and motility. *EMBO J.* 24:3834–3845.

Izumi, Y., Hirose, T., Tamai, Y., et al. (1998). An atypical PKC directly associates and colocalizes at the epithelial tight junction with ASIP, a mammalian homologue of Caenorhabditis elegans polarity protein PAR-3. *J. Cell Biol.* 143:95–106.

Jackson, D.N., and Foster, D.A. (2004). The enigmatic protein kinase Cδ: complex roles in cell proliferation and survival. *FASEB J.* 18:627–636.

Jackson, D., Zheng, Y., Lyo, D., et al. (2005). Suppression of cell migration by protein kinase Cδ. *Oncogene* 24:3067–3072.

Jaken, S., and Parker, P.J. (2000). Protein kinase C binding partners. *Bioessays* 22:245–254.

Jansen, A.P., Dreckschmidt, N.E., Verwiebe, E.G., et al. (2001a). Relation of the induction of epidermal ornithine decarboxylase and hyperplasia to the different skin tumor-promotion susceptibilities of protein kinase C alpha, -delta and -epsilon transgenic mice. *Int. J. Cancer* 93:635–643.

Jansen, A.P., Verwiebe, E.G., Dreckschmidt, N.E., et al. (2001b). Protein kinase C-epsilon transgenic mice: a unique model for metastatic squamous cell carcinoma. *Cancer Res.* 61:808–812.

Johnson, J.E., Giorgione, J., Newton, A.C. (2000). The C1 and C2 domains of protein kinase C are independent membrane targeting modules, with specificity for phosphatidylserine conferred by the C1 domain. *Biochemistry* 39:11360–11369.

Joseloff, E., Cataisson, C., Aamodt, H., et al. (2002). Src family kinases phosphorylate protein kinase C delta on tyrosine residues and modify the neoplastic phenotype of skin keratinocytes. *J. Biol. Chem.* 277:12318–12323.

Joshi, J., Fernandez-Marcos, P.J., Galvez, A., et al. (2008). Par-4 inhibits Akt and suppresses ras-induced lung tumorigenesis. *EMBO J.* 27:2181–2193.

Kajimoto, T., Shirai, Y., Sakai, N., et al. (2004). Ceramide-induced apoptosis by translocation, phosphorylation, and activation of protein kinase Cδ in the Golgi complex. *J. Biol. Chem.* 279:12668–12676.

Kanzaki, M., Mora, S., Hwang, J.B., et al. (2004). Atypical protein kinase C (PKCζ/λ) is a convergent downstream target of the insulin-stimulated phosphatidylinositol 3-kinase and TC10 signaling pathways. *J. Cell Biol.* 164:279–290.

Kashiwagi, M., Ohba, M., Watanabe, H., et al. (2000). PKCeta associates with cyclin E/cdk2/p21 complex, phosphorylates p21 and inhibits cdk2 kinase in keratinocytes. *Oncogene* 19:6334–6341.

Kasper, M., Regl, G., Frischauf, A.M., et al. (2006). GLI transcription factors: mediators of oncogenic Hedgehog signalling. *Eur. J. Cancer* 42:437–445.

Kazanietz, M.G., Areces, L.B., Bahador, A., et al. (1993). Characterization of ligand and substrate specificity for the calcium-dependent and calcium-independent PKC isozymes. *Mol. Pharmacol.* 44:298–307.

Kermorgant, S., Aparicio, T., Dessirier, V., et al. (2001). Hepatocyte growth factor induces colonic cancer cell invasiveness via enhanced motility and protease overproduction. Evidence for PI3 kinase and PKC involvement. *Carcinogenesis* 22:1035–1042.

Keum, E., Kim, Y., Kwon, S., et al. (2004). Syndecan-4 regulates localization, activity and stability of protein kinase C-α. *Biochem. J.* 378:1007–1014.

Kiley, S.C., Clark, K.J., Duddy, S.K., et al. (1999a). Increased protein kinase C delta in mammary tumor cells: relationship to transformation and metastatic progression. *Oncogene* 18:6748–6757.

Kiley, S.C., Clark, K.J., Goodnough, M., et al. (1999b). Protein kinase C δ involvement in mammary tumor metastasis. *Cancer Res.* 59:3230–3238.

Kishimoto, A., Mikawa, K., Hashimoto, K., et al. (1989). Limited proteolysis of protein kinase C subspecies by calcium-dependent neutral protease (calpain). *J. Biol. Chem.* 264:4088–4092.

Koizumi, H., Kohno, Y., Osada, S., et al. (1993). Differentiation-associated localization of nPKCη, a Ca^{++}-independent protein kinase C, in normal human skin and skin diseases. *J. Invest. Dermatol.* 101:858–863.

Konishi, H., Tanaka, M., Takemura, Y., et al. (1997). Activation of protein kinase C by tyrosine phosphorylation in response to H_2O_2. *Proc. Natl. Acad. Sci. U. S. A.* 94:11233–11237.

Kuroki, T., Ikuta, T., Kashiwagi, M., et al. (2000). Cholesterol sulfate, an activator of protein kinase C mediating squamous differentiation: a review. *Mutation Res.* 462:189–195.

Lahn, M., Paterson, B.M., Sundell, K., et al. (2004). The role of protein kinase C-alpha (PKCα) in malignancies of the gastrointestinal tract. *Eur. J. Cancer* 40:10–20.

Lahn, M., and Sundell, K.L. (2004). The role of protein kinase C-alpha (PKCα) in melanoma. *Melanoma Res.* 14:85–89.

Langbein, L., Grund, C., Kuhn, C., et al. (2002). Tight junctions and compositionally related junctional structures in mammalian stratified epithelia and cell cultures derived therefrom. *Eur. J. Cell Biol.* 81:419–435.

Le Good, J.A., and Brindley, D.N. (2004). Molecular mechanisms regulating protein kinase C-zeta turnover and cellular transformation. *Biochem. J.* 378:83–92.

Le Good, J.A., Ziegler, W.H., Parekh, D.B., et al. (1998). Protein kinase C isotypes controlled by phosphoinositide 3-kinase through the protein kinase PDK1. *Science* 281:2042–2045.

Lee, Y.S., Yuspa, S.H., Dlugosz, A.A. (1998). Differentiation of cultured human epidermal keratinocytes at high cell densities is mediated by endogenous activation of the protein kinase C signaling pathway. *J. Invest. Dermatol.* 111:762–766.

Lehel, C., Olah, Z., Jakab, G., et al. (1995a). Protein kinase Cε is localized to the Golgi via its zinc-finger domain and modulates Golgi function. *Proc. Natl. Acad. Sci. U. S. A.* 92:1406–1410.

Lehel, C., Olah, Z., Jakab, G., et al. (1995b). Protein kinase C ε subcellular localization domains and proteolytic degradation sites. A model for protein kinase C conformational changes. *J. Biol. Chem.* 270:19651–19658.

Leinweber, B., Parissenti, A.M., Gallant, C., et al. (2000). Regulation of protein kinase C by the cytoskeletal protein calponin. *J. Biol. Chem.* 275:40329–40336.

Leitges, M., Sanz, L., Martin, P., et al. (2001). Targeted disruption of the zetaPKC gene results in the impairment of the NF-kappaB pathway. *Mol. Cell* 8:771–780.

Leontieva, O.V., and Black, J.D. (2004). Identification of two distinct pathways of protein kinase Cα down-regulation in intestinal epithelial cells. *J. Biol. Chem.* 279:5788–5801.

Li, L., Lorenzo, P.S., Bogi, K., et al. (1999). Protein kinase Cdelta targets mitochondria, alters mitochondrial membrane potential, and induces apoptosis in normal and neoplastic keratinocytes when overexpressed by an adenoviral vector. *Mol. Cell. Biol.* 19:8547–8558.

Li, W., Mischak, H., Yu, J.C., et al. (1994a). Tyrosine phosphorylation of protein kinase C-delta in response to its activation. *J. Biol. Chem.* 269:2349–2352.

Li, W., Yu, J.C., Michieli, P., et al. (1994b). Stimulation of the platelet-derived growth factor β receptor signaling pathway activates protein kinase C-δ. *Mol. Cell. Biol.* 14:6727–6735.

Li, W., Nadelman, C., Gratch, N.S., et al. (2002). An important role for protein kinase C-delta in human keratinocyte migration on dermal collagen. *Exp. Cell Res.* 273:219–228.

Li, Y., Wheeler, D.L., Alters, W., et al. (2005). Early epidermal destruction with subsequent epidermal hyperplasia is a unique feature of the papilloma-independent squamous cell carcinoma phenotype in PKCepsilon overexpressing transgenic mice. *Toxicol. Pathol.* 33:684–694.

Liberman, Z., Plotkin, B., Tennenbaum, T., et al. (2008). Coordinated phosphorylation of insulin receptor substrate-1 by glycogen synthase kinase-3 and protein kinase CβII in the diabetic fat tissue. *Am. J. Endocrinol. Metab.* 294:E1169–E1177.

Lim, S.T., Longley, R., Couchman, J.R., et al. (2003). Direct binding of syndecan-4 cytoplasmic domain to the catalytic domain of protein kinase Cα (PKCα) increases focal adhesion localization of PKCα*. *J. Biol. Chem.* 278:13795–13802.

Liu, Y., Su, W., Thompson, E.A., et al. (2004). Protein kinase CβII regulates its own expression in rat intestinal cells and the colonic epithelium in vivo. *J. Biol. Chem.* 279:45556–45563.

Liu, Y.F., Paz, K., Herschkovitz, A., et al. (2001). Insulin stimulates PKCζ-mediated phosphorylation of insulin receptor substrate-1 (IRS-1). *J. Biol. Chem.* 276:14459–14465.

Lu, Z., Hornia, A., Jiang, Y.W., et al. (1997). Tumor promotion by depleting cells of protein kinase Cδ. *Mol. Cell. Biol.* 17:3418–3428.

Lu, Z., Liu, D., Hornia, A., et al. (1998). Activation of protein kinase C triggers its ubiquitination and degradation. *Mol. Cell. Biol.* 18:839–845.

Luke, C.T., Oki-Idouchi, C.E., Cline, J.M., et al. (2007). RasGRP1 overexpression in the epidermis of transgenic mice contributes to tumor progression during multistage skin carcinogenesis. *Cancer Res.*67:10190–10197.

Mashukova, A., Oriolo, A.S., Wald, F.A., et al. (2009). Rescue of atypical protein kinase C in epithelia by the cytoskeleton and Hsp70 family chaperones. *J. Cell Sci.* 122:2491–2503.

Masso-Welch, P.A., Winston, J.S., Edge, S., et al. (2001). Altered expression and localization of PKC eta in human breast tumors. *Breast Cancer Res. Treat.* 68:211–223.

Masukawa, K., Sakai, N., Ohmori, S., et al. (2006). Spatiotemporal analysis of the molecular interaction between PICK and PKC. *Acta Histochem. Cytochem.* 39:173–181.

Mellor, H., and Parker, P.J. (1998). The extended protein kinase C superfamily. *Biochem. J.* 332:281–292.

Michie, A.M., and Nakagawa, R. (2005). The link between PKCα regulation and cellular transformation. *Immun. Lett.* 96:155–162.

Mills, K.J., Reynolds, S.H., Smart, R.C. (1993). Diacylglycerol is an effector of the clonal expansion of cells containing activated Ha-ras genes. *Carcinogenesis* 14:2645–2648.

Mills, K.J., and Smart, R.C. (1989). Comparison of epidermal protein kinase C activity, ornithine decarboxylase induction and DNA synthesis stimulated by TPA or dioctanoylglycerol in mouse strains with differing susceptibility to TPA-induced tumor promotion. *Carcinogenesis* 10:833–838.

Mineo, C., Ying, Y.S., Chapline, C., et al. (1998). Targeting of protein kinase Cα to caveolae. *J. Cell Biol.* 141:601–610.

Moscat, J., and Diaz-Meco, M.T. (2000). The atypical protein kinase Cs. *EMBO Rep.* 11:399–403.

Moscat, J., Diaz-Meco, M.T., Rennert, P. (2003). NFκB activation by protein kinase C isoforms and B-cell function. *EMBO Rep.* 4:31–35.

Moscat, J., Rennert, P., Diaz-Meco, M.T. (2006). PKCζ at the crossroad of NF-κB and Jak/Stat6 signaling pathways. *Cell Death Differ.* 13:702–711.

Mueller, M.M. (2006). Inflammation in epithelial skin tumours. *Eur. J. Cancer* 42:735–744.

Murray, N.R., Davidson, L.A., Chapkins, R.S., et al. (1999). Overexpression of protein kinase C betaII induces colonic hyperproliferation and increased sensitivity to colon carcinogenesis. *J. Cell Biol.* 145:699–711.

Murray, N.R., Weems, J., Braun, U., et al. (2009). Protein kinase C βII and PKCι/λ: collaborating partners in colon cancer promotion and progression. *Cancer Res.* 69:656–662.

Neill, G.W., Ghali, L.R., Green, J.L., et al. (2003). Loss of protein kinase Cα expression may enhance the tumorigenic potential of Gli1 in basal cell carcinoma. *Cancer Res.* 63:4692–4697.

Newton, A.C. (1997). Regulation of protein kinase C. *Curr. Opin. Cell Biol.* 9:161–167.

Newton, A.C. (2003). Regulation of the ABC kinases by phosphorylation: protein kinase C as a paradigm. *Biochem. J.* 370:361–371.

Ng, T., Shima, D., Squire, A., et al. (1999). PKCα regulates β1 integrin-dependent cell motility through association and control of integrin traffic. *EMBO J.* 18:3909–3923.

Nischt, R., Schmidt, C., Mirancea, N., et al. (2007). Lack of nidogen-1 and -2 prevents basement membrane assembly in skin-organotypic coculture. *J. Invest. Dermatol.* 127:545–554.

Nishikawa, K., Yamamoto, S., Nagumo, H., et al. (1997). Activation of phorbol ester responsive form of protein kinase C zeta in association with Ca(2+)-induced differentiation of primary cultured mouse epidermal cells. *Jpn. J. Pharmacol.* 73:171–174.

Nishizuka, Y. (1988). The molecular heterogeneity of PKC and its implications for cellular regulation. *Nature* 334:661–665.

Nishizuka, Y. (1995). Protein kinase C and lipid signaling for sustained cellular responses. *FASEB J.* 9:484–496.

Novotny-Diermayr, V., Zhang, T., Gu, L., et al. (2002). Protein kinase C delta associates with the interleukin-6 receptor subunit glycoprotein (gp) 130 via Stat3 and enhances Stat3-gp130 interaction. *J. Biol. Chem.* 277:49134–49142.

Ohba, M., Ishino, K., Kashiwagi, M., et al. (1998). Induction of differentiation in normal human keratinocytes by adenovirus-mediated introduction of the eta and delta isoforms of protein kinase C. *Mol. Cell. Biol.* 18:5199–5207.

Ohmori, S., Shirai, Y., Sakai, N., et al. (1998). Three distinct mechanisms for translocation and activation of the delta subspecies of protein kinase C. *Mol. Cell. Biol.* 18:5263–5271.

Oki-Idouchi, C.E., and Lorenzo, P.S. (2007). Transgenic overexpression of RasGRP1 in mouse epidermis results in spontaneous tumors of the skin. *Cancer Res.* 67:276–280.

Osada, S., Hashimoto, Y., Nomura, S., et al. (1993). Predominant expression of nPKCη, a Ca^{2+}-independent isoform of protein kinase C in epithelial tissues, in association with epithelial differentiation. *Cell Growth Differ.* 4:167–175.

Osada, S., Mizuno, K., Saido, T.C., et al. (1992). A new member of the protein kinase C family, nPKC theta, predominantly expressed in skeletal muscle. *Mol. Cell. Biol.* 12:3930–3938.

Oster, H., and Leitges, M. (2006). Protein kinase C α but not PKCζ suppresses intestinal tumor formation in $APC^{Min/+}$ mice. *Cancer Res.* 66:6955–6963.

Papp, H., Czifra, G., Bodo, E., et al. (2004). Opposite roles of protein kinase isoforms in proliferation, differentiation, apoptosis, and tumorigenicity of human HaCaT keratinocytes. *Cell. Mol. Life Sci.* 61:1095–1105.

Paramio, J.M., Segrelles, C., Ruiz, S., et al. (2001). Inhibition of protein kinase B (PKB) and PKCzeta mediates keratin K10-induced cell cycle arrest. *Mol. Cell. Biol.* 21:7449–7459.

Parekh, D.B., Ziegler, W., Parker, P.J. (2000). Multiple pathways control protein kinase C phosphorylation. *EMBO J.* 19:496–503.

Parissenti, A.M., Kirwan, A.F., Kim, S.A., et al. (1998). Inhibitory properties of the regulatory domains of human protein kinase Cα and mouse protein kinase Cε. *J. Biol. Chem.* 273:8940–8945.

Parker, P.J., and Murray-Rust, J. (2004). PKC at a glance. *J. Cell Sci.* 117:131–132.

Parkinson, S.J., Le Good, J.A., Whelan, R.D.H., et al. (2004). Identification of PKCζII: an endogenous inhibitor of cell polarity. *EMBO J.* 23:77–88.

Prevostel, C., Alice, V., Joubert, D., et al. (2000). Protein kinase Cα actively downregulates through caveolae-dependent traffic to an endosomal compartment. *J. Cell Sci.* 113:2575–2584.

Puls, A., Schmidt, S., Grawe, F., et al. (1997). Interaction of protein kinase C ζ with ZIP, a novel protein kinase C-binding protein. *Proc. Natl. Acad. Sci. U. S. A.* 94:6191–6196.

Pysz, M.A., Leontieva, O.V., Bateman, N.W., et al. (2009). PKCα tumor suppression in the intestine is associated with transcriptional inhibition of cyclin D1. *Exp. Cell Res.* 315:1415–1428.

Qi, X., and Mochly-Rosen, D. (2008). PKCδ-Abl complex communicates ER stress to the mitochondria - an essential step in subsequent apoptosis. *J. Cell Sci.* 121:804–813.

Razin, E., Szallasi, Z., Kazanietz, M.G., et al. (1994). Protein kinases C-beta and C-epsilon link the mast cell high-affinity receptor for IgE to the expression of c-fos and c-jun. *Proc. Natl. Acad. Sci. U. S. A.* 91:7722–7726.

Reddig, P.J., Dreckschmidt, N.E., Ahrens, H., et al. (1999). Transgenic mice overexpressing protein kinase C delta in the epidermis are resistant to skin tumor promotion by 12-O-tetradecanoylphorbol-13-acetate. *Cancer Res.* 59:5710–5718.

Reddig, P.J., Dreckschmidt, N.E., Zou, J., et al. (2000). Transgenic mice overexpressing protein kinase C epsilon in their epidermis exhibit reduced papilloma burden but enhanced carcinoma formation. *Cancer Res.* 60:595–602.

Reyland, M.E. (2007). Protein kinase C delta and apoptosis. *Biochem. Soc. Transact.* 35:1001–1004.

Riobo, N.A., Haines, G.M., Emerson, Jr, C.P. (2006). Protein kinase C-δ and mitogen-activated protein/extracellular signal-regulated kinase-1 control GLI activation in hedgehog signaling. *Cancer Res.* 66:839–845.

Riobo, N.A., and Manning, D.R. (2007). Pathways of signal transduction employed by vertebrate Hedgehogs. *Biochem. J.* 403:369–379.

Rosenzweig, T., Braiman, L., Bak, A., et al. (2002). Differential effects of tumor necrosis factor-α on protein kinase C isoforms α and δ mediate inhibition of insulin receptor signaling. *Diabetes* 51:1921–1930.

Rozengurt, E., Rey, O., Waldron, R.T. (2005). Protein kinase D signaling. *J. Biol. Chem.* 280:13205–13208.

Rutberg, S.E., Saez, E., Glick, A., et al. (1996). Differentiation of mouse keratinocytes is accompanied by PKC-dependent changes in AP-1 proteins. *Oncogene* 13:167–176.

Rutberg, S.E., Saez, E., Lo, S., et al. (1997). Opposing activities of c-Fos and Fra-2 on AP-1 regulated transcriptional activity in mouse keratinocytes induced to differentiate by calcium and phorbol esters. *Oncogene* 15:1337–1346.

Rybin, V.O., Guo, J., Gertsberg, Z., et al. (2008). Phorbol 12-myristate 13-acetate-dependent protein kinase Cδ-Tyr^{311} phosphorylation in cardiomyocyte caveolae. *J. Biol. Chem.* 283:17777–17788.

Ryle, C.M., Breitkreutz, D., Stark, H.J., et al. (1989). Density-dependent modulation of synthesis of keratins 1 and 10 in the human keratinocyte line HaCaT and in ras-transfected tumorigenic clones. *Differentiation* 40:42–54.

Sadagurski, M., Yakar, S., Weingarten, G., et al. (2006). Insulin-like growth factor 1 receptor signaling regulates skin development and inhibits skin keratinocyte differentiation. *Mol. Cell. Biol.* 26:2675–2687.

Sanchez, P., de Carcer, G., Sandoval, I.V., et al. (1998). Localization of atypical protein kinase C isoforms into lysosome-targeted endosomes through interaction with p62. *Mol. Cell. Biol.* 18:3069–3080.

Sajan, M.P., Rivas, J., Li, P., et al. (2006). Repletion of atypical protein kinase C following RNA interference-mediated depletion restores insulin-stimulated glucose transport. *J. Biol. Chem.* 281:17466–17473.

Sakaguchi, M., Miyazaki, M., Sonegawa, H., et al. (2004). PKCalpha mediates TGFbeta-induced growth inhibition of human keratinocytes via phosphorylation of S100C/A11. *J. Cell Biol.* 164:979–984.

Saurin, A.T., Durgan, J., Cameron, A.J., et al. (2008). The regulated assembly of PKCε complex controls the completion of cytokinesis. *Nat. Cell Biol.* 10:891–901.

Schechtman, D., and Mochly-Rosen, D. (2001). Adaptor proteins in protein kinase C-mediated signal transduction. *Oncogene* 20:6339–6347.

Segrelles, C., Moral, M., Lara, M.F., et al. (2006). Molecular determinants of Akt-induced keratinocyte transformation. *Oncogene* 25:1174–1185.

Selbie, L.A., Schmitz-Peiffer, C., Sheng, Y., et al. (1993). Molecular cloning and characterization of PKC iota, an atypical isoform of protein kinase C derived from insulin-secreting cells. *J. Biol. Chem.* 268:24296–24302.

Sharkey, N.A., Leach, K.L., Blumberg, P.M. (1984). Competitive inhibition by diacylglycerol of specific phorbol ester binding. *Proc. Natl. Acad. Sci. U. S. A.* 81:607–610.

Shen, S., Alt, A., Wertheimer, E., et al. (2001). PKCdelta activation: a divergence point in the signaling of insulin and IGF-1-induced proliferation of skin keratinocytes. *Diabetes* 50:255–264.

Shimao, Y., Nabeshima, K., Inoue, T., et al. (1999). TPA-enhanced motility and invasion in a highly metastatic variant (L-10) of human rectal adenocarcinoma cell line RCM-1: selective role of PKC-alpha and its inhibition by a combination of PDBu-induced PKC downregulation and antisense oligonucleotides treatment. *Clin. Exp. Metastas.* 17:351–360.

Shtutman, M., Hershko, T., Maissel, A., et al. (2003). PKCeta associates with cyclin E/Cdk2 complex in serum-starved MCF-7 and NIH-3T3 cells. *Exp. Cell Res.* 286:22–29.

Sitailo, L.A., Tibudan, S.S., Denning, M.F. (2004). Bax activation and induction of apoptosis in human keratinocytes by the protein kinase C δ catalytic domain. *J. Invest. Dermatol.* 123:434 – 443.

Sitailo, L.A., Tibudan, S.S., Denning, M.F. (2006). The protein kinase Cδ catalytic fragment targets Mcl-1 for degradation to trigger apoptosis. *J. Biol. Chem.* 281:29703–29710.

Stanwell, C., Denning, M.F., Rutberg, S.E., et al. (1996). Staurosporine induces a sequential program of mouse keratinocyte terminal differentiation through activation of PKC isozymes. *J. Invest. Dermatol.* 106:482–489.

Steinberg, S.F. (2004). Distinctive activation mechanisms and functions for protein kinase Cδ. *Biochem. J.* 384:449–459.

Sumandea, M.P., Rybin, V.O., Hinken, A.C., et al. (2008). Tyrosine phosphorylation modifies protein kinase Cδ-dependent phosphorylation of cardiac troponin I. *J. Biol. Chem.* 283:22680–22689.

Sun, R., Gao, P., Chen, L., et al. (2005). Protein kinase Cζ is required for epidermal growth factor-induced chemotaxis of human breast cancer cells. *Cancer Res.* 65:1433–1441.

Suzuki, A., Hirata, M., Kamimura, K., et al. (2004). aPKC acts upstream of PAR-1b in both the establishment and maintenance of mammalian epithelial polarity. *Curr. Biol.* 14:1425–1435.

Suzuki, A., Ishiyama, C., Hashiba, K., et al. (2002). aPKC kinase activity is required for the asymmetric differentiation of the premature junctional complex during epithelial cell polarization. *J. Cell Sci.* 115:3565–3573.

Suzuki, A., and Ohno, S. (2006). The PAR-aPKC system: lessons in polarity. *J. Cell Sci.* 119:979–987.

Suzuki, A., Yamanaka, T., Hirose, T., et al. (2001). Atypical protein kinase C is involved in the evolutionarily conserved PAR protein complex and plays a critical role in establishing epithelia-specific junctional structures. *J. Cell Biol.* 152:1183–1196.

Svetek, J., Schara, M., Pecar, S., et al. (1995). Spectroscopic characterization of specific phorbol ester binding to PKC-receptor sites in membranes in situ. *Carcinogenesis* 16:2589–2592.

Szalay, J., Bruno, P., Bhati, R., et al. (2001). Associations of PKC isoforms with the cytoskeleton of B16F10 melanoma cells. *J. Histochem. Cytochem.* 49:49–65.

Takahashi, H., Asano, K., Manabe, A., et al. (1998). The alpha and eta isoforms of protein kinase C stimulate transcription of human involucrin gene. *J. Invest. Dermatol.* 110:218–223.

Takai, Y., Kishimoto, A., Iwasa, Y., et al. (1979). Calcium-dependent activation of a multifunctional protein kinase by membrane phospholipids. *J. Biol. Chem.* 254:3692–3695.

Tan, M., Li, P., Sun, M., et al. (2006). Upregulation and activation of PKCα by ErbB2 through Src promotes breast cancer cell invasion that can be blocked by combined treatment with PKCα and Src inhibitors. *Oncogene* 25:3286–3295.

Tapia, J.A., Garcia-Marin, L.J., Jensen, R.T. (2003). Cholecystokinin-stimulated protein kinase C-delta kinase activation, tyrosine phosphorylation, and translocation are mediated by Src tyrosine kinases in pancreatic acinar cells. *J. Biol. Chem.* 278:35220–35230.

Tennenbaum, T., Weiner, A.K., Belanger, A.J., et al. (1993). The suprabasal expression of α6β4 integrins is associated with a high risk for malignant progression in mouse skin carcinogenesis. *Cancer Res.* 53:4803–4810.

Tibudan, S.S., Wang, Y., Denning, M.F. (2002). Activation of protein kinase C triggers irreversible cell cycle withdrawal in human keratinocytes. *J. Invest. Dermatol.* 119:1282–1289.

Tigges, U., Koch, B., Wissing, J., et al. (2003). The F-actin cross-linking and focal adhesion protein filamin A is a ligand and in vivo substrate for protein kinase Cα*. *J. Biol. Chem.* 278:23561–23569.

Tomakidi, P., Mirancea, N., Fusenig, N.E., et al. (1999). Defects of basement membrane and hemidesmosome structure correlate with malignant phenotype and stromal interactions in HaCaT-ras xenografts. *Differentiation* 64:263–275.

Tomakidi, P., Stark, H.J., Herold-Mende, C., et al. (2003). Discriminating expression of differentiation markers evolves in transplants of benign and malignant human skin keratinocytes through stromal interactions. *J. Pathol.* 200:298–307.

Uddin, S., Sassano, A., Deb, D.K., et al. (2002). Protein kinase C-delta (PKCdelta) is activated by type I interferons and mediates phosphorylation of Stat1 on serine 727. *J. Biol. Chem.* 277:14408–14416.

Ueda, E., Ohno, S., Kuroki, T., et al. (1996). The η isoform of protein kinase C mediates transcriptional activation of the human transglutaminase 1 gene. *J. Biol. Chem.* 271:9790–9794.

Ueffing, M., Lovric, J., Philipp, A., et al. (1997). Protein kinase C-ε associates with the Raf-1 kinase and induces the production of growth factors that stimulate Raf-1 activity. *Oncogene* 15:2921–2927.

Urtreger, A.J., Grossoni, V.C., Falbo, K.B., et al. (2005). Atypical protein kinase C-ζ modulates clonogenicity, motility, and secretion of proteolytic enzymes in murine mammary cells. *Mol. Carcinog.* 42:29–39.

Verma, A.K. (1988). Inhibition of tumor promoter 12-O-tetradecanoylphorbol-13-acetate-induced synthesis of epidermal ornithine decarboxylase messenger RNA and diacylglycerol-promoted mouse skin tumor formation by retinoic acid. *Cancer Res.* 48:2168–2173.

Verma, A.K., Pong, R-C., Erickson, D. (1986). Involvement of protein kinase C activation in ornithine decarboxylase gene expression in primay culture of newborn mouse epidermal cells and in skin tumor promotion by 12-O-teradecanoylphorbol-13-acetate. *Cancer Res.* 46:6149–6155.

Verma, A.K., Wheeler, D.L., Aziz, M.H., et al. (2006). Protein kinase Cε and development of squamous cell carcinoma, the nonmelanoma human skin cancer. *Mol. Carcinog.* 45:381–388.

Vetter, M., Blumenthal, S.G., Lindemann, R.K., et al. (2005). Ets1 is an effector of PKCα in cancer cells. *Oncogene* 24:650–661.

Wang, H.B., Yang, C., Leskow, F.C., et al. (2006). Phospholipase Cγ/diacylglycerol-dependent activation of β2-chimaerin restricts EGF-induced Rac signaling. *EMBO J.* 25:2062–2074.

Wang, H.Q., and Smart, R.C. (1999). Overexpression of protein kinase C-alpha in the epidermis of transgenic mice results in striking alterations in phorbol ester-induced inflammation and COX-2, MIP-2 and TNF-alpha expression but not tumor promotion. *J. Cell Sci.* 112:3497–3506.

Wang, Q.J. (2006). PKD at the crossroads of DAG and PKC signaling. *Trends Pharmacol. Sci.* 27:317–323.

Wang, Q.J., Fang, T.W., Yang, D., et al. (2003). Ligand structure-activity requirements for the binding of phorbol esters to protein kinase D. *Mol. Pharmacol.* 64:1342–1348.

Wang, Q.J., Lu, G., Schlapkohl, W.A., et al. (2004). The V5 domain of protein kinase C plays a critical role in determining the isoform-specific localization, translocation, and biological function of protein kinase C-δ and -ε. *Mol. Cancer Res.*2:129–140.

Wang, X.J., Warren, B.S., Beltran, L.M., et al. (1993). Further identification of protein kinase C isozymes in mouse epidermis. *J. Cancer Res. Clin. Oncol.* 119:279–287.

Wang, Y.C., Hsieh, Y.S., Tang, Y.W., et al. (1999). Protein kinase C isoforms in the epidermal tissues of normal and postburn human skin. *Biochem. Mol. Biol. Int.* 47:673–679.

Way, K.J., Chou, E., King, G.L. (2000). Identification of PKC-isoform-specific biological actions using pharmacological approaches. *Trends Pharmacol. Sci.* 21:181–187.

Wheeler, D.L., Li, Y., Verma, A.K. (2005). Protein kinase C epsilon signals ultraviolet light-induced cutaneous damage and development of squamous cell carcinoma possibly through induction of specific cytokines in a paracrine mechanism. *Photochem. Photobiol.* 81:9–18.

Wheeler, D.L., Reddig, P.J., Dreckschmidt, N.E., et al. (2002). Protein kinase Cδ-mediated signal to ornithin decarboxylase induction is independent of skin tumor suppression. *Oncogene* 21:3620–3630.

Wilker, E., Lu, J., Rho, O., et al. (2005). Role of PI3K/Akt signaling in insulin-like growth factor-1 (IGF-1) skin tumor promotion. *Mol. Carcinog.* 44:137–145

Willhauck, M.J., Mirancea, N., Vosseler, S., et al. (2007). Reversion of tumor phenotype in surface transplants of skin SCC cells by scaffold-induced stroma modulation. *Carcinogenesis* 28:595–610.

Yamanaka, T., Horikoshi, Y., Izumi, N., et al. (2006). Lgl mediates apical domain disassembly by suppressing the PAR-3-aPKC-PAR-6 complex to orient apical membrane polarity. *J. Cell Sci.* 119:2107–2118.

Yang, C., Liu, Y., Leskow, F.C., et al. (2005). Rac-GAP-dependent inhibition of breast cancer cell proliferation by β2-chimerin. *J. Biol. Chem.* 280:24363–24370.

Yang, L.C., Ng, D.C., Bikle, D.D. (2003). Role of protein kinase C α in calcium induced differentiation: defective regulation in squamous cell carcinoma. *J. Cell Physiol.* 195:249–259.

Yuspa, S.H., Kilkenny, A.E., Steinert, P.M., et al. (1989). Expression of murine epidermal differentiation markers is tightly regulated by restricted extracellular calcium concentrations in vitro. *J. Cell. Biol.* 109:1207–1217.

Zhang, X.A., Bontrager, A.L., Hemler, M.E. (2001). Transmembrane-4-superfamily proteins associate with activated protein kinase C (PKC) and link PKC to specific beta(1) integrins. *J. Biol. Chem.* 276:25005–25013.

Ziegler, W.H., Tigges, U., Zieseniss, A., et al. (2002). A lipid-regulated docking site on vinculin for protein kinase C. *J. Biol. Chem.* 277:7396–7404.

Chapter 9
Maintaining Energy Balance in Health and Disease: Role of the AMP-Activated Protein Kinase

John W. Scott

Abbreviations

AMPK	AMP-activated protein kinase
ACC	Acetyl CoA carboxylase
AMP	Adenosine 5'-monophosphate
ATP	Adenosine 5'-triphosphate
CaMKK	Ca^{2+}/calmodulin dependent protein kinase kinase
CBS	Cystathionine-β-synthase domain
CBM	Carbohydrate-binding module

9.1 Introduction

Living cells require a constant input of energy to perform useful work such as synthesis of macromolecules (protein, lipids, carbohydrate), active transport of metabolites and ions, and cell movement. Most of these processes use adenosine-5′-triphosphate (ATP) as their energy source, which is the universal currency of energy in virtually all biological systems. Energy is liberated from ATP by hydrolysis of one or more of its phosphoester bonds (yielding ADP or AMP), which is then harnessed to power biochemical reactions that are thermodynamically unfavourable. Cells normally maintain a high ratio of ATP to ADP (or AMP) in order that these cellular processes are kept far from equilibrium; therefore complex mechanisms have evolved to ensure that ATP-production (catabolism) is balanced with ATP-consumption (anabolism). In eukaryotes, the AMP-activated protein kinase (AMPK) is a critical sensor and regulator of cellular energy charge that co-ordinates metabolic processes to balance energy supply with demand. AMPK monitors the cellular AMP to ATP ratio and is activated

J.W. Scott (✉)
St Vincent's Institute and Department of Medicine, University of Melbourne,
41 Victoria Parade, Fitzroy 3065, VIC, Australia
e-mail: jscott@svi.edu.au

C.J. Vidal (ed.), *Post-Translational Modifications in Health and Disease*, Protein Reviews 13,
DOI 10.1007/978-1-4419-6382-6_9,

by metabolic stresses that either inhibit ATP production (hypoxia, fuel deprivation) or stimulate increased ATP consumption (exercise) (Scott et al. 2009). The reason that AMPK senses the AMP/ATP ratio rather than the ADP/ATP ratio is not obvious at first glance, however, the former is a much more sensitive measure of cellular energy charge despite the fact that ADP is the major product of ATP hydrolysis. The explanation for this is due to the action of adenylate kinase, a highly active enzyme found in all eukaryotes that catalyses the reaction 2ADP$\leftrightarrow$AMP+ATP (Hardie and Hawley 2001). Adenylate kinase maintains this reaction close to equilibrium therefore the AMP/ATP ratio tends to vary as the square of the ADP/ATP ratio (for example, a 5-fold rise in the ADP/ATP ratio equates to a 25-fold rise in the AMP/ATP ratio). In other words, the concentration of AMP in the cell rises more dramatically than ADP when ATP levels fall, therefore it stands to reason why natural selection favoured an energy monitoring system that senses the AMP/ATP ratio.

The primary function of AMPK when activated is to conserve cellular ATP by upregulating ATP-producing pathways such as fatty acid oxidation and glycolysis, while inhibiting ATP-consuming processes including fatty acid, sterol, carbohydrate and protein synthesis (Fig. 9.1) (Scott et al. 2009). It achieves this in the main by direct phosphorylation of enzymes that catalyse the rate-limiting steps of these

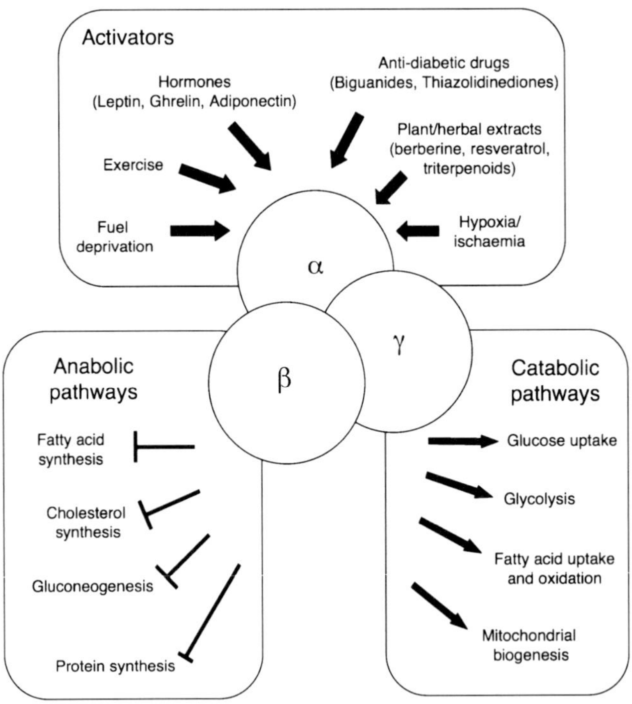

Fig. 9.1 Regulation of energy metabolism by AMPK. Activation of AMPK by metabolic stresses, hormones and various drugs leads to activation and inhibition of catabolic and anabolic pathways, respectively

major metabolic pathways. AMPK also plays an adaptive role to chronic energy challenges, similar to those that occur during exercise training, by reprogramming expression of metabolic genes via phosphorylation of transcription factors and coactivators/regulators (McGee and Hargreaves 2008). It is therefore likely that AMPK is at least partly responsible for some of the beneficial metabolic effects associated with regular physical activity. In addition to physiological and pathological stresses that deplete ATP, AMPK is also activated by drugs used to treat Type 2 diabetes including the biguanides and thiazolidinediones (Zhou et al. 2009). Furthermore, a number of natural plant and herbal compounds such as resveratrol and berberine (present in grapes), and triterpenoids (present in bitter melon), all of which have health-giving properties that include prevention of obesity and insulin resistance, also activate AMPK (Baur et al. 2006; Lee et al. 2006b; Tan et al. 2008). These discoveries have propelled AMPK to the forefront as a potential drug target to treat a range of energy balance disorders and provide a compelling argument for the development of novel, specific activators of AMPK.

AMPK was defined historically as a sensor and regulator of energy balance at the cellular level, and the fact that AMPK orthologues are found in single celled eukaryotes is compatible with this being the function for which it evolved. However, AMPK is now also recognised as a key regulator of energy homeostasis at the whole-body level, responding to appetite controlling hormones such as leptin and ghrelin, as well as other secretory factors such as adiponectin (Kahn et al. 2005). AMPK processes signals from these hormones in both the central nervous system and peripheral tissues in order to control appetite and regulate whole-body energy expenditure. Understanding the role of AMPK in the control of energy homeostasis at both the cellular and whole body levels is crucial to developing new strategies to treat energy balance diseases.

9.2 AMPK Structure and Function

AMPK exists as a $\alpha\beta\gamma$ heterotrimer composed of an α catalytic subunit and regulatory β and γ subunits. In mammals, each subunit is encoded by multiple genes that can also be subject to alternative splicing, therefore at least 12 heterotrimeric combinations are possible giving rise to differences in subcellular localisation, tissue distribution, regulation and function. Orthologues of each subunit occur in all eukaryotes including plants, fungi and even the primitive protist *Giardia lamblia*, which highlights the importance and antiquity of the AMPK system and suggests it emerged very early during eukaryotic evolution.

The mammalian α subunits ($\alpha 1$ and $\alpha 2$) both have conventional Ser/Thr kinase domains at their N-terminus, followed by an autoinhibitory sequence and a C-terminal region that is involved in heterotrimeric formation (Fig. 9.2) (Crute et al. 1998). The kinase domain (PDB ID: 2H6D) adopts a canonical bilobal structure that is typical of the eukaryotic protein kinase superfamily, with the active site located in a cleft between the two lobes (reviewed in Scott et al.

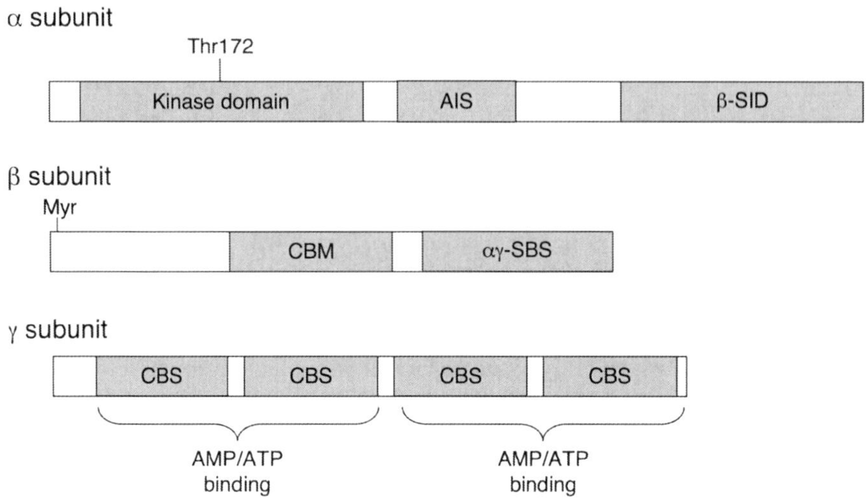

Fig. 9.2 Domain structure of AMPK subunits. *AIS* autoinhibitory sequence, *β-SID* β subunit interacting domain, *Myr* myristoyl moiety, *CBM* carbohydrate-binding module, *αγ-SBS* αγ-subunit binding sequence, *CBS* cystathionine-β-synthase motif

2009). The small N-terminal lobe is predominantly responsible for ATP-binding whereas the large C-terminal lobe contains determinants and structural features that dictate binding of protein substrates. The catalytic loop in the large lobe contains an invariant aspartate residue that functions as the base for catalysis and is absolutely required for catalytic activity. AMPK is also catalytically inactive unless phosphorylated on Thr172 located on the activation loop (or T-loop) of the kinase domain by upstream kinases (Hawley et al. 1996). Protein kinases regulated by phosphorylation of the activation loop, without exception, have an arginine residue immediately preceding the catalytic aspartate, which is required to be neutralised by the presence of a phosphate on the activation loop in order to promote the correct folding and orientation of the ATP-binding site and active site residues (Johnson et al. 1996).

In terms of protein substrate recognition, the kinase domain of AMPK belongs to the basophilic class of protein kinases, requiring the presence of positively charged residues in the sequence N-terminal to the site of phosphorylation on target proteins. The consensus motif for AMPK substrate recognition is H(X,B)XXS/TXXXH where H is a hydrophobic residue, B is a basic residue and S/T serine or threonine residues (Scott et al. 2002). Although the basic residue is critical for recognition, it can be positioned at either the P-3 or P-4 position (the P-/P+ nomenclature denotes the residue position N- or C-terminal to the phosphorylated residue, respectively), whereas there is an absolute requirement for a hydrophobic residue at the P-5 position. The P-3/P-4 basic residue forms an ionic interaction with a conserved acidic patch located at the interface between the small and large lobes of the kinase domain, while the P-5 residue binds in a hydrophobic pocket situated on the large C-terminal lobe. There are a number of other determinants that lie out with the core consensus motif that, although not essential, significantly enhance substrate binding to the kinase domain such as a basic residue at the P-6 and P+3 positions (Scott et al. 2002). Many protein kinases autoregulate their activity through their own structural elements that either directly or indirectly block the catalytic site. Both AMPK α subunit isoforms contain an autoinhibitory sequence immediately C-terminal to the kinase domain, however the autoinhibitory sequence does not block the active site like those found in other ligand regulated kinases such as the Ca^{2+}/calmodulin dependent protein kinase (CaMK) family, but rather binds across the back of the catalytic domain holding the enzyme in an open inactive conformation (Pang et al. 2007). The extreme C-terminus of the α subunit is involved in heterotrimer assembly and binds to the C-terminus of the β subunit (Xiao et al. 2007).

The β subunits (β1 and β2) play a key role in mediating heterotrimeric complex formation, and are also important in mediating subcellular localisation of AMPK to lipid membranes and glycogen (Fig. 9.2). It has also become evident that the β subunits can allosterically regulate AMPK activity in a manner distinct from AMP (McBride et al. 2009; Sanders et al. 2007a; Scott et al. 2008). The β subunits have divergent N-terminal regions of poorly defined function; however the β1 isoform is myristoylated on a glycine residue at the extreme N-terminus (Mitchelhill et al. 1997). The myristoyl group has an autoinhibitory effect on AMPK as removal of the modification by point mutation dramatically increases catalytic activity (Warden et al. 2001). Moreover, removal of the myristoyl group also causes a marked redistribution of β1-containing AMPK complexes from lipid membranes to the cytoplasm. The non-myristoylated AMPK mutants are still activated by AMP, suggesting that the molecular mechanisms by which AMP and myristoylation regulate AMPK activity are distinct. These findings indicate that myristoylation of the β1 subunit confers a secondary level of autoinhibition upon the AMPK heterotrimer, which is removed when the myristoyl group becomes embedded within a lipid bilayer. This mechanism would ensure that β1-containing AMPK complexes only become fully activated when localised to particular membranes or subcellular compartments.

The β subunits also have internal carbohydrate binding modules (CBM) that are related to domains found in enzymes involved in glycogen and starch metabolism. These domains are non-catalytic and are involved in targeting the enzymes in which they are found to polysaccharide structures and in the case of the β subunit CBM, mediates the association of AMPK with glycogen particles (Hudson et al. 2003; Polekhina et al. 2003). The crystal structure of the β1 subunit CBM has been solved in complex with the cyclic sugar β-cyclodextrin, which revealed that binding of sugars to the CBM is largely dependent on hydrophobic interactions mediated by two conserved tryptophan residues (Trp100 and Trp133) that flank the sugar-binding pocket (Polekhina et al. 2005). The β1-CBM preferentially binds oligosaccharides containing between five and seven glucose units connected through an α(1→4) linkage with a single glucose sugar in an α(1→6) branch (Koay et al. 2007). This suggests that AMPK recruitment may be enhanced during periods of glycogen depletion, where α(1→6) branched sugars become exposed after glycogen phosphorylase mediated degradation of α(1→4) linked sugars. In fact, it has been shown that AMPK does not bind to liver α-glycogen particles that are replete with α(1→4) linkages on their surface (Parker et al. 2007). It has become evident that the CBM, rather than just localising AMPK to glycogen, is also a regulatory domain that allosterically inhibits catalytic activity when glycogen is bound (McBride et al. 2009). The extent of branching is important as AMPK is more potently inhibited by highly branched glycogen, which is consistent with the finding that the CBM binds more tightly to branched α(1→6) linked sugars.

Glycogen binding also inhibits phosphorylation and activation of AMPK by upstream kinases but not dephosphorylation by protein phosphatases. This contrasts with the effect of AMP, which inhibits dephosphorylation by protein phosphatases without affecting phosphorylation by the upstream kinases (Suter et al. 2006). The C-terminal domain of the β subunits functions as a scaffold to anchor the α and γ subunits, and is termed the αγ-subunit binding sequence (αγ-SBS) (Iseli et al. 2005). The αγ-SBS forms a network of hydrophobic contacts with the C-terminus of the α subunit whereas the βγ-interface is formed mainly by a hydrogen bonding network between the αγ-SBS and the N-terminus of the γ subunit (Iseli et al. 2008; Xiao et al. 2007). Minimal contacts exist between the α and γ subunits at the heterotrimeric core interface, suggesting that a stable interaction between the α and γ subunits is unlikely in the absence of the β subunit.

The γ subunits (γ1, γ2 and γ3) contain variable N-terminal sequences of unknown function, followed by four tandem repeats of a CBS motif. These motifs invariably occur in tandem pairs and are found in a number of unrelated proteins including IMP dehydrogenase, the CLC chloride channel family and cystathionione-β-synthase (from which the acronym CBS is derived) (Bateman 1997). Pairs of CBS motifs form a discrete structural unit called Bateman domains, which function as adenine nucleotide binding modules (Scott et al. 2004). The γ subunit Bateman domains serve as the allosteric binding sites for AMP and ATP, and are responsible for the energy-sensing properties of AMPK (Fig. 9.2).

The structure of the entire mammalian γ subunit in complex with either AMP or ATP has been solved, which has provided a detailed insight into the molecular basis of nucleotide binding (Xiao et al. 2007). Although there are four potential binding sites for AMP, only three of the sites are able to bind nucleotides as the unoccupied site lacks a conserved aspartate residue that interacts with the 2′ and 3′-hydroxyls of the ribose sugar (Fig. 9.3). Surprisingly, the site equivalent to the unoccupied site in the γ subunit orthologue from *Schizosaccharomyces pombe* can bind ADP by virtue of an interaction between the 2′ and 3′-ribose hydroxyls of ADP and an asparate residue from the β subunit (Jin et al. 2007). Interestingly, there is an aspartate in the corresponding position that is conserved in the sequence of the mammalian β subunit isoforms, raising the possibility that this site in mammalian AMPK may become occupied by a similar mechanism. Two of the three sites bound with AMP can readily exchange with ATP and are therefore the energy-sensing sites, whereas the third molecule of AMP is tightly bound and is non-exchangeable with ATP; however, the function of the non-exchangeable site is unknown (Xiao et al. 2007). The negatively charged

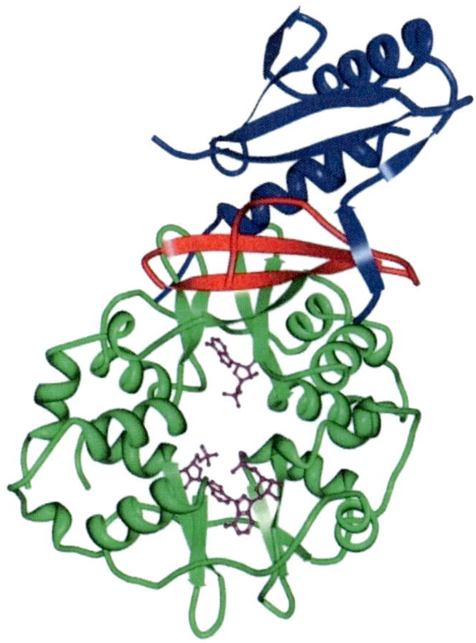

Fig. 9.3 The mammalian αβγ core complex structure. The C-terminal α subunit structure is shown in *blue*, the C-terminal β subunit structure in *red*, and the entire γ subunit in *green*. The three AMP molecules evident in the structure are shown in *magenta*

phosphate moieties of AMP and ATP are held in place by an array of positively charged residues, most of which are located at the mouth of the binding pocket. ATP binding to the exchangeable sites results in only a minor rearrangement of these basic residues despite the fact that ATP has two additional phosphate groups compared with AMP, therefore it is unclear how the occupancy state of the exchangeable binding sites effects allosteric activation of the α catalytic domain.

An interesting feature of the γ subunit nucleotide binding mechanism is that the second CBS motif (CBS2) contributes the majority of the basic residues that interact with the nucleotide phosphate groups. One interpretation of this feature is that CBS2 may be particularly important in the interactions between the α and γ subunits that regulate catalytic activity. Indeed, this possibility is supported by the identification of a pseudosubstrate sequence that occurs in the CBS2 motif of all mammalian γ subunits (Scott et al. 2007). Pseudosubstrate sequences act as competitive inhibitors, blocking access of protein substrates to the active site in the absence of activating ligands. These motifs are usually found adjacent to or overlapping with regulatory ligand-binding sites. There are a number of protein kinases that are regulated by internal pseudosubstrates including cAMP and cGMP-dependent protein kinases (PKA, PKG), the Ca^{2+}/calmodulin dependent protein kinases (CaMK), and the protein kinase C (PKC) family (Kemp et al. 1994). Positively charged residues within the pseudosubstrate of CBS2 that are predicted to be involved in interacting with the substrate-binding site on the kinase domain, are also critical in binding the phosphate group of AMP. Since these two interactions would be unlikely to occur simultaneously, this suggests a simple mechanism whereby in the absence of AMP, the pseudosubstrate would occupy the substrate-binding site on the kinase domain and inhibit catalytic activity. Despite kinetic evidence in favour of this mode of activation, extended crystal structures of the AMPK complex that include the kinase domain will be required to validate this mechanism.

9.3 Regulation of AMPK

AMPK is principally activated by phosphorylation of Thr172 in the activation loop of the α subunit catalytic domain by at least three upstream protein kinases (Fig. 9.4), highlighting the potential for multiple signals to feed into the AMPK system. Conversely, AMPK is inactivated by dephosphorylation of Thr172 by several protein phosphatases including PP2A, PP2C and glycogen bound PP1 in skeletal muscle (Davies et al. 1995; McBride et al. 2009; Paterson et al. 2008). An important question in recent years has been how AMP promotes phosphorylation and activation of AMPK. Initial studies suggested that AMP stimulated phosphorylation of AMPK by two distinct mechanisms: (1) making AMPK a better substrate for activating upstream kinases; (2) making AMPK a worse substrate for inactivating protein phosphatases (Davies et al. 1995; Hawley et al. 2005). However, more recent studies using recombinant AMPK suggest that the former effect is an artefact caused by contamination by protein phosphatases, and that AMP simply protects against dephosphorylation (Suter et al. 2006). In addition to stimulating phosphorylation of Thr172, AMP also directly activates AMPK by an allosteric mechanism, however this effect is relatively modest compared with phosphorylation. Nevertheless, the combination of these two effects of AMP results in a greater than 1,000-fold increase

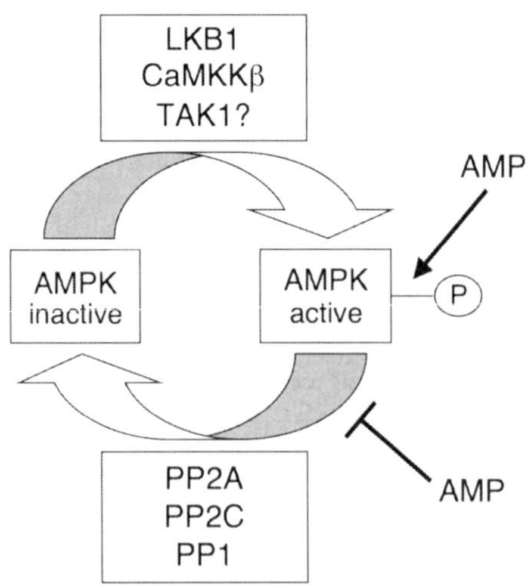

Fig. 9.4 Regulation of AMPK by AMP and upstream kinases. AMPK regulated by reversible phosphorylation of Thr172 by various upstream kinases and protein phosphatases. AMP allosterically activates phosphorylated AMPK as well as inhibiting dephosphorylation of Thr172

in AMPK activity (Suter et al. 2006). Importantly, both effects of AMP are antagonised by high concentrations of ATP, therefore AMPK acts as a sensor of the AMP/ATP ratio rather than AMP in itself (Corton et al. 1995; Davies et al. 1995).

The LKB1 complex is the major upstream AMPK kinase in most mammalian tissues, which exists a heterotrimer consisting of the protein kinase LKB1 and two regulatory subunits STRAD and MO25 (Hawley et al. 2003). STRAD and MO25 are essential for LKB1 to be functional and are also involved in targeting the complex to the cytoplasm (Boudeau et al. 2003). STRAD is termed a pseudokinase as it shares a high degree of sequence homology to the eukaryotic protein kinase superfamily but lacks key residues required for catalytic activity, whereas MO25 is structurally related to Armadillo repeat proteins. LKB1 was first described as a tumour suppressor that is mutated in Peutz-Jeghers syndrome, a rare dominantly inherited disease in humans that is characterised by benign intestinal polyps (or hamartomas) and pigmentation of the skin and mucous membranes (Hemminki et al. 1998). Peutz-Jeghers patients also have significantly increased risk of developing malignant tumours. LKB1 phosphorylates and activates 12 other protein kinases in addition to AMPK, all of which are members of the AMPK-related protein kinase subfamily (Lizcano et al. 2004). These other protein kinases share sequence similarity with the catalytic domain of the AMPK α subunit but are functionally distinct. LKB1 is constitutively active despite the fact that it is phosphorylated by cAMP-dependent protein kinase and p90 ribosomal S6 kinase at multiple sites; however none of these phosphorylation events alters LKB1 activity directly or affects its ability to phosphorylate AMPK (Fogarty and Hardie 2009; Sapkota et al. 2001). Therefore, the current model follows that AMPK is continuously phosphorylated and dephosphorylated by LKB1 and protein phosphatases, respectively, in a futile cycle that is shifted in favour of phosphorylation only when AMP levels rise (Sanders et al. 2007b). Although it would seem counterintuitive for an energy conserving system to wastefully consume ATP in a futile cycle, it does have the advantage of allowing rapid changes in the activity of AMPK to occur in response to perturbations in cellular energy charge.

In addition to changes in cellular energy, increases in intracellular Ca^{2+} also stimulate phosphorylation and activation of AMPK via the Ca^{2+}-calmodulin-dependent protein kinase kinases (CaMKK), regardless of the intracellular AMP concentration (Hawley et al. 2005; Hurley et al. 2005; Woods et al. 2005). There are two CaMKK isoforms (α and β), both of which can phosphorylate AMPK in vitro, however a number of studies suggest that CaMKKβ is the physiological activator of AMPK (Anderson et al. 2008; Stahmann et al. 2006). CaMKKβ differs from CaMKKα mainly in the N-terminal region preceding the catalytic domain, in which there is a regulatory segment in CaMKKβ that renders the enzyme virtually Ca^{2+}-calmodulin independent; although whether this N-terminal region is important for the regulation of AMPK remains to be determined (Tokumitsu et al. 2001). CaMKKβ is predominantly expressed in neuronal tissue but is also found in a small number of other tissues including endothelial and T cells, therefore Ca^{2+}-CaMKK mediated regulation of AMPK is more restricted than LKB1, which is expressed ubiquitously.

Increased cytosolic Ca^{2+} is often accompanied by a surge in demand for ATP, for instance, activation of motor proteins involved in muscle contraction or extrusion of Ca^{2+} from the cytosol, which is predominantly driven by ATP-dependent Ca^{2+} pumps. Under these circumstances, activation of AMPK could be considered as playing a protective role, ensuring that energy metabolism is co-ordinated to meet the increase in demand for ATP that typically follows Ca^{2+} release. In support of this idea, it has been shown that AMPK is activated in response to T-cell receptor (TCR) stimulation in T cells in a Ca^{2+}-CaMKKβ dependent manner (Tamas et al. 2006). TCR activation results in a large influx of Ca^{2+} into the cytosol, which acts as a second messenger to stimulate T cell proliferation and effector function, both of which are energy intensive processes (Fox et al. 2005). Activated T cells rapidly increase their ATP production by promoting glycolysis and oxidative phosphorylation, and it is likely that the CaMKK-AMPK signalling pathway is at least partly responsible for these metabolic effects.

In addition to LKB1 and CaMKKβ, a screen for protein kinases that activate AMPK using a mammalian expression library in yeast identified transforming growth factor-β-activated protein kinase-1 (TAK1) as a potential upstream kinase (Momcilovic et al. 2006). TAK1 is an important component of the signalling pathways that regulate the activities of the nuclear factor-kappa B (NFκB) and activator protein-1 (AP-1) transcription factors in response to cytokines and microbial pathogens (Scheidereit 2006). It acts upstream of several other protein kinases including the IκB kinase (IKK) complex and members of the stress-activated protein kinase family. Although AMPK is activated in response to TGFβ, interleukin-1 (IL-1) and tumour necrosis factor-α (TNFα) (Momcilovic et al. 2006; Suzuki et al. 2005), all of which are known activators of TAK1 signalling, the question of whether TAK1 regulates AMPK in a physiological setting remains to be defined.

9.4 Regulation of Lipid Metabolism by AMPK

As mentioned earlier, the principle function of AMPK in the cell is to co-ordinate metabolic processes to ensure that ATP production remains balanced with ATP consumption. In general, activation of AMPK stimulates ATP production and inhibits ATP consumption primarily by direct phosphorylation of metabolic enzymes that control anabolic (ATP-consuming) and catabolic (ATP-producing) pathways, which allows metabolic flux through these pathways to be rapidly altered. One of the most energy-intensive processes in the cell is the biosynthesis of fatty acids and lipids, and it is now clear that AMPK regulates lipid metabolism at multiple levels not only in an acute manner, but also in a chronic fashion via alterations in the expression pattern of lipogenic genes.

Perhaps the best characterised substrate for AMPK is acetyl-CoA carboxylase (ACC), the rate-limiting enzyme for fatty acid synthesis and an important regulator of fatty acid oxidation (Carling et al. 1989). ACC catalyses the synthesis of malonyl-CoA in an ATP-consuming reaction, and is the first committed step of fatty acid synthesis. There are two ACC isoforms (1 and 2), both of which play different roles and vary in their tissue expression (Abu-Elheiga et al. 1995). ACC1 is predominantly expressed in liver and adipose tissue and is primarily responsible for determining the rate of fatty acid synthesis (Iverson et al. 1990). On the other hand, ACC2 is abundantly expressed in skeletal muscle and, unlike ACC1, is localised to the outer membrane of mitochondria by virtue of an N-terminal targeting sequence (Abu-Elheiga et al. 2000). ACC2 is an important regulator of fatty acid oxidation, and plays a pivotal role in mediating the activity of carnitine palmitoyl transferase-1 (CPT-1), the transporter responsible for shuttling fatty acids into the mitochondria. Allosteric inhibition of CPT-1 by ACC2 derived malonyl-CoA is considered the most important regulatory step in determining the rate of fatty acid oxidation. In fact, genetic deletion of ACC2 results in high levels of fatty acid oxidation and decreased accumulation of triglycerides (Abu-Elheiga et al. 2001). AMPK phosphorylates and inactivates both ACC isoforms, resulting in decreased fatty acid synthesis and increased fatty acid oxidation. Activation of AMPK also promotes fatty acid uptake by stimulating relocalisation of fatty acid translocase (FAT/CD36) from intracellular vesicles to the plasma membrane (Luiken et al. 2003). As well as these acute effects, AMPK also decreases the expression of key lipogenic genes in response to long term energetic challenges. AMPK activation reduces the expression of SREBP1c, which functions as a transcriptional regulator of a number of lipogenic genes including fatty acid synthase (FAS), stearoyl-CoA desaturase-1 (SCD1), and Spot 14 (Foretz et al. 1998; Zhou et al. 2001). This occurs via an indirect mechanism as AMPK does not phosphorylate SREBP1c directly, however the signalling components that link AMPK to SREBP1c are unknown.

AMPK regulates other aspects of lipid metabolism such as triglyceride esterification and hydrolysis particularly in liver and adipose tissue, respectively. The control of these pathways has become an area of intense interest as lipid intermediates in the form of diacylglycerols, ceramides and long-chain fatty acyl-CoA molecules have been reported to activate protein kinases and phosphatases that negatively regulate the insulin signalling pathway (Kraegen and Cooney 2008). This raises the prospect that accumulation of these reactive lipid intermediates in obesity may be a causal factor underlying the development of insulin resistance. Triglyceride esterification is regulated by glycerol-3-phosphate-acyl-transferase (GPAT), which performs the first committed step in triglyceride synthesis and catalyses the formation of lysophosphatidic acid. Although activation of AMPK in the liver reduces GPAT activity and triglyceride esterification, it has yet to be demonstrated whether GPAT is a direct substrate of AMPK (Muoio et al. 1999). The hydrolysis of triglycerides to diglycerides is mediated by hormone sensitive lipase (HSL), which is activated in response to adrenergic factors that stimulate PKA activity (Anthonsen et al. 1998). AMPK phosphorylates and inhibits HSL in adipocytes thereby preventing lipolysis, which contradicts the paradigm that AMPK stimulates catabolic pathways and inhibits anabolic processes (Garton et al. 1989; Garton and Yeaman 1990). However, if fatty acids released by lipolysis are not exported from the adipocyte, they are recycled back into triglyceride in an ATP-consuming manner. Therefore, inhibition of HSL by AMPK appears to be a mechanism to restrict recycling and ensure the rate at which fatty acids are released by lipolysis does not surpass the rate at which they are disposed of by export or mitochondrial oxidation.

3-Hydroxy-3-methylglutaryl-CoA reductase (HMGR) is the rate-limiting enzyme of the cholesterol synthesis (mevalonate) pathway and a substrate for AMPK. Cholesterol is the precursor of steroid hormones such as estradiol, testosterone and the glucocorticoids, whereas geranylgeranyl and farnesyl groups derived from the mevalonate pathway play an important role in providing lipid anchors for many signalling proteins including several members of the small G-protein family (McTaggart 2006). The expression of HMGR is highly regulated at both the transcriptional and translational levels by a negative-feedback control mechanism mediated by sterols and non-sterol metabolites derived from mevalonate (Clarke et al. 1983). In addition, AMPK phosphorylates and inactivates HMGR activity, which results in decreased cholesterol synthesis (Beg et al. 1978; Clarke and Hardie 1990). Interestingly, activation of AMPK by adiponectin in a rodent model of atherosclerosis reduced cholesterol synthesis but also reduced the incidence of atherosclerotic plaques, suggesting that hormonal regulation of HMGR via AMPK may be critical for modulating whole-body cholesterol metabolism and also provide new opportunities for treating cardiovascular disease (Ouchi et al. 2001).

9.5 Regulation of Carbohydrate Metabolism by AMPK

Carbohydrates are an important fuel for virtually all organisms as they provide a rapid source of energy, especially as they are relatively simple to metabolise in comparison with other fuels such as fatty acids and proteins. Glucose is the most important carbohydrate in mammals, and complex cellular and hormonal signalling networks have evolved to ensure that its levels are maintained within narrow limits. The first hint that AMPK was likely to be involved carbohydrate metabolism came from the discovery that its a functional orthologue of the SNF1 protein kinase complex from *Saccharomyces cerevisiae*, which is activated in response to glucose limitation and orchestrates the switch in metabolism that permits growth on alternate carbon sources such as sucrose (Mitchelhill et al. 1994; Woods et al. 1994). Indeed, in mammals, AMPK regulates virtually all aspects of glucose metabolism including glucose uptake, gluconeogenesis and glycogen storage. AMPK also stimulates glycolysis in certain tissues, especially in the heart (see Sect. 9.8), but also in activated monocytes of the immune system (Marsin et al. 2000, 2002).

The uptake of glucose across the plasma membrane is largely mediated by the glucose transporter (GLUT) family of transmembrane proteins, with translocation of GLUT4 from intracellular vesicles to the plasma membrane considered the major mechanism by which glucose transport is stimulated in response to insulin (Sakamoto and Holman 2008). Pharmacological activation of AMPK with AICAR (which is taken into cells and converted to ZMP, an AMP-mimetic) stimulates GLUT4 translocation and glucose uptake independent of the insulin-signalling pathway, a finding that has generated significant interest in developing activators of AMPK as therapeutics to circumvent insulin resistance (Merrill et al. 1997). This concept is reinforced by the fact that genetic deletion of the AMPK α2 subunit (the predominant isoform in skeletal muscle) but not the α1 subunit abolishes AICAR-stimulated glucose uptake in muscle (Jorgensen et al. 2004b). The signalling pathway by which AMPK stimulates GLUT4 translocation is beginning to be unravelled and appears to involve the Rab GTPase-activating proteins (GAP) TBC1D1 and TBC1D4 (AS160), both of which are phosphorylated in response to AMPK activation in skeletal muscle (Kramer et al. 2006; Treebak et al. 2006). Phosphorylation of TBC1D1 and TBC1D4 results in the recruitment of 14-3-3 proteins, which in turn controls the rate of GLUT4 vesicle recycling (Chavez et al. 2008; Chen et al. 2008; Geraghty et al. 2007). Interestingly, TBC1D1 has been identified as a potential candidate gene for severe obesity, suggesting that defects in TBC1D1 signalling may be a contributory factor in obesity-induced insulin resistance (Lee et al. 2006a).

The major site of gluconeogenesis in mammals is the liver, and regulation of hepatic glucose production is crucial for maintaining whole-body glucose homeostasis and is critical for survival. Such is the importance of this process, it is regulated by number of diverse and redundant regulatory cues, which allow for the storage of glucose as glycogen and lipids following a meal and conversely to increase glucose production during a fast or intense exercise. Activation of AMPK in the liver suppresses hepatic glucose production by lowering the expression of gluconeogenic enzymes rather than by directly phosphorylating the enzymes themselves (Bergeron et al. 2001a). AMPK negatively regulates the transcription of the gluconeogenic genes L-type pyruvate kinase (PK), phosphoenolpyruvate carboxykinase (PEPCK), and glucose-6-phosphatase (G6P) by phosphorylation of various transcription factors and coactivators (da Silva Xavier et al. 2000; Leclerc et al. 1998; Lochhead et al. 2000; Woods et al. 2000).

The CREB-regulated transcription coactivator-2 (CRTC2) has emerged as an important regulator of gluconeogenic gene transcription and has been identified as a downstream target of AMPK (Koo et al. 2005). Glucagon stimulates the transcription of gluconeogenic genes when glucose is low via the cAMP-responsive factor CREB (CRE-binding protein) and subsequent recruitment of the coactivator CBP and CRTC2 to the nucleus. This nuclear translocation results in the expression of PGC-1α, which in turn promotes the transcription of PEPCK and G6P. Phosphorylation of CRTC2 by AMPK promotes binding to 14-3-3 proteins in the cytoplasm and prevents relocalisation of CRTC2 to the nucleus, thereby lowering CREB-dependent expression of PEPCK and G6P (Koo et al. 2005). The expression of PK is regulated by the transcription factor hepatic nuclear factor 4α (HNF-4α), and recent evidence suggests that HNF-4α protein is a direct substrate of AMPK, as phosphorylation of HNF-4α by AMPK in vitro reduces its ability to form homodimers that are required for DNA binding (Hong et al. 2003; Leclerc et al. 2001). The importance of AMPK in the regulation of hepatic glucose production is most convincingly demonstrated by findings from rodent models that lack the α2 subunit, which exhibit fasting hyperglycaemia, glucose intolerance and increased hepatic glucose output (Andreelli et al. 2006; Viollet et al. 2003).

Glucose is primarily stored in tissues as glycogen and is the most readily mobilised large-scale source of energy in the cell. The levels of glycogen in the cell are determined by both the rate of production (glycogenesis) and degradation (glycogenolysis), which are mediated by the enzymes glycogen synthase and glycogen phosphorylase, respectively. Glycogen synthesis and degradation are tightly co-ordinated so that glycogen synthase is almost completely inactive when glycogen phosphorylase is maximally active, and vice-versa. It has been known for several years that AMPK phosphorylates and inactivates glycogen synthase in vitro (Carling and Hardie 1989), however the role of AMPK in regulating glycogen synthase activity in vivo is more complex. Paradoxically, activation of AMPK in skeletal muscle actually increases glycogen content and glycogen synthase activity, however this effect is due to stimulation of glucose uptake and the consequent accumulation of glucose-6-phosphate, which allosterically activates glycogen synthase independent of phosphorylation (Aschenbach et al. 2002). However, in support of the initial observation, glycogen synthase has increased activity due to decreased phosphorylation in skeletal muscle lacking the $\alpha2$ subunit of AMPK (Jorgensen et al. 2004a). Therefore, while phosphorylation of glycogen synthase by AMPK does inhibit its activity under basal conditions, stimuli that increase the intracellular levels of glucose-6-phosphate are capable of overriding the inhibitory effects of phosphorylation by AMPK.

9.6 Regulation of AMPK by Hormones and Cytokines

The importance of hormonal factors in the regulation of energy metabolism has been recognised for several decades, however only in recent years has the interplay between these hormones and AMPK come to light. Although AMPK was originally viewed merely as a sensor of cellular energy, it is now regarded as an integral player in processing the signals from these hormones not only in peripheral tissues such as liver, adipose and skeletal muscle, but also in the central nervous system and especially the hypothalamus (Fig. 9.5).

One of the key factors in controlling feeding behaviour and whole-body energy expenditure in mammals is leptin, a protein hormone secreted from adipose tissue that suppresses food intake and prevents lipid accumulation in non-adipose tissue (Kahn et al. 2005). Defects in leptin production or the leptin receptor cause morbid obesity that typically results in the development of Type 2 diabetes (Chen et al. 1996). It is now clear that AMPK mediates many of the systemic metabolic effects of leptin. Classical leptin signalling via the leptin receptor involves recruitment of the JAK2 tyrosine kinase, resulting in phosphorylation of the cytoplasmic tail of the receptor and subsequent binding and activation of STAT proteins (Ghilardi et al. 1996). The STAT proteins are transcriptional activators that translocate to the nucleus when activated and primarily mediate the long term effects of leptin, however the precise signalling pathway that couples the leptin receptor to AMPK has yet to be elucidated. Nevertheless, in skeletal muscle leptin activates AMPK in

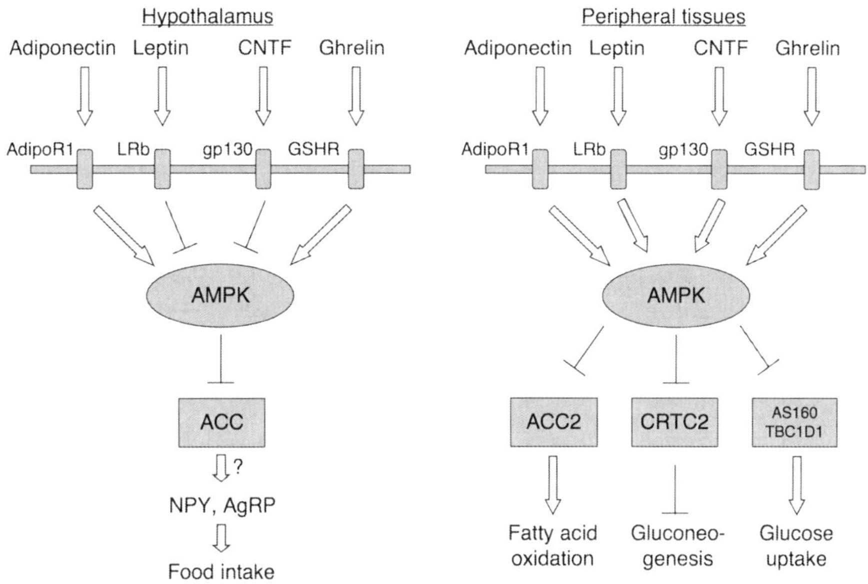

Fig. 9.5 Hormonal control of AMPK. In the hypothalamus, AMPK is bidirectionally regulated by hormones to control appetite. Hormonal control of AMPK in peripheral tissues regulates fatty acid oxidation, glucose uptake and gluconeogenesis

a biphasic manner involving a rapid transient activation, followed by a more sustained activation that lasts a few hours (Minokoshi et al. 2002). The former effect occurs by leptin acting directly at the level of the muscle leptin receptor, whereas the latter effect is mediated through the hypothalamic-sympathetic nervous system culminating in stimulation of muscle α adrenergic receptors. Activation of AMPK in skeletal muscle by leptin results in suppression of ACC2 activity, therefore increasing the rate of fatty acid oxidation. In the longer term, leptin also promotes expression of the AMPK α2 and β2 subunits, both of which are particularly abundant in skeletal muscle, resulting in a long term decrease in intramuscular lipid storage (Steinberg et al. 2003). As well as skeletal muscle, leptin also activates AMPK in adipocytes and, in addition to stimulating fatty acid oxidation, also promotes mitochondrial biogenesis thus increasing the oxidative capacity of this tissue (Wang et al. 2005). Rather surprisingly, leptin has no effect on AMPK activity in the liver despite the fact that the leptin receptor is expressed in this tissue (Brabant et al. 2005; Lee et al. 2002).

Perhaps the most exciting effect of leptin on AMPK occurs in the hypothalamus, where the leptin-AMPK signalling axis plays an important role in controlling appetite. The control of food intake is regulated by distinct neuronal populations in the hypothalamus that process signals from various appetite-regulating hormones and nutrients such as glucose. In contrast to peripheral tissues, leptin inhibits AMPK activity in the periventricular and arcuate nucleus of the hypothalamus, and consequently suppresses food intake by reducing expression of the appetite-stimulating neuropeptides, NPY and agouti-related protein (AgRP) (Minokoshi et al. 2004). Inhibition of AMPK in the hypothalamus by leptin is mediated through melanocortin-4 (MC4) receptor signalling, as agonists for the MC4 receptor mimic the effects of leptin, whereas genetic deletion of the receptor prevents inhibition of AMPK by leptin. Further downstream, the activation state of the AMPK substrate ACC is also important for leptin signalling, as blockade of hypothalamic ACC with pharmacological inhibitors prevents the anorexigenic effects of leptin (Gao et al. 2007). The mechanism by which leptin inhibits AMPK in the hypothalamus while activating AMPK in other tissues is unclear, but it perhaps reflects differences in expression of AMPK subunit isoforms, upstream kinases and phosphatases between tissues. Although leptin has robust effects on AMPK activity and lipid metabolism in skeletal muscle and adipocytes, leptin resistance is common in obesity therefore drugs that activate AMPK, which is downstream of leptin, may be effective in circumventing leptin resistance.

Another adipocyte derived hormone that is an important regulator of whole-body energy expenditure is adiponectin, a protein factor that exists as multimers of various sizes ranging from low molecular mass trimers to high molecular mass dodecamers (Richards et al. 2006). Activation of AMPK by adiponectin is dependent on the AdipoR1 receptor, however the mechanism by which it signals to AMPK is poorly understood although it appears to involve increases in intracellular AMP (Yamauchi et al. 2002, 2007). Adiponectin activates AMPK in skeletal muscle and promotes fatty acid oxidation and glucose uptake, in addition to enhancing insulin sensitivity and promoting mitochondrial biogenesis (Civitarese et al. 2006; Tomas et al. 2002; Wang et al. 2007; Yamauchi et al. 2002). It also activates AMPK in liver and adipocytes resulting in decreased gluconeogenesis and increased glucose uptake, respectively (Wu et al. 2003; Yamauchi et al. 2002). In contrast with leptin, adiponectin activates hypothalamic AMPK to stimulate food intake and this effect is mediated by the low molecular mass trimers and hexamer forms of adiponectin, which can cross the blood-brain barrier unlike the high molecular mass species (Kubota et al. 2007). Crucially, adiponectin is able to reverse the inhibitory effect of leptin on AMPK activity, demonstrating the reciprocal role of these hormones in the control of appetite and whole-body energy expenditure.

Ghrelin is a gut derived hormone that is secreted by the cells lining the fundus of the stomach, which serves to stimulate appetite in response to fasting (Kola et al. 2006). It is a short peptide hormone that binds to the G-protein coupled growth hormone secretagogue (GSH) receptor. While ghrelin has no effect on AMPK in skeletal muscle, it decreases AMPK activity in both liver and adipose tissue (Kola et al. 2005). Similar to adiponectin and in contrast with leptin, ghrelin stimulates AMPK activity in the hypothalamus resulting in increased food intake (Andersson et al. 2004). Cannabinoids also activate hypothalamic AMPK via the GSH receptor to stimulate appetite, which likely explains the phenomenon popularly known as the "munchies" that is associated with cannabis use.

A growing number of other hormones and inflammatory cytokines including tumour necrosis factor α (TNFα) and ciliary neurotrophic factor (CNTF) have all been shown to modulate AMPK activity in a variety of tissues (Steinberg et al. 2009).

The effect of inflammatory cytokines on AMPK activity is particularly relevant, as the expression and release into the circulation of a number of these factors is increased with obesity and Type 2 diabetes. Increased TNFα levels are associated with insulin resistance and genetic deletion of TNFα has been shown to enhance insulin sensitivity (Gonzalez-Gay et al. 2006; Saghizadeh et al. 1996). TNFα decreases AMPK activity chronically in skeletal muscle by upregulating the expression of the deactivating protein phosphatase PP2C, resulting in reduced fatty acid oxidation, which may be connected to decreased insulin sensitivity (Steinberg et al. 2006). CNTF is a cytokine secreted by glial cells that principally acts as a survival factor for motor and sensory neurones (Matthews and Febbraio 2008). The CNTF receptor is expressed in a number of peripheral tissues including skeletal muscle and adipose tissue, highlighting that CNTF can also regulate signalling pathways outwith the central nervous system. CNTF has central and peripheral effects on AMPK activity similar to leptin, however it is still able to modulate AMPK under conditions of leptin resistance, which is significant as bypassing leptin resistance is a major obstacle for treating obesity (Watt et al. 2006).

9.7 AMPK and Exercise

Exercise is arguably the most powerful physiological activator of AMPK, particularly as it stimulates a huge surge in demand for ATP and rapidly increases the fuel requirements of the working muscle. Skeletal muscle is an important tissue with respect to the regulation of whole-body energy metabolism as it makes up a relatively large proportion of body mass, and accounts for a sizeable portion of fuel consumption both during exercise and at rest. It is now well documented that physical inactivity has profound negative effects on whole-body energy metabolism and is a major risk factor for obesity and Type 2 diabetes. Activation of AMPK has been shown to improve the metabolic abnormalities associated with these diseases, therefore it has been postulated that some of the therapeutic metabolic effects of regular physical activity may be attributable to stimulation of AMPK.

AMPK is activated during exercise in an intensity-dependent manner, typically at exercise intensities above 60% of maximal aerobic activity, although it can also be activated at lower intensities if exercise is prolonged (Chen et al. 2003; Fujii et al. 2000; Wojtaszewski et al. 2000, 2002). As mentioned earlier, pharmacological activation of AMPK in skeletal muscle with AICAR stimulates glucose uptake independent of the insulin-signalling pathway (Merrill et al. 1997). Therefore, the role of AMPK in regulating exercise-induced glucose uptake has been an area of intense interest due to the therapeutic potential of stimulating glucose uptake in insulin-resistant muscle. Despite the strong correlation between muscle contraction, AMPK activity, and glucose uptake, the evidence so far paints a complex picture. For instance, contraction-stimulated glucose uptake is normal in skeletal muscle lacking either the α2 or γ3 subunits of AMPK whereas it is impaired in muscle expressing a dominant-negative variant of α2 (Barnes et al. 2004; Jorgensen et al. 2004b; Mu et al. 2001). In spite of these contradictory findings, AMPK is critical for the stimulation of glucose uptake in response to AICAR as this effect is abolished in both α2 and γ3 null skeletal muscle. The effect of AMPK on fatty acid oxidation during exercise also appears to be more complex than anticipated and, similar to the glucose uptake effect, looks likely to involve a number of redundant signalling pathways (Dzamko et al. 2008).

An exciting discovery relating to AMPK and exercise is the observation that chronic activation with AICAR induces a number of metabolic adaptations in skeletal muscle that resemble those that occur with exercise training. These adaptations include increased expression of proteins involved in glucose metabolism, such as GLUT4 and hexokinase, as well as increased mitochondrial biogenesis, which increases the capacity of the muscle to generate ATP by oxidative metabolism (Holmes et al. 1999; Winder et al. 2000). The ability to manipulate mitochondrial biogenesis is important as a reduction in mitochondrial density is thought to be a critical factor contributing to the accumulation of intramuscular lipids in obesity and to the development of insulin resistance. Mitochondrial biogenesis is regulated by various transcription factors and coactivators including nuclear respiratory factors-1 and 2 (NRF-1, NRF-2) and PGC-1α (Mootha et al. 2003; Virbasius and Scarpulla 1994). The expression of NRF-1 and NRF-2 is increased by chronic activation of AMPK, and both proteins are key transcriptional regulators of all five electron transport chain complexes (Bergeron et al. 2001b). On the other hand, PGC-1α is a direct substrate of AMPK and is required for expression of GLUT4 and the mitochondrial proteins cytochrome *c* and uncoupling protein-1 (UCP-1) (Jager et al. 2007).

Although these metabolic effects are observed in skeletal muscle in response to exercise, the question of whether AMPK is necessary is controversial as these responses are unaffected in skeletal muscle lacking either LKB1 or the α2 subunit of AMPK (Jorgensen et al. 2007; Thomson et al. 2007). This is not entirely surprising as activation of calcineurin, protein kinase D (PKD), and p38 MAP kinase have all been shown to be sufficient to increase mitochondrial biogenesis, which highlights the number of redundant signalling pathways involved in regulating oxidative capacity (Akimoto et al. 2005; Kim et al. 2008; Ryder et al. 2003). In spite of this, activation of AMPK is sufficient to induce exercise-type adaptations as evidenced by the fact that skeletal muscle expressing AMPK-activating γ1 or γ3 mutants have increased mitochondrial expression (Garcia-Roves et al. 2008; Nilsson et al. 2006; Rockl et al. 2007). Interestingly, lack of AMPK activity in skeletal muscle is associated with poor exercise performance however whether this is due to defects in fuel delivery or reduced mitochondrial capacity is unknown (Fujii et al. 2007).

9.8 AMPK in the Heart

The heart has perpetually high demands for energy in order to sustain critical processes such as contractile function and ion homeostasis. Control of energy metabolism in the heart is vital as it has a limited capacity to store fuel, therefore it is obligatory that cardiac energy homeostasis is tightly regulated to ensure energy production matches cardiac workload.

In the healthy heart, the large amounts of ATP required to maintain contractile function is derived mainly from oxidative metabolism of lipids in the mitochondria, and to a lesser extent from glucose metabolism (Wisneski et al. 1987). AMPK promotes fatty acid uptake and oxidation in the heart in response to contraction by stimulating translocation of the fatty acid transporter FAT/CD36 from intracellular vesicles to the plasma membrane, as well as enhancing transport of fatty acids into the mitochondria for oxidation via inhibition of ACC2 (Habets et al. 2007; Luiken et al. 2003). However, mitochondrial oxidative metabolism becomes severely attenuated during myocardial ischaemia, therefore under these conditions the heart must quickly switch to non-oxidative sources of ATP (Kloner and Jennings 2001). The steep decline in ATP levels that accompanies cardiac ischaemia rapidly activates AMPK, which orchestrates a shift from oxidative metabolism to anaerobic glucose metabolism by stimulating glucose uptake, as well as increasing flux through glycolysis (Dyck et al. 1999). It enhances the rate of glycolysis in the ischaemic heart by phosphorylating and stimulating phosphofructokinase-2 (PFK2) to produce fructose-2,6-bisphosphate, which in turn allosterically activates phosphofructokinase-1 (PFK1), the rate-limiting enzyme in glycolysis (Marsin et al. 2000). This effect of AMPK also provides an elegant explanation for the so-called Pasteur effect, which is an increased rate of glycolysis in response to hypoxia, a phenomenon first noted by Louis Pasteur.

Although glycolysis is a minor source of ATP in the heart under normal conditions, the small of amounts of ATP that are generated by glycolysis during ischaemia are critical to maintain essential processes such as ion homeostasis via ATP-dependent ion pumps (Xu et al. 1995). Enhancing anaerobic glucose metabolism during myocardial ischaemia is therefore a crucial metabolic response, so much so that inhibition of glucose transport or glycolysis drastically decreases the ability of the heart to withstand ischaemic insults. Although activation of AMPK in response to ischaemia is undoubtedly beneficial for the short term survival of the heart, there is evidence to suggest that it may be harmful to the recovery of the heart during reperfusion (Lopaschuk et al. 1993). This stems from the fact that the AMPK mediated increase in fatty acid oxidation that occurs during reperfusion de-couples glycolysis and glucose oxidation through inhibition of the pyruvate dehydrogenase complex, a phenomenon known as the Randle cycle, thereby causing pyruvate to be converted to the dead end product lactate resulting in lactic acidosis. Sustained increases in lactate can seriously impair cardiac function and efficiency, as it causes a diversion of ATP away from myocardial contraction towards restoring ion balance, which is necessary to counteract the effects of the acidosis and prevent Ca^{2+} overload. In this respect, decreasing fatty acid oxidation through inhibition of malonyl-CoA decarboxylase improves the recovery of contractile function after ischaemia (Dyck et al. 2004). However, future studies are required to definitively answer whether AMPK activation during ischaemia-reperfusion is beneficial or detrimental.

Naturally occurring mutations in the human γ2 subunit of AMPK cause a hypertrophic cardiomyopathy that is characterised by excessive glycogen storage in the cardiomyocyte, which results in the development of a ventricular pre-excitation disorder known as Wolff-Parkinson-White (WPW) syndrome (Akman et al. 2007; Arad et al. 2002; Burwinkel et al. 2005). Excessive glycogen in cardiac myocytes affects the development of the electrically insulating layer between the atria and ventricles, leading to accessory conductance pathways that compromise contractile function. The glycogen storage cardiomyopathy caused by the γ2 subunit mutations is similar to that observed in Pompe's diseases, which also causes WPW syndrome due to impaired glycogenolysis (Bulkley and Hutchins 1978). All of the γ2 subunit mutations occur in the allosteric AMP and ATP binding sites and, with a few exceptions, affect the residues directly involved in nucleotide binding, particularly the positively charged residues that co-ordinate binding of the nucleotide phosphate groups (Xiao et al. 2007). The effect of the γ2 mutations on AMPK activity are complex because, as well as interfering with (and in some cases abolishing) allosteric activation by AMP (a loss of function effect), they also disrupt allosteric inhibition by ATP (a gain of function effect) (Adams et al. 2004; Scott et al. 2004). Therefore, although the mutations prevent activation by AMP, they also increase the basal activity of AMPK in the absence of any stimuli, which explains why these mutations are dominant in nature (Arad et al. 2002; Burwinkel et al. 2005). Constitutive activation of AMPK also provides a simple explanation for the observed glycogen storage phenotype. The increase in AMPK basal activity caused by the mutations stimulates glucose uptake that, in the absence of increased energy demand, causes a buildup of glucose-6-phosphate levels, which is preferentially diverted towards glycogen synthesis (Luptak et al. 2007). The accumulation of glucose-6-phosphate is further exacerbated by the fact that increased AMPK activity also stimulates fatty acid oxidation, which in turn inhibits glucose oxidation as a consequence of the Randle cycle. Although activation of AMPK results in phosphorylation and inhibition of glycogen synthase, this inhibitory effect is overcome by the high levels of glucose-6-phosphate, which is an allosteric activator of glycogen synthase (Jorgensen et al. 2004a). Some of the γ2 mutations also cause elevated glycogen in skeletal muscle where the γ2 subunit is also expressed, however, this does not appear to cause any obvious defects in muscle function. No mutations in γ1 have been identified so far, possibly because they are likely to be more detrimental than mutations in γ2 or γ3 given that the γ1 subunit is the predominant isoform in most tissue types.

9.9 Role of AMPK in Cancer

Several landmark studies in recent years suggest that AMPK signalling is important for the prevention and treatment of a range of human diseases, especially disorders of energy balance such as obesity and Type 2 diabetes. However, evidence is also beginning to emerge that other diseases such as cancer may have underlying defects in energy metabolism, and that activation of AMPK may also be beneficial in treating certain types of cancer. An important characteristic of cancer cells is that they acquire the ability to evade the normal regulatory mechanisms that control cell growth and proliferation, polarity and migration, and programmed cell death (apoptosis) (Hanahan and Weinberg 2000). This transformation to an autonomous state occurs as a result of multiple changes in the genome, typically as a consequence of mutations in oncogenic and tumour suppressor genes. Cancer cells also have fundamental differences in energy metabolism compared with normal cells, for example, cancer cells have a high rate of glucose consumption and produce ATP mainly by metabolising glucose to lactate rather than through oxidative phosphorylation, a phenomenon known as the 'Warburg effect' (Warburg 1956). As AMPK is downstream of the tumour suppressor LKB1, it is now regarded as a potential drug target for the treatment of cancer. As proof of concept, it was recently discovered that Type 2 diabetic patients receiving metformin treatment have a reduced rate of cancer incidence compared with patients on other drug programmes (Evans et al. 2005). Metformin is a widely used anti-diabetic drug that activates AMPK and has been shown to inhibit breast cancer cell growth in an AMPK-dependent manner, primarily via inhibition of the mammalian target of rapamycin (mTOR) pathway (Dowling et al. 2007; Zakikhani et al. 2006).

The mTOR pathway plays an important role in regulating cell growth and is overactivated in a variety of cancers (Schmelzle and Hall 2000). The central component of this pathway is mTOR, which exists within two distinct multiprotein complexes called mTORC1 and mTORC2 (Wullschleger et al. 2006). mTORC1 is composed of mTOR, raptor, PRAS40, mLST8 and is dependent on cellular nutrient availability as withdrawal of glucose, amino acids or oxygen rapidly suppresses mTORC1 activity. On the other hand, the mTORC2 complex consists of mSIN1, PRR5/Protor, mLST8 and rictor and is not responsive to nutrient levels. Several tumour suppressors such as PTEN, TSC1 (hamartin) and TSC2 (tuberin), and LKB1 have all been identified as negative regulators of mTOR signalling (Wullschleger et al. 2006). TSC1 and TSC2 form a heterodimer that is mutated in tuberous sclerosis complex, an inherited genetic disorder characterised by benign tumours in multiple organs that is strikingly similar, but distinct, from Peutz-Jeghers syndrome caused by mutations in LKB1 (van Slegtenhorst et al. 1998). mTORC1 is activated by an upstream pathway involving the TSC1-TSC2 heterodimer and Rheb, a member of the small G-protein family. AMPK inhibits mTORC1 signalling by a dual mechanism involving phosphorylation of TSC2 and raptor (Fig. 9.6). The TSC2 protein has a GTPase-activating protein (GAP) domain that inhibits the ability of Rheb to activate mTORC1, and this inhibitory effect is augmented by AMPK phosphorylation (Corradetti et al. 2004; Inoki et al. 2003). Meanwhile, phosphorylation of raptor by AMPK induces binding to 14-3-3 protein, which suppresses mTORC1 activity by preventing its interaction with downstream substrates 4EBP1 and ribosomal S6 kinase, two key regulators of protein translation (Gwinn et al. 2008). mTORC1 dependent protein translation is known to switch on the expression of a number of important cell growth regulators, including cyclin D1, the hypoxia inducible factor-1α (HIF-1α) transcription factor and c-myc, all of which are highly expressed in many types of cancer (Guertin and Sabatini 2007). In contrast to AMPK, protein kinase B (PKB) phosphorylates the

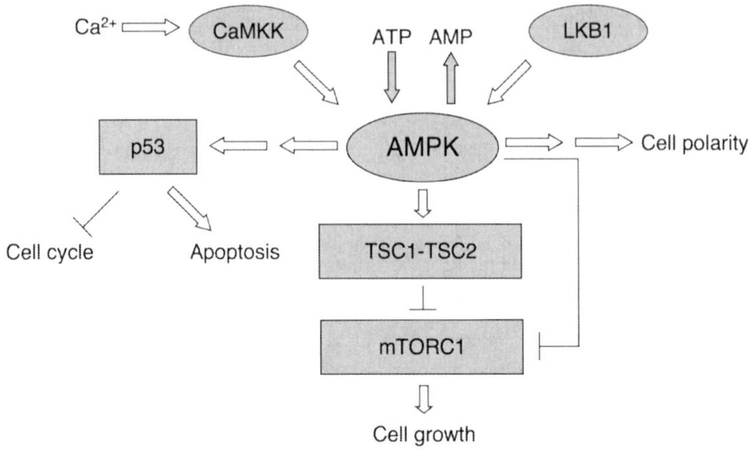

Fig. 9.6 AMPK regulation of cell cycle and growth. AMPK inhibits cell growth via the mTORC1 complex by dual mechanism: direct inhibition and via the TSC1-TSC2 complex. Phosphorylation of p53 by AMPK induces cell cycle arrest

TSC1-TSC2 complex on alternate sites in response to activation of the insulin/insulin-like growth factor-1 (IGF-1) signalling pathway, and inhibits its ability to switch off Rheb (Inoki et al. 2002). Therefore insulin and IGF-1, which indicate the availability of nutrients, and AMPK, which signals a shortage of energy or nutrients, have opposing effects on cell growth.

One of the most commonly mutated tumour suppressors in human cancers is p53, which is frequently referred to as the 'guardian of the genome' (Levine 1997). Expression of p53 is normally very low in the cell however conditions such as DNA damage caused by ionising radiation induce p53 transcription, which leads to the expression of multiple target genes that promote cell cycle arrest or apoptosis. AMPK phosphorylates p53 under conditions of low glucose but not ionising radiation, leading to cell cycle arrest in the G1/S phase and protection from apoptosis, whereas under prolonged glucose starvation AMPK promotes p53-dependent cell senescence (Imamura et al. 2001; Jones et al. 2005). AMPK is also activated by sestrin-1 and sestrin-2, both of which are transcriptionally upregulated by p53 in response to DNA damage and oxidative stress, however the mechanism by which the sestrins promote AMPK phosphorylation and activation is unknown (Budanov and Karin 2008). It has been suggested that the AMPK-p53 signalling axis may function as a metabolic checkpoint for cell growth that ensures energy supplies are adequate for cell cycle progression.

An unexpected role for AMPK that has recently emerged is in the regulation of cell polarity. Establishing and maintaining cell polarity is essential for organism development and tissue function, and loss of cell polarity is frequently associated with tumour invasion, particularly epithelial cell-derived malignant cancers (Williams and Brenman 2008). Clues to the involvement of AMPK in the regulation of cell polarity came from the discovery that LKB1 is required for polarisation of oocytes that determine the anterior-posterior axis of embryos, which likely explains why genetic deletion of the LKB1 gene is embryonic lethal (Martin and St Johnston 2003). LKB1 is also essential for establishing apical-basal polarity in epithelial cells (Baas et al. 2004). Although LKB1 is the upstream kinase for several other protein kinases, some of which have known roles in the establishment of cell polarity (particularly the microtubule-affinity regulating kinases), genetic deletion of AMPK in epithelial cells disrupts polarity as well as increasing cell proliferation, providing direct evidence that AMPK is also involved in these processes (Lee et al. 2007). Maintaining cell polarity is an energy intensive process, therefore it may seem incompatible that AMPK, which is activated by negative energy balance, should promote cell polarity especially during energy shortages. Nevertheless, the maintenance of cell polarity is an essential aspect of normal cell function and, therefore, this may be an instance where AMPK diverts what limited energy is available to a critical survival function.

9.10 AMPK Activators as Therapeutic Drugs

Pharmacological activation of AMPK is now considered a major therapeutic strategy to treat diseases that have underlying defects in the regulation of energy balance. Interest in AMPK as a drug target gained rapid momentum since the discovery that it is activated by the anti-diabetic drug metformin and is largely responsible for mediating its therapeutic effects. Until recently, the number of known activators of AMPK was limited, however a plethora of new activating agents have since been discovered. Unfortunately, the vast majority of these agents activate AMPK indirectly giving rise to multiple off-target effects, however this is expected to be less of a problem with the development of more direct activators. Regardless of their mechanism of action most, if not all, of the activators identified to date elicit positive metabolic effects consistent with activation of AMPK.

The most widely used drug to study the downstream effects of pharmacological activation of AMPK is the riboside AICAR (also known as acadesine) (Fig. 9.7). As mentioned previously, AICAR is a pro-drug that is phosphorylated by adenosine kinase to generate the AMP-mimetic, ZMP (Corton et al. 1995). Although ZMP mimics all the effects of AMP on the AMPK system, it also affects other AMP-sensitive enzymes including fructose-1,6-bisphosphatase and glycogen phosphorylase, both of which are important regulators of glucose metabolism (Vincent et al. 1991; Young et al. 1996). These selectivity problems and poor physicochemical properties have made AICAR and related nucleotide analogues unsuitable leads for further drug development (Dixon et al. 1991).

AMPK is also activated by the glucose lowering drug metformin, which is the most commonly used pharmaceutical to treat Type 2 diabetes (Zhou et al. 2001). The therapeutic effects of metformin are primarily due to repression of gluconeogenesis in the liver and this effect is dependent on the presence of LKB1, which phosphorylates and activates AMPK (Shaw et al. 2005). There is also growing evidence that metformin is a potential anti-cancer drug, which may be related to the fact that LKB1 is a known tumour suppressor (Evans et al. 2005). Despite being able to activate AMPK in intact cells, metformin does not activate or affect the phosphorylation or dephosphorylation of AMPK by upstream kinases or phosphatases in cell-free assays (Hawley et al. 2002). Instead, metformin has been shown to be an inhibitor of complex I of the respiratory chain, which suggests that it activates AMPK by decreasing intracellular ATP and increasing AMP (Owen et al. 2000). Indeed, significant decreases in intracellular ATP have been detected in hepatocytes cultured with metformin, and in other cell types cultured with its more potent analogue phenformin, which likely explains some of their undesirable side-effects such as lactic acidosis (Guigas et al. 2006; Hawley et al. 2005). Inhibition of mitochondrial respiration is likely to be a common mechanism of action for the majority of indirect AMPK activators. In fact, the thiazolidinediones (another class of glucose lowering drug) and a number of natural plant compounds, such as berberine and resveratrol, have all been shown to activate AMPK by inhibiting mitochondrial respiration in much the same manner as metformin (Baur et al. 2006; Brunmair et al. 2004; Turner et al. 2008).

The first direct activator of AMPK to be discovered was the non-nucleoside thienopyridone, A769662 (Fig. 9.7) (Cool et al. 2006). AMPK activation by A769662 is similar to AMP, encompassing not only direct allosteric activation but

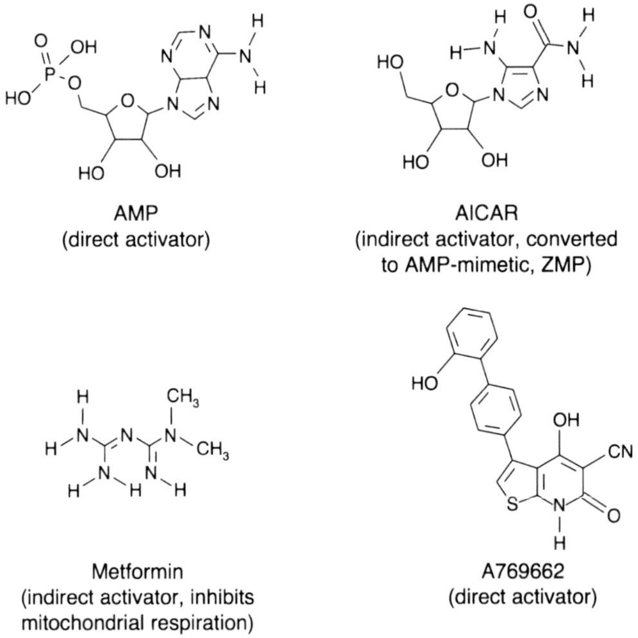

Fig. 9.7 Activators of AMPK. AICAR is a pro-drug that is converted to the AMP-mimetic, ZMP. Metformin indirectly activates AMPK by inhibiting mitochondrial respiration. A769662, like AMP, directly activates AMPK but independently of the AMP-binding sites

also protection against dephosphorylation of Thr172 by protein phosphatases (Goransson et al. 2007; Sanders et al. 2007a). Intriguingly, its mechanism of action appears to be independent of the AMP-binding sites on the γ subunit since it can activate an AMP-insensitive mutant. Furthermore, A769662 selectively activates AMPK complexes containing the β1 subunit isoform but is without effect on β2 complexes (Scott et al. 2008). At present, the molecular basis of this selectivity is unknown, although activation is dependent on the β1 subunit carbohydrate-binding module. These findings are important as it highlights that isoform-specific activation of AMPK is achievable. In terms of metabolic effects, A769662 enhances whole-body fatty acid oxidation, lowers glucose and promotes weight loss predominantly via activation of AMPK in the liver (Cool et al. 2006). Notwithstanding these positive effects, A769662 has poor pharmacokinetic properties therefore it is unlikely to proceed as a potential therapy; however it does provide a useful lead for further drug development.

9.11 Concluding Remarks

Since its discovery over three decades ago, AMPK has ascended from being regarded simply as a sensor of cellular energy, to being recognised as a key player in the regulation of whole-body energy balance, earning the monicker of 'metabolic master switch'. Although it is likely that AMPK emerged as a nutrient sensor in ancient unicellular eukaryotes, it appears to have become adapted during the evolution of multicellular organisms to become not only sensitive to cellular energy status, but also responsive to hormones that regulate whole-body energy balance. The broad spectrum of positive effects that AMPK exerts on both cellular and whole-body energy metabolism has made it an attractive target for new drugs to treat a range of metabolic disorders, particularly obesity and Type 2 diabetes, but perhaps also other diseases such as cancer. Significant strides have been made in the search for novel and direct activators of AMPK, however given that there are 12 possible combinations of AMPK heterotrimer with varying tissue expression and physiological effects, the challenge in the future will be to develop isoform-specific activators that target AMPK in particular tissues. Although this is an ambitious task, the discovery that the thienopyridone A769662 selectively activates AMPK complexes containing the β1 subunit provides proof of concept that the development of isoform specific activators is feasible. Future research in the AMPK field will undoubtedly continue to unravel the intricate mechanisms that regulate energy balance at both the cellular and whole-body levels, and provide critical information to understand and tackle the growing problem of metabolic diseases.

Acknowledgements I would like to thank Prof Bruce Kemp for mentoring and helpful discussions. This work was funded by the Australian Research Council and the National Health and Medical Research Council.

References

Abu-Elheiga, L., Jayakumar, A., Baldini, A., et al. (1995). Human acetyl-CoA carboxylase: characterization, molecular cloning, and evidence for two isoforms. *Proc. Natl. Acad. Sci. USA* 92:4011–4015.

Abu-Elheiga, L., Brinkley, W.R., Zhong, L., et al. (2000). The subcellular localization of acetyl-CoA carboxylase 2. *Proc. Natl. Acad. Sci. USA* 97:1444–1449.

Abu-Elheiga, L., Matzuk, M.M., Abo-Hashema, K.A., et al. (2001). Continuous fatty acid oxidation and reduced fat storage in mice lacking acetyl-CoA carboxylase 2. *Science* 291:2613–2616.

Adams, J., Chen, Z.P., Van Denderen, et al. (2004). Intrasteric control of AMPK via the gamma1 subunit AMP allosteric regulatory site. *Protein Sci.* 13:155–165.

Akimoto, T., Pohnert, S.C., Li, P., Zhang, M., et al. (2005). Exercise stimulates Pgc-1alpha transcription in skeletal muscle through activation of the p38 MAPK pathway. *J. Biol. Chem.* 280:19587–19593.

Akman, H.O., Sampayo, J.N., Ross, F.A., et al. (2007). Fatal infantile cardiac glycogenosis with phosphorylase kinase deficiency and a mutation in the gamma2-subunit of AMP-activated protein kinase. *Pediatr. Res.* 62:499–504.

Anderson, K.A., Ribar, T.J., Lin, F., et al. (2008). Hypothalamic CaMKK2 contributes to the regulation of energy balance. *Cell Metab.* 7:377–388.

Andersson, U., Filipsson, K., Abbott, C.R., et al. (2004). AMP-activated protein kinase plays a role in the control of food intake. *J. Biol. Chem.* 279:12005–12008.

Andreelli, F., Foretz, M., Knauf, C., et al. (2006). Liver adenosine monophosphate-activated kinase-alpha2 catalytic subunit is a key target for the control of hepatic glucose production by adiponectin and leptin but not insulin. *Endocrinology* 147:2432–2441.

Anthonsen, M.W., Ronnstrand, L., Wernstedt, C., et al. (1998). Identification of novel phosphorylation sites in hormone-sensitive lipase that are phosphorylated in response to isoproterenol and govern activation properties in vitro. *J. Biol. Chem.* 273:215–221.

Arad, M., Benson, D.W., Perez-Atayde, et al. (2002). Constitutively active AMP kinase mutations cause glycogen storage disease mimicking hypertrophic cardiomyopathy. *J. Clin. Invest.* 109:357–362.

Aschenbach, W.G., Hirshman, M.F., Fujii, N., et al. (2002). Effect of AICAR treatment on glycogen metabolism in skeletal muscle. *Diabetes* 51:567–573.

Baas, A.F., Kuipers, J., van der Wel, N.N., et al. (2004). Complete polarization of single intestinal epithelial cells upon activation of LKB1 by STRAD. *Cell* 116:457–466.

Barnes, B.R., Marklund, S., Steiler, T.L., et al. (2004). The 5'-AMP-activated protein kinase gamma3 isoform has a key role in carbohydrate and lipid metabolism in glycolytic skeletal muscle. *J. Biol. Chem.* 279:38441–38447.

Bateman, A. (1997). The structure of a domain common to archaebacteria and the homocystinuria disease protein. *Trends Biochem. Sci.* 22:12–13.

Baur, J.A., Pearson, K.J., Price, N.L., et al. (2006). Resveratrol improves health and survival of mice on a high-calorie diet. *Nature* 444:337–342.

Beg, Z.H., Stonik, J.A., Brewer, H.B. Jr. (1978). 3-Hydroxy-3-methylglutaryl coenzyme A reductase: regulation of enzymatic activity by phosphorylation and dephosphorylation. *Proc. Natl. Acad. Sci. USA* 75:3678–3682.

Bergeron, R., Previs, S.F., Cline, G.W., et al. (2001a). Effect of 5-aminoimidazole-4-carboxamide-1-beta-D-ribofuranoside infusion on in vivo glucose and lipid metabolism in lean and obese Zucker rats. *Diabetes* 50:1076–1082.

Bergeron, R., Ren, J.M., Cadman, K.S., et al. (2001b). Chronic activation of AMP kinase results in NRF-1 activation and mitochondrial biogenesis. *Am. J. Physiol. Endocrinol. Metab.* 281:E1340–1346.

Boudeau, J., Baas, A.F., Deak, M., et al. (2003). MO25alpha/beta interact with STRADalpha/beta enhancing their ability to bind, activate and localize LKB1 in the cytoplasm. *EMBO J.* 22:5102–5114.

Brabant, G., Muller, G., Horn, R., et al. (2005). Hepatic leptin signaling in obesity. *FASEB J.* 19:1048–1050.

Brunmair, B., Staniek, K., Gras, F., et al. (2004). Thiazolidinediones, like metformin, inhibit respiratory complex I: a common mechanism contributing to their antidiabetic actions? *Diabetes* 53:1052–1059.

Budanov, A.V. and Karin, M. (2008). p53 target genes sestrin1 and sestrin2 connect genotoxic stress and mTOR signaling. *Cell* 134:451–460.

Bulkley, B.H., and Hutchins, G.M. (1978). Pompe's disease presenting as hypertrophic myocardiopathy with Wolff-Parkinson-White syndrome. *Am. Heart. J.* 96:246–252.

Burwinkel, B., Scott, J.W., Buhrer, C., et al. (2005). Fatal congenital heart glycogenosis caused by a recurrent activating R531Q mutation in the gamma 2-subunit of AMP-activated protein kinase (PRKAG2), not by phosphorylase kinase deficiency. *Am. J. Hum. Genet.* 76:1034–1049.

Carling, D. and Hardie, D.G. (1989). The substrate and sequence specificity of the AMP-activated protein kinase. Phosphorylation of glycogen synthase and phosphorylase kinase. *Biochim. Biophys. Acta* 1012:81–86.

Carling, D., Clarke, P.R., Zammit, V.A., et al. (1989). Purification and characterization of the AMP-activated protein kinase. Copurification of acetyl-CoA carboxylase kinase and 3-hydroxy-3-methylglutaryl-CoA reductase kinase activities. *Eur. J. Biochem.* 186:129–136.

Chavez, J.A., Roach, W.G., Keller, S.R., et al. (2008). Inhibition of GLUT4 translocation by Tbc1d1, a Rab GTPase-activating protein abundant in skeletal muscle, is partially relieved by AMP-activated protein kinase activation. *J. Biol. Chem.* 283:9187–9195.

Chen, H., Charlat, O., Tartaglia, L.A., et al. (1996). Evidence that the diabetes gene encodes the leptin receptor: identification of a mutation in the leptin receptor gene in db/db mice. *Cell* 84:491–495.

Chen, Z.P., Stephens, T.J., Murthy, S., et al. (2003). Effect of exercise intensity on skeletal muscle AMPK signaling in humans. *Diabetes* 52:2205–2212.

Chen, S., Murphy, J., Toth, R., et al. (2008). Complementary regulation of TBC1D1 and AS160 by growth factors, insulin and AMPK activators. *Biochem. J.* 409:449–459.

Civitarese, A.E., Ukropcova, B., Carling, S., et al. (2006). Role of adiponectin in human skeletal muscle bioenergetics. *Cell Metab.* 4:75–87.

Clarke, P.R. and Hardie, D.G. (1990). Regulation of HMG-CoA reductase: identification of the site phosphorylated by the AMP-activated protein kinase in vitro and in intact rat liver. *EMBO J.* 9:2439–2446.

Clarke, C.F., Edwards, P.A., Lan, S.F., et al. (1983). Regulation of 3-hydroxy-3-methylglutaryl-coenzyme A reductase mRNA levels in rat liver. *Proc. Natl. Acad. Sci. USA* 80:3305–3308.

Cool, B., Zinker, B., Chiou, W., et al. (2006). Identification and characterization of a small molecule AMPK activator that treats key components of type 2 diabetes and the metabolic syndrome. *Cell Metab.* 3:403–416.

Corradetti, M.N., Inoki, K., Bardeesy, N., et al. (2004). Regulation of the TSC pathway by LKB1: evidence of a molecular link between tuberous sclerosis complex and Peutz-Jeghers syndrome. *Genes Dev.* 18:1533–1538.

Corton, J.M., Gillespie, J.G., Hawley, S.A., et al. (1995). 5-aminoimidazole-4-carboxamide ribonucleoside. A specific method for activating AMP-activated protein kinase in intact cells? *Eur. J. Biochem.* 229:558–565.

Crute, B.E., Seefeld, K., Gamble, J., et al. (1998). Functional domains of the alpha1 catalytic subunit of the AMP-activated protein kinase. *J. Biol. Chem.* 273:35347–35354.

da Silva Xavier, G., Leclerc, I., Salt, I.P., et al. (2000). Role of AMP-activated protein kinase in the regulation by glucose of islet beta cell gene expression. *Proc. Natl. Acad. Sci. USA* 97:4023–4028.

Davies, S.P., Helps, N.R., Cohen, P.T., et al. (1995). 5'-AMP inhibits dephosphorylation, as well as promoting phosphorylation, of the AMP-activated protein kinase. Studies using bacterially expressed human protein phosphatase-2C alpha and native bovine protein phosphatase-2AC. *FEBS Lett.* 377:421–425.

Dixon, R., Gourzis, J., McDermott, D., et al. (1991). AICA-riboside: safety, tolerance, and pharmacokinetics of a novel adenosine-regulating agent. *J. Clin. Pharmacol.* 31:342–347.

Dowling, R.J., Zakikhani, M., Fantus, I.G., et al. (2007). Metformin inhibits mammalian target of rapamycin-dependent translation initiation in breast cancer cells. *Cancer Res.* 67:10804–10812.

Dyck, J.R., Kudo, N., Barr, A.J., et al. (1999). Phosphorylation control of cardiac acetyl-CoA carboxylase by cAMP-dependent protein kinase and 5'-AMP activated protein kinase. *Eur. J. Biochem.* 262:184–190.

Dyck, J.R., Cheng, J.F., Stanley, W.C., et al. (2004). Malonyl coenzyme a decarboxylase inhibition protects the ischemic heart by inhibiting fatty acid oxidation and stimulating glucose oxidation. *Circ. Res.* 94:e78–84.

Dzamko, N., Schertzer, J.D., Ryall, J.G., et al. (2008). AMPK-independent pathways regulate skeletal muscle fatty acid oxidation. *J. Physiol.* 586:5819–5831.

Evans, J.M., Donnelly, L.A., Emslie-Smith, A.M., et al. (2005). Metformin and reduced risk of cancer in diabetic patients. *BMJ* 330:1304–1305.

Fogarty, S. and Hardie, D.G. (2009). C-terminal phosphorylation of LKB1 is not required for regulation of AMP-activated protein kinase, BRSK1, BRSK2, or cell cycle arrest. *J. Biol. Chem.* 284:77–84.

Foretz, M., Carling, D., Guichard, C., et al. (1998). AMP-activated protein kinase inhibits the glucose-activated expression of fatty acid synthase gene in rat hepatocytes. *J. Biol. Chem.* 273:14767–14771.

Fox, C.J., Hammerman, P.S., Thompson, C.B. (2005). Fuel feeds function: energy metabolism and the T-cell response. *Nat. Rev. Immunol.* 5:844–852.

Fujii, N., Hayashi, T., Hirshman, M.F., et al. (2000). Exercise induces isoform-specific increase in 5'AMP-activated protein kinase activity in human skeletal muscle. *Biochem. Biophys. Res. Commun.* 273:1150–1155.

Fujii, N., Seifert, M.M., Kane, E.M., et al. (2007). Role of AMP-activated protein kinase in exercise capacity, whole body glucose homeostasis, and glucose transport in skeletal muscle -insight from analysis of a transgenic mouse model. *Diabetes Res. Clin. Pract.* 77 Suppl 1:S92–S98.

Gao, S., Kinzig, K.P., Aja, S., et al. (2007). Leptin activates hypothalamic acetyl-CoA carboxylase to inhibit food intake. *Proc. Natl. Acad. Sci. USA* 104:17358–17363.

Garcia-Roves, P.M., Osler, M.E., Holmstrom, M.H., et al. (2008). Gain-of-function R225Q mutation in AMP-activated protein kinase gamma3 subunit increases mitochondrial biogenesis in glycolytic skeletal muscle. *J. Biol. Chem.* 283:35724–35734.

Garton, A.J. and Yeaman, S.J. (1990). Identification and role of the basal phosphorylation site on hormone-sensitive lipase. *Eur. J. Biochem.* 191:245–250.

Garton, A.J., Campbell, D.G., Carling, D., et al. (1989). Phosphorylation of bovine hormone-sensitive lipase by the AMP-activated protein kinase. A possible antilipolytic mechanism. *Eur. J. Biochem.* 179:249–254.

Geraghty, K.M., Chen, S., Harthill, J.E., et al. (2007). Regulation of multisite phosphorylation and 14-3-3 binding of AS160 in response to IGF-1, EGF, PMA and AICAR. *Biochem. J.* 407:231–241.

Ghilardi, N., Ziegler, S., Wiestner, A., et al. (1996). Defective STAT signaling by the leptin receptor in diabetic mice. *Proc. Natl. Acad. Sci. USA* 93:6231–6235.

Gonzalez-Gay, M.A., De Matias, J.M., Gonzalez-Juanatey, C., et al. (2006). Anti-tumor necrosis factor-alpha blockade improves insulin resistance in patients with rheumatoid arthritis. *Clin. Exp. Rheumatol.* 24:83–86.

Goransson, O., McBride, A., Hawley, S.A., et al. (2007). Mechanism of action of A-769662, a valuable tool for activation of AMP-activated protein kinase. *J. Biol. Chem.* 282:32549–32560.

Guertin, D.A. and Sabatini, D.M. (2007). Defining the role of mTOR in cancer. *Cancer Cell* 12:9–22.

Guigas, B., Bertrand, L., Taleux, N., et al. (2006). 5-Aminoimidazole-4-carboxamide-1-beta-D-ribofuranoside and metformin inhibit hepatic glucose phosphorylation by an AMP-activated protein kinase-independent effect on glucokinase translocation. *Diabetes* 55:865–874.

Gwinn, D.M., Shackelford, D.B., Egan, D.F., et al. (2008). AMPK phosphorylation of raptor mediates a metabolic checkpoint. *Mol Cell* 30:214–226.

Habets, D.D., Coumans, W.A., Voshol, P.J., et al. (2007). AMPK-mediated increase in myocardial long-chain fatty acid uptake critically depends on sarcolemmal CD36. *Biochem. Biophys. Res. Commun.* 355:204–210.

Hanahan, D. and Weinberg, R.A. (2000). The hallmarks of cancer. *Cell* 100:57–70.

Hardie, D.G. and Hawley, S.A. (2001). AMP-activated protein kinase: the energy charge hypothesis revisited. *Bioessays* 23:1112–1119.

Hawley, S.A., Davison, M., Woods, A., et al. (1996). Characterization of the AMP-activated protein kinase kinase from rat liver and identification of threonine 172 as the major site at which it phosphorylates AMP-activated protein kinase. *J. Biol. Chem.* 271:27879–27887.

Hawley, S.A., Gadalla, A.E., Olsen, G.S., et al. (2002). The antidiabetic drug metformin activates the AMP-activated protein kinase cascade via an adenine nucleotide-independent mechanism. *Diabetes* 51:2420–2425.

Hawley, S.A., Boudeau, J., Reid, J.L., et al. (2003). Complexes between the LKB1 tumor suppressor, STRAD alpha/beta and MO25 alpha/beta are upstream kinases in the AMP-activated protein kinase cascade. *J. Biol.* 2:28.

Hawley, S.A., Pan, D.A., Mustard, K.J., et al. (2005). Calmodulin-dependent protein kinase kinase-beta is an alternative upstream kinase for AMP-activated protein kinase. *Cell Metab.* 2:9–19.

Hemminki, A., Markie, D., Tomlinson, I., et al. (1998). A serine/threonine kinase gene defective in Peutz-Jeghers syndrome. *Nature* 391:184–187.

Holmes, B.F., Kurth-Kraczek, E.J., and Winder, W.W. (1999). Chronic activation of 5'-AMP-activated protein kinase increases GLUT-4, hexokinase, and glycogen in muscle. *J. Appl. Physiol.* 87:1990–1995.

Hong, Y.H., Varanasi, U.S., Yang, W. et al. (2003). AMP-activated protein kinase regulates HNF4alpha transcriptional activity by inhibiting dimer formation and decreasing protein stability. *J. Biol. Chem.* 278:27495–27501.

Hudson, E.R., Pan, D.A., James, J., et al. (2003). A novel domain in AMP-activated protein kinase causes glycogen storage bodies similar to those seen in hereditary cardiac arrhythmias. *Curr. Biol.* 13:861–866.

Hurley, R.L., Anderson, K.A., Franzone, J.M., et al. (2005). The Ca2+/calmodulin-dependent protein kinase kinases are AMP-activated protein kinase kinases. *J. Biol. Chem.* 280:29060–29066.

Imamura, K., Ogura, T., Kishimoto, A., et al. (2001). Cell cycle regulation via p53 phosphorylation by a 5'-AMP activated protein kinase activator, 5-aminoimidazole-4-carboxamide-1-beta-D-ribofuranoside, in a human hepatocellular carcinoma cell line. *Biochem. Biophys. Res. Commun.* 287:562–567.

Inoki, K., Li, Y., Zhu, T., et al. (2002). TSC2 is phosphorylated and inhibited by Akt and suppresses mTOR signalling. *Nat. Cell Biol.* 4:648–657.

Inoki, K., Zhu, T., and Guan, K.L. (2003). TSC2 mediates cellular energy response to control cell growth and survival. *Cell* 115:577–590.

Iseli, T.J., Walter, M., van Denderen, B.J., et al. (2005). AMP-activated protein kinase beta subunit tethers alpha and gamma subunits via its C-terminal sequence (186-270). *J. Biol. Chem.* 280:13395–13400.

Iseli, T.J., Oakhill, J.S., Bailey, M.F., et al. (2008). AMP-activated protein kinase subunit interactions: beta1:gamma1 association requires beta1 Thr-263 and Tyr-267. *J. Biol. Chem.* 283: 4799–4807.

Iverson, A.J., Bianchi, A., Nordlund, A.C., et al. (1990). Immunological analysis of acetyl-CoA carboxylase mass, tissue distribution and subunit composition. *Biochem. J.* 269:365–371.

Jager, S., Handschin, C., St-Pierre, J., et al. (2007). AMP-activated protein kinase (AMPK) action in skeletal muscle via direct phosphorylation of PGC-1alpha. *Proc. Natl. Acad. Sci. USA* 104:12017–12022.

Jin, X., Townley, R., Shapiro, L. (2007). Structural insight into AMPK regulation: ADP comes into play. *Structure* 15:1285–1295.

Johnson, L.N., Noble, M.E., Owen, D.J. (1996). Active and inactive protein kinases: structural basis for regulation. *Cell* 85:149–158.

Jones, R.G., Plas, D.R., Kubek, S., et al. (2005). AMP-activated protein kinase induces a p53-dependent metabolic checkpoint. *Mol. Cell* 18:283–293.

Jorgensen, S.B., Nielsen, J.N., Birk, J.B., et al. (2004a). The alpha2-5'AMP-activated protein kinase is a site 2 glycogen synthase kinase in skeletal muscle and is responsive to glucose loading. *Diabetes* 53:3074–3081.

Jorgensen, S.B., Viollet, B., Andreelli, F., et al. (2004b). Knockout of the alpha2 but not alpha1 5'-AMP-activated protein kinase isoform abolishes 5-aminoimidazole-4-carboxamide-1-beta-4-ribofuranoside but not contraction-induced glucose uptake in skeletal muscle. *J. Biol. Chem.* 279:1070–1079.

Jorgensen, S.B., Treebak, J.T., Viollet, B., et al. (2007). Role of AMPKalpha2 in basal, training-, and AICAR-induced GLUT4, hexokinase II, and mitochondrial protein expression in mouse muscle. *Am. J. Physiol. Endocrinol. Metab.* 292:E331–E339.

Kahn, B.B., Alquier, T., Carling, D. et al. (2005). AMP-activated protein kinase: ancient energy gauge provides clues to modern understanding of metabolism. *Cell Metab.* 1:15–25.

Kemp, B.E., Parker, M.W., Hu, S., et al. (1994). Substrate and pseudosubstrate interactions with protein kinases: determinants of specificity. *Trends Biochem. Sci.* 19:440–444.

Kim, M.S., Fielitz, J., McAnally, J., et al. (2008). Protein kinase D1 stimulates MEF2 activity in skeletal muscle and enhances muscle performance. *Mol. Cell. Biol.* 28:3600–3609.

Kloner, R.A. and Jennings, R.B. (2001). Consequences of brief ischemia: stunning, preconditioning, and their clinical implications: part 1. *Circulation* 104:2981–2989.

Koay, A., Rimmer, K.A., Mertens, H.D., et al. (2007). Oligosaccharide recognition and binding to the carbohydrate binding module of AMP-activated protein kinase. *FEBS Lett.* 581: 5055–5059.

Kola, B., Hubina, E., Tucci, S.A., et al. (2005). Cannabinoids and ghrelin have both central and peripheral metabolic and cardiac effects via AMP-activated protein kinase. *J. Biol. Chem.* 280:25196–25201.

Kola, B., Boscaro, M., Rutter, G.A., et al. (2006). Expanding role of AMPK in endocrinology. *Trends Endocrinol. Metab.* 17:205–215.

Koo, S.H., Flechner, L., Qi, L., et al. (2005). The CREB coactivator TORC2 is a key regulator of fasting glucose metabolism. *Nature* 437:1109–1111.

Kraegen, E.W. and Cooney, G.J. (2008). Free fatty acids and skeletal muscle insulin resistance. *Curr. Opin. Lipidol.* 19:235–241.

Kramer, H.F., Witczak, C.A., Fujii, N., et al. (2006). Distinct signals regulate AS160 phosphorylation in response to insulin, AICAR, and contraction in mouse skeletal muscle. *Diabetes* 55:2067–2076.

Kubota, N., Yano, W., Kubota, T., et al. (2007). Adiponectin stimulates AMP-activated protein kinase in the hypothalamus and increases food intake. *Cell Metab.* 6:55–68.

Leclerc, I., Kahn, A., and Doiron, B. (1998). The 5'-AMP-activated protein kinase inhibits the transcriptional stimulation by glucose in liver cells, acting through the glucose response complex. *FEBS Lett.* 431:180–184.

Leclerc, I., Lenzner, C., Gourdon, L., et al. (2001). Hepatocyte nuclear factor-4alpha involved in type 1 maturity-onset diabetes of the young is a novel target of AMP-activated protein kinase. *Diabetes* 50:1515–1521.

Lee, Y., Yu, X., Gonzales, F., et al. (2002). PPAR alpha is necessary for the lipopenic action of hyperleptinemia on white adipose and liver tissue. *Proc. Natl. Acad. Sci. USA* 99:11848–11853.

Lee, Y., Naseem, R.H., Park, B.H., et al. (2006a). Alpha-lipoic acid prevents lipotoxic cardiomyopathy in acyl CoA-synthase transgenic mice. *Biochem. Biophys. Res. Commun.* 344:446–452.

Lee, Y.S., Kim, W.S., Kim, K.H., et al. (2006b). Berberine, a natural plant product, activates AMP-activated protein kinase with beneficial metabolic effects in diabetic and insulin-resistant states. *Diabetes* 55:2256–2264.

Lee, J.H., Koh, H., Kim, M., et al. (2007). Energy-dependent regulation of cell structure by AMP-activated protein kinase. *Nature* 447:1017–1020.

Levine, A.J. (1997). p53, the cellular gatekeeper for growth and division. *Cell* 88:323–331.

Lizcano, J.M., Goransson, O., Toth, R., et al. (2004). LKB1 is a master kinase that activates 13 kinases of the AMPK subfamily, including MARK/PAR-1. *EMBO J.* 23:833–843.

Lochhead, P.A., Salt, I.P., Walker, K.S., et al. (2000). 5-aminoimidazole-4-carboxamide riboside mimics the effects of insulin on the expression of the 2 key gluconeogenic genes PEPCK and glucose-6-phosphatase. *Diabetes* 49:896–903.

Lopaschuk, G.D., Wambolt, R.B., and Barr, R.L. (1993). An imbalance between glycolysis and glucose oxidation is a possible explanation for the detrimental effects of high levels of fatty acids during aerobic reperfusion of ischemic hearts. *J. Pharmacol. Exp. Ther.* 264:135–144.

Luiken, J.J., Coort, S.L., Willems, J., et al. (2003). Contraction-induced fatty acid translocase/CD36 translocation in rat cardiac myocytes is mediated through AMP-activated protein kinase signaling. *Diabetes* 52:1627–1634.

Luptak, I., Shen, M., He, H., et al. (2007). Aberrant activation of AMP-activated protein kinase remodels metabolic network in favor of cardiac glycogen storage. *J. Clin. Invest.* 117:1432–1439.

Marsin, A.S., Bertrand, L., Rider, M.H., et al. (2000). Phosphorylation and activation of heart PFK-2 by AMPK has a role in the stimulation of glycolysis during ischaemia. *Curr. Biol.* 10:1247–1255.

Marsin, A.S., Bouzin, C., Bertrand, L. et al. (2002). The stimulation of glycolysis by hypoxia in activated monocytes is mediated by AMP-activated protein kinase and inducible 6-phosphofructo-2-kinase. *J. Biol. Chem.* 277:30778–30783.

Martin, S.G. and St Johnston, D. (2003). A role for Drosophila LKB1 in anterior-posterior axis formation and epithelial polarity. *Nature* 421:379–384.

Matthews, V.B. and Febbraio, M.A. (2008). CNTF: a target therapeutic for obesity-related metabolic disease? *J. Mol. Med.* 86:353–361.

McBride, A., Ghilagaber, S., Nikolaev, A., et al. (2009). The glycogen-binding domain on the AMPK beta subunit allows the kinase to act as a glycogen sensor. *Cell Metab.* 9:23–34.

McGee, S.L. and Hargreaves, M. (2008). AMPK and transcriptional regulation. *Front. Biosci.* 13:3022–3033.

McTaggart, S.J. (2006). Isoprenylated proteins. *Cell. Mol. Life Sci.* 63:255–267.

Merrill, G.F., Kurth, E.J., Hardie, D.G., et al. (1997). AICA riboside increases AMP-activated protein kinase, fatty acid oxidation, and glucose uptake in rat muscle. *Am. J. Physiol.* 273:E1107–E1112.

Minokoshi, Y., Kim, Y.B., Peroni, O.D., et al. (2002). Leptin stimulates fatty-acid oxidation by activating AMP-activated protein kinase. *Nature* 415:339–343.

Minokoshi, Y., Alquier, T., Furukawa, N., et al. (2004). AMP-kinase regulates food intake by responding to hormonal and nutrient signals in the hypothalamus. *Nature* 428:569–574.

Mitchelhill, K.I., Stapleton, D., Gao, G., et al. (1994). Mammalian AMP-activated protein kinase shares structural and functional homology with the catalytic domain of yeast Snf1 protein kinase. *J. Biol. Chem.* 269:2361–2364.

Mitchelhill, K.I., Michell, B.J., House, C.M., et al. (1997). Posttranslational modifications of the 5'-AMP-activated protein kinase beta1 subunit. *J. Biol. Chem.* 272:24475–24479.

Momcilovic, M., Hong, S.P., Carlson, M. (2006). Mammalian TAK1 activates Snf1 protein kinase in yeast and phosphorylates AMP-activated protein kinase in vitro. *J. Biol. Chem.* 281:25336–25343.

Mootha, V.K., Lindgren, C.M., Eriksson, K.F., et al. (2003). PGC-1alpha-responsive genes involved in oxidative phosphorylation are coordinately downregulated in human diabetes. *Nat. Genet.* 34:267–273.

Mu, J., Brozinick, J.T., Jr., Valladares, O., et al. (2001). A role for AMP-activated protein kinase in contraction- and hypoxia-regulated glucose transport in skeletal muscle. *Mol. Cell* 7:1085–1094.

Muoio, D.M., Seefeld, K., Witters, L.A., et al. (1999). AMP-activated kinase reciprocally regulates triacylglycerol synthesis and fatty acid oxidation in liver and muscle: evidence that sn-glycerol-3-phosphate acyltransferase is a novel target. *Biochem. J.* 338:783–791.

Nilsson, E.C., Long, Y.C., Martinsson, S., et al. (2006). Opposite transcriptional regulation in skeletal muscle of AMP-activated protein kinase gamma3 R225Q transgenic versus knock-out mice. *J. Biol. Chem.* 281:7244–7252.

Ouchi, N., Kihara, S., Arita, Y., et al. (2001). Adipocyte-derived plasma protein, adiponectin, suppresses lipid accumulation and class A scavenger receptor expression in human monocyte-derived macrophages. *Circulation* 103:1057–1063.

Owen, M.R., Doran, E., and Halestrap, A.P. (2000). Evidence that metformin exerts its antidiabetic effects through inhibition of complex 1 of the mitochondrial respiratory chain. *Biochem J.* 348:607–614.

Pang, T., Xiong, B., Li, J.Y., et al. (2007). Conserved alpha-helix acts as autoinhibitory sequence in AMP-activated protein kinase alpha subunits. *J. Biol. Chem.* 282:495–506.

Parker, G.J., Koay, A., Gilbert-Wilson, R., et al. (2007). AMP-activated protein kinase does not associate with glycogen alpha-particles from rat liver. *Biochem. Biophys. Res. Commun.* 362:811–815.

Paterson, J., Kelsall, I.R., Cohen, P.T. (2008). Disruption of the striated muscle glycogen-targeting subunit of protein phosphatase 1: influence of the genetic background. *J. Mol. Endocrinol.* 40:47–59.

Polekhina, G., Gupta, A., Michell, B.J., et al. (2003). AMPK beta subunit targets metabolic stress sensing to glycogen. *Curr. Biol.* 13:867–871.

Polekhina, G., Gupta, A., van Denderen, B.J., et al. (2005). Structural basis for glycogen recognition by AMP-activated protein kinase. *Structure* 13:1453–1462.

Richards, A.A., Stephens, T., Charlton, H.K., et al. (2006). Adiponectin multimerization is dependent on conserved lysines in the collagenous domain: evidence for regulation of multimerization by alterations in posttranslational modifications. *Mol. Endocrinol.* 20:1673–1687.

Rockl, K.S., Hirshman, M.F., Brandauer, J., Fujii, N., Witters, L.A., and Goodyear, L.J. (2007). Skeletal muscle adaptation to exercise training: AMP-activated protein kinase mediates muscle fiber type shift. *Diabetes* 56:2062–2069.

Ryder, J.W., Bassel-Duby, R., Olson, E.N., et al. (2003). Skeletal muscle reprogramming by activation of calcineurin improves insulin action on metabolic pathways. *J. Biol. Chem.* 278:44298–44304.

Saghizadeh, M., Ong, J.M., Garvey, W.T., et al. (1996). The expression of TNF alpha by human muscle. Relationship to insulin resistance. *J. Clin. Invest.* 97:1111–1116.

Sakamoto, K. and Holman, G.D. (2008). Emerging role for AS160/TBC1D4 and TBC1D1 in the regulation of GLUT4 traffic. *Am. J. Physiol. Endocrinol. Metab.* 295:E29–E37.

Sanders, M.J., Ali, Z.S., Hegarty, B.D., et al. (2007a). Defining the mechanism of activation of AMP-activated protein kinase by the small molecule A-769662, a member of the thienopyridone family. *J. Biol. Chem.* 282:32539–32548.

Sanders, M.J., Grondin, P.O., Hegarty, B.D., et al. (2007b). Investigating the mechanism for AMP activation of the AMP-activated protein kinase cascade. *Biochem. J.* 403:139–148.

Sapkota, G.P., Kieloch, A., Lizcano, J.M., et al. (2001). Phosphorylation of the protein kinase mutated in Peutz-Jeghers cancer syndrome, LKB1/STK11, at Ser431 by p90(RSK) and cAMP-dependent protein kinase, but not its farnesylation at Cys(433), is essential for LKB1 to suppress cell growth. *J. Biol. Chem.* 276:19469–19482.

Scheidereit, C. (2006). IkappaB kinase complexes: gateways to NF-kappaB activation and transcription. *Oncogene* 25:6685–6705.

Schmelzle, T. and Hall, M.N. (2000). TOR, a central controller of cell growth. *Cell* 103:253–262.

Scott, J.W., Norman, D.G., Hawley, S.A., et al. (2002). Protein kinase substrate recognition studied using the recombinant catalytic domain of AMP-activated protein kinase and a model substrate. *J. Mol. Biol.* 317:309–323.

Scott, J.W., Hawley, S.A., Green, K.A., et al. (2004). CBS domains form energy-sensing modules whose binding of adenosine ligands is disrupted by disease mutations. *J. Clin. Invest.* 113:274–284.

Scott, J.W., Ross, F.A., Liu, J.K. et al. (2007). Regulation of AMP-activated protein kinase by a pseudosubstrate sequence on the gamma subunit. *EMBO J.* 26:806–815.

Scott, J.W., van Denderen, B.J., Jorgensen, S.B., et al. (2008). Thienopyridone drugs are selective activators of AMP-activated protein kinase beta1-containing complexes. *Chem. Biol.* 15:1220–1230.

Scott, J.W., Oakhill, J.S., and van Denderen, B.J. (2009). AMPK/SNF1 structure: a menage a trois of energy-sensing. *Front. Biosci.* 14:596–610.

Shaw, R.J., Lamia, K.A., Vasquez, D., et al. (2005). The kinase LKB1 mediates glucose homeostasis in liver and therapeutic effects of metformin. *Science* 310:1642–1646.

Stahmann, N., Woods, A., Carling, D. et al. (2006). Thrombin activates AMP-activated protein kinase in endothelial cells via a pathway involving Ca2+/calmodulin-dependent protein kinase kinase beta. *Mol. Cell Biol.* 26:5933–5945.

Steinberg, G.R., Rush, J.W., and Dyck, D.J. (2003). AMPK expression and phosphorylation are increased in rodent muscle after chronic leptin treatment. *Am. J. Physiol. Endocrinol. Metab.* 284:E648–E654.

Steinberg, G.R., Michell, B.J., van Denderen, B.J., et al. (2006). Tumor necrosis factor alpha-induced skeletal muscle insulin resistance involves suppression of AMP-kinase signaling. *Cell Metab.* 4:465–474.

Steinberg, G.R., Watt, M.J., and Febbraio, M.A. (2009). Cytokine regulation of AMPK signalling. *Front. Biosci.* 14:1902–1916.

Suter, M., Riek, U., Tuerk, R., et al. (2006). Dissecting the role of 5'-AMP for allosteric stimulation, activation, and deactivation of AMP-activated protein kinase. *J. Biol. Chem.* 281:32207–32216.

Suzuki, A., Kusakai, G., Shimojo, Y., et al. (2005). Involvement of transforming growth factor-beta 1 signaling in hypoxia-induced tolerance to glucose starvation. *J. Biol. Chem.* 280:31557–31563.

Tamas, P., Hawley, S.A., Clarke, R.G., et al. (2006). Regulation of the energy sensor AMP-activated protein kinase by antigen receptor and Ca2+ in T lymphocytes. *J. Exp. Med.* 203:1665–1670.

Tan, M.J., Ye, J.M., Turner, N., et al. (2008). Antidiabetic activities of triterpenoids isolated from bitter melon associated with activation of the AMPK pathway. *Chem. Biol.* 15:263–273.

Thomson, D.M., Porter, B.B., Tall, J.H., et al. (2007). Skeletal muscle and heart LKB1 deficiency causes decreased voluntary running and reduced muscle mitochondrial marker enzyme expression in mice. *Am. J. Physiol. Endocrinol. Metab.* 292:E196–E202.

Tokumitsu, H., Iwabu, M., Ishikawa, Y., et al. (2001). Differential regulatory mechanism of Ca2+/calmodulin-dependent protein kinase kinase isoforms. *Biochemistry* 40:13925–13932.

Tomas, E., Tsao, T.S., Saha, A.K., et al. (2002). Enhanced muscle fat oxidation and glucose transport by ACRP30 globular domain: acetyl-CoA carboxylase inhibition and AMP-activated protein kinase activation. *Proc. Natl. Acad. Sci. USA* 99:16309–16313.

Treebak, J.T., Glund, S., Deshmukh, A., et al. (2006). AMPK-mediated AS160 phosphorylation in skeletal muscle is dependent on AMPK catalytic and regulatory subunits. *Diabetes* 55:2051–2058.

Turner, N., Li, J.Y., Gosby, A., et al. (2008). Berberine and its more biologically available derivative, dihydroberberine, inhibit mitochondrial respiratory complex I: a mechanism for the action of berberine to activate AMP-activated protein kinase and improve insulin action. *Diabetes* 57:1414–1418.

van Slegtenhorst, M., Nellist, M., Nagelkerken, B., et al. (1998). Interaction between hamartin and tuberin, the TSC1 and TSC2 gene products. *Hum. Mol. Genet.* 7:1053–1057.

Vincent, M.F., Marangos, P.J., Gruber, H.E., et al. (1991). Inhibition by AICA riboside of gluconeogenesis in isolated rat hepatocytes. *Diabetes* 40:1259–1266.

Viollet, B., Andreelli, F., Jorgensen, S.B., et al. (2003). The AMP-activated protein kinase alpha2 catalytic subunit controls whole-body insulin sensitivity. *J. Clin. Invest.* 111:91–98.

Virbasius, J.V. and Scarpulla, R.C. (1994). Activation of the human mitochondrial transcription factor A gene by nuclear respiratory factors: a potential regulatory link between nuclear and mitochondrial gene expression in organelle biogenesis. *Proc. Natl. Acad. Sci. USA* 91: 1309–1313.

Wang, M.Y., Orci, L., Ravazzola, M., et al. (2005). Fat storage in adipocytes requires inactivation of leptin's paracrine activity: implications for treatment of human obesity. *Proc. Natl. Acad. Sci. USA* 102:18011–18016.

Wang, C., Mao, X., Wang, L., et al. (2007). Adiponectin sensitizes insulin signaling by reducing p70 S6 kinase-mediated serine phosphorylation of IRS-1. *J. Biol. Chem.* 282:7991–7996.

Warburg, O. (1956). On the origin of cancer cells. *Science* 123:309–314.

Warden, S.M., Richardson, C., O'Donnell, J. Jr., et al. (2001). Post-translational modifications of the beta-1 subunit of AMP-activated protein kinase affect enzyme activity and cellular localization. *Biochem J.* 354:275–283.

Watt, M.J., Dzamko, N., Thomas, W.G., et al. (2006). CNTF reverses obesity-induced insulin resistance by activating skeletal muscle AMPK. *Nat. Med.* 12:541–548.

Williams, T. and Brenman, J.E. (2008). LKB1 and AMPK in cell polarity and division. *Trends Cell Biol.* 18:193–198.

Winder, W.W., Holmes, B.F., Rubink, D.S., et al. (2000). Activation of AMP-activated protein kinase increases mitochondrial enzymes in skeletal muscle. *J. Appl. Physiol.* 88:2219–2226.

Wisneski, J.A., Gertz, E.W., Neese, R.A., et al. (1987). Myocardial metabolism of free fatty acids. Studies with 14C-labeled substrates in humans. *J. Clin. Invest.* 79:359–366.

Wojtaszewski, J.F., Nielsen, P., Hansen, B.F., et al. (2000). Isoform-specific and exercise intensity-dependent activation of 5'-AMP-activated protein kinase in human skeletal muscle. *J. Physiol.* 528:221–226.

Wojtaszewski, J.F., Mourtzakis, M., Hillig, T., et al. (2002). Dissociation of AMPK activity and ACCbeta phosphorylation in human muscle during prolonged exercise. *Biochem. Biophys. Res. Commun.* 298:309–316.

Woods, A., Munday, M.R., Scott, J., et al. (1994). Yeast SNF1 is functionally related to mammalian AMP-activated protein kinase and regulates acetyl-CoA carboxylase in vivo. *J. Biol. Chem.* 269:19509–19515.

Woods, A., Azzout-Marniche, D., Foretz, M., et al. (2000). Characterization of the role of AMP-activated protein kinase in the regulation of glucose-activated gene expression using constitutively active and dominant negative forms of the kinase. *Mol. Cell. Biol.* 20:6704–6711.

Woods, A., Dickerson, K., Heath, R., et al. (2005). Ca2+/calmodulin-dependent protein kinase kinase-beta acts upstream of AMP-activated protein kinase in mammalian cells. *Cell Metab.* 2:21–33.

Wu, X., Motoshima, H., Mahadev, K., et al. (2003). Involvement of AMP-activated protein kinase in glucose uptake stimulated by the globular domain of adiponectin in primary rat adipocytes. *Diabetes* 52:1355–1363.

Wullschleger, S., Loewith, R., and Hall, M.N. (2006). TOR signaling in growth and metabolism. *Cell*, 124:471–484.

Xiao, B., Heath, R., Saiu, P., et al. (2007). Structural basis for AMP binding to mammalian AMP-activated protein kinase. *Nature* 449:496–500.

Xu, K.Y., Zweier, J.L., and Becker, L.C. (1995). Functional coupling between glycolysis and sarcoplasmic reticulum Ca2+ transport. *Circ. Res.* 77:88–97.

Yamauchi, T., Kamon, J., Minokoshi, Y., et al. (2002). Adiponectin stimulates glucose utilization and fatty-acid oxidation by activating AMP-activated protein kinase. *Nat. Med.* 8:1288–1295.

Yamauchi, T., Nio, Y., Maki, T., et al. (2007). Targeted disruption of AdipoR1 and AdipoR2 causes abrogation of adiponectin binding and metabolic actions. *Nat. Med.* 13:332–339.

Young, M.E., Radda, G.K., and Leighton, B. (1996). Activation of glycogen phosphorylase and glycogenolysis in rat skeletal muscle by AICAR- an activator of AMP-activated protein kinase. *FEBS Lett.* 382:43–47.

Zakikhani, M., Dowling, R., Fantus, I.G., et al. (2006). Metformin is an AMP kinase-dependent growth inhibitor for breast cancer cells. *Cancer Res.* 66:10269–10273.

Zhou, G., Myers, R., Li, Y., et al. (2001). Role of AMP-activated protein kinase in mechanism of metformin action. *J. Clin. Invest.* 108:1167–1174.

Zhou, G., Sebhat, I.K., and Zhang, B.B. (2009). AMPK activators. Potential therapeutics for metabolic and other diseases. *Acta Physiol. (Oxf)* 196:175–190.

Chapter 10
Protein Phosphatases in the Brain: Regulation, Function and Disease

Ry Y. Tweedie-Cullen*, C. Sehwan Park*, and Isabelle M. Mansuy

Abbreviations

CaM	Calmodulin
CaMKII	Ca2+/calmodulin-dependent kinase II
CDK	Cyclin-dependent kinase
ChIP	Chromatin immunoprecipitation
ChIP-Seq	Chromatin immunoprecipitation followed by sequencing
CNS	Central nervous system
DSP/DUSP	Dual-specificity phosphatases
HAD	haloacid dehydrogenase
HDAC	Histone deacetylase
I1	Inhibitor 1
KIM	kinase interaction motif
LMPTP/LMWPTP	18 kDa low MW phosphatase
LTP	Long-term potentiation
MKP	MAPK phosphatase
MS	Mass spectrometry
MTM	Myotubularin
MTMR	MTM1-related
NIPP1	Nuclear inhibitor of protein phosphatase 1
NRPTP	Non-receptor PTP
PKA	Protein kinase A
PK	Protein kinase
PP	Protein phosphatase
PP1	Protein phosphatase 1
PP2B	Calcineurin
PPM	Protein phosphatase Mg^{2+}- or Mn^{2+}-dependent

I.M. Mansuy (✉)
Brain Research Institute, Medical Faculty University Zurich/Swiss Federal Institute of Technology Zurich, Winterthurerstrasse 190, CH-8057 Zurich, Switzerland
e-mail: mansuy@hifo.uzh.ch

* Ry Y. Tweedie-Cullen and C.S. Park contributed equally for this chapter

C.J. Vidal (ed.), *Post-Translational Modifications in Health and Disease*, Protein Reviews 13, DOI 10.1007/978-1-4419-6382-6_10, © Springer Science+Business Media, LLC 2011

PPP	Phosphoprotein phosphatase
PTEN	Phosphatase and tensin homolog deleted in chromosome 10
PTK	Protein tyrosine kinase
PTM	Posttranslational modification
PTP	Protein tyrosine phosphatases
PTP-SL	Protein tyrosine phosphatase STEP-like
pSer/pThr/pTyr	Phosphorylated-serine/-threonine/-tyrosine
RLPTP	Receptor-like PTP
ROS	Reactive oxygen species
RPTP	Receptor PTP
Ser/Thr/Tyr	Serine/threonine/tyrosine
SHP	Src-homology domain containing phosphatase
STEP	Striatal-enriched phosphatase

10.1 Introduction

Protein phosphorylation is the most common and important form of reversible protein posttranslational modification (PTM), with up to 30% of all proteins being phosphorylated at any given time (Hunter 1998). Protein kinases (PKs) are the effectors of phosphorylation and catalyse the transfer of a γ-phosphate from ATP to specific amino acids on proteins. Several hundred PKs exist in mammals and are classified into distinct super-families (Ubersax and Ferrell 2007). Proteins are phosphorylated predominantly on serine (Ser), threonine (Thr) and tyrosine (Tyr) residues, which account for 86, 12 and 2% respectively of the phosphoproteome, at least in mammals (Munton et al. 2007; Olsen et al. 2006). In contrast, protein phosphatases (PPs) are the primary effectors of dephosphorylation and can be grouped into three main classes based on sequence, structure and catalytic function. The largest class of PPs is the phosphoprotein phosphatase (PPP) family comprised of PP1, PP2A, PP2B, PP4, PP5, PP6 and PP7, and the protein phosphatase Mg^{2+}- or Mn^{2+}-dependent (PPM) family, comprised primarily of PP2C. The protein Tyr phosphatase (PTP) super-family forms the second group and the aspartate-based protein phosphatases the third (Moorhead et al. 2007).

Protein phosphorylation plays a crucial role in biological functions and controls nearly every cellular process, including metabolism, gene transcription and translation, cell-cycle progression, cytoskeletal rearrangement, protein-protein interactions, protein stability, cell movement, and apoptosis (Manning et al. 2002, Ubersax and Ferrell 2007). These processes depend on the highly regulated and opposing actions of PKs and PPs, through changes in the phosphorylation of key proteins. Histone phosphorylation, along with methylation, ubiquitination, sumoylation and acetylation, also regulates access to DNA through chromatin reorganisation (Jenuwein and Allis 2001).

One of the major switches for neuronal activity is the activation of PKs and PPs by elevated intracellular calcium. The degree of activation of the various isoforms of PKs and PPs is controlled by their individual sensitivities to calcium (Lee 2006).

Furthermore, a wide range of specific inhibitors and targeting partners such as scaffolding, anchoring, and adaptor proteins also contribute to the control of PKs and PPs and recruit them into signalling complexes in neuronal cells (Sim and Scott 1999). Such signalling complexes typically act to bring PKs and PPs in close proximity with target substrates and signalling molecules as well as enhance their selectivity by restricting accessibility to these substrate proteins (Faux and Scott 1996). Phosphorylation events, therefore, are controlled not only by the balanced activity of PKs and PPs but also by their restricted localisation. Regulatory subunits and domains serve to restrict specific proteins to particular subcellular compartments and to modulate protein specificity. These regulators are essential for maintaining the coordinated action of signalling cascades, which in neuronal cells include short-term (synaptic) and long-term (nuclear) signalling. These functions are, in part, controlled by allosteric modification by secondary messengers and reversible protein phosphorylation.

It is thought that around 30% of known PPs are present in all tissues, with the rest showing some level of tissue restriction (Forrest et al. 2006). While protein phosphorylation is a cell-wide regulatory mechanism, recent quantitative proteomics studies have shown that phosphorylation preferentially targets nuclear proteins (Olsen et al. 2006). Many PPs that regulate nuclear events are often enriched in or exclusively present in the nucleus (Moorhead et al. 2007). In neuronal cells, PPs are present in multiple cellular compartments and play a critical role at both pre- and post-synapses, in the cytoplasm and in the nucleus where they regulate gene expression. The various types of PPs are grouped into families which are described below.

10.2 Serine/Threonine PP (PPM/PPP) Families

Protein Ser/Thr phosphatases were originally classified using biochemical assays as either, type 1 (PP1) or type 2 (PP2), and were further subdivided based on metal-ion requirement (PP2A, no metal ion; PP2B, Ca^{2+} stimulated; PP2C, Mg^{2+} dependent) (Moorhead et al. 2007). The protein Ser/Thr phosphatases PP1, PP2A and PP2B of the PPP family, together with PP2C of the PPM family, account for the majority of Ser/Thr PP activity *in vivo* (Barford et al. 1998). In the brain, they are present in different subcellular compartments in neuronal and glial cells, and contribute to different neuronal functions.

10.2.1 PP1

The importance of PP1 in neural systems became apparent after the discovery that high levels of several of its catalytic subunits including α, β, γ-1 and γ-2, are expressed in the brain. In nerve cells, they are associated with different binding proteins that compartmentalise them to varying but distinct subcellular compartments. The catalytic subunit PP1c, for example, is in complex with over 100 different regulatory and targeting subunits (Barford et al. 1998; Faux and Scott 1996; Sim and Scott 1999). The α isoform is found predominantly in the soma of neuronal cells, whereas γ-1, in addition to the soma, is present in large amount in dendrites and presynaptic boutons where it co-localises with CaMKII and synaptophysin (Munton et al. 2004; Strack et al. 1999). PP1 is involved in a large range of cellular functions which are regulated by distinct PP1 holoenzymes.

10.2.2 PP2A, PP4 and PP6

The catalytic subunits for the PP2A-type of Ser/Thr PPs (PP2A, PP4 and PP6) are widely expressed and evolutionarily conserved. PP2A has been implicated in the control of cell growth, proliferation, and differentiation (Mumby 2007). PP2A has been shown in human cells to function as a component of a large multiprotein complex called the striatin-interacting phosphatase and kinase complex where substrate specificity is mediated by several different regulatory subunits (Goudreault et al. 2009). PP4 and PP6 are also thought to function as members of similar multiprotein complexes (Chen and Gingras 2007). Similar to PP1, the diverse functions of PP2A are largely accounted for by the ensemble of regulatory and targeting proteins it associates with.

10.2.3 PP2B and PP7

Calcineurin (PP2B) is characterised by its dependence on Ca^{2+} for activity. It consists of an A-subunit with an N-terminal catalytic domain, a C-terminal regulatory region containing binding sites for the PP2B B-subunit and calmodulin, and an autoinhibitory sequence at the extreme C-terminus. Optimal Ca^{2+} activation of PP2B requires association of Ca^{2+} to both the B-subunit and calmodulin (Barford et al. 1998). In the brain, PP2B is highly abundant and regulates multiple neuronal processes depending on its subcellular localisation and regulatory partners (Mansuy 2003).

10.2.4 PPM

The PPM family, which includes PP2C and pyruvate dehydrogenase phosphatase, are enzymes with Mn^{2+}/Mg^{2+} metal ions that are resistant to classic inhibitors and toxins of the PPP family. Unlike most PPPs, PP2C exists in only one subunit but, like PTPs, it displays a wide variety of structural domains that confer unique functions (Moorhead et al. 2009). In addition, PP2C does not seem to be evolutionarily related to the major family of Ser/Thr PPs and has no sequence homology to ancient PPP enzymes (Moorhead et al. 2009). The current assumption is that PPMs evolved separately from PPPs but converged during evolutionary development (Bellinzoni et al. 2007).

10.2.5 Target Specificity and Structures of Ser/Thr PPs

The ability of PPs to dephosphorylate a specific residue is largely defined by the association of their catalytic subunits with regulatory or targeting proteins (Mansuy and Shenolikar 2006). In contrast to PTPs, PPPs generally consist of small catalytic subunits that bind regulatory or targeting subunits encoded by separate genes to form an enormous variety of distinct holoenzymes with different biological functions (Alonso et al. 2004). The metal coordinating residues (Asp, His, and Asn), which interact with the phosphate group of phosphorylated residues, are invariant between all PPPs and occur within five conserved sequence motifs (Barford et al. 1998). Ser/Thr phosphatases are metalloenzymes that dephosphorylate their substrates in a single reaction step using a metal-activated water molecule or hydroxide group as a nucleophile. The PPM active site is different to that of the PPP family because it contains two Mn^{2+} ions coordinated by four Asp and one Glu residue.

Dephosphorylation is catalysed by metal-activated water molecules acting as nucleophiles in a mechanism similar to that of PPPs (Barford et al. 1998). Differences in the active site are reflected in their different toxin sensitivities. PP2B is specifically and potently inhibited by cyclosporin A and FK506 while PP1 and PP2A are inhibited by a variety of naturally occurring toxins such as okadaic acid and microcystin. The PPM phosphatase, PP2C, however, is only poorly inhibited by the toxins that affect PP1 and PP2A (Barford et al. 1998).

10.3 Protein Tyrosine Phosphatase Family

Protein tyrosine phosphatase (PTPs) are efficient PPs, with constitutive activity causing rapid dephosphorylation of intracellular pTyr residues and disruption of signal transduction pathways. Most PTP family members dephosphorylate Tyr residues and are known as single-specificity PTPs. However, a subset of PTPs can also dephosphorylate Ser and Thr residues (Barford et al. 1998) and are known as dual-specificity phosphatases (DSPs/DUSPs). DSPs may also target messenger RNA, phospholipids and phosphoinositides in addition to phosphoproteins. Tyrosine-specific PTPs are further classified based on their cellular location. Membrane-spanning PTPs are known collectively as receptor-like PTPs (RLPTPs), whereas others lacking membrane-spanning domains are found in the cytoplasm and are known as intracellular or non-receptor PTPs (NRPTPs) (Gee and Mansuy 2005; Paul and Lombroso 2003). While most PTPs are ubiquitously expressed several are brain-specific, including striatal-enriched phosphatase (STEP) and STEP-like (PTP-SL) PTPs (Boulanger et al. 1995; Paul and Lombroso 2003). In general, PTPs are characterised by their sensitivity to vanadate, their ability to hydrolyse p-nitrophenyl phosphate, their insensitivity to okadaic acid and their lack of a requirement for metal ions during catalysis (Paul and Lombroso 2003). The non-catalytic segments of PTPs also possess amino acid sequences that target them

Table 10.1 Classes of PPs and their target specificity

Family	Class	Substrate specificity
Ser/Thr phosphatases		
PPP family	PP1	pSer, pThr
	PP2A	pSer, pThr
	PP4	pSer, pThr
	PP5	pSer, pThr
	PP6	pSer, pThr
	PP2B	pSer, pThr
	PP7	pSer, pThr
PPM family	PP2C	pSer, pThr
PTP superfamily		
Class I Cys-based PTPs		
(a/b) Classical	RPTPs	pTyr
	NRPTPs	pTyr
(c) DSPs	MPKs	pTyr, pThr
	Atypical DSPs	pTyr, pThr, mRNA
	Slingshots	pSer
	PRLs	pTyr
	CDC14s	pSer, pThr
	PTENS	De-phosphoinositides
	Myotubularins	PI(3)P
Class II Cys-based PTPs	LMW-PTP	pTyr
Class III Cys-based PTPs	CDC25	pTyr, pThr
Asp-based PTPs	EyA	pTyr, pSer

to specific intracellular compartments (Paul and Lombroso 2003). Based on the amino acid sequences of their catalytic domains, PTPs can be grouped into four separate families, each with a range of substrate specificities (see Table 10.1) and structures as discussed below (Alonso et al. 2004).

10.3.1 Class I Cys-Based PTPs

Class I PTPs constitute the largest family. They contain the well-known classical receptor (a) and non-receptor PTPs (b), which are strictly tyrosine-specific, and the DSPs (c) which target Ser/Thr as well as Tyr and are the most diverse in terms of substrate specificity (see Table 10.1) (Alonso et al. 2004; Andersen et al. 2004).

10.3.1.1 'Classical': Receptor-Like PTPs

In total, five RPTP sub-types are known (Types I–V) of which all, except type I, are expressed in the CNS (Paul and Lombroso 2003). RPTPs function as an interface between the extracellular environment of a cell and its intracellular signalling components, and typically possess two intracellular phosphatase domains, D1 and D2,

(Tonks 2006) specific for tyrosine. The extracellular regions of RPTPs are highly variable but all contain motifs implicated in cell adhesion. However, most are orphan receptors and their mode of action and their respective functional ligands remain largely unknown. One RPTP whose ligands are well understood is LAR, which can interact with both surface-bound and soluble ligands. These interactions either favour (interaction with protein syndecan) or inhibit (interaction with glycosyl-phosphatidylinositol-anchored protein Dallylike) its phosphatase activity. These ligands modulate the activity of downstream pathways and LAR-associated proteins such as Enabled, ultimately resulting in the regulation of the actin cytoskeleton in synapse morphogenesis and synaptic function (Tonks 2006). Unique to this class is an intracellular helix-turn-helix 'wedge' motif that is thought to mediate dimerisation of two RPTPs for the inhibition of intrinsic phosphatase activity (Majeti et al. 1998, 2000).

10.3.1.2 'Classical': Cytoplasmic/Non-receptor PTPs

NRPTPs can be either tyrosine- or dual-specific, lack a transmembrane domain, and have a single phosphatase domain and multiple variable domains at either the N- and/or C-terminus (Paul and Lombroso 2003). Members of this class include PTP1B, STEP, PTP-SL and src-homology 2 domain containing phosphatases (SHP1 and SHP2) (Paul and Lombroso 2003). PTP1B has been shown to negatively regulate two critical metabolic pathways: the insulin signalling pathway, through dephosphorylation of the insulin receptor, and the leptin-signalling pathway in hypothalamic neurons. The non-catalytic sequence of NRPTPs also controls subcellular distribution, thereby indirectly regulating activity by restricting access to particular substrates at defined subcellular locations (Tonks 2006).

10.3.1.3 Dual-Specificity Phosphatases

Class I DSPs are highly diverse and share reduced sequence identity with each other than do NRPTPs and RPTPs. While DSPs can dephosphorylate all three residues, several do show specificity for one of the three. KAP (cyclin-dependent kinase associated phosphatase) dephosphorylates threonine residues on CDKs and VH1-related DSPs (VHR) preferentially desphosphorylate tyrosine residues on MAPKs. Perhaps the best characterised member of this family, the MAPK phosphatases (MKPs), catalyses the inactivation of MAPKs by dephosphorylation of both the Tyr and Thr sites in the kinase activation loop. The MKPs also display distinct subcellular location and specificity for individual MAPKs (Paul and Lombroso 2003; Tonks 2006). PTEN (Phosphatase and tensin homolog deleted on chromosome 10) and myotubularins (MTM) are unique in that they principally target phosphorylation on D3-phosphorylated inositol phospholipids rather than proteins. As lipid phosphatases, PTEN- and MTM-related (MTMR) proteins dephosphorylate the products of phosphoinositide 3-kinases/AKT signalling, thereby suppressing cell survival and cell proliferation (Wishart and Dixon 2002; Yin and Shen 2008).

10.3.2 Class II: Cys-Based LMW-PTPs

This class of PTPs is represented by a single gene in humans encoding the 18 kDa low MW phosphatase (LM-PTP/LMW-PTP). Related classes are widely distributed in living organisms and were highly conserved through evolution. The preservation of this class of phosphatases and the involvement of LMW-PTPs in many common diseases (Bottini et al. 2002) suggest that they contribute to the regulation of many fundamental processes in cellular physiology (Alonso et al. 2004).

10.3.3 Class III: Cys-Based PTPs

The third class of PTPs contains three cell cycle regulators, CDC25A, CDC25B and CDC25C, which dephosphorylate CDKs at their N-terminal domains, a reaction required to drive progression of the cell cycle. They are themselves regulated by phosphorylation (Honda et al. 1993) and are degraded in response to DNA damage to prevent chromosomal abnormalities.

10.3.4 Class IV: Asp-Based DSPs

The haloacid dehalogenase (HAD) superfamily is a further PP group that uses Asp as a nucleophile and was recently shown to have dual-specificity. These PPs can target both serine and tyrosine (Li et al. 2003; Tootle et al. 2003), but are thought to have greater specificity towards tyrosine (Jemc and Rebay 2007b). A subfamily of HADs, the Eyes Absent Family (EyA), are also transcription factors and can therefore regulate their own phosphorylation and that of transcriptional cofactor/s, and contribute to the control of gene transcription. The combination of these two functions in EyA reveals a greater complexity of transcriptional gene control than previously thought (Jemc and Rebay 2007a, b). A further member of this class is the RNA polymerase II C-terminal domain phosphatase. While this family remains poorly understood, it is known to play important roles in development and nuclear morphology (Alonso et al. 2004; Siniossoglou et al. 1998; Tootle et al. 2003).

10.3.5 Target Specificity and Structure of Tyr Phosphatases

Structurally, most PTPs consist of a combination of modular domains and, in this respect, resemble PTKs but differ markedly from the Ser/Thr PPs (Barford et al. 1998; Manning et al. 2002). In PTKs, protein-protein interactions serve to regulate kinase activity and target the enzyme to substrates or subcellular compartments, which also appears to be the case for PTPs (Alonso et al. 2004). The second significant trait is conservation of the phosphate-binding loop (P- or PTP-loop) in the active site. The conformation of the P-loop is strictly conserved and can be easily superimposed on different PTPs. This structurally conserved arrangement ensures that the catalytic Cys, the nucleophile, and Arg residues involved in phosphate binding remain in close proximity to form a cradle to hold the phosphate group of the substrate for nucleophilic attack (Tabernero et al. 2008). In contrast to PPPs and PPMs which catalyse dephosphorylation in a single step, PTPs catalyse dephosphorylation in two steps by way of a cysteinyl-phosphate enzyme intermediate (Barford et al. 1998). Substrate specificity for pTyr results from the dimension of the catalytic site cleft in which the pTyr recognition loop is located. The tyrosine residue of this motif packs against the phenyl ring of a pTyr substrate and defines the depth of the catalytic site (Barford et al. 1998). Members of the class IV PTP family, such as HADs, use Asp as the nucleophile for catalysis and carry the active-site signature DXDXT/V. This is also identical to that in TFIIF-associating C-terminal domain (CTD) phosphatase-1 (FCP1) and small CTD phosphatase (SCP) enzymes (Moorhead et al. 2007; Zhang et al. 2006).

The structure of the dual-specificity active site is slightly different and allows PTPs to accommodate pSer and pThr in addition to pTyr. DSPs also are not as well conserved and share little sequence similarity beyond the Cys-signature motif (Tonks 2006). *In vitro*, PTPs generally display poor substrate specificity, as they are capable of dephosphorylating most pTyr-containing substrates. However, *in vivo* PTP activity is tightly regulated through compartmentalisation and the binding of inhibitory proteins to ensure appropriate signalling responses (Paul and Lombroso 2003). The majority of PTPs are also modified posttranslationally, the most common modification being phosphorylation. Such modifications have been shown to regulate activity and further demonstrate the tight interplay between kinases and phosphatases (Birle et al. 2002). While much progress has been made on elucidating the structure of PTPs over recent years, the mode of actions and binding partners of many of the PTPs remains undetermined (Barford et al. 1998; Tabernero et al. 2008).

10.4 Regulators of Phosphatases

10.4.1 Regulation of PPPs

The catalytic activity of Ser/Thr PPs and specificity of targeting to their respective substrates is determined by interacting proteins that can regulate and localise PPs to distinct subcellular compartments. With the exception of PP2C, the function of PPs is defined by the regulatory proteins that bind the catalytic subunit or the holoenzyme itself. Many regulators of PP1 have been identified through the presence of a conserved binding motif R/K1-2V/I[X]F/W, where X can be any amino acid except proline, which has allowed for their purification (Moorhead et al. 2007). The diverse localisation of PPs and the various means of their regulation, however, give rise to a multitude of compartment-specific PPP-complexes that function in different signalling pathways (see Table 10.2 for overview of regulatory

Table 10.2 Regulatory mechanisms and their effectors

Family	Type of regulation	Example of effector
Ser/Thr PPPs	Inhibitory regulation	I1, I2, G-substrate, DARPP-32, NIPP1
	Scaffolding	AKAP-like, PDZ-domain containing
	Regulatory subunits	regulatory B, RII, CaM
	Targeting	Repo-Man, PNUTS, ZAP3
	Phosphorylation	PKA, PKC, CaMKII, Aurora kinase
	Substrate binding	Rb1, NFATs
	Stress-responsive	Heat, osmotic pressure (PPM)
PTP Superfamily	Oxidative stress	Insulin, PDGF, UV, B/T-cell activation
	Redox state	Cyclic sulphenamide species
	Ligand-specificity	Modular domains
	Cell-adhesion ligands	Pleiotrophin, cadherin
	Extracellular matrix ligands	Agrin, syndecan, collagen XVIII, dallylike
	Active site occlusion	Dimerisation, wedge motif
	Autoinhibition	Phosphorylated catalytic site, EyA
	Substrate binding	p130cas

mechanisms). In general, PP2B is regulated by modulatory calcineurin-interacting proteins (MCIPs or calcipressins), Ca^{2+}-sensing calmodulin (CaM), and the regulatory subunit B (CNB), which is related in structure and function to CaM (Oliver et al. 2002). At excitatory synapses of neurons, scaffolding proteins such as the A-kinase anchoring protein (AKAP) and microtubule-associated proteins (MAPs) act to stabilise the catalytic subunit A (CNA) of PP2B at the postsynaptic density (Quinlan and Halpain 1996; Sim et al. 2003). AKAP-PP2B is also in a complex with the regulatory subunit (RII) of protein kinase A (PKA) and protein kinase C (PKC) where they can modulate NMDA and AMPA receptor function (Coghlan et al. 1995; Klauck et al. 1996; Oliveria et al. 2003). At presynaptic sites, phosphoproteins including synapsins, synaptotagmin, rabphilin3A, synaptobrevin, and dephosphins interact with PP2B to regulate vesicle endocytosis and exocytosis (Greengard et al. 1993).

PP1 is also anchored to synapses through a member of the AKAP family, yotiao, where antagonises PKA to downregulate NMDA receptor gating kinetics (Lin et al. 1998). Spinophilin, a binding protein with a postsynaptic-density-protein/disc-large/zo1 (PDZ) domain, can sequester PP1 to actin in the synapse where it can associate with D2 dopamine receptors, α-adrenergic receptors, and p70S6 kinase (Burnett et al. 1998). Most regulatory proteins associated with PP1 inhibit its activity, and Inhibitor 1 (I1), G-substrate, and DARPP-32 being examples of particularly effective regulatory inhibitors. I1 variants specific for PP2A, including PP2AI1 and PP2AI2, however, have the opposite effect on PP1 by positively increasing catalytic activity (Katayose et al. 2000). Recent studies have shown direct association between some PPs and PKs, thereby creating autoregulatory modules. In such modules, the kinase can phosphorylate the phosphatase to suppress its function and the phosphatase can counteract the kinase activity through dephosphorylation of amino acids in the activation loop (Bauman and Scott 2002; Shenolikar 2007).

Although PP1 has been studied primarily in the context of cytosolic and synaptic functions, PP1 is also enriched in the nucleus. It has been implicated in gene transcription and, more recently, in chromatin remodelling associated with learning and memory (Canettieri et al. 2003; Miller and Sweatt 2007; Koshibu 2009). Although over 100 regulatory subunits are known for PP1, only a few nuclear interactors of PP1 have been identified. Two of the most abundant of these regulators are the nuclear inhibitor of PP1 (NIPP1/Ard1) and PP1 nuclear targeting subunit (PNUTS/p99). NIPP1 is a small 39 kDa peptide which can bind to PP1, RNA, splicing factors, and transcriptional regulators to control gene transcription, but can be inhibited by PKA and CK2 phosphorylation (Jagiello et al. 1997; Roy et al. 2007; Vulsteke et al. 1997). Binding of PNUTS to RNA through its interaction with PP1 has been implicated in chromosome decondensation, further aiding transcription (Kim et al. 2003; Landsverk et al. 2005). Repo-Man is another putative nuclear targeting subunit of PP1, shown to help maintain chromosome structure during chromosome segregation (Vagnarelli et al. 2006). Interacting protein ZAP3, a polynucleoside kinase, is also known to sequester nuclear PP1 into a complex with various transcriptional factors leading to the modification of RNA (Ulke-Lemee et al. 2007).

10.4.2 Regulation of PTPs

PTP regulation is predominantly dependent on the variable modular domains that define their unique enzymatic activity (Tonks 2006). The classical membrane-bound RPTPs, which are involved in cell-adhesion, are regulated through extracellular ligand-specific binding to their modular domains. To date, studies of RPTPs have shown that phosphatase activity is restricted to the membrane-proximal domain D1 of PTPs with the exception of catalytically dead RPTPs and PTPα, the latter having catalytic activity in both the D1 and the membrane-distal D2 domains (Tonks 2006). Importantly, ligand-induced dimerisation of the D1 domain of RPTPα has been shown to result in a wedge motif, which occludes the active site of the complementary domain (Bilwes et al. 1996; Majeti et al. 1998). In the case of LAR RPTP, however, the D2 domain prevents wedge-mediated dimerisation of D1 and both active sites are accessible and active (Nam et al. 1999). RPTP dimerisation is further regulated by extracellular glycosylation which is dependent on the alternative splicing of three exons, A, B, and C (Xu and Weiss 2002). A representative cell-adhesion ligand is pleiotrophin which inhibits type V RPTPζ activity and results in alterations in phosphorylation of β-catenin, β-adducin, n-cadherin, and p190 Rho GAP thereby affecting cellular cytoarchitecture in the brain (Pariser et al. 2005a, b; Perez-Pinera et al. 2006; Tamura et al. 2006). Type IIb RPTPs such as RPTPμ, a homophilic cell-adhesion molecule highly expressed in neurons and endothelial cells, are regulated entirely through ectodomain-dependent interactions (Brady-Kalnay et al. 1993; Gebbink et al. 1993). RPTPμ dimers, nearly identical in dimension and length to cadherin, enable the receptor to gauge distance and regulate interactions with cadherin/catenin complexes at synapses (Aricescu et al. 2007; Boggon et al. 2002). Two heparan sulphate proteoglycans have also been identified in neurons as high-affinity ligands for type IIa RPTPs (agrin and collagen XVIII for RPTPσ (Aricescu et al. 2002); syndecan and Dallylike for LAR (Johnson et al. 2006)), that mediate synaptic signalling and morphological changes (Johnson et al. 2006).

The other major mechanism of RPTP regulation is through reversible oxidation by reactive oxygen species (ROS) (Lee et al. 1998). The nucleophilic cysteine residue in the active site of PTPs is susceptible to oxidation, which renders the domain inactive and inhibits PTP function (den Hertog et al. 2005; Tonks 2005). Oxidative states play a further role in determining whether the modification is reversible, with higher oxidation states of the nucleophilic cysteine being irreversible and rendering the PTP permanently inactivated. Adding further complexity to their regulation, DSPs and LMW-PTPs contain a second cysteine residue which can form disulphide bonds with adjacent cysteines to prevent oxidation of active-site cysteines (Salmeen and Barford 2005). The phylogenetically distinct D2 domain is also found to have greater sensitivity to oxidation than the D1 domain and can form disulphide bonds thereby stabilising PTP dimers (van der Wijk et al. 2004). With its redox sensor-like role, and oxidation-induced conformational change of the extracellular domains, oxidation of RPTPα D2 is suggested to mediate 'inside-out' retrograde signalling (den Hertog et al. 2005). The multiple levels of regulation, from structural changes to ligand-specificity to ROS signalling, suggest that the diverse family of RPTPs are finely regulated in the course of various cellular processes and physiological functions.

The classical cytoplasmic class of NRPTPs, in comparison, is characterised by regulatory domains that interact either directly with the active site to modulate activity, or indirectly through substrate binding specificity. SHP2 is comprised of two N-terminal SH2 domains that, under basal conditions, occlude the active site of the catalytic domain through intramolecular interactions. Interaction with pTyr ligands on the targeting proteins themselves have been shown to remove autoinhibition of the PTP catalytic site (Neel et al. 2003). In the case of PTP-SL regulation, extracellular signal-regulated kinases 1 and 2 (ERK1/2) binding and subsequent phosphorylation of the kinase interaction motif (KIM) on PTP-SL is necessary for its activation (Pulido et al. 1998). The KIM motif is of significant importance in MAP kinase signalling as it targets KIM-containing PTPs to their substrates, ERK2 and p38. Similarly, LMW-PTPs are also regulated by phosphorylation of Tyr131 and Tyr132 by Lck and Fyn tyrosine kinases in T-cells which, when activated, results in auto-dephosphorylation of the regulatory domains (Bucciantini et al. 1999; Tailor et al. 1997). The highly substrate-specific interaction of PTP-PEST for p130cas has also been attributed to the allosteric binding of the SH3 domain of p130cas which greatly increases its affinity for the substrate and further enhances the specificity of the catalytic domain of PTP-PEST for p130cas (Garton et al. 1997).

10.5 PPs as Regulators of Neuronal Activity

The mechanisms that control neuronal signalling at synapses and in the nucleus are complex and only partially understood. Although some pathways regulated by phosphorylation have been delineated, a global picture of phosphorylation-dependent changes in the brain is lacking, due in part to the lack of suitable technologies in the past to measure protein phosphorylation. In healthy cells, signal transduction pathways depend on the interplay between several classes of enzymes that work synergistically or in opposition to regulate signal transmission (Westphal et al. 1999). PKs and PPs are critical components of these pathways and are required for the modulation of synaptic efficacy and neuronal transmission (Bradshaw et al. 2003; Wang and Kelly 1996), and for memory formation (Lisman et al. 2002; Malleret et al. 2001). In general, PKs and PPs act antagonistically to control signalling pathways through phosphorylation and dephosphorylation of specific targets. Some of these targets, in turn, regulate structural modifications at synaptic contacts (Horne and Dell'Acqua 2007; Oliver et al. 2002; Westphal et al. 1999), and determine whether transcriptional and translational events are initiated to maintain long-term synaptic plasticity and memory (Atkins et al. 2005; Bito et al. 1996; Mansuy and Shenolikar 2006; Winder and Sweatt 2001).

Prominent PKs, including αCaMKII, PKA, MAPK, act as positive regulators of neuronal signalling. They are necessary for mechanisms promoting long-lasting forms of synaptic plasticity such as long term potentiation (LTP), which reflect a strengthening of synaptic connections (Giovannini et al. 2001; Smolen et al. 2006). Phosphorylation of the AMPA receptor subunit GluR1 by CaMKIIα favours the recruitment of new receptors to the membrane and increases the channel's conductance, resulting in long-lasting functional changes (Malenka and Nicoll 1999). In contrast, PPs such as PP1 and PP2B act to weaken synaptic plasticity through dephosphorylation of key proteins required for the induction and maintenance of LTP including NMDA receptor subunits, CaMKIIα, I1 which are substrates of PKs (Strack et al. 2000; Yakel 1997), and PKs themselves (Colbran 2004). PPs are primarily involved in synaptic depression or LTD, a form of synaptic plasticity that reflects a weakening of synapses (Hedou et al. 2008; Jouvenceau and Dutar 2006; Luebke et al. 2004; Mulkey et al. 1994; Norman et al. 2000; Thiels et al. 2000). NMDA receptor signalling, involves differential regulation of the Ras-ERK pathway which also interacts with downstream targets of PTPs (Kim et al. 2005; Pulido et al. 1998). The NR2A and NR2B subunits of the NMDA receptor themselves are phosphorylated by the Src PTKs at multiple tyrosine residues, particularly on Tyr1472 (Tezuka et al. 1999; Zhang et al. 2008), and require specific PTPs for their dephosphorylation and endocytosis (Chen et al. 2003; Snyder et al. 2005; Wang and Salter 1994).

In addition to acting in synaptic terminals to suppress synaptic functions, PPs also contribute to nuclear events and are important regulators of chromatin remodelling. Chromatin remodelling is a complex cellular process regulated by posttranslational modifications of histone proteins, DNA methylation and alterations to chromatin structure. PPs are essential for the regulation of gene expression, and are implicated in multiple brain processes and functions including development (Hsieh and Gage 2005; Levenson and Sweatt 2006), neurogenesis (Tsankova et al. 2007), synaptic plasticity (Allis et al. 2007; Hsieh and Gage 2005), memory (Chwang et al. 2006; Levenson and Sweatt 2005; and Sweatt 2007) and behaviour (Chong and Whitelaw 2004; Richards 2006). PP1 plays a particularly important role in chromatin remodelling because it can form complexes with chromatin regulatory proteins such as histone deacetylases (HDACs) and can thus regulate both histone phosphorylation and acetylation (Koshibu 2009). Both PP1 and PP2A have been implicated in dephosphorylation of Ser10 on histone H3, which has previously been shown to affect chromosome condensation and segregation (Hsu et al. 2000; Nowak et al. 2003). *In vitro*, PP1 was shown to decrease histone phosphorylation and acetylation, leading to inhibition of cAMP-response element binding protein (CREB), and repression of immediate gene response elements necessary for transcription (Canettieri et al. 2003; Ceulemans and Bollen 2004). In contrast to PP1, there is limited evidence for PP2B's involvement in chromatin remodeling in the nucleus, however, shared sequence similarity in the Ser/Thr binding motif suggests that PP2B may also be involved in dephosphorylation of histones. However, studies of PP2B in the nucleus have primarily focused on its role as a phosphatase acting on transcription factors. A recent finding shows that PP2B-dependent dephosphorylation of retinoblastoma protein (Rb) dissociates Rb from a repressor complex, including HDAC and BRG1, to facilitate gene transcription (Qiu and Ghosh 2008). There is further evidence that activity-dependent PP2B dephosphorylation of the myocyte enhancer factor 2 (MEF2) in the nucleus results in enhanced MEF2-dependent transcription of a set of genes which prevents dendritic differentiation and suppresses excitatory synapse number in the brain (Flavell et al. 2006; Shalizi et al. 2006). Recent evidence has also shown that PP2B contributes to the regulation of the immediate early gene Zif268 (Baumgartel et al. 2008), and is therefore, directly associated with transcriptional events. The potential mechanisms and regulation of PP2B in nuclear chromatin remodelling, however, remain to be determined.

10.6 Human Diseases and Therapeutic Targets Based on PPs

Neurodegeneration due to ageing and disease is characterised by a loss of spines and a retraction of dendrites, which reduce both the connectivity and structural complexity of neuronal circuits. The compromised and accelerated decay of synaptic connections with age, in turn, functionally perturbs cellular signalling and synaptic transmission, and reduces the brain's capacity for information processing (Foster 2007; Luebke et al. 2004). On a mechanistic level, multiple molecular processes at both synapses and the nucleus contribute to morphological and functional alterations of neurons in ageing. Since Ser/Thr PP-dependent pathways are negative effectors of neuronal signalling, they are likely to play a significant role in mechanisms associated with age-related cognitive deterioration and neurodegenerative processes. Indeed, the activity and expression of PP1 and PP2B are dysregulated in the aged brain (Foster et al. 2001; Norris et al. 1998; Watson et al. 2002).

A shift in the balance of PKs and PPs in favour of PPs has further been associated with defects in synaptic plasticity and memory in rodents (Wang et al. 1996) and restoration of such a shift by inhibition of PP1 through transgenic expression of I1 in aged mice was shown to restore memory (Genoux et al. 2002). PP1 inhibition can also rescue LTP in adult animals when synapses are either depotentiated or pharmacologically depressed (Jouvenceau and Dutar 2006; Jouvenceau et al. 2006), thus pointing to a general role of PP1 and PP2B in synaptic depression and cognitive deficits. Although a general decrease in phosphatase activity could account for tau pathology in Alzheimer's Disease, a more complex role of PPs is emerging. Increased activation of PP2B has been observed in Alzheimer's Disease patients, and PP2B was found to be involved in Aβ-mediated downregulation of NMDA receptors, increased AMPA receptor internalisation, and spine loss (Knobloch and Mansuy 2008).

In addition to Ser/Thr phosphatases, dysregulation of Tyr phosphatases has also been implicated in the development and pathology of neurodegenerative diseases. Impairments in learning and synaptic plasticity in mice deficient for PTPRZ, an RPTPζ subtype chondroitin sulfate proteoglycan expressed predominantly in the brain, has been attributed to aberrant phosphorylation of p190 Rho GAP which leads to altered Rho-ROCK signalling in hippocampal neurons (Niisato et al. 2005; Tamura et al. 2006). In addition, redox inactivation of MAPK phosphatase 3 (MKP-3) is thought to mediate glutamate excitotoxicity while overexpression of a dominant-negative MKP-3 results in accumulation of phospho-ERK-2 (Levinthal and Defranco 2005). Knockout of PTPσ is also detrimental to the nervous system and results in enhanced neurodegeneration following sciatic nerve injury (McLean et al. 2002). The DSP phosphatidylinositol phosphatase myotubularins (MTMs) have also been directly linked to human diseases such as X-linked myotubular myopathy and mytotonic dystrophy (Buj-Bello et al. 2002; Laporte et al. 1996). Charcot-Marie-Tooth disease is an inherited demyelinating neuropathy resulting from a mutation in the human MTMR2 gene (Bolino et al. 2000; Bolis et al. 2005). Knockout of the DSP laforin or mutations in EPM2A have been implicated in progressive myoclonus epilepsy of Lafora type disease, an inherited neurodegenerative disorder with onset in late childhood and characterised by accumulation of highly phosphorylated glucose polymers (Ganesh et al. 2002, 2006; Serratosa et al. 1999). Deletion of PTEN, initially uncovered as a tumour suppressor PTP, in the brain of mice was shown to result in abnormal social interaction, macrocephaly, and neuronal hypertrophy, characteristics of human autistic spectrum disorder (Kwon et al. 2006).

In the pharmaceutical industry PKs are still the second main drug target after G-protein coupled receptors (Cohen 2002). In contrast, PPs are only just beginning to be recognised as potential therapeutic targets. Because dysregulation of many PPs is implicated in human diseases (Li and Dixon 2000), there is increasing interest in pursuing them (Bialy and Waldmann 2005; Hooft van Huijsduijnen et al. 2004; Tabernero et al. 2008; Tautz et al. 2006). A number of PTPs that have been implicated in oncogenesis and tumour progression are current drug targets for cancer chemotherapy. They include PTP1B, a selective negative regulator of insulin and leptin signalling, which may augment signalling downstream of HER2/Neu; SHP2 essential for growth factor-mediated signalling; Cdc25 phosphatases, which are positive regulators of cell cycle progression; and the phosphatase of regenerating liver (PRL) phosphatases, which promote tumour metastasis (Jiang and Zhang 2008).

Ser/Thr phosphatases are also implicated in several diseases and several are already used as drug targets. PP2B, for instance, is a key component of T-cell signalling pathways and is activated by increased Ca^{2+} resulting from antigen presentation. Inhibition of PP2B by immunosuppressant drugs suppresses T-cell activation, and the microbial products cyclosporin A and FK506 that specifically target PP2B are currently the most effective immunosuppressive agents available. Cyclosporin is widely used to prevent rejection after organ transplantation and in inflammatory diseases such as dermatitis (Ehrchen et al. 2008; Faul et al. 2008; Shenolikar 2007). Specific inhibitors of Ser/Thr phosphatases such as PP5 and PP1γ1 have also been studied as potential anti-tumour drugs. However, developing specific inhibitors has proved difficult (Honkanen and Golden 2002). As PTPs have emerged as drug targets for cancer, a number of strategies are currently being explored for the identification of various classes of PTP inhibitors. These efforts have resulted in many potent and, in some cases, selective inhibitors for PTP1B, SHP2, Cdc25 and PRL phosphatases (Jiang and Zhang 2008). Aspects of PTP biology that may be relevant for cancer research in the future are the regulation of PTPs by oxidation and the possible role of PTPs in angiogeneis (Ostman et al. 2006).

10.7 Current and Future Approaches to Study PPs and Their Functions

The advent of proteomics has opened a new dimension to genomics and has provided novel information about gene expression at the protein level. Its importance is underscored by the recognition that the abundance of mRNA transcripts and the corresponding protein products often do not correlate. This is primarily due to the posttranscriptional regulation of mRNA (Griffin et al. 2002). To date, much of the knowledge gathered about phosphorylation in nerve cells was obtained by studies at the level of single molecules. However, novel proteomic methods are beginning to allow more global characterisations of phosphorylation and analyses of its dynamics (Olsen et al. 2006). Mass spectrometry (MS) is ideally suited to the analysis of protein posttranslational modifications such as phosphorylation, because modern MS techniques allow site-specific *de novo* assignment of phosphorylation and the investigation of signalling networks in a high-throughput discovery-driven manner. Recent MS studies have increased the number of known phosphorylation sites in the mouse by several orders of magnitude (Collins et al. 2005; Munton et al. 2007; Trinidad et al. 2005, 2006, 2008). However, for the majority of these sites the PKs and PPs that modulate them remain unknown. Utilising MS to relate PPs and

PKs to these sites is the logical next step. One such approach is to exploit the common mechanism of PTP catalysis to create substrate-trapping mutants that can be applied to all PTP members to pull-out and determine their substrates by MS (Blanchetot et al. 2005). Similarly, phosphorylation-dependent protein-protein interactions of signalling molecules and their site/s of phosphorylation have also been analysed using MS techniques (Hinsby et al. 2004; Zhou et al. 2007).

Recently, activity-based proteomics has emerged as a powerful method for analysing enzyme activity. This approach applies chemical probes to covalently label active sites of enzymes in an activity-dependent manner, thus providing a direct readout of catalytic activity. The covalently tagged proteins can then be identified by MS analysis (Schmidinger et al. 2006). Quantitative proteomics has also been applied to identify the components of large multi-protein complexes in the CNS. These large complexes act as signal 'transduction machines' and are typically composed of scaffolding molecules, receptors, kinases and PPs such as PP1 and PP2A (Husi et al. 2000; Ranish et al. 2003). The same workflows have been used to delineate the phosphatase interacting partners of the PP1 (Trinkle-Mulcahy et al. 2006), PP2A (Goudreault et al. 2009) and PP4 (Gingras et al. 2005) complexes.

Recent attention has focused on the role of PPs in chromatin remodelling and the use of quantitative proteomics and chromatin immunoprecipitation (ChIP-on-Chip), in particular, has provided unique tools for the analysis of nuclear PPs (Oficjalska-Pham et al. 2006). Recent advances have also furthered genome-wide analyses using techniques such as ChIP-Seq (Schones and Zhao 2008). While quantitative proteomics has the potential to analyse protein modifications in an unbiased fashion it currently lacks the specificity to analyse these changes at the level of the single gene. It promises, however, to be a method of choice in the coming years (Huang and White 2008; Trelle and Jensen 2007; Tweedie-Cullen et al. 2007). In addition, transgenesis and RNAi are complementary methods, which when used in combination with quantitative proteomic and genomic technologies will provide a powerful tool for the analysis of specific phosphatase pathways. Recent studies have investigated the role of PP1 and PP2B in learning and memory by taking advantage of transgenic methodologies in mice (Baumgartel et al. 2007; Genoux et al. 2002; Mansuy 2003). The further generation of such models and their combination with proteomic and genomic methodologies will be instrumental for future studies in the nervous system.

10.8 Conclusions

PPs are critical regulators of protein function and neuronal activity with multiple mechanisms regulating their activity and substrate specificity in vivo. Their importance in higher-order brain function explains why their dysregulation is involved in cognitive decline, neurodegeneration and multiple psychological disorders. It is increasingly clear that the complex signalling networks in which phosphatases participate, must be studied at a systems level; the current 'one pathway – one drug' approach of current drug research is no long applicable in light of current knowledge. The development and increasing maturity of tools available in the fields of genomics and proteomics has opened up an opportunity to study PP function at the systems biology level. These tools can now be exploited and applied to the various animal models already available to elucidate the underlying mechanisms of neuronal signalling and brain functions as mediated by phosphorylation. Their application to the brain and its pathologies will potentially help uncover the mechanisms of many diseases and lead to the development of novel therapies.

References

Allis, C.D., Berger, S.L., Cote, J., *et al.* (2007). New nomenclature for chromatin-modifying enzymes. *Cell* 131:633–636.

Alonso, A., Sasin, J., Bottini, N., *et al.* (2004). Protein tyrosine phosphatases in the human genome. *Cell* 117:699–711.

Andersen, J.N., Jansen, P.G., Echwald, S.M., *et al.* (2004). A genomic perspective on protein tyrosine phosphatases: gene structure, pseudogenes, and genetic disease linkage. *FASEB J.* 18:8–30.

Aricescu, A.R., McKinnell, I.W., Halfter, W., *et al.* (2002). Heparan sulfate proteoglycans are ligands for receptor protein tyrosine phosphatase sigma. *Mol. Cell. Biol.* 22:1881–1892.

Aricescu, A.R., Siebold, C., Choudhuri, K., *et al.* (2007). Structure of a tyrosine phosphatase adhesive interaction reveals a spacer-clamp mechanism. *Science* 317:1217–1220.

Atkins, C.M., Davare, M.A., Oh, M.C., *et al.* (2005). Bidirectional regulation of cytoplasmic polyadenylation element-binding protein phosphorylation by Ca^{2+}/calmodulin-dependent protein kinase II and protein phosphatase 1 during hippocampal long-term potentiation. *J. Neurosci.* 25:5604–5610.

Barford, D., Das, A.K., and Egloff, M.P. (1998). The structure and mechanism of protein phosphatases: insights into catalysis and regulation. *Annu. Rev. Biophys. Biomol. Struct.* 27: 133–164.

Bauman, A.L. and Scott, J.D. (2002). Kinase- and phosphatase-anchoring proteins: harnessing the dynamic duo. *Nat. Cell Biol.* 4:E203–206.

Baumgartel, K., Fernandez, C., Johansson, T., *et al.* (2007). Conditional transgenesis and recombination to study the molecular mechanisms of brain plasticity and memory. *Handb. Exp. Pharmacol.* 3:315–345.

Baumgartel, K., Genoux, D., Welzl, H., *et al.* (2008). Control of the establishment of aversive memory by calcineurin and Zif268. *Nat. Neurosci.* 11:572–578.

Bellinzoni, M., Wehenkel, A., Shepard, W., *et al.* (2007). Insights into the catalytic mechanism of PPM Ser/Thr phosphatases from the atomic resolution structures of a mycobacterial enzyme. *Structure* 15:863–872.

Bialy, L. and Waldmann, H. (2005). Inhibitors of protein tyrosine phosphatases: next-generation drugs? *Angew Chem. Int. Ed. Engl.* 44:3814–3839.

Bilwes, A.M., den Hertog, J., Hunter, T., *et al.* (1996). Structural basis for inhibition of receptor protein-tyrosine phosphatase-alpha by dimerization. *Nature* 382:555–559.

Birle, D., Bottini, N., Williams, S., *et al.* (2002). Negative feedback regulation of the tumor suppressor PTEN by phosphoinositide-induced serine phosphorylation. *J. Immunol.* 169: 286–291.

Bito, H., Deisseroth, K., and Tsien, R.W. (1996). CREB phosphorylation and dephosphorylation: a Ca(2+)- and stimulus duration-dependent switch for hippocampal gene expression. *Cell* 87:1203–1214.

Blanchetot, C., Chagnon, M., Dube, *et al.* (2005). Substrate-trapping techniques in the identification of cellular PTP targets. *Methods* 35:44–53.

Boggon, T.J., Murray, J., Chappuis-Flament, S., *et al.* (2002). C-cadherin ectodomain structure and implications for cell adhesion mechanisms. *Science* 296:1308–1313.

Bolino, A., Muglia, M., Conforti, F.L., *et al.* (2000). Charcot-Marie-Tooth type 4B is caused by mutations in the gene encoding myotubularin-related protein-2. *Nat. Genet.* 25:17–19.

Bolis, A., Coviello, S., Bussini, S., *et al.* (2005). Loss of Mtmr2 phosphatase in Schwann cells but not in motor neurons causes Charcot-Marie-Tooth type 4B1 neuropathy with myelin outfoldings. *J. Neurosci.* 25:8567–8577.

Bottini, N., Borgiani, P., Otsu, A., *et al.* (2002). IL-4 receptor alpha chain genetic polymorphism and total IgE levels in the English population: two-locus haplotypes are more informative than individual SNPs. *Clin. Genet.* 61:288–292.

Boulanger, L.M., Lombroso, P.J., Raghunathan, A., *et al.* (1995). Cellular and molecular characterization of a brain-enriched protein tyrosine phosphatase. *J. Neurosci.* 15:1532–1544.

Bradshaw, J.M., Kubota, Y., Meyer, T., *et al.* (2003). An ultrasensitive Ca2+/calmodulin-dependent protein kinase II-protein phosphatase 1 switch facilitates specificity in postsynaptic calcium signaling. *Proc. Natl. Acad. Sci. USA* 100:10512–10517.

Brady-Kalnay, S.M., Flint, A.J., *et al.* (1993). Homophilic binding of PTP mu, a receptor-type protein tyrosine phosphatase, can mediate cell-cell aggregation. *J. Cell Biol. 122*, 961–972.

Bucciantini, M., Chiarugi, P., Cirri, P., *et al.* (1999). The low Mr phosphotyrosine protein phosphatase behaves differently when phosphorylated at Tyr131 or Tyr132 by Src kinase. *FEBS Lett. 456*, 73–78.

Buj-Bello, A., Furling, D., Tronchere, H., *et al.* (2002). Muscle-specific alternative splicing of myotubularin-related 1 gene is impaired in DM1 muscle cells. *Hum. Mol. Genet. 11*, 2297–2307.

Burnett, P.E., Blackshaw, S., Lai, M.M., *et al.* (1998). Neurabin is a synaptic protein linking p70 S6 kinase and the neuronal cytoskeleton. *Proc. Natl. Acad. Sci. USA* 95:8351–8356.

Canettieri, G., Morantte, I., Guzman, E., *et al.* (2003). Attenuation of a phosphorylation-dependent activator by an HDAC-PP1 complex. *Nat. Struct. Biol.* 10:175–181.

Ceulemans, H. and Bollen, M. (2004). Functional diversity of protein phosphatase-1, a cellular economizer and reset button. *Physiol. Rev.* 84:1–39.

Chen, G.I. and Gingras, A.C. (2007). Affinity-purification mass spectrometry (AP-MS) of serine/threonine phosphatases. *Methods* 42:298–305.

Chen, M., Hou, X., and Zhang, G. (2003). Tyrosine kinase and tyrosine phosphatase participate in regulation of interactions of NMDA receptor subunit 2A with Src and Fyn mediated by PSD-95 after transient brain ischemia. *Neurosci. Lett.* 339:29–32.

Chong, S. and Whitelaw, E. (2004). Epigenetic germline inheritance. *Curr. Opin. Genet. Dev.* 14:692–696.

Chwang, W.B., O'Riordan, K.J., Levenson, J.M., *et al.* (2006). ERK/MAPK regulates hippocampal histone phosphorylation following contextual fear conditioning. *Learn. Mem.* 13:322–328.

Coghlan, V.M., Perrino, B.A., Howard, M., *et al.* (1995). Association of protein kinase A and protein phosphatase 2B with a common anchoring protein. *Science* 267:108–111.

Cohen, P. (2002). Protein kinases – the major drug targets of the twenty-first century? *Nat. Rev. Drug Discov.* 1:309–315.

Colbran, R.J. (2004). Protein phosphatases and calcium/calmodulin-dependent protein kinase II-dependent synaptic plasticity. *J. Neurosci.* 24:8404–8409.

Collins, M.O., Yu, L., Coba, M.P., *et al.* (2005). Proteomic analysis of in vivo phosphorylated synaptic proteins. *J. Biol. Chem.* 280:5972–5982.

den Hertog, J., Groen, A., and van der Wijk, T. (2005). Redox regulation of protein-tyrosine phosphatases. *Arch. Biochem. Biophys.* 434:11–15.

Ehrchen, J., Sunderkotter, C., Luger, T., *et al.* (2008). Calcineurin inhibitors for the treatment of atopic dermatitis. *Expert Opin. Pharmacother.* 9:3009–3023.

Faul, C., Donnelly, M., Merscher-Gomez, S., *et al.* (2008). The actin cytoskeleton of kidney podocytes is a direct target of the antiproteinuric effect of cyclosporine A. *Nat. Med.* 14:931–938.

Faux, M.C. and Scott, J.D. (1996). More on target with protein phosphorylation: conferring specificity by location. *Trends Biochem. Sci.* 21:312–315.

Flavell, S.W., Cowan, C.W., Kim, T.K., *et al.* (2006). Activity-dependent regulation of MEF2 transcription factors suppresses excitatory synapse number. *Science* 311:1008–1012.

Forrest, A.R., Taylor, D.F., Fink, J.L., *et al.* (2006). PhosphoregDB: the tissue and sub-cellular distribution of mammalian protein kinases and phosphatases. *BMC Bioinform* 7:82.

Foster, T.C. (2007). Calcium homeostasis and modulation of synaptic plasticity in the aged brain. *Aging Cell* 6:319–325.

Foster, T.C., Sharrow, K.M., Masse, J.R., *et al.* (2001). Calcineurin links Ca2+ dysregulation with brain aging. *J. Neurosci.* 21:4066–4073.

Ganesh, S., Delgado-Escueta, A.V., Sakamoto, T., *et al.* (2002). Targeted disruption of the Epm2a gene causes formation of Lafora inclusion bodies, neurodegeneration, ataxia, myoclonus epilepsy and impaired behavioral response in mice. *Hum. Mol. Genet.* 11:1251–1262.

Ganesh, S., Puri, R., Singh, S., *et al.* (2006). Recent advances in the molecular basis of Lafora's progressive myoclonus epilepsy. *J. Hum. Genet.* 51:1–8.

Garton, A.J., Burnham, M.R., Bouton, A.H., *et al.* (1997). Association of PTP-PEST with the SH3 domain of p130cas; a novel mechanism of protein tyrosine phosphatase substrate recognition. *Oncogene* 15:877–885.

Gebbink, M.F., Zondag, G.C., Wubbolts, R.W., *et al.* (1993). Cell-cell adhesion mediated by a receptor-like protein tyrosine phosphatase. *J. Biol. Chem.* 268:16101–16104.

Gee, C.E. and Mansuy, I.M. (2005). Protein phosphatases and their potential implications in neuroprotective processes. *Cell. Mol. Life Sci.* 62:1120–1130.

Genoux, D., Haditsch, U., Knobloch, M., Michalon, A., Storm, D., and Mansuy, I.M. (2002). Protein phosphatase 1 is a molecular constraint on learning and memory. Nature 418:970–975.

Gingras, A.C., Caballero, M., Zarske, M., *et al.* (2005). A novel, evolutionarily conserved protein phosphatase complex involved in cisplatin sensitivity. *Mol. Cell. Proteom* 4:1725–1740.

Giovannini, M.G., Blitzer, R.D., Wong, T., *et al.* (2001). Mitogen-activated protein kinase regulates early phosphorylation and delayed expression of Ca2+/calmodulin-dependent protein kinase II in long-term potentiation. *J. Neurosci.* 21:7053–7062.

Goudreault, M., D'Ambrosio, L.M., Kean, M.J., *et al.* (2009). A PP2A phosphatase high density interaction network identifies a novel striatin-interacting phosphatase and kinase complex linked to the cerebral cavernous malformation 3 (CCM3) protein. *Mol. Cell. Proteom* 8:157–171.

Greengard, P., Valtorta, F., Czernik, A.J., *et al.* (1993). Synaptic vesicle phosphoproteins and regulation of synaptic function. *Science* 259:780–785.

Griffin, T.J., Gygi, S.P., Ideker, T., *et al.* (2002). Complementary profiling of gene expression at the transcriptome and proteome levels in Saccharomyces cerevisiae. *Mol. Cell. Proteom* 1:323–333.

Hedou, G.F., Koshibu, K., Farinelli, M., *et al.* (2008). Protein phosphatase 1-dependent bidirectional synaptic plasticity controls ischemic recovery in the adult brain. *J. Neurosci.* 28:154–162.

Hinsby, A.M., Olsen, J.V., and Mann, M. (2004). Tyrosine phosphoproteomics of fibroblast growth factor signaling: a role for insulin receptor substrate-4. *J. Biol. Chem.* 279:46438–46447.

Honda, R., Ohba, Y., Nagata, A., *et al.* (1993). Dephosphorylation of human p34cdc2 kinase on both Thr-14 and Tyr-15 by human cdc25B phosphatase. *FEBS Lett.* 318:331–334.

Honkanen, R.E. and Golden, T. (2002). Regulators of serine/threonine protein phosphatases at the dawn of a clinical era? *Curr. Med. Chem.* 9:2055–2075.

Hooft van Huijsduijnen, R., Sauer, W.H., Bombrun, A., *et al.* (2004). Prospects for inhibitors of protein tyrosine phosphatase 1B as antidiabetic drugs. *J. Med. Chem.* 47:4142–4146.

Horne, E.A. and Dell'Acqua, M.L. (2007). Phospholipase C is required for changes in postsynaptic structure and function associated with NMDA receptor-dependent long-term depression. *J. Neurosci.* 27:3523–3534.

Hsieh, J. and Gage, F.H. (2005). Chromatin remodeling in neural development and plasticity. *Curr. Opin. Cell Biol.* 17:664–671.

Hsu, J.Y., Sun, Z.W., Li, X., Reuben, M., Tatchell, K., Bishop, D.K., Grushcow, J.M., Brame, C.J., Caldwell, J.A., Hunt, D.F., *et al.* (2000). Mitotic phosphorylation of histone H3 is governed by

Ipl1/aurora kinase and Glc7/PP1 phosphatase in budding yeast and nematodes. *Cell* 102:279–291.

Huang, P.H. and White, F.M. (2008). Phosphoproteomics: unraveling the signaling web. *Mol. Cell* 31:777–781.

Hunter, T. (1998). The Croonian Lecture 1997. The phosphorylation of proteins on tyrosine: its role in cell growth and disease. *Philos. Trans. R. Soc. Lond. B Biol. Sci.* 353:583–605.

Husi, H., Ward, M.A., Choudhary, J.S., et al. (2000). Proteomic analysis of NMDA receptor-adhesion protein signaling complexes. *Nat. Neurosci.* 3:661–669.

Jagiello, I., Beullens, M., Vulsteke, V., *et al.* (1997). NIPP-1, a nuclear inhibitory subunit of protein phosphatase-1, has RNA-binding properties. *J. Biol. Chem.* 272:22067–22071.

Jemc, J. and Rebay, I. (2007a). Identification of transcriptional targets of the dual-function transcription factor/phosphatase eyes absent. *Dev. Biol.* 310:416–429.

Jemc, J. and Rebay, I. (2007b). The eyes absent family of phosphotyrosine phosphatases: properties and roles in developmental regulation of transcription. *Annu. Rev. Biochem.* 76:513–538.

Jenuwein, T. and Allis, C.D. (2001). Translating the histone code. *Science* 293:1074–1080.

Jiang, Z.X. and Zhang, Z.Y. (2008). Targeting PTPs with small molecule inhibitors in cancer treatment. *Cancer Metastasis Rev.* 27:263–272.

Johnson, K.G., Tenney, A.P., Ghose, A., *et al.* (2006). The HSPGs Syndecan and Dallylike bind the receptor phosphatase LAR and exert distinct effects on synaptic development. *Neuron* 49:517–531.

Jouvenceau, A. and Dutar, P. (2006). A role for the protein phosphatase 2B in altered hippocampal synaptic plasticity in the aged rat. *J. Physiol. Paris* 99:154–161.

Jouvenceau, A., Hedou, G., Potier, B., *et al.* (2006). Partial inhibition of PP1 alters bidirectional synaptic plasticity in the hippocampus. *Eur. J. Neurosci.* 24:564–572.

Katayose, Y., Li, M., Al-Murrani, S.W., *et al.* (2000). Protein phosphatase 2A inhibitors, I(1)(PP2A) and I(2)(PP2A), associate with and modify the substrate specificity of protein phosphatase 1. *J. Biol. Chem.* 275:9209–9214.

Kim, Y.M., Watanabe, T., Allen, P.B., *et al.* (2003). PNUTS, a protein phosphatase 1 (PP1) nuclear targeting subunit. Characterization of its PP1- and RNA-binding domains and regulation by phosphorylation. *J. Biol. Chem.* 278:13819–13828.

Kim, M.J., Dunah, A.W., Wang, Y.T., *et al.* (2005). Differential roles of NR2A- and NR2B-containing NMDA receptors in Ras-ERK signaling and AMPA receptor trafficking. *Neuron* 46:745–760.

Klauck, T.M., Faux, M.C., Labudda, K., *et al.* (1996). Coordination of three signaling enzymes by AKAP79, a mammalian scaffold protein. *Science* 271:1589–1592.

Knobloch, M. and Mansuy, I.M. (2008). Dendritic spine loss and synaptic alterations in Alzheimer's disease. *Mol. Neurobiol.* 37:73–82.

Koshibu, K., Graff, J., et al. (2009). Protein phosphatase 1 regulates the histone code for long-term memory. *J. Neurosci.* 29(41):13079–13089.

Kwon, C.H., Luikart, B.W., Powell, C.M., *et al.* (2006). Pten regulates neuronal arborization and social interaction in mice. *Neuron* 50:377–388.

Landsverk, H.B., Kirkhus, M., Bollen, M., *et al.* (2005). PNUTS enhances in vitro chromosome decondensation in a PP1-dependent manner. *Biochem. J.* 390:709–717.

Laporte, J., Hu, L.J., Kretz, C., *et al.* (1996). A gene mutated in X-linked myotubular myopathy defines a new putative tyrosine phosphatase family conserved in yeast. *Nat. Genet.* 13:175–182.

Lee, H.K. (2006). Synaptic plasticity and phosphorylation. *Pharmacol. Ther.* 112:810–832.

Lee, S.R., Kwon, K.S., Kim, S.R., *et al.* (1998). Reversible inactivation of protein-tyrosine phosphatase 1B in A431 cells stimulated with epidermal growth factor. *J. Biol. Chem.* 273:15366–15372.

Levenson, J.M. and Sweatt, J.D. (2005). Epigenetic mechanisms in memory formation. *Nat. Rev. Neurosci.* 6:108–118.

Levenson, J.M. and Sweatt, J.D. (2006). Epigenetic mechanisms: a common theme in vertebrate and invertebrate memory formation. *Cell. Mol. Life Sci.* 63:1009–1016.

Levinthal, D.J. and Defranco, D.B. (2005). Reversible oxidation of ERK-directed protein phosphatases drives oxidative toxicity in neurons. *J. Biol. Chem.* 280:5875–5883.

Li, L. and Dixon, J.E. (2000). Form, function, and regulation of protein tyrosine phosphatases and their involvement in human diseases. *Semin. Immunol.* 12:75–84.

Li, X., Oghi, K.A., Zhang, J., Krones, A., *et al.* (2003). Eya protein phosphatase activity regulates Six1-Dach-Eya transcriptional effects in mammalian organogenesis. *Nature* 426:247–254.

Lin, J.W., Wyszynski, M., Madhavan, R., *et al.* (1998). Yotiao, a novel protein of neuromuscular junction and brain that interacts with specific splice variants of NMDA receptor subunit NR1. *J. Neurosci.* 18:2017–2027.

Lisman, J., Schulman, H., and Cline, H. (2002). The molecular basis of CaMKII function in synaptic and behavioural memory. *Nat. Rev. Neurosci.* 3:175–190.

Luebke, J.I., Chang, Y.M., Moore, T.L., *et al.* (2004). Normal aging results in decreased synaptic excitation and increased synaptic inhibition of layer 2/3 pyramidal cells in the monkey prefrontal cortex. *Neuroscience* 125:277–288.

Majeti, R., Bilwes, A.M., Noel, J.P., *et al.* (1998). Dimerization-induced inhibition of receptor protein tyrosine phosphatase function through an inhibitory wedge. *Science* 279:88–91.

Majeti, R., Xu, Z., Parslow, T.G., *et al.* (2000). An inactivating point mutation in the inhibitory wedge of CD45 causes lymphoproliferation and autoimmunity. *Cell* 103:1059–1070.

Malenka, R.C. and Nicoll, R.A. (1999). Long-term potentiation – a decade of progress? *Science* 285:1870–1874.

Malleret, G., Haditsch, U., Genoux, D., *et al.* (2001). Inducible and reversible enhancement of learning, memory, and long-term potentiation by genetic inhibition of calcineurin. *Cell* 104:675–686.

Manning, G., Whyte, D.B., Martinez, R., *et al.* (2002). The protein kinase complement of the human genome. *Science* 298:1912–1934.

Mansuy, I.M. (2003). Calcineurin in memory and bidirectional plasticity. *Biochem. Biophys. Res. Commun.* 311:1195–1208.

Mansuy, I.M. and Shenolikar, S. (2006). Protein serine/threonine phosphatases in neuronal plasticity and disorders of learning and memory. *Trends Neurosci.* 29:679–686.

McLean, J., Batt, J., Doering, L.C., *et al.* (2002). Enhanced rate of nerve regeneration and directional errors after sciatic nerve injury in receptor protein tyrosine phosphatase sigma knock-out mice. *J. Neurosci.* 22:5481–5491.

Miller, C.A. and Sweatt, J.D. (2007). Covalent modification of DNA regulates memory formation. *Neuron* 53:857–869.

Moorhead, G.B., Trinkle-Mulcahy, L., *et al.* (2007). Emerging roles of nuclear protein phosphatases. *Nat. Rev. Mol. Cell. Biol.* 8:234–244.

Moorhead, G.B., De Wever, V., Templeton, G., *et al.* (2009). Evolution of protein phosphatases in plants and animals. *Biochem. J.* 417:401–409.

Mulkey, R.M., Endo, S., Shenolikar, S., *et al.* (1994). Involvement of a calcineurin/inhibitor-1 phosphatase cascade in hippocampal long-term depression. *Nature* 369:486–488.

Mumby, M. (2007). PP2A: unveiling a reluctant tumor suppressor. *Cell* 130:21–24.

Munton, R.P., Vizi, S., and Mansuy, I.M. (2004). The role of protein phosphatase-1 in the modulation of synaptic and structural plasticity. *FEBS Lett.* 567:121–128.

Munton, R.P., Tweedie-Cullen, R., Livingstone-Zatchej, M., *et al.* (2007). Qualitative and quantitative analyses of protein phosphorylation in naive and stimulated mouse synaptosomal preparations. *Mol. Cell. Proteom* 6:283–293.

Nam, H.J., Poy, F., Krueger, N.X., Saito, H., *et al.* (1999). Crystal structure of the tandem phosphatase domains of RPTP LAR. *Cell* 97:449–457.

Neel, B.G., Gu, H., and Pao, L. (2003). The 'Shp'ing news: SH2 domain-containing tyrosine phosphatases in cell signaling. *Trends Biochem. Sci.* 28:284–293.

Niisato, K., Fujikawa, A., Komai, S., *et al.* (2005). Age-dependent enhancement of hippocampal long-term potentiation and impairment of spatial learning through the Rho-associated kinase pathway in protein tyrosine phosphatase receptor type Z-deficient mice. *J. Neurosci.* 25:1081–1088.

Norman, E.D., Thiels, E., Barrionuevo, G., *et al.* (2000). Long-term depression in the hippocampus in vivo is associated with protein phosphatase-dependent alterations in extracellular signal-regulated kinase. *J. Neurochem.* 74:192–198.

Norris, C.M., Halpain, S., and Foster, T.C. (1998). Alterations in the balance of protein kinase/phosphatase activities parallel reduced synaptic strength during aging. *J. Neurophysiol.* 80:1567–1570.

Nowak, S.J., Pai, C.Y., and Corces, V.G. (2003). Protein phosphatase 2A activity affects histone H3 phosphorylation and transcription in Drosophila melanogaster. *Mol. Cell. Biol.* 23:6129–6138.

Oficjalska-Pham, D., Harismendy, O., Smagowicz, W.J., *et al.* (2006). General repression of RNA polymerase III transcription is triggered by protein phosphatase type 2A-mediated dephosphorylation of Maf1. *Mol. Cell* 22:623–632.

Oliver, C.J., Terry-Lorenzo, R.T., Elliott, E., *et al.* (2002). Targeting protein phosphatase 1 (PP1) to the actin cytoskeleton: the neurabin I/PP1 complex regulates cell morphology. *Mol. Cell. Biol.* 22:4690–4701.

Oliveria, S.F., Gomez, L.L., and Dell'Acqua, M.L. (2003). Imaging kinase – AKAP79 – phosphatase scaffold complexes at the plasma membrane in living cells using FRET microscopy. *J. Cell Biol.* 160:101–112.

Olsen, J.V., Blagoev, B., Gnad, F., *et al.* (2006). Global, in vivo, and site-specific phosphorylation dynamics in signaling networks. *Cell* 127:635–648.

Ostman, A., Hellberg, C., and Bohmer, F.D. (2006). Protein-tyrosine phosphatases and cancer. *Nat. Rev. Cancer* 6:307–320.

Pariser, H., Herradon, G., Ezquerra, L., *et al.* (2005a). Pleiotrophin regulates serine phosphorylation and the cellular distribution of beta-adducin through activation of protein kinase C. *Proc. Natl. Acad. Sci. USA* 102:12407–12412.

Pariser, H., Perez-Pinera, P., Ezquerra, L., *et al.* (2005b). Pleiotrophin stimulates tyrosine phosphorylation of beta-adducin through inactivation of the transmembrane receptor protein tyrosine phosphatase beta/zeta. *Biochem. Biophys. Res. Commun.* 335:232–239.

Paul, S. and Lombroso, P.J. (2003). Receptor and nonreceptor protein tyrosine phosphatases in the nervous system. *Cell. Mol. Life Sci.* 60:2465–2482.

Perez-Pinera, P., Alcantara, S., Dimitrov, T., Vega, J.A., and Deuel, T.F. (2006). Pleiotrophin disrupts calcium-dependent homophilic cell-cell adhesion and initiates an epithelial-mesenchymal transition. *Proc. Natl. Acad. Sci. USA* 103:17795–17800.

Pulido, R., Zuniga, A., and Ullrich, A. (1998). PTP-SL and STEP protein tyrosine phosphatases regulate the activation of the extracellular signal-regulated kinases ERK1 and ERK2 by association through a kinase interaction motif. *EMBO J.* 17:7337–7350.

Qiu, Z. and Ghosh, A. (2008). A calcium-dependent switch in a CREST-BRG1 complex regulates activity-dependent gene expression. *Neuron* 60:775–787.

Quinlan, E.M. and Halpain, S. (1996). Postsynaptic mechanisms for bidirectional control of MAP2 phosphorylation by glutamate receptors. *Neuron* 16:357–368.

Ranish, J.A., Yi, E.C., Leslie, *et al.* (2003). The study of macromolecular complexes by quantitative proteomics. *Nat. Genet.* 33:349–355.

Richards, E.J. (2006). Inherited epigenetic variation – revisiting soft inheritance. *Nat. Rev. Genet.* 7:395–401.

Roy, N., Van Eynde, A., Beke, L., *et al.* (2007). The transcriptional repression by NIPP1 is mediated by Polycomb group proteins. *Biochim. Biophys. Acta* 1769:541–545.

Salmeen, A. and Barford, D. (2005). Functions and mechanisms of redox regulation of cysteine-based phosphatases. *Antioxid. Redox Signal.* 7:560–577.

Schmidinger, H., Hermetter, A., and Birner-Gruenberger, R. (2006). Activity-based proteomics: enzymatic activity profiling in complex proteomes. *Amino Acids* 30:333–350.

Schones, D.E. and Zhao, K. (2008). Genome-wide approaches to studying chromatin modifications. *Nat. Rev. Genet.* 9:179–191.

Serratosa, J.M., Gomez-Garre, P., Gallardo, M.E., *et al.* (1999). A novel protein tyrosine phosphatase gene is mutated in progressive myoclonus epilepsy of the Lafora type (EPM2). *Hum. Mol. Genet.* 8:345–352.

Shalizi, A., Gaudilliere, B., Yuan, Z., *et al.* (2006). A calcium-regulated MEF2 sumoylation switch controls postsynaptic differentiation. *Science* 311:1012–1017.

Shenolikar, S. (2007). Analysis of protein phosphatases: toolbox for unraveling cell signaling networks. *Methods Mol. Biol.* 365:1–8.

Sim, A.T. and Scott, J.D. (1999). Targeting of PKA, PKC and protein phosphatases to cellular microdomains. *Cell Calcium* 26:209–217.

Sim, A.T., Baldwin, M.L., Rostas, *et al.* (2003). The role of serine/threonine protein phosphatases in exocytosis. *Biochem J.* 373:641–659.

Siniossoglou, S., Santos-Rosa, H., Rappsilber, J., *et al.* (1998). A novel complex of membrane proteins required for formation of a spherical nucleus. *EMBO J.* 17:6449–6464.

Smolen, P., Baxter, D.A., and Byrne, J.H. (2006). A model of the roles of essential kinases in the induction and expression of late long-term potentiation. *Biophys. J.* 90:2760–2775.

Snyder, E.M., Nong, Y., Almeida, C.G., *et al.* (2005). Regulation of NMDA receptor trafficking by amyloid-beta. *Nat. Neurosci.* 8:1051–1058.

Stone, S.R., Hofsteenge, J., and Hemmings, B.A. (1987). Molecular cloning of cDNAs encoding two isoforms of the catalytic subunit of protein phosphatase 2A. *Biochemistry* 26:7215–7220.

Strack, S., Kini, S., Ebner, F.F., *et al.* (1999). Differential cellular and subcellular localization of protein phosphatase 1 isoforms in brain. *J. Comp. Neurol.* 413:373–384.

Strack, S., McNeill, R.B., and Colbran, R.J. (2000). Mechanism and regulation of calcium/calmodulin-dependent protein kinase II targeting to the NR2B subunit of the N-methyl-D-aspartate receptor. *J. Biol. Chem.* 275:23798–23806.

Tabernero, L., Aricescu, A.R., Jones, E.Y., *et al.* (2008). Protein tyrosine phosphatases: structure-function relationships. *FEBS J.* 275:867–882.

Tailor, P., Gilman, J., Williams, S., *et al.* (1997). Regulation of the low molecular weight phosphotyrosine phosphatase by phosphorylation at tyrosines 131 and 132. *J. Biol. Chem.* 272:5371–5374.

Tamura, H., Fukada, M., Fujikawa, A., *et al.* (2006). Protein tyrosine phosphatase receptor type Z is involved in hippocampus-dependent memory formation through dephosphorylation at Y1105 on p190 RhoGAP. *Neurosci. Lett.* 399:33–38.

Tautz, L., Pellecchia, M., and Mustelin, T. (2006). Targeting the PTPome in human disease. *Expert Opin. Ther. Targets* 10:157–177.

Tezuka, T., Umemori, H., Akiyama, T., *et al.* (1999). PSD-95 promotes Fyn-mediated tyrosine phosphorylation of the N-methyl-D-aspartate receptor subunit NR2A. *Proc. Natl. Acad. Sci. USA* 96:435–440.

Thiels, E., Kanterewicz, B.I., Knapp, L.T., *et al.* (2000). Protein phosphatase-mediated regulation of protein kinase C during long-term depression in the adult hippocampus in vivo. *J. Neurosci.* 20:7199–7207.

Tonks, N.K. (2005). Redox redux: revisiting PTPs and the control of cell signaling. *Cell* 121:667–670.

Tonks, N.K. (2006). Protein tyrosine phosphatases: from genes, to function, to disease. *Nat. Rev. Mol. Cell. Biol.* 7:833–846.

Tootle, T.L., Silver, S.J., Davies, E.L., *et al.* (2003). The transcription factor Eyes absent is a protein tyrosine phosphatase. *Nature* 426:299–302.

Trelle, M.B. and Jensen, O.N. (2007). Functional proteomics in histone research and epigenetics. *Expert Rev. Proteom* 4:491–503.

Trinidad, J.C., Thalhammer, A., Specht, C.G., *et al.* (2005). Phosphorylation state of postsynaptic density proteins. *J. Neurochem.* 92:1306–1316.

Trinidad, J.C., Specht, C.G., Thalhammer, A., *et al.* (2006). Comprehensive identification of phosphorylation sites in postsynaptic density preparations. *Mol. Cell. Proteom* 5:914–922.

Trinidad, J.C., Thalhammer, A., Specht, C.G., *et al.* (2008). Quantitative analysis of synaptic phosphorylation and protein expression. *Mol. Cell. Proteom* 7:684–696.

Trinkle-Mulcahy, L., Andersen, J., Lam, Y.W., *et al.* (2006). Repo-Man recruits PP1 gamma to chromatin and is essential for cell viability. *J. Cell Biol.* 172:679–692.

Tsankova, N., Renthal, W., Kumar, A., *et al.* (2007). Epigenetic regulation in psychiatric disorders. *Nat. Rev. Neurosci.* 8:355–367.

Tweedie-Cullen, R.Y., Wollscheid, B., Livingstone-Zatchej, M., *et al.* (2007). Neuroproteomics and the detection of regulatory phosphosites. *Curr. Proteom* 4:209–222.

Ubersax, J.A. and Ferrell, J.E., Jr. (2007). Mechanisms of specificity in protein phosphorylation. *Nat. Rev. Mol. Cell. Biol.* 8:530–541.

Ulke-Lemee, A., Trinkle-Mulcahy, L., Chaulk, S., *et al.* (2007). The nuclear PP1 interacting protein ZAP3 (ZAP) is a putative nucleoside kinase that complexes with SAM68, CIA, NF110/45, and HNRNP-G. *Biochim. Biophys. Acta* 1774:1339–1350.

Vagnarelli, P., Hudson, D.F., Ribeiro, S.A., *et al.* (2006). Condensin and Repo-Man-PP1 co-operate in the regulation of chromosome architecture during mitosis. *Nat.Cell Biol.* 8:1133–1142.

van der Wijk, T., Overvoorde, J., and den Hertog, J. (2004). H_2O_2-induced intermolecular disulfide bond formation between receptor protein-tyrosine phosphatases. *J. Biol. Chem.* 279: 44355–44361.

Vulsteke, V., Beullens, M., Waelkens, E., *et al.* (1997). Properties and phosphorylation sites of baculovirus-expressed nuclear inhibitor of protein phosphatase-1 (NIPP-1). *J. Biol. Chem.* 272:32972–32978.

Wang, J.H. and Kelly, P.T. (1996). The balance between postsynaptic Ca(2+)-dependent protein kinase and phosphatase activities controlling synaptic strength. *Learn. Mem.* 3:170–181.

Wang, Y.T. and Salter, M.W. (1994). Regulation of NMDA receptors by tyrosine kinases and phosphatases. *Nature* 369:233–235.

Wang, X., Culotta, V.C., and Klee, C.B. (1996). Superoxide dismutase protects calcineurin from inactivation. *Nature* 383:434–437.

Watson, J.B., Khorasani, H., Persson, A., *et al.* (2002). Age-related deficits in long-term potentiation are insensitive to hydrogen peroxide: coincidence with enhanced autophosphorylation of Ca2+/calmodulin-dependent protein kinase II. *J. Neurosci. Res.* 70:298–308.

Westphal, R.S., Tavalin, S.J., Lin, J.W., *et al.* (1999). Regulation of NMDA receptors by an associated phosphatase-kinase signaling complex. *Science* 285:93–96.

Winder, D.G. and Sweatt, J.D. (2001). Roles of serine/threonine phosphatases in hippocampal synaptic plasticity. *Nat. Rev. Neurosci.* 2:461–474.

Wishart, M.J. and Dixon, J.E. (2002). PTEN and myotubularin phosphatases: from 3-phosphoinositide dephosphorylation to disease. *Trends Cell Biol.* 12:579–585.

Xu, Z. and Weiss, A. (2002). Negative regulation of CD45 by differential homodimerization of the alternatively spliced isoforms. *Nat. Immunol.* 3:764–771.

Yakel, J.L. (1997). Calcineurin regulation of synaptic function: from ion channels to transmitter release and gene transcription. *Trends Pharmacol. Sci.* 18:124–134.

Yin, Y. and Shen, W.H. (2008). PTEN: a new guardian of the genome. *Oncogene* 27:5443–5453.

Zhang, Y., Kim, Y., Genoud, N., *et al.* (2006). Determinants for dephosphorylation of the RNA polymerase II C-terminal domain by Scp1. *Mol. Cell* 24:759–770.

Zhang, S., Edelmann, L., Liu, J., *et al.* (2008). Cdk5 regulates the phosphorylation of tyrosine 1472 NR2B and the surface expression of NMDA receptors. *J. Neurosci.* 28:415–424.

Zhou, F., Galan, J., Geahlen, R.L., *et al.* (2007). A novel quantitative proteomics strategy to study phosphorylation-dependent peptide-protein interactions. *J. Proteom Res.* 6:133–140.

Chapter 11
Covalent Protein Modification as a Mechanism for Dynamic Recruitment of Specific Interactors

Nicholas R. Bertos, Veena Sangwan, Xiang-Jiao Yang, and Morag Park

11.1 Introduction

The complement of proteins available to a human cell comprises ca. 20,000–25,000 members (Lander et al. 2004), a number not vastly greater than the number of proteins present in a unicellular organism such as *Saccharomyces cerevisiae* (the genome of which encodes ca. 6,000 different proteins (Goffeau et al. 1996)). An additional level of complexity is generated through the generation of multiple isoforms of many proteins via differential splicing and the use of alternate transcriptional start sites, which increases the repertoire of potential individual protein species. However, the mere presence of a large number of effector molecules does not create the dynamic interchange of information required to mediate the processes of cellular function, or mediate signal transduction to respond to environmental changes.

Additional complexity and flexibility is conferred upon the system by post-translational modifications. These either reversibly or irreversibly alter the configuration of proteins and cause changes in their function, including influences on enzymatic activity, interaction with other proteins, and other characteristics. Reversible modifications include phosphorylation, acetylation and methylation, as well as the covalent attachment of single or multiple moieties of ubiquitin or ubiquitin-related proteins, among others (see Fig. 11.1). The addition of these covalent modifiers can be considered as equivalent to a "gain of function" for the target protein. The modified sites, or combinations thereof, can constitute recognition signals or "codes" for the recruitment of specific docking modules inherently present in single or multiple interaction partners, thus leading to the dynamic

N.R. Bertos (✉)
Rosalind & Morris Goodman Cancer Centre, McGill University, 1160 Ave. des Pins O, Montreal, QC, Canada, H3A 1A3
and Department of Medicine, McGill University Health Centre, Montreal, QC, Canada
e-mail: nicholas.bertos@mcgill.ca

C.J. Vidal (ed.), *Post-Translational Modifications in Health and Disease*, Protein Reviews 13,
DOI 10.1007/978-1-4419-6382-6_11,

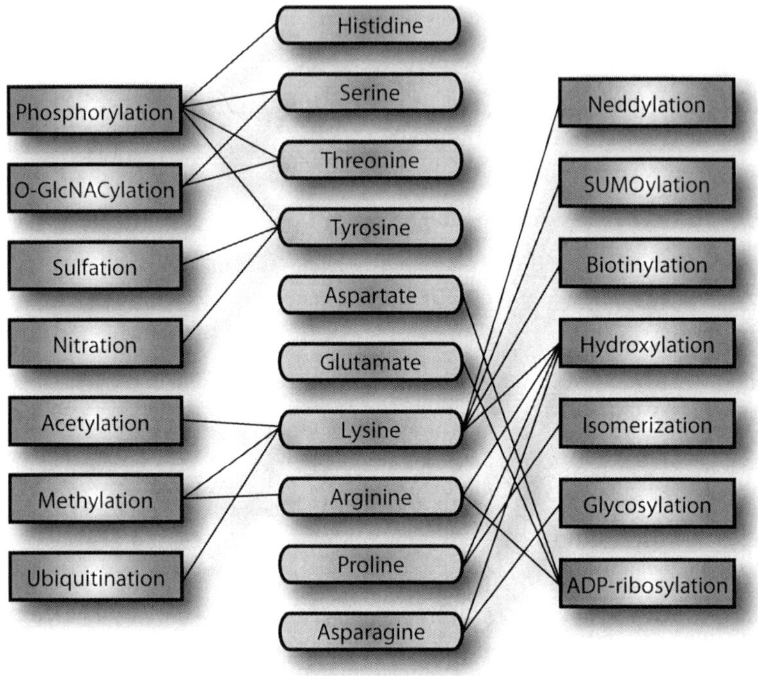

Fig. 11.1 Schematic representation of the most common post-translational modifications and the amino acids at which they can occur, demonstrating the potential for convergence of multiple modifications at specific residues, as well as the range of modifications and corresponding recognition domain-containing effectors that may functionally interact with each other

generation of a specific configuration of protein sets or complexes. Such partnership configurations can be finely controlled with respect to their spatiotemporal features, leading to specific signal transduction to activate or repress downstream functions. The reversibility of many of the covalent interactions involved allows for down-regulation of this activity via removal of the residues required for interactor recruitment when the task is complete, permitting dynamic control of the process.

Additionally, protein modifications can be utilized to not only generate a specific "fingerprint" for the recognition of an element by one or several interaction modules, but to determine which potential interactors are recruited via a particular amino acid. Several of the protein modifications discussed here are targeted to the same amino acid, such that the presence of one modification prevents the establishment of another. For example, acetylation of a lysine residue can act to prevent the addition of a ubiquitin moiety, and vice versa, via same-site competition.

In this chapter, we present an overview of several protein modification interaction domains currently considered to play an important role in the dynamic recruitment of effector or regulatory factors, as well indicate the complexity of the potential interaction space created by interactions between modifications. Where possible, we emphasize how different modifications are linked to cellular signaling networks.

11.2 Phosphorylation-Dependent Interactions

Historically, protein phosphorylation has been the most intensively studied of the covalent modifications, and its role as a key element in signal transduction has long been recognized. Although phosphorylated tyrosine residues were the first to be identified as interaction recruiters, the importance of serine and threonine residues as sites for transient generation of new binding sites is reflected in the number of modules recognized as binding to these motifs. The widespread nature of this modification is demonstrated through phosphoproteomic analyses. 6,600 phosphorylation sites could be identified from ca. 2,200 proteins investigated in HeLa cells; 14% of these changed in abundance within 15 min of exposure to the membrane receptor ligand EGF (*e*pidermal *g*rowth *f*actor), with many proteins containing multiple phosphorylation sites exhibiting different kinetics (Olsen et al. 2006).

11.2.1 Phosphotyrosine-Dependent Interactions

11.2.1.1 SH2 Domain

First identified in 1986 (Sadowski et al. 1986), this domain remains the archetype for recognition of a post-translationally modified amino acid within a sequence-specific context (reviewed in Pawson and Nash 2003). As well as recognizing phosphotyrosine, individual SH2 (*S*rc *h*omology *2*) domains interact with specific sequences of flanking residues, imparting additional specificity to the interactions. SH2 domain-containing proteins are classically considered to act as adapter molecules, coupling tyrosine phosphorylation of proteins such as membrane-localized receptor tyrosine kinases to positive control of downstream signaling events. Additionally, SH2 domains are present in protein tyrosine phosphatases, indicating the flexibility inherent in this interaction motif.

11.2.1.2 PTB Domain

Another class of phosphotyrosine recognition motifs is represented by the PTB (*p*hospho*t*yrosine *b*inding) domain family. These motifs share a low degree of sequence homology, yet adopt similar secondary and tertiary structure conformations (reviewed in Yan et al. 2002). They also exhibit a common binding affinity for phospholipids (reviewed in Uhlik et al. 2005). However, one group of PTB domains, the Dab-like PTBs, exhibits phosphotyrosine-independent binding. This group includes ca. 75% of PTB domains identified (Uhlik et al. 2005), illustrating the heterogeneity in substrate recognition present in this domain family.

11.2.1.3 C2 Domain

Previously recognized as a module responsible for binding phospholipids in a calcium-dependent manner (Davletov and Sudhof 1993; reviewed in Newton and Johnson 1998), this domain has recently been identified as a phosphotyrosine-binding motif (Benes et al. 2005; reviewed in Sondermann and Kuriyan 2005). The interaction is sequence-specific, and mediates the phosphorylation-dependent binding of the C2 domain-containing serine/threonine protein kinase PKCδ (*p*rotein *k*inase *C*δ) to CDCP1 (*C*UB *d*omain-*c*ontaining *p*rotein *1*). In this model, the kinase Src initially phosphorylates and subsequently binds to a tyrosine residue on CDCP1 via the Src SH2 domain, promoting further CDCP1 tyrosine phosphorylation events. This in turn leads to the recruitment of PKCδ via its C2 domain, creating a multimeric signaling complex (Benes et al. 2005).

As well as recruiting modification-dependent binding partners, the interaction of phosphotyrosines with specific recognition motifs can be exploited to create additional signaling pathways. Pyruvate kinase is an important enzyme in the glycolytic metabolic pathway. The *p*yruvate *k*inase *M2* splice isoform (PKM2) represents a variant that promotes the switch to anaerobic glycolysis and is preferentially expressed by cancer cells (Christofk et al., 2008a). PKM2 can interact with phosphotyrosine via a novel motif; this binding competes with that of the allosteric activator fructose-1,6-bisphophate, resulting in decreased PKM2 enzymatic activity (Christofk et al., 2008b). Thus, this motif can act as a sensor of cellular tyrosine phosphorylation levels, coupling this readout to the control of glycolytic metabolism (Christofk et al., 2008a, b).

11.2.2 Phosphoserine/Phosphothreonine-Dependent Interactions

11.2.2.1 The 14-3-3 Protein Family

The seven mammalian members of the 14-3-3 protein family exist in vivo as homo- or hetero-dimers. They play adapter and integrator roles in pathways promoting cell survival and inhibiting apoptosis (reviewed in Morrison 2009), as well as performing many other functional roles, including regulation of the subcellular localization of transcriptional corepressors, i.e., the histone deacetylase HDAC4, in a phosphorylation-dependent manner (Grozinger and Schreiber 2000; McKinsey et al. 2000; Wang et al., 2000a; Nishino et al. 2008; reviewed in Bertos et al. 2001). Their role in driving cell survival positions them as attractive targets for anti-cancer therapies.

11.2.2.2 BRCT Domain

Members of the *BR*CA1 *c*arboxyl-*t*erminal (BRCT) domain family are predominantly found in proteins involved in regulation of the DNA damage response (Koonin et al. 1996). These domains have been found to interact preferentially with phosphoserine-containing peptides (Rodriguez et al. 2003; Yu et al. 2003), and somatic mutations in the BCRT domain of BRCA1 (*br*east *ca*ncer *1*) are linked to an increased risk of early-onset breast and ovarian cancer in affected individuals (Gayther et al. 1995).

11.2.2.3 WD40 Domain

F-box-containing proteins, which play roles in targeting ubiquitinated protein for degradation, mostly also contain WD40 or *l*eucine-*r*ich *r*epeat (LRR) regions that have been linked to the recognition or binding of phosphoproteins, although this interaction has not yet been formally demonstrated (reviewed in Yaffe and Elia 2001).

11.2.2.4 FHA Domain

Proteins containing *f*ork*h*ead-*a*ssociated (FHA) domains include those with functions in DNA damage repair, as well as kinesins, RING-finger proteins, forkhead transcription factors and the proliferation marker Ki-67 (Durocher and Jackson 2002; Mahajan et al. 2008). FHA domains can be divided into 3 subcategories depending on their preferences for specific types of residues surrounding their target phosphoserine residue (Liang and Van Doren 2008). Interestingly, it has been reported that the FHA domain may also be able to bind to phosphotyrosine (Liao et al. 1999; Wang et al., 2000b), although the functional significance of this interaction is not currently clear.

11.2.2.5 Polo-Box Domain

Found in the *P*olo-*l*ike *k*inases (PLKs), the Polo-box domain is essential for proper localization of these important mitotic regulators. The specificity of the interaction of this domain with phosphoserine and phosphothreonine residues is related to the function of PLKs as molecular integrators. Appropriate PLK function requires that prior phosphorylation by other mitotic kinases must have occurred before the PLKs can be properly localized to their target sites, where they proceed to drive progression through the M phase of the cell cycle (reviewed in Lowery et al. 2004). A mass spectrometry-based screen for other Polo-box interactors identified proteins involved in other processes such as translational control, RNA processing and vesicular transport (Lowery et al. 2007), suggesting that this domain may function in additional processes beyond those previously identified.

11.2.2.6 WW Domain

Although generally considered to be modules responsible for interaction with proline-rich surfaces, a subset of WW domain-containing proteins utilize this sequence for recognition of phosphorylated phosphoserine or phosphothreonine residues occurring immediately adjacent to proline (Rodriguez et al. 2003). These include the proline isomerase Pin1 (peptidyl-*p*rolyl cis/trans *i*somerase, *N*IMA-interacting 1) and the ubiquitin ligase Nedd4 (*n*eural precursor cell *e*xpressed, *d*evelopmentally *d*ownregulated *4*); in another illustration of the complexity of the control of such modification-dependent interactions, it has been reported that the WW domain of Pin1 must be phosphorylated to mediate binding to phosphoserine (Lu et al. 2002).

11.2.2.7 FF Domain

First identified as a novel motif often found near WW domains, FF domains contain two highly conserved phenylalanine residues (Bedford and Leder 1999). FF domains are found in the yeast protein Prp40 and the human protein TCERG1 (*t*ran*s*cription *e*longation *r*e*g*ulator *1*, also known as CA150), both of which bind to phosphoserines in the *c*arboxyl-*t*erminal *d*omain (CTD) of RNA polymerase II (Morris and Greenleaf 2000; Goldstrohm et al. 2001), as well as in other proteins known to interact with huntingtin (Faber et al. 1998; Passani et al. 2000). Interestingly, the FF domains of TCERG1 are also able to bind multiple transcription and splicing factors in a phosphorylation-independent manner through multiple weak interactions with motifs comprising negatively charged residues flanked by aromatic amino acids (Smith, et al. 2004).

11.3 Methylation-Dependent Interactions

Methylation of lysine or arginine residues can occur as monomethylation or dimethylation; in the case of lysine, trimethylation is also possible, while symmetric vs. asymmetric dimethylation expands the repertoire of possible configurations for methylated arginine (reviewed in Lee et al. 2005). Although there does not appear to be an obligate requirement for one methylation type vs. another for interaction with methylation-dependent recognition motifs, preferential binding to certain forms exists for specific methylation-specific domain-containing proteins. For example, the chromodomains of the HP1 (*h*eterochromatin *p*rotein *1*) and Polycomb proteins preferentially bind trimethylated lysine-9 of histone H3 (Fischle et al. 2003; Min et al. 2003), while the affinity of the tudor domain is highest for symmetric dimethylarginine (Sprangers et al. 2003).

11.3.1 Chromodomains

Classically found in chromatin-associated proteins, chromodomains exist across a wide variety of organisms, suggesting that they represent an ancient structural motif (reviewed in Eissenberg 2001). In general, their most common binding partners appear to be nucleosomes, and thus it is not surprising that they exhibit an ability to recognize specific methylation patterns on this class of proteins. A screen for binding to specific partner elements conducted using a protein microarray approach, where differentially methylated peptides were used as probes (Kim et al. 2006), established that chromodomains possess specific affinities for methylated lysine residues present in the context of particular histone tail sequences.

11.3.2 The "Royal Family"

Tudor, MBT (malignant brain tumor) and PWWP (proline-tryptophan-tryptophan-proline) Domains. These multiple subfamilies are also predominantly found in chromatin-associated proteins. Due to the structural similarity of some of their members to the chromodomain family, it has been suggested that they may function in a similar manner. In the same binding screen as discussed for chromodomains above (Kim et al. 2006), it was found that tudor domains appeared to be more sensitive to the degree of methylation than to the sequence context of the probe peptide, generally exhibiting a stronger affinity for di- or trimethylated lysine residues, while MBT domains preferentially interact with monomethylated lysines. The affinity of the PWWP domain for methylated lysine residues, on the other hand, has only recently been described (Wang et al. 2009). These differences in partner preference generate an expanded range of specific recognition modules for fine control of effector interaction with methylated proteins.

11.4 Acetylation-Dependent Interactions

11.4.1 Bromodomains

This motif is a ca. 110 amino acid module predominantly found in proteins involved with transcriptional control at the level of chromatin and the nucleosome (reviewed in Jeanmougin et al. 1997). Reversible acetylation of lysine residues is a common modification in the context of the components with which such proteins interact, especially for specific lysine residues in the N-terminal tails of histones H3 and H4. It has been shown that specific interaction of the bromodomain with acetyl-lysine residues occurs via a hydrophobic pocket located between the ZA and BC loops of the four-helix bundle (Dhalluin et al. 1999), thus tethering bromodomain-containing proteins to, e.g., lysine-acetylated histone tail regions.

Although the majority of bromodomain interactions identified to date have been with various acetyllysine residues located on the tail regions of histones H3 and H4, bromodomains have also been found to mediate binding to acetyllysines present on other proteins, such as MyoD (*myo*genic *d*ifferentiation antigen 1) (Polesskaya et al. 2001), Myb (v-myb avian *my*elo*b*lastosis viral oncogene homolog) (Tomita et al. 2000), HIV Tat (*h*uman *i*mmunodeficiency *v*irus *t*rans-*a*ctivator of *t*ranscription) (Col et al. 2001) and p53 (Mujtaba et al. 2004). Furthermore, rather than binding to acetyllysines in general, bromodomains from different proteins also require additional sequence elements to mediate efficient interactions. This combination of specific context-dependent features, along with a "switchable" residue modification, imparts fine control of specificity and permits a wide range of individual interactions to be mediated by multiple bromodomain-containing proteins.

Acetylation and deacetylation of target proteins is mediated through two broad classes of enzymes, historically termed *h*istone *a*cetyl*t*ransferases (HATs) and *h*istone *deac*etylases (HDACs) from the initial substrates identified. Interestingly, several of these proteins, such as the HAT CBP (*C*REB-*b*inding *p*rotein)/p300, also contain bromodomains, which mediate interaction with their enzymatic targets, such as MyoD (Polesskaya et al. 2001). This potentially acts as a positive feedback mechanism potentiating the binding between the two proteins.

11.5 Hydroxylation-Dependent Interactions

Hydroxylation, a potential modification for proline, asparagine, arginine or lysine residues, occurs in the context of multiple proteins, including the hypoxia-inducible transcription factor HIF (*h*ypoxia-*i*nducible *f*actor) and the matrix component collagen. In the latter case, this modification has structural implications (Krane 2008), while in the former case, hydroxylation acts to create a specific recruitment platform for downstream signal modulators.

11.5.1 VHL Domain

The role of this motif in mediating the regulation of HIF (*h*ypoxia-*i*nducible *f*actor) activity, in combination with other post-translational protein modifications governing this central hub of the hypoxic response, illustrates the potential complexity of interactions between these elements, and will therefore be described in detail (see Fig. 11.2). The *v*on *H*ippel-*L*indau (VHL) factor is a crucial element in initiating the response to changes in intracellular oxygen concentrations. This function is mediated through its interaction with the HIF α subunit (reviewed in Ivan and Kaelin 2001). In normoxia, this interaction is promoted, leading to the degradation and subsequent loss of function of HIF via a polyubiquitin-dependent pathway (Huang et al. 1998; Iwai et al. 1999; Lisztwan et al. 1999; Maxwell et al. 1999). Under hypoxic conditions, the abrogation of this interaction leads to a suppression

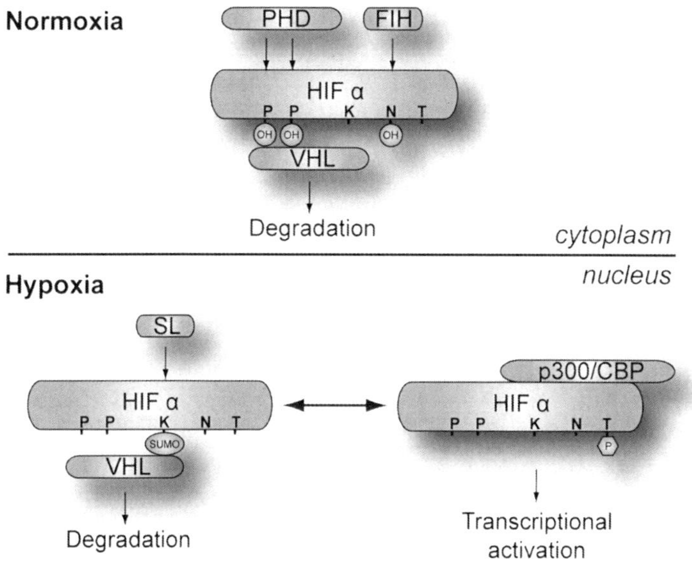

Fig. 11.2 Schematic representation of post-translational modifications involved in regulating HIF α activity under conditions of normal oxygen concentrations vs. hypoxia. PHD, prolyl hydroxylase domain-containing protein; FIH, factor inhibiting HIF; VHL, von Hippel-Lindau factor; p300/CBP, p300/Creb-binding protein; SL, SUMO ligase. Under hypoxic conditions, HIF1 α translocates to the nucleus where it partners with HIF β to exert its transcriptional regulatory function. Note that the role of SUMOylation in promoting HIF1 α degradation is not firmly established at this time

of degradation and thus an increase in the effective concentration of HIF, causing the transcriptional activation of HIF target genes.

Under conditions of normal oxygen concentrations, HIF α is hydroxylated on proline-564 and proline-402 through the action of PHDs (*p*rolyl *h*ydroxylase *d*omain-containing proteins), three of which have been identified so far in humans (Bruick and McKnight 2001; Epstein et al. 2001) (see Fig. 11.2). This process requires the presence of oxygen as a substrate for the hydroxylation reaction. Under hypoxic conditions, the rate of this reaction is reduced, leading to the loss of the hydroxyl groups at these residues. This suggests that the hydroxylation level of the HIF α subunit functions as a direct sensor of oxygen concentrations.

The mechanism through which this protein modification mediates control of HIF α subunit stability occurs via the hydroxylation-dependent recruitment of the von Hippel-Lindau factor, which serves as the recognition component of an ubiquitin ligase (Iwai et al. 1999; Lisztwan et al. 1999) that promotes proteosomal degradation of HIF α (Cockman et al. 2000; Kamura et al. 2000; Ohh et al. 2000; Tanimoto et al. 2000). This interaction is mediated by a specific domain within the von Hippel-Lindau factor, termed the VHL domain, which binds to hydroxylated prolines of the HIF α subunit (Jaakkola et al. 2001). Mutation of this residue in the HIF α protein abrogates this interaction, as does the exogenous addition of a peptide corresponding to the surrounding HIF α sequence bearing a hydroxylated Pro-564 residue, which acts as a competitive inhibitor (Jaakkola et al. 2001).

Interestingly, the ability of prolyl hydroxylases to be regulated by intracellular oxygen concentrations appears to be utilized in a second mode of controlling HIF α activity. An asparagine residue (asparagine-803) located in the C-terminal transcriptional activation domain of HIF α is a target for the asparaginyl hydroxylase FIH (*f*actor *i*nhibiting *H*IF) (Hewitson et al. 2002; Lando et al. 2002a, b). Hydroxylation of this residue suppresses the transcriptional activation function of this domain by preventing its interaction with the transcriptional coactivator p300/CBP (Lando et al. 2002a). Illustrating the concept by which one covalent modification can affect another, the oxygen-independent phosphorylation of threonine-796 in HIF α by casein kinase 2 may be necessary for transcriptional activation (Gradin et al. 2002); phosphorylation at this residue abrogates hydroxylation of asparagine-803 by FIH (Lancaster et al. 2004), suggesting an additional level of control.

A second such instance is provided by the finding that hypoxia also induces the SUMOylation of HIF α, providing an alternative route for enhancement of binding to von Hippel-Lindau factor and subsequent proteosomal degradation (Cheng et al. 2007). This mode of control is depicted in Fig. 11.2; however, previous groups have reported that SUMOylation increases HIF α stability and increases transcriptional activity (Bae et al. 2004), or reduces the transcriptional activity of HIF α without affecting stability (Berta et al. 2007) These different observations may critically depend upon the balance between SUMOylating and deSUMOylating (e.g., SENP1) enzymes (Cheng et al. 2007), which act to modulate this signal. The question of whether different modifications vary with respect to time scale, such that their combination alters the nature of the induction of HIF-dependent transcriptional activity in response to transient vs. long-term decreases in oxygen concentrations, remains open.

11.6 Ubiquitination-Dependent Recognition Motifs

Ubiquitination, initially recognized as a modification involved in targeting proteins for intracellular degradation, has emerged in recent years as an important element in signal transduction. A key difference between the ultimate function of ubiquitination as a marker for degradation vs. as a modifier of function lies in the number of ubiquitin moieties added; while polyubiquitination is often associated with targeting of proteins for proteosomal degradation, monoubiquitination is more closely related to modulation of protein function by generation of a novel protein interaction site (representing a "gain of function"). Interestingly, many ubiquitin-binding domains can mediate autoubiquitination, thus potentially regulating the activity of the domain's "host" proteins.

11.6.1 UBA Domain

The first ubiquitin interaction domain to be described, the UBA domain (Hofmann and Bucher 1996; Bertolaet et al. 2001b) has traditionally been associated with polyubiquitin binding (Wilkinson et al. 2001; Funakoshi et al. 2002; Raasi and Pickart 2003), but is also capable of binding monoubiquitin moieties in vitro, as well as interacting with other UBA domains (Vadlamudi et al. 1996; Bertolaet et al. 2001a; Chen et al. 2001).

Surface plasmon resonance studies of the EDD (*E*3 identified by *d*ifferential *d*isplay) ubiquitin ligase UBA domain show that this module does not exhibit a strong preference for poly- vs. monoubiquitin as a binding partner (Kozlov et al. 2007). NMR (*n*uclear *m*agnetic *r*esonance) titration analysis suggest that the UBA domain of the p62 scaffold protein binds di-ubiquitin with slightly lower affinity than mono-ubiquitin, suggesting that this UBA domain may preferentially interact with extended polyubiquitin chains adopting more open structures (Long et al. 2008). Interestingly, in Paget's disease of bone, the primary defect appears to be mutation or truncation of the p62 UBA domain, which has deleterious effects on the NF-κB (*n*uclear *f*actor *κ-B*) signaling pathway in osteoclasts (reviewed in Layfield and Searle 2008).

The *r*eceptor *t*yrosine *k*inase (RTK) Met plays important roles in cell proliferation and survival, cell migration and epithelial morphogenesis (reviewed in Peschard and Park 2007). Therefore, tight control of its activity is required for normal cell function. One of the mechanisms for its ligand-induced downregulation occurs via phosphotyrosine-dependent recruitment of Cbl E3 ubiquitin ligases, which enhances Met degradation (Peschard et al. 2001, 2004; Abella et al. 2005; Mak et al. 2007).

The c-Cbl UBA domain, is required for both homodimerization and heterodimerization with the Cbl family member Cbl-b (Bartkiewicz et al. 1999; Liu et al. 2003), but does not appear to interact with ubiquitinated lysine residues (Davies et al. 2004; Raasi et al. 2005). Analysis of the crystal structure of this domain reveals that the same surface is used for both homo- and heterodimerization, while site-directed mutagenesis experiments demonstrate the requirement for UBA-mediated dimerization to enable ubiquitin ligase activity directed against the Met RTK (Kozlov et al. 2007). This finding further demonstrates that UBA domains, beyond their role in the recognition of ubiquitinated proteins, can be adapted to serve as protein-protein binding domains, further extending the repertoire of possible protein interactions, and illustrates a general trend where domain archetypes can be adapted to serve as recognition sites for multiple binding motifs.

11.6.2 CUE (*c*oupling of *u*biquitin conjugation to *e*ndoplasmic reticulum degradation) domain

This domain recognizes both mono- and polyubiquitinated residues, with different affinities depending on the "host" protein, as well as acting to promote intramolecular monoubiquitination (Donaldson et al. 2003; Shih et al. 2003).

11.6.3 PAZ/Znf/UBP/HUB Domain

The atypical class II histone deacetylases HDAC6, as well as possessing dual catalytic domains, also contains a zinc finger motif near its carboxyl terminal similar to that found in *u*biquitin-*s*pecific *p*rotease *3* (USP3) and *BR*CA1-*a*ssociated *p*rotein *2* (BRAP2) (reviewed in Bertos et al. 2001) which has been shown to act to interact specifically with ubiquitin (Seigneurin-Berny et al. 2001; Hook et al. 2002; Boyault et al. 2006). This interaction plays an important role in the recognition of misfolded proteins by HDAC6, which acts to transport them to the aggresome for eventual proteasome-independent degradation (Kawaguchi et al. 2003).

11.6.4 UBAN Domain

The NUB (*N*EMO *u*biquitin *b*inding) or UBAN (*u*biquitin *b*inding in *A*BIN and *N*EMO proteins) motif was previously identified as an ubiquitin-binding region present in NEMO (*N*F-κB *e*ssential *mo*dulator), the ABIN family (*A*20 *b*inding and *i*nhibitor of *N*F-κB) and optineurin (Ea et al. 2006; Wu et al. 2006; Zhu et al. 2007; Bloor et al. 2008; Wagner et al. 2008). Although the UBAN domain was previously thought to bind lysine-63-linked ubiquitin chains, recent studies have demonstrated that this motif preferentially interacts with head-to-tail linked linear ubiquitin dimers or multimers (Lo et al. 2009; Rahighi et al. 2009).

NEMO, the regulatory subunit of *I*κB *k*inase (IKK), a complex which plays a key role in regulation of NF-κB, is recruited to polyubiquitinated signaling mediators, leading to IKK recruitment and NF-κB activation. The crystal structure of the ubiquitin-binding region of NEMO reveals that conformational changes occur upon ubiquitin binding (Rahighi et al. 2009), suggesting a possible mechanism for permitting IKK transautophosphorylation and activation. Interestingly, NEMO is itself subject to linear polyubiquitination by the recently described LUBAC (*l*inear *ub*iquitin chain *a*ssembly *c*omplex) ligase complex (Tokunaga et al. 2009), suggesting a possible mechanism for NEMO cis- or trans-interactions (Rahighi et al. 2009).

11.7 Sumoylation-Dependent Interactions

The ubiquitin-related covalent adduct SUMO (*s*mall *u*biquitin-like *mo*difier) exists as four isoforms (SUMO-1, -2, -3 and –4, although the functional role of the last is currently unclear). While SUMO-1 is generally involved in monoSUMOylation, SUMO-2 and -3 are mostly added to their target proteins as polySUMO chains, thus recapitulating the mechanistic differences between mono- and polyubiquitination. The functional consequences of SUMOylation, however, are generally distinct from those of ubiquitination. They are to a large part associated with alterations in protein-protein interactions (reviewed in Geiss-Friedlander and Melchior 2007) rather than targeting of proteins for degradation, although the latter has also been demonstrated to occur (Prudden et al. 2007). SUMOylation can promote protein interactions, e.g., SUMOylated RanGAP1 (*Ran GTPase-activating protein 1*) can interact with RanBP2 (*Ran-binding protein 2*) (Matunis et al. 1996; Mahajan et al. 1997) and SUMOylated p300 interacts with and acetylates HDAC6 (Girdwood et al. 2003); alternately, SUMOylation can lead to the abrogation of previously existing interactions, e.g., SUMOylated CtBP can no longer interact with the PDZ (*postsynaptic density protein-95, disk-large tumor suppressor protein, zonula occludens-1*) domain of nNos (neuronal nitric oxide synthase) (Lin et al. 2003).

11.7.1 SIM Domain

The consensus sequence of SIMs (*SUMO-interacting motifs*) is relatively elastic, although several overall features, such as hydrophobicity and charge clustering, are invariant (reviewed in Kerscher 2007). While multiple ubiquitin-binding domains have been identified, only one SIM has been characterized thus far; it is possible that the variable sequence permits the generation of SIMs with different affinities for the various SUMO isoforms, as well as for mono- vs. polySUMOylated sites.

Yet another level of complexity is generated by the fact that the interaction of some SIMs with their SUMOylated binding partners requires phosphorylation at serine residues adjacent to the hydrophobic core of the SIM domain, generating a phosphorylation-dependent SIM domain-SUMO interaction (Stehmeier and Muller 2009). Thus, the binding affinity of two proteins can be governed by post-translational modification both at the classical modification site, and within the recognition module of the partner protein, illustrating the linkage between SUMO binding and phosphorylation-dependent cellular signaling networks.

11.8 Examples of Multisite Modifications

11.8.1 "Histone Code" or "Chromatin Signature"

As a key integrator upon which multiple pathways converge, and being closely associated with direct control of transcription, it is not surprising that the histone components of chromatin are subject to a wide variety of post-translational modifications. This set of modifications has been proposed to constitute a "histone code", different configurations of which recruit specific combinations of interacting proteins to mediate downstream events (Strahl and Allis 2000). Covalent modifications of residues in the flexible histone amino-terminal tail regions include methylation (where mono-, di- or trimethylation is distinct), acetylation, phosphorylation,

ADP-ribosylation, SUMOylation and ubiquitination (Shiio and Eisenman 2003; reviewed in Bhaumik et al. 2007; Latham and Dent 2007; Munshi et al. 2009).

Multiple examples of the crosstalk between different modifications can be observed in this system. For example, acetylation of histone H4 at lysine-5 and lysine-12 promotes chromatin compaction and thereby gene silencing (Kelly et al. 2000); in the context of existing acetylation at lysine-8 and lysine-16 of the same protein, the same modification is linked to transcriptional activation (reviewed in Yang 2005). Histone H3 serine-10 phosphorylation acts to promote transcriptional activation (DeManno et al. 1999), unless serine-28 is also phosphorylated, in which case this combination constitutes a mark of condensed and therefore transcriptionally inactive chromatin (reviewed in Yang 2005).

As the complexity of the spatiotemporal interactions between different modifications becomes more evident, it is apparent that, beyond the "histone code" that presents a specific temporally limited configuration for recruitment of effectors and interactors, chromatin modifications exist in a state of flux. The resulting regulatory paradigm can be thought of as being similar to a decision tree with multiple interacting branches. Thus, the current output at a specific location would be dependent upon the initial configuration of chromatin modifications as well as the availability and relative abundances of interactors recruited to those modifications, the state of these interactors as determined by their intrinsic modifications, the composition and membership of the interactor complexes themselves, and the dynamics of antagonistic, synergistic and regulatory interactions between these complexes. This output would then generate a novel combination of the factors listed above, while external pathways would also affect several of these parameters.

11.8.2 p53

As a central element in the regulation of the DNA damage response, p53 is subject to multiple levels of post-translational regulation, including a wide variety of post-translational modifications. These include phosphorylation (on serine and threonine residues), acetylation, methylation, ubiquitination, SUMOylation and neddylation (reviewed in Bode and Dong 2004; Yang and Seto 2008). Reminiscent of the regulation of HIF α, a primary mechanism for the control of p53 activity depends on the intrinsic instability of the protein; thus, post-translational modifications targeting this property play important roles in regulating p53 function.

In response to DNA damage, a phosphorylation-acetylation cascade leads to p53 activation; serine phosphorylation and lysine methylation promotes the association of p53 with HATs, which in turn acetylate lysine residues, altering DNA binding properties, creating docking sites for the recruitment of additional interactors, and competitively inhibiting ubiquitination at these sites, leading in turn to decreased degradation (reviewed in Yang and Seto 2008)). Additionally, acetylation of lysine-120 of p53, a site mutated in human cancer, appears to function as a switch determining differential assignments to apoptosis vs. cell cycle control.

Interestingly, transgenic mice in which multiple lysine acetylation sites were deleted by substitution with arginine do not demonstrate a severe phenotype (reviewed in Yang and Seto 2008), suggesting that multiple redundant levels of control may exist to regulate the activity of this key protein. Overall, we believe that multisite modifications constitute a set of still poorly understood regulatory programs for concerted actions in response to different cellular and environmental cues.

11.9 Concluding Remarks

The specificity of many of these interactions, which requires the presence of a covalently modified amino acid or a combination of modifications in the context of defined sequence elements, renders them particularly attractive as targets for the development of exogenous competitive inhibitors. These would have the advantage of blocking a specific interaction or set thereof, in contrast to the broad effects seen following pharmacological blockade of the enzymatic activity of a signaling cascade member. Although methods for delivery of such competitive inhibitors that can be used in the clinical context remain elusive, this approach holds out the promise of eventual specific targeting of therapeutics to abrogate protein-protein interactions.

As additional proteins are studied in detail, a picture is beginning to emerge in which the set of post-translational modifications present on each protein constitutes an additional level of information and control regarding its enzymatic activity, subcellular localization and interaction partners, where these factors may also be interdependent. It may be appropriate to characterize this set of modifications as comprising the "quinary structure" of a protein or protein assembly, which contains the necessary information to fully characterize and identify the role of the effector at a particular point in time.

Regulation via post-reversible post-translational modifications constitutes a dynamic spatiotemporal fine-tuning of cellular element function, thus permitting the cell, and by extension the organism, to continuously adapt and respond to its environment. Modifications at multiple sites can combine in an antagonistic, cooperative or synergistic manner. This further adds to the complexity of understanding the integration of protein modification recognition inputs and functional outputs.

The elucidation of significant portions of the "quinary structure" information set of certain proteins has occurred essentially serendipitously, as a consequence of multiple separate investigations into different aspects of protein function and regulation. The question of whether those proteins that have been identified as bearing multiple layers of functional regulation by a variety of post-translational modifications are indeed more subject to this type of control than others due to their key roles as information integrators, or whether these multiple levels have been revealed precisely because of the intensive investigations directed towards such key proteins, remains open.

Currently, we lack a comprehensive understanding regarding the full extent of post-translational protein modifications, and the degrees of functional interaction between them. This situation likely underlies many of the instances in which conflicting reports arise regarding the effect of a specific modification, since the agonistic or antagonistic effects of other modifications may differ depending on cell type and environment.

A complete catalogue of protein modifications, their temporal relationship and their mutual interactions, as well as of the modules recruited by each moiety in the context of a specific sequence, is likely to be a long-term goal (Yang 2005). However, progress towards the understanding of this fundamental element in the regulation of protein activity is a key element in understanding normal biological processes, and in both elucidating the basic alterations underlying disease processes, and in the design of novel rational therapeutic modalities to target these perturbations.

Acknowledgments Space constraints render it impossible to acknowledge all of the contributions made to this field by the many key individuals and groups who have studied different aspects of this research area in detail, and have made it necessary to refer the reader to reviews in many cases where the primary literature is very large. This work was supported by a post-doctoral fellowship from the Research Institute of the McGill University Health Centre (to N.B.), as well as by operating grants from the Canadian Cancer Society and the Canadian Institutes of Health Research (to X.-J.Y. and M.P.). M.P. holds the Diane and Sal Guerrera Chair in Cancer Genetics at McGill University.

References

Abella, J.V., Peschard, P., Naujokas, M.A., et al. (2005). Met/Hepatocyte growth factor receptor ubiquitination suppresses transformation and is required for Hrs phosphorylation. *Mol. Cell. Biol.* 25:9632–9645.

Bae, S.H., Jeong J.W., Park J.A., et al. (2004). Sumoylation increases HIF-1alpha stability and its transcriptional activity. *Biochem. Biophys. Res. Commun.* 324:394–400.

Bartkiewicz, M., Houghton, A., Baron, R. (1999). Leucine zipper-mediated homodimerization of the adaptor protein c-Cbl. A role in c-Cbl's tyrosine phosphorylation and its association with epidermal growth factor receptor. *J. Biol. Chem.* 274:30887–30895.

Bedford, M.T. and Leder, P. (1999). The FF domain: a novel motif that often accompanies WW domains. *Trends Biochem. Sci.* 24:264–5.

Benes, C.H., Wu, N., Elia, A.E., et al. (2005). The C2 domain of PKCdelta is a phosphotyrosine binding domain. *Cell* 121:271–280.

Berta, M.A., Mazure, N., Hattab, M., et al. (2007). SUMOylation of hypoxia-inducible factor-1alpha reduces its transcriptional activity. *Biochem. Biophys. Res. Commun.* 360:646–652.

Bertolaet, B.L., Clarke, D.J., Wolff, M., et al. (2001a). UBA domains mediate protein-protein interactions between two DNA damage-inducible proteins. *J. Mol. Biol.* 313:955–963.

Bertolaet, B.L., Clarke, D.J., Wolff, M., et al. (2001b). UBA domains of DNA damage-inducible proteins interact with ubiquitin. *Nat. Struct. Biol.* 8:417–422.

Bertos N.R., Wang, A.H., Yang, X.J. (2001). Class II histone deacetylases: structure, function, and regulation. *Biochem. Cell Biol.* 79:243–52.

Bhaumik, S.R., Smith, E., Shilatifard, A. (2007). Covalent modifications of histones during development and disease pathogenesis. *Nat. Struct. Mol. Biol.* 14:1008–1016.

Bloor, S., Ryzhakov, G., Wagner, S., et al. (2008). Signal processing by its coil zipper domain activates IKK gamma. *Proc. Natl. Acad. Sci. USA* 105:1279–1284.

Bode, A.M. and Dong, Z. (2004). Post-translational modification of p53 in tumorigenesis. *Nat. Rev. Cancer.* 4:793–805.

Boyault, C., Gilquin, B., Zhang, Y., et al. (2006). HDAC6-p97/VCP controlled polyubiquitin chain turnover. *EMBO J.* 25:3357–3366.

Bruick, R.K. and McKnight, S.L. (2001). A conserved family of prolyl-4-hydroxylases that modify HIF. *Science* 294:1337–1340.

Chen, L., Shinde, U., Ortolan, T.G., et al. (2001). Ubiquitin-associated (UBA) domains in Rad23 bind ubiquitin and promote inhibition of multi-ubiquitin chain assembly. *EMBO Rep.* 2:933–938.

Cheng, J., Kang, X., Zhang, S., et al. (2007). SUMO-specific protease 1 is essential for stabilization of HIF1alpha during hypoxia. *Cell* 131:584–595.

Christofk, H.R., Vander Heiden, M.G., Harris, M.H., et al. (2008a). The M2 splice isoform of pyruvate kinase is important for cancer metabolism and tumour growth. *Nature* 452:230–233.

Christofk, H.R., Vander Heiden M.G., Wu, N., et al. (2008b). Pyruvate kinase M2 is a phosphotyrosine-binding protein. *Nature* 452:181–186.

Cockman, M.E., Masson, N., Mole, D.R., et al. (2000). Hypoxia inducible factor-alpha binding and ubiquitylation by the von Hippel-Lindau tumor suppressor protein. *J. Biol. Chem.* 275:25733–25741.

Col, E., Caron, C., Seigneurin-Berny, D., et al. (2001). The histone acetyltransferase, hGCN5, interacts with and acetylates the HIV transactivator. Tat. *J. Biol. Chem.* 276:28179–28184.

Davies, G.C., Ettenberg, S.A., Coats, A.O., et al. (2004). Cbl-b interacts with ubiquitinated proteins; differential functions of the UBA domains of c-Cbl and Cbl-b. Oncogene 23:7104–7115.

Davletov, B.A. and Sudhof, T.C. (1993). A single C2 domain from synaptotagmin I is sufficient for high affinity Ca2+/phospholipid binding. *J. Biol. Chem.* 268:26386–26390.

DeManno, D.A., Cottom, J.E., Kline, M.P., et al. (1999). Follicle-stimulating hormone promotes histone H3 phosphorylation on serine-10. *Mol. Endocrinol.* 13:91–105.

Dhalluin, C., Carlson, J.E., Zeng, L., et al. (1999). Structure and ligand of a histone acetyltransferase bromodomain. *Nature* 399:491–496.

Donaldson, K.M., Yin, H., Gekakis, N., et al. (2003). Ubiquitin signals protein trafficking via interaction with a novel ubiquitin binding domain in the membrane fusion regulator, Vps9p. *Curr. Biol.* 13:258–262.

Durocher, D. and Jackson, S.P. (2002). The FHA domain. *FEBS Lett.* 513:58–66.

Ea, C.K., Deng, L., Xia, Z.P., et al. (2006). Activation of IKK by TNFalpha requires site-specific ubiquitination of RIP1 and polyubiquitin binding by NEMO. *Mol. Cell* 22:245–257.

Eissenberg, J.C. (2001). Molecular biology of the chromo domain: an ancient chromatin module comes of age. *Gene* 275:19–29.

Epstein, A.C., Gleadle, J.M., McNeill, L.A., et al. (2001). C. elegans EGL-9 and mammalian homologs define a family of dioxygenases that regulate HIF by prolyl hydroxylation. *Cell* 107:43–54.

Faber, P.W., Barnes, G.T., Srinidhi, J., et al. (1998). Huntingtin interacts with a family of WW domain proteins. *Hum. Mol. Genet.* 7:1463–1474.

Fischle, W., Wang, Y., Jacobs, S.A. et al. (2003). Molecular basis for the discrimination of repressive methyl-lysine marks in histone H3 by Polycomb and HP1 chromodomains. *Genes Dev.* 17:1870–1881.

Funakoshi, M., Sasaki, T., Nishimoto, T., et al. (2002). Budding yeast Dsk2p is a polyubiquitin-binding protein that can interact with the proteasome. *Proc. Natl. Acad. Sci. USA* 99:745–750.

Gayther, S.A., Warren, W., Mazoyer, S., et al. (1995). Germline mutations of the BRCA1 gene in breast and ovarian cancer families provide evidence for a genotype-phenotype correlation. *Nat. Genet.* 11:428–433.

Geiss-Friedlander, R. and Melchior, F. (2007). Concepts in sumoylation: a decade on. *Nat. Rev. Mol. Cell. Biol.* 8:947–956.

Girdwood, D., Bumpass, D., Vaughan, O.A., et al. (2003). P300 transcriptional repression is mediated by SUMO modification. *Mol. Cell* 11:1043–1054.

Goffeau, A., Barrell, B.G., Bussey, H., et al. (1996). Life with 6000 genes. *Science* 274:546, 563–567.

Goldstrohm, A.C., Albrecht, T.R., Sune, C., et al. (2001). The transcription elongation factor CA150 interacts with RNA polymerase II and the pre-mRNA splicing factor SF1. *Mol. Cell. Biol.* 21:7617–7128.

Gradin, K., Takasaki, C., Fujii-Kuriyama, Y., et al. (2002). The transcriptional activation function of the HIF-like factor requires phosphorylation at a conserved threonine. *J. Biol. Chem.* 277:23508–23514.

Grozinger, C.M. and Schreiber, S.L. (2000). Regulation of histone deacetylase 4 and 5 and transcriptional activity by 14-3-3-dependent cellular localization. *Proc. Natl. Acad. Sci. USA* 97:7835–7840.

Hewitson, K.S., McNeill, L.A., Riordan, M.V., et al. (2002). Hypoxia-inducible factor (HIF) asparagine hydroxylase is identical to factor inhibiting HIF (FIH) and is related to the cupin structural family. *J. Biol. Chem.* 277:26351–26355.

Hofmann, K. and Bucher, P. (1996). The UBA domain: a sequence motif present in multiple enzyme classes of the ubiquitination pathway. *Trends Biochem. Sci.* 21:172–173.

Hook, S.S., Orian, A., Cowley, S.M., et al. (2002). Histone deacetylase 6 binds polyubiquitin through its zinc finger (PAZ domain) and copurifies with deubiquitinating enzymes. *Proc. Natl. Acad. Sci. USA* 99:13425–13430.

Huang, L.E., Gu, J., Schau, M., et al. (1998). Regulation of hypoxia-inducible factor 1alpha is mediated by an O2-dependent degradation domain via the ubiquitin-proteasome pathway. *Proc. Natl. Acad. Sci. USA* 95:7987–7992.

Ivan, M. and Kaelin, W.G. Jr. (2001). The von Hippel-Lindau tumor suppressor protein. *Curr. Opin. Genet. Dev.* 11:27–34.

Iwai, K., Yamanaka, K., Kamura, T., et al. (1999). Identification of the von Hippel-lindau tumor-suppressor protein as part of an active E3 ubiquitin ligase complex. *Proc. Natl. Acad. Sci. USA* 96:12436–12441.

Jaakkola, P., Mole, D.R., Tian, Y.M., et al. (2001). Targeting of HIF-alpha to the von Hippel-Lindau ubiquitylation complex by O2-regulated prolyl hydroxylation. *Science* 292:468–472.

Jeanmougin, F., Wurtz, J.M., Le Douarin, B., et al. (1997). The bromodomain revisited. *Trends Biochem. Sci.* 22:151–153.

Kamura, T., Sato, S., Iwai, K., et al. (2000). Activation of HIF1alpha ubiquitination by a reconstituted von Hippel-Lindau (VHL) tumor suppressor complex. *Proc. Natl. Acad. Sci. USA* 97:10430–10435.

Kawaguchi, Y., Kovacs, J.J., McLaurin, A., et al. (2003). The deacetylase HDAC6 regulates aggresome formation and cell viability in response to misfolded protein stress. *Cell* 115:727–738.

Kelly, T.J., Qin, S., Gottschling, D.E., et al. (2000). Type B histone acetyltransferase Hat1p participates in telomeric silencing. *Mol. Cell. Biol.* 20:7051–7058.

Kerscher, O. (2007). SUMO junction-what's your function? New insights through SUMO-interacting motifs. *EMBO Rep.* 8:550–555.

Kim, S., Wong, P., Coulombe, P.A. (2006). A keratin cytoskeletal protein regulates protein synthesis and epithelial cell growth. *Nature* 441:362–365.

Koonin, E.V., Altschul, S.F., Bork, P. (1996). BRCA1 protein products. Functional motifs. *Nat. Genet.* 13:266–268.

Kozlov, G., Nguyen, L., Lin, T., et al. (2007). Structural basis of ubiquitin recognition by the ubiquitin-associated (UBA) domain of the ubiquitin ligase EDD. *J. Biol. Chem.* 282:35787–35795.

Krane, S.M. (2008). The importance of proline residues in the structure, stability and susceptibility to proteolytic degradation of collagens. *Amino Acids* 35:703–710.

Lancaster, D.E., McNeill, L.A., McDonough, M.A., et al. (2004). Disruption of dimerization and substrate phosphorylation inhibit factor inhibiting hypoxia-inducible factor (FIH) activity. *Biochem. J.* 383:429–437.

Lander, E., Linton, L., Birren, B., et al. (2004). Finishing the euchromatic sequence of the human genome. *Nature* 431:931–945.

Lando, D., Peet, D.J., Gorman, J.J., et al. (2002a). FIH-1 is an asparaginyl hydroxylase enzyme that regulates the transcriptional activity of hypoxia-inducible factor. *Genes Dev.* 16:1466–1471.

Lando, D., Peet, D.J., Whelan, D.A., et al. (2002b). Asparagine hydroxylation of the HIF transactivation domain a hypoxic switch. *Science* 295:858–861.

Latham, J.A. and Dent, S.Y. (2007). Cross-regulation of histone modifications. *Nat. Struct. Mol. Biol.* 14:1017–1024.

Layfield, R. and Searle, M.S. (2008). Disruption of ubiquitin-mediated processes in diseases of the brain and bone. *Biochem. Soc. Trans.* 36:469–471.

Lee, D.Y., Teyssier, C., Strahl, B.D., et al. (2005). Role of protein methylation in regulation of transcription. *Endocr. Rev.* 26:147–170.

Liang, X. and Van Doren, S.R. (2008). Mechanistic insights into phosphoprotein-binding FHA domains. *Acc. Chem. Res.* 41:991–999.

Liao, H., Byeon, I.J., Tsai, M.D. (1999). Structure and function of a new phosphopeptide-binding domain containing the FHA2 of Rad53. *J. Mol. Biol.* 294:1041–1049.

Lin, X., Sun, B., Liang, M., et al. (2003). Opposed regulation of corepressor CtBP by SUMOylation and PDZ binding. *Mol. Cell* 11:1389–1396.

Lisztwan, J., Imbert, G., Wirbelauer, C., et al. (1999). The von Hippel-Lindau tumor suppressor protein is a component of an E3 ubiquitin-protein ligase activity. *Genes Dev.* 13:1822–1833.

Liu, J., DeYoung, S.M., Hwang, J.B., et al. (2003). The roles of Cbl-b and c-Cbl in insulin-stimulated glucose transport. *J. Biol. Chem.* 278(38):36754–62.

Lo, Y.C., Lin, S.C., Rospigliosi, C.C., et al. (2009). Structural basis for recognition of diubiquitins by NEMO. *Mol. Cell* 33:602–615.

Long, J., Gallagher, T.R., Cavey, J.R., et al. (2008). Ubiquitin recognition by the ubiquitin-associated domain of p62 involves a novel conformational switch. *J. Biol. Chem.* 283:5427–5440.

Lowery, D.M., Mohammad, D.H., Elia, A.E., et al. (2004). The Polo-box domain: a molecular integrator of mitotic kinase cascades and Polo-like kinase function. *Cell Cycle* 3:128–131.

Lowery, D.M., Clauser, K.R., Hjerrild, M., et al. (2007). Proteomic screen defines the Polo-box domain interactome and identifies Rock2 as a Plk1 substrate. *EMBO J.* 26:2262–2273.

Lu, P.J., Zhou, X.Z., Liou, Y.C., et al. (2002). Critical role of WW domain phosphorylation in regulating phosphoserine binding activity and Pin1 function. *J. Biol. Chem.* 277:2381–2384.

Mahajan, R., Delphin, C., Guan, T., et al. (1997). A small ubiquitin-related polypeptide involved in targeting RanGAP1 to nuclear pore complex protein RanBP2. *Cell* 88:97–107.

Mahajan, A., Yuan, C., Lee, H., et al. (2008). Structure and function of the phosphothreonine-specific FHA domain. *Sci. Signal* 1:re12.

Mak, H.H., Peschard, P., Lin, T., et al. (2007). Oncogenic activation of the Met receptor tyrosine kinase fusion protein, Tpr-Met, involves exclusion from the endocytic degradative pathway. *Oncogene* 26:7213–7221.

Matunis, M.J., Coutavas, E., Blobel, G. (1996). A novel ubiquitin-like modification modulates the partitioning of the Ran-GTPase-activating protein RanGAP1 between the cytosol and the nuclear pore complex. *J. Cell Biol.* 135:1457–1470.

Maxwell, P.H., Wiesener, M.S., Chang, G.W., et al. (1999). The tumour suppressor protein VHL targets hypoxia-inducible factors for oxygen-dependent proteolysis. *Nature* 399:271–275.

McKinsey, T.A., Zhang, C.L., Olson, E.N. (2000). Activation of the myocyte enhancer factor-2 transcription factor by calcium/calmodulin-dependent protein kinase-stimulated binding of 14-3-3 to histone deacetylase 5. *Proc. Natl. Acad. Sci. USA* 97:14400–14405.

Min, J., Zhang, Y., Xu, R.M. (2003). Structural basis for specific binding of Polycomb chromodomain to histone H3 methylated at Lys 27. *Genes Dev.* 17:1823–1828.

Morris, D.P. and Greenleaf, A.L. (2000). The splicing factor, Prp40, binds the phosphorylated carboxyl-terminal domain of RNA polymerase II. *J. Biol. Chem.* 275:39935–39943.

Morrison, D.K. (2009). The 14-3-3 proteins: integrators of diverse signaling cues that impact cell fate and cancer development. *Trends Cell Biol.* 19:16–23.

Mujtaba, S., He, Y., Zeng, L., et al. (2004). Structural mechanism of the bromodomain of the coactivator CBP in p53 transcriptional activation. *Mol. Cell* 13:251–263.

Munshi, A., Shafi, G., Aliya, N., et al. (2009). Histone modifications dictate specific biological readouts. *J. Genet. Genomics* 36:75–88.

Newton, A.C. and Johnson, J.E. (1998). Protein kinase C: a paradigm for regulation of protein function by two membrane-targeting modules. *Biochim. Biophys. Acta* 1376:155–172.

Nishino, T.G., Miyazaki, M., Hoshino, H., et al. (2008). 14-3-3 regulates the nuclear import of class IIa histone deacetylases. *Biochem. Biophys. Res. Commun.* 377:852–856.

Ohh, M., Park, C.W., Ivan, M., et al. (2000). Ubiquitination of hypoxia-inducible factor requires direct binding to the beta-domain of the von Hippel-Lindau protein. *Nat. Cell Biol.* 2:423–427.

Olsen, J.V., Blagoev, B., Gnad, F., et al. (2006). Global, in vivo, and site-specific phosphorylation dynamics in signaling networks. *Cell* 127:635–648.

Passani, L.A., Bedford, M.T., Faber, P.W., et al. (2000). Huntingtin's WW domain partners in Huntington's disease post-mortem brain fulfill genetic criteria for direct involvement in Huntington's disease pathogenesis. *Hum. Mol. Genet.* 9:2175–2182.

Pawson, T. and Nash, P. (2003). Assembly of cell regulatory systems through protein interaction domains. *Science* 300:445–452.

Peschard, P. and Park, M. (2007). From Tpr-Met to Met, tumorigenesis and tubes. *Oncogene* 26:1276–1285.

Peschard, P., Fournier, T.M., Lamorte, L., et al. (2001). Mutation of the c-Cbl TKB domain binding site on the Met receptor tyrosine kinase converts it into a transforming protein. *Mol. Cell* 8:995–1004.

Peschard, P., Ishiyama, N., Lin, T., et al. (2004). A conserved DpYR motif in the juxtamembrane domain of the Met receptor family forms an atypical c-Cbl/Cbl-b tyrosine kinase binding domain binding site required for suppression of oncogenic activation. *J. Biol. Chem.* 279:29565–29571.

Polesskaya, A., Naguibneva, I., Duquet, A., et al. (2001). Interaction between acetylated MyoD and the bromodomain of CBP and/or p300. *Mol. Cell. Biol.* 21:5312–5320.

Prudden, J., Pebernard, S., Raffa, G., et al. (2007). SUMO-targeted ubiquitin ligases in genome stability. *EMBO J.* 26:4089–4101.

Raasi, S. and Pickart, C.M. (2003). Rad23 ubiquitin-associated domains (UBA) inhibit 26S proteasome-catalyzed proteolysis by sequestering lysine 48-linked polyubiquitin chains. *J. Biol. Chem.* 278:8951–8959.

Raasi, S., Varadan, R., Fushman, D., et al. (2005). Diverse polyubiquitin interaction properties of ubiquitin-associated domains. *Nat. Struct. Mol. Biol.* 12:708–714.

Rahighi, S., Ikeda, F., Kawasaki, M., et al. (2009). Specific recognition of linear ubiquitin chains by NEMO is important for NF-kappaB activation. *Cell* 136:1098–1109.

Rodriguez, M., Yu, X., Chen, J., et al. (2003). Phosphopeptide binding specificities of BRCA1 COOH-terminal (BRCT) domains. *J. Biol. Chem.* 278:52914–52918.

Sadowski, I., Stone, J.C., Pawson, T. (1986). A noncatalytic domain conserved among cytoplasmic protein-tyrosine kinases modifies the kinase function and transforming activity of Fujinami sarcoma virus P130gag-fps. *Mol. Cell. Biol.* 6:4396–4408.

Seigneurin-Berny, D., Verdel, A., Curtet, S., et al. (2001). Identification of components of the murine histone deacetylase 6 complex: link between acetylation and ubiquitination signaling pathways. *Mol. Cell. Biol.* 21:8035–8044.

Shih, S.C., Prag, G., Francis, S.A., et al. (2003). A ubiquitin-binding motif required for intramolecular monoubiquitylation, the CUE domain. *EMBO J.* 22:1273–1281.

Shiio, Y. and Eisenman, R.N. (2003). Histone sumoylation is associated with transcriptional repression. *Proc. Natl. Acad. Sci. USA* 100:13225–13230.

Smith, M.J., Kulkarni, S., Pawson, T. (2004). FF domains of CA150 bind transcription and splicing factors through multiple weak interactions. *Mol. Cell. Biol.* 24:9274–9285.

Sondermann, H. and Kuriyan, J. (2005). C2 can do it, too. *Cell* 121:158–160.

Sprangers, R., Groves, M.R., Sinning, I., et al. (2003). High-resolution X-ray and NMR structures of the SMN Tudor domain: conformational variation in the binding site for symmetrically dimethylated arginine residues. *J. Mol. Biol.* 327:507–520.

Stehmeier, P. and Muller, S. (2009). Phospho-regulated SUMO interaction modules connect the SUMO system to CK2 signaling. *Mol. Cell* 33:400–409.

Strahl, B.D. and Allis, C.D. (2000). The language of covalent histone modifications. *Nature* 403:41–45.

Tanimoto, K., Makino, Y., Pereira, T., et al. (2000). Mechanism of regulation of the hypoxia-inducible factor-1 alpha by the von Hippel-Lindau tumor suppressor protein. *EMBO J.* 19:4298–4309.

Tokunaga, F., Sakata, S., Saeki, Y., et al. (2009). Involvement of linear polyubiquitylation of NEMO in NF-kappaB activation. *Nat. Cell Biol.* 11:123–132.

Tomita, A., Towatari, M., Tsuzuki, S., et al. (2000). c-Myb acetylation at the carboxyl-terminal conserved domain by transcriptional co-activator p300. *Oncogene* 19:444–451.

Uhlik, M.T., Temple, B., Bencharit, S., et al. (2005). Structural and evolutionary division of phosphotyrosine binding (PTB) domains. *J. Mol. Biol.* 345:1–20.

Vadlamudi, R.K., Joung, I., Strominger, J.L., et al. (1996). p62, a phosphotyrosine-independent ligand of the SH2 domain of p56lck, belongs to a new class of ubiquitin-binding proteins. *J. Biol. Chem.* 271:20235–20237.

Wagner, S., Carpentier, I., Rogov, V., et al. (2008). Ubiquitin binding mediates the NF-kappaB inhibitory potential of ABIN proteins. *Oncogene* 27:3739–3745.

Wang, A.H., Kruhlak, M.J., Wu, J., et al. (2000a). Regulation of histone deacetylase 4 by binding of 14-3-3 proteins. *Mol. Cell. Biol.* 20:6904–6912.

Wang, P., Byeon, I.J., Liao, H., et al. (2000b). II. Structure and specificity of the interaction between the FHA2 domain of Rad53 and phosphotyrosyl peptides. *J. Mol. Biol.* 302:927–940.

Wang, Y., Reddy, B., Thompson, J., et al. (2009). Regulation of Set9-mediated H4K20 methylation by a PWWP domain protein. *Mol. Cell* 33:428–437.

Wilkinson, C.R., Seeger, M., Hartmann-Petersen, R., et al. (2001). Proteins containing the UBA domain are able to bind to multi-ubiquitin chains. *Nat. Cell Biol.* 3:939–943.

Wu, C.J., Conze, D.B., Li, T., et al. (2006). Sensing of Lys 63-linked polyubiquitination by NEMO is a key event in NF-kappaB activation. *Nat. Cell Biol.* 8:398–406.

Yaffe, M.B. and Elia, A.E. (2001). Phosphoserine/threonine-binding domains. *Curr. Opin. Cell Biol.* 13:131–138.

Yan, K.S., Kuti, M., Zhou, M.M. (2002). PTB or not PTB – that is the question. *FEBS Lett.* 513:67–70.

Yang, X.J. (2005). Multisite protein modification and intramolecular signaling. *Oncogene* 24:1653–1662.

Yang, X.J. and Seto, E. (2008). Lysine acetylation: codified crosstalk with other posttranslational modifications. *Mol. Cell* 31:449–461.

Yu, X., Chini, C.C., He, M., et al. (2003). The BRCT domain is a phospho-protein binding domain. *Science* 302:639–642.

Zhu, G., Wu, C.J., Zhao, Y., et al. (2007). Optineurin negatively regulates TNFalpha-induced NF-kappaB activation by competing with NEMO for ubiquitinated RIP. *Curr. Biol.* 17: 1438–1443.

Chapter 12
Regulation of Gene Expression by the Ubiquitin–Proteasome System and Implications for Neurological Disease

Lisa Lukaesko and Robert Meller

Abbreviations

APC	Adenomatous polyposis coli
ARD1	Arrest-defective 1 N-acetyltransferase
BARD-1	BRCA-associated RING domain protein
BMP	Bone morphogenetic protein
BRCA	Breast cancer susceptibility gene 1
CBP	CREB binding protein
CDK9	Cell cycle arrest associated protein 9
GSK-3β	Glycogen synthase kinase-3β
HAF	Hypoxia-associated factor
HAT	Histone acetyltransferase
HIF-1	Hypoxia-inducible factor-1
HOS	Homologue of slimb
MAD	Mothers against decapentaplegic
MDM2	Mouse double minute 2 protein
NES	Nuclear export signal
PHD	Prolyl hydroxylases
Pin1	Peptidylprolyl cis/trans isomerase
PML-NB	Promyelocytic leukemia nuclear bodies
PPIase	Peptidyl-prolyl cis–trans isomerase
pVHL	von Hippel Lindau tumor suppressor
RACK	Receptor of activated protein kinase
SAG	Sensitive to Apoptosis Gene
SCF	Skp-fbox-cullin
SUMO	Small ubiquitin related modifier
TGF	Transforming growth factor

R. Meller (✉)
RS Dow Neurobiology Laboratories and Cancer Research Laboratories,
Legacy Research, 1225 Ne 2nd Ave, Portland, OR 97232, USA
e-mail: rmeller@downeurobiology.org

C.J. Vidal (ed.), *Post-Translational Modifications in Health and Disease*, Protein Reviews 13,
DOI 10.1007/978-1-4419-6382-6_12,

TRRAP	Transcriptional histone acetyltransferase cofactor
UCP	E2-EPF Ubiquitin carrier protein
UPS	Ubiquitin–proteasome system
VDU2	p-VHL-interacting deubiquitinating enzyme 2
VEGF	Vascular endothelial growth factor
βTrCP	Beta transduction repeat containing protein

12.1 Introduction

This chapter will focus on the regulation of transcription factor function by the ubiquitin–proteasome system and its implications in neurological diseases. Many transcription factors and transcription co-factors are regulated by their rapid degradation by the ubiquitin–proteasome system, to enhance or inhibit gene expression. Following their prior phosphorylation, proteins, such as transcription factors, bind E3-ligases resulting in their ubiquitination. As such, transcription factors are the convergence point of multiple intracellular signaling pathways enabling the complex regulation of gene expression to various biological stimuli.

Many transcription factors are ubiquitinated in the cytoplasm following their export from the nucleus. In addition, the enrichment of ubiquitin and proteasome subunits in the promyelocytic leukemia nuclear bodies (PML-NB) sub-compartment of the nucleus suggests some transcription factors are subject to local ubiquitination and degradation in the nucleus. PML-NBs are dynamic regions that are in intimate contact with the cellular chromatin (for more information the reader is referred to some recent reviews (Fedorova and Zink 2008; Krieghoff-Henning and Hofmann 2008; Zimber et al. 2004)). The PML-NBs contain other post-translational modification enzymes, such as acetylases and protein kinases. The PML-NBs have an organized structure consisting of a capsule enriched in PML protein, small ubiquitin related modifier1 (SUMO1) and CREB binding protein (CBP), whereas the central lattice contains the proteasome subunits. As such, there is a concentration of nuclear ubiquitin–proteasome system components in close proximity to sites of gene expression and regulation.

12.2 Overview of the Ubiquitin–Proteasome System

The following is a brief overview of the ubiquitin–proteasome system to address key details (for more details the reader is directed to the previous chapters in this book). The degradation of proteins by the ubiquitin–proteasome system is a two-stage process (Fig. 12.1). First proteins are conjugated to ubiquitin, an 8–9 kDa protein present in all cells, by the sequential action of E1-, E2- and E3-ligase proteins. Second, the ubiquitinated protein is enzymatically degraded by the 26S proteasome, a multi-catalytic protease complex. The biological importance of the proteasome is

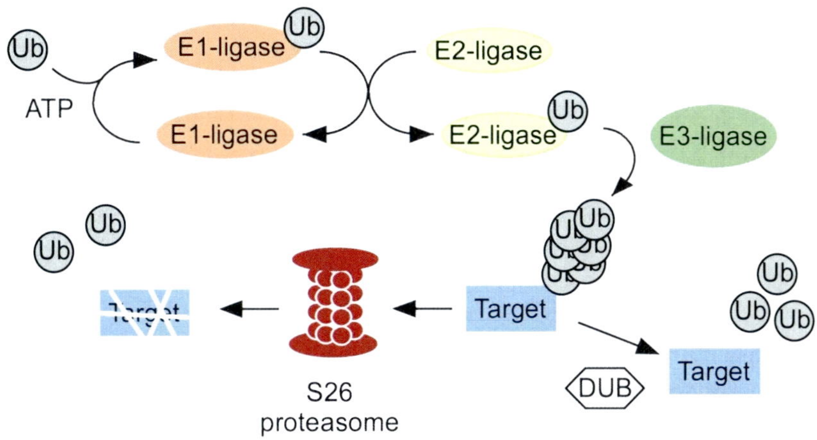

Fig. 12.1 Overview of protein ubiquitination and proteasomal degradation. The following is an overview of the common components of the ubiquitin–proteasome system. Ubiquitin (*gray*) is added to an E1-ligase protein (*orange*) in a reaction requiring ATP. The ubiquitin is transferred to the E2-ligase (*yellow*) and then transferred to the substrate protein (*blue box*) by an E3-ligase (*green*). Poly-ubiquitinated proteins are transferred to the S26 proteasome for degradation. De-ubiquitinating enzyme (DUB, *white hexagon*) removes ubiquitin modifications from substrate proteins. The colour and shapes of these components are consistent throughout Figs. 12.2–12.6

defined by the observation that deletion of 13/14 genes which make up the core 20S complex results in lethal phenotypes (Heinemeyer et al. 1991; Orlowski 1999). Protein degradation by 26S proteasome requires ATP. A chain of at least five ubiquitin molecules attached to the substrate is required for 26S proteasomal degradation (Thrower et al. 2000).

The specificity of the ubiquitination reaction is regulated by the E3-ligase which catalyses the transfer of the ubiquitin from the E2-ligase to either a Lys residue on the substrate protein or the N-terminus of the protein (Orlowski 1999; Pickart 2001). Two classes of E3-ligases have been defined; those containing either a HECT or a RING domain. RING E3-ligases can be single proteins, containing the substrate recognition site and the catalytic ligase domain in the single protein, or a multi-protein complex, whereby the recognition/binding protein is distinct from the catalytic protein (i.e. the SKIP/cullin/Fbox family; SCF). In addition, the E4-ligase regulates the poly-ubiquitination of a previously mono-ubiquitinated target substrate (Grossman et al. 2003). The different mechanisms of ubiquitin transfer to a target protein have been considered in the following review (Hochstrasser 2006).

The existence of six lysine residues within the ubiquitin amino acid sequence enables the formation of complex poly-ubiquitin chains with different topologies and functions (Varadan et al. 2002, 2004). The Lys48 poly-ubiquitin chain modification is associated with the proteasomal targeting of substrate proteins. In contrast, mono-ubiquitination regulates protein–protein interactions and signaling. The Lys63 poly-ubiquitin modification is associated with lysosomal degradation of proteins and the formation of signaling complexes (Ikeda and Dikic 2008). Because of the multiple biological roles of ubiquitination, when considering the evidence for a role of the ubiquitin–proteasome system in regulating gene expression, one must consider the evidence for the role of the proteasome (usually via proteasome inhibitors) as well as the type of modification added to the target protein. This is important for neurological studies because high levels or prolonged exposure to proteasome inhibitors induces neuronal cell death (Qiu et al. 2000).

12.3 Regulation of Gene Expression by the Ubiquitination of Histones

Eukaryotic DNA is tightly packaged into chromatin. The DNA is wound around nucleosomes, which consist of octamers of four histones (H2A, H2B, H3, H4). In addition to providing an efficient mechanism to store DNA, the structure of the nucleosomes also represses transcription by blocking promoter regions and preventing the binding of RNA polymerases. Ubiquitination of histones, which affects packing efficiency, has a significant impact on transcriptional regulation. Since the subject of ubiquitination modifications of histones has been reviewed elsewhere in this volume, we will give only a brief review of the subject.

Histones are modified through both mono-ubiquitination and poly-ubiquitination which regulate chromatin winding and integrate other post-translational modification events. The E3-ligases, Bre1, mouse double minute 2 (MDM2) and breast cancer suppressor gene 1 (BRCA1), have all been implicated in the ubiquitination of histones. First identified in *S. cerevisiae,* the E2-ligase Rad6 works in conjunction with Bre1 and MDM2 to ubiquitinate histones. Two potential mammalian homologues of Rad6, HR6A and HR6B, have also been identified (Baarends et al. 1999; Koken et al. 1991).

The most studied of the histone E3-ligases is the *S. cerevisiae* Bre1 (named Brl1 in *S. Pombe* and RNF20/40 in humans), a RING-finger type E3-ligase for H2B. Mono-ubiquitination of H2B by Bre/Rad1 is required for disassembly of nucleosomes during the RNA polymerase II-mediated elongation of transcripts. Bre/Rad1 is recruited to histones through a multi-step process. The histone chaperone FACT, a factor required for displacement of the H2A/H2B dimer from the core nucleosomes, recruits the transcription initiation complex Paf1 to chromatin. Chromatin bound Paf1 recruits Rad6 and Bre1 to promote ubiquitination of H2B on Lys120, which is required for efficient elongation by RNA polymerase II. Other studies have found that RNA polymerase II is required for the recruitment of Rad5 and Bre1 to histones promoting transcriptional elongation (Henry et al. 2003; Hwang et al. 2003; Kao et al. 2004; Robzyk et al. 2000; Wood et al. 2003; Xiao et al. 2005).

Murine double minute 2 (MDM2), an E3-ligase best known for its role in the ubiquitination of transcription factor p53, also represses transcription of p53-responsive genes through its modification of histones. MDM2 promotes ubiquitination of H2A and H2B in vitro and H2B in vivo in close proximity to p53 promoters, thereby inhibiting transcription of p53-responsive genes. This transcriptional repression was demonstrated to be independent of the role of MDM2 in ubiquitin-mediated degradation of p53 (Minsky and Oren 2004). In addition, BRCA1 promotes ubiquitination of H2A,

H2AX, and H2B in vitro (Chen et al. 2002; Mallery et al. 2002). However, the effects these modifications have on histone assembly and transcription are not clear.

Ubiquitination also regulates histone methylation to modify histone function. Ubiquitination of H2B by Rad6/Bre1 promotes H3 methylation on Lys4 and Lys79 (Cerretti et al. 1992; Weake and Workman 2008). H2B ubiquitination regulates di- and tri-methylation on these Lys residues on H3 and not mono-methylation (Dehe et al. 2005; Schneider et al. 2005; Shahbazian et al. 2005). Recent evidence has demonstrated that ubiquitination of H2A regulates the binding of the Cps35 subunit of a methylation complex, COMPASS (Lee et al. 2007). Recruitment of Cps35 to chromatin was impaired in Rad-deficient *S. cerevisiae* strains and methylation of Lys4 on H3 was reduced (Lee et al. 2007).

Deubiquitination of histones by the deubiquitinating enzyme (DUB) Ubp8 has also been shown to be important for regulating methylation of H2B (Wyce et al. 2007). De-ubiquitination of H2B promotes the recruitment of the kinase Ctk1 resulting in phosphorylation of RNA polymerase II (Wyce et al. 2007). This phosphorylation promotes the recruitment of the methyltransferase, Set 2 (Wyce et al. 2007). The regulation of histones is clearly a complex system, whereby multiple post translational modification events regulate access of transcription factors to the chromatin, thereby affecting gene expression patterns and rates.

12.4 Regulation of Transcription Machinery by the Ubiquitin–Proteasome System

12.4.1 Regulation of RNA Polymerase by the Ubiquitin–Proteasome System

The recruitment of RNA polymerase to the chromatin is a critical step in RNA synthesis. RNA polymerase is a large multi-subunit complex (labeled rpb 1–8 in yeast), which is regulated by additional protein co-factors. Specifically, RNA polymerase II is ubiquitinated by multiple E3-ligases, although not all ubiquitination events result in proteasomal degradation. This may be a result of different poly-ubiquitin linkage topologies, for example Lys48 and Lys63 (Varadan et al. 2004). Some common features of RNA polymerase II ubiquitination have been identified. The ubiquitination and proteasomal degradation of RNA polymerase II results in a reduction in RNA synthesis. First, ubiquitin is predominantly added to the largest polymerase subunit, rpb1, on the C-terminal domain. Up to six lysine residues in this region are substrates for ubiquitination (Li et al. 2007a). Second, phosphorylation of the Ser5 residue in the C-terminal domain is a prerequisite for ubiquitination (Daulny et al. 2008; Mikhaylova et al. 2008; Starita et al. 2005; Yasukawa et al. 2008). In addition, the ubiquitination of RNA polymerase II subunits, rpb1 and rpb 2, result in rpd4/7 dissociating from the RNA polymerase complex, further decreasing transcription (Daulny et al. 2008).

The E3-ligase Wwp2 may ubiquitinate rpb1 in pluri-potent stem cells, but in a non-phospho-Ser5 dependent manner (Li et al. 2007a). Wwp2 is a member of the

Nedd-4 family of HECT E3-ligases and interacts with the C-terminus domain of rpb1 similar to other E3-ligases that regulate damage-induced rpb1 ubiquitination (Li et al. 2007a). Nedd 4 E3-ligases either bind their substrates directly via a WW-domain interaction with a PPxY motif, or via linker proteins for WW-domain independent interactions (Foot et al. 2008). Wwp2 is also associated with the regulation of membrane ion channels (Foot et al. 2008). This suggests that a single E3-ligase, Wwp2, has multiple substrates with diverging cellular location and functions.

RNA polymerase II is a point of convergence of multiple signaling pathways following UV radiation-induced DNA damage and transcription arrest. BRCA-1 associates with BRCA1-associated RING domain protein (BARD-1) to ubiquitinate a number of proteins associated with DNA damage and repair. Mutations in this gene are associated with a high susceptibility to develop breast cancer. Following DNA damage BRCA1 ubiquitinates the major RNA polymerase subunit rpb1 (Starita et al. 2005). Ubiquitination of rpb1 is dependent on Ser5 phosphorylation (Starita et al. 2005). In addition, BRCA1 also ubiquitinates rpb8 to promote cell survival following DNA damage (Wu et al. 2007).

Additional E3-ligases have been associated with rpb1 ubiquitination following UV-light induced DNA damage and transcription arrest. Nedd4 was identified as a direct acting Rpb1 E3-ligase (Anindya et al. 2007), however, in the same assay BRCA1 did not induce Rpb ubiquitination. The mammalian Elongin complex has also been shown to regulate Rpb1 ubiquitination following UV exposure. Following damage, Elongin associates with the cul5/rbx2 module to promote rpb1 ubiquitination (Yasukawa et al. 2008). This seems counter to the established role of Elongin complex in reducing the pausing of RNA synthesis, thus enhancing transcription rates. However, further experiments show that Elongin deficient cells suppress UV-stimulated rpb1 ubiquitination (Yasukawa et al. 2008). Hence, Elongin may have dual roles: enabling efficient transcription under "normal" cellular circumstances, but assisting in transcription arrest following cell damage.

The von Hippel Lindau tumor suppressor (pVHL) is associated with regulation of HIF-1α (see Sect. 12.5.6 of this chapter). However, studies show that pVHL also ubiquitinates rpb1, dependent on Ser5 phosphorylation (Mikhaylova et al. 2008). Interestingly, this effect is also dependent on the prior hydroxylation of a proline residue by Prolin-hydroxylase (PHD1). PHD1 activity is enhanced under conditions of low oxidative stress. PHD1 hydroxylates Pro1465, enabling phosphorylation of the Ser5 residue and the subsequent binding of pVHL to rpb1 (Mikhaylova et al. 2008). pVHL-mediated rpb1 ubiquitination is not associated with degradation. This would suggest that pVHL-mediated poly-ubiquitination is via a Lys63-linked or another non-degradation associated poly-ubiquitin modification. Indeed, the pVHL–mediated ubiquitination of rpb1 enhances the recruitment of rpb1 to chromatin (Mikhaylova et al. 2008).

What is unclear from these studies on the role of RNA polymerase ubiquitination is whether the system has redundancy, or whether specific E3-ligases were identified because of the test system used to investigate the biology. However, these studies suggest that the regulation of RNA polymerase II by E3-ligases is biologically complex, depending on the context of the cell cycle timing and the biological stimuli.

12.4.2 Regulation of the Transcription Co-factors CBP/p300 by the Ubiquitin–Proteasome System

The CREB binding protein (CBP) and p300 are critical regulators of transcription by forming a bridging complex between chromatin-bound transcription factors, RNA polymerase and RNA helicase. CBP and p300 integrate signals from multiple transcription factors such as Smad, STAT, CREB, AP1 and c-myc transcription factors. Hence, their levels and functions are very tightly controlled.

CBP has both E3- and E4-ligase properties, but is also a target of E3-ligases, which regulate its function and degradation. In addition to their ubiquitin ligase roles, CBP and p300 acetylate histones, to relax chromatin and promote transcription, as well as non-histone proteins, to regulate their function (a general review of CBP has been provided by Barco and Kandel 2006). CBP functions are tightly regulated by phosphorylation by multiple protein kinases, some of which regulate E3-ligase targeting of CBP for ubiquitination.

CBP is regulated by the E3-ligase MDM2 (Sanchez-Molina et al. 2006). Following stimulation of PDGF receptors, ras activation leads to an intracellular signaling cascade resulting in ERK and Akt activation and the subsequent degradation of CBP. The decrease of CBP was hypothesized to alter transcription rates by increasing the local competition for CBP between several transcription factors (Sanchez-Molina et al. 2006). A number of phosphorylatable residues on CBP have been shown to regulate its function (Impey et al. 2002). However, it is unclear from this study whether the direct CBP phosphorylation regulated its ubiquitination. The knockdown of MDM2 with siRNA lowered MDM2 levels and caused a reciprocal enhancement of CBP levels and function. In addition, an MDM2 mutant (C246A) was unable to promote the degradation of CBP (Sanchez-Molina et al. 2006). The identification of MDM2 as an E3-ligase for CBP is interesting given that these two proteins interact to regulate p53 transcription (see Sect. 12.5.4).

p300 was identified as a substrate for the SCF Fbx3 E3-ligase in a proteomics analysis of K652 cells and confirmed by immunoprecipitation (Shima et al. 2008). Loss of Fbx3 by siRNA resulted in stabilization of p300 and HIPK2 (another transcription cofactor). Interestingly, p300 associates with PML protein in nuclear bodies, resulting in the stabilization of p300. The effect on HIPK2 was not via a blockade in ubiquitination, but rather preventing the proteasomal degradation of the ubiquitinated HIPK2 in the nuclear bodies. In contrast, the non-nuclear body-associated p300 and HIPK2 were degraded by the proteasome (Shima et al. 2008). The loss of PML blocks p53 activation, suggesting that PML blocks p300 degradation thereby enabling it to regulate p53-mediated transcription.

p300 has E4-ligase function resulting in the poly-ubiquitination of a previously mono-ubiquitinated protein, thereby promoting its proteasomal degradation (Grossman et al. 2003). In addition to having intrinsic E3/E4-ligase properties CBP and p300 have been shown to interact with other nuclear E3-ligases to help regulate their function. The anaphase promoting complex (APC) is a multi-subunit regulatory protein which has E3-ligase properties and requires CBP/p300 (Turnell et al. 2005).

The APC regulates the transition of G_1 phase in the cell cycle. The APC E3-ligase has a similar organization to the SCF multi-subunit E3-ligases, in that APC11 consists of the catalytic RING domain and APC2 forms the bridge between the substrate and catalytic domain. The substrate is brought to the APC via either cdh1 or cdc20, depending on the temporal context of the cell cycle. These substrate-binding proteins recognize overlapping targets (Turnell et al. 2005). Cdc20 functions in mitosis to promote ubiquitination of Securin, a scaffold protein, and cyclins. In contrast, cdh1 is utilized in late mitosis and early G_1 to regulate cyclin, cyclin kinase and other G_1 regulatory protein degradation. A loss of CBP in cells results in slower degradation of cyclin B and slow cell cycle progression (Turnell et al. 2005).

The APC however has other functions in regulating gene transcription, and this is brought about via its interaction with CBP/p300. Loss of CBP/p300 binding to the APC results in the loss of the p21 response to UV light-induced DNA damage, a p53-mediated response (Turnell et al. 2005). The effect of the APC is not via a direct p53-mediated mechanism but rather the regulation of transcription events (Turnell et al. 2005). In addition, association of the APC with p300 increases its histone 4 acetylase activity. Hence, CBP/p300 directly regulates protein ubiquitination and also regulates the ability of the APC to ubiquitinate targets.

The p300/CBP associated factor (PCAF) has an E3-ligase RING domain in its amino terminus (Linares et al. 2007). PCAF (HDM2) of mouse targets the human homologue of MDM2 for ubiquitination and degradation. In addition, PCAF acetylates HDM2. Acetylation of the HDM2 homologue MDM2 results in an enhancement of p53 function. However, MDM2 is not acetylated by PCAF, but rather by p300 and CBP (Wang et al. 2004b). The significance of this difference in substrates between the species is not yet clear. To further complicate the system, it has been shown that MDM2 ubiquitinates and promotes the degradation of PCAF (Jin et al. 2004). The unraveling of this system shows a highly complex whereby transcription co-factors, such as CBP/p300, are under very tight regulation. The regulation of CBP/p300 by MDM2 and the converse regulation of MDM2 by CBP/p300 suggest that these proteins function under tight regulation in response to various biological stimuli. This regulation involves integration and convergence of multiple post-translational modifications of CBP/p300 including acetylation, ubiquitination and phosphorylation.

12.5 Regulation of Transcription Factors by the Ubiquitin–Proteasome System

There are many examples of transcription factors falling under the regulation of the ubiquitin–proteasome system. In general, their function is reduced due to their rapid degradation. However some transcription factors use multiple E3-ligases to regulate their function, thereby integrating multiple signaling pathways. Recent studies are beginning to unravel the additional complexities of non-degradation signaling by the ubiquitin–proteasome system, whereby ubiquitination may affect the function of the transcription factor by regulating the binding of co-factors, or via the regulation of the cellular location of the transcription factor. Here we will focus on some specific transcription factor families and highlight the mechanisms whereby the ubiquitin–proteasome system may regulate their function.

12.5.1 Regulation of the Forkhead Transcription Factors

The Forkhead (winged helix) transcription factor family is defined by a large 100 amino acid binding domain. There are currently over 40 members encoded in the human genome (Katoh and Katoh 2004), which are the homolog of the DAF transcription factors described in *C. elegans*. Since their original identification, the nomenclature for the Forkhead proteins has changed, which can lead to some confusion in the literature: the common Forkhead proteins are FOXO-1 (FKHR), FOXO3a (FKHRL1), and FOXO4 (AFX). Gene deletion studies show that FOXO1 knockout mice are not viable past embryonic day 10, showing the importance of this transcription factor in development (Hosaka et al. 2004). In contrast, FOXO3a and FOXO4 knockouts are viable, and with no apparent defects (Arden 2008; Hosaka et al. 2004).

The Forkhead transcription factors are involved in the response of a cell to stress, nutrient deficiency and prevention of tumor progression. Activation of Forkhead transcription factor results in G1 arrest of cells due to the expression of cell cycle arrest proteins, such as p27kip1 (Stahl et al. 2002). In response to cell stressors, such as oxidative stress, seizures and ischemia, Forkhead regulates the expression of cell death mediators such as Fas ligand, TRAIL and the pro-apoptotic protein Bim (Gilley et al. 2003; Shinoda et al. 2004; Stahl et al. 2002). Members of the Forkhead family of transcription factors are regulated by multiple signaling pathways, resulting in the phosphorylation, ubiquitination, methylation and acetylation of Forkhead, each of which regulate Forkhead-mediated gene expression patterns. The reader is referred to some alternative reviews for further details of other post translational modifications of Forkhead (Vogt et al. 2005).

Forkhead transcription factors are activated following their phosphorylation by JNK in response to mild oxidative stress (Essers et al. 2004) (Fig. 12.2). The phosphorylation of Forkhead by JNK is on different residues to those phosphorylated by Akt, which turns off Forkhead-mediated gene expression. The stress activated GTPase ral activates JNK resulting in Thr447/451 phosphorylation of Forkhead (Essers et al. 2004). JNK-mediated phosphorylation of Forkhead induces gene expression and appears not to promote ubiquitination of Forkhead. When in the nucleus, the binding of active FOXO to CBP/p300 promotes acetylation of histone proteins, but may result in acetylation of FOXO. Acetylated FOXO recruits the deacetylase SIRT resulting in the preferential expression of GADD45, cell cycle and glucogenic genes, rather than pro-apoptotic genes such as Bim and Fas-ligand (Frescas et al. 2005).

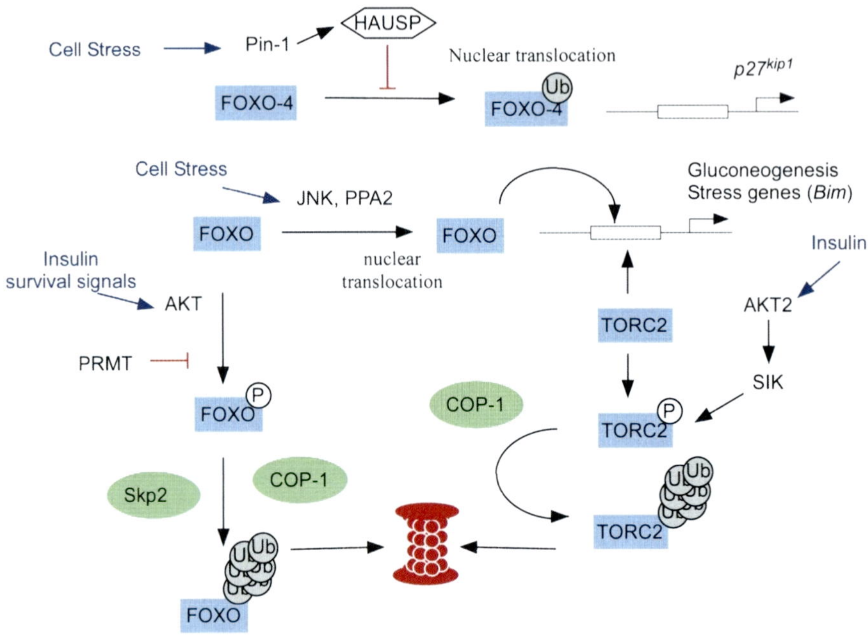

Fig. 12.2 Overview of forkhead transcription factor regulation by the ubiquitin–proteasome system. Forkhead transcription factors regulate gene expression in response to cell stress and also promote the expression of gluconeogenesis genes when glucose levels are low. Forkhead is normally repressed following its phosphorylation by Akt, which promotes its nuclear export and ubiquitination by Skp2. Forkhead is activated either by dephosphorylation by protein phosphatases or via phosphorylation of Forkhead by Jun N-terminal kinase (JNK) promoting its nuclear import. In response to conditions of high glucose availability Forkhead mediated-gluconeogenesis gene expression is rapidly turned off by the activation of Akt and the ubiquitination of Forkhead and its co-factor TORC2 by the E3-ligase Cop1. Forkhead proteins are also regulated by mono-ubiquitination, which affects their nuclear localization. The de-ubiquitinating enzyme HAUSP removes mono-ubiquitin modifications from Forkhead preventing the nuclear translocation of Forkhead (FOXO-4) and reducing gene expression

Forkhead FOXO-1 is inactivated by the phosphorylation of Ser253 by Akt (Brunet et al. 1999), where as IKKβ regulates FOXO3a phosphorylation (Hu et al. 2004). Phosphorylated Forkhead binds to 14-3-3 protein and chromosome region maintenance protein 1 and is exported from the nucleus (Kau et al. 2003). Phosphorylation blocks the nuclear location sequence in the Forkhead protein. In addition, Forkhead can be activated following its de-phosphorylation by protein phosphatase 2A (PP2A) (Yan et al. 2008). Forkhead is released from 14-3-3 revealing a nuclear localization sequence (NLS) motif resulting in translocation to the nucleus, where it can bind to its promoter region. Interestingly, Akt phosphorylation of Forkhead is prevented by methylation of two Arg residues (Arg 248/250) close to the Ser 253 site by PRMT revealing yet another layer of regulatory complexity of this transcription factor (Yamagata et al. 2008).

In addition to exporting Forkhead from the nucleus, Akt-induced phosphorylation of Forkhead promotes its degradation (Matsuzaki et al. 2003) (Fig. 12.2). Cytoplasmic FOXO-1 is ubiquitinated by Skp2, a member of the skp1/cullin/F-box E3-ligase family of proteins, which may play a role in tumorogenesis (Huang et al. 2005). Whether 14-3-3 facilitates ubiquitination or the assembly of the ubiquitination E3-ligase complex is not yet known. Alternatively 14-3-3 may protect Forkhead from the E3-ligase, preserving some of the cells Forkhead complement.

In response to cell survival signals, FOXO3a is ubiquitinated by MDM2 resulting in the proteasomal degradation of FOXO3a. Phosphorylation of FOXO3a on Ser294, Ser344 and Ser425 residues mediates the interaction between MDM2 and FOXO3a (Yang et al. 2008). Loss of these residues stabilizes the FOXO3a protein.

Forkhead proteins also play a role in the expression of genes associated with gluconeogenesis, and these studies have revealed an additional Forkhead E3-ligase, Cop1 (Fig. 12.2). In response to insulin, the E3-ligase Cop1 is expressed and this serves as an E3-ligase for FOXO-1, thereby reducing the expression of gluconeogenesis genes (Kato et al. 2008). For Forkhead to be ubiquitinated, it must be first phosphorylated on Ser256 by Akt. Akt2-mediated phosphorylation of FOXO, following insulin stimulation, results in FOXO ubiquitination by Cop1 (Kato et al. 2008). In addition to regulating Forkhead ubiquitination, Cop1 also regulates the ubiquitination of the Forkhead co-factor TORC2 (Dentin et al. 2007). Activation of Akt by insulin results in the phosphorylation and activation of the Ser/Thr protein kinase SIK (also known as Probable serine/threonine-protein kinase SNF1LK) on Ser378. Activated SIK phosphorylates TORC2 on Ser171 enhancing its Cop1-mediated ubiquitination and proteasomal degradation. This interesting example shows how a single E3-ligase may be responsible for the ubiquitination of multiple proteins to selectively turn down a specific gene expression pattern.

The two examples of E3-ligase identified for Forkhead highlight some of the challenges in identifying E3-ligases for a given protein substrate. Multiple E3-ligases can regulate protein ubiquitination when activated by different biological stimuli. For example Skp2 was identified in cancer cells and may regulate an anti-oncogenic phenotype and in contrast Cop1 was activated by insulin. However, care must be taken when interpreting such studies, as it is not yet clear what degree of redundancy E3-ligases may have, nor how truly selective an E3-ligase is for a given substrate protein.

Ubiquitination does not just promote degradation of Forkhead proteins. Mono-ubiquitination may also regulate Forkhead function. AFX/FOXO4 mono-ubiquitination results in its nuclear accumulation. Following cell stress, peptidylproline cis/trans isomerase 1 (Pin1) is activated and interacts with the deubiquitinating enzyme HAUSP/USP6 to attenuate FOXO-4 mono-ubiquitination and its nuclear accumulation (Brenkman et al. 2008). Blocking nuclear translocation of FOXO4 results in an attenuation of $p27^{Kip1}$ transcription. Interestingly, the FOXO-1 E3-ligase Skp2 also ubiquitinates the gene target of FOXO-4, $P27^{Kip1}$, as well as other cell cycle arrest associated proteins (CDK9) (Kiernan et al. 2001; Wang et al. 2003). Hence, a single E3-ligase can regulate a transcription factor family member and a family member gene product.

Taken together these examples show the rich crosstalk between intracellular signaling pathways which regulate the function of the Forkhead transcription factor family. Clearly much still remains to be investigated with respect to how different E3-ligases regulate differential gene expression depending on the biological stimuli and how Forkhead mitochondrial gene expression may also be regulated by the ubiquitin–proteasome system (Jacobs et al. 2008).

12.5.2 Regulation of bZIP Transcription Factors by the UPS

The AP1 transcription factors, also known as leucine zipper family, are regulated by ubiquitin-dependent proteasomal degradation. The AP1 family consists of the transcription factors Fos, Jun, CREB and ATFs. These transcription factors act as either homo- or heterodimers to regulate gene expression responses to multiple stimuli, including cytokines, growth factors, stress, and infections (for a more detailed review of AP1 transcription factors see Hai and Hartman 2001).

Fos is the protein encoded by the *c-fos* gene. Fos is known as an immediate-early gene due to its ability to be rapidly expressed and translated following stimuli. Fos protein is regulated by the E3-ligase Ubr1 in a signal transducer and activator of transcription protein (STAT)-dependent manner (Sasaki et al. 2006). The nuclear export and ubiquitination of Fos is regulated following the phosphorylation of Thr232 in the nuclear localization sequence NLS by Erk5 (Sasaki et al. 2006). Interestingly, in one of the earlier works studying the ubiquitin–proteasome system, a non-Ubr E3-ligase was also shown to ubiquitinate Fos protein (Stancovski et al. 1995). Ubr is an E3-ligase associated with N-end ubiquitination. This more exotic ubiquitination modification does not result in lysine residue modification, but is associated with the regulation of degron signals on proteins. Fos is an inherently unstable protein due to two degron motifs at the N- and C-termini, which promotes its degradation in a ubiquitin-independent manner (Basbous et al. 2007, 2008). Fos can also be ubiquitinated creating a second faster mechanism for its degradation by the proteasome and hence turning off Fos signaling. The ubiquitin-independent mechanism is inhibited by Erk1/2 phosphorylation of the C terminal domain on Ser363 and Ser374 (Basbous et al. 2007). One interesting hypothesis suggests that the 20S proteasome could partner with different 19S, 11S and an alternative 11S proteasome cap sub-units to form homogeneous or heterogeneous proteasome complexes, which may result in different proteasomal activities and requirements for substrates (Basbous et al. 2008; Rechsteiner and Hill 2005). It is generally believed that the 19S cap is required for Lys48-linked poly-ubiquitin-mediated degradation. Hence, alternative configurations of proteasome cap and core subunits may enable the degradation of non-K48 linked poly-ubiquitinated proteins and non-ubiquitinated proteins.

Fos is not the only AP1 transcription factor regulated by N-end rule ubiquitination. ATF5, which regulates suppression of gene expression, is also regulated by N-end rule ubiquitination (Wei et al. 2008). A number of E3-ligases have been associated with N-end ubiquitination, especially the Ubr class of E3-ligases. The chemotoxic agent cisplatin induces cell death via a number of mechanisms, but has been shown to stabilize ATF5 protein levels via the reduced ubiquitination of ATF5 (Wei et al. 2008). Interestingly cisplatin treatment resulted in a relocation of cdc34 (aka UbcH3) from the nucleus to the cytoplasm, thereby depleting the E2-ligase necessary for ATF5 ubiquitination (Wei et al. 2008). This novel mechanism of action of cisplatin opens up the possibility that E2-ligases may also be molecular/ drug targets for regulating ubiquitin–proteasome system-regulated cellular events.

Three E3-ligases have been identified for Jun: Fbw7, hCOP1/hDET and Itch (Bianchi et al. 2003; Gao et al. 2004; Nateri et al. 2004). Fbw7 and Itch are down-regulated by UV light, resulting in stabilization of c-jun levels (Anzi et al. 2008). In addition to ubiquitinating c-jun the Fbw7/hCdc4 SCF E3-ligase complex also ubiquitinates c-Myc, cyclin E and Notch transcription factors (Nateri et al. 2004; Welcker et al. 2004). Fbwx7-mediated c-jun degradation results in a loss in apoptotic signaling in neurons, and conversely the loss of Fbw7 results in stabilization of c-jun and neuronal apoptosis (Nateri et al. 2004). The effect of Fbxw7 knock-down is mitigated by overexpression of the inhibitory JNK interacting protein JIP (Nateri et al. 2004). To date it is not clear which residues are phosphorylated to regulate Jun degradation, both JNK and CSK3 (COOH-terminal Src kinase) phosphorylate Jun, thus promoting its ubiquitination (Nateri et al. 2004; Wei et al. 2005; Zhu et al. 2006).

Cyclic AMP mediated transcription events are mediated by CRE-response element binding protein (CREB). The transcriptional activity of CREB is enhanced following its phosphorylation on a critical Ser133 residue, which promotes its association with CBP (Barco and Kandel 2006; Johannessen et al. 2004; Mayr and Montminy 2001). Other residues on CREB are also phosphorylated resulting in enhanced or repressed transcriptional activity (Kornhauser et al. 2002). Activation of CREB results in the transcription of many genes, but interestingly it also drives the expression of its inhibitory proteins cAMP response element modulator (CREM) and inducible cAMP early repressor (ICER) (Barco and Kandel 2006). Both CREB and its repressor proteins are ubiquitinated by E3-ligases promoting their rapid degradation.

It is not yet clear which E3-ligase ubiquitinates CREB, but CREB ubiquitination and degradation is reported following hypoxia and PDGF-BB treatment of pulmonary smooth muscle cells (Garat et al. 2006). This effect is regulated by the Akt-mediated phosphorylation of Ser103 and Ser107 residues. Phosphorylation of this region results in nuclear export of CREB and its subsequent degradation by the proteasome. The export of CREB from the nucleus is blocked by Leptomycin B, which prevents the degradation of CREB (Garat et al. 2006). Akt is associated with the activation of CREB via the phosphorylation of the Ser133 residue (Perkinton et al. 2002; Walton and Dragunow 2000). In addition, brief ischemic conditions in brain cells activate both Akt and CREB, resulting in new gene expression and protection (Meller et al. 2005; Noshita et al. 2001; Tanaka et al. 1999; Walton et al. 1996). It is not clear how Akt signaling switches from promoting CREB activation to regulating the nuclear export and proteasomal degradation of CREB.

Functionally, the proteasome regulation of CREB may play a role in the development of long-term potentiation (LTP), a cellular model of learning. Following brief tetanic stimulation, the response of a neuron to a given stimulation is increased. Many complex molecular pathways regulate LTP. Proteasome inhibitors have been shown to enhance the induction, but impair the maintenance of late phase LTP in the hippocampus (Dong et al. 2008). Proteasome inhibitors appear to block CREB function in LTP by preventing the degradation of the CREB repressor ATF4, reducing expression of BDNF (Dong et al. 2008). In the Aplysia model of LTP, serotonin induces the degradation of CREB repressors via the proteasome in a protein kinase C-dependent manner (Upadhya et al. 2004). These two studies suggest that under conditions of enhanced CREB function, its repressors are rapidly degraded.

The CREB ICER has been shown to be under tight regulation by the ubiquitin–proteasome system. Activation of MAPK has been shown to drive ICER phosphorylation (Ser41) and its ubiquitination in JEG-3 and mouse pituitary AtT20 cells (Yehia et al. 2001). MAPK also regulates CREB transcription via phosphorylation of CREB, CBP and affecting CBP recruitment to the CRE (Johannessen et al. 2004; Meller et al. 2005). ICER phosphorylation was blocked by MAPK inhibitors (PD98059) and cAMP (Yehia et al. 2001). Consistent with phosphorylation of ICER regulating the stability of this protein, loss of the Ser41 site on ICER prolongs the half life of the ICER protein (Yehia et al. 2001).

The SCF (skp1-fbox-cullin) complex has been identified as an ICER E3-ligase in yeast studies and utilizes the E2-ligases rad6 (UbcH2) and cdc34 (UbcH3) (Pati et al. 1999). Interestingly, loss of these two specific E2 ligases resulted in elevated levels of ICER, suggesting that both cdc34 and rad6 are necessary for ICER ubiquitination. This study shows how E2-ligases, in addition to E3-ligases, may regulate the target specificity of the ubiquitination reaction.

12.5.3 Regulation of Smad Transcription Factors by the UPS

The Smad family of transcription factors are activated by transforming growth factor (TGF) and bone morphogenic protein (BMP). The mammalian Smad proteins are homologues of the drosophila mother against decapentaplegic (MAD) and the *C. elegans* SMA transcription factors. These transcription factors are broken down into three different subgroups depending on their function. The receptor Smads (rSmads) interact with activated TGF receptors (Smad2 and 3) or BMP receptors (Smad1, 5 and 8). Following activation of their respective receptors, the receptor associated Smads (rSmads) are phosphorylated, bind to Smad4 and move to the nucleus. Once in the nucleus they bind to the Smad promoter and recruit additional transcription factors and cofactors such as CBP/p300 to initiate TGF and BMP-mediated gene expression (Shen et al. 1998; Yang et al. 2000; Zhang et al. 1998). Smad signaling is blocked by inhibitor Smads (iSmads) 6 and 7, which compete for the binding site at the receptor. In addition, the nuclear actions of active Smads are inhibited by SnoN and Ski (He et al. 2003). For a more detailed review, the reader is referred to the following review articles (Inoue and Imamura 2008; Ross and Hill 2008; Schmierer and Hill 2007).

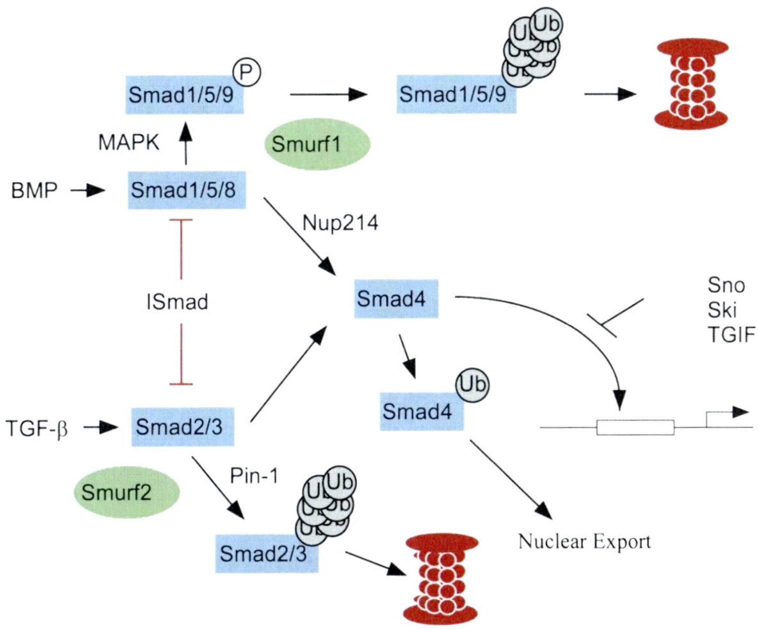

Fig. 12.3 Overview of Smad transcription factor regulation by the ubiquitin–proteasome system. Activation of receptor associated Smads by either TGF (Smad 2 and 3) or BMP receptors (Smads 1, 5 and 8) results in binding to a co-activator Smad4 and gene expression. Smad gene expression is inhibited by inhibitory co-factors Sno, Ski and TGIF. Smads are ubiquitinated by Smurf1 and Smurf2. Smurf1 ubiquitinates MAPK phosphorylated Smad1 associated with BMP signaling. Smurf2 ubiquitinates Smad2 and requires a co-factor Pin-1. In addition, mono-ubiquitination of SMAD4 results in its nuclear export

The Smad system is highly regulated by the ubiquitin–proteasome system (Inoue and Imamura 2008) (Fig. 12.3). Smad proteins are ubiquitinated by the HECT E3-ligase Smad ubiquitination regulatory factor1 (Smurf1), Smurf2 and a multi-subunit RING-type E3-ligase, ROC1-SCF (Fbw1a) (Fukuchi et al. 2001; Li et al. 2004). The E3-ligase CHIP has been shown to ubiquitinate Smad1 and Smad4 independent of TGF receptor activation (Li et al. 2005a). Additional E3-ligases associated with the poly-ubiquitination and degradation of Smad proteins are WWP1, Itch, Arkadia, SCFbwi1, SCF skp2, APC and ectodermin (Inoue and Imamura 2008). Here we will outline the role of some of these E3-ligases, but clearly the large number of E3-ligases with the potential to regulate Smad signaling emphasizes the complex regulation of this developmentally important pathway.

The E3-ligase Smurf1 ubiquitinates Smad1, 5 and 7, whereas Smurf2 ubiquitinates Smad1 and 2 as well as the inhibitor protein SnoN (Inoue and Imamura 2008). In addition, Smurf1 prevents nuclear import of active Smad1 by blocking its interaction with the nuclear import factor NUP214 (Sapkota et al. 2007). Ubiquitination of Smad proteins is associated with their degradation, thereby attenuating either TGF or BMP signaling.

Smurf binds to Smads promoting their ubiquitination and preventing signaling from upstream receptors (Fig. 12.3). Smurf-mediated Smad ubiquitination is regulated by prior phosphorylation of the substrate protein, similar to many protein ubiquitination pathways. The Smurf WW2 domain has been shown to bind to a PPXY motif in the linker region of Smads (Sangadala et al. 2007). In addition, Pin1, a peptidyl-prolyl cis–trans isomerase (PPIase) appears to be a necessary co-factor to regulate Smad ubiquitination by Smurf (Nakano et al. 2009). Pin-1 binds to Smad2 and 3, but not 4, due to their phosphorylation on a ST-P motif. Phosphorylation of this motif increases the association of Smads with Smurf (Nakano et al. 2009). Since Pin-1 enhances Smad ubiquitination, this suggests that Pin-1 may function as a regulator of protein ubiquitination, not only for Smads, but other proteins whose ubiquitination is regulated by phosphorylation.

Smad3 is ubiquitinated by ROC1-SCF, promoting Smad3 degradation in HaCaT human keratinocyte cells (Fukuchi et al. 2001). Smad3 ubiquitination was enhanced following TGF receptor stimulation, and resulted in a direct interaction between ROC1and the C-terminal MH2 domain of Smad3. p300 has biphasic effects on TGF signaling and Smad function. Binding of CBP to the Smad3 complex promotes Smad3-mediated gene expression. However, binding of Smad3 to CBP also promotes the ubiquitination of Smad3 by Roc1-SCF (Fukuchi et al. 2001). This may be part of a mechanism to "turn off" the TGF signal.

Smurfs ubiquitinate the trans-membrane receptors that activate Smad signaling. Both Smurf1 and 2 associate with Smad7 to promote the ubiquitination of both Smad 7 and the bound TGF receptor (Ebisawa et al. 2001; Kavsak et al. 2000). Smurfs are localized in the nucleus, but binding of the E3-ligase to Smad7 promotes their export to the membrane surface. Smad7 binding to Smurf may also enhance its ability to bind and utilize the E2-ligase UbcH7 (Ogunjimi et al. 2005). Smurfs can also facilitate Smad signaling via the degradation of the inhibitory factor SnoN. Smad2 activation promotes Smurf to ubiquitinate the inhibitory co-repressor SnoN (Bonni et al. 2001). Activated Smad3 also targets SnoN for degradation, but via the APC E3-ligase (Stroschein et al. 2001).

Additional E3-ligases shown to regulate Smad2 function are Nedd4-2 and Itch. Nedd4-2 ubiquitinates Smad2, reducing Smad2 mediated gene expression. Nedd 4-2 also ubiquitinates the TGF receptor 1 when bound to Smad2, and Smad4 when bound to Smad7 (Kuratomi et al. 2005). The protein kinase Itch, which is associated with NF-κB signaling, also regulates Smad2 function. Itch induces phosphorylation of Smad2, which increases its interaction with TGF receptor 1 and Smad7 to decrease TGF signaling (Bai et al. 2004).

Smad7 may play an important role in integrating Wnt/β-catenin signaling into the Smad signaling pathway. The Wnt signaling protein Axin binds to the E3-ligase Arkadia and Smad7 to promote Smad7 ubiquitination, thereby enhancing Smad signaling (Liu et al. 2006). Reducing Axin or Arkadia expression increases the stability of Smad7. Axin is inhibited by Wnt1 and Axin overexpression attenuates Smad7 degradation (Liu et al. 2006). In addition, it has been shown that Arkadia can also ubiquitinate Sno and Ski (Nagano et al. 2007). Interestingly, the binding of Axin to Smad7 competes with β-catenin (Tang et al. 2008). Smad7 binding to

Axin also displaces GSK3β and Smurf2 from the Axin complex, which reduces β-catenin phosphorylation and degradation (Tang et al. 2008). However the Smad7 stabilized β-catenin is preferentially transferred to the cell membrane, rather than the nucleus, where it interacts with cadherin proteins increasing cell-cell adhesion signals (Tang et al. 2008). As such, these examples show the complex regulation between transcription factor systems which are regulated by the ubiquitin–proteasome system.

In addition to regulating Smad function by poly-ubiquitination, mono-ubiquitination may also regulate Smad4 biological activity. Smad can be mono-ubiquitinated in the C-terminus domain (Lys507) promoting its binding to rSmads and enhancing transcriptional activity (Moren et al. 2003). In contrast to Lys507, Lys519 mono-ubiquitination has been shown to inhibit Smad4 function by blocking its interaction with phospho-Smad2. FAM/USP9x was identified by siRNA screen of Smad4 function, and ectodermin/Tif1γ is a potential mono-ubiquitin E3-ligase (Dupont et al. 2009). What is unclear is whether one form of mono-ubiquitination is dominant over the other form. In addition the E3-ligase which regulates the Lys507 modification has not yet been identified. These studies suggest that mono-ubiquitination of residues may have similarity to phosphorylation of residues on a protein, whereby seemingly adjacent phosphorylation sites have opposing effects on the function of the protein.

12.5.4 Regulation of the p53 Pathway by the UPS

The p53 pathway is a critical regulator of the response to cellular stressors (i.e. DNA damage, chromosomal aberrations, telomere erosion, hypoxia, or oncogenic responses) (Michael and Oren 2003). In unstressed cells, the p53 is shut off. In response to cellular stressors, the p53 pathway is activated thereby shutting down the multiplication of cells through cell cycle arrest, apoptosis, and as more recently shown, necrosis (Tu et al. 2009). An essential component of this response is the regulation of the levels and the activity of the transcription factor, p53. Upon activation of the pathway, p53 organizes into a tetramer, translocates into the nucleus where it upregulates and downregulates genes that are essential mediators of cell cycle arrest and apoptosis (for current reviews of p53 responsive genes, see references Riley et al. 2008; Bunz et al. 1998).

The levels of p53 are kept low in unstressed cells, but upon the introduction of stressors the levels of p53 dramatically increase (Kastan et al. 1991; Maltzman and Czyzyk 1984). The p53 pathway uses the ubiquitin–proteasome system to regulate p53 and a number of E3-ligases have been implicated in this process, including Pirh2, COP1, TOPORS, ARF-BP1, and MDM2 (Chen et al. 2005; Dornan et al. 2004; Honda et al. 1997; Leng et al. 2003; Rajendra et al. 2004). The most studied of these is the MDM2 which provides many examples of how ubiquitin–proteasome system is used to regulate transcription factors (Fig. 12.4). MDM2 regulates p53 levels through a negative feedback loop where p53 induces the expression of

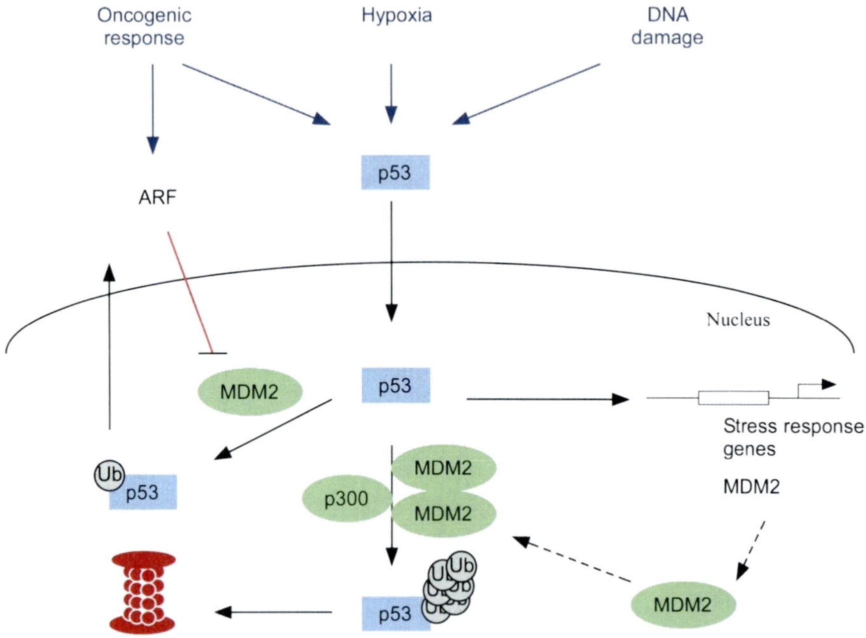

Fig. 12.4 Overview of p53 transcription factor regulation by the ubiquitin–proteasome system. p53-mediated gene expression is activated in response to oncogenic stressors, hypoxia and DNA damage. p53 regulates the expression of its own E3-ligase MDM2. When MDM2 levels are high in conjunction with p300, p53 is poly-ubiquitinated and degraded by the proteasome. When levels of MDM2 are low, p53 is mono-ubiquitinated which drives its export from the nucleus. The p53-regulated gene ARF inhibits MDM2 function

MDM2 (Barak et al. 1993; Wu et al. 1993). MDM2 then binds to p53 and promotes inactivation and proteasome-mediated degradation of p53. The level of p53 reduces the overall transcriptional activity of p53 and hence the levels of MDM2 transcribed. Continual p53 promoted expression is likely required as MDM2 is a very short-lived protein due to its self-ubiquitination and degradation (Chang et al. 1998). Interestingly, an in vitro assay using recombinant proteins showed that MDM2 modifies p53 through covalent attachment of mono-ubiquitin to multiple lysines of p53, although modification through a chain of Lys48-linked ubiquitin is associated with proteasomal-mediated degradation (Fang et al. 2000; Honda and Yasuda 2000). However, it is possible that in vivo MDM2 could be modified by p53 through poly-ubiquitination by an E4-enzyme.

The regulation of p53 export out of the nucleus may also be regulated by MDM2. Co-expression of MDM2 along with p53 promoted nuclear export of p53 (Boyd et al. 2000; Geyer et al. 2000). Furthermore, an expressed ubiquitin–p53 fusion protein promoted nuclear export of p53. It has been proposed this ubiquitination of p53 by MDM2 causes a conformation change in p53 and exposes or activates the nuclear export signal (NES) in p53, thereby promoting nuclear export of p53 (Gu et al. 2001; Lohrum et al. 2001). This nuclear shuttling of p53 by MDM2 is highly dependent on the type of ubiquitin modification. Expression of varying levels of MDM2 along with p53 in the p53-null human lung carcinoma cell line revealed that low levels of MDM2 promoted mono-ubiquitination and rapid translocation of p53 out of the nucleus, whereas high levels of MDM2 promoted poly-ubiquitination and nuclear degradation of p53 (Li et al. 2003). The necessity of having a two-step process for ubiquitination of p53 is unknown, but translocation of p53 to the cytoplasm could prevent any extraneous transcriptional activity by p53 in a rapid and reversible fashion (Brooks et al. 2004).

The interactions between p53 and MDM2 are modulated by a number of proteins. One known modulator of this interaction is the tumor suppressor, Arf (Arf is known as p19Arf in mouse and p14Arf in humans). Arf is induced in response to oncogenic signals, such as overexpression of the oncogenes E1A, Myc and Ras (de Stanchina et al. 1998; Palmero et al. 1998; Zindy et al. 1998). Inactivation of Arf is known to commonly occur during cancer development and its inactivation is associated with upregulation of oncogenes such as retinoblastoma (Chang et al. 2007; Matheu et al. 2008). Arf prevents MDM2-mediated degradation of p53 by binding to MDM2 thereby inhibiting MDM2 activity as well as preventing export of the p53-MDM2 complex out of the nucleus (Honda et al. 1997; Midgley et al. 2000; Zhang and Xiong 1999).

Another important modifier of MDM2 activity is the transcriptional co-activator p300/CBP which has E3-ligase activity (Fig. 12.4). In vitro, MDM2 alone can only promote the mono-ubiquitination of p53, but co-expression of MDM2 with p300 promotes poly-ubiquitination of p53 (Grossman et al. 2003; Lai et al. 2001). Although it was not examined whether the proteasome-targeted K48 linked polyubiquitin chains were formed, it has been speculated that p300 is necessary for promoting proteasome-mediated degradation of p53 (Grossman et al. 2003). However, work using the p53 null fibroblasts found that the expression of p300 was able to promote ubiquitination of p53, but failed to promote degradation of p53 (Zhu et al. 2001).

12.5.5 Regulation of the Wnt/β-Catenin Pathway by the UPS

Wnt-1 signaling pathway is a highly conserved pathway that plays a critical role in cell adhesion, cellular proliferation, cellular differentiation, and stem cell maintenance, thereby affecting many developmental processes such as neurogenesis, hematopoiesis, and body axis formation (Clevers 2006). Three major cascades of Wnt have been identified: one pathway commonly referred to as the canonical Wnt signaling and two later identified non-canonical Wnt signaling pathways. The transcriptional factor complex of β-catenin/Tcf has been identified as a major effector of the canonical Wnt signaling pathway and will be the focus of this review (Widelitz 2005).

Stimulation of transcriptional activity in Wnt signaling is essentially regulated by the levels of cytoplasmic β-catenin. In the absence of Wnt signaling, cytoplasmic

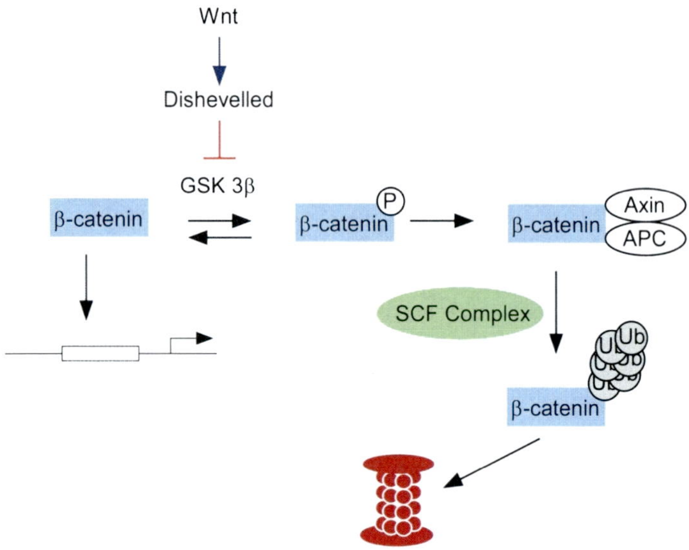

Fig. 12.5 *Overview of β-catenin transcription factor regulation by the ubiquitin–proteasome system*. β-catenin levels are normal low in the cell, due to the phosphorylation of β-catenin by GSK-3β, which increases the binding of β-catenin to axin and APC. Once bound to these co-factors β-catenin is ubiquitinated by a SCF E3-ligase and degraded. Following activation of Wnt signaling, dishevelled blocks GSK-3β function thus stabilizing β-catenin protein levels and enabling the nuclear translocation of β-catenin and promoting gene expression

levels of β-catenin are kept low through ubiquitin-mediated degradation (Aberle et al. 1997) (Fig. 12.5). Ubiquitin-mediated degradation of β-catenin is initiated through phosphorylation of β-catenin by glycogen synthase kinase-3β (GSK-3β) (Hedgepeth et al. 1997; Stambolic et al. 1996). This phosphorylation promotes the association of β-catenin with a degradation complex consisting of adenomatous polyposis coli (APC), Axin, Casein kinase-1α, and GSK-3β (Hart et al. 1998; Kishida et al. 2001; Sakanaka et al. 1998; Price 2006). The formation of this complex is necessary for subsequent ubiquitination of β-catenin by the SCF E3-ligase complex, which consists of Skp1, Cullin1, and Roc1/Rbx1 (Liu et al. 2004b; Winston et al. 1999). The recruitment of β-catenin to the SCF complex is mediated through F box motif ubiquitin ligase receptors, Slimb, Beta transduction repeat containing protein (β TrCP), and Homologue of Slimb (HOS) (Fuchs et al. 1999; Latres et al. 1999; Jiang and Struhl 1998). In the presence of Wnt, Dishevelled inhibits GSK-3β and β-catenin remains unphosphorylated (Wagner et al. 1997). β-Catenin can then form a transcriptional complex with β-catenin/Tcf/Lef and promote transcriptional transactivation of Wnt-responsive genes (Behrens et al. 1996).

Recent work has shed light on the purpose of the association of β-catenin to the APC/Axin/GSK-3β degradation complex. The association of phosphorylated β-catenin with the APC protein is critical for their recruitment to the SCF E3-ligase complex. Furthermore, the association β-catenin with APC is required to preserve the phosphorylation of β-catenin (Su et al. 2008). This work would suggest that APC plays a critical protective role to ensure that β-catenin is ubiquitinated. The importance of the regulation of β-catenin levels in Wnt signaling is illustrated by the fact that β-catenin levels are not only regulated in the cytoplasm by the SCF complex, but also in the nucleus. The histone acetyltransferase (HAT) complex component, Transcriptional Histone Acetyltransferase Cofactor (TRRAP), recruits Skp1 to β-catenin residing on its target promoter in chromatin and promotes β-catenin ubiquitiation and degradation (Finkbeiner et al. 2008).

Ubiquitination plays a role in the Wnt pathway by not only regulating β-catenin levels, but also ubiquitin ligase receptors as well. The Wnt pathway was demonstrated to regulate the mRNA levels of ubiquitin ligase receptor, βTrCP. The expression of a dominant negative Tcf decreased the levels of βTrCP1 mRNA whereas overexpression of stabilized β-catenin in HEK293T cells provided elevated levels of βTrCP mRNA. β-catenin/Tcf affect mRNA levels by inducing the expression of an RNA-binding protein, CRD-BP, which stabilizes βTrCP1 mRNA (Noubissi et al. 2006). Elevated levels of βTrCP resulted in enhanced β-catenin degradation (Latres et al. 1999). This would suggest that β-catenin/Tcf through induction of CRD-BP form a negative feedback loop to regulate the expression levels of β-catenin.

12.5.6 Regulation of the HIF-1α Pathway by the UPS

Hypoxia-inducible factor-1 (HIF-1) is a basic helix-loop helix-PAS domain transcription factor that regulates the cellular response to oxygen deprivation. Through induction of over 100 target genes, HIF-1 is able to promote adaptation of tissue to hypoxic conditions (i.e. increased oxygen delivery to tissues, increased angiogenesis and increased anaerobic glycolysis) (Yang et al. 1998) (Koh et al. 2008b). HIF-1 consists of two subunits: an oxygen-regulated subunit, HIF-α and a constitutively expressed subunit, HIF-1β (Wang et al. 1995). Currently, there are known three isoforms of HIF-α: HIF-1α, HIF-2α and HIF-3α. HIF-1α is expressed ubiquitiously, whereas HIF-2α shows tissue-specific expression. HIF-α/HIF-1β heterodimer translocates to the nucleus from the cytoplasm during hypoxic conditions where it binds to HIF binding sequence within the hypoxia-response element (HRE) (Wang and Semenza 1993; Wood et al. 1998).

Under normal oxygen levels (normoxia), HIF-α levels are kept low through continual ubiquitination and proteasome-mediated degradation (Fig. 12.6). The initiation of ubiquitination of HIF-α requires its hydroxylation on proline residues located within the oxygen-dependent domain (ODD) by specific proline hydroxylases (PHD1, PHD2 and PHD3s) (Ivan et al. 2001; Jaakkola et al. 2001). PHD activity depends on oxygen, so under hypoxic conditions PHDs are inhibited and incapable of hydroxylating HIF-α (Fong and Takeda 2008). Hydroxylation by PHDs under normoxic conditions allows HIF-α to bind to von Hippel-Lindau tumor suppressor gene (pVHL) (Jaakkola et al. 2001; Maxwell et al. 1999). HIF-α

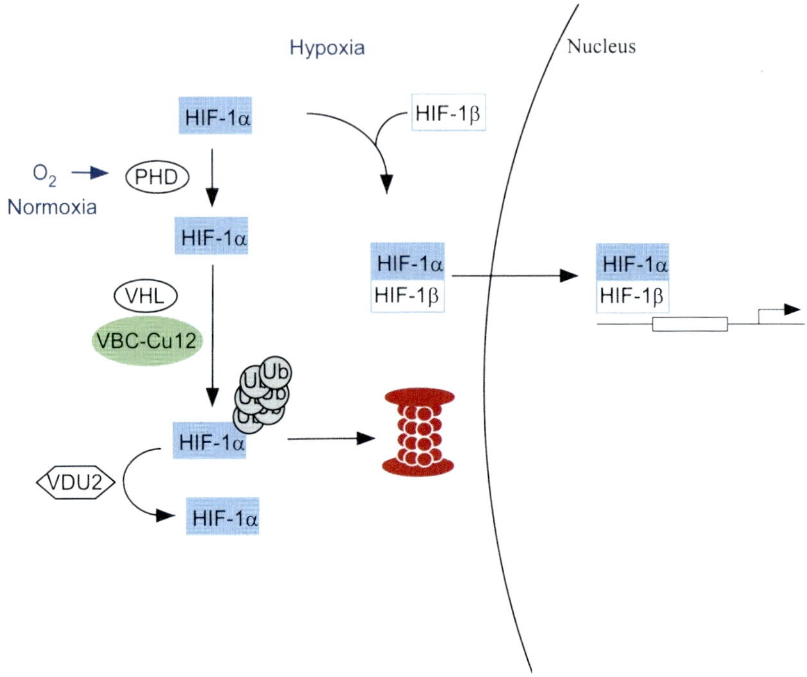

Fig. 12.6 Overview of HIF-1α transcription factor regulation by the ubiquitin–proteasome system. Under normal (normoxic conditions) HIF-1α levels are kept low. HIF-1α is hydroxylated on proline residues located within the oxygen-dependent domain (ODD) by specific prolyl hydroxylases (PHD) and then acetylated by arrest-defective 1 N-acetyltransferase. Acetylated HIF-1α binds to von Hippel–Lindau tumor suppressor gene (pVHL), is ubiquitinated by VBC-cu12 and then rapidly degraded by the proteasome. HIF-1α can be de-ubiquitinated by the DUB VDU2. Under hypoxic conditions, PHD is inhibited, which stabilizes HIF-1α and promotes its interaction with HIF-1β, leading to nuclear import and gene expression

and pVHL then form a complex with a member of the cullin-RING E3-ligase, the Elongin B/C-cullin2-Ring box 1 ubiquitin ligase (VBC-Cul2) (Kibel et al. 1995).

Other E3-ligases complexes can also ubiquitinate HIF-α. Thus, overexpression of SAG (sensitive to apoptosis gene), a member of the ROC1/RBX1 family, in the presence of pVHL can promote ubiquitination and degradation of HIF1-α. This effect is inhibited by a small interfering RNA against SAG. Like ROC1/RBX1, SAG forms a complex with the E3-ligase complex proteins, Cul-5 and pVHL (Tan et al. 2008). Receptor of activated protein kinase (RACK) is also involved in E3-ligase complex formation. RACK competes with Hsp90 of HIF-1α leading to instability. RACK then binds with Elongin B and C to promote formation of the E3-ligase complex leading to ubiquitination of HIF-1α (Liu et al. 2007). Hypoxia-associated factor (HAF), an E3-ligase expressed in proliferating cells, is able to promote ubiquitination irrespective of oxygen levels which results in a decrease in HIF-1α levels (Koh et al. 2008a).

Currently, p-VHL-interacting deubiquitinating enzyme 2 (VDU2), is the only DUB known for HIF-1α. VDU2 was found to interact with HIF-1α leading to the deubiquitination and stabilization of HIF-1α. Consequently, there is an increased level of expression of the HIF-1α responsive gene, vascular endothelial growth factor (VEGF) (Li et al. 2002, 2005b). The levels of pVHL are also known to be regulated by ubiquitination by its interaction with the E2-EPF ubiquitin carrier protein (UCP), a protein that is often overexpressed in liver and gastric cancers (Jung et al. 2006; Ohh 2006). pVHL and UCP were found to interact in HEK293T cells through co-immunoprecipitation experiments. Overexpression of UCP lead to decrease in pVHL levels in the presence of MG132, suggesting that UCP mediates proteasomal degradation of pVHL. As a result, HIF-1α levels were elevated (Jung et al. 2006).

These studies show the regulation of HIF1 and its regulatory protein by multiple ubiquitin ligases and demonstrate the necessity to control levels of these proteins during normoxic conditions. It would be interesting to determine the effect of ischemia and other cell stress conditions on the regulation of these E3-ligases.

12.6 Therapeutic Implications for Neurological Disease

Given the critical role of the ubiquitin–proteasome system in regulating cell function, it is perhaps not surprising that evidence for dysfunction of this system may be implicated in neuropathological disease. Experimental evidence supports a dysfunctional ubiquitin–proteasome system in Huntington's disorder, Alzheimer's and Parkinson's diseases. In addition, acute neurological disorders, such as stroke and seizure, also affect ubiquitin-proteasome function. While many studies may implicate the ubiquitin–proteasome system in the etiology of the disease, very few have identified an effect of ubiquitin–proteasome system dysfunction on gene expression regulation. Given the regulation of multiple transcription factors by the ubiquitin–proteasome system, this is clearly an area for further investigation.

12.6.1 Acute Neurological Conditions

Prolonged ischemic conditions result in neuronal cell death (stroke). Following harmful ischemia, ubiquitinated proteins accumulate in the brain due to proteasomal inhibition. The accumulation of ubiquinated proteins, or their subsequent proteasomal or autophagic processing may contribute to additional cellular stress (Hayashi et al. 1991; Keller et al. 2000; Liu et al. 2004a, 2005). Interestingly, the proteasome inhibitors MLN519 and bortezomib protect against ischemia induced brain damage (Henninger et al. 2006; Williams et al. 2003, 2006). However further studies suggest that the protective effect of proteasome inhibitor MNL519 may be due to the inhibition of neuro-inflammation associated NF-κB signaling, rather than preventing global proteasomal function (Williams et al. 2006).

Ischemic cell death is a delayed event and requires novel protein synthesis. The transcription factor HIF-1α is stabilized under hypoxic conditions where it mediates a program of gene expression (see Sect. 12.5.6). However, shorter periods of ischemia may help prevent cell death in a process called ischemic tolerance. Following brief ischemia (preconditioning) a different pattern of gene expression is observed compared to that occurring following harmful ischemia (Stenzel-Poore et al. 2003). Interestingly, the prior exposure to brief ischemia can re-program the genomic response to a harmful ischemic insult (Stenzel-Poore et al. 2003). Recent proteomic studies have started to identify proteins ubiquitinated following brief periods of ischemia. Of the proteins which were identified by proteomics some are associated with gene expression regulation (Meller et al. 2008). Hence, the ubiquitin–proteasome system may regulate gene expression following brain ischemia.

A role for the ubiquitin–proteasome system in seizures and epilepsy is less clear. Two studies have recently identified mutations in E3-ligase proteins in familial forms of epilepsy. For example a mutation in the FBX25 gene, which is a member of the SCF E3-ligase family was identified in a patient with familial epilepsy and mental retardation (Hagens et al. 2006). In a recent study of Angelman syndrome, a neurological disorder characterized by mental retardation and seizures, a mutation in the HECT family of E3-ligases UBE3A was observed in 10% of cases (Lalande and Calciano 2007). To date it is not clear whether these mutations are associated with specific types of epilepsy or with mental retardation. In addition, it is not yet clear what the consequence these mutations have on protein ubiquitination.

12.6.2 Progressive Neurological Disease

Alzheimers disease is characterized by the accumulation of β-amyloid proteins which form plaques and tangles in the brain. A shift in reading of the β-amyloid and ubiquitin genes results in the mutant expression of ubiquitin-B and β-amyloid (van Leeuwen et al. 1998). Brain plaques are enriched with microtubule proteins including tau. Interestingly the E3-ligase CHIP has been identified as a potential therapeutic target to degrade tau-enriched inclusions (Dickey et al. 2007). It is not yet fully clear whether the plaques that form in Alzheimer's disease are a self survival attempt by the cell or directly contribute to neurotoxicity.

The accumulation of ubiquitinated proteins results in the ER stress response. It has recently been shown that mutations in the SEL-1L gene which mediates ER stress gene expression may confer susceptibility to Alzheimer's disease (Saltini et al. 2006). To date, no studies have shown how ubiquitination with the mutant ubiquitin (+1) directly affects the function of transcription factors. Ubiquitin-mediated Wnt signaling may be aberrant in Alzheimers Disease. The phosphorylated form of β-catenin has been shown to accumulate in Alzheimer's patients. Furthermore, there appears to be enhanced levels of ubiquitinated form of β-catenin in these patients (Ghanevati and Miller 2005). Other groups report that tau hyperphosphorylation results in a decrease in β-catenin phosphorylation and increased function (Li et al. 2007b). Hence, whether β-catenin is dysfunctional in Alzheimer's disease and its effect on disease progression is not yet clear.

Aberrant proteasome function has been noted in Parkinson's disease (Mandel et al. 2005). Multiple genes associated with the ubiquitin proteasome including E3-ligases (skp1A) and proteasome subunits show decreased expression in sporadic Parkinson's disease (Mandel et al. 2005; Zhang et al. 2005). In addition, there are 10 Parkinson's disease related genes, of which one is an E3-ligase (Parkin2) and one is a deubiquitinating enzyme (Parkin 5) (Le and Appel 2004). The Parkin2 gene, which encodes the E3-ligase, functions with CHIP and LIM-kinase 1 to ubiquitinate proteins (Imai et al. 2002; Lim et al. 2007). To date, no studies have shown a definitive dysfunction of a gene expression patterns mediated by the ubiquitin proteasome as a cause of Parkinson's Disease.

A better understanding of the impact of an aberration in the ubiquitin–proteasome system on gene expression regulation can be shown in Huntington's disease. Huntington's disease is due to a large poly-glutamine (polyQ) expansion in the Huntington gene (Htt). There is a number of poly-glutamine expansion diseases, such as spinocerebellar ataxia, and some non-coding expansions as observed in fragile X syndromes. Huntington's disease results in neuronal cell damage and inclusion bodies in the brain. The cells show abnormal phenotypes and one common observation is a decrease in proteasome activity in animal models of Huntington's. This has recently been confirmed by mass spectrometric experiments of brain samples from patients with Huntington's disease, which show an enhanced content of poly-ubiquitinated proteins in diseased brains, thus strongly suggesting a dysfunction in the proteasome (Bennett et al. 2007). In addition, proteasome activators reduce cell death in Huntington's disease models (Seo et al. 2007).

CRE-mediated gene expression is reduced in cell culture models of Huntington's, whereby Htt is overexpressed in PC12 cells. Microarray experiments show a reduction in CRE-mediated gene expression; whether this is due to reduced CREB levels or enhanced CREB inhibition is not clear (Wyttenbach et al. 2001). Overexpression of the mutant poly Q expanded Htt gene results in the formation of nuclear aggregates in hippocampal cells, which are enriched in CBP and then subject to ubiquitination and proteasomal degradation (Jiang et al. 2003). This observation is consistent with the observation of reduced CRE-mediated gene expression in Huntington's disease models (Wyttenbach et al. 2001). Given the importance of CREB-mediated gene expression for neuronal survival, this is likely to be detrimental to neurons (Walton and Dragunow 2000). In addition, the loss of CBP will increase competition for the remaining CBP thereby affecting the rates of transcription mediated by other transcription factors.

To further support this view, it has been shown that histone mono-ubiquitination is altered by mutant Htt (Kim et al. 2008). The decreased interaction between Htt and the E3-ligase Bmi results in a reduction in H2A mono-ubiquitination. Mono-ubiquitination of H2A is associated with reduced gene expression (Cao et al. 2005; Wang et al. 2004a; Zhou et al. 2008). Hence, this would suggest that Htt disrupts normal gene expression to change the pattern of genes regulated by histones.

This interesting example of so called "epigenetic re-modeling" of transcription suggests that the mutation of Htt would have a large global consequence on gene expression patterns and has implications for planned therapies for this disease.

12.7 Concluding Remarks

Many neurological diseases show dysfunction in their ubiquitin–proteasome systems either due to the selective loss of proteins, or a general downregulation of function. Since gene expression is under such tight regulation by the ubiquitin–proteasome system, a strong case for regulation of gene expression via transcription factor degradation may be made for multiple neurological diseases. However, the recent finding that the mono-ubiquitination of histones is altered in Huntington's disease strongly suggests that ubiquitin-mediated epigenetic reprogramming mechanisms may have profound effects in neurological diseases. Clearly, this identifies a novel area of biology, which requires further understanding in order to extrapolate from these observations into viable therapeutic options.

Acknowledgements The authors would like to acknowledge the generous funding support of the National Institute for Health (NINDS) grants NS054023 & NS059588 (awarded to Robert Meller) and NS024728 & NS039016 (awarded to Roger Simon). Local funding was provided by the Legacy Good Samaritan Foundation. In addition the authors would like to thank the following Lab members for their assistance with the preparation of this manuscript: Dr. Roger P. Simon, Dr. Simon Thompson, Andrea Ordonez, Michelle Ashley, Kristin McCarthy, Phebe Matson and Veronica Jessick.

References

Aberle, H., Bauer, A., Stappert, J., et al. (1997). Beta-catenin is a target for the ubiquitin-proteasome pathway. *EMBO J.* 16:3797–3804.

Anindya, R., Aygun, O., Svejstrup, J.Q. (2007). Damage-induced ubiquitylation of human RNA polymerase II by the ubiquitin ligase Nedd4, but not Cockayne syndrome proteins or BRCA1. *Mol. Cell* 28:386–397.

Anzi, S., Finkin, S., Shaulian, E. (2008). Transcriptional repression of c-Jun's E3 ubiquitin ligases contributes to c-Jun induction by UV. *Cell Signal.* 20:862–871.

Arden, K.C. (2008). FOXO animal models reveal a variety of diverse roles for FOXO transcription factors. *Oncogene* 27:2345–2350.

Baarends, W.M., Hoogerbrugge, J.W., Roest, H.P., et al. (1999). Histone ubiquitination and chromatin remodeling in mouse spermatogenesis. *Dev. Biol.* 207:322–333.

Bai, Y., Yang, C., Hu, K., et al. (2004). Itch E3 ligase-mediated regulation of TGF-beta signaling by modulating smad2 phosphorylation. *Mol. Cell* 15:825–831.

Barak, Y., Juven, T., Haffner, R., et al. (1993). mdm2 expression is induced by wild type p53 activity. *EMBO J.* 12:461–468.

Barco, A. and Kandel, E.R. (2006). The role of CREB and CBP in brain function. In: *Transcription Factors in the Nervous System* (Thiel, P.D.G., ed). Wiley, Weinheim, Germany, pp. 206–241.

Basbous, J., Chalbos, D., Hipskind, R., et al. (2007). Ubiquitin-independent proteasomal degradation of Fra-1 is antagonized by Erk1/2 pathway-mediated phosphorylation of a unique C-terminal destabilizer. *Mol. Cell. Biol.* 27:3936–3950.

Basbous, J., Jariel-Encontre, I., Gomard, T., et al. (2008). Ubiquitin-independent versus ubiquitin-dependent proteasomal degradation of the c-Fos and Fra-1 transcription factors: is there a unique answer? *Biochimie* 90:296–305.

Behrens, J., von Kries, J.P., Kuhl, M., et al. (1996). Functional interaction of beta-catenin with the transcription factor LEF-1. *Nature* 382:638–642.

Bennett, E.J., Shaler, T.A., Woodman, B., et al. (2007). Global changes to the ubiquitin system in Huntington's disease. *Nature* 448:704–708.

Bianchi, E., Denti, S., Catena, R., et al. (2003). Characterization of human constitutive photomorphogenesis protein 1, a RING finger ubiquitin ligase that interacts with Jun transcription factors and modulates their transcriptional activity. *J. Biol. Chem.* 278:19682–19690.

Bonni, S., Wang, H.R., Causing, C.G., et al. (2001). TGF-beta induces assembly of a Smad2-Smurf2 ubiquitin ligase complex that targets SnoN for degradation. *Nat. Cell Biol.* 3:587–595.

Boyd, S.D., Tsai, K.Y., Jacks, T. (2000). An intact HDM2 RING-finger domain is required for nuclear exclusion of p53. *Nat. Cell Biol.* 2:563–568.

Brenkman, A.B., de Keizer, P.L., van den Broek, N.J., et al. (2008). The peptidyl-isomerase Pin1 regulates p27kip1 expression through inhibition of Forkhead box O tumor suppressors. *Cancer Res.* 68:7597–7605.

Brooks, C.L., Li, M., Gu, W. (2004). Monoubiquitination: the signal for p53 nuclear export? *Cell Cycle* 3:436–438.

Brunet, A., Bonni, A., Zigmond, M.J., et al. (1999). Akt promotes cell survival by phosphorylating and inhibiting a Forkhead transcription factor. *Cell* 96:857–868.

Bunz, F., Dutriaux, A., Lengauer, C., et al. (1998). Requirement for p53 and p21 to sustain G2 arrest after DNA damage. *Science* 282:1497–1501.

Cao, R., Tsukada, Y., Zhang, Y. (2005). Role of Bmi-1 and Ring1A in H2A ubiquitylation and Hox gene silencing. *Mol. Cell* 20:845–854.

Cerretti, D.P., Kozlosky, C.J., Mosley, B., et al. (1992). Molecular cloning of the interleukin-1 beta converting enzyme. *Science* 256:97–100.

Chang, D.L., Qiu, W., Ying, H., et al. (2007). ARF promotes accumulation of retinoblastoma protein through inhibition of MDM2. *Oncogene* 26:4627–4634.

Chang Y.C., Lee, Y.S., Tejima, T., et al. (1998). mdm2 and bax, downstream mediators of the p53 response, are degraded by the ubiquitin-proteasome pathway. *Cell Growth Differ.* 9:79–84

Chen, A., Kleiman, F.E., Manley, J.L., et al. (2002). Autoubiquitination of the BRCA1*BARD1 RING ubiquitin ligase. *J. Biol. Chem.* 277:22085–22092.

Chen, D., Kon, N., Li, M., et al. (2005). ARF-BP1/Mule is a critical mediator of the ARF tumor suppressor. *Cell* 121:1071–1083.

Clevers, H. (2006). Wnt/beta-catenin signaling in development and disease. *Cell* 127:469–480

Daulny, A., Geng, F., Muratani, M., et al. (2008). Modulation of RNA polymerase II subunit composition by ubiquitylation. *Proc. Natl. Acad. Sci. USA* 105:19649–19654.

de Stanchina, E., McCurrach, M.E., Zindy, F., et al. (1998). E1A signaling to p53 involves the p19(ARF) tumor suppressor. *Genes Dev.* 12:2434–2442.

Dehe, P.M., Pamblanco, M., Luciano, P., et al. (2005). Histone H3 lysine 4 mono-methylation does not require ubiquitination of histone H2B. *J. Mol. Biol.* 353:477–484.

Dentin, R., Liu, Y., Koo, S.H., et al. (2007). Insulin modulates gluconeogenesis by inhibition of the coactivator TORC2. *Nature* 449:366–369.

Dickey, C.A., Kamal, A., Lundgren, K., et al. (2007). The high-affinity HSP90-CHIP complex recognizes and selectively degrades phosphorylated tau client proteins. *J. Clin. Invest.* 117:648–658.

Dong, C., Upadhya, S.C., Ding, L., et al. (2008). Proteasome inhibition enhances the induction and impairs the maintenance of late-phase long-term potentiation. *Learn Memory* 15:335–347.

Dornan, D., Wertz, I., Shimizu, H., et al. (2004). The ubiquitin ligase COP1 is a critical negative regulator of p53. *Nature* 429:86–92.

Dupont, S., Mamidi, A., Cordenonsi, M., et al. (2009). FAM/USP9x, a deubiquitinating enzyme essential for TGFbeta signaling, controls Smad4 monoubiquitination. *Cell* 136:123–135.

Ebisawa, T., Fukuchi, M., Murakami, G., et al. (2001). Smurf1 interacts with transforming growth factor-beta type I receptor through Smad7 and induces receptor degradation. *J. Biol. Chem.* 276:12477–12480.

Essers, M.A., Weijzen, S., de Vries-Smits, A.M., et al. (2004). FOXO transcription factor activation by oxidative stress mediated by the small GTPase Ral and JNK. *EMBO J.* 23: 4802–4812.

Fang, S., Jensen, J.P., Ludwig, R.L., et al. (2000). Mdm2 is a RING finger-dependent ubiquitin protein ligase for itself and p53. *J. Biol. Chem.* 275:8945–8951.

Fedorova. E. and Zink, D. (2008). Nuclear architecture and gene regulation. *Biochim. Biophys. Acta* 1783:2174–2184

Finkbeiner, M.G., Sawan, C., Ouzounova, M., et al. (2008). HAT cofactor TRRAP mediates beta-catenin ubiquitination on the chromatin and the regulation of the canonical Wnt pathway. *Cell Cycle* 7:3908–3914

Fong, G.H. and Takeda, K. (2008). Role and regulation of prolyl hydroxylase domain proteins. *Cell Death Differ.* 15:635–641.

Foot, N.J., Dalton, H.E., Shearwin-Whyatt, L.M., et al. (2008). Regulation of the divalent metal ion transporter DMT1 and iron homeostasis by a ubiquitin-dependent mechanism involving Ndfips and WWP2. *Blood* 112:4268–4275.

Frescas, D., Valenti, L., Accili, D. (2005). Nuclear trapping of the forkhead transcription factor FoxO1 via Sirt-dependent deacetylation promotes expression of glucogenetic genes. *J. Biol. Chem.* 280:20589–20595.

Fuchs, S.Y., Chen, A., Xiong, Y., et al. (1999). HOS, a human homolog of Slimb, forms an SCF complex with Skp1 and Cullin1 and targets the phosphorylation-dependent degradation of IkappaB and beta-catenin. *Oncogene* 18:2039–2046.

Fukuchi, M., Imamura, T., Chiba, T., et al. (2001). Ligand-dependent degradation of Smad3 by a ubiquitin ligase complex of ROC1 and associated proteins. *Mol. Biol. Cell* 12:1431–1443.

Gao, M., Labuda, T., Xia, Y., et al. (2004). Jun turnover is controlled through JNK-dependent phosphorylation of the E3 ligase Itch. *Science* 306:271–275.

Garat, C.V., Fankell, D., Erickson, P.F., et al. (2006). Platelet-derived growth factor BB induces nuclear export and proteasomal degradation of CREB via phosphatidylinositol 3-kinase/Akt signaling in pulmonary artery smooth muscle cells. *Mol. Cell. Biol.* 26:4934–4948.

Geyer, R.K., Yu, Z.K., Maki, C.G. (2000). The MDM2 RING-finger domain is required to promote p53 nuclear export. *Nat. Cell Biol.* 2:569–573.

Ghanevati, M. and Miller, C.A. (2005). Phospho-beta-catenin accumulation in Alzheimer's disease and in aggresomes attributable to proteasome dysfunction. *J. Mol. Neurosci.* 25:79–94

Gilley, J., Coffer, P.J., Ham, J. (2003). FOXO transcription factors directly activate bim gene expression and promote apoptosis in sympathetic neurons. *J. Cell Biol.* 162:613–622.

Grossman, S.R., Deato, M.E., Brignone, C., Chan, H.M., et al. (2003). Polyubiquitination of p53 by a ubiquitin ligase activity of p300. *Science* 300:342–344.

Gu, J., Nie, L., Wiederschain, D., et al. (2001). Identification of p53 sequence elements that are required for MDM2-mediated nuclear export. *Mol. Cell Biol.* 21:8533–8546.

Hagens, O., Minina, E., Schweiger, S., et al. (2006). Characterization of FBX25, encoding a novel brain-expressed F-box protein. *Biochim. Biophys. Acta* 1760:110–118.

Hai, T. and Hartman, M.G. (2001). The molecular biology and nomenclature of the activating transcription factor/cAMP responsive element binding family of transcription factors: activating transcription factor proteins and homeostasis. *Gene* 273:1–11.

Hart, M.J., de los Santos, R., Albert, I.N., et al. (1998). Downregulation of beta-catenin by human Axin and its association with the APC tumor suppressor, beta-catenin and GSK3 beta. *Curr. Biol.* 8:573–581.

Hayashi, T., Takada, K., Matsuda, M. (1991). Changes in ubiquitin and ubiquitin-protein conjugates in the CA1 neurons after transient sublethal ischemia. *Mol. Chem. Neuropathol.* 15:75–82.

He, J., Tegen, S.B., Krawitz, A.R., et al. (2003). The transforming activity of Ski and SnoN is dependent on their ability to repress the activity of Smad proteins. *J. Biol. Chem.* 278: 30540–30547.

Hedgepeth, C.M., Conrad, L.J., Zhang, J., et al. (1997). Activation of the Wnt signaling pathway: a molecular mechanism for lithium action. *Dev Biol* 185:82–91.

Heinemeyer, W., Kleinschmidt, J.A., Saidowsky, J., et al. (1991). Proteinase yscE, the yeast proteasome/multicatalytic-multifunctional proteinase: mutants unravel its function in stress induced proteolysis and uncover its necessity for cell survival. *EMBO J.* 10:555–562.

Henninger, N., Sicard, K.M., Bouley, J., et al. (2006). The proteasome inhibitor VELCADE reduces infarction in rat models of focal cerebral ischemia. *Neurosci. Lett.* 398:300–305.

Henry, K.W., Wyce, A., Lo, W.S., et al. (2003). Transcriptional activation via sequential histone H2B ubiquitylation and deubiquitylation, mediated by SAGA-associated Ubp8. *Genes Dev.* 17:2648–2663.

Hochstrasser, M. (2006). Lingering mysteries of ubiquitin-chain assembly. *Cell* 124:27–34.

Honda, R., Tanaka, H., Yasuda, H. (1997). Oncoprotein MDM2 is a ubiquitin ligase E3 for tumor suppressor p53. *FEBS Lett.* 420:25–27.

Honda, R. and Yasuda, H. (2000). Activity of MDM2, a ubiquitin ligase, toward p53 or itself is dependent on the RING finger domain of the ligase. *Oncogene* 19:1473–1476.

Hosaka, T., Biggs, W.H. 3rd, Tieu. D., et al. (2004). Disruption of forkhead transcription factor (FOXO) family members in mice reveals their functional diversification. *Proc. Natl. Acad. Sci. USA* 101:2975–2980.

Hu, M.C., Lee, D.F., Xia, W., et al. (2004). IkappaB kinase promotes tumorigenesis through inhibition of forkhead FOXO3a. *Cell* 117:225–237.

Huang, H., Regan, K.M., Wang, F., et al. (2005). Skp2 inhibits FOXO1 in tumor suppression through ubiquitin-mediated degradation. *Proc. Natl. Acad. Sci. USA* 102:1649–1654.

Hwang, W.W., Venkatasubrahmanyam, S., Ianculescu, A.G., et al. (2003). A conserved RING finger protein required for histone H2B monoubiquitination and cell size control. *Mol. Cell* 11:261–266

Ikeda, F. and Dikic, I. (2008). Atypical ubiquitin chains: new molecular signals. 'Protein Modifications: Beyond the Usual Suspects' review series. *EMBO Rep.* 9:536–542.

Imai, Y., Soda, M., Hatakeyama, S., et al. (2002). CHIP is associated with Parkin, a gene responsible for familial Parkinson's disease, and enhances its ubiquitin ligase activity. *Mol. Cell* 10:55–67.

Impey, S., Fong, A.L., Wang, Y., et al. (2002). Phosphorylation of CBP mediates transcriptional activation by neural activity and CaM kinase IV. *Neuron* 34:235–244.

Inoue, Y. and Imamura, T. (2008). Regulation of TGF-beta family signaling by E3 ubiquitin ligases. *Cancer Sci.* 99:2107–2112.

Ivan, M., Kondo, K., Yang, H., et al. (2001). HIFalpha targeted for VHL-mediated destruction by proline hydroxylation: implications for O_2 sensing. *Science* 292:464–468.

Jaakkola, P., Mole, D.R., Tian, Y.M., et al. (2001). Targeting of HIF-alpha to the von Hippel-Lindau ubiquitylation complex by O2-regulated prolyl hydroxylation. *Science* 292:468–472.

Jacobs, K.M., Pennington, J.D., Bisht, K.S., et al. (2008). SIRT3 interacts with the daf-16 homolog FOXO3a in the mitochondria, as well as increases FOXO3a dependent gene expression. *Int. J. Biol. Sci.* 4:291–299.

Jiang, H., Nucifora, F.C. Jr., Ross, C.A., et al. (2003). Cell death triggered by polyglutamine-expanded huntingtin in a neuronal cell line is associated with degradation of CREB-binding protein. *Hum. Mol. Genet.* 12:1–12.

Jiang, J. and Struhl, G. (1998). Regulation of the Hedgehog and Wingless signalling pathways by the F-box/WD40-repeat protein Slimb. *Nature* 391:493–496.

Jin, Y., Zeng, S.X., Lee, H., et al. (2004). MDM2 mediates p300/CREB-binding protein-associated factor ubiquitination and degradation. *J. Biol. Chem.* 279:20035–20043.

Johannessen, M., Delghandi, M.P., Moens, U. (2004). What turns CREB on? *Cell Signal.* 16:1211–1227.

Jung, C.R., Hwang, K.S., Yoo, J., et al. (2006). E2-EPF UCP targets pVHL for degradation and associates with tumor growth and metastasis. *Nat. Med.* 12:809–816.

Kao, C.F., Hillyer, C., Tsukuda, T., et al. (2004). Rad6 plays a role in transcriptional activation through ubiquitylation of histone H2B. *Genes Dev.* 18:184–195.

Kastan, M.B., Onyekwere, O., Sidransky, D., et al. (1991). Participation of p53 protein in the cellular response to DNA damage. *Cancer Res.* 51:6304–6311.

Kato, S., Ding, J., Pisck, E., et al. (2008). COP1 functions as a FoxO1 ubiquitin E3 ligase to regulate FoxO1-mediated gene expression. *J. Biol. Chem.* 283:35464–35473.

Katoh, M. and Katoh, M. (2004). Human FOX gene family. *Int. J. Oncol.* 25:1495–1500.

Kau, T.R., Schroeder, F., Ramaswamy, S., et al. (2003). A chemical genetic screen identifies inhibitors of regulated nuclear export of a Forkhead transcription factor in PTEN-deficient tumor cells. *Cancer Cell* 4:463–476.

Kavsak, P., Rasmussen, R.K., Causing, C.G., et al. (2000). Smad7 binds to Smurf2 to form an E3 ubiquitin ligase that targets the TGF beta receptor for degradation. *Mol. Cell* 6:1365–1375.

Keller, J.N., Huang, F.F., Zhu, H., et al. (2000). Oxidative stress-associated impairment of proteasome activity during ischemia-reperfusion injury. *J. Cereb. Blood Flow Metab.* 20: 1467–1473.

Kibel, A., Iliopoulos, O., DeCaprio, J.A., et al. (1995). Binding of the von Hippel-Lindau tumor suppressor protein to Elongin B and C. *Science* 269:1444–1446.

Kiernan, R.E., Emiliani, S., Nakayama, K., et al. (2001). Interaction between cyclin T1 and SCF(SKP2) targets CDK9 for ubiquitination and degradation by the proteasome. *Mol. Cell. Biol.* 21:7956–7970

Kim, M.O., Chawla, P., Overland, R.P., et al. (2008). Altered histone monoubiquitylation mediated by mutant huntingtin induces transcriptional dysregulation. *J. Neurosci.* 28:3947–3957.

Kishida, M., Hino, S., Michiue, T., et al. (2001). Synergistic activation of the Wnt signaling pathway by Dvl and casein kinase Iepsilon. *J. Biol. Chem.* 276:33147–33155.

Koh, M.Y., Darnay, B.G., Powis, G. (2008a). Hypoxia-associated factor, a novel E3-ubiquitin ligase, binds and ubiquitinates hypoxia-inducible factor 1alpha, leading to its oxygen-independent degradation. *Mol. Cell Biol.* 28:7081–7095.

Koh, M.Y., Spivak-Kroizman, T.R., Powis, G. (2008b). HIF-1 regulation: not so easy come, easy go. *Trends Biochem. Sci.* 33:526–534.

Koken, M.H., Reynolds, P., Jaspers-Dekker, I., et al. (1991). Structural and functional conservation of two human homologs of the yeast DNA repair gene RAD6. *Proc. Natl. Acad Sci. USA* 88:8865–8869.

Kornhauser, J.M., Cowan, C.W., Shaywitz, A.J., et al. (2002). CREB transcriptional activity in neurons is regulated by multiple, calcium-specific phosphorylation events. *Neuron* 34: 221–233.

Krieghoff-Henning, E. and Hofmann, T.G.. (2008). Role of nuclear bodies in apoptosis signalling. *Biochim. Biophys. Acta* 1783:2185–2194.

Kuratomi, G., Komuro, A., Goto, K., et al. (2005). NEDD4-2 (neural precursor cell expressed, developmentally down-regulated 4-2). negatively regulates TGF-beta (transforming growth factor-beta) signalling by inducing ubiquitin-mediated degradation of Smad2 and TGF-beta type I receptor. *Biochem. J.* 386:461–470.

Lai, Z., Ferry, K.V., Diamond, M.A., et al. (2001). Human mdm2 mediates multiple monoubiquitination of p53 by a mechanism requiring enzyme isomerization. *J. Biol. Chem.* 276:31357–31367.

Lalande, M. and Calciano, M.A. (2007). Molecular epigenetics of Angelman syndrome. *Cell. Mol. Life Sci.* 64:947–960.

Latres, E., Chiaur, D.S., Pagano, M. (1999). The human F box protein beta-Trcp associates with the Cul1/Skp1 complex and regulates the stability of beta-catenin. *Oncogene* 18:849–854.

Le, W. and Appel, S.H. (2004). Mutant genes responsible for Parkinson's disease. *Curr. Opin. Pharmacol.* 4:79–84.

Lee, J.S., Shukla, A., Schneider, J., et al. (2007). Histone crosstalk between H2B monoubiquitination and H3 methylation mediated by COMPASS. *Cell* 131:1084–1096.

Leng, R.P., Lin, Y., Ma, W., et al. (2003). Pirh2, a p53-induced ubiquitin-protein ligase, promotes p53 degradation. *Cell* 112:779–791.

Li, H., Zhang, Z., Wang, B., et al. (2007a). Wwp2-mediated ubiquitination of the RNA polymerase II large subunit in mouse embryonic pluripotent stem cells. *Mol. Cell. Biol.* 27:5296–5305.

Li, H.L., Wang, H.H., Liu, S.J., et al. (2007b). Phosphorylation of tau antagonizes apoptosis by stabilizing beta-catenin, a mechanism involved in Alzheimer's neurodegeneration. *Proc. Natl. Acad. Sci. USA* 104:3591–3596.

Li, L., Xin, H., Xu, X., et al. (2004). CHIP mediates degradation of Smad proteins and potentially regulates Smad-induced transcription. *Mol. Cell. Biol.* 24:856–864.

Li, M., Brooks, C.L., Wu-Baer, F., et al. (2003). Mono- versus polyubiquitination: differential control of p53 fate by Mdm2. *Science* 302:1972–1975.

Li, R.F., Zhang, F., Lu, Y.J., et al. (2005a). Specific interaction between Smad1 and CHIP: a surface plasmon resonance study. *Colloid Surf. B Biointerfaces* 40:133–136.

Li, Z., Na, X., Wang, D., et al. (2002). Ubiquitination of a novel deubiquitinating enzyme requires direct binding to von Hippel-Lindau tumor suppressor protein. *J. Biol. Chem.* 277: 4656–4662.

Li, Z., Wang, D., Messing, E.M., et al. (2005b). VHL protein-interacting deubiquitinating enzyme 2 deubiquitinates and stabilizes HIF-1alpha. *EMBO Rep.* 6:373–378.

Lim, M.K., Kawamura, T., Ohsawa, Y., et al. (2007). Parkin interacts with LIM Kinase 1 and reduces its cofilin-phosphorylation activity via ubiquitination. *Exp. Cell Res.* 313:2858–2874.

Linares, L.K., Kiernan, R., Triboulet, R., et al. (2007). Intrinsic ubiquitination activity of PCAF controls the stability of the oncoprotein Hdm2. *Nat. Cell Biol.* 9:331–338.

Liu, C.L., Ge, P., Zhang, F., et al. (2005). Co-translational protein aggregation after transient cerebral ischemia. *Neuroscience* 134:1273–1284.

Liu, C.L., Martone, M.E., Hu, B.R. (2004a). Protein ubiquitination in postsynaptic densities after transient cerebral ischemia. *J. Cereb. Blood Flow Metab.* 24:1219–1225.

Liu, J., Stevens, J., Matsunami, N., et al. (2004b). Targeted degradation of beta-catenin by chimeric F-box fusion proteins. *Biochem. Biophys. Res. Commun.* 313:1023–1029.

Liu, W., Rui, H., Wang, J., et al. (2006). Axin is a scaffold protein in TGF-beta signaling that promotes degradation of Smad7 by Arkadia. *EMBO J.* 25:1646–1658.

Liu, Y.V., Baek, J.H., Zhang, H., et al. (2007). RACK1 competes with HSP90 for binding to HIF-1alpha and is required for O(2)-independent and HSP90 inhibitor-induced degradation of HIF-1alpha. *Mol. Cell* 25:207–217.

Lohrum, M.A., Woods, D.B., Ludwig, R.L., et al. (2001). C-terminal ubiquitination of p53 contributes to nuclear export. *Mol. Cell. Biol.* 21:8521–8532.

Mallery, D.L., Vandenberg, C.J., Hiom, K. (2002). Activation of the E3 ligase function of the BRCA1/BARD1 complex by polyubiquitin chains. *EMBO J.* 21:6755–6762.

Maltzman, W. and Czyzyk, L. (1984). UV irradiation stimulates levels of p53 cellular tumor antigen in nontransformed mouse cells. *Mol. Cell. Biol.* 4:1689–1694.

Mandel, S., Grunblatt, E., Riederer, P., et al. (2005). Gene expression profiling of sporadic Parkinson's disease substantia nigra pars compacta reveals impairment of ubiquitin-proteasome subunits, SKP1A, aldehyde dehydrogenase, and chaperone HSC-70. *Ann. NY Acad. Sci.* 1053:356–375.

Matheu, A., Maraver, A., Serrano, M. (2008). The Arf/p53 pathway in cancer and aging. *Cancer Res.* 68:6031–6034.

Matsuzaki, H., Daitoku, H., Hatta, M., et al. (2003). Insulin-induced phosphorylation of FKHR (Foxo1) targets to proteasomal degradation. *Proc. Natl. Acad. Sci. USA* 100:11285–11290.

Maxwell, P.H., Wiesener, M.S., Chang, G.W., et al. (1999). The tumour suppressor protein VHL targets hypoxia-inducible factors for oxygen-dependent proteolysis. *Nature* 399:271–275.

Mayr, B. and Montminy, M. (2001). Transcriptional regulation by the phosphorylation-dependent factor CREB. *Nat. Rev. Mol. Cell. Biol.* 2:599–609.

Meller, R., Minami, M., Cameron, J.A., et al. (2005). CREB-mediated Bcl-2 protein expression after ischemic preconditioning. *J. Cereb. Blood Flow Metab.* 25:234–246.

Meller, R., Thompson, S.J., Lusardi, T.A., et al. (2008). Ubiquitin proteasome-mediated synaptic reorganization: a novel mechanism underlying rapid ischemic tolerance. *J. Neurosci.* 28:50–59.

Michael, D. and Oren, M. (2003). The p53-Mdm2 module and the ubiquitin system. *Semin. Cancer Biol.* 13:49–58.

Midgley, C.A., Desterro, J.M., Saville, M.K., et al. (2000). An N-terminal p14ARF peptide blocks Mdm2-dependent ubiquitination in vitro and can activate p53 in vivo. *Oncogene* 19:2312–2323.

Mikhaylova, O., Ignacak, M.L., Barankiewicz, T.J., et al. (2008). The von Hippel-Lindau tumor suppressor protein and Egl-9-Type proline hydroxylases regulate the large subunit of RNA polymerase II in response to oxidative stress. *Mol. Cell. Biol.* 28:2701–2717.

Minsky, N. and Oren, M. (2004). The RING domain of Mdm2 mediates histone ubiquitylation and transcriptional repression. *Mol. Cell* 16:631–639.

Moren, A., Hellman, U., Inada, Y., et al. (2003). Differential ubiquitination defines the functional status of the tumor suppressor Smad4. *J. Biol. Chem.* 278:33571–33582.

Nagano, Y., Mavrakis, K.J., Lee, K.L., et al. (2007). Arkadia induces degradation of SnoN and c-Ski to enhance transforming growth factor-beta signaling. *J. Biol. Chem.* 282:20492–20501.

Nakano, A., Koinuma, D., Miyazawa, K., et al. (2009). Pin1 downregulates TGF-beta signaling by inducing degradation of Smad proteins. *J. Biol. Chem.* 284:6109–6115.

Nateri, A.S., Riera-Sans, L., Da Costa, C., et al. (2004). The ubiquitin ligase SCFFbw7 antagonizes apoptotic JNK signaling. *Science* 303:1374–1378.

Noshita, N., Lewen, A., Sugawara, T., et al. (2001). Evidence of phosphorylation of Akt and neuronal survival after transient focal cerebral ischemia in mice. *J. Cereb. Blood Flow Metab.* 21:1442–1450.

Noubissi, F.K., Elcheva, I., Bhatia, N., et al. (2006). CRD-BP mediates stabilization of betaTrCP1 and c-myc mRNA in response to beta-catenin signalling. *Nature* 441:898–901.

Ogunjimi, A.A., Briant, D.J., Pece-Barbara, N., et al. (2005). Regulation of Smurf2 ubiquitin ligase activity by anchoring the E2 to the HECT domain. *Mol. Cell* 19:297–308.

Ohh, M. (2006). pVHL's kryptonite: E2-EPF UCP. *Cancer Cell* 10:95–97.

Orlowski, R.Z. (1999). The role of the ubiquitin-proteasome pathway in apoptosis. *Cell Death Differ.* 6:303–313.

Palmero, I., Pantoja, C., Serrano, M. (1998). p19ARF links the tumour suppressor p53 to Ras. *Nature* 395:125–126.

Pati, D., Meistrich, M.L., Plon, S.E. (1999). Human Cdc34 and Rad6B ubiquitin-conjugating enzymes target repressors of cyclic AMP-induced transcription for proteolysis. *Mol. Cell. Biol.* 19:5001–5013.

Perkinton, M.S., Ip, J.K., Wood, G.L., et al. (2002). Phosphatidylinositol 3-kinase is a central mediator of NMDA receptor signalling to MAP kinase (Erk1/2), Akt/PKB and CREB in striatal neurones. *J. Neurochem.* 80:239–254.

Pickart, C.M. (2001). Mechanisms underlying ubiquitination. *Annu. Rev. Biochem.* 70:503–533

Price, M.A. (2006). CKI, there's more than one: casein kinase I family members in Wnt and Hedgehog signaling. *Genes Dev.* 20:399–410.

Qiu, J.H., Asai, A., Chi, S., et al. (2000). Proteasome inhibitors induce cytochrome c-caspase-3-like protease-mediated apoptosis in cultured cortical neurons. *J. Neurosci.* 20:259–265.

Rajendra, R., Malegaonkar, D., Pungaliya, P., et al. (2004). Topors functions as an E3 ubiquitin ligase with specific E2 enzymes and ubiquitinates p53. *J. Biol. Chem.* 279:36440–36444.

Rechsteiner, M. and Hill, C.P. (2005). Mobilizing the proteolytic machine: cell biological roles of proteasome activators and inhibitors. *Trends Cell Biol.* 15:27–33.

Riley, T., Sontag, E., Chen, P., et al. (2008). Transcriptional control of human p53-regulated genes. *Nat. Rev. Mol. Cell. Biol.* 9:402–412.

Robzyk, K., Recht, J., Osley, M.A. (2000). Rad6-dependent ubiquitination of histone H2B in yeast. *Science* 287:501–504.

Ross, S. and Hill, C.S. (2008). How the Smads regulate transcription. *Int. J. Biochem. Cell Biol.* 40:383–408.

Sakanaka, C., Weiss, J.B., Williams, L.T. (1998). Bridging of beta-catenin and glycogen synthase kinase-3beta by axin and inhibition of beta-catenin-mediated transcription. *Proc. Natl. Acad. Sci. USA* 95:3020–3023.

Saltini, G., Dominici, R., Lovati, C., et al. (2006). A novel polymorphism in SEL1L confers susceptibility to Alzheimer's disease. *Neurosci. Lett.* 398:53–58.

Sanchez-Molina, S., Oliva, J.L., Garcia-Vargas, S., et al. (2006). The histone acetyltransferases CBP/p300 are degraded in NIH 3T3 cells by activation of Ras signalling pathway. *Biochem. J.* 398:215–224.

Sangadala, S., Metpally, R.P., Reddy, B.V. (2007). Molecular interaction between Smurf1 WW2 domain and PPXY motifs of Smad1, Smad5, and Smad6--modeling and analysis. *J. Biomol. Struct. Dyn.* 25:11–23.

Sapkota, G., Alarcon, C., Spagnoli, F.M., et al. (2007). Balancing BMP signaling through integrated inputs into the Smad1 linker. *Mol. Cell* 25:441–454.

Sasaki, T., Kojima, H., Kishimoto, R., et al. (2006). Spatiotemporal regulation of c-Fos by ERK5 and the E3 ubiquitin ligase UBR1, and its biological role. *Mol. Cell* 24:63–75.

Schmierer, B. and Hill, C.S. (2007). TGFbeta-SMAD signal transduction: molecular specificity and functional flexibility. *Nat. Rev. Mol. Cell. Biol.* 8:970–982.

Schneider, J., Wood, A., Lee, J.S., et al. (2005). Molecular regulation of histone H3 trimethylation by COMPASS and the regulation of gene expression. *Mol. Cell* 19:849–856.

Seo, H., Sonntag, K.C., Kim, W., et al. (2007). Proteasome activator enhances survival of huntington's disease neuronal model cells. *PLoS ONE* 2:e238.

Shahbazian, M.D., Zhang, K., Grunstein, M. (2005). Histone H2B ubiquitylation controls processive methylation but not monomethylation by Dot1 and Set1. *Mol. Cell* 19:271–277.

Shen, X., Hu, P.P., Liberati, N.T., et al. (1998). TGF-beta-induced phosphorylation of Smad3 regulates its interaction with coactivator p300/CREB-binding protein. *Mol. Biol. Cell* 9:3309–3319.

Shima, Y., Shima, T., Chiba, T., et al. (2008). PML activates transcription by protecting HIPK2 and p300 from SCFFbx3-mediated degradation. *Mol. Cell. Biol.* 28:7126–7138.

Shinoda, S., Schindler, C.K., Meller, R., et al. (2004). Bim regulation may determine hippocampal vulnerability after injurious seizures and in temporal lobe epilepsy. *J. Clin. Invest.* 113:1059–1068.

Stahl, M., Dijkers, P.F., Kops, G.J., et al. (2002). The forkhead transcription factor FoxO regulates transcription of p27Kip1 and Bim in response to IL-2. *J. Immunol.* 168:5024–5031.

Stambolic, V., Ruel, L., Woodgett, J.R. (1996). Lithium inhibits glycogen synthase kinase-3 activity and mimics wingless signalling in intact cells. *Curr. Biol.* 6:1664–1668.

Stancovski, I., Gonen, H., Orian, A., et al. (1995). Degradation of the proto-oncogene product c-Fos by the ubiquitin proteolytic system in vivo and in vitro: identification and characterization of the conjugating enzymes. *Mol. Cell. Biol.* 15:7106–7116.

Starita, L.M., Horwitz, A.A., Keogh, M.C., et al. (2005). BRCA1/BARD1 ubiquitinate phosphorylated RNA polymerase II. *J. Biol. Chem.* 280:24498–24505.

Stenzel-Poore, M.P., Stevens, S.L., Xiong, Z., et al. (2003). Effect of ischaemic preconditioning on genomic response to cerebral ischaemia: similarity to neuroprotective strategies in hibernation and hypoxia-tolerant states. *Lancet* 362:1028–1037.

Stroschein, S.L., Bonni, S., Wrana, J.L., et al. (2001). Smad3 recruits the anaphase-promoting complex for ubiquitination and degradation of SnoN. *Genes Dev.* 15:2822–2836.

Su, Y., Fu, C., Ishikawa, S., et al. (2008). APC is essential for targeting phosphorylated beta-catenin to the SCFbeta-TrCP ubiquitin ligase. *Mol. Cell* 32:652–661.

Tan, M., Gu, Q., He, H., et al. (2008). SAG/ROC2/RBX2 is a HIF-1 target gene that promotes HIF-1 alpha ubiquitination and degradation. *Oncogene* 27:1404–1411.

Tanaka, K., Nogawa, S., Nagata, E., et al. (1999). Temporal profile of CREB phosphorylation after focal ischemia in rat brain. *Neuroreport* 10:2245–2250.

Tang, Y., Liu, Z., Zhao, L., et al. (2008). Smad7 stabilizes beta-catenin binding to E-cadherin complex and promotes cell-cell adhesion. *J. Biol. Chem.* 283:23956–23963.

Thrower, J.S., Hoffman, L., Rechsteiner, M., et al. (2000). Recognition of the polyubiquitin proteolytic signal. *EMBO J.* 19:94–102.

Tu, H.C., Ren, D., Wang, G.X., et al. (2009). The p53-cathepsin axis cooperates with ROS to activate programmed necrotic death upon DNA damage. *Proc. Natl. Acad. Sci. USA* 106:1093–1098.

Turnell AS, Stewart GS, Grand RJ, Rookes SM, Martin A, Yamano H, Elledge SJ, Gallimore PH. (2005). The APC/C and CBP/p300 cooperate to regulate transcription and cell-cycle progression. *Nature* 438:690–695.

Upadhya, S.C., Smith, T.K., Hegde, A.N. (2004). Ubiquitin-proteasome-mediated CREB repressor degradation during induction of long-term facilitation. *J. Neurochem.* 91:210–219.

van Leeuwen F.W., de Kleijn, D.P., van den Hurk, H.H., et al. (1998). Frameshift mutants of beta amyloid precursor protein and ubiquitin-B in Alzheimer's and Down patients. *Science* 279:242–247.

Varadan, R., Assfalg, M., Haririnia, A., et al. (2004). Solution conformation of Lys63-linked diubiquitin chain provides clues to functional diversity of polyubiquitin signaling. *J. Biol. Chem.* 279:7055–7063.

Varadan, R., Walker, O., Pickart, C., et al. (2002). Structural properties of polyubiquitin chains in solution. *J. Mol. Biol.* 324:637–647.

Vogt, P.K., Jiang, H., Aoki, M. (2005). Triple layer control: phosphorylation, acetylation and ubiquitination of FOXO proteins. *Cell Cycle* 4:908–913.

Wagner, U., Brownlees, J., Irving, N.G., et al. (1997). Overexpression of the mouse dishevelled-1 protein inhibits GSK-3beta-mediated phosphorylation of tau in transfected mammalian cells. *FEBS Lett.* 411:369–372.

Walton, M., Sirimanne, E., Williams, C., et al. (1996). The role of the cyclic AMP-responsive element binding protein (CREB). in hypoxic-ischemic brain damage and repair. *Brain Res. Mol. Brain Res.* 43:21–29.

Walton, M.R. and Dragunow, I. (2000). Is CREB a key to neuronal survival? *Trends Neurosci.* 23:48–53.

Wang, G.L., Jiang, B.H., Rue, E.A., et al. (1995). Hypoxia-inducible factor 1 is a basic-helix-loop-helix-PAS heterodimer regulated by cellular O2 tension. *Proc. Natl. Acad. Sci. USA* 92:5510–5514.

Wang, G.L. and Semenza, G.L. (1993). General involvement of hypoxia-inducible factor 1 in transcriptional response to hypoxia. *Proc. Natl. Acad. Sci. USA* 90:4304–4308.

Wang, H., Wang, L., Erdjument-Bromage, H., et al. (2004a). Role of histone H2A ubiquitination in polycomb silencing. *Nature* 431:873–878.

Wang, W., Ungermannova, D., Chen, L., et al. (2003). A negatively charged amino acid in Skp2 is required for Skp2-Cks1 interaction and ubiquitination of p27Kip1. *J. Biol. Chem.* 278:32390–32396.

Wang, X., Taplick, J., Geva, N., et al. (2004b). Inhibition of p53 degradation by Mdm2 acetylation. *FEBS Lett.* 561:195–201.

Weake, V.M. and Workman, J.L. (2008). Histone ubiquitination: triggering gene activity. *Mol. Cell* 29:653–663.

Wei, W., Jin, J., Schlisio, S., et al. (2005). The v-Jun point mutation allows c-Jun to escape GSK3-dependent recognition and destruction by the Fbw7 ubiquitin ligase. *Cancer Cell* 8:25–33.

Wei, Y., Jiang, J., Liu, D., et al. (2008). Cdc34-mediated degradation of ATF5 is blocked by cisplatin. *J. Biol. Chem.* 283:18773–18781.

Welcker, M., Orian, A., Grim, J.E., et al. (2004). A nucleolar isoform of the Fbw7 ubiquitin ligase regulates c-Myc and cell size. *Curr. Biol.* 14:1852–1857.

Widelitz, R. (2005). Wnt signaling through canonical and non-canonical pathways: recent progress. *Growth Factors* 23:111–116.

Williams, A.J., Dave, J.R., Tortella, F.C. (2006). Neuroprotection with the proteasome inhibitor MLN519 in focal ischemic brain injury: relation to nuclear factor kappaB (NF-kappaB), inflammatory gene expression, and leukocyte infiltration. *Neurochem. Int.* 49:106–112.

Williams, A.J., Hale, S.L., Moffett, J.R., et al. (2003). Delayed treatment with MLN519 reduces infarction and associated neurologic deficit caused by focal ischemic brain injury in rats via antiinflammatory mechanisms involving nuclear factor-kappaB activation, gliosis, and leukocyte infiltration. *J. Cereb. Blood Flow Metab.* 23:75–87.

Winston, J.T., Strack, P., Beer-Romero, P., et al. (1999). The SCFbeta-TRCP-ubiquitin ligase complex associates specifically with phosphorylated destruction motifs in IkappaBalpha and beta-catenin and stimulates IkappaBalpha ubiquitination in vitro. *Genes Dev.* 13:270–283.

Wood, A., Krogan, N.J., Dover, J., et al. (2003). Bre1, an E3 ubiquitin ligase required for recruitment and substrate selection of Rad6 at a promoter. *Mol. Cell* 11:267–274.

Wood, S.M., Wiesener, M.S., Yeates, K.M., et al. (1998). Selection and analysis of a mutant cell line defective in the hypoxia-inducible factor-1 alpha-subunit (HIF-1alpha).. Characterization of hif-1alpha-dependent and -independent hypoxia-inducible gene expression. *J. Biol. Chem.* 273:8360–8368.

Wu, W., Nishikawa, H., Hayami, R., et al. (2007). BRCA1 ubiquitinates RPB8 in response to DNA damage. *Cancer Res.* 67:951–958.

Wu, X., Bayle, J.H., Olson, D., et al. (1993). The p53-mdm-2 autoregulatory feedback loop. *Genes Dev.* 7:1126–1132.

Wyce, A., Xiao, T., Whelan, K.A., et al. (2007). H2B ubiquitylation acts as a barrier to Ctk1 nucleosomal recruitment prior to removal by Ubp8 within a SAGA-related complex. *Mol. Cell* 27:275–288.

Wyttenbach, A., Swartz, J., Kita, H., et al. (2001). Polyglutamine expansions cause decreased CRE-mediated transcription and early gene expression changes prior to cell death in an inducible cell model of Huntington's disease. *Hum. Mol. Genet.* 10:1829–1845.

Xiao, T., Kao, C.F., Krogan, N.J., et al. (2005). Histone H2B ubiquitylation is associated with elongating RNA polymerase II. *Mol. Cell. Biol.* 25:637–651.

Yamagata, K., Daitoku, H., Takahashi, Y., et al. (2008). Arginine methylation of FOXO transcription factors inhibits their phosphorylation by Akt. *Mol. Cell* 32:221–231.

Yan, L., Lavin, V.A., Moser, L.R., et al. (2008). PP2A regulates the pro-apoptotic activity of FOXO1. *J. Biol. Chem.* 283:7411–7420.

Yang, J.Y., Zong, C.S., Xia, W., et al. (2008). ERK promotes tumorigenesis by inhibiting FOXO3a via MDM2-mediated degradation. *Nat. Cell Biol.* 10:138–148.

Yang, X., Ji, X., Shi, X., et al. (2000). Smad1 domains interacting with Hoxc-8 induce osteoblast differentiation. *J. Biol. Chem.* 275:1065–1072.

Yang, Y., Shuaib, A., Li, Q., et al. (1998). Neuroprotection by delayed administration of topiramate in a rat model of middle cerebral artery embolization. *Brain Res.* 804:169–176.

Yasukawa, T., Kamura, T., Kitajima, S., et al. (2008). Mammalian Elongin A complex mediates DNA-damage-induced ubiquitylation and degradation of Rpb1. *EMBO J.* 27:3256–3266.

Yehia, G., Schlotter, F., Razavi, R., et al. (2001). Mitogen-activated protein kinase phosphorylates and targets inducible cAMP early repressor to ubiquitin-mediated destruction. *J. Biol. Chem.* 276:35272–35279.

Zhang, Y., Feng, X.H., Derynck, R. (1998). Smad3 and Smad4 cooperate with c-Jun/c-Fos to mediate TGF-beta-induced transcription. *Nature* 394:909–913.

Zhang, Y., James, M., Middleton, F.A., et al. (2005). Transcriptional analysis of multiple brain regions in Parkinson's disease supports the involvement of specific protein processing, energy metabolism, and signaling pathways, and suggests novel disease mechanisms. *Am. J. Med. Genet. B Neuropsychiatr. Genet.* 137B:5–16.

Zhang, Y. and Xiong, Y. (1999). Mutations in human ARF exon 2 disrupt its nucleolar localization and impair its ability to block nuclear export of MDM2 and p53. *Mol. Cell* 3:579–591.

Zhou, W., Zhu, P., Wang, J., et al. (2008). Histone H2A monoubiquitination represses transcription by inhibiting RNA polymerase II transcriptional elongation. *Mol. Cell* 29:69–80.

Zhu, F., Choi, B.Y., Ma, W.Y., et al. (2006). COOH-terminal Src kinase-mediated c-Jun phosphorylation promotes c-Jun degradation and inhibits cell transformation. *Cancer Res.* 66:5729–5736.

Zhu, Q., Yao, J., Wani, G., et al. (2001). Mdm2 mutant defective in binding p300 promotes ubiquitination but not degradation of p53: evidence for the role of p300 in integrating ubiquitination and proteolysis. *J. Biol. Chem.* 276:29695–29701.

Zimber, A., Nguyen, Q.D., Gespach, C. (2004). Nuclear bodies and compartments: functional roles and cellular signalling in health and disease. *Cell Signal.* 16:1085–1104.

Zindy, F., Eischen, C.M., Randle, D.H., et al. (1998). Myc signaling via the ARF tumor suppressor regulates p53-dependent apoptosis and immortalization. *Genes Dev.* 12:2424–2433.

Chapter 13
Small Ubiquitin-Like Modifiers and Other Ubiquitin-Like Proteins

Martijn van Hagen and Alfred C.O. Vertegaal

Abbreviations

Aos1	Activation of Smt3p
ATG8	Autophagy-related protein 8
ATG12	Autophagy-related protein 12
E1	SUMO activating enzyme
E2	SUMO protein carrier protein
E3	SUMO ligase
FAT10	Ubiquitin-like protein FAT10
FUBI	Ubiquitin-like protein FUBI
GMP1	GAP-modifying protein 1
HDAC	Histone deacetylase
HECT	Homologous to the E6-AP carboxyl terminus
HIF	Hypoxia-inducible factor
HUB1	Ubiquitin-like modifier HUB1
IκBα	NFκB inhibitor α
IR	Internal repeats
ISG15	Interferon-stimulated gene 15
Mdm2	Mouse double minute 2
NDSM	Negatively charged amino acid-dependent sumoylation motif
NEDD8	Neural precursor cell expressed developmentally down-regulated protein 8
NFκB	Nuclear factor κB
NPC	Nuclear pore complex
Pc2	Polycomb protein 2
PDSM	Phosphorylation-dependent sumoylation motif
PIAS	Protein inhibitor of activated STAT
PIC1	PML-interacting clone 1

A.C.O. Vertegaal (✉)
Department of Molecular Cell Biology, Leiden University Medical Center,
P.O. Box 9600, Mailzone S-1-P, 2300 RC Leiden, The Netherlands
e-mail: vertegaal@lumc.nl

C.J. Vidal (ed.), *Post-Translational Modifications in Health and Disease*, Protein Reviews 13, DOI 10.1007/978-1-4419-6382-6_13, © Springer Science+Business Media, LLC 2011

PML	Promyelocytic leukemia protein
PUP	Prokaryotic ubiquitin-like protein
RanBP	Ran-binding protein
RanGAP	Ran GTPase-activating protein
RING	Really interesting new gene
RNF4	RING finger protein 4
SAE	SUMO activating enzyme
SENP	SUMO protease
SILAC	Stable-isotope labeling by amino acids in cell culture
SIM	SUMO interaction motif
Siz	SAP and Miz-finger domain-containing protein 1
SMT3	Suppressor of mif two 3
STAT	Signal transducer and activator of transcription
SUMO	Small ubiquitin-like modifier
TAF	TBP-associated factor
TBP	TATA box binding protein
TDG	Thymine-DNA glycosylase
Uba2	Ubiquitin-activating enzyme E1-like
Ubc9	Ubiquitin carrier protein 9
UBL1	Ubiquitin-like protein 1
UFM1	Ubiquitin-fold modifier 1
Ulp	Ubiquitin-like-specific protease
URM1	Ubiquitin-related modifier 1

13.1 Introduction

Small ubiquitin-like modifiers (SUMOs) are ubiquitin family members that are covalently conjugated to target proteins to regulate their functions (Geiss-Friedlander and Melchior 2007; Meulmeester and Melchior 2008; Ulrich 2009). Several groups independently identified SUMO-1, also known as PIC1, GMP1, SMT3, UBL1 and sentrin (Boddy et al. 1996; Mahajan et al. 1997; Matunis et al. 1996; Okura et al. 1996; Shen et al. 1996). The conjugation and deconjugation pathways of SUMOs and ubiquitin are remarkably similar.

RanGAP1 was one of the first target proteins identified for SUMO-1 (Mahajan et al. 1997, 1998; Matunis et al. 1996, 1998). Sumoylation is required for the binding of RanGAP1 to RanBP2, a nuclear pore component, thereby regulating the subcellular partitioning of RanGAP1. Sumoylation does not appear to affect the stability of RanGAP1. These initial studies on the sumoylation of RanGAP1 have provided a useful framework for functional sumoylation studies of other target proteins with an emphasis on the regulation of protein-protein interactions and subcellular localization by sumoylation.

Currently, hundreds of SUMO target proteins and non-covalent SUMO interacting proteins have been identified. The functional analysis of these proteins will provide detailed insight into how sumoylation affects different cellular processes. Similar studies will help us to understand how other ubiquitin-like proteins regulate cellular processes. It is clear that these small proteins are covalent post-translational regulators of extensive sets of target proteins with a major impact on virtually all aspects of cellular life.

13.2 Sumoylation Machinery

13.2.1 The SUMO Family

SUMO and ubiquitin only share approximately 16% sequence identity with each other (Table 13.1). Nevertheless, the three-dimensional structures of SUMOs and ubiquitin are remarkably similar (Bayer et al. 1998; Ding et al. 2005). The two most striking differences between ubiquitin and SUMO are the surface charge distributions and SUMOs extended and flexible N-terminal tail which is absent in ubiquitin (Bayer et al. 1998). A single SUMO gene has been found in *S. cerevisiae*, *C. elegans* and *D. melanogaster*. Larger SUMO families have been uncovered in vertebrates and in plants. In vertebrates the SUMO protein family consists of three different isoforms, SUMO-1, -2 and -3 (Saitoh and Hinchey 2000). Although a gene that may encode a fourth SUMO family member has been identified, this might be a pseudogene and the existence of a SUMO-4 protein has yet to be proven. SUMOs are synthesized as precursor proteins that are processed by SUMO proteases to generate the mature form that can be conjugated to target proteins via C-terminal diglycine motifs. Mature SUMO-1 shares approximately 50% sequence identity

Table 13.1 Ubiquitin family including% homology to ubiquitin

Ubiquitin-like protein	% Local identity with ubiquitin	% Identity with full length mature ubiquitin (76 AA)	% Local positive with ubiquitin	% Positive with full length mature ubiquitin (76 AA)
SUMO-1	20	17	51	43
SUMO-2	16	14	44	39
SUMO-3	16	14	44	39
SMT3 (yeast)	23	16	48	33
NEDD8	57	57	77	77
FUBI	35	35	55	55
HUB1	26	21	43	34
ISG15	31/36	29/36	50/57	46/57
FAT10	30/41	39/34	56/61	53/50
URM1	–	–	–	–
UFM1	–	–	–	–
ATG8	–	–	–	–
ATG12	–	–	–	–
PUP	–	–	–	–

with the other two vertebrate SUMO family members while mature SUMO-2 and -3 share 95% sequence identity with each other. Given their similarity, they are often named together as SUMO-2/3. SUMO-1 and SUMO-2/3 behave differently in cells (Saitoh and Hinchey 2000). SUMO-1 is nearly completely present in a conjugated form. SUMO-2/3 on the other hand, is present in both conjugated form and as a relatively large pool of free, unconjugated SUMO. SUMO-2/3 conjugation has been shown to be increased by various cellular stresses, implying a role for their conjugation in modulating stress responses. Proteomics studies have shown that a subset of SUMO substrates shows a preference for either SUMO-1 or SUMO-2/3 (Rosas-Acosta et al. 2005b; Vertegaal et al. 2006). Other substrates can be modified equally well by both SUMO isoforms, suggesting partial redundancy. Consistently SUMO-2/3 can compensate for the loss of SUMO-1 (Evdokimov et al. 2008). Interestingly, SUMO-2/3 is able to form chains via internal sumoylation sites (Matic et al. 2008; Tatham et al. 2001) (further discussed in Sect. 13.6).

13.2.2 The Sumoylation Conjugation Machinery

Similar to ubiquitin, SUMO proteins are conjugated to target proteins by E1, E2 and E3 enzymes (Fig. 13.1). The SUMO conjugation and deconjugation machinery is specific for SUMO and is unable to conjugate ubiquitin, and vice versa with the

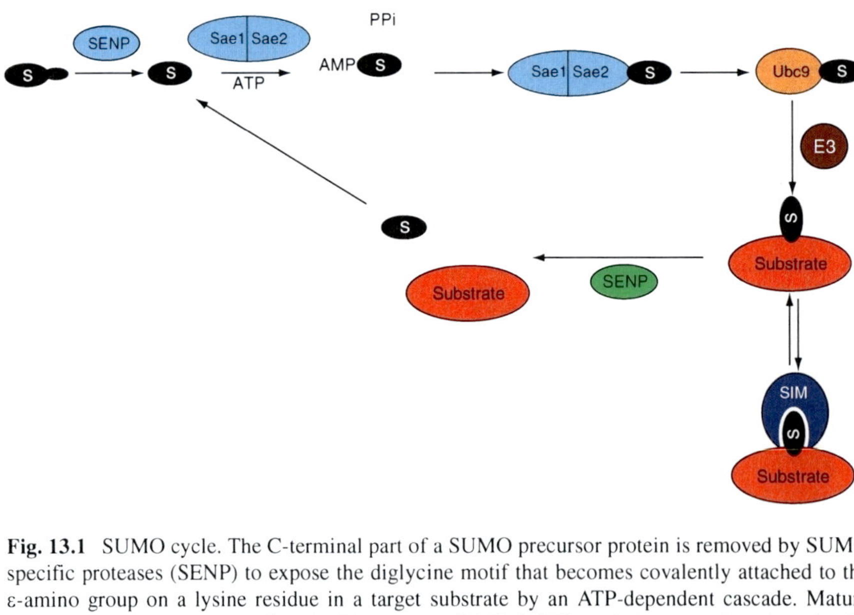

Fig. 13.1 SUMO cycle. The C-terminal part of a SUMO precursor protein is removed by SUMO specific proteases (SENP) to expose the diglycine motif that becomes covalently attached to the ε-amino group on a lysine residue in a target substrate by an ATP-dependent cascade. Mature SUMO is activated by the dimeric E1 enzyme (SAE1-SAE2) and then transferred to the E2 enzyme (Ubc9). E3 factors catalyse the conjugation of SUMOs to target proteins. SUMO binding proteins interact with sumoylated proteins via a SUMO interaction motif (SIM). SUMOs can be removed from target proteins by SUMO specific proteases

exception of the E3 enzyme TOPORS (Johnson et al. 1997; Johnson and Blobel 1997). The first step in the conjugation process is the activation of SUMO by a single E1 enzyme in an ATP-dependent step. Next, a thioester bond is created between the SUMO protein and a cysteine residue in the E1 enzyme (Johnson et al. 1997). The activated SUMO is then passed on to a single E2 enzyme, the conjugating enzyme Ubc9. The action of recombinant E1 and E2 enzyme is sufficient to conjugate SUMOs to target proteins in vitro (Desterro et al. 1999; Okuma et al. 1999). Efficient sumoylation in vivo requires the additional action of several E3 enzymes (Johnson and Gupta 2001).

13.2.2.1 The E1 Enzyme

The E1 enzyme was first identified in *S. cerevisiae*. It is a heterodimer composed of Aos1 and Uba2. Both proteins are required for the formation of the thioester bond between SMT3 and the active site cysteine in Uba2. Deletion of either Aos1 or Uba2 in *S. cerevisiae* is lethal, indicating that there are no other SMT3 E1 enzymes present in yeast (Dohmen et al. 1995; Johnson et al. 1997). The mammalian homologues of Aos1 and Uba2 were identified shortly thereafter and termed SUMO activating enzyme 1 and 2 (SAE1 and SAE2), respectively (Desterro et al. 1999; Okuma et al. 1999). In contrast to the heterodimeric SUMO E1, the ubiquitin E1 is a monomer. Interestingly, SAE1 and Aos1 resemble the N terminus of the ubiquitin E1, while SAE2 and Uba2 resemble the C terminus.

13.2.2.2 The E2 Enzyme

A large family of E2 enzymes is responsible for ubiquitination. In contrast, a single E2 (Ubc9) is responsible for sumoylation. Ubc9 is an essential protein for cell viability; conditional *S. cerevisiae* Ubc9 mutants were unable to complete cell division and arrested in the G_2/M stage (Johnson and Blobel 1997; Seufert et al. 1995). Mice deficient for Ubc9 die at the early postimplantation stage due to major chromosome condensation and segregation defects (Nacerddine et al. 2005).

Based on its sequence similarity with other Ubc proteins, Ubc9 was originally classified as an ubiquitin conjugating enzyme (Seufert et al. 1995). However, Ubc9 was found to specifically form thioesters with SUMO/SMT3 instead of with ubiquitin (Desterro et al. 1997; Gong et al. 1997; Johnson and Blobel 1997) and to be required for the sumoylation of RanGAP1 (Lee et al. 1998; Saitoh et al. 1998).

Ubc9 does not discriminate between the different SUMO family members and catalyses the conjugation of all of them. The E2 enzyme has a limited capacity for target protein recognition. It is able to recognize and bind to the sumoylation consensus sequence that is found in most target proteins (Bernier-Villamor et al. 2002; Sampson et al. 2001). Thus, the E2 enzyme has a limited control over the specificity of the conjugation process. Specificity is furthermore the responsibility of E3 enzymes.

Interestingly, Ubc9 itself has also been shown to be a target for sumoylation. Knipscheer et al. reported an increased activity of SUMO-modified Ubc9 towards

Sp100 (Knipscheer et al. 2008). This increased activity depended on the presence of a SUMO interaction motif (SIM; further discussed in Sect. 13.5) found in Sp100 but did not extend to all SIM domain containing proteins tested. The authors proposed a model where the Ubc9-conjugated SUMO could provide an extra interface to interact with SIM domain-containing target proteins. Sumoylation of a subset of SIM domain-containing targets could then be enhanced possibly based on the positioning of the SIM domain and the SUMO acceptor lysine in the target.

13.2.2.3 The E3 Enzymes

After being activated by the E1 enzyme and transferred to the E2 enzyme, sumoylation requires the action of a third enzyme for the final step of conjugation, the E3 enzyme. Unlike the unique E1 and E2 enzymes, a number of E3 enzymes have been identified. They are subdivided into four different types: the Siz/PIAS family, the nucleoporin RanBP2, polycomb protein 2 (PC2) and TOPORS. The E3 enzymes are able to bind both Ubc9 and target proteins and likely enhance sumoylation by stabilizing their interaction.

PIAS/Siz

The first identified SUMO E3 ligases were Siz1 and Siz2 in *S. cerevisiae*. These two enzymes are responsible for the bulk of SUMO conjugation in this organism. Deleting both genes greatly reduced the total amount of SUMO conjugates, but was not lethal (Johnson and Gupta 2001). Siz proteins are found in all eukaryotes. The members of the PIAS protein family are the mammalian homologues of the Siz proteins (Sachdev et al. 2001). Five PIAS proteins are present in humans, PIAS1, PIAS3, PIASy, PIASxα and PIASxβ; the last two are splice variants encoded by the same gene. Significant sequence similarity exists between the different PIAS proteins, all of which contain an SP-RING domain that is essential for their ligase activity. Mice deficient for PIASy or PIASx had mild phenotypes and no noticeable difference in the overall sumoylation pattern, indicating that PIAS proteins might act in a redundant manner (Roth et al. 2004; Santti et al. 2003; Wong et al. 2004).

Unlike ubiquitin E3 ligases, PIAS proteins are much less specific in targeting substrates. One substrate can be modified by two or more different PIAS proteins and the PIAS proteins themselves are able to recognize and modify multiple substrates. PIAS target specificity could be enhanced through binding partners, subnuclear localization or tissue specific expression. Additionally, little is known about SUMO isoform selectivity by the PIAS proteins (Rosas-Acosta et al. 2005a). Other SP-RING containing SUMO E3 ligases are ZIP3 in yeast (Cheng et al. 2006) and MMS21, an E3 ligase involved in DNA repair (Andrews et al. 2005; Branzei et al. 2006; Potts and Yu 2005, 2007; Zhao and Blobel 2005).

RanBP2

Another SUMO E3 ligase is the nuclear pore complex (NPC) component RanBP2 (Pichler et al. 2002). RanBP2 localizes to the cytoplasmic filaments of the NPC and interacts with SUMO-modified RanGAP1 and the GTPase Ran. The E3 activity of RanBP2 has been mapped to an unusual domain. So far, two classes of ubiquitin ligases have been defined. The first class is formed by the RING finger E3 ligases that function by forming a bridge between the E2 enzyme and the substrate. The second comprises the HECT E3 ligases, which act by forming a thioester intermediate with ubiquitin on a cysteine residue before transferring ubiquitin to a substrate. Interestingly, RanBP2 does not contain a RING finger domain nor does its activity depend upon free cysteine side chains (Pichler et al. 2002, 2004). The catalytic activity in RanBP2 has been mapped to two ~50-residue internal repeats (IR1 and IR2) separated by a middle domain (M). A fragment containing these domains efficiently catalyzed sumoylation of a subset of SUMO substrates. This suggests that in addition to its SUMO ligase activity, RanBP2 also contains innate substrate selectivity (Pichler et al. 2004). Sumoylation of topoisomerase IIα by RanBP2 in vivo is critical for the resolution of sister centromeres (Dawlaty et al. 2008).

Pc2 and TOPORS

The polycomb group (PcG) protein Pc2 is a third type of SUMO ligase. PcG proteins form large multimeric nuclear bodies that function in modulating transcriptional repression. Pc2 recruits Ubc9 and the corepressor CtBP to PcG bodies and enhances its sumoylation (Kagey et al. 2003). Recently, the nuclear RING finger protein TOPORS has also been shown to exhibit SUMO E3 ligase activity (Weger et al. 2005). Interestingly, this is the first SUMO E3 ligase that also exhibits ubiquitin ligase activity.

13.2.3 The SUMO Deconjugation Machinery

SUMO conjugation is a reversible process and, once conjugated, SUMO and substrate can be separated and recycled with the aid of SUMO-specific isopeptidases (Mukhopadhyay and Dasso 2007). SUMO deconjugating enzymes have two essential functions in the cell. They cleave conjugated SUMO proteins from their substrates and they are required for processing newly synthesized SUMO proteins in order to expose the diglycine motif.

All SUMO deconjugating enzymes contain a catalytic domain at their C-terminus. This domain shows significant sequence similarity between different homologues and is distinct from the catalytic domain found in ubiquitin deconjugating enzymes.

The N-termini of the deconjugating enzymes show much more variability. This is thought to be important in determining the subcellular localization of these enzymes, thereby restricting their activity to distinct sets of targets.

In *S. cerevisiae*, two SUMO deconjugating enzymes were found, ubiquitin-like modifier protease 1 (Ulp1) and Ulp2. Each enzyme appears to have a distinct set of target proteins and studies using mutant yeast have shown that they cannot functionally compensate for each others' loss. Ulp1 localizes to the nuclear periphery, is involved in the C-terminal processing of SUMO precursor proteins and is essential for viability (Li and Hochstrasser 1999; Mossessova and Lima 2000; Takahashi et al. 2000). Ulp2, on the other hand, cannot process the SUMO precursor and is not essential in yeast, even though Δ*ulp2* cells are more vulnerable to stress and show defects in genome maintenance (Li and Hochstrasser 2000).

In mammals, seven genes encode proteins that contain sequence similarity with the C-terminal catalytic domain of the Ulps. At least one of these (SENP8) turned out to be a NEDD8-specific deconjugating enzyme, rather than a SUMO deconjugating enzyme (Mendoza et al. 2003). Similar to the yeast deconjugating enzymes, mammalian SENPs have distinct functions and subcellular localizations (Bailey and O'Hare 2002; Gong et al. 2000; Hang and Dasso 2002; Mukhopadhyay and Dasso 2007; Nishida et al. 2000; Zhang et al. 2002). SENP1 and SENP2 are involved in SUMO-1 and SUMO-2 precursor processing, respectively, while they show limited activity towards SUMO-3 (Reverter and Lima 2004; Xu and Au 2005). SUMO-3, in turn, is processed by SENP5 (Gong and Yeh 2006). SENP6 is involved in SUMO-2/3 chain remodeling (Mukhopadhyay et al. 2006). Interestingly, SENP1 and SENP2 are essential for mammalian development; a key target protein for SENP1 is sumoylated HIF1α and a key target protein for SENP2 is sumoylated Mdm2 (Cheng et al. 2007; Chiu et al. 2008).

13.3 SUMO Target Proteins

13.3.1 Sumoylation Consensus Site

After the identification of several SUMO target proteins including RanGAP1, IκBα, PML and p53 it was realized that the SUMO attachment sites in these proteins were situated in a small consensus motif ψKxE, with ψ being a large hydrophobic amino acid V, I, L, M or F (Desterro et al. 1998; Duprez et al. 1999; Kamitani et al. 1998; Mahajan et al. 1998; Matunis et al. 1998; Rodriguez et al. 1999; Rodriguez et al. 2001). This consensus motif is transferrable and combined with a nuclear localization signal it enables the sumoylation of substrates in cells. Occasionally, the acidic residue downstream of a SUMO attachment site was found to be an aspartic acid instead of a glutamic acid. These kinds of consensus motifs are not available for ubiquitin targets. The sumoylation consensus motif has been very helpful to rapidly identify candidate SUMO target proteins and to obtain sumoylation-deficient mutant proteins for functional studies.

Two potential pitfalls should be mentioned with respect to this sumoylation consensus motif. Firstly, the sequence is very short and the presence of such a short sequence in a protein does not necessarily indicate that the protein is being sumoylated. The presence of a nuclear localization signal in addition to a sumoylation consensus site increases the chance that the candidate protein is indeed sumoylated (Rodriguez et al. 2001), although cytoplasmic proteins can also be sumoylated. Moreover, amino acids flanking the sumoylation consensus site will determine whether the consensus site is situated in a part of the protein that is accessible for the SUMO conjugation machinery. Secondly, the absence of a sumoylation consensus motif does not necessarily indicate that the protein is not being sumoylated. Several proteins have been found that are conjugated to SUMOs on alternative lysines (Denison et al. 2005; Hay 2005; Johnson 2004; Vertegaal et al. 2006; Zhou et al. 2004) including the ubiquitin E2-25K protein that is sumoylated on a non-consensus site that is part of an α-helix (Pichler et al. 2005). Interestingly, sumoylation consensus sites are also present in the E2-25K protein, but they are not conjugated in the context of the protein. However, the same consensus sites can be sumoylated when they are part of small unstructured peptides, highlighting the relevance of secondary structure elements for sumoylation.

13.3.2 Transcription Factors

The largest functional group of SUMO targets are transcriptional regulators including specific transcription factors, basal transcription factors, chromatin remodelers, co-activators and co-repressors (Gill 2005; Girdwood et al. 2004). The emerging picture is that transcription factors are generally inhibited by sumoylation via several different mechanisms.

Sumoylation can inhibit a transcription factor via steric hindrance by blocking the DNA-binding domain of the protein. Two examples of transcription factors that are inhibited via this mechanism are the transcription factor SATB2 and the basal transcription factor complex TFIID. Sumoylation of SATB2 influences the expression of immunoglobulin μ gene via diminished binding of SATB2 to matrix attachment regions in cells (Dobreva et al. 2003). The TFIID complex directly interacts with DNA and is composed of the TATA box binding protein (TBP) and TBP-associated factors (TAFs). TAF5 and TAF12 are targets for SUMO-1 and the sumoylation of TAF5 interferes with DNA binding of the TFIID complex (Boyer-Guittaut et al. 2005).

Recruitment of sumoylated transcription factors to nuclear bodies is a second mechanism of transcriptional inhibition by sumoylation. The transcription factor LEF1 is a target for the SUMO E3 ligase PIASy, which is able to recruit LEF1 to nuclear bodies, thereby inactivating the protein (Sachdev et al. 2001). Sumoylated Sp3 was reported to also localize to nuclear bodies and furthermore to the nuclear periphery, correlating with transcriptional inhibition of Sp3 by SUMO (Ross et al. 2002).

A third mechanism of transcriptional inhibition by SUMOs involves the recruitment of histone deacetylases (HDACs) or other transcriptional repressors. The transcription factor Elk1 is negatively regulated by sumoylation (Yang et al. 2003). Interestingly, sumoylated Elk1 recruits histone deacetylases to reduce the acetylation of histones and repress transcription of Elk1 target genes (Yang and Sharrocks 2004). The p300 co-activator contains a repressor domain that functions in a sumoylated state and recruits HDAC6 (Girdwood et al. 2003). p300 mediated repression in cells is counteracted by treatment with HDAC inhibitors or HDAC6 RNAi.

An alternative mechanism of transcription factor inhibition occurs in the NFκB pathway. The NFκB inhibitor IκBα was one of the first transcriptional regulators that was identified as a SUMO target (Desterro et al. 1998). Activation of NFκB involves the controlled degradation of IκB family members via site-specific phosphorylation-mediated ubiquitination. Upon proteasomal degradation of ubiquitinated IκB, NFκB translocates to the nucleus to activate κB sites containing promoters. SUMO-1 and ubiquitin can compete for the same acceptor lysine in IκBα; sumoylation of IκBα occurs via the same lysine that is normally ubiquitinated to activate NFκB. Since sumoylation and ubiquitination of IκBα are mutually exclusive, sumoylation prevents the degradation of IκBα and inhibits NFκB activation.

13.3.3 DNA Repair

Thymine DNA glycosylase (TDG) is a key SUMO substrate and functions as a base excision repair enzyme that removes thymine or uracil from DNA strands when they are mismatched with guanine (Hardeland et al. 2002; Steinacher and Schar 2005). TDG has been shown to bind very tightly to DNA and in vitro it was unable to release itself from a site of repair. In order to gain access to the repair site, newly recruited repair enzymes have to displace TDG. One of the downstream enzymes (AP)-endonuclease 1 (APE1) stimulates the sumoylation of TDG. SUMO-modified TDG has a greatly decreased affinity for DNA due to conformational changes induced by SUMO in a covalent and non-covalent manner (Baba et al. 2005). As sumoylation is a reversible and dynamic modification, it could act as a switch on TDG by enabling its release from the repair site when other downstream enzymes are recruited. Desumoylation by the SENP proteins would then prepare TDG for another round of mismatch repair.

PCNA is another important SUMO target protein involved in DNA replication and repair (Ulrich 2005). Homotrimeric PCNA forms a ring-like structure around DNA and slides along the DNA with the replication fork. During replication of the DNA, it functions as a moving scaffold for many different replication and/or repair enzymes. PCNA is regulated by mono- and poly-ubiquitination and by sumoylation in *S. cerevisiae* (Hoege et al. 2002; Stelter and Ulrich 2003). Interestingly, all three modifications take place on the same target lysine, K164, although K127 can also act as a minor site for sumoylation. PCNA modification takes place whenever the replication fork encounters errors in the DNA or stalls during replication. Mono-ubiquitination of PCNA initiates an error-prone DNA repair mechanism, whereas poly-ubiquitination is responsible for starting an error-free repair mechanism.

In yeast, sumoylation of PCNA takes place in S phase. The helicase Srs2 specifically binds sumoylated PCNA to protect the replicating DNA from undergoing recombination (Papouli et al. 2005; Pfander et al. 2005).

13.3.4 Target Protein Proteomics

The identification of a consensus site for sumoylation enabled initial rapid growth of the SUMO field based on single substrate projects. After 2003, the number of identified SUMO substrates significantly increased based on unbiased proteomics sumoylation projects with some studies reporting hundreds of SUMO targets (Wilson and Heaton 2008). Challenges that need to be met in these studies include the low abundance of many SUMO targets and the substoichiometric nature of the modification (Andersen et al. 2009). It is estimated that less than 1% of the total amount of many target proteins in cells is actually sumoylated at a given time point, although exceptions exist such as RanGAP1 (Mahajan et al. 1997; Matunis et al. 1996). Moreover, SUMO proteases are very potent enzymes and will remove SUMOs from target proteins in most buffers. Therefore, it is critical to employ denaturing buffers that cause the inactivation of SUMO proteases and furthermore limit the copurification of proteins that bind to SUMOs in a non-covalent manner.

In general, single or tandem epitope tagged versions of SUMOs have been used in proteomics studies. A key issue in these studies is the challenge to distinguish purified sumoylated proteins from fortuitously co-purifying contaminants. Due to the non-quantitative nature of most studies, it has been difficult to distinguish background proteins from true SUMO targets and verification of sumoylation is recommended prior to a detailed study of a chosen substrate. The collection of target proteins for the *S. cerevisiae* SUMO family member SMT3 resulting from proteomics studies is especially large, although there are significant differences in the results obtained in different studies (Denison et al. 2005; Hannich et al. 2005; Panse et al. 2004; Wohlschlegel et al. 2004; Wykoff and O'Shea 2005; Zhou et al. 2004). Other studies have provided collections of target proteins for mammalian SUMO family members (Gocke et al. 2005; Li et al. 2004; Manza et al. 2004; Rosas-Acosta et al. 2005b; Schimmel et al. 2008; Vertegaal et al. 2004; Zhao et al. 2004b). The largest functional group of sumoylated proteins consists of transcriptional regulators. In addition, these proteomics studies have uncovered the broad impact of SUMOs on virtually all cellular processes including transcription, metabolism, cell cycle regulation, glycosylation, DNA repair, pre-mRNA splicing and RNA editing.

Most proteomics studies have not attempted to directly identify sumoylation sites due to the complex tandem mass spectrometry spectra involved. This is particularly true for mammalian SUMOs that contain very long C-terminal tryptic fragments (Matic et al. 2008). The C-terminal tryptic fragment of SMT3 is much smaller, explaining why it has been possible to identify consensus and non-consensus sumoylation sites directly by mass spectrometry (Denison et al. 2005).

Recently, quantitative proteomics techniques have become available that are very helpful to overcome the challenges associated with sumoylation proteomics

(Mann 2006). We have applied SILAC technology to establish target protein preferences of SUMO-1 and SUMO-2 (Vertegaal et al. 2006). SILAC technology is furthermore particularly suitable to study conditional sumoylation; dynamic alterations in the sumoylated proteome can be established using this technique (Mann 2006; Schimmel et al. 2008).

13.4 SUMO Binding Proteins

Non-covalent interactions of proteins with covalently ubiquitinated proteins are important to determine the fate of these ubiquitinated proteins. The non-covalent interactors have been named ubiquitin-binding proteins. At least twenty different ubiquitin binding domains exist that bind mono-ubiquitinated or poly-ubiquitinated proteins (Ikeda and Dikic 2008). Similarly, a single SUMO interaction motif (SIM) has been identified in proteins that interact with sumoylated proteins in a non-covalent manner (Hannich et al. 2005; Hecker et al. 2006; Minty et al. 2000; Song et al. 2004). This domain consists of a hydrophobic core, frequently flanked by acidic residues or serines. This SIM interacts with a specific groove in SUMOs (Hecker et al. 2006; Song et al. 2004). Database searches reveal hundreds of SIM-containing proteins, but whether all these proteins are able to bind sumoylated proteins is currently unclear. Nevertheless, the identification of SIM containing proteins is an exciting finding that should help us to comprehend more deeply how sumoylation controls cellular processes.

Interestingly, SIMs can display a limited preference for different SUMO family members and can direct the covalent SUMO conjugation of lysines in the SIM containing protein USP25 (Meulmeester et al. 2008). The presence of multiple closely spaced SIMs enables RNF4 to preferentially bind to SUMO chains compared to SUMO monomers in vitro (Tatham et al. 2008). Surprisingly, a single SIM in CENP-E also preferentially interacts with SUMO chains (Zhang et al. 2008). The sumoylated protein PML can also interact with sumoylated proteins via SIMs to recruit sumoylated proteins to nuclear bodies (Boddy et al. 1996; Shen et al. 2006). Other examples of SIM containing proteins are p73α (Minty et al. 2000), TDG (Takahashi et al. 2005) and Daxx (Lin et al. 2006). Furthermore, an extensive collection of non-covalent SMT3 interactors has been uncovered by proteomics (Makhnevych et al. 2009).

13.5 SUMOs in Chains

An important feature of the ubiquitin system is the formation of ubiquitin chains via all seven internal lysines (Peng et al. 2003). Different types of chains mediate different fates of target proteins, ranging from proteasomal proteolysis to kinase activation (Ikeda and Dikic 2008). Interestingly, SUMO-2, SUMO-3 and the single SUMO SMT3 in budding yeast contain internal sumoylation consensus sites that are located in their N-terminal flexible domains (Ulrich 2008; Vertegaal 2007). SUMO-2 and SUMO-3 have been shown to multimerize in vitro (Tatham et al. 2001). Although SUMO-1 is lacking an internal sumoylation site, it appears that SUMO-1 is also able to form multimers in vitro (Pedrioli et al. 2006) but SUMO-1 polymers have not been found in cells. Using an in vitro to in vivo strategy, we have recently found that SUMO-2 and SUMO-3 are able to form multimers in cells at endogenous levels (Matic et al. 2008). SUMO-1 can also be present on SUMO-2/3 chains in cells at the distal site of the chain, thereby limiting SUMO chain extension.

Mechanistic insight into SUMO chain formation was obtained via structural studies (Capili and Lima 2007; Knipscheer et al. 2007). Ubc9, the SUMO E2 enzyme, not only interacts with SUMO via thioester formation but also in a non-covalent manner to stimulate SUMO chain formation in vitro. These SUMO chains can thus be formed on Ubc9 and can be transferred en-bloc to a target protein. In addition to the E2 enzyme, several E3 enzymes including RanBP2/NUP358 and Siz1 have been shown to enhance SUMO/SMT3 chain formation in vitro (Cooper et al. 2005; Johnson and Gupta 2001; Pichler et al. 2002). SUMO proteases including Ulp2 in yeast (Bylebyl et al. 2003) and SENP6 in mammalian cells (Mukhopadhyay et al. 2006) limit the amount of SUMO chains in cells. SENP6 has a preference for SUMO chains compared to SUMO monomers attached to a target protein.

We are limited in our understanding of the functional relevance of SUMO chains. It was shown that yeast cells expressing a lysine-deficient SMT3 mutant are viable. However, SMT3 polymers were shown to play a role during meiosis (Cheng et al. 2006). Furthermore, SUMO-2/3 chains accumulate upon proteasome inhibition, but polymer formation appears to be dispensable for the proteasomal degradation of a subset of SUMO-2/3 conjugates (Schimmel et al. 2008). Exceptions may exist, such as PML and the PML-RARα protein (Lallemand-Breitenbach et al. 2008; Tatham et al. 2008) but the use of SUMO mutants that are unable to form chains in cells is still needed to establish this more rigorously. Specific SUMO polymer binding proteins exist such as ZIP1 (Cheng et al. 2006), RNF4 (Tatham et al. 2008) and CENP-E (Zhang et al. 2008) that have a preference for SUMO multimers. The identification of other SUMO chain binding proteins will be important to increase our functional insight in SUMO polymerization.

13.6 Crosstalk Between Sumoylation and Other Post-translational Modifications

13.6.1 Crosstalk Between Sumoylation and Phosphorylation

Post-translational regulation of proteins includes many different types of modifications and these modifications can act in concert to control target proteins (Hunter 2007). We are still limited in our understanding of crosstalk, but it is now recognized that sumoylation can cooperate with phosphorylation, acetylation and ubiquitination.

The group of Sistonen found that phosphorylation events downstream of sumoylation sites can influence sumoylation efficiency (Hietakangas et al. 2006). The phosphorylation-dependent sumoylation motif PDSM that was identified is ψKxExxSP. This motif is present in several transcriptional regulators and proline directed phosphorylation increases sumoylation of the upstream lysine within the motif. Later, the group of Sharrocks found that acidic amino acids downstream of sumoylation sites in a so called NDSM (negatively charged amino acid-dependent sumoylation motif) also increase sumoylation efficiency (Yang et al. 2006).

13.6.2 Crosstalk Between Sumoylation and Acetylation

Crosstalk has also been observed between sumoylation and acetylation. Sumoylation negatively regulates transcription via the modification of histones in *S. cerevisae* and serves as a potential block to histone acetylation, a modification of histones that positively regulates transcription (Nathan et al. 2006). Transcriptional inhibition via this mechanism can occur locally via the recruitment of the histone deacetylase HDAC2 by the sumoylated ETS domain transcription factor Elk1 (Yang and Sharrocks 2004). Interplay between sumoylation and acetylation has furthermore been shown to regulate MEF2 and HIC1 (Gregoire and Yang 2005; Shalizi et al. 2006; Stankovic-Valentin et al. 2007). For MEF2, crosstalk between sumoylation and acetylation occurs in a phosphorylation-dependent manner (Shalizi et al. 2006).

13.6.3 Crosstalk Between Sumoylation and Ubiquitination

Different kinds of crosstalk have been observed between sumoylation and ubiquitination. Ubiquitin conjugating and deconjugating enzymes are sumoylated, affecting their activity (Meulmeester et al. 2008; Pichler et al. 2005). SUMO can also directly compete with ubiquitin for an acceptor lysine in a target protein (Desterro et al. 1998) or modify the same acceptor lysine in a target protein in a sequential manner (Huang et al. 2003). Recently, it has been uncovered that a subset of sumoylated proteins can be ubiquitinated and degraded via SUMO-targeted ubiquitin E3 ligases (Fig. 13.2) (Lallemand-Breitenbach et al. 2008; Prudden et al. 2007; Sun et al. 2007; Tatham et al. 2008). This affects a significant number of SUMO-2/3 target proteins (Schimmel et al. 2008).

13.6.4 Other Ubiquitin Family Members

In addition to SUMOs, the ubiquitin family comprises ISG15, NEDD8, ATG8, ATG12, FUBI, URM1, UFM1, FAT10, and HUB1 (Herrmann et al. 2007; Kerscher

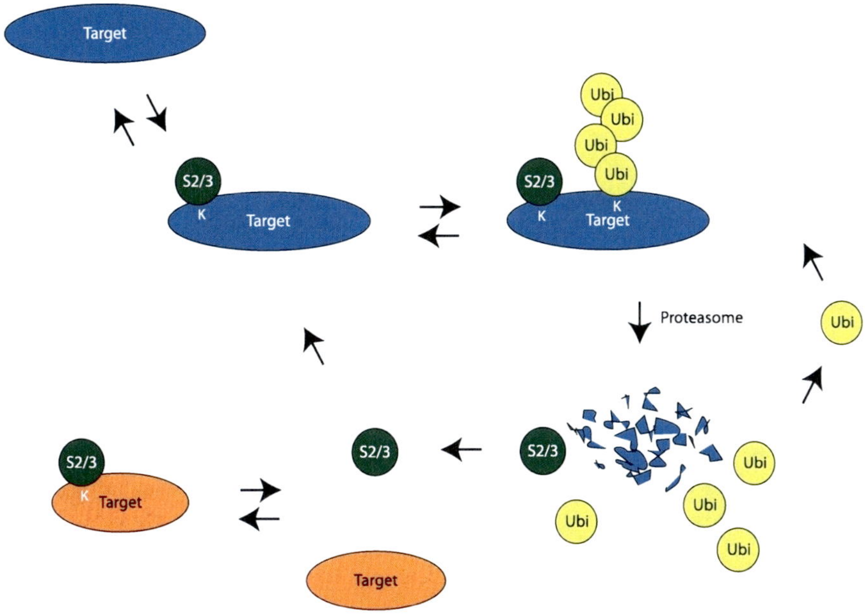

Fig. 13.2 SUMO ubiquitin crosstalk. The ubiquitin-proteasome system is a component of the SUMO-2/3 cycle. A subset of SUMO-2/3 conjugates are subsequently ubiquitinated and degraded by the proteasome. Inhibition of the proteasome leads to accumulation of a subset of SUMO-2/3 conjugates and a depletion of free SUMO-2/3, indicating that crosstalk between sumoylation and ubiquitination is important for SUMO-2/3 recycling. Sumoylation of a second subset of SUMO-2/3 conjugates is decreased upon proteasome-inhibition due to a lack of recycled SUMO-2/3 (Schimmel et al. 2008 (© <2008> The American Society for Biochemistry and Molecular Biology)

et al. 2006) (Table 13.1). The family members URM1, UFM1, ATG8, ATG12 and PUP have no detectable amino acid sequence conservation with ubiquitin, nevertheless all family members share the characteristic ubiquitin fold. Most mature ubiquitin-like proteins contain a C-terminal diglycine motif that is required for covalent conjugation to target proteins. Most family members employ similar biochemical mechanisms for conjugation to target proteins. They are synthesized as precursor proteins that require processing to make the C-terminal glycine available for conjugation to target proteins employing E1, E2 and E3 enzymes. They are also removed from target proteins by specific proteases.

13.6.4.1 ISG15 and FAT10

ISG15 and FAT10 have specific immunological functions (Liu et al. 2005; Lukasiak et al. 2008; Ritchie and Zhang 2004). ISG15 was identified as an Interferon Stimulated Gene that contains a tandem ubiquitin-like domain (Haas et al. 1987; Reich et al. 1987). Analyzing the conjugation machinery of ISG15 yielded a surprise, the ubiquitin E2 enzyme UbcH8 turned out to also be responsible for ISG15 conjugation to target proteins, indicating unexpected enzyme sharing between the ubiquitin and ISG15 system (Durfee et al. 2008; Zhao et al. 2004a). In non-stimulated cells, the amount of ISG15 is low and UbcH8 will mainly conjugate ubiquitin whereas in interferon stimulated cells, UbcH8 will also conjugate the large amounts of ISG15 present. Interferon-induced E3 enzymes (Dastur et al. 2006; Wong et al. 2006; Zou and Zhang 2006) mediate the conjugation of ISG15 and ubiquitin. ISG15 also utilizes the deconjugating machinery of ubiquitin, including USP2, USP5, USP13 and USP14 (Catic et al. 2007). Specific components of the ISG15 pathway are the E1 enzyme UBE1L (Pitha-Rowe et al. 2004) and USP18 (Malakhov et al. 2002). ISG15 target proteins include the signal-transducers ERK1, JAK1 and STAT1 (Malakhov et al. 2003).

13.6.4.2 NEDD8

NEDD8 was originally identified as a gene that shows developmentally down-regulated expression in the mouse brain (Kumar et al. 1992; Rabut and Peter 2008). The main target proteins for NEDD8 are the cullin subunits of SCF ubiquitin E3 ligases (Hori et al. 1999; Osaka et al. 1998; Pan et al. 2004; Singer et al. 1999). Neddylation of cullins facilitates the recruitment of E2 enzymes to the SCF complex. This is important for the ubiquitination of SCF complex substrates including the cell cycle regulators cyclin E (Singer et al. 1999) and p27 (Morimoto et al. 2000). Other NEDD8 targets are p53 (Xirodimas et al. 2004) and ribosomal proteins (Xirodimas 2008). The conjugation machinery and deconjugation components NEDP1 (Mendoza et al. 2003; Wu et al. 2003) and the COP9 signalosome subunit CSN5 are specific for NEDD8 whereas the deconjugation components UCHL1, UCHL3, PfUCH54, USP21 and Ataxin-3 remove both NEDD8 and ubiquitin from targets and the RING finger protein Mdm2 mediates both the ubiquitination and neddylation of p53 (Xirodimas et al. 2004).

13.6.4.3 ATG8 and ATG12

Autophagy is a cellular pathway that degrades cytoplasmic components employing vesicles that fuse with lysosomes to degrade the contents of these vesicles. Dissecting the authophagy pathway in yeast, two ubiquitin-like proteins were identified, ATG8 and ATG12 that are critical components of the autophagy pathway required for proper membrane docking and membrane fusion of autophagy vesicles with lysomes (Mizushima et al. 1998a, b; Ohsumi and Mizushima 2004).

13.6.4.4 PUP

Ubiquitin-like protein systems were thought to be restricted to eukaryotes, but recently, an ubiquitin-like system was also found in prokaryotes (Pearce et al.

2008). The protein PUP was shown to be involved in proteasomal degradation in *M. tuberculosis*. PUP was found to be covalently attached to a lysine residue in a target protein, but the conjugation chemistry differed from other ubiquitin-like systems since the C-terminal residue attached to a target protein was the deamidated glutamine rather than the preceding glycine residue.

13.7 Perspective

The sumoylation field has rapidly matured and sumoylation is now recognized as a critical post-translational modifier of an extensive set of target proteins. Hundreds of SUMO target proteins have been identified in proteomics screens. A significant amount of work is still needed to understand precisely how all identified target proteins are regulated by sumoylation.

Most SUMO target proteins were identified in screens using regular growth conditions of yeast and mammalian cells. Our understanding of the dynamic regulation of target proteins by specific stimuli or treatments is still limited. Previously, it has been shown that sumoylation levels can increase due to specific stress stimuli (Saitoh and Hinchey 2000). Another example of the dynamic regulation of protein sumoylation is the degradation of a subset of SUMO-2 conjugates by the ubiquitin-proteasome system. We anticipate that quantitative proteomics will greatly improve our insight into conditional sumoylation in the future.

Large scale studies are furthermore needed to improve our insight into target protein control by individual SUMO E3 ligases and SUMO proteases. The identification of the extensive subsets of SUMO target proteins that are regulated by these enzymes will enable us to understand in detail how these ligases and proteases affect cellular processes.

It has now convincingly been shown that SUMO-2 and SUMO-3 are able to form SUMO chains in cells (Matic et al. 2008). Our understanding of the functional relevance of SUMO chains is still limited. The identification of SUMO chain conjugates and SUMO chain binding proteins will be key objectives.

A decade of sumoylation research has provided major fundamental insight into protein sumoylation. Nevertheless, our understanding of the clinical significance of sumoylation is still limited. Due to the substoichiometric nature of sumoylation, it is difficult to study the sumoylation status of key target proteins in clinical samples. This is currently hampering progress and new tools are urgently needed here. Compared to more mature fields such as protein phosphorylation and protein ubiquitination, the protein sumoylation field is still young and we have only started to unravel its significance.

Acknowledgments We are grateful to Joost Schimmel, Ivo Hendriks and Sarah Sparks for critically reading the manuscript. We thank Ivo Hendriks for preparing Table 13.1. We apologise to colleagues whose work was not cited due to space constrains. Research in the laboratory of ACOV is supported by the Netherlands Organisation for Scientific Research (NWO).

References

Andersen, J.S., Matic, I., and Vertegaal, A.C. (2009) Identification of SUMO target proteins by quantitative proteomics. *Methods Mol. Biol.* 497:19–31.

Andrews, E.A., Palecek, J., Sergeant, J., et al. (2005) Nse2, a component of the Smc5-6 complex, is a SUMO ligase required for the response to DNA damage. *Mol. Cell. Biol.* 25:185–196.

Baba, D., Maita, N., Jee, J.G., et al. (2005) Crystal structure of thymine DNA glycosylase conjugated to SUMO-1. *Nature* 435:979–982.

Bailey, D. and O'Hare, P. (2002) Herpes simplex virus 1 ICP0 co-localizes with a SUMO-specific protease. *J. Gen. Virol.* 83:2951–2964.

Bayer, P., Arndt, A., Metzger, S., et al. (1998) Structure determination of the small ubiquitin-related modifier SUMO-1. *J. Mol. Biol.* 280:275–286.

Bernier-Villamor, V., Sampson, D.A., and Matunis, M.J. (2002) Structural basis for E2-mediated SUMO conjugation revealed by a complex between ubiquitin-conjugating enzyme Ubc9 and RanGAP1. *Cell* 108:345–356.

Boddy, M.N., Howe, K., Etkin, L.D., et al. (1996) PIC 1, a novel ubiquitin-like protein which interacts with the PML component of a multiprotein complex that is disrupted in acute promyelocytic leukaemia. *Oncogene* 13:971–982.

Boyer-Guittaut, M., Birsoy, K., Potel, C., et al. (2005) SUMO-1 modification of human transcription factor (TF) IID complex subunits: inhibition of TFIID promoter-binding activity through SUMO-1 modification of hsTAF5. *J. Biol. Chem.* 280:9937–9945.

Branzei, D., Sollier, J., Liberi, G., et al. (2006) Ubc9- and mms21-mediated sumoylation counteracts recombinogenic events at damaged replication forks. *Cell* 127:509–522.

Bylebyl, G.R., Belichenko, I., and Johnson, E.S. (2003) The SUMO isopeptidase Ulp2 prevents accumulation of SUMO chains in yeast. *J. Biol. Chem.* 278:44113–44120.

Capili, A.D. and Lima, C.D. (2007) Structure and analysis of a complex between SUMO and Ubc9 illustrates features of a conserved E2-Ubl interaction. *J. Mol. Biol.* 369:608–618.

Catic, A., Fiebiger, E., Korbel, G.A., et al. (2007) Screen for ISG15-crossreactive deubiquitinases. *PLoS ONE* 2: e679.

Cheng, C.H., Lo, Y.H., Liang, S.S., et al. (2006) SUMO modifications control assembly of synaptonemal complex and polycomplex in meiosis of Saccharomyces cerevisiae. *Genes Dev.* 20:2067–2081.

Cheng, J., Kang, X., Zhang, S. et al. (2007) SUMO-specific protease 1 is essential for stabilization of HIF1alpha during hypoxia. *Cell* 131:584–595.

Chiu, S.Y., Asai, N., Costantini, F. et al. (2008) SUMO-specific protease 2 is essential for modulating p53-Mdm2 in development of trophoblast stem cell niches and lineages. *PLoS Biol* 6: e310.

Cooper, H.J., Tatham, M.H., Jaffray, E., et al. (2005) Fourier transform ion cyclotron resonance mass spectrometry for the analysis of small ubiquitin-like modifier (SUMO) modification: identification of lysines in RanBP2 and SUMO targeted for modification during the E3 auto-SUMOylation reaction. *Anal. Chem.* 77:6310–6319.

Dastur, A., Beaudenon, S., Kelley, M., et al. (2006) Herc5, an interferon-induced HECT E3 enzyme, is required for conjugation of ISG15 in human cells. *J. Biol. Chem.* 281:4334–4338.

Dawlaty, M.M., Malureanu, L., Jeganathan, K.B., et al. (2008) Resolution of sister centromeres requires RanBP2-mediated SUMOylation of topoisomerase IIalpha. *Cell* 133:103–115.

Denison, C., Rudner, A.D., Gerber, S.A., et al. (2005) A proteomic strategy for gaining insights into protein sumoylation in yeast. *Mol. Cell. Proteomics* 4:246–254.

Desterro, J.M., Thomson, J., and Hay, R.T. (1997) Ubch9 conjugates SUMO but not ubiquitin. *FEBS Lett.* 417:297–300.

Desterro, J.M., Rodriguez, M.S., and Hay, R.T. (1998) SUMO-1 modification of IkappaBalpha inhibits NF-kappaB activation. *Mol. Cell* 2:233–239.

Desterro, J.M., Rodriguez, M.S., Kemp, G.D. et al. (1999) Identification of the enzyme required for activation of the small ubiquitin-like protein SUMO-1. *J. Biol. Chem.* 274:10618–10624.

Ding, H., Xu, Y., Chen, Q., et al. (2005) Solution structure of human SUMO-3 C47S and its binding surface for Ubc9. *Biochemistry* 44:2790–2799.

Dobreva, G., Dambacher, J., and Grosschedl, R. (2003) SUMO modification of a novel MAR-binding protein, SATB2, modulates immunoglobulin mu gene expression. *Genes Dev.*17:3048–3061.

Dohmen, R.J., Stappen, R., McGrath, J.P., et al. (1995) An essential yeast gene encoding a homolog of ubiquitin-activating enzyme. *J. Biol. Chem.* 270:18099–18109.

Duprez, E., Saurin, A.J., Desterro, J.M., et al. (1999) SUMO-1 modification of the acute promyelocytic leukaemia protein PML: implications for nuclear localisation. *J. Cell Sci.* 112 (Pt 3):381–393.

Durfee, L.A., Kelley, M.L., and Huibregtse, J.M. (2008) The basis for selective E1-E2 interactions in the ISG15 conjugation system. *J. Biol. Chem.* 283:23895–23902.

Evdokimov, E., Sharma, P., Lockett, S.J., et al. (2008) Loss of SUMO1 in mice affects RanGAP1 localization and formation of PML nuclear bodies, but is not lethal as it can be compensated by SUMO2 or SUMO3. *J. Cell Sci.* 121:4106–4113.

Geiss-Friedlander, R. and Melchior, F. (2007) Concepts in sumoylation: a decade on. *Nat. Rev. Mol. Cell. Biol.* 8:947–956.

Gill, G. (2005) Something about SUMO inhibits transcription. *Curr. Opin. Genet. Dev.* 15:536–541.

Girdwood, D., Bumpass, D., Vaughan, O.A., et al. (2003) P300 transcriptional repression is mediated by SUMO modification. *Mol.Cell* 11:1043–1054.

Girdwood. D.W., Tatham, M.H., and Hay, R.T. (2004) SUMO and transcriptional regulation. *Semin. Cell Dev. Biol.* 15:201–210.

Gocke, C.B., Yu, H., and Kang, J. (2005) Systematic identification and analysis of mammalian small ubiquitin-like modifier substrates. *J. Biol. Chem.* 280:5004–5012.

Gong, L. and Yeh, E.T. (2006) Characterization of a family of nucleolar SUMO-specific proteases with preference for SUMO-2 or SUMO-3. *J. Biol. Chem.* 281:15869–15877.

Gong, L., Kamitani, T., Fujise, K., et al. (1997) Preferential interaction of sentrin with a ubiquitin-conjugating enzyme, Ubc9. *J. Biol. Chem.* 272: 28198–28201.

Gong, L., Millas, S., Maul, G.G., et al. (2000) Differential regulation of sentrinized proteins by a novel sentrin-specific protease. *J. Biol. Chem.* 275:3355–3359.

Gregoire, S. and Yang, X.J. (2005) Association with class IIa histone deacetylases upregulates the sumoylation of MEF2 transcription factors. *Mol. Cell. Biol.* 25:2273–2287.

Haas, A.L., Ahrens, P., Bright, P.M. et al. (1987) Interferon induces a 15-kilodalton protein exhibiting marked homology to ubiquitin. *J. Biol. Chem.* 262:11315–11323.

Hang, J. and Dasso, M. (2002) Association of the human SUMO-1 protease SENP2 with the nuclear pore. *J. Biol. Chem.* 277:19961–19966.

Hannich, J.T., Lewis, A., Kroetz, M.B., et al. (2005) Defining the SUMO-modified proteome by multiple approaches in *Saccharomyces cerevisiae*. *J. Biol. Chem.* 280:4102–4110.

Hardeland, U., Steinacher, R., Jiricny, J. et al. (2002) Modification of the human thymine-DNA glycosylase by ubiquitin-like proteins facilitates enzymatic turnover. *EMBO J.* 21:1456–1464.

Hay, R.T. (2005) SUMO: a history of modification. *Mol. Cell* 18:1–12.

Hecker, C.M., Rabiller, M., Haglund, K., et al. (2006) Specification of SUMO1- and SUMO2-interacting motifs. *J. Biol. Chem.* 281:16117–16127.

Herrmann, J., Lerman, L.O., and Lerman, A. (2007) Ubiquitin and ubiquitin-like proteins in protein regulation. *Circ. Res.* 100:1276–1291.

Hietakangas, V., Anckar, J., Blomster, H.A., et al. (2006) PDSM, a motif for phosphorylation-dependent SUMO modification. *Proc. Natl. Acad. Sci. USA* 103:45–50.

Hoege, C., Pfander, B., Moldovan, G.L., et al. (2002) RAD6-dependent DNA repair is linked to modification of PCNA by ubiquitin and SUMO. *Nature* 419:135–141.

Hori, T., Osaka, F., Chiba, T., et al. (1999) Covalent modification of all members of human cullin family proteins by NEDD8. *Oncogene* 18:6829–6834.

Huang, T.T., Wuerzberger-Davis, S.M., Wu, Z.H., et al. (2003) Sequential modification of NEMO/IKKgamma by SUMO-1 and ubiquitin mediates NF-kappaB activation by genotoxic stress. *Cell* 115:565–576.

Hunter, T. (2007) The age of crosstalk: phosphorylation, ubiquitination, and beyond. *Mol. Cell* 28:730–738.

Ikeda, F. and Dikic, I. (2008) Atypical ubiquitin chains: new molecular signals. 'Protein Modifications: Beyond the Usual Suspects' review series. *EMBO Rep.* 9:536–542.

Johnson, E.S. (2004) Protein modification by SUMO. *Annu. Rev. Biochem.* 73:355–382.

Johnson, E.S. and Blobel, G. (1997) Ubc9p is the conjugating enzyme for the ubiquitin-like protein Smt3p. *J. Biol. Chem.* 272:26799–26802.

Johnson, E.S. and Gupta, A.A. (2001) An E3-like factor that promotes SUMO conjugation to the yeast septins. *Cell* 106:735–744.

Johnson, E.S, Schwienhorst, I., Dohmen, R.J., et al. (1997) The ubiquitin-like protein Smt3p is activated for conjugation to other proteins by an Aos1p/Uba2p heterodimer. *EMBO J.* 16:5509–5519.

Kagey, M.H., Melhuish, T.A., and Wotton, D. (2003) The polycomb protein Pc2 is a SUMO E3. *Cell* 113:127–137.

Kamitani, T., Kito, K., Nguyen, H.P., et al. (1998) Identification of three major sentrinization sites in PML. *J. Biol. Chem.* 273:26675–26682.

Kerscher, O., Felberbaum, R., and Hochstrasser, M. (2006) Modification of proteins by ubiquitin and ubiquitin-like proteins. *Annu. Rev. Cell Dev. Biol.* 22:159–180.

Knipscheer, P., van Dijk, W.J., Olsen, J.V., et al. (2007) Noncovalent interaction between Ubc9 and SUMO promotes SUMO chain formation. *EMBO J.* 26:2797–2807.

Knipscheer, P., Flotho, A., Klug, H., et al. (2008) Ubc9 sumoylation regulates SUMO target discrimination. *Mol. Cell* 31:371–382.

Kumar, S., Tomooka, Y., and Noda, M. (1992) Identification of a set of genes with developmentally down-regulated expression in the mouse brain. *Biochem. Biophys. Res. Commun.* 185:1155–1161.

Lallemand-Breitenbach, V., Jeanne, M., Benhenda, S., et al. (2008) Arsenic degrades PML or PML-RARalpha through a SUMO-triggered RNF4/ubiquitin-mediated pathway. *Nat. Cell Biol.* 10:547–555.

Lee, G.W., Melchior, F., Matunis, M.J., et al. (1998) Modification of Ran GTPase-activating protein by the small ubiquitin-related modifier SUMO-1 requires Ubc9, an E2-type ubiquitin-conjugating enzyme homologue. *J. Biol. Chem.* 273:6503–6507.

Li, S.J. and Hochstrasser, M. (1999) A new protease required for cell-cycle progression in yeast. *Nature* 398:246–251.

Li, S.J. and Hochstrasser, M. (2000) The yeast ULP2 (SMT4) gene encodes a novel protease specific for the ubiquitin-like Smt3 protein. *Mol. Cell. Biol.* 20:2367–2377.

Li, T., Evdokimov, E., Shen, R.F., et al. (2004) Sumoylation of heterogeneous nuclear ribonucleoproteins, zinc finger proteins, and nuclear pore complex proteins: a proteomic analysis. *Proc. Natl. Acad. Sci. USA* 101:8551–8556.

Lin, D.Y., Huang, Y.S., Jeng, J.C., et al. (2006) Role of SUMO-interacting motif in Daxx SUMO modification, subnuclear localization, and repression of sumoylated transcription factors. *Mol. Cell* 24:341–354.

Liu, Y.C., Penninger, J., and Karin, M. (2005) Immunity by ubiquitylation: a reversible process of modification. *Nat. Rev. Immunol.* 5:941–952.

Lukasiak, S., Schiller, C., Oehlschlaeger, P., et al. (2008) Proinflammatory cytokines cause FAT10 upregulation in cancers of liver and colon. *Oncogene* 27:6068–6074.

Mahajan, R., Delphin, C., Guan, T., et al. (1997) A small ubiquitin-related polypeptide involved in targeting RanGAP1 to nuclear pore complex protein RanBP2. *Cell* 88: 97–107.

Mahajan, R., Gerace, L., and Melchior, F. (1998) Molecular characterization of the SUMO-1 modification of RanGAP1 and its role in nuclear envelope association. *J. Cell Biol.* 140:259–270.

Makhnevych, T., Sydorskyy, Y., Xin, X., et al. (2009) Global map of SUMO function revealed by protein-protein interaction and genetic networks. *Mol. Cell* 33:124–135.

Malakhov, M.P., Malakhova, O.A., Kim, K.I., et al. (2002) UBP43 (USP18) specifically removes ISG15 from conjugated proteins. *J. Biol. Chem.* 277:9976–9981.

Malakhov, M.P., Kim, K.I., Malakhova, O.A., et al. (2003) High-throughput immunoblotting. Ubiquitin-like protein ISG15 modifies key regulators of signal transduction. *J. Biol. Chem.* 278:16608–16613.

Mann, M. (2006) Functional and quantitative proteomics using SILAC. *Nat. Rev. Mol. Cell. Biol.* 7:952–958.

Manza, L.L., Codreanu, S.G., Stamer, S.L., et al. (2004) Global shifts in protein sumoylation in response to electrophile and oxidative stress. *Chem. Res. Toxicol.* 17: 1706–1715.

Matic, I., van Hagen, M., Schimmel, J., et al. (2008) In vivo identification of human small ubiquitin-like modifier polymerization sites by high accuracy mass spectrometry and an in vitro to in vivo strategy. *Mol. Cell. Proteomics* 7:132–144.

Matunis, M.J., Coutavas, E., and Blobel, G. (1996) A novel ubiquitin-like modification modulates the partitioning of the Ran-GTPase-activating protein RanGAP1 between the cytosol and the nuclear pore complex. *J. Cell Biol.* 135: 1457–1470.

Matunis, M.J., Wu, J., and Blobel, G. (1998) SUMO-1 modification and its role in targeting the Ran GTPase-activating protein, RanGAP1, to the nuclear pore complex. *J. Cell Biol.*140:499–509.

Mendoza, H.M., Shen, L.N., Botting, C., et al. (2003) NEDP1, a highly conserved cysteine protease that deNEDDylates Cullins. *J. Biol. Chem.* 278:25637–25643.

Meulmeester, E. and Melchior, F. (2008) Cell biology: SUMO. *Nature* 452:709–711.

Meulmeester, E., Kunze, M., Hsiao, H.H., et al. (2008) Mechanism and consequences for paralog-specific sumoylation of ubiquitin-specific protease 25. *Mol. Cell* 30:610–619.

Minty, A., Dumont, X., Kaghad, M., et al. (2000) Covalent modification of p73alpha by SUMO-1. Two-hybrid screening with p73 identifies novel SUMO-1-interacting proteins and a SUMO-1 interaction motif. *J. Biol. Chem.* 275:36316–36323.

Mizushima, N., Noda, T., Yoshimori, T., et al. (1998a) A protein conjugation system essential for autophagy. *Nature* 395:395–398.

Mizushima, N., Sugita, H., Yoshimori, T., et al. (1998b) A new protein conjugation system in human. The counterpart of the yeast Apg12p conjugation system essential for autophagy. *J. Biol. Chem.* 273:33889–33892.

Morimoto, M., Nishida, T., Honda, R., et al. (2000) Modification of cullin-1 by ubiquitin-like protein Nedd8 enhances the activity of SCF(skp2) toward p27(kip1). *Biochem. Biophys. Res. Commun.* 270:1093–1096.

Mossessova, E. and Lima, C.D. (2000) Ulp1-SUMO crystal structure and genetic analysis reveal conserved interactions and a regulatory element essential for cell growth in yeast. *Mol. Cell* 5:865–876.

Mukhopadhyay, D. and Dasso, M. (2007) Modification in reverse: the SUMO proteases. *Trends Biochem. Sci.* 32:286–295.

Mukhopadhyay, D., Ayaydin, F., Kolli, N., et al. (2006) SUSP1 antagonizes formation of highly SUMO2/3-conjugated species. *J. Cell Biol.* 174:939–949.

Nacerddine, K., Lehembre, F., Bhaumik, M., et al. (2005) The SUMO pathway is essential for nuclear integrity and chromosome segregation in mice. *Dev. Cell* 9:769–779.

Nathan, D., Ingvarsdottir, K., Sterner, D.E., et al. (2006) Histone sumoylation is a negative regulator in *Saccharomyces cerevisiae* and shows dynamic interplay with positive-acting histone modifications. *Genes Dev.* 20:966–976.

Nishida, T., Tanaka, H., and Yasuda, H. (2000) A novel mammalian Smt3-specific isopeptidase 1 (SMT3IP1) localized in the nucleolus at interphase. *Eur. J. Biochem.* 267:6423–6427.

Ohsumi, Y. and Mizushima, N. (2004) Two ubiquitin-like conjugation systems essential for autophagy. *Semin. Cell Dev. Biol.* 15:231–236.

Okuma, T., Honda, R., Ichikawa, G., et al. (1999) In vitro SUMO-1 modification requires two enzymatic steps, E1 and E2. *Biochem. Biophys. Res. Commun.* 254:693–698.

Okura, T., Gong, L., Kamitani, T., et al. (1996) Protection against Fas/APO-1- and tumor necrosis factor-mediated cell death by a novel protein, sentrin. *J. Immunol.* 157:4277–4281.

Osaka. F., Kawasaki, H., Aida, N., et al. (1998) A new NEDD8-ligating system for cullin-4A. *Genes Dev.* 12:2263–2268.

Pan, Z.Q., Kentsis, A., Dias, D.C., et al. (2004) Nedd8 on cullin: building an expressway to protein destruction. *Oncogene* 23:1985–1997.

Panse, V.G., Hardeland, U., Werner, T., et al. (2004) A proteome-wide approach identifies sumoylated substrate proteins in yeast. *J. Biol. Chem.* 279:41346–41351.

Papouli, E., Chen, S., Davies, A.A., et al. (2005) Crosstalk between SUMO and ubiquitin on PCNA is mediated by recruitment of the helicase Srs2p. *Mol. Cell* 19:123–133.

Pearce, M.J., Mintseris, J., Ferreyra, J., et al. (2008) Ubiquitin-like protein involved in the proteasome pathway of *Mycobacterium tuberculosis*. *Science* 322:1104–1107.

Pedrioli, P.G., Raught, B., Zhang, X.D., et al. (2006) Automated identification of SUMOylation sites using mass spectrometry and SUMmOn pattern recognition software. *Nat. Methods* 3:533–539.

Peng, J., Schwartz, D., Elias, J.E., et al. (2003) A proteomics approach to understanding protein ubiquitination. *Nat. Biotechnol.* 21:921–926.

Pfander, B., Moldovan, G.L., Sacher, M., et al. (2005) SUMO-modified PCNA recruits Srs2 to prevent recombination during S phase. *Nature* 436:428–433.

Pichler, A., Gast, A., Seeler, J.S., et al. (2002) The nucleoporin RanBP2 has SUMO1 E3 ligase activity. *Cell* 108:109–120.

Pichler, A., Knipscheer, P., Saitoh, H., et al. (2004) The RanBP2 SUMO E3 ligase is neither. *Nat. Struct. Mol. Biol.* 11:984–991.

Pichler, A., Knipscheer, P., Oberhofer, E., et al. (2005) SUMO modification of the ubiquitin-conjugating enzyme E2-25K. *Nat. Struct. Mol. Biol.* 12:264–269.

Pitha-Rowe, I., Hassel, B.A., and Dmitrovsky, E. (2004) Involvement of UBE1L in ISG15 conjugation during retinoid-induced differentiation of acute promyelocytic leukemia. *J. Biol. Chem.* 279:18178–18187.

Potts, P.R. and Yu, H. (2005) Human MMS21/NSE2 is a SUMO ligase required for DNA repair. *Mol. Cell Biol.* 25:7021–7032.

Potts, P.R. and Yu, H. (2007) The SMC5/6 complex maintains telomere length in ALT cancer cells through SUMOylation of telomere-binding proteins. *Nat. Struct. Mol. Biol.* 14:581–590.

Prudden, J., Pebernard, S., Raffa, G., et al. (2007) SUMO-targeted ubiquitin ligases in genome stability. *EMBO J.* 26:4089–4101.

Rabut, G. and Peter, M. (2008) Function and regulation of protein neddylation. 'Protein modifications: beyond the usual suspects' review series. *EMBO Rep.* 9:969–976.

Reich, N., Evans, B., Levy, D., et al. (1987) Interferon-induced transcription of a gene encoding a 15-kDa protein depends on an upstream enhancer element. *Proc. Natl. Acad. Sci. USA* 84:6394–6398.

Reverter. D. and Lima, C.D. (2004) A basis for SUMO protease specificity provided by analysis of human Senp2 and a Senp2-SUMO complex. *Structure* 12:1519–1531.

Ritchie, K.J. and Zhang, D.E. (2004) ISG15: the immunological kin of ubiquitin. *Semin. Cell Dev. Biol.* 15: 237–246.

Rodriguez, M.S., Desterro, J.M., Lain, S., et al. (1999) SUMO-1 modification activates the transcriptional response of p53. *EMBO J.* 18:6455–6461.

Rodriguez, M.S., Dargemont, C., and Hay, R.T. (2001) SUMO-1 conjugation in vivo requires both a consensus modification motif and nuclear targeting. *J. Biol. Chem.* 276:12654–12659.

Rosas-Acosta, G., Langereis, M.A., Deyrieux, A., et al. (2005a) Proteins of the PIAS family enhance the sumoylation of the papillomavirus E1 protein. *Virology* 331:190–203.

Rosas-Acosta, G., Russell, W.K., Deyrieux, A., et al. (2005b) A universal strategy for proteomic studies of SUMO and other ubiquitin-like modifiers. *Mol. Cell. Proteomics* 4:56–72.

Ross, S., Best, J.L., Zon, L.I., et al. (2002) SUMO-1 modification represses Sp3 transcriptional activation and modulates its subnuclear localization. *Mol. Cell* 10:831–842.

Roth, W., Sustmann, C., Kieslinger, M., et al. (2004) PIASy-deficient mice display modest defects in IFN and Wnt signaling. *J. Immunol.* 173:6189–6199.

Sachdev, S., Bruhn, L., Sieber, H., et al. (2001) PIASy, a nuclear matrix-associated SUMO E3 ligase, represses LEF1 activity by sequestration into nuclear bodies. *Genes Dev.* 15:3088–3103.

Saitoh, H. and Hinchey, J. (2000) Functional heterogeneity of small ubiquitin-related protein modifiers SUMO-1 versus SUMO-2/3. *J. Biol. Chem.* 275:6252–6258.

Saitoh, H., Sparrow, D.B., Shiomi, T., et al. (1998) Ubc9p and the conjugation of SUMO-1 to RanGAP1 and RanBP2. *Curr. Biol.* 8:121–124.

Sampson, D.A., Wang, M., and Matunis, M.J. (2001) The small ubiquitin-like modifier-1 (SUMO-1) consensus sequence mediates Ubc9 binding and is essential for SUMO-1 modification. *J. Biol. Chem.* 276:21664–21669.

Santti, H., Mikkonen, L., Hirvonen-Santti, S., et al. (2003) Identification of a short PIASx gene promoter that directs male germ cell-specific transcription in vivo. *Biochem. Biophys. Res. Commun.* 308:139–147.

Schimmel, J., Larsen, K.M., Matic, I., et al. (2008) The ubiquitin-proteasome system is a key component of the SUMO-2/3 cycle. *Mol. Cell. Proteomics* 7:2107–2122.

Seufert, W., Futcher, B., and Jentsch, S. (1995) Role of a ubiquitin-conjugating enzyme in degradation of S- and M-phase cyclins. *Nature* 373:78–81.

Shalizi, A., Gaudilliere, B., Yuan, Z., et al. (2006) A calcium-regulated MEF2 sumoylation switch controls postsynaptic differentiation. *Science* 311:1012–1017.

Shen, Z., Pardington-Purtymun, P.E., Comeaux, J.C., et al. (1996) UBL1, a human ubiquitin-like protein associating with human RAD51/RAD52 proteins. *Genomics* 36: 271–279.

Shen, T.H., Lin, H.K., Scaglioni, P.P., et al. (2006) The mechanisms of PML-nuclear body formation. *Mol. Cell* 24:331–339.

Singer, J.D., Gurian-West, M., Clurman, B., et al. (1999) Cullin-3 targets cyclin E for ubiquitination and controls S phase in mammalian cells. *Genes Dev.* 13:2375–2387.

Song, J., Durrin, L.K., Wilkinson, T.A., et al. (2004) Identification of a SUMO-binding motif that recognizes SUMO-modified proteins. *Proc. Natl. Acad. Sci. USA* 101:14373–14378.

Stankovic-Valentin, N., Deltour, S., Seeler, J., et al. (2007) An acetylation/deacetylation-SUMOylation switch through a phylogenetically conserved psiKXEP motif in the tumor suppressor HIC1 regulates transcriptional repression activity. *Mol. Cell. Biol.* 27:2661–2675.

Steinacher, R. and Schar, P. (2005) Functionality of human thymine DNA glycosylase requires SUMO-regulated changes in protein conformation. *Curr. Biol.* 15:616–623.

Stelter, P. and Ulrich. H.D. (2003) Control of spontaneous and damage-induced mutagenesis by SUMO and ubiquitin conjugation. *Nature* 425:188–191.

Sun, H., Leverson, J.D., and Hunter, T. (2007) Conserved function of RNF4 family proteins in eukaryotes: targeting a ubiquitin ligase to SUMOylated proteins. *EMBO J.* 26:4102–4112.

Takahashi, Y., Mizoi, J., Toh, E., et al. (2000) Yeast Ulp1, an Smt3-specific protease, associates with nucleoporins. *J. Biochem.* 128:723–725.

Takahashi, H., Hatakeyama, S., Saitoh, H., et al. (2005) Noncovalent SUMO-1 binding activity of thymine DNA glycosylase (TDG) is required for its SUMO-1 modification and colocalization with the promyelocytic leukemia protein. *J. Biol. Chem.* 280:5611–5621.

Tatham, M.H., Jaffray, E., Vaughan, O.A., et al. (2001) Polymeric chains of SUMO-2 and SUMO-3 are conjugated to protein substrates by SAE1/SAE2 and Ubc9. *J. Biol. Chem.* 276:35368–35374.

Tatham, M.H., Geoffroy, M.C., Shen, L., et al. (2008) RNF4 is a poly-SUMO-specific E3 ubiquitin ligase required for arsenic-induced PML degradation. *Nat. Cell Biol.* 10:538–546.

Ulrich, H.D. (2005) Mutual interactions between the SUMO and ubiquitin systems: a plea of no contest. *Trends Cell Biol.* 15:525–532.

Ulrich, H.D. (2008) The fast-growing business of SUMO chains. *Mol. Cell* 32:301–305.

Ulrich, H.D. (2009) The SUMO system: an overview. *Methods Mol. Biol.* 497:3–16.

Vertegaal, A.C. (2007) Small ubiquitin-related modifiers in chains. *Biochem Soc. Trans.* 35:1422–1423.

Vertegaal, A.C., Ogg, S.C., Jaffray, E., et al. (2004) A proteomic study of SUMO-2 target proteins. *J. Biol. Chem.* 279: 33791–33798.

Vertegaal, A.C., Andersen, J.S., Ogg, S.C., et al. (2006) Distinct and overlapping sets of SUMO-1 and SUMO-2 target proteins revealed by quantitative proteomics. *Mol. Cell. Proteomics* 5:2298–2310.

Weger, S., Hammer, E., and Heilbronn, R. (2005) Topors acts as a SUMO-1 E3 ligase for p53 in vitro and in vivo. *FEBS Lett.* 579:5007–5012.

Wilson, V.G. and Heaton, P.R. (2008) Ubiquitin proteolytic system: focus on SUMO. *Expert Rev. Proteomics* 5:121–135.

Wohlschlegel, J.A., Johnson, E.S., Reed, S.I., et al. (2004) Global analysis of protein sumoylation in Saccharomyces cerevisiae. *J. Biol. Chem.* 279:45662–45668.

Wong, K.A., Kim, R., Christofk, H., et al. (2004) Protein inhibitor of activated STAT Y (PIASy) and a splice variant lacking exon 6 enhance sumoylation but are not essential for embryogenesis and adult life. *Mol. Cell. Biol.* 24:5577–5586.

Wong. J.J., Pung, Y.F., Sze, N.S., et al. (2006) HERC5 is an IFN-induced HECT-type E3 protein ligase that mediates type I IFN-induced ISGylation of protein targets. *Proc. Natl. Acad. Sci. USA* 103:10735–10740.

Wu, K., Yamoah, K., Dolios, G., et al. (2003) DEN1 is a dual function protease capable of processing the C terminus of Nedd8 and deconjugating hyper-neddylated CUL1. *J. Biol. Chem.* 278:28882–28891.

Wykoff, D.D. and O'Shea, E.K. (2005) Identification of sumoylated proteins by systematic immunoprecipitation of the budding yeast proteome. *Mol. Cell. Proteomics* 4:73–83.

Xirodimas, D.P. (2008) Novel substrates and functions for the ubiquitin-like molecule NEDD8. *Biochem. Soc. Trans.* 36:802–806.

Xirodimas, D.P., Saville, M.K., Bourdon, J.C., et al. (2004) Mdm2-mediated NEDD8 conjugation of p53 inhibits its transcriptional activity. *Cell* 118:83–97.

Xu, Z. and Au, S.W. (2005) Mapping residues of SUMO precursors essential in differential maturation by SUMO-specific protease, SENP1. *Biochem. J.* 386:325–330.

Yang, S.H. and Sharrocks, A.D. (2004) SUMO promotes HDAC-mediated transcriptional repression. *Mol. Cell* 13:611–617.

Yang, S.H., Jaffray, E., Hay, R.T., et al. (2003) Dynamic interplay of the SUMO and ERK pathways in regulating Elk-1 transcriptional activity. *Mol. Cell* 12:63–74.

Yang, S.H., Galanis, A., Witty, J., et al. (2006) An extended consensus motif enhances the specificity of substrate modification by SUMO. *EMBO J.* 25:5083–5093.

Zhang, H., Saitoh, H., and Matunis, M.J. (2002) Enzymes of the SUMO modification pathway localize to filaments of the nuclear pore complex. *Mol. Cell. Biol.* 22:6498–6508.

Zhang, X.D., Goeres, J., Zhang, H., et al. (2008) SUMO-2/3 modification and binding regulate the association of CENP-E with kinetochores and progression through mitosis. *Mol. Cell* 29:729–741.

Zhao, X. and Blobel, G. (2005) A SUMO ligase is part of a nuclear multiprotein complex that affects DNA repair and chromosomal organization. *Proc. Natl. Acad. Sci. USA* 102:4777–4782.

Zhao, C., Beaudenon, S.L., Kelley, M.L., et al. (2004a) The UbcH8 ubiquitin E2 enzyme is also the E2 enzyme for ISG15, an IFN-alpha/beta-induced ubiquitin-like protein. *Proc. Natl. Acad. Sci. USA* 101:7578–7582.

Zhao, Y., Kwon, S.W., Anselmo, A., et al. (2004b) Broad spectrum identification of cellular small ubiquitin-related modifier (SUMO) substrate proteins. *J. Biol. Chem.* 279: 20999–21002.

Zhou, W., Ryan, J.J., and Zhou, H. (2004) Global analyses of sumoylated proteins in Saccharomyces cerevisiae. Induction of protein sumoylation by cellular stresses. *J. Biol. Chem.* 279:32262–32268.

Zou, W. and Zhang, D.E. (2006) The interferon-inducible ubiquitin-protein isopeptide ligase (E3) EFP also functions as an ISG15 E3 ligase. *J. Biol. Chem.* 281:3989–3994.

Chapter 14
ER-associated Degradation and Its Involvement in Human Disease: Insights from Yeast

Nathalie Campagnolo and Michel Ghislain

14.1 Introduction

Proteins destined for the secretory pathway are synthesized on cytosolic ribosomes and transported, either co- or post-translationally, to the endoplasmic reticulum (ER) (Brodsky 1996). The ER is endowed with a quality control system that monitors newly synthesised proteins for correct chain folding and modification, as well as assembly into functional macromolecular complexes (Kopito 1997; Ellgaard and Helenius 2003; Sitia and Braakman 2003). Aberrant and misfolded proteins are retro-translocated to the cytosol, where they are ubiquitylated and eliminated by the 26S proteasome (Werner et al. 1996; Plemper and Wolf 1999; Hampton 2002; Kostova and Wolf 2003; McCracken and Brodsky 2003; Meusser et al. 2005). ER-associated protein degradation (ERAD) comprises at least three steps: (i) the recognition of a misfolded or unassembled polypeptide by the quality control system, (ii) its retro-translocation (or dislocation) into the cytosol across the ER membrane and (iii) ubiquitin-mediated degradation by the proteasome (Fig. 14.1). A second mechanism for dealing with misfolded proteins is termed the unfolded protein response (UPR), which increases the ER folding capacity and may trigger apoptosis (Ron and Walter 2007). A third mechanism is the sequestration of misfolded proteins in special subcellular compartments and autophagy (Klionsky 2007).

This review focuses on the budding yeast *S. cerevisiae*, a unicellular eukaryote that is highly amenable to genetic and biochemical analyses (Sommer and Jentsch 1993; Hampton et al. 1996; Hiller et al. 1996; McCracken and Brodsky 1996). Studies using yeast and mammalian cells have shown that the basic events of the ERAD pathway are conserved in eukaryotes (Ward et al. 1995; Lenk et al. 2002; Gnann et al. 2004). Hereafter, genes and proteins that are conserved between yeast and mammals, but have different names, will be prefixed by *y-* or *m-* (Table 14.1).

M. Ghislain (✉)
Université Catholique de Louvain, Institut des Sciences de la Vie, Unité de Biochimie Physiologique, Croix du Sud, 4-5, boîte 15, B-1348, Louvain-la-Neuve, Belgium
e-mail: michel.ghislain@uclouvain.be

C.J. Vidal (ed.), *Post-Translational Modifications in Health and Disease*, Protein Reviews 13, DOI 10.1007/978-1-4419-6382-6_14, © Springer Science+Business Media, LLC 2011

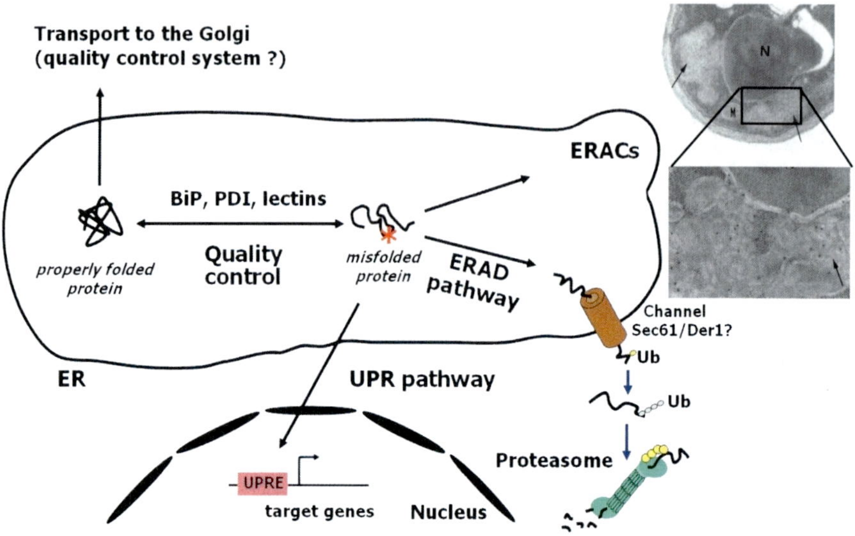

Fig. 14.1 Overview of ER protein quality control and proteasome mediated degradation. Newly synthesised secretory and membrane proteins are monitored for non-native conformation by a quality control system consisting of various ER folding sensors (y-Kar2p/m-BiP, PDI and lectin-like chaperones). Unfolded or misfolded proteins are transported across the ER membrane into the cytosol, where they are marked for proteasomal elimination by E3 ligase-mediated conjugation to ubiquitin (Ub). The accumulation of unfolded and misfolded proteins in the ER induces UPR, resulting in the activation of genes coding for ER chaperones and proteins involved in the secretory and ERAD pathways. Another mechanism for overcoming the toxicity linked to elevated amounts of misfolded proteins is their sequestration into ER-associated compartments (ERACs). The thin-section electron micrograph shows ERACs (*arrows*) in a yeast cell expressing a mislocalised variant of the plasma membrane H^+-ATPase (Pma1-D378N). An antibody conjugated to gold spheres (*black dots*) has been used to locate Kar2p molecules. The nucleus (N), mitochondria (M) and vacuoles can be seen (Courtesy of André Goffeau, 1995)

14.2 Recognition of Aberrant Proteins in the ER

Newly synthesised proteins undergo several post-translational modifications in the ER, including the removal of the signal sequence, disulphide bond formation, *N*-glycosylation and the addition of glycosylphosphatidylinositol. Protein folding involves ER-resident molecular chaperones of the heat shock protein (Hsp) 70 family, such as *y*-Kar2p/*m*-BiP (Kozutsumi et al. 1989; Normington et al. 1989; Rose et al. 1989) and its nucleotide exchange factor, *y*-Lhs1/*m*-ORP150 (Saris et al. 1997; Takeuchi 2006). Disulphide-bond formation is catalysed by protein disulphide isomerases (PDIs), including *y*-Pdi1p, 4 other non-essential Pdi1-related proteins (Norgaard et al. 2001) and *y*-Ero1p (Fassio and Sitia 2002; Gross et al. 2006). These oxidoreductases ensure that irreversible formation of disulphide bonds does not occur until protein folding is complete. However, no significant effects on ERAD are seen in various *PDI*-deleted strains (Norgaard et al. 2001). The attachment of

Table 14.1 Mammalian counterparts of the yeast ERAD components and associated diseases

Activity/function	Yeast	Mammalian	Disease/disorder
E2s	Ubc6	UBE2J1, UBE2J2	Cystic fibrosis
	Ubc7	UBE2G1, UBEG2	
E3s	Hrd1-Hrd3	Synoviolin/HRD1-SEL1	Net sodium imbalance, rheumatoid arthritis, cystic fibrosis, accumulation of Pael-R
	Doa10	TEB4[a]	PFIC II
		RMA1, RFP2, Kf-1	Cystic fibrosis, PFIC II
Cdc48-Ufd1-Npl4 complex	Cdc48	p97 or VCP	Inclusion body myopathy with Paget disease of bone and frontotemporal dementia (Weihl et al. 2006)
Cdc48 receptor	Ubx2	VIMP[b]	
Dislocation pore?	Der1	DERLIN	Cystic fibrosis, ALS
	Usa1	HERP	Enhancement of β-amyloid protein generation (Sai et al. 2002)
Substrate adaptors for the proteasome	Dsk2	Ubiquilin1	Alzheimer's disease
	Rad23	HHR23A and B	Machado-Joseph disease

[a]Sequence identity found in the N-terminal RING-CH domain and an internal 130-residue block, the TEB4-Doa10 (TD) domain, that might be involved in retro-tranlocation (Wang et al. 2008)
[b]Sequence homology is restricted to the UBX domain

N-linked carbohydrates is also essential for the maturation of secretory and membrane proteins. The added glycans stabilize the protein conformation by interacting with the hydrophobic protein surface and also act as substrates for *m*-calnexin (*y*-Cne1p) and *m*-calreticulin, ER lectins that assist the folding of some glycoproteins (Song et al. 2001; Helenius and Aebi 2004). Note that evidence for a canonical calnexin cycle in yeast is lacking. It is also not known which of the several yeast peptidyl-prolyl isomerases (cyclophilins and FK506-binding proteins) are responsible for protein folding in the ER, as none of these are required for cell viability (Dolinski et al. 1997). The mammalian enzymes involved in the secretory protein factory have recently been reviewed (Anelli and Sitia 2008).

As folding sensors, ER luminal Hsp70s, lectins and PDIs are key components of the quality control system (Bukau et al. 2006). These molecular chaperones monitor the thermodynamic stability of non-native proteins and the presence of unpaired cysteine residues or immature glycans (Sitia and Braakman 2003; Krebs et al. 2004). Folding or assembly intermediates are retained in the ER and allowed to undergo further folding attempts. It is still not clear how molecules that have not had the time to fold are distinguished from those that have failed to fold after many attempts and must therefore be eliminated. According to the mannose timer hypothesis (Jakob et al. 1998; Nakatsukasa et al. 2001; Ellgaard and Helenius 2003), the terminal mannose residue of the $Man_9GlcNAC_2$ oligosaccharide on misfolded glycosylated proteins is removed by α1,2-mannosidase I (*y*-Mns1p), producing the sugar moiety $Man_8GlcNAC_2$. The mannosidase reaction is slow, so only terminally misfolded proteins that spend an excessive amount of time in the ER will be trimmed. The Man_8-species are recognized by α-mannosidase-like lectins so they may be trimmed more extensively. However, as all folded glycoproteins also contain a Man_8 oligosaccharide, it seems likely that the signal for ERAD consists of a polypeptide determinant in addition to the glycan structure (Byrd et al. 1982; Spear and Ng 2005). The mannose timer hypothesis has been challenged in the fission yeast *Schizosaccharomyces pombe*, as ER mannose trimming does not appear to be required for ERAD (Movsichoff et al. 2005).

14.2.1 Model ERAD Substrates

Several model proteins have been used to identify the ERAD components in yeast. Hydroxy-methylglutaryl-CoA reductase (*y*-Hmg2p) catalyses the synthesis of mevalonate, a precursor of activated isoprenes and sterols. Following the accumulation of sterols and other isoprenoids, *y*-Hmg2p undergoes rapid ER-associated degradation (Hampton et al. 1996). Other ERAD substrates are mutant forms of secretory or membrane proteins that fail to fold properly, as judged by an altered trypsin digestion pattern. Transmembrane chimeras have also been constructed that contain misfolded segments exposed on the cytosolic side of the ER membrane, the luminal side or both. Two different ERAD pathways can be distinguished according to the topology of the misfolded lesion (Fig. 14.2). ER luminal proteins and membrane proteins with lesions in a luminal domain are degraded via the so-called ERAD-L pathway, whereas membrane proteins with cytosolic lesions are degraded via the ERAD-C pathway (Ahner and Brodsky 2004).

14.2.1.1 Soluble Luminal ERAD Substrates

A mutant form of vacuolar carboxypeptidase Y, named CPY*, is retained in the ER, as shown by the lack of Golgi-specific carbohydrate modification, and is rapidly degraded by the ubiquitin-proteasome system (Finger et al. 1993; Hiller et al. 1996). Retrograde transport of CPY* through the Sec61 translocon depends on Sc-Kar2p and the Hsp40 co-chaperones *y*-Jem1p and *y*-Scj1p (the role of *y*-Sec63p is controversial) (Plemper et al. 1997; Nishikawa et al. 2001). These may play an important role in degradation by remodelling the aberrant proteins into a partially unfolded or reduced form, more amenable for retro-translocation. A non-glycosylated version of CPY* is not degraded, suggesting a role for *N*-linked glycans in entry into the ERAD pathway (Knop et al. 1996). CPY* degradation is impaired by mutation of glucosidases I and II (Jakob et al. 1998; Hitt and Wolf 2004a) or of an α-mannosidase-like lectin, named Mnl1p or Htmp1 in yeast (Jakob et al. 2001; Nakatsukasa et al. 2001) or EDEM (ER degradation enhancing α-mannosidase-like protein) in mammals (Hosokawa et al. 2001; Molinari et al. 2003). Mutagenesis analysis has revealed that a single, specific *N*-linked glycan at the C-terminus of CPY is required for sorting

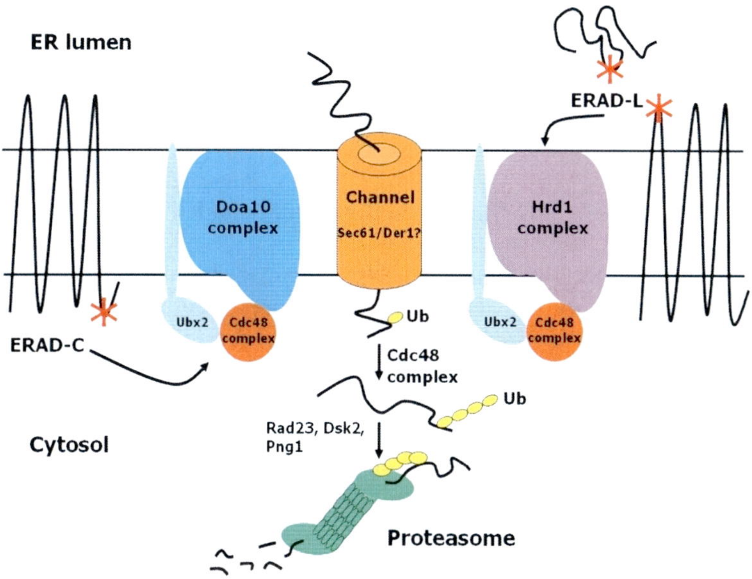

Fig. 14.2 Proteins with luminal and cytosolic lesions are targeted for proteasome-mediated degradation through distinct protein-ubiquitin ligase complexes. Membrane proteins with misfolded cytosolic domains are substrates of the ubiquitin ligase *y*-Doa10p (ERAD-C pathway), whereas membrane and soluble proteins with luminal misfolded domains are substrates of the ubiquitin ligase *y*-Hrd1p (ERAD-L pathway). A third checkpoint that screens transmembrane domains (ERAD-M) and can activate *y*-Hrd1p is not represented. Ubiquitylated substrates are extracted from the ER membrane by the Cdc48-Ufd1p-Npl4 complex, which is recruited to the ER membrane via its interaction with *y*-Ubx2p, and are delivered to the 26S proteasome by *y*-Rad23p and *y*-Dsk2p. The identity of the dislocation pore (*y*-Der1 or Sec61) is still a matter of debate

into the ERAD pathway, supporting the bipartite nature of the ERAD signal (Spear and Ng 2005).

Studies on mammalian cells have suggested that *m*-EDEM1 helps misfolded glycoproteins leave the calnexin/calreticulin cycle, where they are attempting to fold, and maintains substrate solubility for retro-translocation (Oda et al. 2003; Kanehara et al. 2007). *m*-EDEM associates with ER-resident *m*-BiP (Hsp70) and *m*-ERdj5 (Hsp40), a co-chaperone which cleaves the disulphide bonds of misfolded proteins (Ushioda et al. 2008). The related proteins *m*-EDEM2 and *m*-EDEM3 accelerate ERAD of a misfolded α1-antitrypsin variant (Mast et al. 2005; Hirao et al. 2006). In yeast, Htm1p functions as an α1,2-specific exo-mannosidase that generates the $Man_7GlcNAc_2$ oligosaccharide with a terminal α1,6-linked mannosyl residue on degradation substrates. Htm1p activity requires processing of the *N*-glycan by glucosidase I, glucosidase II and mannosidase I, resulting in a sequential order of specific *N*-glycan structures that reflect the folding status of the glycoprotein (Clerc et al. 2009). A second class of lectin-like ERAD factors consists of mammalian OS-9 and erlectin/XTP3-B proteins and yeast Yos9p, which contain a mannose-6-phosphate receptor homology domain (Buschhorn et al. 2004;

Bhamidipati et al. 2005; Kim et al. 2005; Szathmary et al. 2005). It has been proposed that *y*-Yos9p has a proof-reading function in scanning for the Man_8 determinant in substrates ready to be retro-translocated (Denic et al. 2006; Gauss et al. 2006a; Hirsch et al. 2006).

14.2.1.2 Membrane ERAD Substrates with a Misfolded Cytosolic Domain

y-Ste6p is an ATP-binding cassette (ABC) transporter that functions at the plasma membrane to export the **a**-factor mating pheromone from the cell (McGrath and Varshavsky 1989). Ste6p is slowly targeted for degradation in the vacuole through ubiquitin-mediated endocytosis (Kolling and Hollenberg 1994). However, a C-terminally truncated variant, Ste6-Q1249-X, is retained in the ER and degraded through the ERAD pathway (Loayza et al. 1998). *y*-Kar2p is dispensable, whereas the cytosolic Hsp70 *y*-Ssa1p chaperone and the Hsp40 co-chaperones *y*-Ydj1p and *y*-Hlj1p are not (Huyer et al. 2004b). *y*-Ydj1p is tethered to the ER membrane by prenylation, whereas *y*-Hlj1p is a C-terminally anchored ER membrane protein. These results indicate that ERAD of Ste6-Q1249-X employs different machinery from that of the soluble luminal substrate CPY*. *y*-Ssa1p and *y*-Ydj1p (but not *y*-Hlj1p) are required for the degradation of Ura3p-CL1, a cytosolic soluble protein with a transplantable degradation signal (Metzger et al. 2008). As its degradation also requires the ER-localized ubiquitin conjugating and ligase enzymes, it is likely that the cytosolic face of the ER membrane serves as a platform for the degradation of Ura3p-CL1 and other cytosolic misfolded proteins (Metzger et al. 2008).

y-Pma1p is a plasma membrane H^+-ATPase of the P-type family, which pumps protons out of the cell at the expense of ATP hydrolysis. The generated proton chemical gradient drives the uptake of nutrients and ions (Serrano et al. 1986). The Pma1-D378N variant is poorly folded and accumulates in ER-associated compartments (ERACs) prior to ER-associated degradation (Harris et al. 1994; Nakamoto et al. 1998). The mutant pump binds *y*-Esp1p, an ER membrane PDI that functions in protein quality control. The loss of *y*-Esp1 allows Pma1-D378N to travel to the plasma membrane (Wang and Chang 1999). No stabilisation of Ste6-Q1249-X is seen in cells lacking *y*-Esp1, indicating that *y*-Esp1p is a specific recognition factor (Huyer et al. 2004b).

14.2.1.3 Membrane-Associated ERAD Substrates with a Luminal Lesion

y-Pdr5p is a plasma membrane multidrug ABC transporter which mediates the efflux of a large variety of cytotoxic compounds (Balzi et al. 1994; Bissinger and Kuchler 1994). A mutant form, Pdr5-C1427Y (Pdr5*), is retained in the ER membrane and targeted for proteasomal degradation through the ERAD pathway (Egner et al. 1998; Plemper et al. 1998). There is no evidence that Pdr5* is misfolded, although its glycosylation pattern differs from that of intact *y*-Pdr5p (Plemper et al. 1998). It has been proposed that replacement of the cysteine residue affects the formation of disulphide bonds with other cysteine residues facing the ER lumen

(Egner et al. 1998). The chaperone lectin *y*-Htm1p, but not *y*-Kar2p, is involved in the degradation process (Plemper et al. 1998; Jakob et al. 2001; Gnann et al. 2004).

The CTG* chimera, consisting of CPY* fused to the green fluorescent protein (GFP) via a transmembrane segment from *y*-Pdr5p, is degraded through the ERAD pathway (Taxis et al. 2003). *y*-Kar2p is dispensable, whereas the cytosolic chaperone proteins *y*-Ssa1p, *y*-Hlj1p, *y*-Cwc23p and *y*-Jid1p are not. This indicates that the recognition of the misfolded CPY* moiety in CTG* is not the primary step targeting the protein chimera for ERAD. However, the misfolded CPY* domain might be responsible for a signalling event that recruits the cytosolic chaperones onto the stable GFP moiety for the unfolding of this domain to occur before proteasomal degradation (Ahner and Brodsky 2004).

14.2.1.4 Transmembrane Proteins Exposing Misfolded Domains to the ER Lumen and Cytosol

The use of different ERAD substrates in yeast has revealed the existence of the ERAD-C and ERAD-L checkpoints that monitor the folding of the cytosolic and luminal domains of newly synthesised proteins (see Fig. 14.2). What is the fate of a transmembrane protein with two lesions, one exposed to the ER lumen and the other to the cytosol? This question was addressed by analysing the fate of the KSS chimera, consisting of Ste6-Q1249-X fused to an ER luminal misfolded artificial domain. KSS turned out to be degraded as efficiently in cells lacking *y*-Mnl1p as in wild-type cells, even though the misfolded ER luminal domain confers Mnl1p-dependent degradation when transplanted to another stable transmembrane protein (Vashist and Ng 2004).

The use of modular ERAD substrates has shown that nascent transmembrane proteins are monitored by a two-step quality control system. The first, the ERAD-C checkpoint, is located at the cytosolic face of the ER membrane. If a lesion is detected, the membrane protein is ubiquitylated and degraded by the proteasome. If the cytosolic domains are found to be correctly folded, the transmembrane protein will pass to the second checkpoint, ERAD-L, which checks the luminal domains and targets misfolded proteins for degradation through a pathway shared by soluble secretory proteins (Vashist and Ng 2004). The existence of different ERAD pathways in mammalian cells has been recently analyzed using the plasma membrane V2R vasopressin receptor and three mutant forms that contain a misfolded cytosolic, luminal or intramembrane domain and are retained in different intracellular compartments (Schwieger et al. 2008). Unlike the wild-type protein, each variant is targeted for proteasomal degradation through a seemingly similar ERAD pathway involving *m*-DERLINs (see below) (Schwieger et al. 2008).

14.2.2 ER-Associated Ubiquitylation Machinery

Conjugation of ubiquitin to ERAD substrates occurs during their retro-translocation and is mediated by the E2 ubiquitin-conjugating (Ubc) enzymes *y*-Ubc6p, *y*-Ubc7p

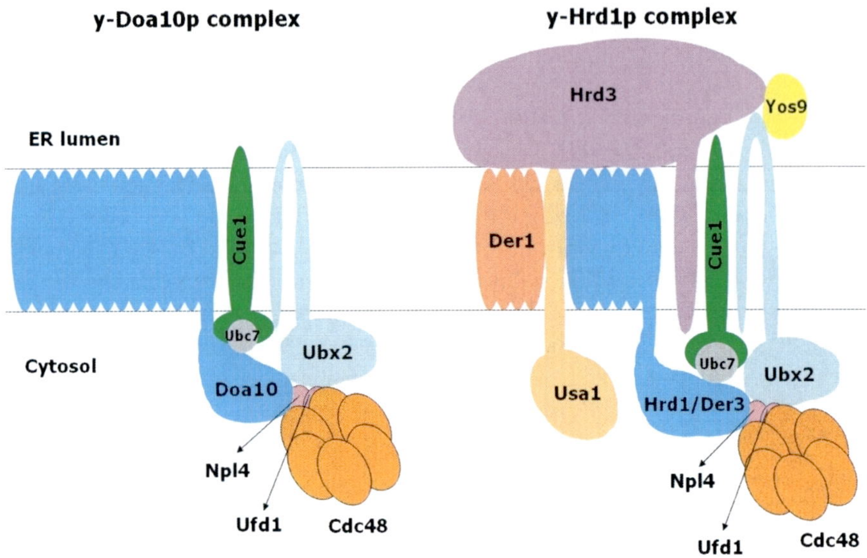

Fig. 14.3 Yeast ER-associated ubiquitin ligase complexes. The Doa10 complex of the ERAD-C pathway is composed of the ubiquitin ligase Doa10p, the ubiquitin-conjugating enzyme Ubc7p and its membrane-anchoring factor *y*-Cue1p. The Hrd1 complex of the ERAD-L pathway contains a large ER lumenal region consisting of *y*-Hrd3p and *y*-Yos9 that may serve as the receptor site for ERAD-L substrates. *y*-Usa1p interacts with *y*-Der1p, a putative constituent of the retro-translocation pore. Note that the ER membrane ubiquitin conjugating enzyme Ubc6p is not found in these complexes. *y*-Ubx2 acts as a membrane anchor of the Cdc48-Ufd1-Npl4 complex, which is involved in the extraction of ubiquitylated substrates from the ER membrane (Carvalho et al. 2006; Denic et al. 2006)

and *y*-Ubc1p (Hiller et al. 1996; Friedlander et al. 2000). The E3 ubiquitin ligases *y*-Hrd1p and *y*-Doa10p belong to the RING zinc finger domain family (Hampton et al. 1996; Plemper et al. 1998; Huyer et al. 2004b). They act as separate scaffolds for the assembly of components involved in substrate recognition, extraction and ubiquitylation for the ERAD-L and-C pathways (Fig. 14.3). In the case of some ERAD substrates, elongation of the polyubiquitin chain requires *y*-Ufd2p (Koegl et al. 1999; Richly et al. 2005; Nakatsukasa et al. 2008).

14.2.2.1 The y-Doa10 Complex

y-Doa10p (Degradation of soluble Matα2 transcription repressor), alias y-Ssm4p, is a 14-transmembrane segment protein of the ER/nuclear envelope, both termini facing the cytosol (Swanson et al. 2001; Kreft et al. 2006). *y*-Doa10p is part of a complex that includes both *y*-Ubc7p, which is tethered to the ER membrane via its association with *y*-Cue1p (Biederer et al. 1997), and the Cdc48-Ufd1-Npl4 complex, which is recruited to the membrane via *y*-Ubx2p (Schuberth et al. 2004; Neuber et al. 2005). The second ERAD E2 enzyme, *y*-Ubc6p, is anchored to the ER membrane through its C-terminal hydrophobic tail. *y*-Doa10 is able to target cytosolic (Ura3p-CL1), nuclear (Matα2p) and ER membrane (Ste6-Q1249-X) proteins for proteasomal degradation (Metzger et al. 2008). The Cdc48-Ufd1-Npl4 complex facilitates the degradation of membrane-embedded Doa10 substrates, but is not required for soluble Doa10 substrates. Thus, while *y*-Doa10p ubiquitinates both membrane and soluble proteins, the mechanisms of the subsequent proteasome targeting differ (Ravid et al. 2006).

14.2.2.2 The y-Hrd1 Complex

y-Hrd1p (HmgCoA reductase degradation), alias *y*-Der3p (Degradation in the ER), is a 551 amino acid protein with six transmembrane spans and both termini facing the cytosol (Gardner et al. 2000; Deak and Wolf 2001). *y*-Hrd1p assembles with *y*-Ubc7p-Cue1p in a RING finger-dependent manner (Hampton et al. 1996; Friedlander et al. 2000). Physical association with *y*-Ubc1p and *y*-Ubc6p has been reported (Bays et al. 2001a). *y*-Hrd1p also interacts with the Cdc48 complex and *y*-Hrd3p, a membrane protein with a large ER-luminal domain that functions as a substrate receptor (Hampton et al. 1996; Gardner et al. 2000). In the absence of *y*-Hrd3p, *y*-Hrd1p is rapidly degraded by the ubiquitin-proteasome system (Plemper et al. 1999; Gardner et al. 2000). The interaction between *y*-Hrd1p and the Cdc48-Ufd1-Npl4 complex requires the membrane protein *y*-Ubx2p, which contains an N-terminal UBA domain, known to bind to the ubiquitin chain of ERAD substrates, and a C-terminal ubiquitin-like (UBX) domain, which binds Cdc48 (Decottignies et al. 2004; Schuberth et al. 2004; Neuber et al. 2005).

The transmembrane protein *y*-Der1p is linked to *y*-Hrd1p via *y*-Hrd3p and *y*-Usa1p, the latter of which has no known function (Carvalho et al. 2006; Gauss et al. 2006b). Deletion of the *USA1* gene is synthetically lethal with deletion of *IRE1*, which encodes a key player in the UPR (Carvalho et al. 2006). The lectin-like *y*-Yos9p physically and mechanistically interacts with *y*-Hrd1p by binding to a luminal domain of *y*-Hrd3 and to the chaperone *y*-Kar2p (Carvalho et al. 2006; Denic et al. 2006; Gauss et al. 2006a). The Kar2p-Yos9p-Hrd3p luminal complex is not disrupted by the loss of *y*-Hrd1p, but these proteins no longer interact with the Cdc48p complex (Carvalho et al. 2006; Denic et al. 2006).

14.2.3 Retro-translocation into the Cytosol and Delivery to the 26S Proteasome

The proteasomal degradation of ER luminal and membrane proteins requires their retrograde transport out of the ER back to the cytosol (see Fig. 14.2). The identity of the protein components responsible for retro-translocation is still debated, the potential candidates being the Sec61p translocation channel, Der1p and Doa10p (Nakatsukasa and Brodsky 2008).

Sec61p is an obvious candidate for the dislocation channel because of its role in the translocation of newly synthesised proteins in the ER. In mammalian cells, the Sec61 complex associates with MHC class I heavy chains that are targeted for ERAD by the human cytomegalovirus-encoded glycoprotein US2 (Wiertz et al. 1996). Studies in yeast also indicate that the Sec61 channel functions in the export of a misfolded secretory protein (CPY* and derivatives) or membrane protein (Pdr5*) for proteasomal degradation (Pilon et al. 1997; Plemper et al. 1997, 1998; Willer et al. 2008). Instrumental in these studies were the isolation of *sec61* alleles that are defective for protein export, but not for import, at an appropriate temperature (Pilon et al. 1997) and the modification of the translocation properties of model ERAD substrates (Willer et al. 2008). Formation of a disulphide-linked intermediate complex between Sec61p and a transmembrane ERAD substrate has been recently reported (Scott and Schekman 2008). These authors proposed that Sec61p acts directly in the retro-translocation of ERAD substrates or indirectly in remodelling membrane proteins before they are transferred to a separate ER channel for retro-tranlocation.

A role for Derlin-1 as the ER retro-translocation channel is based on its interaction with US11, a virally-encoded ER protein that targets MHC1 class I heavy chains for retro-translocation, and with VIMP, a membrane protein that recruits the cytosolic Cdc48-Ufd1-Npl4 complex to the ER membrane (Lilley and Ploegh 2004; Ye et al. 2004). Derlin-1 also facilitates the retro-translocation of cholera toxin (Bernardi et al. 2008). Real-time fluorescence analysis of pro-α factor retro-translocation in an *in vitro* mammalian system has shown a requirement for Derlin-1, but not Sec61α (Wahlman et al. 2007). The yeast Derlin-1 homolog (*y*-Der1p) is required for the efficient degradation of CPY* (Knop et al. 1996; Hitt and Wolf 2004b), but not of CTG* and other membrane ERAD substrates (Taxis et al. 2003; Huyer et al. 2004b; Vashist and Ng 2004). ERADication of these Der1p- and Sec61p-independent substrates requires *y*-Doa10p (Walter et al. 2001; Vashist and Ng 2004; Kreft et al. 2006), and it has been proposed that a subset of the Doa10 transmembrane segments might form a retro-translocation channel (Kreft et al. 2006).

The driving force for substrate retro-translocation can be derived from three non-mutually exclusive sources: chaperone-mediated extraction, (poly)ubiquitin-mediated ratcheting and proteasome-mediated retro-translocation (Tsai et al. 2002; McCracken and Brodsky 2003). Single-spanning membrane proteins, such as MHC class I heavy chain and the α subunit of the T cell receptor, may be completely dislocated from the membrane to the cytosol before their degradation by the proteasome. Polytopic membrane proteins may be processively dislocated and degraded from either the N- or the C-terminus by the proteasome. It is also possible that cytoplasmic loop(s) of the substrates may first be "clipped" by the endoproteolytic activity of the proteasome, then dislocated or extracted from the ER membrane (Nakatsukasa and Brodsky 2008). Retrograde transport of soluble proteins through the protein-conducting channel is conceptually more difficult. During import into the ER the N-terminal signal peptide targets these proteins to the ER membrane and into the opening in the translocon. In the ER lumen, the signal peptide is removed and the opening of the channel from the luminal side must therefore be triggered by a mechanism different from that used during protein import (Romisch 1999).

Ubiquitylated ERAD substrates are released from the ER by a soluble ubiquitin-specific chaperone complex consisting of Cdc48p, Ufd1p and Npl4p (Bays et al. 2001b; Hitchcock et al. 2001; Ye et al. 2001; Braun et al. 2002; Jarosch et al. 2002; Rabinovich et al. 2002). Yeast Cdc48p, named p97 or valosin-containing protein (VCP) in mammals, has two ATP-binding domains and two Second Regions of Homology (SRH) motifs (Frohlich et al. 1991). X-ray crystallographic studies showed six subunits assembled into a ring, the overall shape of which changes during the ATPase catalytic cycle (De LaBarre and Brunger 2003; Dreveny et al. 2004). A nucleotide-dependent conformational switch may apply tension to bound proteins and thereby allow polypeptide unfolding (Bukau et al. 2006). Cdc48p/p97 binds to the C-terminal domain of Ufd1p and to the N-terminus of the zinc-binding motif protein *y*-Npl4p (Meyer et al. 2000; Rape et al. 2001; Bruderer et al. 2004). The N-terminal domain of *y*-Ufd1, which is structurally similar to that of p97, and the Npl4 zinc finger contain binding sites for ubiquitin (Alam et al. 2004; Park et al. 2005). The yeast Cdc48-Ufd1-Npl4 complex is recruited to the ER membrane by the UBX domain-containing protein *y*-Ubx2p and associates with *y*-Der1p (Carvalho et al. 2006). Ubx2p also contains a ubiquitin-associated (UBA) domain that binds ubiquitin chains and may transport the ERAD substrates to Cdc48p (Schuberth et al. 2004; Neuber et al. 2005).

After their retro-translocation to the cytosol, the multi-ubiquitylated substrates are delivered to the proteasome by *y*-Rad23p and *y*-Dsk2p (Chen and Madura 2002; Funakoshi et al. 2002; Medicherla et al. 2004). These proteins have one or more UBA domains and a ubiquitin-like (UBL) domain, which is recognized by the proteasome subunit *y*-Rpn1 (Elsasser et al. 2002). Both the 26S proteasome and the E4 ubiquitin chain elongation factor *y*-Ufd2p associate with the *y*-Rad23 UBL domain in a mutually exclusive manner *in vitro* (Kim et al. 2004). *y*-Rad23p and Cdc48p have been shown to interact with a conserved peptide *N*-glycanase, Pgn1p, which is required for the deglycosylation of misfolded glycoproteins during proteasome-dependent degradation (Suzuki et al. 2000; Hirsch et al. 2004; Nita-Lazar and Lennarz 2005).

The multi-ubiquitylated versus de-ubiquitylated state of an ERAD substrate and its subsequent targeting to the proteasome are determined by a subtle balance between *y*-Ufd2p and *y*-Doa1p (a WD40 repeat protein controlling the cellular ubiquitin concentration) and y-Otu1p (a de-ubiquitinating enzyme). *y*-Doa1p and *y*-Otu1p bind to *y*-Cdc48p, enhancing their inhibitory effect on *y*-Ufd2p (Richly et al. 2005; Rumpf and Jentsch 2006). As the C-terminal end of *y*-Doa1p interacts with the UBX domain, it is possible that the binding of *y*-Ubx2p to *y*-Cdc48p is also inhibited by *y*-Doa1p (Decottignies et al. 2004).

HHR23, the human homolog of *y*-Rad23p, interacts with the second ubiquitin interaction motif of the human S5a subunit (*y*-Rpn10) of the 26S proteasome (Hiyama et al. 1999). It also interacts with the protease domain of ataxin-3, a de-ubiquitinating enzyme involved in the development of the neurodegenerative Machado-Joseph disease (Wang et al. 2000). Ubiquilins 1–4 are human homologues of *y*-Dsk2 that bind presenilins and are found in Lewy bodies and neurofibrillar tangles (Mah et al. 2000). It has been proposed that, under conditions of high levels of protein aggregation and subsequent proteasome overloading, ubiquilin-1 interacts with Eps15, an essential component of the clathrin-mediated endocytosis pathway, to promote the trafficking of protein aggregates to the aggresome (Madsen et al. 2007).

14.3 Accumulation of Unfolded Proteins in Subcellular Compartments

14.3.1 The Unfolded Protein Response

The accumulation of unfolded proteins in the lumen of the ER induces a coordinated adaptive program called the UPR (Shamu et al. 1994; Chapman et al. 1998; Sidrauski et al. 1998) (see Fig. 14.1). *y*-Ire1p is a conserved membrane kinase and contains a luminal domain responsible for sensing misfolded proteins and cytosolic kinase and ribonuclease domains (Hampton 2003). In yeast, under normal growing conditions, the monomeric protein is bound to Kar2p in an inactive complex. In the presence of high amounts of unfolded or misfolded proteins, the Kar2p-Ire1p complex dissociates and Ire1p assembles into an oligomeric complex with kinase activity (Shamu and Walter 1996).

Autophosphorylation and activation of the C-terminal endoribonuclease domain of *y*-Ire1p promotes the splicing of the pre-mRNA for the *y*-Hac1 (*m*-XBP-1) transcription factor with the help of the tRNA ligase *y*-Rgl1p. y-Hac1p activates the expression of target genes by binding to a conserved Unfolded Protein Response Element (UPRE) in their promoters (Cox and Walter 1996). The UPR alleviates stress by upregulating the expression of ER chaperones (*y*-Kar2p) and components of the ERAD pathway, such as *y*-Ubc7p, *y*-Hrd1p and *y*-Dfm1p (Der1-like-family-member) (Travers et al. 2000). *y*-Dfm1p binds to *y*-Cdc48p and has been identified in a protein complex containing the Cdc48p cofactors *y*-Ubx1p and *y*-Ubx7p (Hitt and Wolf 2004b). This suggests that the Der1p complex is involved in ERAD and the Dfm1p complex is involved in ER stress and homeostasis (Sato and Hampton 2006).

In addition to the IRE1 kinase, mammals utilize two other major transducers for sensing ER stress, the PERK eIF2a kinase, which attenuates protein translation in response to ER stress, and the ER transmembrane transcription factor ATF6. The accumulation of misfolded proteins allows ATF6 to reach the Golgi, where transmembrane proteases release a cytosolic transcriptionally active form, which enters the nucleus and induces the transcription of target genes (Wu and Kaufman 2006).

14.3.2 Proliferation of Subcellular Compartments

A second means of coping with non-native proteins in the ER is to sequester them in special sub-compartments (see Fig. 14.1). Overexpression of the ER membrane

3-hydroxy 3-methylglutaryl coenzyme A (HMG-CoA) reductase in yeast leads to the formation of stacked membrane pairs surrounding the nuclear envelope, called karmellae (Wright et al. 1988). A second type of ER sub-compartment, known as the ER-associated compartment (ERAC), consists of a network of tubulo-vesicular structures (Huyer et al. 2004a). ERAC formation is induced by the yeast plasma membrane H^+-ATPase variant Pma1-D378N and some, but not all, mutants of the pheromone ABC-transporter Ste6p (Wright et al. 1988; Supply et al. 1993; Harris et al. 1994; de Kerchove d'Exaerde et al. 1995; Loayza et al. 1998). The increased volume of the yeast ER may help to accommodate newly synthesized ER enzymes and inhibit the aggregation of unfolded proteins by reducing their concentration. The recently identified juxta-nuclear quality control (JUNQ) compartment serves as a temporary storage site for misfolded ubiquitylated proteins that cannot be folded or degraded because of the limited capacity of the ubiquitin-proteasome system under certain stress conditions (Kaganovich et al. 2008).

The term "macroautophagy" designates a cellular process leading to the nonselective sequestration of cytoplasmic components in double-membrane vesicles, called autophagosomes, for degradation in the vacuole/lysosome (Levine and Klionsky 2004). Activation of the UPR in yeast induces the formation of autophagosomes that selectively include ER membranes, their delimiting double membranes also being partially derived from the ER (Bernales et al. 2007). This so-called ER-phagy can remove damaged parts of the ER and participate in reducing the size of the ER back to normal once the folding stress is gone. ER-phagy could therefore represent an important degradative function of the UPR and be an integral player in achieving homeostatic control (Bernales et al. 2007). *y*-ATG19, a member of the cytoplasm-to-vacuole targeting (Cvt) pathway, increases the degradation of the misfolded Pma1-D378T form of the plasma membrane H^+-ATPase. Therefore, the efficient degradation of Pma1-D378T requires the cooperation of the ERAD and Cvt/autophagy pathways (Mazon et al. 2007).

14.4 ERAD Substrates Linked to Human Diseases

14.4.1 Regulation of Net Sodium Balance and Hypertension

Net sodium balance in humans is maintained through various ion transporters expressed along the entire nephron. Of these ion transporters, epithelial sodium channels (ENaCs), located along the aldosterone-sensitive distal nephron, play a pivotal role in the homeostasis of sodium balance. An increased abundance and activity of ENaCs in the plasma membrane causes hereditary hypertension, such as Liddle syndrome. The E3 ubiquitin ligase Nedd4-2 binds to the PY motif in the C-terminus of ENaC and catalyzes the ubiquitination of the NH_2 terminus for subsequent degradation (Staub and Rotin 2006). ENaC and Nedd4–2 are phosphorylation substrates for the short-lived serum- and glucocorticoid-induced kinase (Sgk1), the expression of which is stimulated by aldosterone (Chen et al. 1999). Phosphorylation of Nedd4–2 results in the recruitment of 14-3-3 proteins to the phosphorylated sites, leading to a reduced interaction between Nedd4–2 and ENaC and causing reduced ubiquitylation of ENaC and its accumulation at the plasma membrane. Sgk1 is ubiquitylated by the action of UBE2J1-2 (UBC6), UBE2G1-2 (UBC7) and the E3 ligase HRD1 (synoviolin) (Arteaga et al. 2006). Sgk1 may be a key component of the cellular stress response, and HRD1 is upregulated to protect the cell against ER stress-induced apoptosis by degrading unfolded proteins accumulated in the ER (Arteaga et al. 2006). To be functional *in vitro*, HRD1/synoviolin requires UBE2G2, although its cognate E2 enzyme in cells has not been identified (Nadav et al. 2003; Kikkert et al. 2004; Omura et al. 2006).

14.4.2 Rheumatoid Arthritis

Rheumatoid arthritis is a disease associated with painful joints that affects approximately 1% of the population worldwide and for which no specific cure is available. Rheumatoid synovial cells produce large amounts of various proteins, such as cytokines and proteases, which might confer an autonomous proliferation property on the cells. The overexpression of HRD1/synoviolin results in a 'hyper-ERAD' state, allowing the cells to deal with the accumulation of unfolded proteins (about 30% of all newly synthesized, ER-sorted proteins) (Yagishita et al. 2008). Systemic overexpression of HRD1 in the mouse causes spontaneous arthropathy, with marked joint swelling, at 20 weeks of age in 30% of the mice analysed. Analysis of the affected joints shows bone destruction and severe synovial cell hyperplasia. HRD1-overexpressing mice exhibit pathologic features similar to those of rheumatoid arthritis. It has been hypothesized that the anti-apoptotic effects of HRD1 cause arthropathy by triggering synovial cell outgrowth (Amano et al. 2003).

14.4.3 Cystic Fibrosis

Cystic fibrosis arises from the misfolding and premature degradation of the CFTR chloride channel, an ABC-transporter located in the plasma membrane of epithelial cells (Ward et al. 1995). An ER membrane-associated ubiquitin ligase complex containing the E3 RMA1, the E2 UBE2J1 and Derlin-1 cooperates with the cytosolic HSC70-CHIP E4 complex to promote the proteasomal degradation of CFTRΔF508, a mutant form in which phenylalanine 508 is deleted (Meacham et al. 2001; Younger et al. 2006). RMA1 can recognize folding defects in the mutated transporter coincidentally with translation, whereas CHIP appears to act post-translationally. CHIP reduces the ATPase activity of HSC70 and HSP70 and inhibits the HSC70-HSP70 substrate-binding cycle (Meacham et al. 2001; Younger et al. 2006). CHIP activity is dependent on the C-terminal U box, a domain that shares similarity with yeast Ufd2p (Hoppe 2005).

14.4.4 Progressive Familial Intrahepatic Cholestasis Type II

Bile secretion is mediated by several ABC-transporters located in the canalicular membrane of the hepatocyte (Alrefai and Gill 2007). Progressive familial intrahepatic cholestasis type II (PFIC II) is associated with mutations in the bile salt export pump Bsep. Rat Bsep variants carrying PFIC II mutations are degraded by the proteasome through the ERAD pathway. The E3 ubiquitin ligases Rma1 and TEB4 contribute to the degradation of the G238V variant containing a cytosolic lesion, whereas HRD1 contributes to the degradation of a mutant lacking the luminal glycosylation domain (Wang et al. 2008).

14.4.5 Drug Efflux Pumps and Xenobiotic Metabolism

The human ABCG2 protein is a half ABC transporter bearing a single ATP-binding fold at the NH_2-terminus and containing six transmembrane domains. It is located in the plasma membrane, where it forms homodimers bound through disulphide-bonded cysteine residues. This efflux pump is suggested to be responsible for protecting the body against toxic xenobiotics and metabolites. The formation of an intra-molecular disulphide bond between Cys592 and Cys608 and *N*-glycosylation at Asn596 are critical check points for the stability and degradation of the *de novo* synthesized ABCG2 protein (Wakabayashi-Nakao et al. 2009). Non-synonymous single nucleotide polymorphisms, such as Q141K, F208S, and S441N, have also been found to greatly affect the stability of ABCG2 in the ER and to enhance the protein degradation rate via ubiquitination and proteasomal proteolysis (Wakabayashi-Nakao et al. 2009). The identity of the E3 ligase involved is unknown.

14.4.6 Accumulation of a G-Protein-Coupled Transporter and Neurodegeneration

Parkin, the gene product of *PARK2*, is an E3 ubiquitin ligase that is required for the degradation of several substrates, including Parkin-associated endothelin receptor-like receptor (Pael-R) (Imai et al. 2001). Accumulation of Pael-R in the ER of dopaminergic neurons induces ER stress leading to neurodegeneration (Imai et al. 2001). CHIP facilitates Parkin-mediated Pael-R ubiquitylation by promoting the dissociation of HSP70 from Parkin and Pael-R. CHIP also enhances the ability of

Parkin to inhibit cell death induced by Pael-R (Imai et al. 2002). Unfolded Pael-R interacts with HRD1, via its proline-rich region, and the disruption of endogenous *HRD1* by small interfering RNA induces Pael-R accumulation and caspase-3 activation (Omura et al. 2006). ATF6 overexpression, which induces *HRD1* through the UPR pathway, increases Pael-R degradation. These results suggest that in addition to Parkin, HRD1 is also involved in the degradation of Pael-R (Omura et al. 2006).

14.4.7 SOD1 and Amyotrophic Lateral Sclerosis

Human copper/zinc superoxide dismutase (SOD1) is responsible for destroying free superoxide radicals in the body. SOD1 is a soluble cytoplasmic protein that functions as a homodimer to convert naturally-occurring, but harmful, superoxide radicals into molecular oxygen and hydrogen peroxide. Mutations in the *SOD1* gene have been implicated as one of the causes of familial amyotrophic lateral sclerosis (ALS). In mouse motor neurons and human embryonic kidney cells expressing the G93A variant, mutant SOD1 protein (SOD^{mut}) interacts with the C-terminal cytoplasmic region of Derlin-1 and triggers ER stress by causing dysfunction of ER-associated degradation (Nishitoh et al. 2008). SOD^{mut} activates apoptosis signal-regulating kinase 1 (ASK1) by triggering ER stress-induced IRE1 activation in mouse motor neurons. Dissociation of SOD^{mut} from Derlin-1 protects motor neurons from SOD^{mut}-induced cell death. Furthermore, deletion of *ASK1* partially mitigates motor neuron loss *in vitro* and extends the life-span of SOD^{mut} transgenic mice. These results indicate that the interaction of SOD^{mut} with Derlin-1 is crucial for disease progression in familial ALS (Nishitoh et al. 2008).

14.5 Concluding Remarks

Eukaryotic cells have several quality control systems to ensure that only correctly folded proteins are transported along the secretory pathway. The use of modular ERAD substrates has shown that newly-synthesized ER-transmembrane proteins are monitored by a two-step quality control and targeted to the proteasome for degradation, through distinct pathways. The molecular mechanisms underlying retrotranslocation and the protein composition of the channel remain to be determined. Analysis of additional substrates and the identification of new factors that are required for regulation of ER-associated degradation will help to define pharmaceutical strategies to control the degradation of membrane transporters that are associated with human disease.

Acknowledgements We apologize to those whose work we were unable to cite due to space constraints. N. C is supported by a grant from the Fonds pour la formation à la Recherche dans l'Industrie et dans l'Agriculture (F.R.I.A.).

References

Ahner, A., and Brodsky, J.L. (2004). Checkpoints in ER-associated degradation: excuse me, which way to the proteasome? *Trends Cell Biol.* 14:474–478.

Alam, S.L., Sun, J., Payne, M., et al. (2004). Ubiquitin interactions of NZF zinc fingers. *EMBO J.* 23:1411–1421.

Alrefai, W.A., and Gill, R.K. (2007). Bile acid transporters: structure, function, regulation and pathophysiological implications. *Pharm. Res.* 24:1803–1823.

Amano, T., Yamasaki, S., Yagishita, N., et al. (2003). Synoviolin/Hrd1, an E3 ubiquitin ligase, as a novel pathogenic factor for arthropathy. *Genes Dev.* 17:2436–2449.

Anelli, T., and Sitia, R. (2008). Protein quality control in the early secretory pathway. *EMBO J.* 27:315–327.

Arteaga, M.F., Wang, L., Ravid, T., et al. (2006). An amphipathic helix targets serum and glucocorticoid-induced kinase 1 to the endoplasmic reticulum-associated ubiquitin-conjugation machinery. *Proc. Natl. Acad. Sci. USA.* 103:11178–11183.

Balzi, E., Wang, M., Leterme, S., et al. (1994). PDR5, a novel yeast multidrug resistance conferring transporter controlled by the transcription regulator PDR1. *J. Biol. Chem.* 269:2206–2214.

Bays, N.W., Gardner, R.G., Seelig, L.P., et al. (2001a). Hrd1p/Der3p is a membrane-anchored ubiquitin ligase required for ER-associated degradation. *Nat. Cell Biol.* 3:24–29.

Bays, N.W., Wilhovsky, S.K., Goradia, A., et al. (2001b). HRD4/NPL4 is required for the proteasomal processing of ubiquitinated ER proteins. *Mol. Biol. Cell* 12:4114–4128.

Bernales, S., Schuck, S., Walter, P. (2007). ER-phagy: selective autophagy of the endoplasmic reticulum. *Autophagy* 3:285–287.

Bernardi, K.M., Forster, M.L., Lencer, W.I. et al. (2008). Derlin-1 facilitates the retro-translocation of cholera toxin. *Mol. Biol. Cell* 19:877–884.

Bhamidipati, A., Denic, V., Quan, E.M. et al. (2005). Exploration of the topological requirements of ERAD identifies Yos9p as a lectin sensor of misfolded glycoproteins in the ER lumen. *Mol. Cell* 19:741–751.

Biederer, T., Volkwein, C., Sommer, T. (1997). Role of Cue1p in ubiquitination and degradation at the ER surface. *Science* 278:1806–1809.

Bissinger, P.H., and Kuchler, K. (1994). Molecular cloning and expression of the *Saccharomyces cerevisiae* STS1 gene product. A yeast ABC transporter conferring mycotoxin resistance. *J. Biol. Chem.* 269:4180–4186.

Braun, S., Matuschewski, K., Rape, M., et al. (2002). Role of the ubiquitin-selective CDC48 (UFD1/NPL4) chaperone (segregase) in ERAD of OLE1 and other substrates. *EMBO J.* 21:615–621.

Brodsky, J.L. (1996). Post-translational protein translocation: not all hsc70s are created equal. *Trends Biochem. Sci.* 21:122–126.

Bruderer, R.M., Brasseur, C., Meyer, H.H. (2004). The AAA ATPase p97/VCP interacts with its alternative co-factors, Ufd1-Npl4 and p47, through a common bipartite binding mechanism. *J. Biol. Chem.* 279:49609–49616.

Bukau, B., Weissman, J., Horwich, A. (2006). Molecular chaperones and protein quality control. *Cell* 125:443–451.

Buschhorn, B.A., Kostova, Z., Medicherla, B. et al. (2004). A genome-wide screen identifies Yos9p as essential for ER-associated degradation of glycoproteins. *FEBS Lett.* 577:422–426.

Byrd, J.C., Tarentino, A.L., Maley, F., et al. (1982). Glycoprotein synthesis in yeast. Identification of Man8GlcNAc2 as an essential intermediate in oligosaccharide processing. *J. Biol. Chem.* 257:14657–14666.

Carvalho, P., Goder, V., Rapoport, T.A. (2006). Distinct ubiquitin-ligase complexes define convergent pathways for the degradation of ER proteins. *Cell* 126:361–373.

Chapman, R., Sidrauski, C., Walter, P. (1998). Intracellular signaling from the endoplasmic reticulum to the nucleus. *Annu. Rev. Cell Dev. Biol.* 14:459–485.

Chen, L., and Madura, K. (2002). Rad23 promotes the targeting of proteolytic substrates to the proteasome. *Mol. Cell Biol.* 22:4902–4913.

Chen, S.Y., Bhargava, A., Mastroberardino, L., et al. (1999). Epithelial sodium channel regulated by aldosterone-induced protein sgk. *Proc. Natl. Acad. Sci. USA*. 96:2514–2519.

Clerc, S., Hirsch, C., Oggier, D.M., et al. (2009). Htm1 protein generates the N-glycan signal for glycoprotein degradation in the endoplasmic reticulum. *J. Cell Biol.* 184:159–172.

Cox, J.S., and Walter, P. (1996). A novel mechanism for regulating activity of a transcription factor that controls the unfolded protein response. *Cell* 87:391–404.

de Kerchove d'Exaerde, A., Supply, P., Dufour, J.P., et al. (1995). Functional complementation of a null mutation of the yeast *Saccharomyces cerevisiae* plasma membrane H(+)-ATPase by a plant H(+)-ATPase gene. *J. Biol. Chem.* 270:23828–23837.

Deak, P.M., and Wolf, D.H. (2001). Membrane topology and function of Der3/Hrd1p as a ubiquitin-protein ligase (E3) involved in endoplasmic reticulum degradation. *J. Biol. Chem.* 276:10663–10669.

Decottignies, A., Evain, A., Ghislain, M. (2004). Binding of Cdc48p to a ubiquitin-related UBX domain from novel yeast proteins involved in intracellular proteolysis and sporulation. *Yeast* 21:127–139.

De LaBarre, B., and Brunger, A.T. (2003). Complete structure of p97/valosin-containing protein reveals communication between nucleotide domains. *Nat. Struct. Biol.* 10:856–863.

Denic, V., Quan, E.M., Weissman, J.S. (2006). A luminal surveillance complex that selects misfolded glycoproteins for ER-associated degradation. *Cell* 126:349–359.

Dolinski, K., Muir, S., Cardenas, M., et al. (1997). All cyclophilins and FK506 binding proteins are, individually and collectively, dispensable for viability in *Saccharomyces cerevisiae*. *Proc. Natl. Acad. Sci. USA*. 94:13093–13098.

Dreveny, I., Kondo, H., Uchiyama, K., et al. (2004). Structural basis of the interaction between the AAA ATPase p97/VCP and its adaptor protein p47. *EMBO J.* 23:1030–1039.

Egner, R., Rosenthal, F.E., Kralli, A., Sanglard, D. et al. (1998). Genetic separation of FK506 susceptibility and drug transport in the yeast Pdr5 ATP-binding cassette multidrug resistance transporter. *Mol. Biol. Cell* 9:523–543.

Ellgaard, L., and Helenius, A. (2003). Quality control in the endoplasmic reticulum. *Nat. Rev. Mol. Cell Biol.* 4:181–191.

Elsasser, S., Gali, R.R., Schwickart, M., et al. (2002). Proteasome subunit Rpn1 binds ubiquitin-like protein domains. *Nat. Cell Biol.* 4:725–730.

Fassio, A., and Sitia, R. (2002). Formation, isomerisation and reduction of disulphide bonds during protein quality control in the endoplasmic reticulum. *Histochem. Cell Biol.* 117:151–157.

Finger, A., Knop, M., Wolf, D.H. (1993). Analysis of two mutated vacuolar proteins reveals a degradation pathway in the endoplasmic reticulum or a related compartment of yeast. *Eur. J. Biochem.* 218:565–574.

Friedlander, R., Jarosch, E., Urban, J., et al. (2000). A regulatory link between ER-associated protein degradation and the unfolded-protein response. *Nat. Cell Biol.* 2:379–384.

Frohlich, K.U., Fries, H.W., Rudiger, M., et al. (1991). Yeast cell cycle protein CDC48p shows full-length homology to the mammalian protein VCP and is a member of a protein family involved in secretion, peroxisome formation, and gene expression. *J. Cell Biol.* 114:443–453.

Funakoshi, M., Sasaki, T., Nishimoto, T. et al. (2002). Budding yeast Dsk2p is a polyubiquitin-binding protein that can interact with the proteasome. *Proc. Natl. Acad. Sci. USA*. 99:745–750.

Gardner, R.G., Swarbrick, G.M., Bays, N.W., et al. (2000). Endoplasmic reticulum degradation requires lumen to cytosol signaling. Transmembrane control of Hrd1p by Hrd3p. *J. Cell Biol.* 151:69–82.

Gauss, R., Jarosch, E., Sommer, T. et al. (2006a). A complex of Yos9p and the HRD ligase integrates endoplasmic reticulum quality control into the degradation machinery. *Nat. Cell Biol.* 8:849–854.

Gauss, R., Sommer, T., Jarosch, E. (2006b). The Hrd1p ligase complex forms a linchpin between ER-lumenal substrate selection and Cdc48p recruitment. *EMBO J.* 25:1827–1835.

Gnann, A., Riordan, J.R., Wolf, D.H. (2004). Cystic fibrosis transmembrane conductance regulator degradation depends on the lectins Htm1p/EDEM and the Cdc48 protein complex in yeast. *Mol. Biol. Cell* 15:4125–4135.

Gross, E., Sevier, C.S., Heldman, N., et al. (2006). Generating disulfides enzymatically: reaction products and electron acceptors of the endoplasmic reticulum thiol oxidase Ero1p. *Proc. Natl. Acad. Sci. USA*. 103: 299–304.

Hampton, R.Y., Gardner, R.G., Rine, J. (1996). Role of 26S proteasome and HRD genes in the degradation of 3-hydroxy-3-methylglutaryl-CoA reductase, an integral endoplasmic reticulum membrane protein. *Mol. Biol. Cell* 7:2029–2044.

Hampton, R.Y. (2002). ER-associated degradation in protein quality control and cellular regulation. *Curr. Opin. Cell Biol.* 14:476–482.

Hampton, R.Y. (2003). IRE1: a role in UPREgulation of ER degradation. *Dev. Cell* 4:144–146.

Harris, S.L., Na, S., Zhu, X., et al. (1994). Dominant lethal mutations in the plasma membrane H(+)-ATPase gene of *Saccharomyces cerevisiae*. *Proc. Natl. Acad. Sci. USA*. 91:10531–10535.

Helenius, A., and Aebi, M. (2004). Roles of N-linked glycans in the endoplasmic reticulum. *Annu. Rev. Biochem.* 73:1019–1049.

Hiller, M.M., Finger, A., Schweiger, M., et al. (1996). ER degradation of a misfolded luminal protein by the cytosolic ubiquitin-proteasome pathway. *Science* 273:1725–1728.

Hirao, K., Natsuka, Y., Tamura, T., et al. (2006). EDEM3, a soluble EDEM homolog, enhances glycoprotein endoplasmic reticulum-associated degradation and mannose trimming. *J. Biol. Chem.* 281:9650–9658.

Hirsch, C., Misaghi, S., Blom, D., et al. (2004). Yeast N-glycanase distinguishes between native and non-native glycoproteins. *EMBO Rep.* 5:201–206.

Hirsch, C., Gauss, R., Sommer, T. (2006). Coping with stress: cellular relaxation techniques. *Trends Cell Biol.* 16:657–663.

Hitchcock, A.L., Krebber, H., Frietze, S., et al. (2001). The conserved npl4 protein complex mediates proteasome-dependent membrane-bound transcription factor activation. *Mol. Biol. Cell* 12:3226–3241.

Hitt, R., and Wolf, D.H. (2004a). DER7, encoding alpha-glucosidase I is essential for degradation of malfolded glycoproteins of the endoplasmic reticulum. *FEMS Yeast Res.* 4:815–820.

Hitt, R., and Wolf, D.H. (2004b). Der1p, a protein required for degradation of malfolded soluble proteins of the endoplasmic reticulum: topology and Der1-like proteins. *FEMS Yeast Res.* 4:721–729.

Hiyama, H., Yokoi, M., Masutani, C., et al. (1999). Interaction of hHR23 with S5a. The ubiquitin-like domain of hHR23 mediates interaction with S5a subunit of 26 S proteasome. *J. Biol. Chem.* 274:28019–28025.

Hoppe, T. (2005). Multiubiquitylation by E4 enzymes: 'one size' doesn't fit all. *Trends Biochem. Sci.* 30:183–187.

Hosokawa, N., Wada, I., Hasegawa, K., et al. (2001). A novel ER alpha-mannosidase-like protein accelerates ER-associated degradation. *EMBO Rep.* 2:415–422.

Huyer, G., Longsworth, G.L., Mason, D.L., et al. (2004a). A striking quality control subcompartment in Saccharomyces cerevisiae: the endoplasmic reticulum-associated compartment. *Mol. Biol. Cell* 15:908–921.

Huyer, G., Piluek, W.F., Fansler, Z., et al. (2004b). Distinct machinery is required in *Saccharomyces cerevisiae* for the endoplasmic reticulum-associated degradation of a multispanning membrane protein and a soluble luminal protein. *J. Biol. Chem.* 279:38369–38378.

Imai, Y., Soda, M., Inoue, H., et al. (2001). An unfolded putative transmembrane polypeptide, which can lead to endoplasmic reticulum stress, is a substrate of Parkin. *Cell* 105:891–902.

Imai, Y., Soda, M., Hatakeyama, S., et al. (2002). CHIP is associated with Parkin, a gene responsible for familial Parkinson's disease, and enhances its ubiquitin ligase activity. *Mol. Cell* 10:55–67.

Jakob, C.A., Burda, P., Roth, J., et al. (1998). Degradation of misfolded endoplasmic reticulum glycoproteins in *Saccharomyces cerevisiae* is determined by a specific oligosaccharide structure. *J. Cell Biol.* 142:1223–1233.

Jakob, C.A., Bodmer, D., Spirig, U., et al. (2001). Htm1p, a mannosidase-like protein, is involved in glycoprotein degradation in yeast. *EMBO Rep.* 2:423–430.

Jarosch, E., Taxis, C., Volkwein, C., et al. (2002). Protein dislocation from the ER requires polyubiquitination and the AAA-ATPase Cdc48. *Nat. Cell Biol.* 4:134–139.

Kaganovich, D., Kopito, R., Frydman, J. (2008). Misfolded proteins partition between two distinct quality control compartments. *Nature* 454:1088–1095.

Kanehara, K., Kawaguchi, S., Ng D.T. (2007). The EDEM and Yos9p families of lectin-like ERAD factors. *Semin. Cell Dev. Biol.* 18:743–750.

Kikkert, M., Doolman, R., Dai, M., et al. (2004). Human HRD1 is an E3 ubiquitin ligase involved in degradation of proteins from the endoplasmic reticulum. *J. Biol. Chem.* 279:3525–3534.

Kim, I., Mi, K., Rao, H. (2004). Multiple interactions of rad23 suggest a mechanism for ubiquitylated substrate delivery important in proteolysis. *Mol. Biol. Cell* 15:3357–3365.

Kim, W., Spear, E.D., Ng, D.T. (2005). Yos9p detects and targets misfolded glycoproteins for ER-associated degradation. *Mol. Cell* 19:753–764.

Klionsky, D.J. (2007). Autophagy: from phenomenology to molecular understanding in less than a decade. *Nat. Rev. Mol. Cell. Biol.* 8:931–937.

Knop, M., Finger, A., Braun, T., et al. (1996). Der1, a novel protein specifically required for endoplasmic reticulum degradation in yeast. *EMBO J.* 15:753–763.

Koegl, M., Hoppe, T., Schlenker, S., et al. (1999). A novel ubiquitination factor, E4, is involved in multiubiquitin chain assembly. *Cell* 96:635–644.

Kolling, R., and Hollenberg, C.P. (1994). The ABC-transporter Ste6 accumulates in the plasma membrane in a ubiquitinated form in endocytosis mutants. *EMBO J.* 13:3261–3271.

Kopito, R.R. (1997). ER quality control: the cytoplasmic connection. *Cell* 88:427–430.

Kostova, Z., and Wolf, D.H. (2003). For whom the bell tolls: protein quality control of the endoplasmic reticulum and the ubiquitin-proteasome connection. *EMBO J.* 22:2309–2317.

Kozutsumi, Y., Normington, K., Press, E., et al. (1989). Identification of immunoglobulin heavy chain binding protein as glucose-regulated protein 78 on the basis of amino acid sequence, immunological cross-reactivity, and functional activity. *J. Cell Sci. Suppl.* 11:115–137.

Krebs, M.P., Noorwez, S.M., Malhotra, R., et al. (2004). Quality control of integral membrane proteins. *Trends Biochem. Sci.* 29:648–655.

Kreft, S.G., Wang, L., Hochstrasser, M. (2006). Membrane topology of the yeast endoplasmic reticulum-localized ubiquitin ligase Doa10 and comparison with its human ortholog TEB4 (MARCH-VI). *J. Biol. Chem.* 281:4646–4653.

Lenk, U., Yu, H., Walter, J., et al. (2002). A role for mammalian Ubc6 homologues in ER-associated protein degradation. *J. Cell Sci.* 115:3007–3014.

Levine, B., and Klionsky, D.J. (2004). Development by self-digestion: molecular mechanisms and biological functions of autophagy. *Dev. Cell* 6:463–477.

Lilley, B.N., and Ploegh, H.L. (2004). A membrane protein required for dislocation of misfolded proteins from the ER. *Nature* 429:834–840.

Loayza, D., Tam, A., Schmidt, W.K., et al. (1998). Ste6p mutants defective in exit from the endoplasmic reticulum (ER) reveal aspects of an ER quality control pathway in *Saccharomyces cerevisiae*. *Mol. Biol. Cell* 9:2767–2784.

Madsen, L., Schulze, A., Seeger, M., et al. (2007). Ubiquitin domain proteins in disease. *BMC Biochem.* 8(Suppl 1): S1.

Mah, A.L., Perry, G., Smith, M.A., et al. (2000). Identification of ubiquilin, a novel presenilin interactor that increases presenilin protein accumulation. *J. Cell Biol.* 151:847–862.

Mast, S.W., Diekman, K., Karaveg, K., et al. (2005). Human EDEM2, a novel homolog of family 47 glycosidases, is involved in ER-associated degradation of glycoproteins. *Glycobiology* 15:421–436.

Mazon, M.J., Eraso, P., Portillo, F. (2007). Efficient degradation of misfolded mutant Pma1 by endoplasmic reticulum-associated degradation requires Atg19 and the Cvt/autophagy pathway. *Mol. Microbiol.* 63:1069–1077.

McCracken, A.A., and Brodsky, J.L. (1996). Assembly of ER-associated protein degradation in vitro: dependence on cytosol, calnexin, and ATP. *J. Cell Biol.* 132:291–298.

McCracken, A.A., and Brodsky, J.L. (2003). Evolving questions and paradigm shifts in endoplasmic-reticulum-associated degradation (ERAD). *Bioessays* 25:868–877.
McGrath, J.P., and Varshavsky, A. (1989). The yeast STE6 gene encodes a homologue of the mammalian multidrug resistance P-glycoprotein. *Nature* 340:400–404.
Meacham, G.C., Patterson, C., Zhang, W., et al. (2001). The Hsc70 co-chaperone CHIP targets immature CFTR for proteasomal degradation. *Nat. Cell Biol.* 3:100–105.
Medicherla, B., Kostova, Z., Schaefer, A., et al. (2004). A genomic screen identifies Dsk2p and Rad23p as essential components of ER-associated degradation. *EMBO Rep.* 5:692–697.
Metzger, M.B., Maurer, M.J., Dancy, B.M., et al. (2008). Degradation of a cytosolic protein requires endoplasmic reticulum-associated degradation machinery. *J. Biol. Chem.* 283:32302–32316.
Meusser, B., Hirsch, C., Jarosch, E., et al. (2005). ERAD: the long road to destruction. *Nat. Cell Biol.* 7:766–772.
Meyer, H.H., Shorter, J.G., Seemann, J., et al. (2000). A complex of mammalian ufd1 and npl4 links the AAA-ATPase, p97, to ubiquitin and nuclear transport pathways. *EMBO J.* 19:2181–2192.
Molinari, M., Calanca, V., Galli, C., et al. (2003). Role of EDEM in the release of misfolded glycoproteins from the calnexin cycle. *Science* 299:1397–1400.
Movsichoff, F., Castro, O.A., Parodi, A.J. (2005). Characterization of *Schizosaccharomyces pombe* ER alpha-mannosidase: a reevaluation of the role of the enzyme on ER-associated degradation. *Mol. Biol. Cell* 16:4714–4724.
Nadav, E., Shmueli, A., Barr, H., et al. (2003). A novel mammalian endoplasmic reticulum ubiquitin ligase homologous to the yeast Hrd1. *Biochem. Biophys. Res. Commun.* 303:91–97.
Nakamoto, R.K., Verjovski-Almeida, S., Allen, K.E., et al. (1998). Substitutions of aspartate 378 in the phosphorylation domain of the yeast PMA1 H+-ATPase disrupt protein folding and biogenesis. *J. Biol. Chem.* 273:7338–7344.
Nakatsukasa, K., Nishikawa, S., Hosokawa, N., et al. (2001). Mnl1p, an alpha-mannosidase-like protein in yeast *Saccharomyces cerevisiae*, is required for endoplasmic reticulum-associated degradation of glycoproteins. *J. Biol. Chem.* 276:8635–8638.
Nakatsukasa, K., and Brodsky, J.L. (2008). The recognition and retrotranslocation of misfolded proteins from the endoplasmic reticulum. *Traffic* 9:861–870.
Nakatsukasa, K., Huyer, G., Michaelis, S., et al. (2008). Dissecting the ER-associated degradation of a misfolded polytopic membrane protein. *Cell* 132:101–112.
Neuber, O., Jarosch, E., Volkwein, C., et al. (2005). Ubx2 links the Cdc48 complex to ER-associated protein degradation. *Nat. Cell Biol.* 7:993–998.
Nishikawa, S.I., Fewell, S.W., Kato, Y., et al. T (2001). Molecular chaperones in the yeast endoplasmic reticulum maintain the solubility of proteins for retrotranslocation and degradation. *J. Cell Biol.* 153:1061–1070
Nishitoh, H., Kadowaki, H., Nagai, A., et al. (2008). ALS-linked mutant SOD1 induces ER stress- and ASK1-dependent motor neuron death by targeting Derlin-1. *Genes Dev.* 22:1451–1464.
Nita-Lazar, M., and Lennarz, W.J. (2005). Pkc1p modifies CPY* degradation in the ERAD pathway. *Biochem. Biophys. Res. Commun.* 332:357–361.
Norgaard, P., Westphal, V., Tachibana, C., et al. (2001). Functional differences in yeast protein disulfide isomerases. *J. Cell Biol.* 152:553–562.
Normington, K., Kohno, K., Kozutsumi, Y., et al. (1989). *S. cerevisiae* encodes an essential protein homologous in sequence and function to mammalian BiP. *Cell* 57:1223–1236.
Oda, Y., Hosokawa, N., Wada, I., et al. (2003). EDEM as an acceptor of terminally misfolded glycoproteins released from calnexin. *Science* 299:1394–1397.
Omura, T., Kaneko, M., Okuma, Y., et al. (2006). A ubiquitin ligase HRD1 promotes the degradation of Pael receptor, a substrate of Parkin. *J. Neurochem.* 99:1456–1469.
Park, S, Isaacson, R., Kim, H.T., et al. (2005). Ufd1 exhibits the AAA-ATPase fold with two distinct ubiquitin interaction sites. *Structure* 13:995–1005.
Pilon, M., Schekman, R., Romisch, K. (1997). Sec61p mediates export of a misfolded secretory protein from the endoplasmic reticulum to the cytosol for degradation. *EMBO J.* 16: 4540–4548.

Plemper, R.K., Bohmler, S., Bordallo, J., et al. (1997). Mutant analysis links the translocon and BiP to retrograde protein transport for ER degradation. *Nature* 388: 891–895.

Plemper, R.K., Egner, R., Kuchler, K., et al. (1998). Endoplasmic reticulum degradation of a mutated ATP-binding cassette transporter Pdr5 proceeds in a concerted action of Sec61 and the proteasome. *J. Biol. Chem.* 273: 32848–32856.

Plemper, R.K., Bordallo, J., Deak, P.M., et al. (1999). Genetic interactions of Hrd3p and Der3p/Hrd1p with Sec61p suggest a retro-translocation complex mediating protein transport for ER degradation. *J. Cell Sci.* 112 (Pt 22):4123–4134.

Plemper, R.K., and Wolf, D.H. (1999). Endoplasmic reticulum degradation. Reverse protein transport and its end in the proteasome. *Mol. Biol. Rep.* 26:125–130.

Rabinovich, E., Kerem, A., Frohlich, K.U., et al. (2002). AAA-ATPase p97/Cdc48p, a cytosolic chaperone required for endoplasmic reticulum-associated protein degradation. *Mol. Cell Biol.* 22:626–634.

Rape, M., Hoppe, T., Gorr, I., et al. (2001). Mobilization of processed, membrane-tethered SPT23 transcription factor by CDC48 (UFD1/NPL4), a ubiquitin-selective chaperone. *Cell* 107: 667–677.

Ravid, T., Kreft, S.G., Hochstrasser, M. (2006). Membrane and soluble substrates of the Doa10 ubiquitin ligase are degraded by distinct pathways. *EMBO J.* 25:533–543.

Richly, H., Rape, M., Braun, S., et al. (2005). A series of ubiquitin binding factors connects CDC48/p97 to substrate multiubiquitylation and proteasomal targeting. *Cell* 120:73–84.

Romisch, K. (1999). Surfing the Sec61 channel: bidirectional protein translocation across the ER membrane. *J. Cell Sci.* 112 (Pt 23):4185–4191.

Ron, D., and Walter, P. (2007). Signal integration in the endoplasmic reticulum unfolded protein response. *Nat. Rev. Mol. Cell. Biol.* 8:519–529.

Rose, M.D., Misra, L.M., Vogel, J.P. (1989). KAR2, a karyogamy gene, is the yeast homolog of the mammalian BiP/GRP78 gene. *Cell* 57:1211–1221.

Rumpf, S., and Jentsch, S. (2006). Functional division of substrate processing cofactors of the ubiquitin-selective Cdc48 chaperone. *Mol. Cell* 21:261–269.

Sai, X., Kawamura, Y., Kokame, K., et al. (2002). Endoplasmic reticulum stress-inducible protein, Herp, enhances presenilin-mediated generation of amyloid beta-protein. *J. Biol. Chem.* 277:12915–12920.

Saris, N., Holkeri, H., Craven, R.A., et al. (1997). The Hsp70 homologue Lhs1p is involved in a novel function of the yeast endoplasmic reticulum, refolding and stabilization of heat-denatured protein aggregates. *J. Cell Biol.* 137:813–824.

Sato, B.K., and Hampton, R.Y. (2006). Yeast Derlin Dfm1 interacts with Cdc48 and functions in ER homeostasis. *Yeast* 23:1053–1064.

Schuberth, C., Richly, H., Rumpf, S., et al. (2004). Shp1 and Ubx2 are adaptors of Cdc48 involved in ubiquitin-dependent protein degradation. *EMBO Rep.* 5:818–824.

Schwieger, I., Lautz, K., Krause, E., et al. (2008). Derlin-1 and p97/valosin-containing protein mediate the endoplasmic reticulum-associated degradation of human V2 vasopressin receptors. *Mol. Pharmacol.* 73:697–708.

Scott, D.C., and Schekman, R. (2008). Role of Sec61p in the ER-associated degradation of short-lived transmembrane proteins. *J. Cell Biol.* 181:1095–1105.

Serrano, R., Kielland-Brandt, M.C., Fink, G.R. (1986). Yeast plasma membrane ATPase is essential for growth and has homology with (Na+, K+), K+- and Ca2+-ATPases. *Nature* 319:689–693.

Shamu, C.E., Cox, J.S., Walter, P. (1994). The unfolded-protein-response pathway in yeast. *Trends Cell Biol.* 4:56–60.

Shamu, C.E., and Walter, P. (1996). Oligomerization and phosphorylation of the Ire1p kinase during intracellular signaling from the endoplasmic reticulum to the nucleus. *EMBO J.* 15:3028–3039.

Sidrauski, C., Chapman, R., Walter, P. (1998). The unfolded protein response: an intracellular signalling pathway with many surprising features. *Trends Cell Biol.* 8:245–249.

Sitia, R., and Braakman, I. (2003). Quality control in the endoplasmic reticulum protein factory. *Nature* 426:891–894.

Sommer, T., and Jentsch, S. (1993). A protein translocation defect linked to ubiquitin conjugation at the endoplasmic reticulum. *Nature* 365:176–179.

Song, Y., Sata, J., Saito, A., et al. (2001). Effects of calnexin deletion in *Saccharomyces cerevisiae* on the secretion of glycosylated lysozymes. *J. Biochem.* 130:757–764.

Spear, E.D., and Ng, D.T. (2005). Single, context-specific glycans can target misfolded glycoproteins for ER-associated degradation. *J. Cell Biol.* 169:73–82.

Staub, O., and Rotin, D. (2006). Role of ubiquitylation in cellular membrane transport. *Physiol. Rev.* 86:669–707.

Supply, P., Wach, A., Thines-Sempoux, D., et al. (1993). Proliferation of intracellular structures upon overexpression of the PMA2 ATPase in *Saccharomyces cerevisiae. J. Biol. Chem.* 268:19744–19752.

Suzuki, T., Park, H., Hollingsworth, N.M., et al. (2000). PNG1, a yeast gene encoding a highly conserved peptide: N-glycanase. *J. Cell Biol.* 149:1039–1052.

Swanson, R., Locher, M., Hochstrasser, M. (2001). A conserved ubiquitin ligase of the nuclear envelope/endoplasmic reticulum that functions in both ER-associated and Matalpha2 repressor degradation. *Genes Dev.* 15:2660–2674.

Szathmary, R., Bielmann, R., Nita-Lazar, M., et al. (2005). Yos9 protein is essential for degradation of misfolded glycoproteins and may function as lectin in ERAD. *Mol. Cell* 19:765–775.

Takeuchi, S. (2006). Molecular cloning, sequence, function and structural basis of human heart 150 kDa oxygen-regulated protein, an ER chaperone. *Protein J.* 25:517–528.

Taxis, C., Hitt, R., Park, S.H., et al. (2003). Use of modular substrates demonstrates mechanistic diversity and reveals differences in chaperone requirement of ERAD. *J. Biol. Chem.* 278:35903–35913.

Travers, K.J., Patil, C.K., Wodicka, L., et al. (2000). Functional and genomic analyses reveal an essential coordination between the unfolded protein response and ER-associated degradation. *Cell* 101:249–258.

Tsai, B., Ye, Y., Rapoport, T.A. (2002). Retro-translocation of proteins from the endoplasmic reticulum into the cytosol. *Nat. Rev. Mol. Cell. Biol.* 3:246–255.

Ushioda, R., Hoseki, J., Araki, K., et al. (2008). ERdj5 is required as a disulfide reductase for degradation of misfolded proteins in the ER. *Science* 321:569–572.

Vashist, S., and Ng, D.T. (2004). Misfolded proteins are sorted by a sequential checkpoint mechanism of ER quality control. *J. Cell Biol.* 165:41–52.

Wahlman, J., DeMartino, G.N., Skach, W.R., et al. (2007). Real-time fluorescence detection of ERAD substrate retrotranslocation in a mammalian *in vitro* system. *Cell* 129:943–955.

Wakabayashi-Nakao, K., Tamura, A., Furukawa, T., et al. (2009). Quality control of human ABCG2 protein in the endoplasmic reticulum: ubiquitination and proteasomal degradation. *Adv. Drug Deliv. Rev.* 61:66–72.

Walter, J., Urban, J., Volkwein, C., et al. (2001). Sec61p-independent degradation of the tail-anchored ER membrane protein Ubc6p. *EMBO J.* 20:3124–3131.

Wang, G., Sawai, N., Kotliarova, S., et al. (2000). Ataxin-3, the MJD1 gene product, interacts with the two human homologs of yeast DNA repair protein RAD23, HHR23A and HHR23B. *Hum. Mol. Genet.* 9:1795–1803.

Wang, L., Dong, H., Soroka, C.J., et al. (2008). Degradation of the bile salt export pump at endoplasmic reticulum in progressive familial intrahepatic cholestasis type II. *Hepatology* 48:1558–1569.

Wang, Q., and Chang, A. (1999). Eps1, a novel PDI-related protein involved in ER quality control in yeast. *EMBO J.* 18:5972–5982.

Ward, C.L., Omura, S., Kopito, R.R. (1995). Degradation of CFTR by the ubiquitin-proteasome pathway. *Cell* 83: 121–127.

Weihl, C.C., Dalal, S., Pestronk, A., et al. (2006). Inclusion body myopathy-associated mutations in p97/VCP impair endoplasmic reticulum-associated degradation. *Hum. Mol. Genet.* 15:189–199.

Werner, E.D., Brodsky, J.L., McCracken, A.A. (1996). Proteasome-dependent endoplasmic reticulum-associated protein degradation: an unconventional route to a familiar fate. *Proc. Natl. Acad. Sci. USA.* 93:13797–13801.

Wiertz, E.J., Tortorella, D., Bogyo, M., et al. (1996). Sec61-mediated transfer of a membrane protein from the endoplasmic reticulum to the proteasome for destruction. *Nature* 384: 432–438.

Willer, M., Forte, G.M., Stirling, C.J. (2008). Sec61p is required for ERAD-L: genetic dissection of the translocation and ERAD-L functions of Sec61P using novel derivatives of CPY. *J. Biol. Chem.* 283:33883–33888.

Wright, R., Basson, M., D'Ari, L., et al. (1988). Increased amounts of HMG-CoA reductase induce "karmellae": a proliferation of stacked membrane pairs surrounding the yeast nucleus. *J. Cell Biol.* 107:101–114.

Wu, J., and Kaufman, R.J. (2006). From acute ER stress to physiological roles of the Unfolded Protein Response. *Cell Death Differ.* 13:374–384.

Yagishita, N., Yamasaki, S., Nishioka, K., et al. (2008). Synoviolin, protein folding and the maintenance of joint homeostasis. *Nat. Clin. Pract. Rheumatol.* 4:91–97.

Ye, Y., Meyer, H.H., Rapoport, T.A. (2001). The AAA ATPase Cdc48/p97 and its partners transport proteins from the ER into the cytosol. *Nature* 414:652–656.

Ye, Y., Shibata, Y., Yun, C., et al. (2004). A membrane protein complex mediates retro-translocation from the ER lumen into the cytosol. *Nature* 429:841–847.

Younger, J.M., Chen. L., Ren, H.Y., et al. (2006). Sequential quality-control checkpoints triage misfolded cystic fibrosis transmembrane conductance regulator. *Cell* 126:571–582.

Chapter 15
Regulation of Chromatin Structure and Transcription Via Histone Modifications

Kajan Ratnakumar, Avnish Kapoor, and Emily Bernstein

15.1 Introduction

Eukaryotes package their genome into a highly organized structure, known as chromatin, composed of DNA and its intimately associated proteins. The nucleosome is the basic fundamental repeating unit of chromatin, a spherical-shaped macromolecule composed of approximately 146 base pairs of DNA wrapped twice around an octamer of four histone proteins – H2A, H2B, H3 and H4 (Fig. 15.1; Luger et al. 1997). Nucleosome assembly involves the association of an H3/H4 heterotetramer with DNA, and subsequent addition of two H2A/H2B heterodimers, facilitated by proteins known as histone chaperones (Park and Luger 2008). Chromatin is generally classified into either euchromatin or heterochromatin, depending on its level of compaction.

Euchromatin is 'open' and poised for gene expression, while heterochromatin is compact and refractory to transcription. Euchromatin is best described by the 'beads on a string' model, which is thought to represent the lowest level of chromatin compaction (10 nm fiber). Heterochromatin is formed by the addition of linker histone H1 and various non-histone proteins, which further compact nucleosomes into higher order structures (30 nm fiber and beyond). Finally, chromatin reaches its most condensed state during mitosis (Horn and Peterson 2002).

While chromatin plays a structural role, its regulation is highly dynamic. At least three critical factors contribute to chromatin dynamics – the post-translational modification of histones (see below), ATP-dependent chromatin remodeling, and the incorporation of specialized histone variants into chromatin. ATP-driven chromatin remodeling complexes rearrange or mobilize nucleosomes during cellular processes such as transcription, and their recruitment to chromatin can depend on the PTM-status of particular loci (Wu et al. 2009). Histone variants differ from canonical histones in amino acid sequence and therefore, are subject to distinct

E. Bernstein (✉)
Department of Oncological Sciences, Mount Sinai School of Medicine, New York, NY 10029, USA
e-mail: emily.bernstein@mssm.edu

C.J. Vidal (ed.), *Post-Translational Modifications in Health and Disease*, Protein Reviews 13, DOI 10.1007/978-1-4419-6382-6_15, © Springer Science+Business Media, LLC 2011

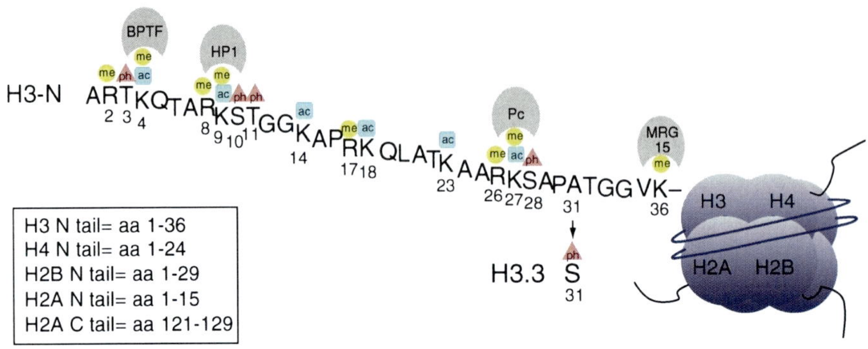

Fig. 15.1 The canonical nucleosome consists of 146bp of DNA wrapped around the histone octamer (two of each H2A, H2B, H3 and H4). The histone tails protrude beyond the nucleosomal core, are unstructured in nature, and highly post-translationally modified. See histone H3 tail as an example. Modified residues are numbered below the H3 tail sequence and PTMs are shown above; triangles represent phosphorylation (ph), squares represent acetylation (ac) and circles depict methylation (me). A single residue can be modified by different PTMs under different biological conditions (e.g. K9ac in transcription vs. K9me3 in transcriptional silencing). Representative readers of lysine methylation are shown above the methylation sites (see text for details). The H3.3 S31ph site is shown below the canonical H3 sequence (which contains an alanine at position 31). Lengths of histone tails are listed in the box below. Although not depicted here, PTMs are also found in the nucleosomal core, such as H3K79me

PTMs, resulting in unique biological outcomes (Bernstein and Hake 2006). Several key cellular pathways are regulated by histone variant PTMs, and these will be described where relevant.

PTMs are particularly abundant on the N-terminal histone tails, which protrude from the nucleosomal core. However, PTMs also exist within the core and on certain C-terminal tails, for example, H2A. Histone PTMs can result in 'on' or 'off' chromatin states regarding transcriptional status, including phosphorylation, acetylation, methylation and ubiquitination - the focus of this chapter. In order to simplify histone PTM terminology, we follow the nomenclature presented in Table 15.1 (Turner 2005).

The "histone code" hypothesis has been proposed to explain the complex and combinatorial pattern of PTMs and their biological consequences. This hypothesis states that PTMs can act through two mechanisms: (1) by structurally changing the chromatin fiber through internucleosomal contacts, thus regulating DNA accessibility, and (2) by serving as docking sites for effector molecules (generally referred to as 'readers') that initiate distinct biological processes (Strahl and Allis 2000; Turner 1993; 2000). As discussed throughout the chapter, histone PTMs are placed by enzymes referred to as 'writers' (e.g. methyltransferases), recognized and bound by 'readers' (e.g. HP1) and finally, removed by 'erasers' (e.g. demethylases), summarized in Table 15.2. Table 15.3 provides a list of the PTM-recognition domains of the readers discussed throughout the chapter. Finally, while the focus of the chapter is mammalian histone PTMs, groundbreaking studies in other organisms

Table 15.1 Histone posttranslational modification nomenclature

PTM	Abbreviation	Example
Phosphorylation	S/T/Y#ph	H3S10ph
Acetylation	K#ac	H4K16ac
Arginine methylation	R#me	
monomethyl-	me1	H4R3me1
dimethyl symmetric-	me2s	H4R3me2s
dimethyl asymmetric-	me2a	H4R3me2a
Lysine methylation	K#me	
monomethyl-	me1	H3K27me1
dimethyl-	me2	H3K27me2
trimethyl-	me3	H3K27me3
Ubiquitylation	K#ub	
monoubiquityl-	ub1	H2AK119ub1
polyubiquityl-	ubn	H2BK123ubn[a]

This nomenclature was put forth in 2005 (Turner 2005) and is used to describe PTMs throughout the chapter. The # represents the amino acid that is modified; examples are on right side of the table
[a]Polyubiquitylation of H2B was recently described in *S. cerevisae*; this K residue is conserved in mammals

have significantly contributed to our understanding of chromatin regulation and are mentioned where applicable.

15.2 Histone Phosphorylation

Of all PTMs found on cellular proteins in general, phosphorylation is perhaps the most widely documented and well characterized. The ATP-dependent addition of a phosphate moiety to serine, threonine or tyrosine residues is mediated by kinases and removed by phosphatases. Historically, the interaction of phosphorylated proteins with downstream binding effectors has been recognized as an important step in controlling signal transduction cascades, for example, in the activation of receptor tyrosine kinases. Histone proteins are no exception to the action of kinases. Phosphorylation is found on all four core histones, the linker histone H1, as well as histone variants, and is carried out by a variety of kinases. Histone phosphorylation can broadly be divided into three classes – mitotic, gene transcription-induced, and apoptotic and DNA damage-induced phosphorylation. Interestingly, several individual phosphorylation sites play a role in more than one biological process.

15.2.1 Mitotic Phosphorylation

Histone phosphorylation occurs in a cell cycle-dependent manner and has long been recognized to play a role in mitotic chromosome condensation, however, the mechanisms remain unclear. Mitotic phosphorylation is well characterized on several

Table 15.2 Identified writers, eraser, and readers of the histone code and their effects on transcriptional readout

PTM	Writer(s)	Eraser(s)	Reader(s)	Transcriptional readout
Phosphorylation				
H3S10ph, H3S28ph	Aurora B (M); MSK1, 2, RSK2 (I)	[PP1] (M)	14-3-3 (I)	ON (I)
H3.3 S31ph	N/D	N/D	N/D	N/A
H2A.X S139ph	ATM/ATR, DNA PK	PP2C, PP4	MDC1	N/A
H2A.X Y142ph	WSTF	EYA	N/D	N/A
macroH2A S137ph	CDKs	N/D	N/D	[ON] (I)
Acetylation				
H3K9ac	CBP, p300	SIRT1, 2, 3	BAF180	ON
H3K14ac	GCN5, PCAF	HDACs	Brg1, BAF180, PCAF, TAFII250	ON
H4K16ac	GCN5, hMOF	SIRT1, 2, 3	GCN5	ON
Arginine methylation				
H3R2me1/2	CARM1	PAD4/JMJD6	N/D	ON/OFF
H4R3me1/2	PRMT1, PRMT5	PAD4/JMJD6	DNMT3A	OFF
Lysine methylation				
H3K4me1/2/3	MLL1-4 SET1A, B, SET7/9	LSD1/JARID1A, B, C, D	WDR5/CHD1, BPTF, ING2, JMJD2A	ON
H3K9me1/2/3	SUV39H1, 2, G9a, GLP, ESET, RIZ1	LSD1/JHDM2A, B, JMJD2A, B, C, D	HP1, CDY, CDYL2	OFF
H3K27me1/2/3	EZH1, EZH2	UTX, JMJD3	Pc	OFF
H3K36me1/2/3	SET2	JHDM1 A, B/JMJD2A, B, C	MRG15	ON
H3K79me1/2/3	DOT1L	N/D	53BP1	ON
H4K20me1/2/3	PR-SET7/SUV4-20H1, 2	N/D	L3MBTL1, L3MBT2, 53BP1/ JMJD2A	OFF
Ubiquitylation				
H2AK119ub1	RING1A,B	2A-DUB, UBP-M, USP21	N/D	OFF
H2BK120ub1	RNF20	ATXN7L3, USP22, and ENY2	N/D	ON

This table categorizes many of the histone PTMs, readers, writers and erasers discussed in detail throughout the chapter. Proteins listed are all human homologs. Slashes (/) represent separation of different methylation states of the PTM and the readers, writers and erasers. Proteins or transcriptional readout in [] are likely candidates or functions, but not yet proven experimentally in humans. (M) represents mitosis and (I) represents interphase. N/D is not determined, N/A represents not applicable to transcriptional readout. While all possible histone regulatory molecules could not be discussed in the text, many other readers, writers, and erasers have been added to the table for completeness

Table 15.3 Domains responsible for histone code 'reading'

PTM	Reader domain(s)
Phosphorylation	14-3-3, Tandem BRCT domains
Acetylation	Bromodomain (BD)
Arginine methylation	
me0	WD40 repeats
me1	N/D
me2s	PHD finger
me2a	N/D
Lysine methylation	
me0	Plant homeodomain finger (PHD finger)
me1	Malignant brain tumor repeats (MBT)
me2	WD40 repeats, tandem tudor, MBT, PHD finger, chromo barrel
me3	Chromodomain (CD), double CD, chromo barrel, PHD finger, double tudor
Ubiquitylation	N/D

The readers of the histone code contain specific PTM-binding domains. These are summarized based on type of modification and methylation state. It is clear that many lysine methyl-binding domains have been discovered, while reading domains of other PTMs are less well characterized. While not discussed in the text, me0 is the unmodified state. Recently, PHD fingers have been shown to bind me0 histone tails (Taverna et al. 2007). N/D is not determined. See text for details

key residues of nucleosomal H3, including S10ph and S28ph (Cerutti and Casas-Mollano 2009). Linker H1 phosphorylation also peaks during mitosis, as does phosphorylation of S31 on the H3 variant H3.3 (van Holde 1988; Hake et al. 2005). Interestingly, S31 is one of five residues that differ between H3.3 and the canonical H3.1, suggesting unique histone variant PTM regulation (Fig. 15.1; Hake et al. 2005). Finally, other mitotic phosphorylation sites on the H3 tail include, T3ph and T11ph. Unlike S10ph and S28ph, T3ph and T11ph peak just prior to mitosis (Cerutti and Casas-Mollano 2009).

The enzymatic writers of H3S10ph and S28ph are the Aurora kinases (A and B), which are necessary for kinetochore assembly and microtubule attachment (Crosio et al. 2002). Aurora B is a component of the chromosome passenger complex (CPC), which includes the inner centromere protein, INCENP and survivin (Vader et al. 2006). This complex is required for proper mitotic execution. While the mammalian eraser of H3S10ph has yet to be reported, a study in the budding yeast, *S.cerevisiae*, identified a Protein Phosphatase 1 (PP1) homolog as the H3S10 eraser, implicating PP1 as the potential eraser in mammals (Hsu et al. 2000).

Interestingly, a link between H3S10ph and the mitotic loss of the heterochromatin protein HP1β, which binds the adjacent methylated K9 (Fig. 15.1), from chromatin has been demonstrated in mammals (Fischle et al. 2005). This is referred to as a 'methyl/phos switch', and suggests that an additional role of histone phosphorylation is to regulate the recruitment of effector proteins to chromatin. A recent study in the fission yeast, *S. pombe*, also reported this 'methyl/phos switch' and demonstrated that condensin, a factor required for chromosome condensation during mitosis, was present on chromosomes concomitant with H3S10ph (Chen et al. 2008). This implicates condensin, or factors required for its loading, as potential S10ph readers.

Finally, the cell cycle-dependent phosphorylation of the H2A variant, macroH2A was recently reported. MacroH2A is an unusually large histone variant, containing an H2A domain and an evolutionarily conserved macro domain. MacroH2A is phosphorylated at serine 137 (S137ph), and while this phosphorylation is detected throughout the cell cycle, it peaks during mitosis (Bernstein et al. 2008). In vitro kinase assays have implicated the cell cycle-regulated Cdk/cyclin complexes in writing this PTM (Bernstein et al. 2008). The reader(s) of this modification is currently unknown.

15.2.2 Phosphorylation in Gene Activation

Histone phosphorylation events exhibit complex relationships during the cell cycle. For example, while H3S10 and S28 are phosphorylated during mitosis, these residues are also phosphorylated during periods of immediate-early gene induction. Thus, one PTM can serve two opposing roles and suggests that PTMs should be studied in a context-dependent manner.

The phosphorylation of H3 was first noted to occur as a rapid 'nucleosome response' to external mitogenic stimulation by growth factors such as EGF or FGF. Immediate early genes, c-fos and c-jun, are transcribed following growth factor administration and their promoters acquire H3S10ph (Mahadevan et al. 1991). There is an extensive list of writers for H3S10ph during gene activation, and their activity largely depends upon the stimulus received by the cell. Among those characterized include MAP kinases MSK1 and MSK2, which respond to tumor promoting factor phorbol ester TPA, and Ribosomal S6 Kinase (RSK2), which responds to EGF (Bode and Dong 2005).

The readers of H3S10ph during gene activation are the phospho-serine binding 14-3-3 proteins (Winter et al. 2008; Taverna et al. 2007). 14-3-3's localize to H3 tails following phosphorylation of S10, but the biological significance of this interaction is unknown. The stability of 14-3-3's association with H3 tails bearing S10ph is enhanced when the H3 tail simultaneously bears acetylation on K9 or K14 (Winter et al. 2008). This phospho-acetyl motif serves to enhance the association of 14-3-3 and can overcome repressive chromatin marks that inhibit transcription. Thus, different histone PTMs can work together, and therefore, often need to be addressed in a combinatorial fashion.

15.2.3 Apoptotic and DNA Damage Induced-Histone Phosphorylation

The histone variant H2A.X is central to the cell's response to DNA double strand breaks (DSBs). In response to DSB-inducing ionizing radiation, H2A.X becomes rapidly phosphorylated at serine 139 (S139ph) in mammalian cells (Rogakou et al. 1998).

This modified form of H2A.X, referred to as γH2A.X, persists until DNA damage has been repaired and the cell is released from the damage-induced checkpoint. γH2A.X is a hallmark of DNA damage and can be used to assess DSB-induced genomic instability.

The writers of γH2A.X include two members of the phosphatidylinositol-3 kinase-like family (PI3KK), ATM (Ataxia Telangiectasia Mutated) and ATR (ATM and Rad3-related), as well as DNA PK (Motoyama and Naka 2004). These kinases recognize the highly conserved SQEX motif (X represents any amino acid) present on the C terminal tail of H2A.X. γH2A.X is highly conserved; in budding yeast, the predominant form of H2A is also phosphorylated in response to DNA damage.

Upon the induction of DNA damage, γH2A.X provides a template for factors involved in the DNA damage response, resulting in 'γH2A.X foci', which can extend megabases from the actual break sites. While numerous factors are found in γH2A.X foci, peptide association studies determined that ATM, the MRN complex (Mre11, Rad50, NBS1-responsible for repair via homologous recombination) and MDC1 (Mediator of DNA damage Checkpoint 1) directly bind to S139ph (Stucki et al. 2005). Among these potential readers, MDC1 binding was strongest and occurred via its tandem BRCT (BRCA1 C-Termini) domains. The factors present at γH2A.X foci, whether directly bound or recruited by readers, function to repair damaged DNA and activate the checkpoint in order to prevent cells from entering the next phase of the cell cycle prior to recovery from damage. Upon repair completion, dephosphorylation of H2A.X occurs. The erasers responsible for removing the phosphate moiety from S139 are the mammalian protein phosphatase 2A and 4 complexes (PP2A and PP4) (Bonner et al. 2008).

In addition to S139ph, phosphorylation of the very C-terminal residue of H2A.X, tyrosine 142 (Y142), was recently reported (Xiao et al. 2009). Interestingly, S139ph correlates with the loss of Y142ph and thus, appear to be mutually exclusive during the DNA damage response. The Y142ph kinase is WTSF (Williams-Beuren Syndrome Transcription Factor) and the eraser of this mark is the tyrosine phosphatase EYA (Eyes Absent) (Cook et al. 2009). An interesting interplay between these two modifications was suggested in the delineation between the DNA damage repair and apoptotic pathways. In the absence of Y142 desphosphorylation, the pro-apopotic stress response kinase, Jnk1, is recruited preferentially to break sites. Concomitantly, the pro-repair factor MDC1 is lost from break sites, suggesting that the chromatin signature at DSB sites can determine whether the cell responds to damage through repair mechanisms or by programmed cell death. Of interest, Y142ph is the first report of histone tyrosine phosphorylation (Xiao et al. 2009).

15.3 Histone Acetylation

While most histone PTMs result in single-site specific effects, acetylation is a more global affair with multiple lysine residues becoming acetylated at once, mainly on H3 and H4 (Fig. 15.1). The first indication that histone acetylation positively regulated transcription came from incubation studies of chromatin with acetic anhydride, which provides an acetyl moiety. This resulted in increased global histone acetylation and transcriptional activity (Csordas et al. 1984). Enzymes of the histone acetyltransferase (HAT) family add an acetyl group from acetyl-CoA to the ε-amino group of lysines, which in turn, are removed by an enzyme family known as histone deacetylases (HDACs) (Roth et al. 2001). This section focuses on histone acetylation, HAT and HDAC activity, and their functional readout on histones.

Histone acetylation is broadly classified as a charge-altering modification. The structure of the nucleosome is largely based upon the physical attraction between the negatively charged DNA phosphate backbone and the positively charged histones. Acetylation adds a negative charge to positively charged lysines, reducing the strong interaction between DNA and the histone octamer, thus facilitating the unwinding of DNA from the octamer. The protein machinery that mediates fundamental cellular processes such as transcription, DNA replication and repair, require access to naked DNA, and acetylation plays a critical role during these events. On a global scale, histone acetylation and deacetylation are involved in the dynamic structural alterations of the chromatin template. Acetylation is found mainly in euchromatin and deacetylation predominates in condensed chromatin, contributing to open and closed chromatin states, respectively (Roth et al. 2001).

15.3.1 Histone Acetyltransferases

HATs can be divided into two main classes – Class A HATs acetylate lysine residues on nucleosomal histones and generally function during transcriptional activation, whereas Class B HATs acetylate free histones, which are thought to become incorporated into the nucleosome during DNA replication (Sterner and Berger 2000; Roth et al. 2001). Due to the focus of this chapter on transcriptional control, we discuss Class A HATs below.

While much of the histone acetylation work has been done in *S.cerevisiae*, the first report of HAT activity came from studies in the ciliated protozoan *Tetrahymena thermophila* in 1995 (Brownell and Allis 1995; Shahbazian and Grunstein 2007). This activity was subsequently attributed to Gcn5, a factor originally identified in *S.cerevisiae* to positively regulate transcription (Georgakopoulos and Thireos 1992; Brownell et al. 1996). Gcn5 has specificity for histones H3 and H4 (H3K14, H4K8 and H4K16) and optimal activity requires additional subunits, which are components of the yeast SAGA and ADA HAT complexes (Roth et al. 2001).

Class A HATs consists of the GNAT (*G*cn5-related *N* *a*cetyl*t*ransferase) and MYST (named after *MOZ*, *Y*bf2/Sas3, *S*as2 and *T*ip60) families (Sterner and Berger 2000). Both contain a HAT domain, but have variable domains, contributing to their unique function. For example, GNAT members contain a lysine acetyl-binding motif at their C-termini, known as a bromodomain (BD) (Table 15.3). This domain allows GNAT members to read the modifications they write, suggesting a mechanism by which HATs remain at sites of their activity for prolonged function.

MYST family members, on the other hand, often contain a zinc finger motif (ZF) and/or methyl-lysine binding chromodomains (CD) (Roth et al. 2001).

The well-characterized mammalian Class A HATs include p300, CBP (*C*REB-*B*inding *P*rotein), PCAF (*p*300/*C*BP *A*ssociated *F*actor) and GCN5. Interestingly, p300 and CBP contain both a BD and a ZF motif, which are peculiar to both the GNAT and MYST families (Roth et al. 2001). Both p300 and CBP can acetylate all core histones in the context of the nucleosome, and have overlapping substrate specificities with equal activity on H3K9, H3K14 and H4K5. However, p300 acetylates H4K8 to a higher degree than CBP, while CBP has greater activity on H4K12 (McManus and Hendzel 2003). PCAF is a GCN5 ortholog and like GCN5, it physically interacts with p300 and CBP. Both PCAF and GCN5 have been shown to acetylate both free and nucleosome-bound H3K14 (Table 15.2; Yang et al. 1996).

Mouse knockout experiments have demonstrated that the loss of p300, CBP or GCN5 results in embryonic lethality, indicating that histone acetylation plays a crucial role during mammalian development (Roth et al. 2001). Moreover, it suggests that despite the similarity between HATs and their targets, they are not functionally redundant. Mutations in CBP or p300 have been implicated in Rubinstein-Taybi Syndrome, a developmental disorder characterized by mental retardation (Roth et al. 2001). Genetic loss-of-function mutations and chromosomal translocations of CBP and p300 have also been detected in various tumors (Wang et al. 2008).

In the context of transcription, HAT complexes are recruited to gene promoters by transcriptional activators. For example, GCN5 interacts with the transcriptional activator hADA2 to regulate the transcription factor Pax5, which determines B cell lymphopoiesis and midbrain patterning (Barlev et al. 2003). AML1/RUNX1, a transcription factor required for hematopoietic gene expression, physically interacts with p300 and CBP to stimulate transcription (Aikawa et al. 2006). It is thought that these interactions allow the specific acetylation of target gene promoters.

15.3.2 Histone Deacetylases

HDACs are crucial to the formation and maintenance of heterochromatin. In mammals there are five classes of HDACs (Class I, IIa, IIb, III and Class IV), and while all classes contain a conserved HDAC domain, they differ in structure, function and tissue specific expression patterns. Mammals contain a class of NAD-dependent HDACs, known as the Sirtuins or Class III HDACs, which have been implicated in aging (Haberland et al. 2009).

Class I HDACs (HDAC1, 2, 3 and 8) are homologs of the yeast HDAC, Rpd3. They are relatively simple in structure; all have a conserved deacetylase domain, and with the exception of HDAC8, are subunits of repressor complexes. For example, HDAC1 and 2 are components of the Sin3 complex (Sin3A/Sin3B), which has various roles, including gene repression during development (Cunliffe 2008). Class IIa HDACs (HDAC4, 5, 7 and 9) have an extended conserved N-terminal region involved in protein-protein interactions. They have minimal catalytic activity and the mechanism by which they function is relatively unknown (Haberland et al. 2009). Class IIb and Class IV HDACs make up much smaller classes, with only two members in Class IIb (HDAC6 and 10) and one known member in Class IV (HDAC11). The precise roles of these members are still unclear; however, HDAC6 functions mainly in the cytoplasm and has two deacetylase domains, distinguishing it from other HDACs.

The role of HDACs in the formation of heterochromatin is well characterized at the telomeres of *S.cerevisiae*. Telomeres are heterochromatic and their formation begins in late S phase following replication of the genome. The key regulators of this process are the Silent information regulator (Sir) proteins including the HDAC, Sir2. Sir2 is an NAD-dependent enzyme that is recruited to telomeres by telomeric DNA-binding proteins, and subsequently deacetylates K16 on H4 (Shahbazian and Grunstein 2008). This deacetylation event serves to allow other Sir proteins to bind to chromatin, increasing the levels of chromatin compaction.

The recruitment of HDACs occurs in a similar fashion to that of HATs. HDACs are recruited to gene promoters by transcriptional repressors where they deacetylate histones and allow chromatin compaction. For example, the tumor suppressor Rb interacts with HDAC1 targeting it to promoters of genes required for cell cycle progression (Luo et al. 1998). Additionally, the yeast histone deacetylase 1 (Hda1) is directed to chromatin by the transcriptional repressor Tup1, while Ume6, a sequence specific repressor, recruits Rpd3. Deletion of either Hda1 or Rpd3 results in the hyperacetylation of chromatin and aberrant gene expression (Shahbazian and Grunstein 2007). Interestingly, it has also been observed that transcriptional activators can recruit HDACs to chromatin. In the case of growth factor stimulation in mammalian cells, which leads to the expression of immediate early genes, Sin3A gets recruited to these gene promoters by the transcriptional activator Elk1 (Yang et al. 2001). This mediates the rapid repression of genes following activation and is crucial to the temporal regulation of gene expression.

15.4 Histone Methylation

Histones are modified via the enzymatic addition of methyl groups through the donor, S-adenosylmethionine (SAM). This reaction occurs on the ε-nitrogen of lysines and arginines, and is mainly found on histones H3 and H4 (Shilatifard 2006; Ruthenburg et al. 2007). Mass spectrometry and biochemical studies have confirmed that lysines can be mono-, di-, or trimethylated in vivo (Ruthenburg et al. 2007; Taverna et al. 2007). Similarly, methyl groups can be added to arginine residues in order to generate monomethyl, symmetrical dimethyl or asymmetric dimethyl states (Wysocka et al. 2006). Symmetrical and asymmetrical dimethylation refers to the addition of either one methyl group to each nitrogen or two methyl groups to one nitrogen of the guanidinium group, respectively.

Importantly, in contrast to acetylation, methylation does not influence the net charge of the affected residues. This suggests that methylation primarily serves as an 'information-storage' mark. In support of this hypothesis, studies indicate that both the degree of methylation and lysine residue in question, are differentially "read" by specific binding proteins (Fig. 15.1). This in turn leads to particular downstream effects, including gene activation and repression (Taverna et al. 2007). In recent years, considerable progress has been made in identifying and characterizing the enzymatic machinery involved in adding and removing histone methylation (Shilatifard 2006; Ruthenburg et al. 2007; Shi and Whetstine 2007). The distinct proteins responsible for 'writing', 'reading' and 'erasing' histone methylation will be discussed (see Table 15.2).

15.5 Reading, Writing and Erasing Histone Arginine Methylation

Histone methyltransferases (HMTs) are grouped into three classes: PRMTs (Protein Arginine Methyltransferases), SET domain-containing lysine HMTs (KMTs) and non-SET KMTs. PRMTs catalyze arginine methylation and to date, at least four mammalian PRMTs are known to catalyze histone methylation: PRMT1, PRMT5, PRMT6 and CARM1. Type I PRMTs include PRMT1, PRMT6, and CARM1 (cofactor associated arginine methyltransferase), which generate Rme1 and Rme2a marks. PRMT1 methylates H4R3 (me2a), PRMT6 methylates H3R2, while CARM1 methylates H3R2, H3R17 and H3R26 (Wysocka et al. 2006; Litt et al. 2009). Type II PRMTs include PRMT5, which generate Rme1 and Rme2s; specifically PRMT5 methylates H3R8 and H4R3 (me2s) (Wysocka et al. 2006).

While histone arginine methylation is involved in multiple chromatin events, its role as a transcriptional coactivator through nuclear hormone receptors (NR) has been particularly explored (Xu et al. 2003). For example, CARM1 cooperates with PRMT1 and p300/CBP in NR-mediated transcriptional activation (Koh et al. 2001). PRMT5 and 6, on the other hand, have been shown to be associated with transcriptional repression. PRMT5-mediated H4R3me2s is required for subsequent DNA methylation in mammals. In fact, DNMT3A, a *de novo* DNA methyltransferase, reads H4R3me2s via its PHD (Plant Homeodomain) finger domain, providing a mechanism by which this repressive PTM and DNA methylation are coordinated (Zhao et al. 2009). PRMT6-catalyzed H3R2me prevents transcriptional activation by abrogating H3K4me3 readers from binding (see below; Litt et al. 2009). A reader for PRMT6-mediated H3R2 methylation has yet to be identified.

Removal of arginine methylation is performed by at least two possible mechanisms. In the first, peptidylarginine deiminases (PADs) convert methyl-arginine to citrulline while releasing methylamine, known as demethylimination (Wysocka et al. 2006). Mammals encode four PADs, of which the nuclear PAD4, was shown to carry out demethylimination with broad specificity (H3R2, R8, R17 and R26 and H4R3) (Wysocka et al. 2006). Like PRMT1 and CARM1, PAD4 is recruited to hormone-induced gene promoters where its presence correlates with loss of arginine methylation and disengagement of RNA polymerase II (Wysocka et al. 2006).

However, PAD4 only converts the mono-methyl state into citrulline in vitro. This, combined with its broad specificity, leaves several questions unanswered, including how the removal of the dimethyl state is carried out. Recently, JMJD6 was identified to have *bone fide* demethylation activity (the second mechanism) with specificity for H3R2me2 and H4R3me2, and additional arginine demethylases will likely continue to be identified (Litt et al. 2009).

15.5.1 Reading, Writing and Erasing Histone Lysine Methylation

Like arginine methylation, lysine methylation contributes to both active and repressive chromatin functions and varies considerably depending on the modified lysine residue (Shilatifard 2006; Taverna et al. 2007). Lysine methylation is likely the most complex modification of histones. In most cases, the biochemical machinery devoted to each individual methylation site is unique and methylation of lysines only a few residues apart, can have distinct biological readouts. The methylation of H3K4, 9, 27, 36 and H4K20 catalyzed by SET-domain containing KMTs, and H3K79me written by a non-SET domain containing KMT, will be discussed.

15.5.1.1 Transcriptional Activation by Lysine Methylation

Methylation of residues H3K4, H3K36 and H3K79 are associated with transcriptionally active chromatin. In the cases of H3K4 and H3K36 methylation, the responsible KMTs are associated with RNA polymerase II during transcriptional initiation and elongation, resulting in methylation of promoter and coding regions, respectively (Shilatifard 2006; Ruthenburg et al. 2007). Importantly, the methylation state (me1, me2, me3) also plays a key role in transcriptional readout. Analyses of the distribution of different H3K4 methylation states using high-resolution genome-wide studies indicate that H3K4me2 and H3K4me3 are found predominantly at active loci, however H3K4me2 can also be present on poised inactive genes (Ruthenburg et al. 2007). In *S. cerevisiae*, a single KMT, Set1, is responsible for all three states of H3K4 methylation, while humans encode multiple H3K4 KMTs including MLL (mixed lineage leukemia) 1-4, hSET1A/B, SET 7/9, among others (Table 15.2; Shilatifard 2006; Ruthenburg et al. 2007). These KMTs exist in multi-subunit complexes that often contain H3K4me binding proteins (Ruthenburg et al. 2007).

Considerable attention has been given to proteins that read H3K4me2/3 (Ruthenburg et al. 2007; Taverna et al. 2007). At least four factors associated with transcriptionally active genes have been reported to interact with H3K4me2/3. This includes the CD-containing CHD1 (Chromo-helicase/ATPase-DNA binding protein 1), a core component of ATP-dependent chromatin remodeling complexes, and WDR5, a component of MLL complexes. WDR5 interacts with H3K4me2 through its WD40 repeats to promote H3K4me3, and has therefore been referred to as a 'presenter' rather than a reader (Table 15.3; Ruthenburg et al. 2007; Taverna et al. 2007). Another well-characterized chromatin remodeling complex, NURF, associates with H3K4me3 through its subunit BPTF (Bromodomain PHD finger Transcription Factor; Fig. 15.1). BPTF recognizes H3K4me3 via its PHD finger to assist NURF-mediated chromatin remodeling for transcriptional activation (Ruthenburg et al. 2007; Taverna et al. 2007). The ING (Inhibitor of Growth) family of tumor suppressors has also been identified to bind H3K4me3 through its PHD fingers (Ruthenburg et al. 2007; Taverna et al. 2007). ING2 binds H3K4me3 upon DNA damage, and stabilizes the Sin3A-HDAC complex at the promoters of proliferation genes (Ruthenburg et al. 2007).

Other methylation sites associated with gene activation include H3K36me and H3K79me. Set2 is responsible for K36 methylation in *S. cerevisiae* and is associated with elongating RNAPII (Shilatifard 2006). Set2 differs from Set1 as it remains associated with RNAPII throughout the body of the transcribed gene. Specifically, the presence of H3K36me2 within ORFs correlates with the 'on' state of transcription. A subunit of the yeast Rpd3 HDAC complex, Eaf3, reads H3K36me2/3 in yeast. Its human homolog MRG15, which associates with Sin3A-HDAC complexes, also reads H3K36me2/3 through its chromo barrel domain, which has similar structure to CDs (Sun et al. 2008).

DOT1 proteins are responsible for H3K79me, and are the only KMTs that lack an identifiable SET domain (Shilatifard 2006). In contrast to other methylated lysines, no specific function has been attributed to the different methylation states of H3K79 in mammals, which all seem to localize along the length of active genes, suggesting overlapping functions (Frederiks et al. 2008). Mice deficient for DOT1L display a global loss of H3K79me. Interestingly, reduced levels of heterochromatin marks at centromeres and telomeres were observed, accompanied by aneuploidy, telomere elongation and proliferation defects (Jones et al. 2008). Thus, the role of H3K79me is thought to prevent binding of heterochromatic proteins in euchromatin. When H3K79me is lost, heterochromatic proteins can bind euchromatin, resulting in a reduction and dilution of these factors in heterochromatin and a concomitant loss of silencing (Shilatifard 2006; Jones et al. 2008).

15.5.1.2 Transcriptional Silencing and Heterochromatin Formation by Lysine Methylation

Of the histone methylation sites found primarily in transcriptionally silent chromatin, H3K9, H3K27 and H4K20 are best characterized. The first SET-domain containing KMT was identified in *Drosophila*, Su(var)3-9; its human homolog, SUV39H1, was later shown to have specificity for H3K9 (Tschiersch et al. 1994; Rea et al. 2000). The role of SUV39H1 and H3K9me in heterochromatin formation was indicated through its association with Heterochromatin Protein, HP1. Subsequent studies demonstrated that H3K9me3 provides a binding site for the CD of HP1 (Fig. 15.1; Maison and Almouzni 2004). H3K9me3 is specifically enriched in pericentric heterochromatin, while H3K9me1 and H3K9me2 are often found in euchromatic regions (Maison and Almouzni 2004; Rice et al. 2003). While SUV39H1 is responsible for heterochromatic methylation, mouse knock out studies suggest that G9a and G9a related protein (GLP) are the primary H3K9 KMTs in euchromatin (Tachibana et al. 2005).

Methylation of H3K27 is linked to the silencing of Hox genes, X-chromosome inactivation (Xi) in female mammals and genomic imprinting (Cao and Zhang 2004; Whitcomb et al. 2007; Reik 2007). Originally discovered in *Drosophila* as being essential for proper body patterning, the Polycomb Group (PcG) family of proteins lies at the heart of this silencing system (Ringrose and Paro 2004). Biochemical and genetic evidence has identified at least two major complexes of PcG proteins that are important for H3K27me-dependent silencing (Ringrose and Paro 2004; Whitcomb et al. 2007). These are referred to as PRC1 and PRC2 (Polycomb Repressive Complex). *Drosophila* PRC2 contains Enhancer of Zeste (E(Z); human homologs EZH1 and EZH2), which catalyzes H3K27me3, Extra Sex Combs (ESC; human homolog EED) and Suppressor of Zeste-12 (SUZ12) (Whitcomb et al. 2007). Both EED and SUZ12 are essential for efficient H3K27 methylation and the integrity/stability of the complex (Whitcomb et al. 2007). Analogous to HP1 binding of H3K9me3, H3K27 is bound by the CD-containing protein Polycomb (Pc) which itself is a component of PRC1 (Fig. 15.1; Cao and Zhang 2004; Whitcomb et al. 2007). PRC1 also contains Ring1B, which possesses ubiquitin E3 ligase activity specific for H2A (discussed below), linking H3K27me and H2A ubiquitylation (Wang et al. 2004).

Methylation of H4K20 also marks mammalian heterochromatin. H4K20 methylation states also demarcate different genomic regions, with H4K20me3 found in constitutive heterochromatin and H4K20me2 or H4K20me1 occurring in non-overlapping fashion throughout the euchromatic regions (Yang and Mizzen 2008). SUV4-20H1 and SUV4-20H2, two SET-domain containing proteins are responsible for di- and trimethylation, while PR-Set7 is mainly responsible for H4K20me1 (Yang and Mizzen 2008). H4K20me1 has also been linked to transcriptional repression and X inactivation (Kohlmaier et al. 2004; Karachentsev et al. 2005). Intriguingly, H3K9me3 is required for the induction of H4K20me3 at constitutive heterochromatin and like SUV39H1, the SUV4-20H KMTs also interact with HP1 isoforms (Schotta et al. 2004).

Recent studies have uncovered a link between H4K20me and the DNA damage checkpoint - H4K20me is required for localization of the repair protein, 53BP1, to DNA DSBs. The tandem tudor domains of 53BP1 read H4K20me2, but not H4K20me3 (Taverna et al. 2007). Readers for HK20me1 include transcriptional repressors L3MBTL1 and L3MBTL2, which contain three tandem MBT (Table 15.3; Malignant Brain Tumor) repeats (Yang and Mizzen 2009; Guo et al. 2009).

15.5.1.3 Histone Lysine Demethylation

The identification of enzymes responsible for removing lysine methylation has lagged behind the discovery of histone methyltransferases until recently (Klose and

Zhang 2007; Shi and Whetstine 2007). For many years, it was believed that histone methylation was static and could only be removed by histone exchange or by cleavage of the methylated histone tail, which, in fact, has recently been shown for mammalian H3 (Duncan et al. 2008). The tremendous amount of effort put into identifying lysine demethylases (KDMs) has led to discovery of novel enzymes including families of amine oxidases and hydroxylases (Klose and Zhang 2007; Shi and Whetstine 2007).

The first KDM identified was LSD1 (lysine-specific demethylase 1), which acts on H3K4 methylation. LSD1 is a FAD-dependent amine oxidase that produces hydrogen peroxide and formaldehyde in its reaction (Shi and Whetstine 2007). Since the reaction mechanism requires a protonated nitrogen to initiate demethylation, LSD1's activity is limited to me1 and me2. Interestingly, LSD1 can function both as an activator or repressor, depending on the complex it associates with. When LSD1 associates with the transcriptional repressor complex Co-REST, it acts as an H3K4me1/me2 demethylase contributing to transcriptional repression. However, LSD1 association with the androgen receptor converts it into an H3K9me1/2 demethylase, allowing it to function as a transcription activator (Shi and Whetstine 2007).

The Jumonji family of proteins containing the JmjC domain has also been shown to function as KDMs. JmjC is similar to the bacterial AlkB catalytic domain and is capable of carrying out hydroxylation on methylated lysines (Shi and Whetstine 2007). There are 27 JmjC domain-containing genes within the human genome, 15 of which whose protein products demethylate specific lysines or arginines in the H3 tail (Klose et al. 2006). The first JmjC domain protein identified as a demethylase, JHDM1, targets H3K36me1/2 (Shi and Whetstine 2007). The apparent inability to demethylate H3K36me3, despite using a different reaction mechanism than LSD1, was surprising. JHMD2, like JHMD1, was also specifically shown to demethylate me1 and me2 states of H3K9, but not H3K9me3. The JHDM3 family was subsequently shown to demethylate both H3K9me3 and H3K36me3, demonstrating the reversibility of the trimethyl mark (Klose and Zhang 2007). These enzymes require a JmjN domain in addition to JmjC domains to efficiently catalyze trimethyl demethylation. Their C-termini often contain lysine methyl-binding PHD and Tudor domains, which likely contribute to their activity and specificity in chromatin (Shi and Whetstine 2007). Finally, the JARID1 subfamily of JmjC proteins is capable of demethylating H3K4me2 and me3, but fails to initiate demethylation of H3K4me1 (Secombe and Eisenman 2007). Thus, the counterpart of JARID1 in full K4 demethylation is LSDI.

KDM activity for H3K27 methylation has been attributed to UTX and JMJD3. Their activity is specific for H3K27me2/3 and interestingly, UTX is found in MLL complexes allowing for concomitant H3K4 methylation and H3K27 demethylation (Lee et al. 2007). JMJD3 plays a role in neural lineage commitment by regulating the expression of neurogenesis genes (Burgold et al. 2008). The demethylases for H3K79me and H4K20me have yet to be identified and it remains plausible that these marks are relatively static or alternative mechanisms are utilized to erase them (Klose and Zhang 2007).

15.6 Histone Ubiquitylation

Ubiquitin (ub) is a 76 amino acid protein and since its discovery 34 years ago, a variety of ubiquitin-dependent cellular processes have been identified (Hochstrasser 1996). These include protein degradation, cell-cycle regulation, protein trafficking and transcriptional regulation. Depending on the nature of ubiquitylation, it can either be a signal for specific downstream effects (mono-ubiquitylation) or result in protein degradation (poly-ubiquitylation). Several years after the discovery of ubiquitin, it became clear that a single ubiquitin moiety is conjugated to both H2A and H2B in mammals (Osley 2006). However, it took many years to decipher the role and regulators of these PTMs. Consistent with other PTMs, monoubiquitylation of histones is reversible and can be removed from target histones by a class of thiol proteases known as ubiquitin specific proteases (called UBPs in yeast and plants; USPs in mammals) (Weake and Workman 2008).

15.6.1 H2B Ubiquitylation

H2B is mono-ubiquitylated at K120 in mammals (H2BK120ub1), which corresponds to K123 in *S. cerevisiae*, K119 in *S. pombe* and K143 in *Arabidopsis* (Osley 2006). H2Bub is a mark of transcriptional activation and accumulating evidence has demonstrated a crosstalk between H2B ubiquitylation and H3 methylation, including H3K4me3 and K79me3 (Shilatifard 2006). For example, in the absence of H2Bub, H3K4 KMT complexes are only able to monomethylate H3K4.

Rad6 is the E2 ubiquitin conjugase that together with the E3 ubiquitin ligase Bre1, is required for H2B monoubiquitylation in yeast (Weake and Workman 2008). Rad6/Bre1 mediated H2B ubiquitylation is a prerequisite for H3K4 and H3K79 methylation (Shilatifard 2006). Humans have two sequence homologs of Rad6: HR6A and HR6B that might act redundantly, and the Bre1 homolog RNF20 regulates H2Bub in vivo (Weake and Workman 2008).

Deubiquitylation of H2Bub is carried out by two UBPs in *S. cerevisiae*: UBP8 and UBP10. UBP8 is a component of the SAGA complex and deubiquitylates H2B in vivo. UBP10, functions independently of SAGA and appears to affect distinct pools of H2Bub in vivo (Weake and Workman 2008). Recently, the *Drosophila* homolog of UBP10, *scrawny*, was shown to be required in germline, epithelium, and intestinal stem cells to repress the premature expression of key differentiation genes (Buszczak et al. 2009). Loss of the *Arabidopsis* homolog, UBP26, results in reactivation of transgenes and transposons by altering DNA methylation and H3K4 and H3K9 levels at these loci (Sridhar et al. 2007).

15.6.2 H2A Ubiquitylation

H2A ubiquitylation is catalyzed by the E3 ubiquitin ligase, Ring1B, a component of PRC1 (Table 15.2, Wang et al. 2004). PRC1 contains three RING domain-containing subunits Ring1A, Ring1B and Bmi-1, with only Ring1B possessing in vitro E3 ligase activity towards H2A (Wang et al. 2004). RNAi-mediated knockdown of Ring1B and its homolog, Ring1A, depletes H2Aub from the Xi in mice, consistent with its role in silencing (Whitcomb et al. 2007). It has been suggested that H2Aub might directly contribute to transcriptional repression by regulating higher order chromatin structure (Weake and Workman 2008). Of note, the corresponding lysine residues in the H2A variants H2A.Z and macroH2A1 are also subjected to monoubiquitylation (Weake and Workman 2008).

Specific deubiquitylases for H2Aub have been identified including Ubp-M, USP21 and 2A-DUB (Weake and Workman 2008). H2A deubiquitylation by Ubp-M is required for Hox gene expression and cell-cycle progression, while USP21 relieves the repression of transcription initiation caused by H2Aub-mediated inhibition of H3K4me. Finally, 2A-DUB interacts with the HDAC, PCAF, and preferentially deubiquitylates hyperacetylated nucleosomes in vitro (Zhu et al. 2007).

While histone monoubiquitylation was discovered decades ago, only two sites on H2A and H2B have been well characterized – H2Aub1 in transcriptional repression and H2Bub1 in transcriptional activation. Both are involved in crosstalk with other modifications including methylation and acetylation on the H3 tail. Of note, recent evidence suggests that H2B in yeast can be polyubiquitylated; however, its function is currently unknown (Geng and Tansey 2008). Future work focused on the discovery and characterization of novel sites of histone ubiquitylation, as well as the role of histone polyubiquitylation, will be of interest.

15.7 Concluding Remarks

While the sole purpose of the nucleosome was once thought to be the packaging of DNA, it is now clear that the nucleosome and its associated histone PTM profile play an active role in numerous DNA-templated cellular processes. Chromatin is an extremely dynamic template and we are just at the brink of understanding the complexity of histone PTMs. PTMs are numerous and diverse, they crosstalk with one another, and interact with various histone modifying complexes to bring about additional PTMs, remove PTMs and remodel chromatin. However, this complexity must be tightly regulated. It is becoming increasingly apparent that chromatin-modifying factors are involved in disease states, which has led to great interest in targeting these molecules for therapeutic purposes. In fact, the inhibition of HDACs (HDACi) is a current focus of cancer therapy (Haberland et al. 2008).

While beyond the scope of this chapter, it should be noted that histone PTMs are only one means of altering the chromatin template. Other factors include DNA methylation, histone variants, and non-coding RNAs. This chapter has focused on well-characterized PTMs, however, other regulatory PTMs exist, including sumoylation, ADP-ribosylation, proline isomerization, formylation, glycosylation and biotinylation (van Holde 1988; Kouzarides 2007). There is no doubt that this list will continue to grow and that new mechanisms for regulating histone PTMs, their contribution to the histone code, and downstream biological effects, will come into light – just like discovery of the histone demethylases, only a few years ago. It will be exciting to watch the world of chromatin 'unfold'.

Acknowledgements The authors wish to thank Matthew Goldberg, Luis Duarte, Andrew Xiao and Sandra Hake for discussions and critical reading of this chapter. We apologize to those whose work could not be cited due to space limitations. This work was supported by The Ellison Medical Foundation and the American Skin Association to E.B.

References

Aikawa, Y., Nguyen, L.A., Isono, K., et al. (2006). Roles of HIPK1 and HIPK2 in AML1- and p300-dependent transcription, hematopoiesis and blood vessel formation. *EMBO J.* 25:3955–3965.

Barlev, N.A., Emelyanov, A.V., Castagnino, P., et al. (2003). A novel human Ada2 homologue functions with Gcn5 or Brg1 to coactivate transcription. *Mol. Cell. Biol.* 23:6944–6957.

Bernstein, E., and Hake, S.B. (2006). The nucleosome: a little variation goes a long way. *Biochem. Cell Biol.* 84:505–517.

Bernstein, E., Muratore-Schroeder, T.L., Diaz, R.L., et al. (2008). A phosphorylated subpopulation of the histone variant macroH2A1 is excluded from the inactive X chromosome and enriched during mitosis. *Proc. Natl. Acad. Sci U.S.A* 105:1533–1538.

Bode, A.M., and Dong, Z. (2005). Inducible covalent posttranslational modification of histone H3. *Sci. STKE* 2005:re4.

Bonner, W.M., Redon, C.E., Dickey, J.S., et al. (2008). γH2A.X and cancer. *Nat. Rev. Cancer.* 8:957–967.

Brownell, J.E., and Allis, C.D. (1995). An activity gel assay detects a single, catalytically active histone acetyltransferase subunit in *Tetrahymena macronuclei. Proc. Natl. Acad. Sci. U.S.A.* 92:6364–6368.

Brownell, J.E., Zhou, J., Ranalli, T., et al. (1996). Tetrahymena histone acetyltransferase A: a homolog to yeast Gcn5p linking histone acetylation to gene activation. *Cell* 84:843–851.

Burgold, T., Spreafico, F., De Santa, F., et al. (2008). The histone H3 lysine 27-specific demethylase Jmjd3 is required for neural commitment. *PLoS One* 3:e3034.

Buszczak, M., Paterno, S., Spradling, A.C. (2009). Drosophila stem cells share a common requirement for the histone H2B ubiquitin protease scrawny. *Science* 323:248–251.

Cao, R., and Zhang, Y. (2004). The functions of E(Z)/EZH2-mediated methylation of lysine 27 in histone H3. *Curr. Opin. Genet. Dev.* 14:155–164.

Cerutti, H., and Casas-Mollano, J.A. (2009). Histone H3 phosphorylation: universal code or lineage specific dialects? *Epigenetics* 4:71–75.

Chen, E.S., Zhang, K., Nicolas, E., et al. (2008). Cell cycle control of centromeric repeat transcription and heterochromatin assembly. *Nature* 451:734–7.

Cook, P.J., Ju, B.G., Telese, F., et al. (2009). Tyrosine dephosphorylation of H2A.X modulates apoptosis and survival decisions. *Nature* 458:591–596.

Crosio, C., Fimia, G.M., Loury, R., et al. (2002). Mitotic phosphorylation of histone H3: spatio-temporal regulation by mammalian Aurora kinases. *Mol. Cell. Biol.* 22:874–885.

Csordas, A., Multhaup, I., Grunicke, H. (1984). Transcription of chemically acetylated chromatin with homologous RNA polymerase B. *Biosci. Rep.* 4:155–163.

Cunliffe, V.T. (2008). Eloquent silence: developmental functions of Class I histone deacetylases. *Curr. Opin. Genet. Dev.* 18:404–410.

Duncan, E.M., Muratore-Schroeder, T.L., Cook, R.G., et al. (2008). Cathepsin L proteolytically processes histone H3 during mouse embryonic stem cell differentiation. *Cell* 7:284–294.

Fischle, W., Tseng, B.S., Dormann, H.L., et al. (2005). Regulation of HP1-chromatin binding by histone H3 methylation and phosphorylation. *Nature* 438:1116–1122.

Frederiks, F., Tzouros, M., Oudgenoeg, G., et al. (2008). Nonprocessive methylation by Dot1 leads to functional redundancy of histone H3K79 methylation states. *Nat. Struct. Mol. Biol.* 15:550–557.

Geng, F., and Tansey, W.P. (2008). Polyubiquitylation of histone H2B. *Mol. Biol. Cell.* 19:3616–3624.

Georgakopoulos, T., and Thireos, G. (1992). Two distinct yeast transcriptional activators require the function of the GCN5 protein to promote normal levels of transcription. *EMBO J.* 11:4145–4152.

Guo, Y., Nady, N., Qi, C., et al. (2009). Methylation-state-specific recognition of histones by the MBT repeat protein L3MBTL2. *Nucleic Acids Res.* 37:2204–2210.

Haberland, M., Montgomery, R.L., Olson, E.N. (2009). The many roles of histone deacetylases in development and physiology: implications for disease and therapy. *Nat Rev Genet.* 10:32–42.

Hake, S.B., Garcia, B.A., Kauer, M., et al. (2005). Serine 31 phosphorylation of histone variant H3.3 is specific to regions bordering centromeres in metaphase chromosomes. *Proc. Natl. Acad. Sci. U.S.A.* 102:6344–6349.

Hochstrasser, M. (1996). Ubiquitin-dependent protein degradation. *Annu. Rev. Genet.* 30:405–439.

Horn, P.J., and Peterson, C.L. (2002). Chromatin higher order folding-wrapping up transcription. *Science* 297:1824–1827

Hsu, J.Y., Sun, Z.W., Li, X., et al. (2000). Mitotic phosphorylation of histone H3 is governed by Ipl1/aurora kinase and Glc7/PP1 phosphatase in budding yeast and nematodes. *Cell* 102:279–291.

Jones, B., Su, H., Bhat, A., et al. (2008). The histone H3K79 methyltransferase Dot1L is essential for mammalian development and heterochromatin structure. *PLoS Genet.* 12:1–11.

Karachentsev, D., Sarma, K., Reinberg, D., et al. (2005). PR-Set7-dependent methylation of histone H4 Lys 20 functions in repression of gene expression and is essential for mitosis. *Genes Dev.* 19:431–435.

Klose, R.J., Kallin, E.M., Zhang, Y. (2006). JmjC-domain-containing proteins and histone demethylation. *Nat. Rev. Genet.* 7:715–727.

Klose, R.J., and Zhang, Y. (2007). Regulation of histone methylation by demethylimination and demethylation. *Nat. Rev. Mol. Cell Biol.* 8:307–318.

Koh, S.S., Chen, D., Lee, Y.H., et al. (2001). Synergistic enhancement of nuclear receptor function by p160 coactivators and two coactivators with protein methyltransferase activities. *J. Biol. Chem.* 276:1089–1098.

Kohlmaier, A., Savarese, F., Lachner, M., et al. (2004). A chromosomal memory triggered by Xist regulates histone methylation in X inactivation. *PLoS Biol.* 2:E171

Kouzarides, T. (2007). Chromatin modifications and their function. *Cell* 128:693–705.

Lee, M.G., Villa, R., Trojer, P., et al. (2007). Demethylation of H3K27 regulates polycomb recruitment and H2A ubiquitination. *Science* 318:447–450.

Litt, M., Qiu, Y., Huang, S. (2009). Histone arginine methylations: their roles in chromatin dynamics and transcriptional regulation. *Biosci. Rep.* 29:131–141.

Luger, K., Mader, A. W., Richmond, R. K., et al. (1997). Crystal structure of the nucleosome core particle at 2.8 A resolution. *Nature* 38:251–260.

Luo, R.X., Postigo, A.A., Dean, D.C. (1998). Rb interacts with histone deacetylase to repress transcription. *Cell* 92:463–473.

Mahadevan, L.C., Willis, A.C., Barratt, M.J. (1991). Rapid histone H3 phosphorylation in response to growth factors, phorbol esters, okadaic acid, and protein synthesis inhibitors. *Cell* 65:775–783.

Maison, C., and Almouzni, G. (2004). HP1 and the dynamics of heterochromatin maintenance. *Nat. Rev. Mol. Cell. Biol.* 5:296–304.

McManus, K.J., and Hendzel, M.J. (2003). Quantitative analysis of CBP- and P300-induced histone acetylations in vivo using native chromatin. *Mol. Cell. Biol.* 23:7611–7627.

Motoyama, N., and Naka, K. (2004). DNA damage tumor suppressor genes and genomic instability. *Curr. Opin. Genet. Dev.* 14:11–6.

Osley, A. (2006). Regulation of histone H2A and H2B ubiquitylation, *Brief. Funct. Genomic Proteomic* 5:179–189.

Park, Y.J., and Luger, K. (2008). Histone chaperones in nucleosome eviction and histone exchange. *Curr. Opin. Struct. Biol.* 18:282–289.

Rea, S., Eisenhaber, F., O'Carroll, D., et al. (2000). Regulation of chromatin structure by site-specific histone H3 methyltransferases. *Nature* 406:593–599.

Reik, W. (2007). Stability and flexibility of epigenetic gene regulation in mammalian development. *Nature* 24:425–432.

Rice, J.C., Briggs, S.D., Ueberheide, B., et al. (2003). Histone methyltransferases direct different degrees of methylation to define distinct chromatin domains. *Mol. Cell* 12:1591–1598.

Ringrose, L., and Paro, R. (2004). Epigenetic regulation of cellular memory by the Polycomb and Trithorax group proteins. *Annu. Rev. Genet.* 38:413–443.

Rogakou, E.P., Pilch, D.R., Orr, A.H., et al. (1998). DNA double-stranded breaks induce histone H2A.X phosphorylation on serine 139. *J. Biol. Chem.* 273:5858–5868.

Roth, S.Y., Denu, J.M., Allis, C.D. (2001). Histone acetyltransferases. *Annu. Rev. Biochem.* 70:81–120.

Ruthenburg, A.J., Allis, C.D., Wysocka, J. (2007). Methylation of lysine 4 on histone H3: intricacy of writing and reading a single epigenetic mark. *Mol. Cell.* 12:15–30.

Schotta, G., Lachner, M., Sarma, K., et al. (2004). A silencing pathway to induce H3-K9 and H4-K20 trimethylation at constitutive heterochromatin. *Genes Dev.* 18:1251–1262.

Secombe, J., and Eisenman, R.N. (2007). The function and regulation of the JARID1 family of histone H3 lysine 4 demethylases: the Myc connection. *Cell Cycle* 6:1324–1328.

Shahbazian, M.D., and Grunstein, M. (2007). Functions of site-specific histone acetylation and deacetylation. *Annu. Rev. Biochem.*76:75–100.

Shi, Y., and Whetstine, J.R. (2007). Dynamic regulation of histone lysine methylation by demethylases. *Mol. Cell.* 12:1–14.

Shilatifard, A. (2006). Chromatin modifications by methylation and ubiquitination: implications in the regulation of gene expression. *Annu. Rev. Biochem.* 75:243–269.

Sridhar, V.V., Kapoor, A., Zhang, K., et al. (2007). Control of DNA methylation and heterochromatic silencing by histone H2B deubiquitination. *Nature* 447:735–738.

Sterner, D.E., and Berger, S.L. (2000). Acetylation of histones and transcription-related factors. *Microbiol. Mol. Biol. Rev.* 64:435–459.

Strahl, B. D., and Allis, C. D. (2000). The language of covalent histone modifications. *Nature* 403:41–45.

Stucki, M., Clapperton, J.A., Mohammad, R., et al. (2005). MDC1 directly binds phosphorylated histone H2A.X to regulate cellular responses to DNA double-strand breaks. *Cell* 123:1213–1226.

Sun, B., Hong, J., Zhang, P., et al. (2008). Molecular basis of the interaction of *Saccharomyces cerevisiae* Eaf3 chromo domain with methylated H3K36. *J. Biol. Chem.* 283:36504–36512.

Tachibana, M., Ueda, J., Fukuda, M., et al. (2005). Histone methyltransferases G9a and GLP form heteromeric complexes and are both crucial for methylation of euchromatin at H3-K9. *Genes Dev.* 19:815–826.

Taverna, S.D., Li, H., Ruthenburg, A.J., et al. (2007). How chromatin-binding modules interpret histone modifications: lessons from professional pocket pickers. *Nat. Struct. Mol. Biol.* 14:1025–1040.

Tschiersch, B., Hofmann, A., Krauss, V., et al. (1994). The protein encoded by the Drosophila position-effect variegation suppressor gene Su(var)3–9 combines domains of antagonistic regulators of homeotic gene complexes. *EMBO J.* 13:3822–3831.

Turner, B. M. (1993). Decoding the nucleosome. *Cell* 75:5–8.

Turner, B.M. (2000). Histone acetylation and an epigenetic code. *Bioessays* 22:836–845.

Turner, B.M. (2005). Reading signals on the nucleosome with a new nomenclature for modified histones. *Nat. Struct. Mol. Biol.* 12:110–112.

Vader, G., Kauw, J.J., Medema, R.H., et al. (2006). Survivin mediates targeting of the chromosomal passenger complex to the centromere and midbody. *EMBO Rep.* 7:85–92.

van Holde, K. E. (1988). *Chromatin*. Springer, New York.

Wang, H., Wang, L., Erdjument-Bromage, H., et al. (2004). Role of histone H2A ubiquitination in Polycomb silencing. Nature 14:873–878.

Wang, L., Tang, Y., Cole, P.A., et al. (2008). Structure and chemistry of the p300/CBP and Rtt109 histone acetyltransferases: implications for histone acetyltransferase evolution and function. *Curr. Opin. Struct. Biol.* 18:741–747.

Weake, V.M., and Workman, J.L. (2008). Histone ubiquitination: triggering gene activity. *Mol. Cell* 29:653–663.

Whitcomb, S.J., Basu, A., Allis, C.D., et al. (2007). Polycomb Group proteins: an evolutionary perspective. *Trends Genet.* 23:494–502.

Winter, S., Simboeck, E., Fischle, W., et al. (2008). 14-3-3 proteins recognize a histone code at histone H3 and are required for transcriptional activation. *EMBO J.* 27:88–99.

Wu, J.I., Lessard, J., Crabtree, G.R. (2009). Understanding the words of chromatin regulation. *Cell* 136:200–206.

Wysocka, J., Allis, C.D., Coonrod, S. (2006). Histone arginine methylation and its dynamic regulation. *Front. Biosci.* 11:344–355.

Xiao, A., Li, H., Shechter, D., et al. (2009). WSTF regulates the H2A.X DNA damage response via a novel tyrosine kinase activity. *Nature* 457:57–62.

Xu, W., Cho, H., Evans, R.M. (2003). Acetylation and methylation in nuclear receptor gene activation. *Methods Enzymol.* 364:205–323.

Yang, X.J., Ogryzko, V.V., Nishikawa, J., et al. (1996). A p300/CBP-associated factor that competes with the adenoviral oncoprotein E1A. *Nature* 382:319–324.

Yang, S.H., Vickers, E., Brehm, A., et al. (2001). Temporal recruitment of the mSin3A-histone deacetylase corepressor complex to the ETS domain transcription factor Elk-1. *Mol. Cell. Biol.* 21:2802–2814.

Yang, H., and Mizzen, C.A. (2009). The multiple facets of histone H4-lysine 20 methylation. *Biochem. Cell Biol.* 87:151–161.

Zhao, Q., Rank, G., Tan, Y.T., et al. (2009). PRMT5-mediated methylation of histone H4R3 recruits DNMT3A, coupling histone and DNA methylation in gene silencing. *Nat. Struct. Mol. Biol.* 16:304–311.

Zhu, P., Zhou, W., Wang, J., et al. (2007). A histone H2A deubiquitinase complex coordinating histone acetylation and H1 dissociation in transcriptional regulation, *Mol. Cell* 27:609–621.

Chapter 16
Chromatin: The Entry to and Exit from DNA Repair

Anastas Gospodinov and Zdenko Herceg

16.1 Introduction

DNA is under constant threat from DNA damaging (genotoxic) agents (Fig. 16.1) and DNA lesions are continuously produced during the life of a cell. Repair of DNA lesions occurs in the context of chromatin, a highly condensed structure composed of DNA and specialized proteins (histones). Usually, DNA compaction in chromatin is believed to represent a hindrance to the processes of DNA metabolism and thus DNA repair requires a mechanism to deal with compacted chromatin structure. In line with this thinking, early observations of nuclease sensitivity kinetics of newly synthesized DNA in UV irradiated cells suggested that reversible changes of chromatin accessibility are a part of the repair process (Smerdon and Lieberman 1978). This led to the "access, repair, restore" model for DNA repair. While this may be valid for most repair mechanisms, accumulating data indicates that chromatin changes have a more complex role, such as providing interfaces for the recruitment of signaling and repair proteins. This appears to be particularly important in the process of double-strand break (DSB) repair. In nucleotide excision repair (NER), on the other hand, in addition to local changes at the sites of damage, histone acetylation seems to facilitate relaxation on genome-wide scale to improve accessibility to DNA (Rubbi and Milner 2003). Finally it is the restoration of chromatin structure to its initial state that signals the end of the repair process (Keogh et al. 2006; Groth et al. 2007).

As it is likely that both DNA repair processes and those that control chromatin structure have evolved together, changes in histone modifications and ATP dependent chromatin remodeling seem to be as much an intrinsic part of the repair process as lesion recognition and removal. Here we will review the current understanding of the role of chromatin modification/remodeling in the repair of different DNA lesions.

Z. Herceg (✉)
Epigenetics Group, International Agency for Research on Cancer (IARC),
150 Cours Albert Thomas, 69008 Lyon, France
e-mail: herceg@iarc.fr

C.J. Vidal (ed.), *Post-Translational Modifications in Health and Disease*, Protein Reviews 13, DOI 10.1007/978-1-4419-6382-6_16,

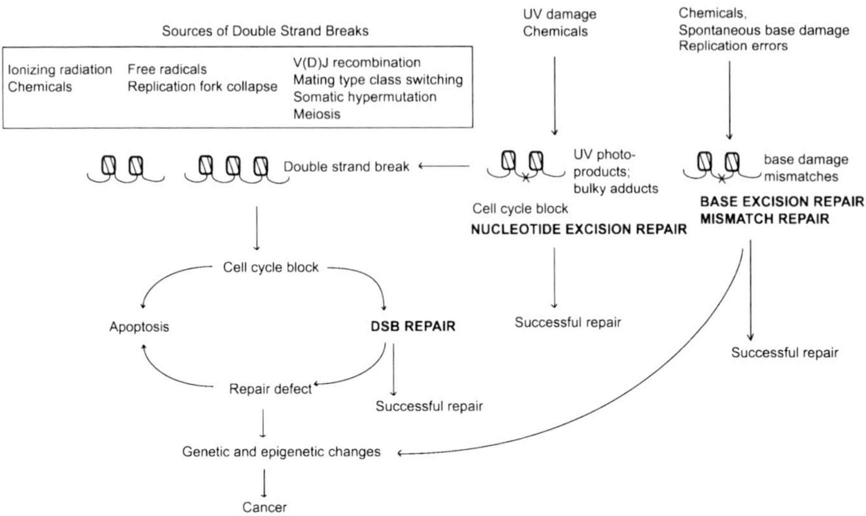

Fig. 16.1 Causes and consequences of DNA damage. DNA of cells is under constant threat of both endogenous and exogenous genotoxic agents. Double strand breaks result from ionizing radiation or radiomimetic chemicals, free radicals resulting from cellular metabolism or from processes of controlled genome rearrangements specific to some cell types. DNA breaks can be either repaired by HR or NHEJ, lead to cell suicide or result in genetic changes that may ultimately initiate cancer. DNA lesions specific to NER are generated mostly by physical or chemical agents exogenous to the cell. They are usually rapidly repaired, but in S-phase may stall DNA polymerases and be converted to double strand breaks. If left unrepaired, base damage or replication errors result in accumulation of mutations that may lead to cancer

16.2 Repair of Double Strand Breaks

16.2.1 Ways to Repair a Double Strand Break

Double strand breaks (DSBs) are the most dangerous type of damage. A single DSB in yeast, if left unrepaired, results in cell death (Resnick and Martin 1976). DNA breaks result from endogenous factors – replication fork collapse, V(D)J recombination, yeast mating type switching and meiosis as well as from exogenous agents such as ionizing radiation, crosslinking agents that stall polymerases, topoisomerase poisons, etc. (Fig. 16.1). In higher eukaryotes, failure to repair the break, may lead to loss of genetic material or to tumorigenic chromosomal rearrangements. In addition to DSBs being a cause of cancer, agents that induce them are used as treatment for cancer to activate cell death pathways (Helleday et al. 2008) and their efficiency is influenced by the repair capacity of cells.

A central place in cellular response to DNA damage play ATM (ataxia telangiectasia mutated) and ATR (ATM and Rad3-related) protein kinases, which are recruited to the sites of DNA damage (Falck et al. 2005) by the Mre11-Rad50-Nbs1 (MRN) complex (Lee and Paull 2005), and interaction with ATRIP (Cortez et al. 2001;

Falck et al. 2005; Zou and Elledge 2003), respectively. These kinases have multiple targets (Shiloh 2003), including protein kinases Chk1 and Chk2, which in turn target other proteins to induce cell-cycle arrest and facilitate DNA repair.

DSBs can be subjected to two repair pathways – nonhomologous end joining (NHEJ) and homologous recombination (HR) repair. These distinct types of repair are carried out by distinct sets of factors, although they have some shared molecular components. NHEJ is initiated by Ku70 and Ku80 proteins, which form a heterodimer with affinity for DSB ends and recruit DNA-dependent protein kinase catalytic subunit (DNA-PKcs), which phosphorylates itself and its target proteins (Burma and Chen 2004). The final step in rejoining is mediated by ligase IV (Lees-Miller and Meek 2003), which is associated with a dimer of XRCC4 (Modesti et al. 2003) and XLF (Cernunnos) (Ahnesorg et al. 2006). As NHEJ is not guided by a DNA template it can be error prone.

HR repair is considered an error-free mechanism for repair as it uses homologous sequences somewhere else in the genome to repair the break. HR uses mostly the sister chromatid as repair template and functions in late S and G2 phases. For HR repair, broken ends are processed to produce long 3' ssDNA tails that would be able to invade the homologous duplex. MRN complex was implicated in the generation of a ssDNA stretch (Ira et al. 2004; Nakada et al. 2004), but it was later found that *S. pombe* Ctp1 is essential for efficient formation of RPA-coated single-strand DNA adjacent to DSBs, indicating that it functions together with the MRN complex in $5' \rightarrow 3'$ resection (Limbo et al. 2007). More recent data has suggested a two-step mechanism for DSB processing – first Mre11 and Sae2 (*S. cerevisiae* Ctp1 homolog) remove a small oligonucleotide from DNA ends and then Exo1 and Sgs1 process this intermediate to generate extensive tracts of ssDNA that serve as Rad51 substrate (Mimitou and Symington 2008). This mechanism seems to be conserved from yeast to mammals (Gravel et al. 2008).

On ssDNA, Rad51 replaces RPA forming a nucleoprotein filament capable of searching for, invading, and transferring strands with a homologous duplex. In human cells, Rad51 nucleoprotein filament formation is mediated by Rad51 paralogs – Rad51B, Rad51C, Rad51D, XRCC2, and XRCC3 (Paques and Haber 1999; Thompson and Schild 2001) and Rad52 (New et al. 1998; Shinohara et al. 1998; Shen et al. 1996). Strand invasion is stimulated by Rad54 (Petukhova et al. 1998; Sigurdsson et al. 2002), which also promotes branch migration (Bugreev et al. 2006). In most instances the homologous sequence is that on the sister chromatid, but the HR machinery has the capacity to find donor sequences even if it is located on a different chromosome (Richardson et al. 1998). Once a joint molecule is formed invading strand is extended by DNA polymerases and branch migration to restoration of the genetic information spanning the break. Finally, resolution of the Holiday junction intermediates (West 2003; Ip et al. 2008) and ligation completes the process.

In eukaryotes, the repair of DSBs is influenced by the fact that it is carried out in the context of the chromatin structure. Nonrandom arrangement of genomes in the nucleus is a determinant of chromosome translocations as a consequence of illegitimate DSB repair and a factor in tumor development (Soutoglou and Misteli 2008b). Recent data demonstrated the relative immobility of a DSB in mammalian cells, which depends on Ku80 protein, but surprisingly is independent of the cohesion complex, H2AX phosphorylation, and MRN complex (Soutoglou et al. 2007). While positional stability of severed DNA provides the physical proximity of the ends – an obvious requirement for repair, another such requirement is that DNA ends are made accessible to repair complexes. Multiple mechanisms seem to cooperate to provide damage accessibility during DSB repair. In an elegant recent study in living cells it was reported that chromatin undergoes an energy-dependent local expansion immediately after introduction of a DSB. The change corresponded to a 30–40% reduction in the density of chromatin fibers in the vicinity of the DSB (Kruhlak et al. 2006). ATM kinase has a key role in increasing accessibility by phosphorylating chromatin proteins. One of ATM substrates KAP-1 is phosphorylated at the sites of damage, and as a result spreads throughout chromatin, promoting global relaxation and rapid genome surveillance (Ziv et al. 2006). Goodarzi and co-workers (Goodarzi et al. 2008) have provided evidence that ATM has a specialized role in repair of DSBs in tightly compacted heterochromatic parts of the genome. Removal of factors that promote compaction such as KAP-1, HP1 and histone deacetylases bypassed the requirement for ATM in repair of this subset of breaks, supporting the notion that ATM mediates chromatin accessibility.

Yet, simple increase of chromatin accessibility is not sufficient for repair. It has been found that proteins involved in DNA damage signaling and repair show complex subcompartmentalization at DSBs. Most available data indicate that DSB repair requires a dynamic sequence of interrelated chromatin changes induced by chromatin modification and remodeling complexes. The following section focuses on the chromatin modifications occurring at DSBs and providing interfaces for entry of DNA damage surveillance proteins, as well as on the functions of ATP-dependent chromatin remodelers in DSB repair.

16.2.2 Histone Modifications in DSB Repair

16.2.2.1 Phosphorylation

By far the most studied repair-related posttranslational histone modification is the phosphorylation of histone H2AX, a histone H2A isoform, at serine 139. It is one of the earliest events after break induction (Rogakou et al. 1999) and it serves to facilitate the accumulation of repair proteins, recruitment of chromatin modifiers and as a signal of incomplete repair. In mammals, the principal kinase that phosphorylates H2AX is ATM (Burma et al. 2001) (Fig. 16.2a). DNA-PK can function redundantly, but only in the absence of ATM. (Ward and Chen 2001; Stiff et al. 2004). During S phase, ATR mediates H2AX phosphorylation in response to replication stress (Ward and Chen 2001). In yeast cells H2AX variant is missing, thus phosphorylation is carried out on H2A S129 by Mec1 and Tel1 kinases which are homologues of ATR and ATM, respectively (Downs et al. 2000; Shroff et al. 2004). For reasons of simplicity, in this review yeast H2A will be referred to as H2AX.

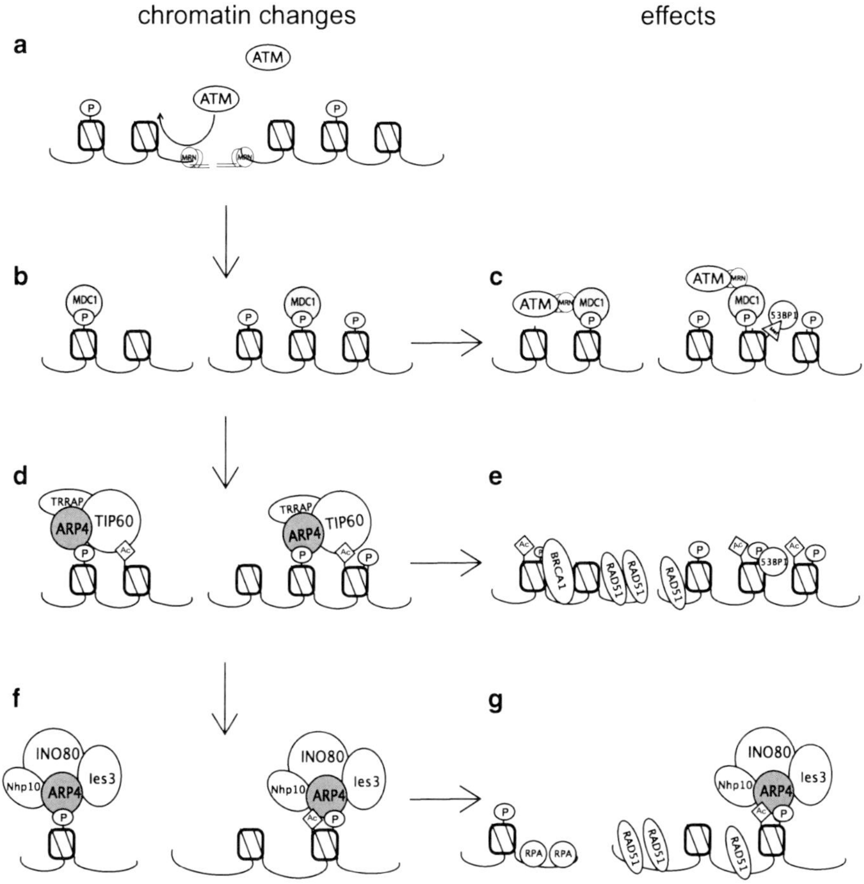

Fig. 16.2 Chromatin changes and their effects during double-strand break repair. (**a**) Upon break induction MRN complex binds DSB ends, allowing recruitment of ATM kinase that phosphorylates H2AX serine139. Phosphorylated H2AX is bound by MDC1 (**b**), which recruits MRN and consequently ATM, thus establishing a feed-back loop for H2AX phosphorylation (**c**), which extends several megabase pairs from the break site. Conformational changes in chromatin that result from DSB formation expose H3K79me2, which facilitates 53BP1 recruitment. H2AX phosphorylation further provides a binding interface for attachment of Arp4 (**d**) and Arp4-contining complexes (**d**, **f**). H4 acetylation mediated by TRRAP-TIP60 complex facilitates the access of RAD51, BRCA1 and 53BP1 to the sites of damage. INO80 chromatin remodeler is recruited through the interaction between ARP4 subunit and phosphorylated H2AX, but also needs additional subunits and possibly histone acetylation (**f**). Subsequent loss of histones in the vicinity of the break is thought to stimulate ssDNA formation (depicted by RPA binding) and/or recruitment of repair proteins

In both yeast and mammalian cells phosphorylated H2AX encompasses large regions around the break – several tens of kilobases in yeast (Shroff et al. 2004) and megabase sized regions in mammalian cells (Rogakou et al. 1999). Accumulation of γ-H2AX could be visualized as discrete nuclear foci following immunofluorescent staining. Available data suggest that phosphorylation of H2AX creates an epigenetic signal that is recognized by several proteins known to participate in DNA damage repair and signaling. Initial studies showed that foci formation of 53BP1 (Rappold et al. 2001), BRCA1, Rad50, and Rad51 (Paull et al. 2000) (Fig. 16.2) depended on H2AX phosphorylation, suggesting that γ-H2AX plays a role in recruitment of repair proteins. Yet, a microirradiation approach revealed that in H2AX(−/−) cells Nbs1, 53BP1 and Brca1 localized to irradiated tracks immediately following DNA damage induction, but they were lost from DNA damage sites at later times (Celeste et al. 2003). Thus, although dispensable for initial association with DSB chromatin, H2AX phosphorylation is necessary for the prolonged retention of repair proteins (Kobayashi et al. 2002; Downs et al. 2004; Stucki and Jackson 2006; Toh et al. 2006).

MDC1 plays a central role in the interaction of upstream repair factors with γ-H2AX. Changes to C-terminal tail of H2AX that prevent the interaction with MDC1, completely abolished NBS1 and 53BP1 foci formation, although phosphorylation and foci formation of H2AX remained intact (Stucki et al. 2005; Bekker-Jensen et al. 2005). It has been suggested that γ-H2AX-bound MDC1 (Stewart et al. 2003) (Fig. 16.2b) establishes a positive feed-back loop spreading H2AX phosphorylation to megabase distances away from the break (Kinner et al. 2008). Binding of NBS1 to MDC1 stabilizes the former at DSBs (Lukas et al. 2004), promotes ATM activation (Falck et al. 2005; You et al. 2005) which in turn increases H2AX phosphorylation (Fig. 16.2c). This model may also explain size differences of phosphorylated H2AX domains between mammalian and yeast cells as the latter lack MDC1 to serve as signal amplifier. The model is supported by the structure of irradiation induced foci (IRIF) (Bekker-Jensen et al. 2006), where ATM, MRN complex proteins, MDC1, 53BP1 and BRCA1 were localized within the entire region of γ-H2AX chromatin. The role of γ-H2AX was further confirmed by a recent study which demonstrated that immobilization of repair factors to chromatin triggers DNA damage response in the absence of DNA damage. Targeting of LacR fused NBS1, MRE11, MDC1 or ATM to a stably integrated LacO array led to H2AX phosphorylation. In H2AX deficient cells, accumulation of NBS1 and MRE11 by either another (immobilized) MRN component or by MDC1 was not affected. However, upon immobilization of NBS1 or MRE11, the accumulation of the downstream factors, MDC1 and 53BP1, was strongly impaired in the absence of H2AX, underscoring the importance of H2AX phosphorylation in maintaining these factors at the sites of DNA damage (Soutoglou and Misteli 2008a).

Phosphorylated H2AX also serves as the entry point for enzyme complexes that induce further post-translational modifications of histones and chromatin remodeling. Key to this role is Arp4 – a protein that binds phosphorylated H2AX and is a subunit of NuA4 histone acetylase complex, as well as of SWR1 and INO80 chromatin remodeling complexes (Shen et al. 2000; Krogan et al. 2003; Kobor et al. 2004; Mizuguchi et al. 2004) (Fig. 16.2d). The importance of Arp4 mediated recruitment is underscored by the fact that yeast Arp4 mutants were more sensitive to DNA damage than unphosphorylatable H2AX mutants (Downs et al. 2004). NuA4 complex is recruited to DSBs by Arp4 almost concomitantly with the appearance of γ-H2AX (Downs et al. 2004) and association of INO80 and SWR1 is thought to occur later on. Recruitment of chromatin remodelers requires NuA4 function (Downs et al. 2004) as well as additional subunits like Nhp10 (Morrison et al. 2004) and Ies3 (van Attikum et al. 2004) in INO80 complex. Available data suggest that Arp4-containing complexes bind in a step-wise manner at DSB sites to promote dynamic chromatin reconfiguration during repair.

It is important to note that phosphorylated H2AX serves as the primary signal of incomplete repair and that its dephosphorylation is a prerequisite to switching off checkpoint signaling at the "exit" from the repair process (Keogh et al. 2006). It has been shown that in mammalian cells, H2AX phosphorylation reached its peak within 1–2 h after damage followed by a steady decrease over a course of 12 h that correlated with kinetics of DNA repair (Paull et al. 2000). γ-H2AX is dephosphorylated by PP2A phosphatase (Chowdhury et al. 2005), while in yeast a related protein Pph3 is a part of a three protein complex (HTP-C) involved in H2A dephosphorylation (Keogh et al. 2006). Mammalian PP2A colocalizes with γ-H2AX foci, which indirectly implies that dephosphorylation might occur on chromatin (Chowdhury et al. 2005). However, several factors have been described to function in the removal of phosphorylated H2AX from the sites of damage. Data from J. Workman laboratory showed that *D. melanogaster* TIP60 acetylates phosphorylated H2Av (the analogue of H2AX) and drives its exchange for unmodified histones (Kusch et al. 2004). In HeLa cells, TIP60 acetylates γ-H2AX at K5, promoting its ubiquitination by Ubc13 and release (Ikura et al. 2007). Another study found that H4 acetylation by TIP60 complex is required prior to dephosphorylation of H2AX (Jha et al. 2008). It is possible that histone loss by chromatin remodeling may also contribute to the release of γ-H2AX, although the region where histone loss occurs (several kb in yeast) is not comparable to the size of γ-H2AX domains (several tens of kb in the same organism).

Additional histone phosphorylation events have been described to occur during DSB repair. In yeast, mutation of serine 122 of H2A showed stronger repair defect than γ-H2AX mutant (Wyatt et al. 2003). Histone H2B is phosphorylated on S14 in response to DNA damage and this may be linked with chromatin condensation and apoptosis (Fernandez-Capetillo et al. 2004). In response to DSBs casein kinase 2 phosphorylates histone H4 serine 1, and this modification was found to be inhibitory to NuA4-mediated acetylation of H4 (Cheung et al. 2005; Utley et al. 2005).

16.2.2.2 Acetylation

A number of recent studies have demonstrated that histone acetylation is an important prerequisite for DSB repair. Mutation of the four acetylatable N-terminal residues of histone H4 (lysines 5, 8, 12 and 16) abolish repair of both DSB and UV lesions and the defect was linked to the NuA4 acetylase complex (Bird et al. 2002). Histone H3 acetylation seems similarly important as mutants with histone H3 N-terminal deletion are more sensitive than histone H4 N-terminus deleted cells. GCN5, the enzyme that catalyses acetylation of N-tail lysines in H3

(K9, K14, K18, K23, K27, K36) was shown to be recruited to DSBs and yeast cells deleted for GCN5 gene lose viability following induction of a single DSB (Tamburini and Tyler 2005).

In mammalian cells, it was found that mutated TIP60 (mammalian homologue of yeast NuA4) causes defects in DSB repair and apoptosis (Ikura et al. 2000). Evidence from our group has revealed that TIP60 is recruited to the sites of DSB in a TRRAP-dependent manner, which leads to hyperacetylation of histone H4 and relaxation of the chromatin structure. Histone acetylation by TRRAP/TIP60 HAT is important for the accumulation of repair proteins to sites of DSBs and acetylation defects compromise homologous recombination repair (Murr et al. 2006). Deletion of TRRAP affected the loading of 53BP1, Rad51 and BRCA1 but not MDC1, indicating that some repair proteins are recruited in an acetylation independent way, while others require chromatin relaxation mediated by TRRAP/TIP60 (Murr et al. 2006) (Fig. 16.2e). Knock-down of TIP49, another subunit of TIP60, impaired Rad51 redistribution to chromatin and nuclear foci formation in response to damage, also suggesting that histone acetylation modulates the access of the repair factors to the sites of lesions (Gospodinov et al. 2009).

Taken together, data from yeast and mammalian studies clearly indicate that acetylation by NuA4/TIP60 complex of histone H4 N-terminal tail is required for efficient repair of DSBs. In this respect, an important question that deserves further investigation is how NuA4/TIP60 complexes are recruited to the sites of damage. Although Arp4 is necessary for the association of NuA4, INO80 and SWR1 complexes to phosphorylated H2AX at the sites of damage, it has not been shown that mutation of kinases or of serine 129 of H2AX affect NuA4 binding or acetylation. Presently, there is no direct evidence that Arp4 homologue BAF53 has the same function in recruitment of TIP60 and mammalian chromatin remodeling complexes to γ-H2AX domains. A direct interaction between TRRAP and NBS1 has been demonstrated (Robert et al. 2006), however the complex lacks acetylase activity.

Histone deacetylation is also important for DSB repair and histone deacetylases Rpd3, Sir2, and Hst1 are recruited to the HO lesion (Tamburini and Tyler 2005). Acetylation status of histone H3 and H4 lysines was shown to change during homologous recombination repair, and these changes appear to be triggered by the HR repair process (Tamburini and Tyler 2005), pointing to the transient nature of this mark.

There are two not mutually exclusive possibilities how histone acetylation promotes repair: 1. by conferring negative charge to histone tails leading to chromatin relaxation; and 2. by providing binding surface to participants in the repair process – via the bromodomains found in these proteins (Huyen et al. 2004). In addition, histone acetyl marks may have signaling roles in DNA damage response and DNA repair (Tamburini and Tyler 2005). It is possible that histone acetylation and phosphorylation of H2AX cooperate to bring and retain repair factors – the former facilitating access, while the latter provide binding interfaces. An additional facet of the role of histone acetylation in DNA damage response has come from a recent study showing that H3K14 is acetylated globally after induction of DSBs by ionizing radiation (IR) in an HMGN1 dependent manner. ATM localization to chromatin before and after IR and its activation in response to damage depended on HMGN1, suggesting that this mark participates in the regulation of ATM activation by its intranuclear organization (Kim et al. 2009).

16.2.2.3 Methylation

The role of histone methylation in the DNA damage response and DNA repair is less clear, although the involvement of lysine methylation has recently received considerable attention. Structural studies showed that binding of human 53BP1 to methylated H3K79 occurs via 53BP1 two tandem Tudor folds (Huyen et al. 2004). As levels of methylated H3K79 do not change following DNA damage a model was put forward in which DSBs induce changes in the higher-order chromatin structure and thus expose methylated H3K79 (Fig. 16.2c), which is then bound by 53BP1 (Huyen et al. 2004). Consistent with data in human cells, activation of Rad53 in yeast DOT1 and H3K79 mutants was impaired and Rad9 was not recruited to DSBs (Wysocki et al. 2005). A recent study has identified a role for H3K79 methylation in the control of 5′–3′ strand resection and checkpoint activation caused by DSBs or uncapped telomeres. The authors showed that Rad9 binding to methylated H3K79 inhibited resection and deletion of DOT1 methylase resulted in a more rapid accumulation of ssDNA and enhanced activation of Mec1 kinase, suggesting a model in which Rad9 bound to methylated H3K79 serves as a functional or physical barrier to resection (Lazzaro et al. 2008).

Methylation of H4K20 was also shown to have a role in repair in both yeast and mice. Deletion of Set9 methyltransferase in *S. pombe*, resulted in impaired ability of Crb2 (Rad9 homolog) to localize to the sites of damage (Sanders et al. 2004). While complete loss of methylation at H4K20 in mouse cells is not compatible with life, transition to monomethylation of H4K20, following knock-out of Suv4-20h methyltransferase resulted in increased sensitivity to damaging stress. Similarly to yeast, these defects may be at least in part due to reduced recruitment of 53BP1 to sites of damage in these cells (Schotta et al. 2008). The essential nature of H4K20 methylation might be explained by its contribution to maintenance of genome stability during DNA replication (Tardat et al. 2007). RP-Set7, which maintains H4K20 mono- and tri-metylation associates with the replication foci and its depletion leads to massive induction of DNA breaks followed by checkpoint activation (Tardat et al. 2007; Schotta et al. 2008).

Available data firmly establish the role of different histone marks as the means for entry of DSB repair proteins. The dynamic nature and interdependencies of different histone marks that provide the necessary chromatin environment during the steps of DSB repair are beginning to be elucidated (Tamburini and Tyler 2005). In a recent study in human cells the kinetics of appearance and removal of several histone modifications in response to IR was followed (Falk et al. 2007). TIP60 HAT colocalized with γ-H2AX foci and acetylated H4 increased rapidly, reaching maximum 20 min post-irradiation. The increase of H4K5 acetylation was accompanied by decreased H3K9 dimethylation. These changes, characteristic of decondensed chromatin were replaced by 40 min postirradiation by histone modification typical of condensed chromatin – decrease of H4K5 acetylation and increased metylation at H3K9. Future studies will undoubtedly further improve our knowledge regarding the spatial organization and the dynamic nature of chromatin modifications in DSB repair pathways.

16.2.3 Chromatin Remodeling in DSB Repair

In addition to histone modification, the other key mechanism that alters chromatin structure is ATP-dependent chromatin remodeling. Chromatin remodelers are large multi-subunit complexes that weaken the interaction between histones and DNA using the energy from ATP hydrolysis. At least four distinct ATP-dependent chromatin remodeling complexes – INO80, RSC, SWI/SNF and SWR1 are directly involved in DSB repair (Huang et al. 2005). INO80 is the most studied chromatin remodeling complex with respect to DNA repair. Strains deleted for Ino80, or Arp5 and Arp8 – subunits specific for the INO80 complex are hypersensitive to DSB-inducing agents (Shen et al. 2000; Shen et al. 2003). Knock-down of INO80 subunits dramatically reduced homologous recombination repair in HeLa cells (Wu et al. 2007) and INO80 was required for homologous recombination in *Arabidopsis thaliana* (Fritsch et al. 2004).

INO80 complex is recruited to a single HO-induced DSB in yeast by binding to phosphorylated H2AX (Morrison et al. 2004; van Attikum et al. 2004; Tsukuda et al. 2005) (Fig. 16.2f) and was shown to promote nucleosome eviction in a 5–6 kb region surrounding the DSB (Tsukuda et al. 2005; van Attikum et al. 2007). Although a general agreement exists that INO80 complex participates in HR repair, there are conflicting results regarding the exact step in which INO80 is involved. Data from one group showed that strains lacking INO80 chromatin remodeling activity are less efficient in 5′–3′ resection at DSB ends and observed consistent defects in Mre11 and Mec1 recruitment and checkpoint activation (van Attikum et al. 2004). Other authors found that 5′–3′ resection in Arp8 (INO80 complex specific subunit) deletion strain was normal but loading of repair proteins Rad51 and Rad52 was defective (Tsukuda et al. 2005). In the model favored by these authors, chromatin remodeling by INO80 complex would expose DNA at recipient locus and aid recruitment of proteins involved in homologous recombination (Fig. 16.2g).

Recent data point to a role of INO80 complex in checkpoint adaptation – a process in which after prolonged arrest cells resume cell cycle progression in the presence of unrepaired DNA breaks. Ino80Δ yeast cells failed to escape checkpoint arrest and this correlated with decreased levels of γ-H2AX and increased Htz1 (H2AZ) incorporation at the DSB site. The data showed that INO80 antagonizes Htz1 incorporation mediated by the SWR1 complex and thus maintains γ-H2AX levels in the vicinity of a DSB (Papamichos-Chronakis et al. 2006).

Similar to INO80 complex, yeast SWR1 chromatin remodeler is characterized by a split ATPase domain in the core ATPase subunit and the presence of Rvb proteins

(Bao and Shen 2007a,b) and is involved in transcriptional regulation and DNA repair. The homolog of SWR1 complex in higher eukaryotes is TIP60, which appears to be a fusion of yeast SWR1 remodeling and NuA4 acetylase complexes (Bao and Shen 2007a,b). SWR1 participates in the exchange of H2A with H2AZ variant (Mizuguchi et al. 2004), at promoters of inactive genes (Li et al. 2005) and heterochromatin boundaries (Meneghini et al. 2003). SWR1 complex contains Arp4 and is recruited to phosphorylated H2AX chromatin regions (Downs et al. 2004; van Attikum et al. 2007).Kinetics of SWR1 recruitment to sites of DSBs (Downs et al. 2004) and the fact that proper H2AZ deposition depends on NuA4 and Gcn5 acetylases (Kobor et al. 2004) has led to the suggestion that SWR1 might be necessary to further modify the chromatin environment in the vicinity of DSBs (Bao and Shen 2007a, b). SWR1 may have other roles at these sites as it was found to facilitate binding of yeast Ku80 protein and mediate NHEJ repair (van Attikum et al. 2007).

SWI/SNF is another chromatin remodeler recruited to HO-endonuclease induced DNA break site in yeast (Chai et al. 2005). Its deficiency in both yeast and mammalian cells led to DSB hypersensitivity (Chai et al. 2005; Klochendler-Yeivin et al. 2006). SWI/SNF ATPase subunit BRG1 interacts with BRCA1 suggesting participation in HR repair (Bochar et al. 2000; Moynahan et al. 1999). Both in vivo and in vitro evidence point to a role of human SWI/SNF in V(D)J recombination (Kwon et al. 2000; Morshead et al. 2003; Patenge et al. 2004). Existing data indicate that the remodeler participates in recombinational repair and may facilitate HR repair by remodeling nucleosomes at the donor locus (Chai et al. 2005). While SWI/SNF is needed before synapsis, another chromatin remodeler – RSC is required following synapsis to complete the repair event (Chai et al. 2005). RSC (remodels the structure of chromatin) is a conserved and abundant 15-subunit chromatin remodeling complex, essential for viability in yeast (Wong et al. 2006). It is recruited very early in the DSB response to sites of damage in a MRX (yeast MRN complex) dependent way. Although it is the first chromatin remodeler that is recruited to the break sites, it seems to be required only late in the repair process – during ligation. Efficiency of this step in Rsc2Δ mutants was 20% of that of the wild type (Chai et al. 2005). Additional functions of RSC have been described – yeast Rsc1 and Rsc2 subunits were shown to interact with Mre11 and Ku80, and these interactions were critical for survival following genotoxic stress (Shim et al. 2005). It has also been shown that RSC participates in the generation of a histone (H3) free region of several hundred bp near the break that is needed for access of repair factors (Shim et al. 2007). In addition RSC is necessary for full H2AX phosphorylation as it facilitates Tel1 and Mec1 recruitment to the break site and for cohesin loading (Liang et al. 2007).

How precisely do chromatin remodelers facilitate repair and the steps at which each one of them is required are not clear. It is likely that they function non-redundantly and it is tempting to speculate that INO80 may be required to promote DSB end processing and/or to facilitate Rad51 nucleoprotein filament formation, then SWI/SNF to catalyze a step necessary for synapsis, while RSC functions later on to complete repair, but more experimental data in yeast and in other organisms is required to justify such a model. Further studies are also needed to understand the extent of conservation of repair related chromatin remodeling in yeast and higher eukaryotes.

16.3 Nucleotide Excision Repair

16.3.1 NER and Histone Modifications

Nucleotide excision repair is conserved repair mechanism that deals with a broad spectrum of lesions which represent a chemical modification to DNA and structurally distort the double helix (Wood 1999; Batty and Wood 2000) such as cyclobutane pyrimidine dimers (CPD), 6–4 photoproducts (6–4PP) and bulky adducts. NER is carried out by excision of a short DNA fragment containing the lesion by a complex of XP (Xeroderma pigmentosum) proteins and TFIIH, followed by re-synthesis of the missing stretch using the undamaged strand as a template (Sancar et al. 2004). Based on differences in lesion recognition, NER is divided into two subpathways – transcription coupled repair (TC-NER), which is involved in the repair of the transcribed strand of active genes (Bohr et al. 1985; Mellon et al. 1987), where lesions are repaired much more efficiently and global genome repair (GG-NER), responsible for repair of all other parts of the genome (Svejstrup 2002). Lesion recognition in GG-NER is initiated by XPC, aided by XPA as well as RPA (Volker et al. 2001; You et al. 2003). In vivo it is strongly stimulated by a two protein complex called UV-DDB (Wakasugi et al. 2002), which serves to facilitate sensing of poorly recognized lesions such as CPD. In TC-NER, lesion-blocked RNA polymerase II, stimulates accumulation of downstream repair factors (Svejstrup 2002) and serves as a damage sensor. In TC-NER XPC protein is dispensable, but two additional factors – CSA and CSB (Hanawalt 2002; Ljungman and Lane 2004) are required.

Chromatin structure exerts inhibitory effect on NER. An early study hinted that nucleosome structure might present a barrier to repair by showing that cleavage of DNA in permeabilized cells by an enzyme recognizing CPDs was more efficient if cells were salt extracted to remove histones (Wilkins and Hart 1974). Following studies revealed that NER is more efficient in naked DNA than in chromatinized DNA (Evans and Linn 1984) and that repair is markedly more efficient in linker DNA than in DNA wrapped around the nucleosome particle (Lan and Smerdon 1985).

Evidence from more than 2 decades of studies points to a role of histone acetylation in maximizing NER efficiency. Early research from Smerdon group showed that total protein acetylation increases in response to UV irradiation (Ramanathan and Smerdon 1986) and that repair DNA synthesis is enhanced in hyperacetylated nucleosomes (Ramanathan and Smerdon 1989), suggesting that histone acetylation makes DNA more accessible to repair factors. Available data indicate that histone acetyltransferases modify chromatin both locally at the lesion sites and that in response to damage (repaired by NER), overall chromatin accessibility is increased as a result of histone acetylation.

Yeast GCN5 affected NER of MFA2 – a gene whose transcription is GCN5 dependent (Teng et al. 2002) and K9 and K14 of histone H3 at the MFA2 promoter were hyperacetylated in response to UV irradiation (Yu et al. 2005). As both repair of the transcribed and the non-transcribed strand were affected, it was concluded that GCN5 facilitates repair via changes to chromatin structure (Teng et al. 2002) as opposed to mediating transcription and TC-NER as a consequence, which would have affected repair of the transcribed strand only. At the same time repair of RPB2, a gene whose expression is not GCN5 dependent was not influenced. This suggests that local chromatin changes or presence of remodelers during transcription influences NER, and that a particular chromatin modifier could affect repair of only a subset of genes (Teng et al. 2002).

A study on HeLa cells demonstrated that following UV irradiation histone acetylation was increased. The same study found that human TBP-free TAF(II) complex (TFTC) complex (which contains GCN5 acetyltransferase) was targeted to UV damaged sites and preferentially acetylated nucleosomes assembled on UV damaged DNA templates (Brand et al. 2001). Another GCN5 containing acetylase complex – human STAGA associated with DDB1 protein (Martinez et al. 2001), part of UV-DDB. Interaction with the UV-DDB complex was also shown for p300, a HAT involved in transcriptional regulation, consistent with a role in facilitating access of repair machinery (Datta et al. 2001). Additional support for the role of p300 in NER came from a study showing that p300 interacted with PCNA and colocalized with sites of newly synthesized DNA after UV irradiation (Hasan et al. 2001).

The global increase in chromatin accessibility mediated by histone acetylation following DNA damage is likely to play a critical role in maximizing repair efficiency. UV irradiation triggers genome wide hyperacetylation of histone H3 and H4 (Yu et al. 2005). Rubbi and Milner demonstrated that local UV damage, led to p53 dependent increase of acetylation of H3K9 and global chromatin relaxation (Rubbi and Milner 2003). Acetylation of H3K14 was also increased in this process (Allison and Milner 2004). Chromatin relaxation was independent of proteins involved in lesion recognition and the requirement for p53 could be circumvented by histone deacetylase inhibitor Trichostatin A. Microinjection of an antibody against p300 inhibited NER and recruitment of p300 to NER sites was p53 dependent (Rubbi and Milner 2003). Treatment with RNA Pol II blocking agent α-amanitin induced chromatin relaxation, suggesting that lesion-stalled Pol II triggers global chromatin relaxation. Thus, in addition to its role in TC-NER, RNA Pol II might facilitate GG-NER as well, by mediating global increase in chromatin accessibility following DNA damage (Rubbi and Milner 2003).

ING proteins are a group of chromatin acetylation regulators that associate with HATs and HDACs and were shown to affect NER (Doyon et al. 2006). Interaction between p53 and p300 acetyltransferase occurs through their binding to ING2 (Wang et al. 2006a). p33(ING1), another member of ING group, was shown to enhance repair of UV lesions. Cells overexpressing p33(ING1) exhibited an enhanced repair capacity following exposure to UV irradiation (Cheung et al. 2001). Interestingly, knockdown of a splice variant of ING1b resulted in NER deficiency. ING1b was shown to be required for global histone H4 acetylation and UV-induced global chromatin relaxation (Kuo et al. 2007). Like ING1 protein, ING2 was shown to promote hyperacetylation of H4 upon UV irradiation and induce chromatin relaxation (Wang et al. 2006b).

Data regarding the importance of other histone modifications in NER is limited. It appears that, similarly to DSB repair, H3K79 methylation has a role in facilitating NER (Bostelman et al. 2007; Evans et al. 2008).

16.3.2 Chromatin Remodeling in NER

A number of experiments carried out mostly in vitro point to a role of chromatin remodeling in NER. Using an in vitro system with purified components and dinucleosome substrates, Ura and co-workers found that a dinucleosome assembly presented a hindrance to 6–4 photoproduct (6–4PP) excision, suggesting that human NER machinery requires space greater than the size of linker DNA to excise the lesion efficiently. Dual incision in this system was facilitated by ACF chromatin remodeling factor specifically in linker DNA, but not at the centre of nucleosome cores (Ura et al. 2001). Other investigators showed that removal of an acetylaminofluorene-guanine (AAF-G) adduct placed near the centre of twofold symmetry of the nucleosome core was stimulated by SWI/SNF chromatin remodeling activity. They also noted that NER factors XPA, XPC and RPA stimulated SWI/SNF chromatin remodeling activity, which in turn stimulated adduct removal (Hara and Sancar 2002). SWI/SNF complex stimulated removal of AAF-G adducts and 6–4PP, but not removal of CPD (Hara and Sancar 2003), suggesting that greater distortion of DNA caused by 6–4PP somehow influences chromatin remodeling.

Although it is technically difficult to assess the effects of chromatin remodeling on NER in vivo as the lesions which are subject to this pathway cannot be targeted to a precise genomic location, in vitro results strongly suggest that in addition to histone acetylation, chromatin remodeling participates in NER. There are however data that do not fit with this assumption. In vivo SWI/SNF was found to increase accessibility of chromatin at the MFA2 promoter in response to UV irradiation, yet swi2Δ mutant yeast exhibited normal repair efficiency (Yu et al. 2005). Bucceri and colleagues assessed the kinetics of UV damage removal in yeast under conditions of photolyase overexpression. The observed rates in most parts of the genome were extremely high with 50% of CPD removed within 5 s and 80% in 90 s (Bucceri et al. 2006). These results imply that induction of damage may result in rapid nucleosome unwrapping similar to chromatin expansion observed upon DSB induction (Kruhlak et al. 2006), which may be sufficient for photolyase entry. It may also be a result of the lower spatial requirements of a single enzyme to access DNA, compared to the multi-protein NER complex.

16.4 Other Types of Repair

Small base alterations do not distort the DNA helix and are generally repaired by base excision repair (BER) pathway. BER is a multistep process initiated by damage-specific DNA glycosylases, which release the damaged base. The abasic site is subsequently processed by an AP (apurinic/apyrimidinic) endonuclease or glycosylase-associated AP-lyase activity creating free 3′ end for DNA polymerase β synthesis. AP site can be repaired by either short patch repair, which replaces a single nucleotide before ligation or by long patch repair in which de novo synthesis of 2–13 nucleotides and displacement (by an endonuclease) of the lesion-containing strand occurs (Fortini and Dogliotti 2007). Presently, data about the necessity of chromatin changes during BER is largely limited. Some investigators found that chromatin structure did not inhibit BER or the function of its component proteins. For example, it was reported that Flap endonuclease 1 (FEN1) which participates in long patch BER could bind and cleave its substrates with similar efficiency irrespective whether presented with naked DNA or one assembled in nucleosomes (Huggins et al. 2002). Contrary to that in an in vitro system with BER components it was found that uracil DNA glycosylase (UDG) and apurinic/apyrimidinic endonuclease (APE) have 10 times reduced efficiency on nucleosomes substrates compared with naked DNA and synthesis activity of pol β on nucleosomal substrates was completely inhibited (Beard et al. 2003). In agreement with this observation, 8-oxoguanine DNA glycosylase, APE and pol β were strongly inhibited on nucleosomes, and SWI/SNF remodeling stimulated processing by these enzymes to the levels observed in naked DNA (Menoni et al. 2007). Acetylation may have a role in BER as a study showed that thymine DNA glycosylase associated with p300 and the resulting complexes were competent for both excision and acetylation (Tini et al. 2002). Future investigations are needed to conclusively determine the chromatin modification/remodeling requirements in BER.

Mismatch repair (MMR) is responsible for correcting base insertion-deletions that occur as errors during the replication process. Currently, there is no direct evidence to support the notion that MMR requires chromatin modifications and/or remodeling. A proteomic examination of the interaction partners of human MutL homologues failed to identify molecules associated with modifying chromatin (Cannavo et al. 2007). It has been suggested that MMR may utilize chromatin changes that occur during DNA replication (Flores-Rozas et al. 2000; Kleczkowska et al. 2001; Escargueil et al. 2008).

16.5 Concluding Remarks

As a result of the experimental efforts of many researchers during the last two decades it is now established that chromatin changes in response to DNA damage are required for efficient DNA repair and the mechanistic understanding of how this happens is constantly growing. Still, more remains to be understood. For example, information is scarce as to what particular chromatin changes are required at a given stage of DNA repair and how they relate with the other chromatin modification/remodeling, signaling and repair events.

A straight forward conclusion, from the involvement of chromatin modification/remodeling factors in DNA repair is that with respect to this process, different parts of the genome are unequal, much more so than it was suggested by the discovery of TC-NER. This is supported by direct experimental data (Teng et al. 2002;

Goodarzi et al. 2008). Thus, not only are the complexes that control epigenetic information involved in repairing DNA, but by directly affecting repair epigenetic patterns may be able to influence the genetic dimension of inheritance.

The knowledge on chromatin modification and remodeling in DNA repair has clear health implications, especially in light of the current efforts to modify activity of histone deacetylases as well as DNA methyltransferases in so called "epigenetic therapy". Histone modifiers, most notably acetyltransferases have been directly linked to oncogenesis (Yang 2004). Deregulation of HATs may contribute to cancer development and progression through impaired transcription, but it is likely that aberrant regulation of DNA repair by epigenetic factors is involved in oncogenesis as well. However, the way histone modifications are disrupted in cancer and how it relates to DNA repair, remains largely unknown (Herceg 2007). It also seems possible that small changes in efficiency of histone modifiers by agents present in the environment may influence mutation rates and lead to development of cancer. Due to the subtle and cumulative nature of these effects this will be hard to prove (Herceg 2007).

Yet, what is already known about histone modifications and repair may be clinically useful. For example, human H2AX gene maps to a region that frequently exhibits mutations and copy number changes in cancer and it has been suggested that it may be used as a susceptibility marker in lymphomas, leukemias and other cancers (Bonner et al. 2008). Whatever future discoveries are made in this exciting field, they will not only improve our knowledge of the mechanisms underlying diseases, but also provide benefits to human health.

Acknowledgements We thank to Drs Anupam Paliwal, Stoyno Stoynov, and Boyka Anachkova for critical reading of the manuscript. A.G. is supported by a Special Training Fellowship from the International Agency for Research on Cancer (IARC), Lyon, France. Work in our laboratory is supported by the International Agency for Research on Cancer (IARC), Lyon (France), and grants from National Institute of Health/National Cancer Institute, USA, the Association pour la Recherche sur le Cancer (ARC), France, la Ligue Nationale (Française) contre le Cancer (France), Institut National du Cancer, (France), Swiss Bridge Award, and Agence Nationale de Recherhe Contre le Sida et Hépatites Virales (ANRS, France) (to Z.H.).

References

Ahnesorg, P., Smith, P., Jackson, S.P., *et al.* (2006). XLF interacts with the XRCC4-DNA ligase IV complex to promote DNA nonhomologous end-joining. *Cell* 124:301–313.

Allison, S.J., and Milner, J. (2004). Remodelling chromatin on a global scale: a novel protective function of p53. *Carcinogenesis* 25:1551–1557.

Bao, Y., and Shen, X. (2007a). Chromatin remodeling in DNA double-strand break repair. *Curr. Opin. Genet. Dev.* 17:126–131.

Bao, Y., and Shen, X. (2007b). INO80 subfamily of chromatin remodeling complexes. *Mutat. Res.* 618:18–29.

Batty, D.P., and Wood, R.D. (2000). Damage recognition in nucleotide excision repair of DNA. *Gene* 241:193–204.

Beard, B.C., Wilson, S.H., Smerdon, M.J. (2003). Suppressed catalytic activity of base excision repair enzymes on rotationally positioned uracil in nucleosomes. *Proc. Natl. Acad. Sci. USA.* 100:7465–7470.

Bekker-Jensen, S., Lukas, C., Melander, F., *et al.* (2005). Dynamic assembly and sustained retention of 53BP1 at the sites of DNA damage are controlled by Mdc1/NFBD1. *J. Cell Biol.* 170:201–211.

Bekker-Jensen, S., Lukas, C., Kitagawa, R., *et al.* (2006). Spatial organization of the mammalian genome surveillance machinery in response to DNA strand breaks. *J. Cell Biol.* 173: 195–206.

Bird, A.W., Yu, D.Y., Pray-Grant, M.G., *et al.* (2002). Acetylation of histone H4 by Esa1 is required for DNA double-strand break repair. *Nature* 419:411–415.

Bochar, D.A., Wang, L., Beniya, H., *et al.* (2000). BRCA1 is associated with a human SWI/SNF-related complex: linking chromatin remodeling to breast cancer. *Cell* 102:257–265.

Bohr, V.A., Smith, C.A., Okumoto, D.S., *et al.* (1985). DNA repair in an active gene: removal of pyrimidine dimers from the DHFR gene of CHO cells is much more efficient than in the genome overall. *Cell* 40:359–369.

Bonner, W.M., Redon, C.E., Dickey, J.S., *et al.* (2008). GammaH2AX and cancer. *Nat. Rev. Cancer* 8:957–967.

Bostelman, L.J., Keller, A.M., Albrecht A.M., *et al.* (2007). Methylation of histone H3 lysine-79 by Dot1p plays multiple roles in the response to UV damage in Saccharomyces cerevisiae. *DNA Repair (Amst)* 6:383–395.

Brand, M., Moggs, J.G., Oulad-Abdelghani, M., *et al.* (2001). UV-damaged DNA-binding protein in the TFTC complex links DNA damage recognition to nucleosome acetylation. *EMBO J.* 20:3187–3196.

Bucceri, A., Kapitza, K., Thoma, F. (2006). Rapid accessibility of nucleosomal DNA in yeast on a second time scale. *EMBO J.* 25:3123–3132.

Bugreev, D.V, Mazina, O.M, Mazin, A.V., *et al.* (2006). Rad54 protein promotes branch migration of Holliday junctions. *Nature* 442:590–593.

Burma, S., and Chen, D.J. (2004). Role of DNA-PK in the cellular response to DNA double-strand breaks. *DNA Repair (Amst)* 3:909–918.

Burma, S., Chen, B.P., Murphy, M., *et al.* (2001). ATM phosphorylates histone H2AX in response to DNA double-strand breaks. *J. Biol. Chem.* 276:42462–42467.

Cannavo, E., Gerrits, B., Marra, G., *et al.* (2007). Characterization of the interactome of the human MutL homologues MLH1, PMS1, and PMS2. *J. Biol. Chem.* 282:2976–2986.

Celeste, A., Fernandez-Capetillo, O., Kruhlak, M.J., *et al.* (2003). Histone H2AX phosphorylation is dispensable for the initial recognition of DNA breaks. *Nat. Cell Biol.* 5:675–679.

Chai, B., Huang, J., Cairns, B.R., *et al.* (2005). Distinct roles for the RSC and Swi/Snf ATP-dependent chromatin remodelers in DNA double-strand break repair. *Genes. Dev.* 19:1656–1661.

Cheung, K. J. Jr., Mitchell, D., Lin, P., *et al.* (2001). The tumor suppressor candidate p33(ING1) mediates repair of UV-damaged DNA. *Cancer Res.* 61:4974–4977.

Cheung, W.L, Turner, F.B., Krishnamoorthy T., *et al.* (2005). Phosphorylation of histone H4 serine 1 during DNA damage requires casein kinase II in S. cerevisiae. *Curr. Biol.* 15:656–660.

Chowdhury, D., Keogh, M.C., Ishii, H., *et al.* (2005). Gamma-H2AX dephosphorylation by protein phosphatase 2A facilitates DNA double-strand break repair. *Mol. Cell* 20:801–809.

Cortez, D., Guntuku, S., Qin, J., *et al.* (2001). ATR and ATRIP: partners in checkpoint signaling. *Science* 294:1713–1716.

Datta, A., Bagchi, S., Nag, A., *et al.* (2001). The p48 subunit of the damaged-DNA binding protein DDB associates with the CBP/p300 family of histone acetyltransferase. *Mutat. Res.* 486:89–97.

Downs. J.A., Lowndes, N.F., Jackson, S.P., *et al.* (2000). A role for Saccharomyces cerevisiae histone H2A in DNA repair. *Nature* 408:1001–1004.

Downs, J. A., Allard, S., Jobin-Robitaille, O., *et al.* (2004). Binding of chromatin-modifying activities to phosphorylated histone H2A at DNA damage sites. *Mol. Cell* 16:979–990.

Doyon. Y., Cayrou, C., Ullah, M., *et al.* (2006). ING tumor suppressor proteins are critical regulators of chromatin acetylation required for genome expression and perpetuation. *Mol. Cell* 21:51–64.

Escargueil, A.E., Soares, D.G, Salvador, M., *et al.* (2008). What histone code for DNA repair? *Mutat. Res.* 658:259–270.

Evans, D.H., and Linn, S. (1984). Excision repair of pyrimidine dimers from simian virus 40 minichromosomes in vitro. *J. Biol. Chem.* 259:10252–10259.

Evans, M.L., Bostelman, L. J., Albrecht, A.M., *et al.* (2008). UV sensitive mutations in histone H3 in Saccharomyces cerevisiae that alter specific K79 methylation states genetically act through distinct DNA repair pathways. *Curr. Genet.* 53:259–274.

Falck, J., Coates, J., Jackson, S.P. (2005). Conserved modes of recruitment of ATM, ATR and DNA-PKcs to sites of DNA damage. *Nature* 434:605–611.

Falk, M., Lukasova, E., Gabrielova, B., *et al.* (2007). Chromatin dynamics during DSB repair. *Biochim. Biophys. Acta* 1773:1534–1545.

Fernandez-Capetillo, O., Allis, C.D., Nussenzweig, A. (2004). Phosphorylation of histone H2B at DNA double-strand breaks. *J. Exp. Med.* 199:1671–1677.

Flores-Rozas, H., Clark, D., Kolodner, R.D. (2000). Proliferating cell nuclear antigen and Msh2p-Msh6p interact to form an active mispair recognition complex. *Nat. Genet.* 26:375–378.

Fortini, P., and Dogliotti, E. (2007). Base damage and single-strand break repair: mechanisms and functional significance of short- and long-patch repair subpathways. *DNA Repair (Amst)* 6:398–409.

Fritsch, O., Benvenuto, G. Bowler, C., *et al.* (2004). The INO80 protein controls homologous recombination in Arabidopsis thaliana. *Mol. Cell* 16:479–485.

Goodarzi, A.A., Noon, A.T., Deckbar, D., *et al.* (2008). ATM signaling facilitates repair of DNA double-strand breaks associated with heterochromatin. *Mol. Cell* 31:167–177.

Gospodinov, A., Tsaneva, I., Anachkova, B. (2009). RAD51 foci formation in response to DNA damage is modulated by TIP49. *Int. J. Biochem. Cell. Biol.* 41:925–933.

Gravel, S., Chapman, J.R. Magill, C., *et al.* (2008). DNA helicases Sgs1 and BLM promote DNA double-strand break resection. *Genes Dev.* 22:2767–2772.

Groth, A., Rocha, W., Verreault, A., *et al.* (2007). Chromatin challenges during DNA replication and repair. *Cell* 128:721–733.

Hanawalt, P. C. (2002). Subpathways of nucleotide excision repair and their regulation. *Oncogene* 21:8949–8956.

Hara, R. and Sancar, A. (2002). The SWI/SNF chromatin-remodeling factor stimulates repair by human excision nuclease in the mononucleosome core particle. *Mol. Cell. Biol.* 22:6779–6787.

Hara, R., and Sancar, A. (2003). Effect of damage type on stimulation of human excision nuclease by SWI/SNF chromatin remodeling factor. *Mol. Cell. Biol.* 23:4121–4125.

Hasan, S., Hassa, P.O., Imhof, R., *et al.* (2001). Transcription coactivator p300 binds PCNA and may have a role in DNA repair synthesis. *Nature* 410:387–391.

Helleday, T., Petermann, E., Lundin, C., *et al.* (2008). DNA repair pathways as targets for cancer therapy. *Nat. Rev. Cancer* 8:193–204.

Herceg, Z. (2007). Epigenetics and cancer: towards an evaluation of the impact of environmental and dietary factors. *Mutagenesis* 22:91–103.

Huang, J., Liang, B., Qiu, J., *et al.* (2005). ATP-dependent chromatin-remodeling complexes in DNA double-strand break repair: remodeling, pairing and (re)pairing. *Cell Cycle* 4:1713–1715.

Huggins, C.F., Chafin, D.R., Aoyagi, S., *et al.* (2002). Flap endonuclease 1 efficiently cleaves base excision repair and DNA replication intermediates assembled into nucleosomes. *Mol. Cell* 10:1201–1211.

Huyen, Y., Zgheib, O., DiTullio, R.A., *et al.* (2004). Methylated lysine 79 of histone H3 targets 53BP1 to DNA double-strand breaks. *Nature* 432:406–411.

Ikura, T., Ogryzko, V.V., Grigoriev, M., *et al.* (2000). Involvement of the TIP60 histone acetylase complex in DNA repair and apoptosis. *Cell* 102:463–473.

Ikura, T., Tashiro, S., Kakino, A., *et al.* (2007). DNA damage-dependent acetylation and ubiquitination of H2AX enhances chromatin dynamics. *Mol. Cell. Biol.* 27:7028–7040.

Ip, S.C., Rass, U., Blanco, M.G., *et al.* (2008). Identification of Holliday junction resolvases from humans and yeast. *Nature* 456:357–361.

Ira, G., Pellicioli, A., Balijja, A., *et al.* (2004). DNA end resection, homologous recombination and DNA damage checkpoint activation require CDK1. Nature 431:1011–1017.

Jha, S., Shibata, E., Dutta, A., *et al.* (2008). Human Rvb1/Tip49 is required for the histone acetyltransferase activity of Tip60/NuA4 and for the downregulation of phosphorylation on H2AX after DNA damage. *Mol. Cell. Biol.* 28:2690–2700.

Keogh, M.C., Kim, J.A., Downey, M., *et al.* (2006). A phosphatase complex that dephosphorylates gammaH2AX regulates DNA damage checkpoint recovery. *Nature* 439:497–501.

Kim, Y.C., Gerlitz, G., Furusawa, T., *et al.* (2009). Activation of ATM depends on chromatin interactions occurring before induction of DNA damage. *Nat. Cell Biol.* 11:92–96.

Kinner, A., Wu, W., Iliakis, G. (2008). Gamma-H2AX in recognition and signaling of DNA double-strand breaks in the context of chromatin. *Nucleic Acids Res.* 36:5678–5694.

Kleczkowska, H.E., Marra, G., Lettieri, T., *et al.* (2001). hMSH3 and hMSH6 interact with PCNA and colocalize with it to replication foci. *Genes Dev.* 15:724–736.

Klochendler-Yeivin, A., Picarsky, E., Yaniv, M. (2006). Increased DNA damage sensitivity and apoptosis in cells lacking the Snf5/Ini1 subunit of the SWI/SNF chromatin remodeling complex. *Mol. Cell. Biol.* 26:2661–2674.

Kobayashi, J., Tauchi, H., Sakamoto, S., *et al.* (2002). NBS1 localizes to gamma-H2AX foci through interaction with the FHA/BRCT domain. *Curr. Biol.* 12:1846–1851.

Kobor, M.S., Venkatasubrahmanyam, S., Moneghini, M.D., *et al.* (2004). A protein complex containing the conserved Swi2/Snf2-related ATPase Swr1p deposits histone variant H2A.Z into euchromatin. *PLoS Biol.* 2:E131.

Krogan, N.J., Keogh, M.C., Datta, N., *et al.* (2003). A Snf2 family ATPase complex required for recruitment of the histone H2A variant Htz1. *Mol. Cell* 12:1565–1576.

Kruhlak, M.J., Celeste, A., Dellaire, G., *et al.* (2006). Changes in chromatin structure and mobility in living cells at sites of DNA double-strand breaks. *J. Cell Biol.* 172:823–834.

Kuo, W.H., Wang, Y., Wong, R.P., *et al.* (2007). The ING1b tumor suppressor facilitates nucleotide excision repair by promoting chromatin accessibility to XPA. *Exp. Cell Res.* 313:1628–1638.

Kusch, T., Florens, L., Macdonald, W.H., *et al.* (2004). Acetylation by Tip60 is required for selective histone variant exchange at DNA lesions. *Science* 306:2084–2087.

Kwon, J., Morshead, K.B., Guyon, J.R., *et al.* (2000). Histone acetylation and hSWI/SNF remodeling act in concert to stimulate V(D)J cleavage of nucleosomal DNA. *Mol. Cell* 6:1037–1048.

Lan, S.Y., and Smerdon, M.J. (1985). A nonuniform distribution of excision repair synthesis in nucleosome core DNA. *Biochemistry* 24:7771–7783.

Lazzaro, F., Sapountzi, V., Granata, M., *et al.* (2008). Histone methyltransferase Dot1 and Rad9 inhibit single-stranded DNA accumulation at DSBs and uncapped telomeres. *EMBO J.* 27:1502–1512.

Lee, J.H., and Paull, T.T. (2005). ATM activation by DNA double-strand breaks through the Mre11-Rad50-Nbs1 complex. *Science* 308:551–554.

Lees-Miller, S.P., and Meek, K. (2003). Repair of DNA double strand breaks by non-homologous end joining. *Biochimie* 85:1161–1173.

Li, B., Pattenden, S.G., Lee, D., *et al.* (2005). Preferential occupancy of histone variant H2AZ at inactive promoters influences local histone modifications and chromatin remodeling. *Proc. Natl. Acad. Sci. USA.* 102:18385–18390.

Liang, B., Qiu, J., Ratnakumar, K., *et al.* (2007). RSC functions as an early double-strand-break sensor in the cell's response to DNA damage. *Curr. Biol.* 17:1432–1437.

Limbo, O., Chahwan, C., Yamada, Y., *et al.* (2007). Ctp1 is a cell-cycle-regulated protein that functions with Mre11 complex to control double-strand break repair by homologous recombination. *Mol. Cell* 28:134–146.

Ljungman, M., and Lane, D.P. (2004). Transcription - guarding the genome by sensing DNA damage. *Nat. Rev. Cancer* 4:727–737.

Lukas, C., Melander, F., Stucki, M., *et al.* (2004). Mdc1 couples DNA double-strand break recognition by Nbs1 with its H2AX-dependent chromatin retention. *EMBO J.* 23:2674–2683.

Martinez, E., Palhan, V.B., Tjernberg, A., *et al.* (2001). Human STAGA complex is a chromatin-acetylating transcription coactivator that interacts with pre-mRNA splicing and DNA damage-binding factors in vivo. *Mol. Cell. Biol.* 21:6782–6795.

Mellon, I., Spivak, G., Hanawalt, P.C. (1987). Selective removal of transcription-blocking DNA damage from the transcribed strand of the mammalian DHFR gene. *Cell* 51:241–249.

Meneghini, M.D, Wu, M., Medhani, H.D. (2003). Conserved histone variant H2A.Z protects euchromatin from the ectopic spread of silent heterochromatin. *Cell* 112:725–736.

Menoni, H., Gasparutto, D., Hamiche, A., *et al.* (2007). ATP-dependent chromatin remodeling is required for base excision repair in conventional but not in variant H2A.Bbd nucleosomes. *Mol. Cell. Biol.* 27:5949–5956.

Mimitou, E.P., and Symington, L.S. (2008). Sae2, Exo1 and Sgs1 collaborate in DNA double-strand break processing. *Nature* 455:770–774.

Mizuguchi, G., Shen, X., Landry, J., *et al.* (2004). ATP-driven exchange of histone H2AZ variant catalyzed by SWR1 chromatin remodeling complex. *Science* 303:343–348.

Modesti, M., Junop, M.S., Ghirlando, R., *et al.* (2003). Tetramerization and DNA ligase IV interaction of the DNA double-strand break repair protein XRCC4 are mutually exclusive. *J. Mol. Biol.* 334:215–228.

Morrison, A.J., Highland, J., Krogan, N.J., *et al.* (2004). INO80 and gamma-H2AX interaction links ATP-dependent chromatin remodeling to DNA damage repair. *Cell* 119:767–775.

Morshead, K.B., Ciccone, D.N., Taverna, S.D., *et al.* (2003). Antigen receptor loci poised for V(D)J rearrangement are broadly associated with BRG1 and flanked by peaks of histone H3 dimethylated at lysine 4. *Proc. Natl. Acad. Sci. USA.* 100:11577–11582.

Moynahan, M.E., Chiu, J.W., Koller, B.H., *et al.* (1999). Brca1 controls homology-directed DNA repair. *Mol. Cell* 4:511–518.

Murr, R., Loizou, J.I., Yang, Y.G., *et al.* (2006). Histone acetylation by Trrap-Tip60 modulates loading of repair proteins and repair of DNA double-strand breaks. *Nat. Cell. Biol.* 8:91–99.

Nakada, D., Hirano, Y., Sugimoto, K., *et al.* (2004). Requirement of the Mre11 complex and exonuclease 1 for activation of the Mec1 signaling pathway. *Mol. Cell. Biol.* 24:10016–10025.

New, J.H., Sugiyama, T., Zaitseva, E., *et al.* (1998). Rad52 protein stimulates DNA strand exchange by Rad51 and replication protein A. *Nature* 391:407–410.

Papamichos-Chronakis, M., Krebs, J.E., Peterson, C.L., *et al.* (2006). Interplay between Ino80 and Swr1 chromatin remodeling enzymes regulates cell cycle checkpoint adaptation in response to DNA damage. *Genes Dev.* 20:2437–2449.

Paques, F., and Haber, J.E. (1999). Multiple pathways of recombination induced by double-strand breaks in Saccharomyces cerevisiae. *Microbiol. Mol. Biol. Rev.* 63:349–404.

Patenge, N., Elkin, S.K., Oettinger, M.A. (2004). ATP-dependent remodeling by SWI/SNF and ISWI proteins stimulates V(D)J cleavage of 5 S arrays. *J. Biol. Chem.* 279:35360–35367.

Paull T.T., Rogakou, E.P, Yamazaki, V., *et al.* (2000). A critical role for histone H2AX in recruitment of repair factors to nuclear foci after DNA damage. *Curr. Biol.* 10:886–895.

Petukhova, G., Stratton, S., Sung, P. (1998). Catalysis of homologous DNA pairing by yeast Rad51 and Rad54 proteins. *Nature* 393:91–94.

Ramanathan, B., and Smerdon, M.J. (1986). Changes in nuclear protein acetylation in u.v.-damaged human cells. *Carcinogenesis* 7:1087–1094.

Ramanathan, B. and Smerdon, M.J. (1989). Enhanced DNA repair synthesis in hyperacetylated nucleosomes. *J. Biol. Chem.* 264:11026–11034.

Rappold, I., Iwabuchi. K., Date, T., *et al.* (2001). Tumor suppressor p53 binding protein 1 (53BP1) is involved in DNA damage-signaling pathways. *J. Cell Biol.* 153:613–620.

Resnick. M.A., and Martin, P. (1976). The repair of double-strand breaks in the nuclear DNA of Saccharomyces cerevisiae and its genetic control. *Mol. Gen. Genet.* 143:119–129.

Richardson, C., Moynahan, M.E., Jasin, M. (1998). Double-strand break repair by interchromosomal recombination: suppression of chromosomal translocations. *Genes Dev.* 12:3831–3842.

Robert, F., Hardy, S., Nagy, Z., *et al.* (2006). The transcriptional histone acetyltransferase cofactor TRRAP associates with the MRN repair complex and plays a role in DNA double-strand break repair. *Mol. Cell. Biol.* 26:402–412.

Rogakou, E.P., Boon, C., Redon, C., *et al.* (1999). Megabase chromatin domains involved in DNA double-strand breaks in vivo. *J. Cell Biol.* 146:905–916.
Rubbi, C.P., and Milner, J. (2003). p53 is a chromatin accessibility factor for nucleotide excision repair of DNA damage. *EMBO J.* 22:975–986.
Sancar, A., Lindsey-Boltz, L.A., Unsal-Kaçmaz, K. *et al.* (2004). Molecular mechanisms of mammalian DNA repair and the DNA damage checkpoints. *Annu. Rev. Biochem.* 73:39–85.
Sanders, S.L., Portoso, M., Mata, J., *et al.* (2004). Methylation of histone H4 lysine 20 controls recruitment of Crb2 to sites of DNA damage. *Cell* 119:603–614.
Schotta, G., Sengupta, R., Kubicek, S., *et al.* (2008). A chromatin-wide transition to H4K20 monomethylation impairs genome integrity and programmed DNA rearrangements in the mouse. *Genes Dev.* 22:2048–2061.
Shen, Z., Cloud, K.G, Chen, D.J., *et al.* (1996). Specific interactions between the human RAD51 and RAD52 proteins. *J. Biol. Chem.* 271:148–152.
Shen, X., Mizuguchi, G., Hamiche, A., *et al.* (2000). A chromatin remodelling complex involved in transcription and DNA processing. *Nature* 406:541–544.
Shen, X., Ranallo, R., Choi, E., *et al.* (2003). Involvement of actin-related proteins in ATP-dependent chromatin remodeling. *Mol. Cell* 12:147–155.
Shiloh, Y. (2003). ATM and related protein kinases: safeguarding genome integrity. *Nat. Rev. Cancer* 3:155–168.
Shim, E.Y., Ma, J.L., Oum J.H., *et al.* (2005). The yeast chromatin remodeler RSC complex facilitates end joining repair of DNA double-strand breaks. *Mol. Cell. Biol.* 25:3934–3944.
Shim, E.Y., Hong, S.J., Oum J.H., *et al.* (2007). RSC mobilizes nucleosomes to improve accessibility of repair machinery to the damaged chromatin. *Mol. Cell. Biol.* 27:1602–1613.
Shinohara, A., Shinohara, M., Ohta, T., *et al.* (1998). Rad52 forms ring structures and co-operates with RPA in single-strand DNA annealing. *Genes Cells* 3:145–156.
Shroff, R., Arbel-Eden, A., Sattler, U., *et al.* (2004). Distribution and dynamics of chromatin modification induced by a defined DNA double-strand break. *Curr. Biol.* 14:1703–1711.
Sigurdsson, S., Van Komen, S., Petukhova, G., *et al.* (2002). Homologous DNA pairing by human recombination factors Rad51 and Rad54. *J. Biol. Chem.* 277:42790–42794.
Smerdon, M.J., and Lieberman, M.W. (1978). Nucleosome rearrangement in human chromatin during UV-induced DNA- reapir synthesis. *Proc. Natl. Acad. Sci. USA.* 75:4238–4241.
Soutoglou, E., and Misteli, T. (2008a). Activation of the cellular DNA damage response in the absence of DNA lesions. *Science* 320:1507–1510.
Soutoglou, E., and Misteli, T. (2008b). On the contribution of spatial genome organization to cancerous chromosome translocations. *J. Natl. Cancer Inst. Monogr.* (39):16–19.
Soutoglou, E., Dorn, J.F., Sengupta, K., *et al.* (2007). Positional stability of single double-strand breaks in mammalian cells. *Nat. Cell Biol.* 9:675–682.
Stewart, G.S., Wang, B., Bignell, C.R., *et al.* (2003). MDC1 is a mediator of the mammalian DNA damage checkpoint. *Nature* 421:961–966.
Stiff, T., O'Driscoll, M., Rief, N., *et al.* (2004). ATM and DNA-PK function redundantly to phosphorylate H2AX after exposure to ionizing radiation. *Cancer Res.* 64:2390–2396.
Stucki, M., and Jackson, S.P. (2006). gammaH2AX and MDC1: anchoring the DNA-damage-response machinery to broken chromosomes. *DNA Repair (Amst)* 5:534–543.
Stucki, M., Clapperton, J.A., Mohammad, D., *et al.* (2005). MDC1 directly binds phosphorylated histone H2AX to regulate cellular responses to DNA double-strand breaks. *Cell* 123: 1213–1226.
Svejstrup, J.Q. (2002). Mechanisms of transcription-coupled DNA repair. *Nat. Rev. Mol. Cell. Biol.* 3:21–29.
Tamburini, B.A., and Tyler, J.K. (2005). Localized histone acetylation and deacetylation triggered by the homologous recombination pathway of double-strand DNA repair. *Mol. Cell. Biol.* 25:4903–4913.
Tardat, M., Murr, R., Herceq, Z., *et al.* (2007). PR-Set7-dependent lysine methylation ensures genome replication and stability through S phase. *J. Cell Biol.* 179:1413–1426.

Teng, Y., Yu, Y., Waters, R., *et al.* (2002). The Saccharomyces cerevisiae histone acetyltransferase Gcn5 has a role in the photoreactivation and nucleotide excision repair of UV-induced cyclobutane pyrimidine dimers in the MFA2 gene. J. Mol. Biol. 316:489–499.

Thompson, L.H., and Schild, D. (2001). Homologous recombinational repair of DNA ensures mammalian chromosome stability. *Mutat. Res.* 477:131–153.

Tini, M., Benecke, A., Um, S.J., *et al.* (2002). Association of CBP/p300 acetylase and thymine DNA glycosylase links DNA repair and transcription. *Mol. Cell* 9:265–277.

Toh, G.W., O'Shaughnessy, A.M., Jimeno, S.J., *et al.* (2006). Histone H2A phosphorylation and H3 methylation are required for a novel Rad9 DSB repair function following checkpoint activation. *DNA Repair (Amst)* 5:693–703.

Tsukuda, T., Fleming, A.B., Nickoloff, J.A., *et al.* (2005). Chromatin remodelling at a DNA double-strand break site in Saccharomyces cerevisiae. *Nature* 438:379–383.

Ura, K., Araki, M., Saeki, H., *et al.* (2001). ATP-dependent chromatin remodeling facilitates nucleotide excision repair of UV-induced DNA lesions in synthetic dinucleosomes. *EMBO J.* 20:2004–2014.

Utley, R.T., Lacoste, N., Jobin-Robitaille, O., *et al.* (2005). Regulation of NuA4 histone acetyltransferase activity in transcription and DNA repair by phosphorylation of histone H4. *Mol. Cell. Biol.* 25:8179–8190.

van Attikum, H., Fritsch, O., Hohn, B., *et al.* (2004). Recruitment of the INO80 complex by H2A phosphorylation links ATP-dependent chromatin remodeling with DNA double-strand break repair. *Cell* 119:777–788.

van Attikum, H., Fritsch, O., Gasser, S.M. (2007). Distinct roles for SWR1 and INO80 chromatin remodeling complexes at chromosomal double-strand breaks. *EMBO J.* 26:4113–41125.

Volker, M., Mone, M.J., Karmakar, P., *et al.* (2001). Sequential assembly of the nucleotide excision repair factors in vivo. *Mol. Cell* 8: 213–224.

Wakasugi, M., Kawashima, A., Morioka, H., *et al.* (2002). DDB accumulates at DNA damage sites immediately after UV irradiation and directly stimulates nucleotide excision repair. *J. Biol. Chem.* 277:1637–1640.

Wang, J., Chin, M.Y., Li, G. (2006a). The novel tumor suppressor p33ING2 enhances nucleotide excision repair via inducement of histone H4 acetylation and chromatin relaxation. *Cancer Res.* 66:1906–1911.

Wang, Y., Wang, J., Li, G. (2006b). Leucine zipper-like domain is required for tumor suppressor ING2-mediated nucleotide excision repair and apoptosis. *FEBS Lett.* 580: 3787–3793.

Ward, I.M., and Chen, J. (2001). Histone H2AX is phosphorylated in an ATR-dependent manner in response to replicational stress. *J. Biol. Chem.* 276: 47759–47762.

West, S.C. (2003). Molecular views of recombination proteins and their control. *Nat. Rev. Mol. Cell. Biol.* 4:435–445.

Wilkins, R.J. and Hart, R.W. (1974). Preferential DNA repair in human cells. *Nature* 247: 35–36.

Wong, L.Y, Recht, J., Laurent, B.C. (2006). Chromatin remodeling and repair of DNA double-strand breaks. *J. Mol. Histol.* 37:261–269.

Wood, R.D. (1999). DNA damage recognition during nucleotide excision repair in mammalian cells. *Biochimie* 81:39–44.

Wu, S., Shi, Y., Mulligan, P., *et al.* (2007). A YY1-INO80 complex regulates genomic stability through homologous recombination-based repair. *Nat. Struct. Mol. Biol.* 14:1165–1172.

Wyatt, H.R., Liaw, H., Green, G.R., *et al.* (2003). Multiple roles for Saccharomyces cerevisiae histone H2A in telomere position effect, Spt phenotypes and double-strand-break repair. *Genetics* 164:47–64.

Wysocki, R., Javaheri, A., Allard, S., *et al.* (2005). Role of Dot1-dependent histone H3 methylation in G1 and S phase DNA damage checkpoint functions of Rad9. *Mol. Cell. Biol.* 25:8430–8443.

Yang, X.J. (2004). The diverse superfamily of lysine acetyltransferases and their roles in leukemia and other diseases. *Nucleic Acids Res.* 32:959–976.

You, J.S., Wang, M., Lee, S.H.. (2003). Biochemical analysis of the damage recognition process in nucleotide excision repair. *J. Biol. Chem.* 278:7476–7485.

You, Z., Chahwan, C., Bailis, J., *et al.* (2005). ATM activation and its recruitment to damaged DNA require binding to the C terminus of Nbs1. *Mol. Cell. Biol.* 25:5363–5379.

Yu,Y., Teng, Y., Liu, H., *et al.* (2005). UV irradiation stimulates histone acetylation and chromatin remodeling at a repressed yeast locus. *Proc. Natl. Acad. Sci. USA.* 102:8650–8655.

Ziv, Y., Bielopolski, D., Galanty, Y., *et al.* (2006). Chromatin relaxation in response to DNA double-strand breaks is modulated by a novel ATM- and KAP-1 dependent pathway. *Nat. Cell Biol.* 8:870–876.

Zou, L., and Elledge, S.J. (2003). Sensing DNA damage through ATRIP recognition of RPA-ssDNA complexes. *Science* 300:1542–1548.

Chapter 17
Poly(ADP-Ribosyl)ation of Chromosomal Proteins, Epigenetic Regulation and Human Genomic Integrity in Health and Disease

Rafael Alvarez-Gonzalez

17.1 Introduction

The accurate hereditary transmission of genetic information, or "blue print", which makes up a human being, from generation to generation occurs with a high degree of fidelity. Molecular accuracy occurs not only due to the wealth of information stored in the long deoxynucleotide sequence of DNA of 6×10^9 billion units per diploid genome of somatic cells (primary structure). Instead, as it has become clear recently, the intrinsic ability of the DNA double helix (secondary structure) to interact with a multitude of proteins at the molecular level (epigenetic information) is pivotal for high binding specificity. According to the rapidly emerging field of epigenetics, the selectivity of DNA-protein interactions is occasionally dictated by electronegativity of DNA and intrinsic structural properties of the properly folded polypeptides, which may operate as single monomeric molecules, as in the case of histone H1 (Suganuma and Workman 2008; Izzo et al. 2008) or DNA metabolizing enzymes such as DNA polymerase β (Beard and Wilson 2006). More frequently however, DNA-protein complexes develop as highly organized oligomeric macromolecular scaffolds, including the nucleosomal histone octamers or "cores" (Tremethik 2007), which comprise 50% of the total nuclear protein, or as DNA replication (Tye and Swayer 2000) and transcription multi-subunit molecular machines (Malecka et al. 2009). More importantly in the context of this volume, these DNA-bound oligomeric protein structures are also found amongst post-translational modification enzymes, e.g. PARP-1 itself (Kawaichi et al. 1981; Mendoza-Alvarez and Alvarez-Gonzalez 1993). The organization of these mega-unit protein-nucleic acid complexes adopts very sophisticated, elaborate and highly regulated "biochemical engines" which orchestrate the individual molecular events that modulate chromosomal dynamics (Tremethik

R. Alvarez-Gonzalez (✉)
Graduate School of Biomedical Sciences, Institute for Cancer Research, University of North Texas Health Science Center at Fort Worth, 3500 Camp Bowie Blvd., Fort Worth, TX 76107, USA
e-mail: Rafael.Alvarez-Gonzalez@unthsc.edu

C.J. Vidal (ed.), *Post-Translational Modifications in Health and Disease*, Protein Reviews 13, DOI 10.1007/978-1-4419-6382-6_17,

2007; Suganuma and Workman 2008). Indeed, the temporal and topographic distribution of these biochemical engines, ultimately determines the right patterns of gene expression. Not surprisingly, when epigenetic mechanisms are out of synchrony, cells and tissues may loose growth control, as it occurs in carcinogenesis (*vide infra*) or over time, it may also lead to altered genomic homeostatic systems that may result in the development of chronic disease.

17.2 Protein-Poly(ADP-Ribosyl)ation and Human Genomic Integrity

The proper maintenance of genomic integrity in individual human cells, and by inference, the proper balance between health and disease, requires flawless molecular integration between nucleosomal histone proteins (Althaus 1992) and complex oligomeric DNA metabolizing enzymes, e.g., PARP-1 (Mendoza-Alvarez and Alvarez-Gonzalez 1993), which epigenetically coordinate poly-functional proteins, such as the tumor suppressor protein known as p53 (for a recent review see Alvarez-Gonzalez 2007), a key transcription factor with DNA nick sensing ability. The dynamics of this elaborated genetic expression system seems to be orchestrated by these two "guardian angels" (Chatteerjee et al. 1999; Lane 1992; Tong et al. 2001) of genomic integrity which quickly detect, protect and facilitate the repair of nicks and double strand breaks on DNA (Althaus 1992) Interestingly, the molecular mechanism appears to involve the homodimerization of PARP-1 on a DNA nick (Mendoza-Alvarez and Alvarez-Gonzalez 1993), coupled with either the homodimerization and/or the tetramerization of sequence dependent-DNA binding of p53. Not surprisingly, the disruption of accurate protein-protein and protein-DNA interactions of the p53/PARP-1 macromolecular complex seems to either turn on (Wieler et al. 2003) or turn off (Conde et al. 2001) pivotal control genetic switches that dramatically increase tumor latency in murine experimental models (*vide infra*). In other words, PARP-1 and p53 molecular interplay on DNA appear to control the toggle switch between cell survival, following repair of genotoxic damage, or programmed cell death (apoptosis) after a threshold of un-repairable DNA-damage level has been reached (Kumari et al. 1998).

Nonetheless, not all molecular details of the scenario described above have completely been elucidated. For example, even though both protein guardians of genomic integrity become covalently poly(ADP-ribosyl)ated in the process (Kumari et al. 1998; Mendoza-Alvarez and Alvarez-Gonzalez 2001; Simbulan-Rosenthal et al. 2001), we do not know whether PARP-1/p53 multimeric complexes involve either heterodimer formation or alternatively, a heterodimer of homodimers as a plausible protein-poly(ADP-ribosyl)ation enzymatic intermediate. Regardless of the complexity of p53/PARP-1 protein oligomers on DNA, one conclusion is undisputable: PARP-1 and p53 work in tandem, as DNA sensors and cell cycle checkpoints, respectively, and together facilitate either the repair of DNA-damage to promote cell survival (Bouchard et al. 2003), or launch the apoptotic program

(Scovassi and Poirier 1999; Alvarez-Gonzalez et al. 1999) that eliminates highly mutated cells, thus preventing the development of tissue-specific malignancies (Conde et al. 2001), or alternatively, the unstoppable process of chronic disease (*vide infra*).

17.3 Protein-Poly(ADP-Ribosyl)ation and Epigenetics

Of the epigenetic pathways listed above, reversible protein-poly(ADP-ribosyl)ation is known to be a transient enzymatic regulatory cycle initiated by a family of enzymes known as ADP-ribose polymerases, of which the best understood, and most abundant example, is poly(ADP-ribose) polymerase-1 or "PARP-1" for short (Alvarez-Gonzalez 2007). It has been reported that up to 2% of the total nuclear proteins of interphase chromatin may correspond to PARP-1 (Alvarez-Gonzalez 2007), a constitutive enzyme that is present throughout the cell cycle, in most somatic and germinal cells (Atorino et al. 2000), except mature erythrocytes. This fact alone implies that, leaving the histone proteins out, the next protein in line which structurally and functionally maintains the integrity of the human genome may very well be PARP-1 itself.

This chapter will focus on discussing how this ubiquitous eukaryotic enzyme and its partner catabolic protein, responsible for the hydrolytic degradation of protein-bound ADP-ribose polymers, namely poly(ADP-ribose) glycohydrolase (PARG) (Fig. 17.1), may directly (or indirectly) activate (or inactivate) the epigenetic balance between health and disease. Two of the main hallmarks of an epigenetic pathway are: (i) the information is not directly related to the primary structure of DNA, e.g., the overall genomic sequence; and (ii) the information is properly transmitted from generation to generation with a high degree of fidelity, just like the DNA replication pathway. Indeed, the covalent ADP-ribose polymerization of chromatin proteins (*vide supra*) fulfills both criteria and can thus be classified as an epigenetic pathway.

Another major characteristic of epigenetic mechanisms is the reversibility of the system. This implies the participation of two enzymatically catalyzed steps, one to transfer a given chemical moiety from a donor substrate onto a protein acceptor, and its counterpart, a different enzyme responsible for the removal of the protein-bound modifier as depicted on Fig. 17.1 for protein-poly(ADP-ribosyl)ation (*vide infra*). Thus, a classic biochemical cycle of post-translational protein functional regulation is typically characterized for having both an activating and a de-activating arm. Therefore, when this process occurs in the cell nucleus, and the protein subjected to specific chemical modification reactions is a DNA-binding protein, the biochemical cycle may also be classified as an epigenetic pathway.

As it might be expected, any malfunctions on either side of the epigenetic pathway may lead to loss of genomic integrity and cancer, or alterations of cell physiology via over- or under-activation of specific tissue- and cell-specific gene expression patterns, and therefore different kinds of genetic diseases and/or chronic diseases, including

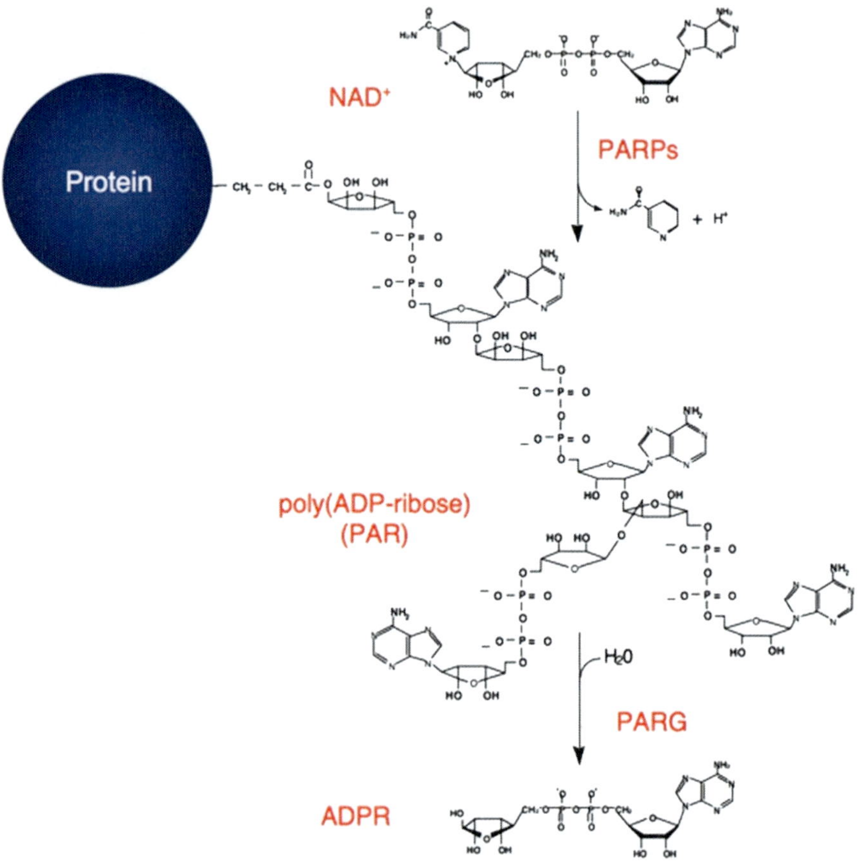

Fig. 17.1 Enzymatic cycle of protein-poly(ADP-ribosyl)ation. A family of post-translational modification enzymes known as ADP-ribose polymerases, e.g., PARP-1 and PARP-2 (PARPs) utilize the oxidized form of β-nicotinamide adenine dinucleotide (NAD^+) as an ADP-ribose donor (Loetscher et al. 1987) to assemble covalently bound polymers of ADP-ribose (PAR) of more than 240 ADP-ribose units onto DNA-binding proteins, including PARP-1 and PARP-2 themselves. PARPs efficiently catalyze four kinetically distinct chemical reactions: (i) ADP-ribose chain initiation (Mendoza-Alvarez and Alvarez-Gonzalez 1999); (ii) ADP-ribose chain elongation (Alvarez-Gonzalez 1988); (iii) ADP-ribose polymer branching (Alvarez-Gonzalez and Jacobson 1987); and (iv) chain termination. The enzyme responsible for the degradation of these chromatin-bound polymers is poly(ADP-ribose) glycohydrolase(s) (PARG) which generates large amounts of free ADP-ribose or "ADPR" (Pacheco-Rodriguez and Alvarez-Gonzalez 1999)

aging. This is the main reason why, there has been a recent expansion of research efforts to better understand epigenetic pathways, with the resulting explosion of literature reports in the field of epigenetics in the last few years, including three chapters in this volume, one on histone changes and gene expression (Bernstein and co-workers, Chap. 15), another one on histone modifications and DNA repair (Gospodinov and Herceg, Chap. 16), and this one on protein-poly(ADP-ribosyl)ation.

Obviously, due to the multi-factorial origin of many human pathological conditions, a common molecular origin for the substantially different patterns of gene expression observed when comparing normal cells and tissues with those typical of chronic disease, an epigenetic abnormality may provide a target for effective pharmacological intervention. Therefore, a thorough understanding of the protein molecules, their chemical modifications and the enzymes involved should provide with a more transparent window of observation into human health and disease. However, connecting the dots between the development of chronic diseases over time and key molecular events that actually tilt the homeostatic balance between health and disease seems to be rather elusive at best.

In the following segments of this review, a small but significant number of human ailments that have been reported to be related to protein-poly(ADP-ribosyl) ation abnormalities will be briefly reviewed. It is also important to say that in an attempt to keep them all somewhat related, the topics selected for discussion have a common point of relationship, and that is that they all involve a step of specific tissue or organ inflammation, early in the process of pathological development.

17.4 Protein-Poly(ADP-Ribosyl)ation Abnormalities and Chronic Disease

All diseases listed below will connect the epigenetic pathway of protein-poly(ADP-ribosyl)ation with: (i) acute phase responses and inflammation, at the molecular level; and (ii) the ultimate outcome of chronic disease as an imbalance between cell survival and cell death (apoptosis and/or necrosis).

During the last few years, a dizzying wave of literature reports connecting the field of protein-poly(ADP-ribosyl)ation to a large number of chronic human diseases others than cancer, has steadily been accumulating in the literature (for a recent example see Pacher and Szabo 2008). However, due to the plethora of molecular pathways that appear to be regulated by this reversible protein modification cycle, and the fact that most reports have just touched upon the surface of significance PAR metabolism, and a cohesive and detailed metabolic integration of the depth and breath of ailments connected to this unique and transient biochemical pathway is quickly becoming a web that is quite complex and thus very difficult to untangle.

As a result, in this review, an attempt is made to strictly focus on the significance of the molecular and epigenetic properties of PARP-1 and its protein partners in a group of selected chronic diseases, including arthritis, cancer, chronic heart failure and myocardial infarction, diabetes, and stroke, from the protein chemistry point of view.

17.5 Protein-Poly(ADP-Ribosyl)ation and Aging

The time-dependent accumulation of genotoxic damage, and of the molecular wear and tear of particularly proteins and enzymes in most human tissues, typically manifests as a combination of chronic diseases. In a given individual, these clinical ailments together with a decreased efficiency of specific biochemical pathways resulting from genetic mutations or not, are generally associated with the process of aging. While it may sound too pretentious, to even relate any specific biochemical pathway to aging, one must ideally focus on a few potentially important molecular events as key contributors to this rather complex phenomenon.

As it might also be expected, the observation of unusual levels of protein-poly(ADP-ribosyl)ation activity, as a function of aging, in peripheral blood samples from animal species with a characteristic lifespan (Grube and Burkle 1992) has been reported. Interestingly, in these studies, it was clear that the steady state levels of PARP activity correlated with the life span of each species tested with rodents at the lower end of the spectrum and humans at the opposite end (Grube and Burkle 1992). In fact, Homo sapiens displayed a fivefold higher level of PARP activity when compared to rats. Even though several explanations have been proposed to explain this difference (Beneke and Burkle 2007), a full mechanistic explanation for this dramatic difference remains to be established.

While it is recognized that the process of aging is obviously very complex, a discussion to understand the physiology and environmental factors that contribute to the unavoidable time-dependent reduction of physiological efficiency in mammals is well beyond the scope of this chapter. Nonetheless, it must be noted that it has recently been reported that tissue inflammation and endothelial dysfunction may also significantly contribute to the acceleration of aging manifestations (Csiszar et al. 2008). Thus, identifying the molecular connection of these interesting observations to protein-poly(ADP-ribose) metabolism and genetic integrity (El-Domyati et al. 2008) is very attractive. Furthermore, unraveling how PARP-activity and its role in the maintenance of genomic integrity (Alvarez-Gonzalez 2007) contributes to the genetic decision making process between cell survival (Perkin and Gilmore 2006) and p53-dependent apoptosis (Pietsch et al. 2008) may shed light on the molecular steps of the aging process.

17.6 Protein-Poly(ADP-Ribosyl)ation and Arthritis

Amongst the most common examples of chronic diseases, that may be associated with protein-poly(ADP-ribosyl)ation, enzymatic anomalies, and tissue inflammation, we may identify several autoimmune diseases (Negri et al. 1990), including systemic lupus erythematosus (SLE) (Jeoung et al. 2004), glomerulonephritis (Messmer et al. 2000) and Sjogren's syndrome (Rosen and Casciola-Rosen 2004). These arthritic or rheumatoid diseases are characterized for the accumulation of autoantigens, autoantibodies, or a high concentration of antigen-antibody immunocomplexes, within the tissue involved. Indeed, one of the most frequent antigenic culprits observed in autoimmune diseases is either PARP-1 or PAR [polymers of (ADP-ribose)] (Fig. 17.1), or sometimes, both.

It must be noted however, that these do not represent the only protein complexes detected, and thus, may or may not represent the main immune elicitors. Interestingly, little has been done to properly distinguish between two distinct (ADP-ribosyl)ation antigenic scenarios with regards to the presence of a high antibody titer to PAR in these patients. It is not clear whether these antigens represent protein-free PAR molecules or protein-bound ADP-ribose chains.

Clearly, since protein-free PAR's have not been detected in healthy cells and tissues, one may argue that ADP-ribose chains must be exclusively present as protein-bound polymers, e.g., poly(ADP-ribosyl)ated-chromatin adducts. However, given the wide spectrum of proteins that can be subjected to this post-translational covalent modification reaction, we still do not know the oligomeric structure of most chromatin protein targets, and more importantly, how chromatin protein adducts play an immunogenic role in tissue inflammation at the molecular level. A very likely possibility is that abundant polypeptides produced by immune cell rupture and release of nucleoplasmic components during the inflammatory response results in the release of chromatin components, especially the most abundant proteins in chromatin, the histone proteins (Kouzarides 2007; Tremethik 2007), which become effective immunogens. This becomes even more relevant as lymphocytes and macrophages recruited to swollen tissues via diapedesis, rupture as a result of triggering the apoptotic program (Kumari et al. 1998; Alvarez-Gonzalez et al. 1999; Messmer et al. 2000).

Once immune cells have fulfilled their physiological function as acute phase response recruiters and the immunological defense system of the injured individual mounts an attack on micro-environmental microbial opportunists, an autoimmune reaction against self nucleoplasmic components, e.g., chromatin molecular components, including PARP-1 and protein-bound PAR's, also concomitantly ensues, leading to arthritic pain and associated symptoms.

17.7 Protein-Poly(ADP-Ribosyl)ation and Cancer

It has become evident that the complex process of carcinogenesis and the protein-poly(ADP-ribosyl)ation cycles were definitely related at the molecular level (for a recent review on this topic see Alvarez-Gonzalez 2007). Ever since it was reported in the early 1980s that PARP is activated by strand breaks on DNA (Benjamin and Gill 1980), which when occurring in intact cells and tissues, as a result of carcinogen exposure, causes a dramatic increase in the levels of chromatin-bound PAR's (Juarez-Salinas et al. 1979).

Over the last 30 years, literally thousands of reports have scrutinized the relationship between DNA-damage, DNA repair and poly(ADP-ribose) metabolism from most biomedical angles, from the molecular and genetic level to organ physiology, cancer biology and clinical chemotherapy regimens. Since 1980, protein-poly(ADP-ribosyl)ation has undoubtedly been a very active area of research as far as cancer is concerned. The efforts in this area have been quite substantial, and literature reports are too numerous to list here. For readers wishing to obtain further information on this topic, a reference is listed below (Alvarez-Gonzalez 1999) to a book published just before the end of the twentieth century. Now, in the dawn of the twenty-first century, as a result of the tireless efforts of dozens of protein-poly(ADP-ribosyl)ation experts, whom have published world-class experimental work performed during the last 45 years, in leading Universities and medical research-intensive institutions across the globe, from Canada and the USA to Japan and also to France, Germany, Italy, Spain, Switzerland, UK, etc., we have apparently reached a climax of unparalleled success with the recent application of highly specific competitive inhibitors of PARP-1 and PARP-2 to enhance the efficacy of standard cancer chemotherapy regimens in patients with different kinds of tumors (Donawho et al. 2007; Plummer et al. 2008). Nonetheless, in spite of the promise and excitement in this arena, this is still the early phase of the currently ongoing clinical trials with these cancer chemotherapeutic agents, which rely mostly on their potency as PARP inhibitors in vitro, and it is still possible that they may affect other enzymes with similar biochemical activity, e.g., using NAD^+ as a substrate, and thus, may affect other physiological phenomena with serious side effects.

Obviously, we must continue our experimental efforts with diligence until we identify all proteins, genetic markers and all molecular details of epigenetic control, to fully elucidate the various poly(ADP-ribosyl)ation epigenetic cycles and completely unravel how a given protein-poly(ADP-ribosyl)ation cycle participates in the different phases of cancer initiation, promotion, progression, en even metastasis. One outcome is almost certain, we should witness the unveiling of key poly(ADP-ribosyl)ated proteins as important players in the maintenance of genomic integrity and cancer, such as p53 (Kumari et al. 1998; Mendoza-Alvarez and Alvarez-Gonzalez 2001; Alvarez-Gonzalez 2007).

Again, in order to keep this chapter focused as it pertains to the role of protein-bound PAR metabolism in human health and disease, the brief discussion in this section will center on the relationship of this post-translational modification reaction with cancer via the epigenetic regulation of transcription (p53) and tissue inflammation via NF-kappa B regulation of gene expression (see below).

17.8 Protein-Poly(ADP-Ribosyl)ation in Heart Failure and Myocardial Infarction

The primordial role of cell death and apoptosis in heart failure has also gained a lot of attention in the last 12 years (Bromme and Holtz 1996). The hypoxic or ischemic shock caused by a sudden decrease in blood flow, and thus oxygen homeostasis, through the heart, may trigger localized vascular inflammation to avoid significant damage. Alternatively, massive hypoxic conditions may elicit the endothelial and myocyte cell death programs (de Boer et al. 2000). Thus, the epigenetic role of protein-poly(ADP-ribosyl)ation in modulating the various phases of the apoptotic response (Alvarez-Gonzalez et al. 1999) makes PARP-1 a potential target for cardio pharmacological modulators that may help avoid the dramatic consequences of tissue damage during chronic, or even acute, for that matter, myocardial infarction.

In the year 2000, a solid attempt was made to find the connection between the biochemical activation of PARP-1 and the epigenetic inactivation of this pivotal pathway in homeostatic heart physiology during chronic heart failure (de Boer et al. 2000). In that report, it was clear that there was a molecular mechanistic connection between protein-poly(ADP-ribosyl)ation and myocardial infarction, although the link was more towards the end of the pathway leading to massive heart damage (Pillai et al. 2005).

More recently, most industrialized nations have reported cardiovascular and heart disease as the number one cause of unexpected deaths. Because of this, there is currently a lot of interest worldwide to develop preventive and therapeutic measures to turn the tide in favor of human health. As a result, many investigators studying various genetic and epigenetic factors that contribute to heart disease are actively pursuing all kinds of molecular leads to ameliorate the population morbidity and mortality due to this particular problem. Obviously, protein-poly(ADP-ribosyl)ation is not the exception to this rule (Molnar et al. 2006).

A sudden burst in the investigation of PARP inhibitors as potential effective drugs in treating heart disease has been sparked (Booz 2007; Bartha et al. 2008) in the last 2 years. Nonetheless, in spite of the undeniable potential of PARP inhibitors as efficient therapeutic agents to reduce the consequences of heart failure, one should keep in mind that not all details of the exact molecular mechanism have been elucidated. Thus, it is anticipated that a significant body of experimental evidence describing the exact genetic and epigenetic role of protein-poly(ADP-ribosyl)ation pathway in heart disease will become available within the next 10 years.

17.9 Protein-Poly(ADP-Ribosyl)ation and Diabetes

An indirect connection between the covalent poly(ADP-ribosyl)ation or chromatin proteins and the pathogenesis of diabetes has been considered since the 1960s (Lindall and Lazarow, 1964) via NAD metabolism and function, as a prosthetic or co-enzyme factor, that facilitates the balance between oxidative catabolism (NAD/NADP) and reductive anabolism (NADH/NADPH).

Based on this information, as well as key data obtained with the application of powerful affinity chromatography purification techniques (Alvarez-Gonzalez et al. 1983) to study NAD^+ metabolism and protein-poly(ADP-ribosyl)ation together, it was concluded that there is actually a molecular signaling pathway that takes information from bio-energetic pathways in the NAD/NADH-dependent mitochondrial synthesis of ATP in the mitochondrion and the epigenetic poly(ADP-ribosyl)ation of chromatin proteins in the nucleoplasm (Loetscher et al. 1987). Interestingly, this link does not only involve the epigenetic regulation of key proteins involved in endocrine transcription (Valdor et al. 2008), but also the PARP-1 gene itself (Masutani et al. 1999).

To further illustrate the complexity of the direct and indirect metabolic links between protein-poly(ADP-ribosyl)ation, pyridine nucleotide metabolism, and diabetes in pancreatic tissue, the Japanese group of Okamoto and collaborators reported in 1999 that the CD38-cyclic ADP-ribose signaling system plays a pivotal role in insulin secretion (Okamoto 1999) and thus channels overall NAD consumption to the cytoplasm, away from the nucleoplasm. Therefore, even though this pathway does not represent post-translational modification of proteins directly, it undermines the efficiency of protein-poly(ADP-ribosyl)ation in the nucleoplasm by depleting the intracellular NAD+pool as well.

Interestingly, in spite of the obvious link between protein-poly(ADP-ribosyl) ation and the development of diabetes as one of the culprits (the other one is heart disease, *vide supra*) of what is now called metabolic syndrome, we still lack fundamental information about the molecular role that the biochemical pathway of chromatin poly(ADP-ribosyl)ation plays in pancreatic physiology, especially as a "stepping stone" into the genetic and epigenetic development of this increasingly more prevalent chronic disease.

17.10 Protein-Poly(ADP-Ribosyl)ation and Stroke

The localized and many times irreversible brain cell damage that occurs as a result of ischemia and hypoxic shock within minutes of blood vessel clogging is a well recognized clinical manifestation of a stroke episode. Also, as it was mentioned above, there seems to be a direct connection between constitutive PARP enzymatic activity and structural integrity of this protein to insure cell survival (Alvarez-Gonzalez 2001; Bouchard et al. 2003), when possible. However, massive ischemia may rapidly induce the cell death program or apoptosis (Scovassi and Poirier 1999) which results in the neuronal proteolytic degradation of PARP-1 (Joashi et al. 1999). Not surprisingly, about a decade ago, a search for a direct connection between the protein structure and function of PARP-1 and stroke was launched (Choi 1997; Endres et al. 1997).

After 12 years of research into stroke, hypotoxic neuronal damage and apoptosis, over 100 literature reports that dig deeper into our understanding of how exactly PARP may be used as a chemotherapeutic target to reduce the irreversible damage of brain function after a stroke, at least with experimental animal models (Haddad et al. 2008), indicates that prevention of PARP-1 enzymatic activation with competitive inhibitors, and its apparently consequential susceptibility to proteolytic degradation by caspases, probably represents a very interesting therapeutic possibility. Even though, we still lack a lot of information regarding the biochemical and epigenetic role of PARP-1 in neuronal homeostasis and health, the potential benefits of PARP-1 specific inhibition were elegantly summarized (Moroni and Chiarugi 2009) as this chapter was being written.

17.11 Concluding Remarks

Undoubtedly, the notion that protein-poly(ADP-ribosyl)ation cycles play a pivotal role in the developing stages of various chronic diseases of drastic pathological, physiological and anatomical differences is a very difficult biomedical concept to accept. However, it is likely that all mechanisms of disease that operate at the genetic, biochemical and molecular levels coalesce at a central pivotal point, such as the concept of epigenetic control of tissue-specific gene expression, both temporally and topographically, within specific chromosomal domains at different stages of cell homeostasis and differentiation. Such a point may be at the level of the decision making process between cell survival and cell death (Alvarez-Gonzalez 2001; Bouchard et al. 2003). Alternatively, the important bridge in this cause effect relationship may be at the level of either maintenance of structural chromosomal and genomic integrity (Althaus 1992; Chatteerjee et al. 1999; Conde et al. 2001; Tong et al. 2001; Alvarez-Gonzalez 2007) or the ubiquitously characteristic inflammatory responses of tissues to all kinds of environmental insults, including microbial infections, via NF-kappa B dependent regulation of gene expression (Oliver et al. 1999; Chang and Alvarez-Gonzalez 2001; Hassa et al. 2003) or both.

Finally the undisputable link between chromatin protein-poly(ADP-ribosyl) ation with either the genetics of tissue inflammation via NF-kappa B-dependent gene expression and the four different stages of the apoptotic (Alvarez-Gonzalez et al. 1999) cell-death program (condemnation, commitment, execution and demolition) strongly suggests that other chronic ailments such as pulmonary and neurodegenerative diseases may also represent important pathological conditions arising over time as a result of abnormal ADP-ribose polymer metabolism. Further research into the molecular and epigenetic mechanisms of these diseases will either confirm or refute such an interesting possibility.

In closing, it can be concluded that, after almost 50 years since the discovery of chromatin-bound ADP-ribose polymers (Chambon et al. 1963), our scientific journey into either the biochemical, genetic, molecular, protein chemistry or epigenetic realms of chromosomal-poly(ADP-ribosyl)ation cycles, should provide us with thorough mechanistic details about the physiological function(s) of PARP enzymes, and their catalytic products, in balancing human health and disease. More importantly, the knowledge generated in this highly sophisticated, but very interesting field of protein structure and function, should helps us develop more effective, and disease-specific, therapeutic approaches to alleviate most chronic ailments in the near future.

References

Althaus, F.R. (1992). Poly ADP-ribosylation: a histone shuttle mechanism in DNA excision repair. *J. Cell Sci.* 102:663–670.

Alvarez-Gonzalez, R., Juarez-Salinas, H., Jacobson, E.L., et al. (1983). Evaluation of immobilized boronates for studies of adenine and pyridine nucleotide metabolism. *Anal. Biochem.* 135:69–77.

Alvarez-Gonzalez, R., and Jacobson, M.K. (1987). Characterization of polymers of adenosine diphosphate ribose generated *in Vitro* and *in Vivo*. *Biochemistry* 26:3218–3224.

Alvarez-Gonzalez, R. (1988). 3′-deoxyNAD as a substrate for poly(ADP-ribose) polymerase and the reaction mechanism of poly(ADP-ribose) elongation. *J. Biol. Chem.* 263: 17690–17696.

Alvarez-Gonzalez, R. (ed) (1999). ADP-ribosylation reactions: from bacterial pathogenesis to cancer. Kluwer, Dordrecht/Boston/London.

Alvarez-Gonzalez, R., Spring, H., Muller, M., et al. (1999). Selective loss of poly(ADP-ribose) and the 85 kDa fragment of poly(ADP-ribose) polymerase in nucleoli during alkylation-induced apoptosis in HeLa cells. *J. Biol. Chem.* 274:32122–32126.

Alvarez-Gonzalez, R. (2001). PARP-regulation of eukaryotic gene expression. Survival or Death? *Trends Genet.* 17:607–608.

Alvarez-Gonzalez, R. (2007). Genomic maintenance: the p53 poly(ADP-ribosyl)ation connection. *Sci. STKE* (415):pe68.

Atorino, L., Alvarez-Gonzalez, R., Cardone, A., et al. (2000). Metabolic changes in the poly(ADP-ribosyl)ation pathway of differentiating rat germinal cells. *Arch. Biochem. Biophys.* 381:111–118.

Bartha, E., Kiss, G.N., Kalman, E., et al. (2008). Effect of L-2286, a poly(ADP-ribose) polymerase inhibitor and enalapril on myocardial remodeling and heart failure. *J. Cardiovasc. Pharmacol.* 52:253–261.

Beneke, S., and Burkle, A. (2007). Poly(ADP-ribosyl)ation in mammalian aging. *Nucleic Acids Res.* 35:7456–7465.

Benjamin, R.C., and Gill, D.M. (1980). ADP-ribosylation in mammalian cell ghosts. Dependence of poly(ADP-ribose) synthesis on strand breakage on DNA. *J. Biol. Chem.* 255: 10493–10501.

Beard, W.A., and Wilson, S.H. (2006). Structure and mechanism of DNA polymerase beta. *Chem. Rev.* 106:361–382.

Booz, G.W. (2007). PARP inhibitors and heart failure. Translational medicine caught in the act. *Congestive Heart Fail.* 13:105–112.

Bouchard, V.J., Rouleau, M., Poirier, G.G. (2003). PARP-1, a determinant of cell survival in response to DNA damage. *Exp. Hematol.* 31:446–454.

Bromme, H.J., and Holtz, J. (1996). Apoptosis in the heart: when and why. *Mol. Cell. Biochem.* 163–164:261–275.

Chambon, P., Weill, J.D., Mandel, P. (1963). Nicotinamide mononucleotide activation of a new DNA-dependent polyadenylic acid synthesizing enzyme. *Biochem. Biophys. Res. Commun.* 11:9–43.

Chang, W.J., and Alvarez-Gonzalez, R. (2001). The sequence specific DNA-binding of NF-κB is reversibly regulated by the automodification reaction of poly(ADP-ribose) polymerase-1. *J. Biol. Chem.* 276:47664–47670.

Chatteerjee, S., Berger, S.J., Berger, N.A. (1999). Poly(ADP-ribose) polymerase: a guardian of the genome that facilitates DNA repair by protecting against DNA recombination. *Mol. Cell. Biochem.* 193:23–30.

Choi, D.W. (1997). At the scene of ischemic brain injury. Is PARP a perp. *Nature (Med)* 3:1073–1074.

Conde, C., Mark, M., Oliver, F.J., Huber, A., et al. (2001). Loss of poly(ADP-ribose) polymerase 1 causes increased tumor latency in p53 deficient mice. *EMBO J.* 20:3535–3543.

Csiszar, A., Wang, M., Lakkata, E.G., et al. (2008). Inflammation and endothelial dysfunction during aging: role of NF-kappa B. *J. Appl. Physiol.* 105:1333–1341.

De Boer, R.A., van Veldhuisen, D.J., van der Wijk, J., et al. (2000). Additional use of immunostaining for active caspase 3 and cleaved actin and PARP fragments to detect apoptosis in patients with chronic heart failure. *J. Card. Fail.* 6:330–337.

Donawho, C.K., Luo, Y., Penning, T.D., et al. (2007). ABT-888, an orally active poly(ADP-ribose) polymerase inhibitor that potentiates DNA-damaging agents in pre-clinical tumor models. *Clin. Cancer Res.* 13:2728–2737.

El-Domyati, M.M., Al-Din, A.B., Barakat, M.T., et al. (2008). Deoxyribonucleic acid repair and apoptosis in testicular germ cells of aging fertile men: the role of the poly(adenosine diphosphate ribosyl)ation pathway. *Fertil. Steril.* 91(Suppl 5):2221–2229

Endres, M., Wang, Z.Q., Namura, S., et al. (1997). Ischemic brain injury is mediated by the activation by poly(ADP-ribose) polymerase. *J. Cereb. Blood Flow Metab.* 17:1143–1151.

Gospodinov, A., and Herceg, Z. (2009). Chromatin. The entry and exit from DNA repair. In: Vidal, C.J. (ed) *Post-translational Modifications of Proteins*, Springer, New York.

Grube, K., and Burkle, A. (1992). Poly(ADP-ribose) polymerase activity in mononuclear leukocytes of 13 mammalian species correlates with species-specific life span. *Proc. Natl. Acad. Sci. U.S.A.* 89:11759–11763.

Haddad, M., Beray-Berthat, V., Coqueran, B., et al. (2008). Reduction of hemorrhagic transformation by PJ34, a poly(ADP-ribose) polymerase inhibitor after permanent focal cerebral ischemia in mice. *Eur. J. Pharmacol.* 588:52–57.

Hassa, P.O., Buerki, C., Lombardi, C., et al. (2003). Transcriptional co-activation of nuclear factor-κB-dependent gene expression by p300 is regulated by poly(ADP-ribose) polymerase-1. *J. Biol. Chem.* 278:45145–45153.

Izzo, A., Kamieniarz, K., Schneider, R. (2008). The histone H1 family: specific members, specific functions? *Biol. Chem.* 389:333–343.

Jeoung, D., Lim, Y., Lee, E.B., et al. (2004). Identification of autoantibody against poly(ADP-ribose) polymerase (PARP) fragment as a serological marker in systemic lupus erythematosus. *J. Autoimmunity* 22: 87–94.

Joashi, U.C., Greenwood, K., Taylor, D.L., et al. (1999). Poly(ADP-ribose) polymerase cleavage precedes neuronal cell death in the hippocampus and cerebellum following injury to the development rat forebrain. *Eur. J. Neurosci.* 11:91–100.

Juarez-Salinas, H., Sims, J.L., Jacobson, M.K. (1979). Poly(ADP-ribose) levels in carcinogen-treated cells. *Nature (London)* 282:740–741.

Kawaichi, K., Ueda, K., Hayaishi, O. (1981). Multiple autopoly(ADP-ribosyl)ation of rat liver poly(ADP-ribose) synthetase: Mode of modification and properties of automodified synthetase. *J. Biol. Chem.* 256:9483–9489.

Kouzarides, T. (2007). Chromatin modifications and their function. *Cell* 128:693–705.

Kumari, S.R., Alvarez-Gonzalez, R., Mendoza-Alvarez, H. (1998). Functional interactions of p53 with poly(ADP-ribose) polymerase (PARP) during apoptosis following DNA-damage: covalent poly(ADP-ribosyl)ation by exogenous PARP and non-covalent binding of p53 to Mr 85,000 proteolytic fragment. *Cancer Res.* 58:5075–5078.

Lane, D.P. (1992). Cancer. p53- Guardian of the genome. *Nature (London)* 358:15–16.

Lindall, A.W., and Lazarow, L. (1964). A critical study of pyridine nucleotide concentrations in normal fed, normal fasted and diabetic rat liver. *Metabolism* 13:259–271.

Loetscher, P., Alvarez-Gonzalez, R., Althaus, F.R. (1987). Poly(ADP-ribose) may signal changing metabolic conditions to the chromatin of mammalian cells. *Proc. Natl. Acad. Sci. U.S.A.* 84: 1286–1289.

Malecka, K.A., Ho, W.C., Marmorstein, R. (2009). Crystal structure of a p53 core tetramer bound to DNA. *Oncogene* 28:325–333.

Masutani, M., Suzuki, H., Kamada, N., et al. (1999). Poly(ADP-ribose) polymerase gene disruption conferred mice resistant to streptosotocyn-induced diabetes. *Proc. Natl. Acad. Sci. U.S.A.* 96:2301–2304.

Mendoza-Alvarez, H., and Alvarez-Gonzalez, R. (1993). Poly(ADP-ribose) polymerase is a catalytic dimer and the automodification reaction is intermolecular. *J. Biol. Chem.* 268: 22575–22580.

Mendoza-Alvarez, H., and Alvarez-Gonzalez, R. (2001). Regulation of p53 sequence-specific binding by covalent poly(ADP-ribosyl)ation. *J. Biol. Chem.* 276:36425–36430.

Mendoza-Alvarez, H., and Alvarez-Gonzalez, R. (1999). Biochemical characterization of mono(ADP-ribosyl)ated poly(ADP-ribose) polymerase. *Biochemistry* 38:3948–3953.

Messmer, U.K., Winkel, G., Briner, G.A., et al. (2000). Suppression of apoptosis by glucocorticoids in glomerular endothelial cells: effects on proapoptotic pathways. *Br. J. Pharmacol.* 29:1673–1683.

Molnar. A., Toth, A., Bagi, Z., et al. (2006). Activation of the poly(ADP-ribose) polymerase pathway in human heart failure. *Mol. Med.* 12:143–152.

Moroni, F., and Chiarugi, A. (2009). Post-ischemic brain damage: targeting PARP-1 within the ischemic neurovascular units as a realistic avenue to stroke treatment. *FEBS J.* 276:36–45.

Negri, C., Scovassi, A.I., Cerino, A., et al. (1990). Autoantibodies to poly(ADP-ribose) polymerase in autoimmune diseases. *Autoimmunity* 6:203–209.

Oliver, F.J., Menissier-de Murcia, J., Nacci, C., et al. (1999). Resistance to endotoxic shock as a consequence to defective NF-κB activation in poly(ADP-ribose) polymerase-1 deficient mice. *EMBO J.* 18:4446–4454.

Okamoto, H. (1999). The CD38-cyclic ADP-ribose signaling system in insulin secretion. *Mol. Cell. Biochem.* 193:115–118.

Pacheco-Rodriguez, G., and Alvarez-Gonzalez, R. (1999). Measurement of poly(ADP-ribose) glycohydrolase activity by high resolution polyacrylamide gel electrophoresis: specific inhibition by histones and nuclear matrix proteins. *Mol. Cell. Biochem.* 193:13–18.

Pacher, P., and Szabo, C. (2008). Role of peroxynitrate-poly(ADP-ribose) polymerase pathway in human disease. *Am. J. Pathol.* 173:1–13.

Perkin, N.D., and Gilmore, T.D. (2006). Good cop, bad cop: the different faces of NF-kappa B. *Cell Death Differ.* 13:759–772.

Pietsch, E.C., Sykes, S.M., MacMahon, S.B., et al. (2008). The p53 family and programmed cell death. *Oncogene* 27:6507–6521.

Pillai, J.B., Isbatan, A., Imai, S.I., et al. (2005). Poly(ADP-ribose) polymerase 1-dependent cardiac myocyte cell death during heart failure is mediated by NAD + cell depletion and reduced Sir2α deacetylase activity. *J. Biol. Chem.* 280:43121–43130.

Plummer, R., Jones, C., Middleton, M., et al. (2008). Phase I study of the poly(ADP-ribose) polymerase inhibitor, AG-14699, in combination with temozolomide in patients with advanced solid tumors. *Clin. Cancer. Res.* 14:7917–7923.

Rosen, A., and Casciola-Rosen, L. (2004). Altered autoantigen structure in Sjogren's syndrome: implications for the pathogenesis of autoimmune tissue damage. *Crit. Rev. Oral Biol. Med.* 15:156–164.

Scovassi, A.I., and Poirier, G.G. (1999). Poly(ADP-ribosyl)ation and apoptosis. *Mol. Cell. Biochem.* 199:125–137.

Simbulan-Rosenthal, C.M., Rosenthal, D.S., Luo, R. B., et al. (2001). Poly(ADP-ribosyl)ation of p53 in vitro and in vivo modulates binding to its DNA consensus sequence. *Neoplasia* 3:179–188.

Suganuma, T., and Workman, J.L. (2008). Crosstalk among histone modifications. *Cell* 135:604–607.

Tong, W.M., Cortes, U., Wang, Z.Q. (2001). Poly(ADP-ribose) polymerase: a guardian angel protecting the genome and suppressing tumorigenesis. *Biochim. Biophys. Acta* 1552:27–37.

Tremethik, D.J. (2007). Higher-ordered structures of chromatin: the elusive 30 nm fiber. *Cell* 128:651–654.

Tye, B.K., and Swayer, S. (2000). The hexameric eukaryotic MCM helicase: building symmetry from non-identical parts. *J. Biol. Chem.* 275:34833–34836.

Valdor, R., Schreiber, V., Saenz, J., et al. (2008). Regulation of NFAT by poly(ADP-ribose) polymerase in T cells. *Mol. Immunol.* 24:1863–1871.

Wieler, S., Cagne, J.P., Vaziri, H., et al. (2003). Poly(ADP-ribose) polymerase-1 is a positive regulator of the p53-mediated G1 cycle arrest response following ionizing radiation. *J. Biol. Chem.* 278:18914–18921.

Chapter 18
Post-translational Proteolytic Processing on Intracellular Proteins by Cathepsins and Cystatins

Nobuhiko Katunuma, Masae Takahashi, and Tadashi Tezuka

18.1 Antigen Processing and Proteolytic Modification of Biologically Active Proteins by Cathepsins in Health and Diseases

18.1.1 The Importance of Sequential Limited Proteolysis in the Post-translational Processing of Proteins

Almost all proteins are synthesized from pre-pro-proteins and then processed to biologically active mature proteins. This processing is performed in specific intracellular locations by specific proteinases and, therefore, specific cathepsins play important roles in this respect. Here we show an interesting abnormal processing of serum albumin, called abnormal proalbuminemia Tokushima (Matsuda 1986). This inborn error of metabolism was found in a specific family in Tokushima. The metabolic disorder patients showed a weak resistance for heavy metal intoxication. This abnormal proalbumina shows Arg·Gly·Val·Phe·*His*·Arg/Asp·Ala – , while normal proalbumin shows Arg·Gly·Val·Phe·*Arg*·Arg/Asp·Ala – as shown in Fig. 18.1. Therefore, the processing protease can not cleave off the proalbumin part and, therefore, this abnormal proalbumin variant is secreted into the serum. Interestingly, the same abnormal proalbuminemia was found in Lille in France, suggesting a genealogical connection.

Also, we found that the intracellular degradation of cathepsins themselves are initiated from the nicking of a special ordered bond in the lysosome, such as the peptide bond after the 47th amino acid in the case of mature cathepsin B, the peptide bond after the 177th amino acid in mature cathepsin H, and the peptide bond after

N. Katunuma (✉)
Institute for Health Sciences, Tokushima Bunri University, 180 Nishihamabouji, Yamashiro-cho, Tokushima 770-8514, Japan
e-mail: katunuma@tokushima.bunri-u.ac.jp

C.J. Vidal (ed.), *Post-Translational Modifications in Health and Disease*, Protein Reviews 13, 425
DOI 10.1007/978-1-4419-6382-6_18, © Springer Science+Business Media, LLC 2011

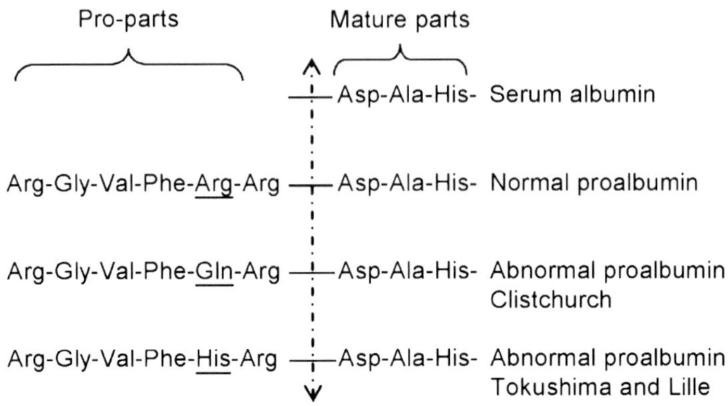

Fig. 18.1 Abnormal amino acid sequence of the pro-parts of albumin in abnormal proalbuminemia

the 178th in mature cathepsin L. Therefore, the intracellular degradation of biologically active proteins and enzymes occurs via ordered limited proteolysis processes in relevant working organelles.

18.1.2 Functional Shear of Individual Cathepsins from Their Morphological Aspects

From the enzyme chemical aspects, each cathepsin has a different cleavage bond to make different product peptides. Furthermore, cathepsins B, H, and L are located in different parts of hepatocytes; these differences are apparent when hepatocytes are stained using antibodies to each cathepsin, as shown in Fig. 18.2. Each cathepsin is located in different lysosome particles, as shown by double immunocytochemical staining with different sizes of gold particles, using an electron microscopic technique (Fig. 18.3). Cathepsin B and cathepsin H in the islet cells of the pancreas are located in separate lysosomes, so a selective incorporation mechanism of substrate proteins is very important. The mechanisms of post-translocational proteolytic processing are not simple and the selective incorporation mechanism of substrate proteins is still unknown.

18.1.3 Antigen Processing by Lysosomal Cathepsins

Allergies and immunity are expressed through the antigen-antibody reaction. Antibody production is initiated by antigen processing by cathepsins in antigen-presenting cells, such as macrophages and dendritic cells. Proteins as an

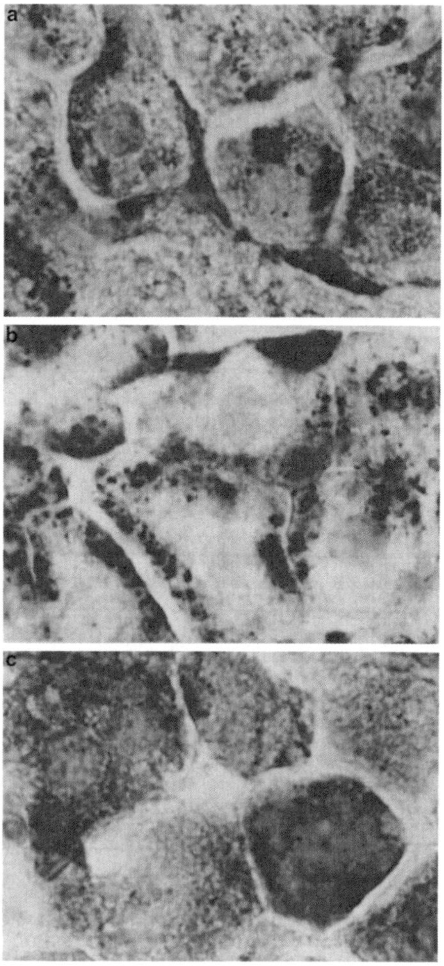

Fig. 18.2 Different intracellular localizations of cathepsin B (*upper*), cathepsin H (*center*), and cathepsin L (*lower*) in hepatocytes, determined with antibodies

antigen are processed to a 15-mer antigenic peptide (epitope) by various cathepsins in the lysosomes and then the processed signal peptides (epitope) are presented to MHC class II in the cells. These signals are further transduced to helper T lymphocytes and then their signals are transduced to B-lymphocytes to produce the antibodies. The classes of antibodies to be produced are determined by the signals of the presented epitopes. Therefore, lysosomal cathepsins play an important role in the class-switching of antibody production. The type of cathepsin determines the kind of antibodies to be formed (Katunuma 1997, 2003).

Cysteine cathepsins are a protease family located in lysosomes; eleven kinds of cysteine cathepsins are presently registered in the human genome database (Turk

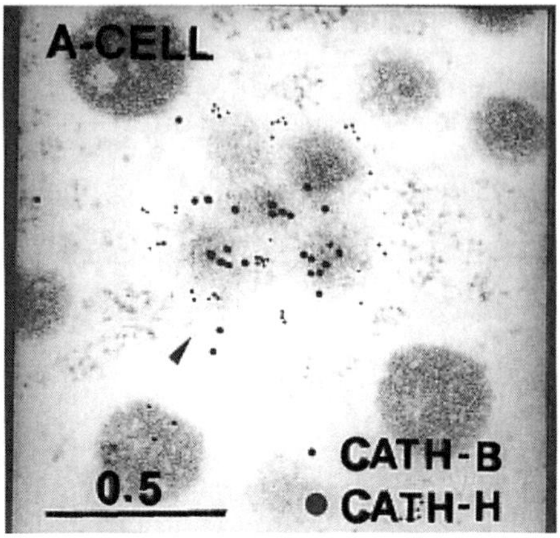

Fig. 18.3 Electron-microscopy showing double immunostaining of cathepsins B and H labeled with gold particles in islet cells of the pancreas

et al. 2001). Lysosomal cathepsins play a role not only in general protein catabolism, but also in the production of bioactive proteins and peptides by their limited proteolysis. One of their most important roles is antigen processing by limited proteolysis.

Each cathepsin has an individual susceptible bond on the substrate proteins; different cathepsins produce different individual biologically active peptides. As a result, the different expressions of cathepsins induce individual physiological or pathological conditions. Therefore, the development of specific inhibitors for individual cathepsins or the development of knockout mice lacking genes for individual cathepsins are essential to analyze the individual roles of cathepsins and to clarify the pathogenesis of the specific diseases mediated by the abnormal expression of cathepsins.

18.1.4 Development of Specific Inhibitors for Individual Cathepsins to Analyze and Regulate the Post-translational Processing

The development of specific inhibitors for the cathepsin family began with the discovery of two kinds of natural cathepsin inhibitors from soil bacteria (Pedolobacteria). As Fig. 18.4 shows, the first group contains amino-acid derivatives of aliphatic aldehydes, such as leupeptin and antipain, that make a thioether bond with the SH-Cys active site of cysteine proteases (cathepsins); those were developed by Umezawa's group (Umezawa 1972). The second group contains amino-acid derivatives of epoxysuccinate, such as E-64, that make a thiohemiacetal bond with the SH-Cys active site of cathepsins; this was discovered by Hanada et al. (1978) and developed as a cathepsin inhibitor by Katunuma's group (Hashida et al. 1980; Hara et al. 1988). These inhibitors were specific for the cysteine protease

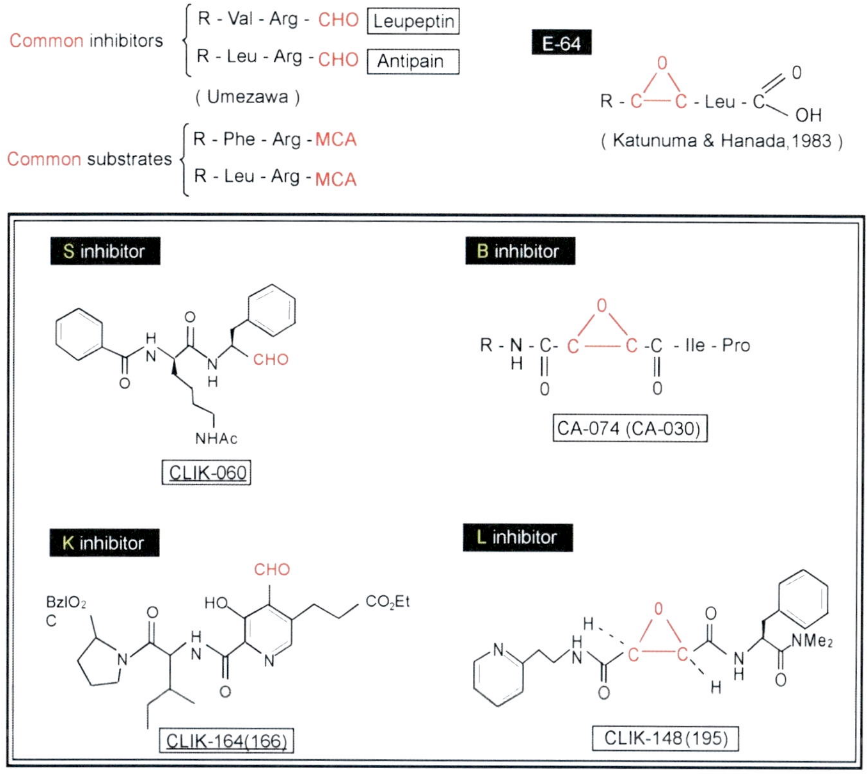

Fig. 18.4 Design of specific inhibitors for individual cysteine cathepsins. Epoxysuccinate or aldehyde serves as the active site to be bound with the active thiol group of cathepsins

group, but unfortunately, not specific for individual cathepsins. Using these two frame compounds as the active site of inhibitors, we designed compounds inhibitors that exhibit specificity for individual cathepsins, based on their different tertiary structures of substrate binding pockets of cathepsins, determined using X-ray crystallographic analysis by Turk et al. (1995). These designed inhibitors were then chemically synthesized and their specificities were tested by our group.

As Fig. 18.4 shows, individual cathepsin-specific inhibitors were developed. The epoxysuccinate derivatives CA-074 and CA-030 were designed as specific inhibitors for cathepsin B (Katunuma 1995) and CLIK-148 and CLIK-195 (CLIK: Cathepsin L Inhibitor Katunuma) were designed as specific inhibitors for cathepsin L (Katunuma 1999). The aldehyde derivative CLIK-60 was designed as a cathepsin S-specific inhibitor. As pyridoxal propionate (aromatic aldehyde) derivatives, CLIK-163 and CLIK-166 have been developed. These compounds were designed based on the structures of their substrate binding pocket (Katunuma 1995, 1999, 2000).

These inhibitors easily pass through the cell membrane and are effective not only in vitro, but also in vivo. These inhibitors showed 100% inhibition at concentrations of 10^{-6}–10^{-7} M, both in vitro and in vivo, as shown in Table 18.1.

Table 18.1 Inhibition specificities for various cathepsins

	10^{-x} M	B	L	S	K
CA-30	−7	100	0	0	0
CA-74	−7	100	0	0	0
CLIK-148	−6	0	100	30	0
	−7	0	63	0	0
CLIK-195	−6	0	100	25	0
	−7	0	85	0	0
CLIK-60	−6	25	30	100	10
	−7	0	0	86	0
CLIK-164	−5	0	20	60	100
	−6	0	0	20	60
CLIK-166	−4	0	0	17	100
	−5	0	0	0	100

The mechanisms to show the cathepsin B specificity by CA·030 were clarified by X-ray crystallography (Musil et al. 1991).

18.1.5 The Role of Special Cathepsins for Antigen Processing and Presentation

Immunological events are started from the phagocytosis of antigens. The antigenic proteins are then processed to antigenic 15-mer epitopes by various cathepsins to present to MHC class II genes. There are many unknowns regarding the selective incorporation of antigen proteins to be incorporated into relevant lysosomes. The resulting antibodies and cytokines are determined by signals of the epitopes that are processed and the signals are transduced to the helper T cells (Th-1 or Th-2). For instant, ovalbumin as an antigen is able to produce not only IgE and INFα, but also IgG1 and IL-4 by the different epitopes produced by distinct cathepsins.

The immune response to rabies vaccine was used as a model for antigen processing by cathepsin B. CA-074 as a cathepsin B-specific inhibitor, the rabies vaccine, and its epitope $ER_{281-299,}$ (EECLDALESTMTTKSVSFR) derived from rabies glycoprotein, were used to challenge 2C5 strain or B6 strain cells, which respond specifically to rabies vaccine. As shown in Fig. 18.5, when rabies vaccine or antigenic epitope $ER_{281-299}$ was incubated with antigen-presenting cells (2C5 strain or B6 strain), $[H^3]$-thymidine incorporation by rabies vaccine was suppressed by the addition of CA-074 in a dose-dependent manner, but the proliferation response by the processed antigenic peptide $ER_{281-299}$ was not inhibited by CA-074.

Importantly, the C-terminus sequence of the antigenic peptide has an FR (phenylalanine-arginine) structure in the C-terminus that is susceptible to cleavage by cathepsin B. This response of the 2C5 clone to rabies vaccine was also suppressed by the F(ab)′ of cathepsin B-specific antibody or by Z-RR-MCA (Z-Arg·Arg-MCA), a specific substrate for cathepsin B, but methylcoumarinamide (MCA) itself showed no effect, as shown in Fig. 18.6. This further confirmed that

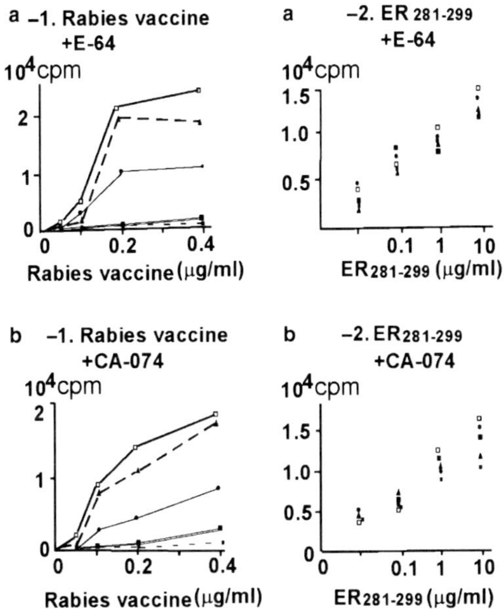

Fig. 18.5 Inhibitory effects of E-64d and CA-074 on the proliferative response of human T cell clones by re-challenge with a rabies vaccine or antigenic peptide. The 3×10^4 cells of 2C5 clone were incubated for 72 h with rabies vaccine [(**a**)-1, (**b**)-1] or with the synthetic peptide ER·281–299 [(**a**)-2, (**b**)-2] in the presence of 1×10^5 cells of Rx-PBMC plus various concentrations of cathepsin B inhibitors E-64d or CA-074. E-64d was used in (**a**)-1 and (**a**)-2 and CA-074 was used in (**b**)-1 and (**b**)-2. (▲···▲) Pre-pulsed Rx-PBMC + 5 μg/ml of inhibitor; (●——●) Rx-PBMC + antigen + 1.25 μg/ml of inhibitor;(■===■) Rx-PBMC + antigen + 2.5 μg/ml of inhibitor;(▪▪) Rx-PBMC + antigen + 5 μg/ml of inhibitor;(□——□) Rx-PBMC + antigen

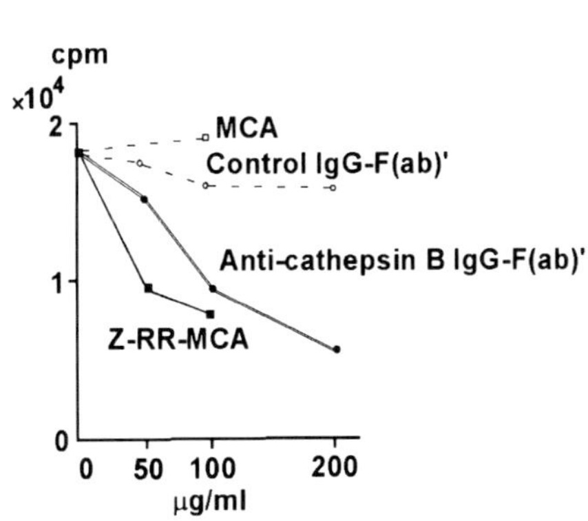

Fig. 18.6 Inhibition of T cell proliferation by the antibody fragment of anti-cathepsin B antibody and the specific substrate of cathepsin B, Z-RR-MCA. Cells (3×10^4) of 2C5 clone were incubated for 72 h with 1×10^5 cells of Rx-PBMC and 0.2 μg/ml of rabies vaccine in the presence of various concentrations of the F(ab)' of anti-cathepsin B antibody IgG. The cells were pulsed with 1 mCi/well of [^{3}H] thymidine for the last 12 h and harvested

			57						63			
MHC-II, β-chain	R	P	V	A	E	S	W	N	S	Q	K	D
Cathepsin B	W	L	V	A	N	S	W	N	T	D	W	G
			217						223			
Cathepsin L	W	L	V	K	N	S	W	G	K	E	W	G
Cathepsin II	W	I	V	K	N	S	W	G	S	N	W	G
Cathepsin S	W	L	V	K	N	S	W	G	L	H	F	G
Cathepsin J	W	I	V	K	N	S	W	G	S	Q	W	G
Papain	I	L	I	K	N	S	W	G	T	G	W	G

Fig. 18.7 One of the active sites of cathepsin B shares homology with a part of the antigenic peptide binding domain (desetope) of MHC class II, but no homology with the other cathepsins

cathepsin B is the determinant protease responsible for the antigen processing of vaccines of rabies to be presented to the MHC class II region. Furthermore, a part of the active site of cathepsin B, V_{217}-N_{223} (VANSWNT), shows high homology with a part of the binding domain (desetope) of the MHC class II β-chain, V_{57}-N_{63} (VAESWNS), because the alanine and the asparagine are the same amino acids on cathepsin B and the MHC class II region, although the alanine and the asparagine in the same active domain of the other cathepsins are commonly substituted by lysine and glycine, respectively, as shown in Fig. 18.7. This is also one important reason why only cathepsin B specifically makes antigenic peptides bind with the MHC class II region. We conclude that the antigenic fragments of hepatitis B surface antigen and rabies vaccines are processed by cathepsin B specifically and show common affinity to bind to the desetope of the MHC class II β-chain (Matsunaga et al. 1993).

18.1.6 Class-Switching of Antibody Formation by Different Cathepsins

Different cathepsins produce different antigenic epitopes (peptides) to present to the MHC class II region, and then the different signals are transduced to the different helper T lymphocytes (Th-1 or Th-2). Because the cathepsins participate in antigen processing, the inhibitors of individual cathepsins regulate the class-switching of antibody formation through the formation of different signal peptides. For instance, cathepsin B participates in antigen processing for the signal transduction to helper T2 type (Th-2) cells; as a result, IgG2a and IFNα formation are stimulated and IgE, IgG1, and IL-4 productions are suppressed. Cathepsin L participates in expression to helper Th1 type (Th-1) signal transductions to produce IgE, IgG1, and IL-4. Therefore, the classes of immune responses were switched between Th-1 and Th-2 types by various cathepsin inhibitors as follows:

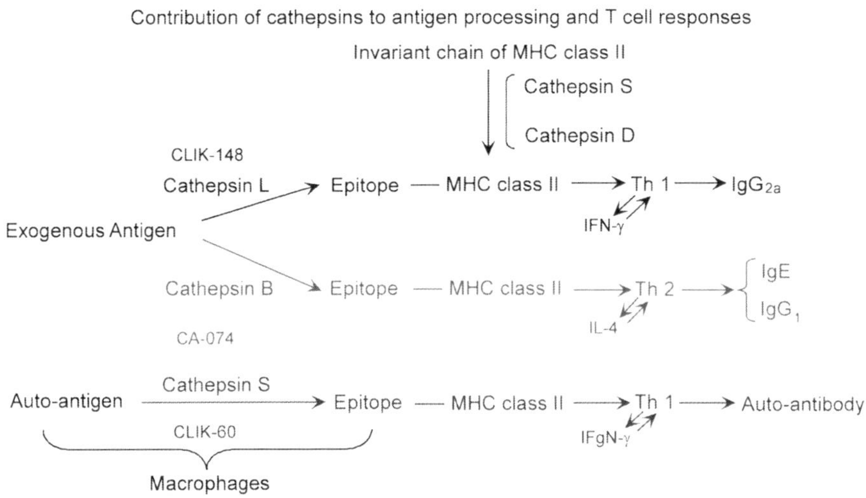

Scheme

Class-switching of antibody formation to ovalbumin was observed by administration of a cathepsin B-specific inhibitor, such as CA-074, in vivo. The Th-2 type of immunoglobulins and cytokines, such as IgG1 and IgE, the passive cutaneous anaphylaxis reaction, and IL-4 levels were decreased and IgG2a and IFN-α levels were elevated in the CA-074-treated ovalbumin-immunized rabbits, as shown in Figs. 18.8 and 18.9. The immune responses were opposite in rabbits treated with CLIK-148, which is a cathepsin L-specific inhibitor (Katunuma et al. 2003).

18.1.7 Auto-Antigen α-Fodrin was Processed by Cathepsin S in Sjögren's Disease

Autoimmune diseases are expressed by specific auto-antigens processed by specific cathepsins. An inhibitor of one specific cathepsin (in this case, thc cathepsin S inhibitor CLIK-60) prevented the expression of Sjögren's syndrome. The auto-antigen α-fodrin of Sjögren's syndrome is processed by cathepsin S to make autoantibody in the mouse model of Sjögren's syndrome, as shown in Fig. 18.10 and Table 18.2.

The acceleration of [H^3] thymidine incorporation into splenocyte T cells of the mouse model of Sjögren's syndrome was assayed. The [H^3] thymidine incorporation was stimulated by α-fodrin, but not by ovalbumin. Thymectomized mice with Sjögren's syndrome were used, as shown in Fig. 18.10. The [H^3] thymidine incorporation was suppressed completely by CLIK-60, which is a cathepsin S-specific inhibitor and the other cathepsin inhibitors had no effect. The pathological signs

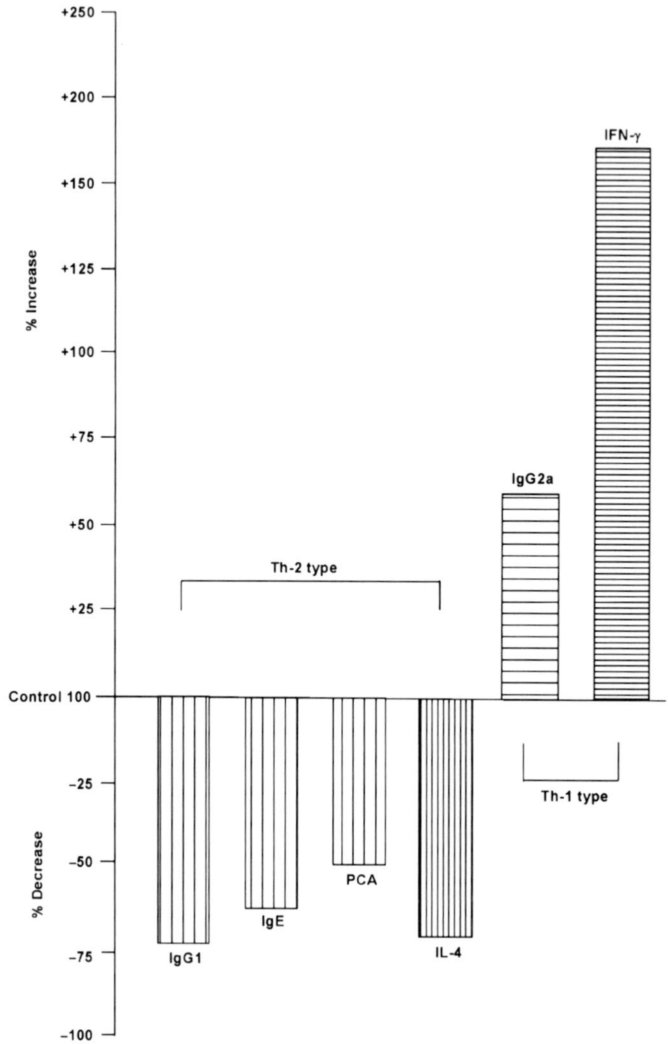

Fig. 18.8 Changes in the ovalbumin-dependent production of cytokines and immunoglobulins by CA-074 administration

in mice with Sjögren's syndrome and their prevention by CLIK-60 are shown in Table 18.2. Pathological lesions and increased secretion volumes of saliva and tears were almost completely prevented by treatment with CLIK-60 (Maekawa et al. 1998; Saegusa et al. 2002; Ishimaru et al. 2004).

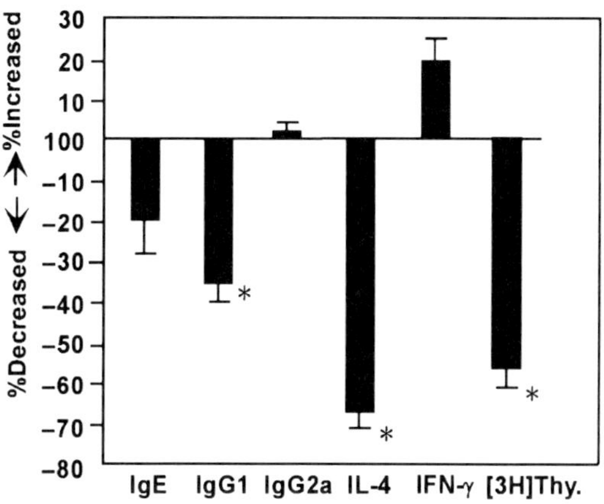

Fig. 18.9 Changes in ovalbumin-dependent immune responses in mice by peroral administration of pyridoxine in a high-protein diet. Percent changes in productions of immunoglobulins and cytokines by 6.0-mg pyrodoxine diet in 70% casein were compared with 0.58-mg pyridoxine diet in 70% casein. Each group consisted of eight mice. A statistically significant difference was found between the standard and excess pyridoxine diets. $^*P<0.05$ (for details, See Katunuma et al. 2000)

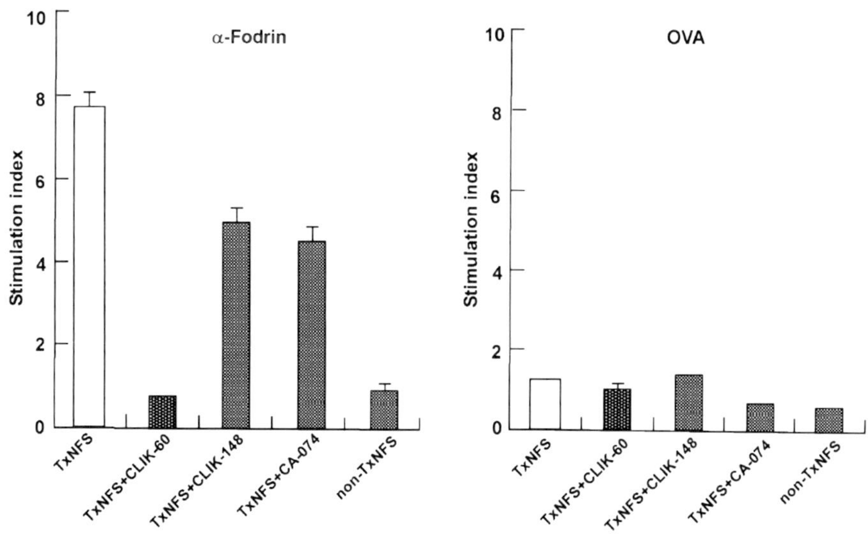

Fig. 18.10 Suppressed expression of Sjögren's syndrome in the mouse model of Sjögren's syndrome by the cathepsin S inhibitor CLIK-060. Response to autoantigen α-fodrin (*left panel*) and ovalbumin (*right panel*). TxNFS: thymectomized mice with Sjögren's syndrome 3 days after birth. Stimulation indices: accelerated [^{3}H] thymidine incorporation in T cells of the spleen from mice with Sjogren's syndrome

Table 18.2 Pathological symptoms of Sjogren's syndrome model mice and their suppression by CLIK-60

	Grade of legions	
Pathological legion grade	Lacrimal G.	Submandibular
Non-Tx SS	0.3	0.3
Tx	3.8	3.0
SS+CA-074	2.8	2.5
SS+CLIK-148	3.5	2.8
SS+CLIK-060	0.9	1.0
Secretion volume of saliva and tears		
	Saliva	Tears
Non-Tx SS	9.0	3.0
Tx	2.0	1.0
SS+CLIK-060	7.0	2.9

Txx-SS: Thymectomized Sjogren's model mice. Secretion volume given in μl/20 min.

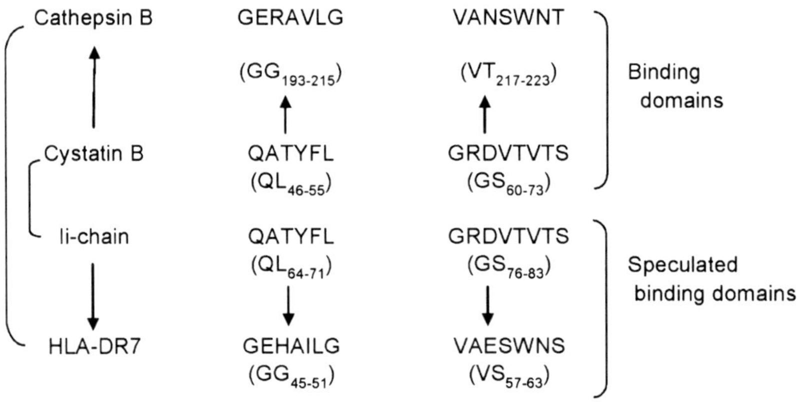

Fig. 18.11 Possible binding domains between Ii-chain and HLA·DR7. The domains $GG_{193-215}$ and $VT_{217-223}$ of cathepsin B that correspond to the second and the third active sites of cathepsin B are exposed on the surface of the substrate binding section, as shown by X-ray crystallography. It is well known that the two domains of cystatin B, QL_{46-55} and GS_{60-73}, which show the highest homology with the corresponding two domains of the Ii-chain, are binding domains with two active sites of cathepsin B. The domain VS_{57-63} of HLA-DR7 is generally considered to be the desetope that binds with the antigenic peptide

18.1.8 *Immunological Significance of the Relationship Between the Invariant Chain of the MHC Class II Region and Cathepsins*

The primary structure of p31 of the invariant chain (Ii-chain) shows about 50% homology with those of the cystatin family, as shown in Fig. 18.11. At a concentration of 3.8×10^{-7} M, the Ii-chain inhibited 75% of cathepsin L and 55% of cathepsin H activities, while cathepsin B activity was not inhibited at all, as shown in Fig. 18.12.

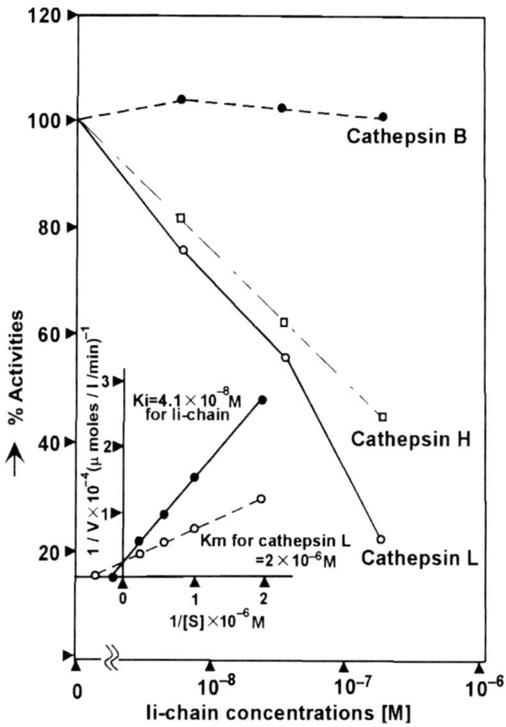

Fig. 18.12 Inhibition profiles of cathepsins L, H, and B by purified human Ii-chain. The reactions were carried out in 50 mM acetate buffer, pH 5.5, with 10^{-3} M of L-cysteine for 30 min and the released methylcoumarinamide was fluorometrically assayed. The molar concentrations of the Ii-chain added were calculated as 60-kDa dimer form. The K_m for cathepsin L was 1.97×10^{-6} and the K_i value for Ii-chain was 4.1×10^{-8} M

The inhibition mode of cathepsin L by the Ii-chain shows typical competitive kinetics and the Ki value of the Ii-chain for cathepsin L was 4.1×10^{-8} M. The antigen processing by cathepsin B is apparently not deterred by the released Ii-chain; in addition, further degradation of processed antigenic peptides by cathepsin L or H is protected by the Ii-chain (Zhang et al. 2000).

18.1.9 Cathepsin L Activity Controls Adipogenesis and Glucose Tolerance

The important role that cathepsin L activity plays in the control of adipogenesis and glucose tolerance has been recently elucidated by Guo-Ping Shi and his group (Yang et al. 2007). They have been demonstrated in vivo a role for cathepsin L in the degradation of essential proteins for adipogenesis and glucose metabolism, such as the matrix protein fibronectin, as well as the insulin receptor (IR) and the insulin-like growth factor 1 receptor (IGF-IR). Cathepsin L inhibition by CLIK-195 leads to the reduction of preadipocyte adipogenesis or lipid accumulation, protection of fibronectin from degradation, accumulation of IR and IGF-IR-β subunits, and an increase in glucose uptake, as shown in Figs. 18.13–18.16. Cathepsin

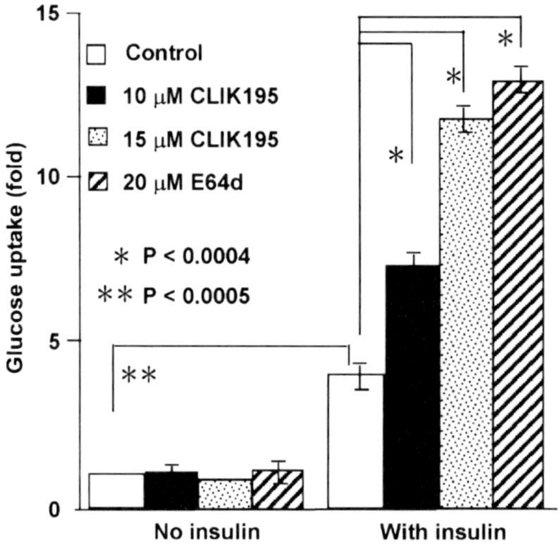

Fig. 18.13 Cathepsin L inhibition affects 3T3-L1 cell adipogenesis and insulin receptor proteolysis. Cathepsin L inhibition with CLIK-195- or E-64d enhanced insulin-induced glucose uptake in 3T3-L1 adipocytes. Data are presented as the increase in glucose uptake (cpm) relative to the counts per minute from untreated cells (mean ± s.e.m. of quadruplicate experiments)

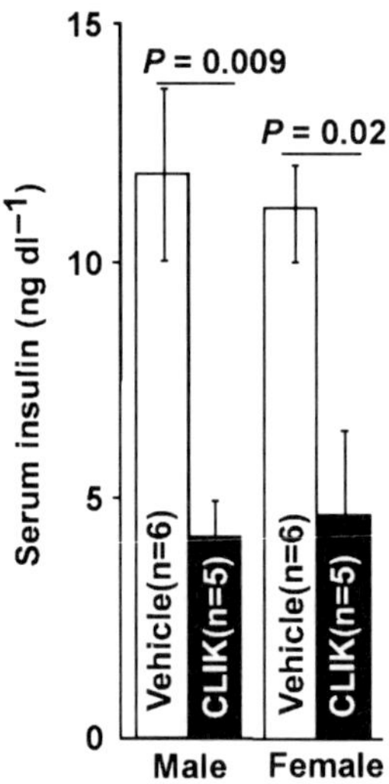

Fig. 18.14 Pharmacological syndrome inhibition of cathepsin L reduces body weight gain and glucose intolerance in CLIK-195-treated male and female *ob/ob* mice

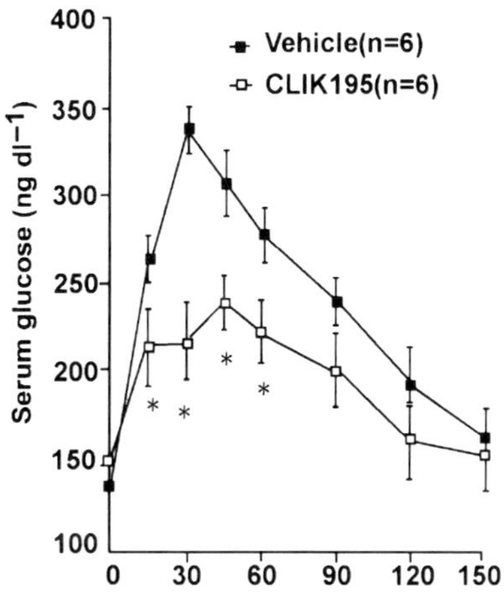

Fig. 18.15 CLIK-195 reduced glucose intolerance in male *ob/ob* mice

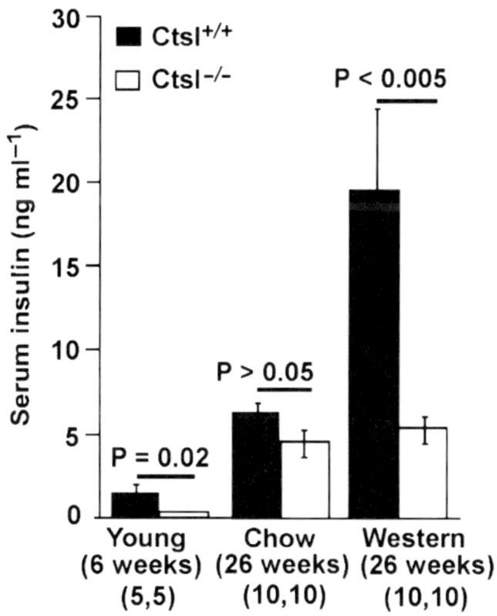

Fig. 18.16 Cathepsin L deficiency increased glucose tolerance. Serum insulin level was significantly lower in $Ctsl^{-/-}$ mice compared with $Ctsl^{+/+}$ mice at a young age and after 26 weeks of consuming a Western diet

L-deficient mice have reduced levels of serum glucose and insulin, but increased levels of muscle IR β-subunits, fibronectin, and glucose transporter (Glut 4) in muscle (Figs. 18.17 and 18.18). The inhibition of cathepsin L demonstrated reduced bodyweight gain and serum insulin levels and increased glucose tolerance, and increased levels of muscle IR β-subunits, fibronectin, and Glut 4. Cathepsin L is a novel target for diabetes therapy (Yang et al. 2007).

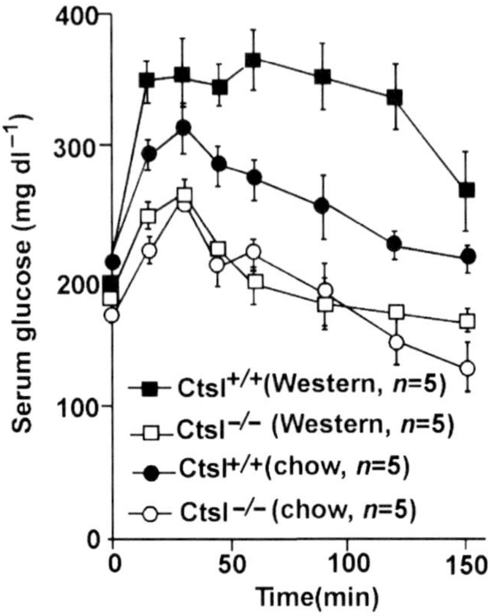

Fig. 18.17 Cathepsin L deficiency increases glucose tolerance. (**a**) Glucose tolerance testing showed increased glucose tolerance in *Ctsl*$^{-/-}$ mice that had consumed either chow or Western diets

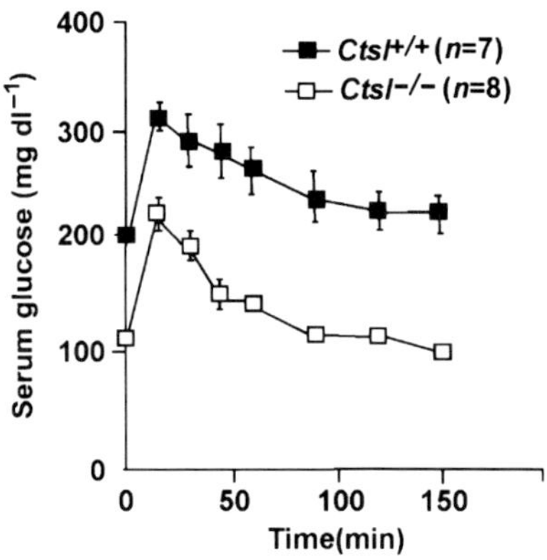

Fig. 18.18 Reduced glucose intolerance

18.1.10 Cleavage of Retinoid X Receptors α by Lysosomal Cathepsin L

Retinoid X receptors (RXRs) belong to the steroid/thyroid hormone receptor superfamily and their endogenous ligand has been shown to be 9-cis-retinoic acid (9-cisRA). We characterized a protease responsible for the cleavage of the 9-cisRA receptor α (RXRα) in two human delivered cell lines, HepG2 and JEC-3. The presence of the protease in the cytoplasm was confirmed by incubating full-length S^{35}RXRα with each fraction. The cytoplasmic fraction cleaved RXRα into small pieces with molecular sizes of 45, 43, and 31 kDa. To characterize the RXRα cleaving protease, a series of protease inhibitors was added in the reaction of S^{35}RXRα with the cytoplasm of JEG-3 cells. The proteolytic cleavage was inhibited by cysteine protease inhibitors, but not by serine protease inhibitors, aspartic protease inhibitors, metalloprotease inhibitors, or members of the calpain family. As shown in Fig. 18.19, RXRα was specifically inhibited by cathepsin L inhibitors, such as CLIK-088, -112, and -121. In addition, Nagaya et al. (1998) reported the intracellular proteolytic cleavage of RXRα by cathepsin L using CLIK-148 and suggested the potentiality of this process for modulating thyroid hormone action. Inhibition of cathepsin L activity in the monolayer of hepatocytes resulted in increased nuclear RXRα protein and augmentation of T3-dependent induction of spot-14 mRNA.

18.1.11 Short Summary and Discussions

All proteins and biologically active peptides are synthesized as their pre-pro forms, therefore to make mature active forms proteolytic processing is indispensable processes. Intracellular cathepsins play a major role in the intracellular processing of proteins and peptides, not only intracellular proteins catabolism. We developed the specific inhibitors for individual cathepsins using the structure based new inhibitor design. Using these specific inhibitors, we clarified the mechanisms of processing

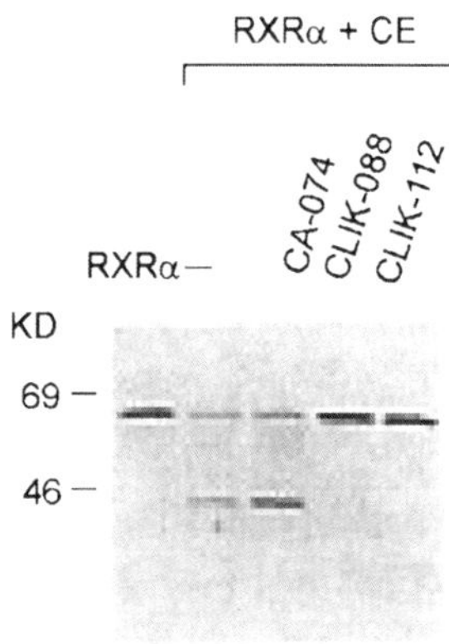

Fig. 18.19 Characterization of a cytoplasmic protease in hepatocytes cultured as a monolayer. When cathepsin B (CA-074) or L-type-specific inhibitors (CLIK-088 and CLIK-112) were added to the incubation reaction, cathepsin L-type-specific inhibitors CLIK-088 and CLIK-112, but not cathepsin B-specific CA-074, inhibit proteolytic cleavage of RXRα by cytoplasmic extract, indicating that cathepsin L-type protease is the enzyme that cleaves RXRα at its amino terminus

of various proteins (enzymes) and biologically active peptides. Also these inhibitors participate to the elucidation of pathogenesis of various diseases and therapy for the special diseases. Especially, we put the focus on the antigen processing mechanisms and autoantigen processing in autoimmune diseases. The most important question to be clarified at the present is the following problem; The individual cathepsin (B, L, H) are located in different particles (lysosomes) from their electron-microscopic pictures (labeled by gold). The selection incorporation mechanisms of the objective substrate proteins into target lysosomes are unknown. These recognition mechanisms of target substrate proteins have been entirely unknown.

18.2 Post-translational Covalent Modifications of Cystatin Family

18.2.1 Intracellular Cathepsin Inhibitors Such as Cystatins

The cysteine protease family (cathepsins) plays an important role in intracellular protein metabolism and cathepsin activities are regulated by cystatins, which are endogenous cysteine protease inhibitors. Among them, more than ten kinds of endogenous cysteine cathepsin inhibitors have been reported at the present. Cystatin α (A) and cystatin β (B) are well known endogenous cysteine protease inhibitors with molecular weights of 10–15 kDa; cystatin α is located only in the epidermis and cystatin β is located in all parenchymal cells ubiquitously.

18.2.2 History of Cystatin Studies

In 1970, endogenous cysteine protease inhibitors of the cystatin family began to be studied as cathepsin inhibitors in various mammalian organs by Järvinen et al. (1976) and Lenny et al. (1979). Then Katunuma and co-workers in 1982, and Turk and colleagues in 1983 reported the presence of two kinds of endogenous cysteine protease inhibitors in rats and humans, respectively (Kominami et al. 1982 Machleidt et al. 1983). They were named "cystatin α" and "cystatin β" in the case of rats by the Katunuma's group, and cystatin A and cystatin B in the human case by the Turk's group. Cystatin α and cystatin A or cystatin β and cystatin B are the same kinds of cysteine protease inhibitors and their amino acid sequences show strong homology.

More than ten kinds of cystatins have been reported in many organs and secretory fluids, and the individual cystatins show different inhibitory specificities against various cathepsins. We focused on the regulation of the inhibitory activities by their covalent modification of these cystatins at the molecular level. The inhibition mechanism and the regulation mechanism of cathepsin activity by cystatins in situ are very important for the regulation of protein catabolism; however, little is known.

We found that cystatin α is phosphorylated by protein kinase C and then targeted in the cornified envelope in skin; then the phosphorylated cystatin α is conjugated with a filaggrin linker segment peptide mediated by transglutaminase. The phosphorylated cystatin α in skin inhibits bacterial growth and plays an important role in the protection against bacterial infection. On the other hand, cystatin β is distributed in all parenchymal cells of various animal organs and plays a role in the regulation of intralysosomal cathepsin activities. The cysteine residue located in the third position of the N-terminus of the cystatin β molecule is reversibly modified to make mixed disulfate derivatives with glutathione and the glutathionated cystatin β. The changes between the glutathionated form and the deglutathionated form of cystatin β and also their dimers are regulated by the redox potential in the cells (Wakamatsu et al. 1984; Tsukahara et al. 1987). Therefore, the intracellular cathepsin activities should be regulated by the redox potential in the cells.

18.2.3 Classification of Cystatin Family and Inhibition Mechanisms of Cystatins

As shown in Fig. 18.20, the cystatin family is divided into two groups; one is located inside cells and the other is secreted into various body fluids. The intracellular group includes cystatin α and β and the secreted group includes cystatin S in saliva (Isimura et al. 1984), cystatin C and γ-trace in cerebrospinal fluid (Barrett et al. 1984), and egg white cystatin (Turk et al. 1983). Another secreted group is a high molecular weight cystatin group, including kininogen (Ohkubo et al. 1984). Cystatins are classified expediently into three groups: family 1, including cystatin α and β (A and B); family 2, including cystatin S and egg white cystatin; and family 3, which are high molecular weight inhibitors like kininogen.

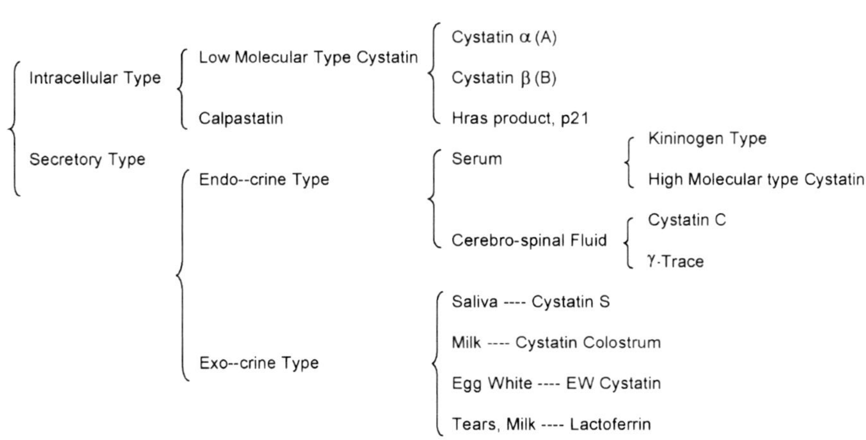

Fig. 18.20 Biological classification of endogenous cysteine protease inhibitors

Each cystatin shows a different affinity for individual cathepsins. For example, egg white cystatin and cystatin α do not inhibit cathepsin B activity, but strongly inhibit cathepsin L activity. In other words, various cystatins share inhibitory functions against individual cathepsins. Turk and co-workers (1983, 1985, 1991) explained the inhibition mechanism of cystatin at the molecular level as follows. The tertiary structure of cystatin C, determined using X-ray crystallography, is shown in Fig. 18.21. The pink loop and green loop in the Fig. 18.21 structure have homologous amino acid sequences in all cystatins and are the common binding sites with cathepsins. The low molecular weight cystatin family has three binding domains showing homologous amino acid sequences, illustrated in Fig. 18.22 with

Fig. 18.21 Tertiary structure of egg white cystatin (ribbon-like presentation). The upper left (*red part*) end is the N-terminus. The upper right two loops (*pink* and *green parts*) are binding domains with cathepsins

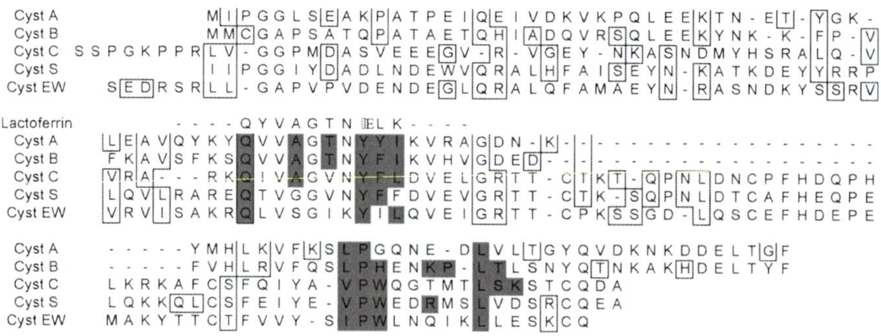

Fig. 18.22 Sequence homology of the cystatin family. The highlighted domains are the common binding domains with cathepsins

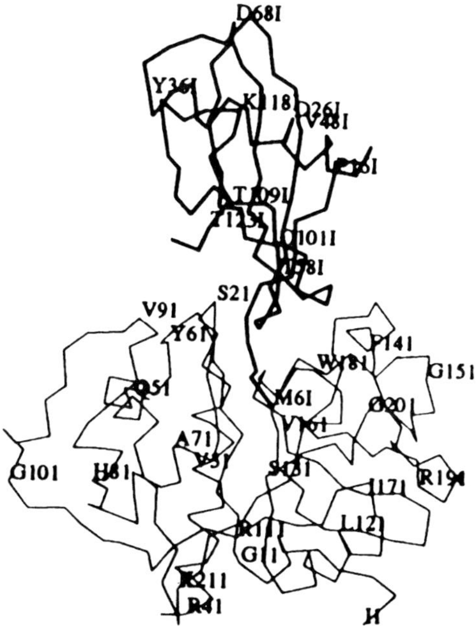

Fig. 18.23 X-ray co-crystallography of cystatin C with cathepsin B; the N-terminus of cystatin is inserted into the substrate binding pocket of cathepsins

red and green as the common binding sites with cathepsin. The N-terminus of cystatins, indicated by a red symbol in Fig. 18.21 is inserted into the substrate binding pocket of the cathepsins. X-ray co-crystallography of the complex between cystatin C and cathepsin B was demonstrated by Turk (1995), as shown in Fig. 18.23.

18.2.4 Inhibition Mechanism of Cathepsin by Cystatins In Vivo

The contribution of the cystatin family to the inhibition of cathepsin activities in situ is very important from the aspect of regulation of protein catabolism. The intracellular localizations of cystatins are shown in Figs. 18.24–18.26. The immunohistochemical examinations of cystatin α and cystatin β in skin are shown in Fig. 18.24 using anti-cystatin α and anti-cystatin β antibodies. Cystatin α is located only in the epidermis. In contrast, cystatin β is ubiquitously detected not only in epidermal cells, but also in hair follicular cells and dermal cells. As shown in Fig. 18.26, anti-cystatin β antibodies were labeled by large gold particles and the anti-insulin antibody wasv labeled by small gold particles. Both sizes of gold particles were observed in the β-cell granules of the islets of Langerhans of the pancreas by immuno-electron microscopy. When cystatin α was injected intravenously, it was incorporated into lysosomes of kidney cells after 30 min. and the intralysosomal cathepsin H activity was reciprocally inhibited by incorporated cystatin α, as shown in Fig. 18.25.

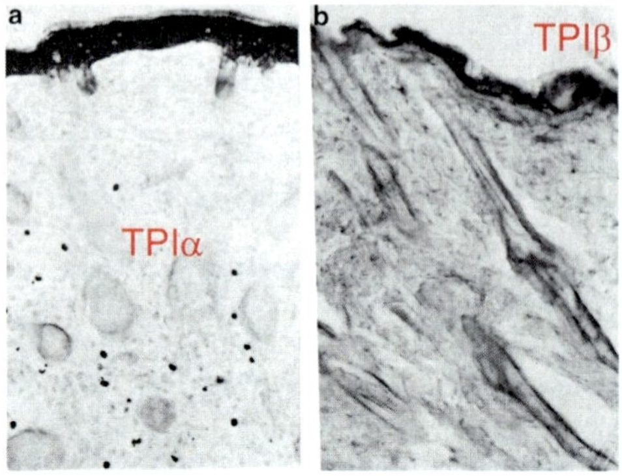

Fig. 18.24 Immunohistochemical staining of cystatin α (TPI-α) and cystatin β (TPI-β) in skin

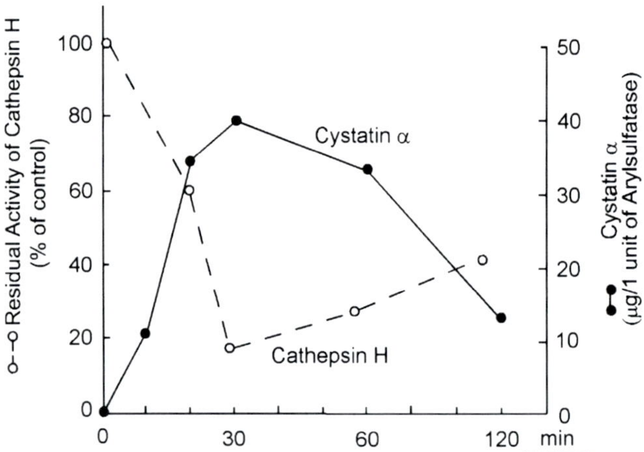

Fig. 18.25 Incorporation of cystatin α into kidney lysosomes and inhibition of intra-lysosomal cathepsin H after intravenous injection of cystatin α

Therefore, as Fig. 18.27 shows, the cystatin β located originally in secretory granules is secreted outside of the cells and then incorporated into lysosomes by exophagy, and the cathepsin activities in the lysosomes are inhibited. We determined the cystatin α gene expression, as shown in Fig. 18.28. When the cystatin α gene plasmid was transfected to cancer cells, surprisingly, not only cystatin α was expressed, but the expression of cathepsin B was also increased. This result indicates the coordinated relationship between the expression of both the cystatin α and cathepsin B genes.

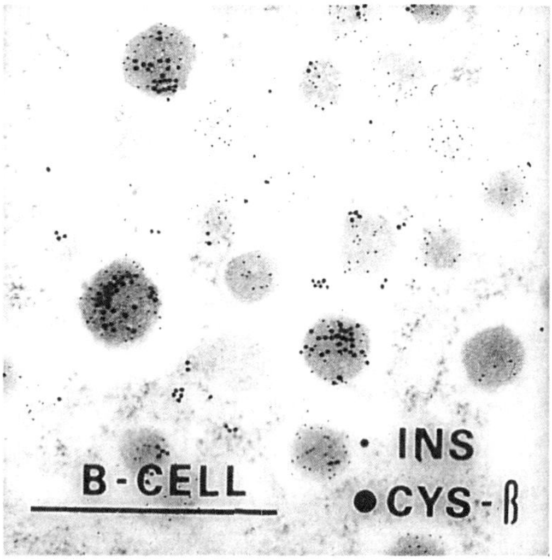

Fig. 18.26 Localization of cystatin β and insulin in β-cells of the island of Langerhans in the pancreas. Cystatin β is labeled by big gold particles and insulin is labeled by small gold particles. Cystatin β is located in secretory granules

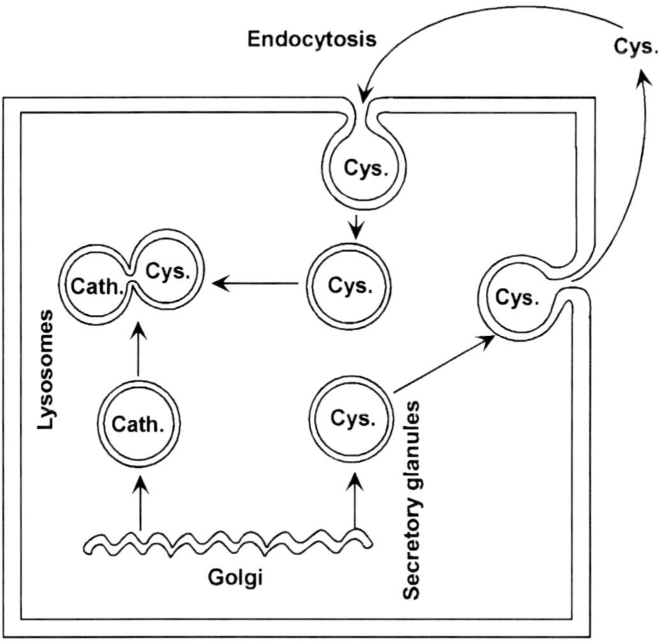

Fig. 18.27 Hypothetical mechanism of lysosomal cathepsin inhibition by cystatin β in situ. The inhibition mechanism of lysosomal cathepsins by cystatins was originally located in secretory granules

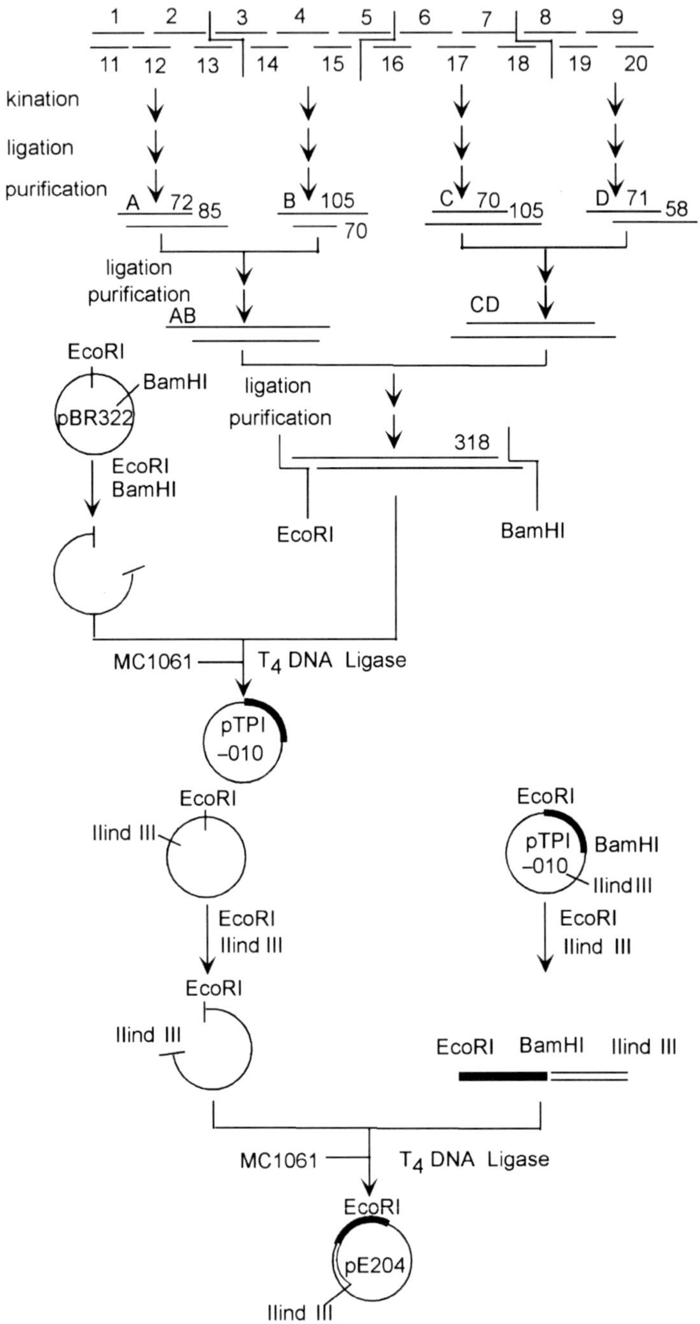

Fig. 18.28 Structures of the cystatin α gene and the expression plasmid of the cystatin α gene

18.2.5 Post-translational Covalent Modification of Cystatin α

When skin was stained by immunohistochemistry using anticystatin α antibody, only the cornified envelope of the skin was stained. The sphingosine treatment of newborn rat skin resulted in the suppression of the targeting of cystatin α into the cornified envelope, as shown in Fig. 18.29. Therefore, cystatin·α was stained in the cornified envelope in the case of untreated skin using anti-cystatin α antibody, because sphingosine is a powerful inhibitor of protein kinase C (PKC) and cystatin α was not phosphorylated.

We found that the hematoxylin stainable protein in newborn rat epidermis was a phosphorylated cystatin α. Alkaline phosphatase treatment of the hematoxylin stainable protein in keratohyalin granules resulted in the release of cystatin α. A threonine residue located in near C-terminus of cystatin α is phosphorylated by PKC (Protein Kinase C) and the phosphorylated cystatin α is then incorporated into the cornified envelope.

The participation of PKC in the phosphorylation of cystatin α was confirmed. The specific inhibitor of PKC, H-7, inhibited the incorporation of ^{32}P into cystatin α, as shown in Table 18.3. The cystatin α was then conjugated with the filaggrin linker segment peptide, which is rich in glutamine residues, mediated by epidermal transglutaminase in the presence of calcium, to yield a high molecular weight protein, as shown in Fig. 18.30 (Takahashi et al. 1994; Takahashi et al. 1999).

The phosphorylated cystatin α possesses the capacity to inhibit the cysteine protease activity of bacteria and viruses. Furthermore, the cornified envelope, which contains cystatin α, shows inhibitory activity to cathepsins, but the cornified envelope from the skin treated with sphingosine lost its inhibitory activity, because

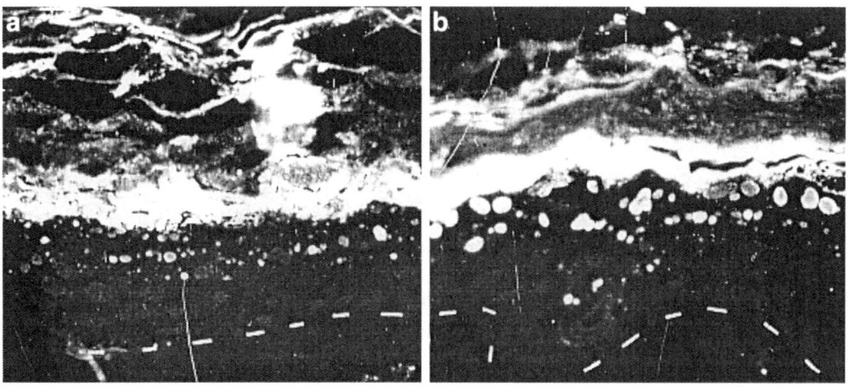

Fig. 18.29 Inhibition of cystatin α incorporation into keratohyaline particles with cathepsin B. Keratohyaline granules of skin detected by sphingosine. Indirect immuno-fluorescence staining of newborn rat skin by using anti-cystatin α antibodies. (**a**) Sphingosine-treated skin. (**b**) Normal control skin

Table 18.3 Incorporation of 32P into cystatin α

Substrate	Inhibitor	32P incorporated into protein fraction (cpm)	32P incorporated into cystatin α or histone (cpm)
–	–	523.9	
–	+H-7	448.0	
Cystatin α (86 μg)	–	1,146.4	622.3
Cystatin α (86 μg)	+H-7	711.4	263.3
Cystatin α (172 μg)	–	1,363.1	839.9
Cystatin α (172 μg)	+H-7	789.1	344.1
Histone (86 μg)	–	6,820.4	6,296.9
Histone (86 μg)	+H-7	792.3	344.3

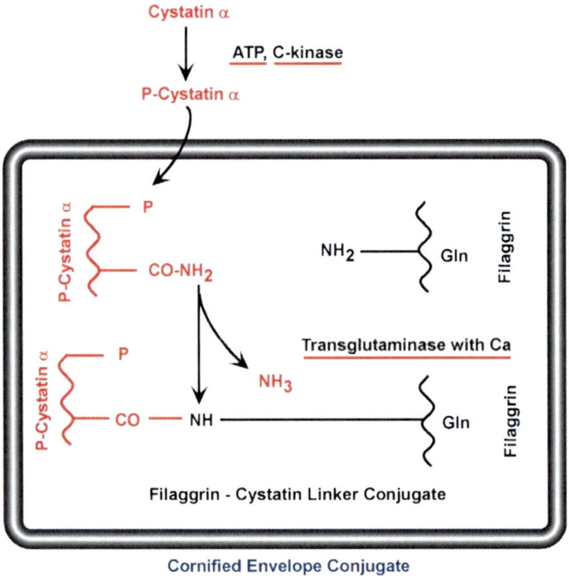

Fig. 18.30 Fate of cystatin α to make filaggrin conjugates in skin

by the treatment, the cystatin α was not incorporated into the envelope. As shown in Fig. 18.31, phosphorylated cystatin α strongly inhibited the activity of cathepsin L, but did not inhibit the activities of cathepsin B or cathepsin H; therefore, cystatin α in the cornified envelope showed a specific inhibition for cathepsin L.

Cathepsin L is an important protease for the survival of bacteria and viruses. In 1994 Katunuma's group reported that cystatin α strongly suppressed the bacterial growth of cultured *Staphylococcus aureus* V8. Thus, when *Staphylococcus aureus* V8 was inoculated onto the sphingosine-treated skin, the incorporation of cystatin α was suppressed and many more colonies were obtained compared with that of normal skin (Takahashi 1994). This anti-bacterial action of cystatin α and its blocking effect on poliovirus proliferation (Korant et al. 1985) led authors to propose that

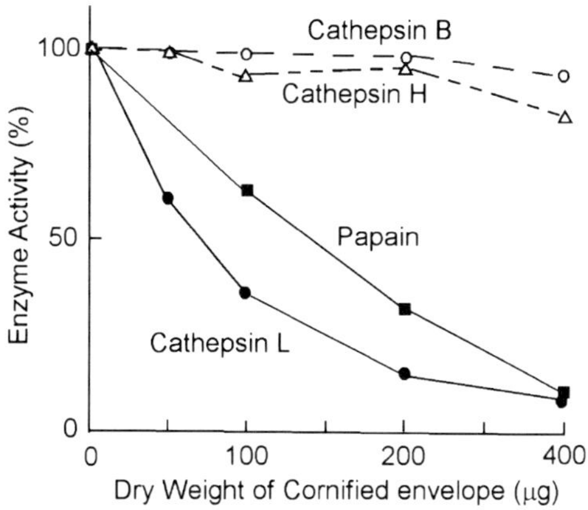

Fig. 18.31 Inhibition of various cathepsins by purified cornified envelope extract

skin cystatin α plays an important role in the protection against bacterial and viral infections. In this context fell our observations regarding the anti-bacteral action of lactoferrin, a member of the cystatin family. We found that lactoferrin strongly inhibits cathepsin L activity (Ki = 10^{-8} M) and, therefore, that the intake of lactoferrin in milk suppresses the growth of *Staphylococcus epidermis* in the small intestine.

18.2.6 Post-translational Covalent Modification of Cystatin β

Cystatin β is located ubiquitously in all cells and regulates proteolysis in their lysosomes (Fig. 18.26). Cystatin β possesses at the third position of the N-terminus a cysteine residue (Fig. 18.32) with which glutathione reacts to form a mixed disulfate complex, as shown in Fig. 18.33 (Katunuma 1985). Since the glutathionated cystatin β did not bind with cathepsins, it loses its inhibitory activity. As mentioned earlier, Turk's group (1995) made it clear using X-ray co-crystallography of cystatin with cathepsin B that while the N-terminus of cystatin was inserted into the binding pocket of cathepsin, the glutathionated N-terminus of cystatin β was unable to bind with the binding pocket. The changes in coefficient of the oxidized and reduced forms of glutathione affected the binding of cystatin β with cathepsins. The coefficient of the oxidized and reduced forms of glutathione regulates the inhibitory activity of cystatin β, as shown in Fig. 18.33. When the oxidized form of glutathione was increased by adding menadione (vitamin K_3) to a cultured macrophages system, the binding of cystatin β with glutathione to make an inactive form of cystatin β was increased, as Figs. 18.33 and 18.34 show (Tsukahara et al. 1987, 1984). Therefore, the activities of cathepsins were regulated by the intracellular redox potentials through the changes of the inhibitory activity of cystatin β, as shown in Fig. 18.34.

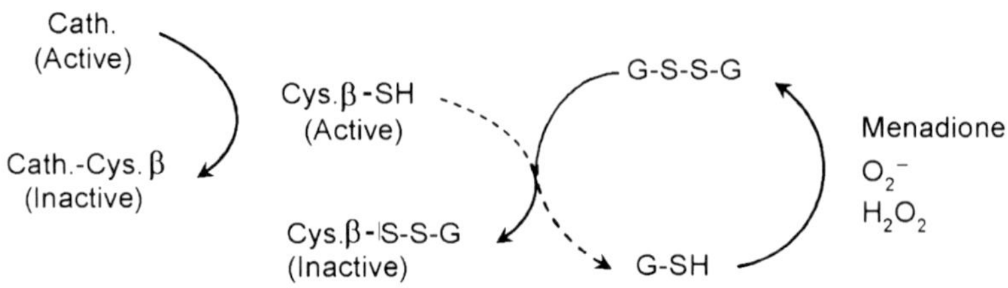

Fig. 18.32 Regulation of cathepsin activity through covalent modification of cystatin β with glutathione. The arrow shows the reaction when menadione is added

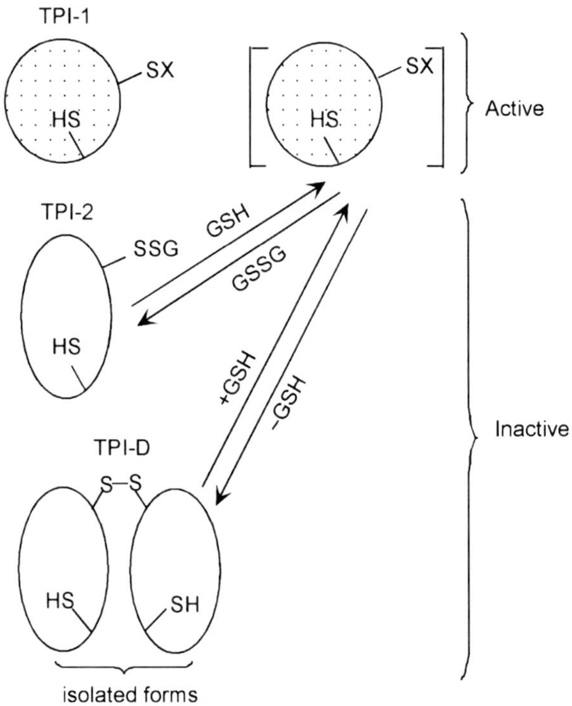

Fig. 18.33 Interconversion of three forms of cystatin β by redox potential

The inhibitory activity of cystatin α did not change under reducing conditions and cystatin α is able to make conjugates with cathepsin H in either condition. In contrast, cystatin β was available to make conjugates with cathepsin H only under reducing conditions; glutathionated cystatin β under oxidizing conditions could not bind with cathepsin H. Therefore, the inhibitory activity of cystatin β depends on the intracellular ratio of oxidized form and reduced form of glutathione. As a result, cathepsin activities are regulated by the level of intracellular redox potential through the changes in cystatin β activities.

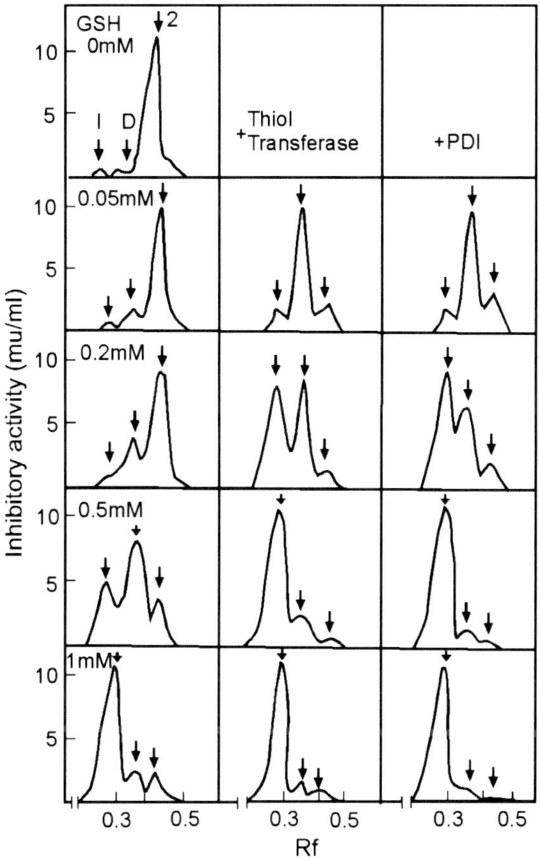

Fig. 18.34 Enzymatic dethiolation of cystatin β. Purified cystatin β (mixed disulfide with glutathione, 10 μg) was incubated with the indicated concentrations of glutathione and thiol transferase (0.2 unit) or protein disulfide isomerase (0.1 unit) in 50 μl of 0.1 M sodium phosphate buffer, pH 7.5, containing 5 mM EDTA for 1 h at 37°C. Then, aliquots of the reaction mixtures were promptly subjected to polyacrylamide gel electrophoresis without sodium dodecyl sulfate at pH 8.0. After electrophoresis, gels were cut into slices and the papain inhibitory activity of their extracts was examined in the presence of 8 mM cysteine

18.2.7 Short Summary and Discussion

At the present, about ten kinds of cystatins are reported. Cystatin β is distributed in all organs, on the contrary, other cystatins are located in special organs or body fluids. Cathepsins are originally located in lysosomes, but the other cystatins are originally located in cytoplasm or special body fluids or serum. Therefore, how to contact between cathepsins and cystatin family in situ is a big question about in general. Cystatin α and cystatin family in milk or serum play a role in the protection from the bacterial infection. Invariant chain is a kind of cystatin family and might play a role in protection from cathepsin action in antigen presentation.

References

Barrett AJ, Davies ME and Grubb A (1984) The place of human gamma – trace (cystatin C) amongst the cysteine proteinase inhibitors. Biochem Biophys Res Commun 120:631–636.

Hanada K, Tamai M, Yamagishi M et al. (1978) Isolation and characterization of E-64, a new thiol protease inhibitor. Agric Biol Chem 42:523.

Hara K, Kominami E and Katunuma N. (1988) Effect of proteinase inhibitors on intracellular processing of cathepsin B, H and L in rat macrophages. FEBS Lett 231:229–231.

Hashida S, Towatari T, Kominami E and Katunuma N (1980) Inhibition by E-64 derivatives of rat liver cathepsin B and cathepsin L in vitro and in vivo. J Biochem 88:1805–1811.

Ike Y., Yamato M, Kominami E and Katunuma N (1989) Total synthesis of cystatin – α gene and its expression in *E. Coli*. In: Katunuma N and Kominami E (eds) Intracellular Proteolysis. Mechanisms and Regulations, pp. 391–393.

Isemura S, Saitoh E, Isemura M and Sanada K (1984) Cystatin S: A cysteine proteinase inhibitor of human saliva. J Biochem 96:1311–1314.

Ishimaru N, Arakaki R, Katunuma N and Hayashi Y (2004) Critical role of cathepsin-inhibitors for autoantigen processing and autoimmunity. Advan Enzyme Regul 44:309–320. Pergamon Press, Oxford.

Järvinen M (1976) Purification and properties of two protease inhibitors from rat skin inhibiting papain and other SH – proteases. Acta Chem Scand-B 30:933–940.

Järvinen MJ and Rinne A (1982) Human spleen cysteineproteinase inhibitor. Purification, fractionation into isoelectric variants and some properties of the variants. Biochim Biophys Acta 708:210–217.

Katunuma N (1997) New aspects on antigen presentation mechanism for immuno-responses and allergy expression. Medical aspects of proteases and protease inhibitors. pp. 153–172. Ios Press Inc, Tokyo.

Katunuma N and Kominami E (1985) Molecular basis of intracellular regulation of thiol proteinase inhibitors. Curr Top Cell Regul 27:345–360.

Katunuma N and Kominami E. (1995) Structure, properties, mechanisms, and assays of cysteine protease inhibitors: Cystatins and E-64 derivatives. Meth Enzymol 251:382–397. Academic Press, Inc.

Katunuma N, Yamato M, Kominami E and Ike Y (1988a) Total synthesis of the cystatin α gene and its expression in *E. coli*. FEBS Lett 238:116–118.

Katunuma N, Matsunaga Y, Matsui A et al. (1998b) Novel physiological functions of cathepsins B and L on antigen processing and osteoclastic bone resorption Advan Enzyme Regul 38:235–251. Pergamon Press, Oxford.

Katunuma N, Murata E, Kakegawa H, et al. (1999) Structure based development of novel specific inhibitors for cathepsin L and cathepsin S *in vitro* and *in vivo*. FEBS Lett 458:6–10.

Katunuma N, Matsui A, Endo K et al. (2000) Inhibition of intracellular cathepsin activities and suppression of immune responses mediated by helper T lymphocyte type-2 by peroral or intraperitoneal administration of vitamin B_6. Biochem Biophys Res Commun 272:151–155.

Katunuma N, Matsui A, Inubushi T et al. (2000) Structure-based development of pyridoxal propionate derivatives as specific inhibitors of cathepsin K *in vitro* and *in vivo*. Biochem Biophys Res Commun 267:850–854.

Katunuma N, Matsunaga Y, Himeno K and Hayashi Y (2003) Insights into the roles of cathepsins in antigen processing and presentation revealed by specific inhibitors. Biol Chem 384:883–890.

Kominami E, Wakamatsu N and Katunuma N (1982) Purification and characterization of thiol proteinase inhibitors from rat serum and liver. J Biol Chem 257:14648–14652.

Korant BD, Brzin J and Turk V (1985) Cystatin, a protein inhibitor of cysteine protease alters viral protein cleavage in infected human cells. Biochem Biophys Res Commun 127:1072–1076.

Lenny JF, Tolan JR, Sugai WJ and Lee AG (1979) Thermostable endogenous inhibitors of cathepsin B and H. Eur J Biochem. 101:153–161.

Machleidt W, Borchart U, Fritz H et al. (1983) Protein inhibitors of cysteine proteinases.II. Primary structure of stefin, a cytosolic protein inhibitor of cysteine proteinases from human polymorphonuclear granulocytes. Hoppe Seylers Z Physiol Chem 264:1481–1486.

Maekawa Y, Himeno K, Ishikawa H et al. (1998) Switch of CD4+T cell differentiation from Th2 to Th1 by treatment with cathepsin B inhibitor in experimental leishmaniasis. J Immunol 161:2120–2127.

Matsuda Y, Ogushi F, Ogawa K and Katunuma N (1986) Structure and properties of albumin Tokushima and its proteolytic processing by cathepsin B in vitro. J Biochem (Tokyo) 100:375–379.

Matsunaga Y, Saibara T, Kido H and Katunuma N (1993) Participation of cathepsin B in processing of antigen presentation to MHC Class II. FEBS Lett 324:325–330.

Musil D, Zucic D, Turk, et al. (1991) The refined 2.15A X-ray crystal structure of human liver cathepsin B: the structural basis for its specificity. EMBO J 10:2321–2330.

Nagaya T, Murata Y, Yamaguchi S et al. (1998) Intracellular proteolytic cleavage of 9-cis-retinoic acid receptor α by cathepsin L-type protease is a potential mechanism for modulating thyroid hormone action. J Biol Chem 273: 33166–33173.

Ohkubo I, Kurachi K, Takasawa T et al. (1984) Isolation of a human cDNA for alpha-2-thiol proteinase inhibitor and its identity with low molecular weight kininogen. Biochemistry 23:5691–5697.

Saegusa K, Ishimaru N, Yanagi K et al. (2002) Cathepsin S inhibitor prevents autoantigen presentation and autoimmunity. J Clin Invest 110:361–369.

Takahashi M, Tezuka T, Towatari T and Katunuma N (1991) Identification of hematoxylin-stainable protein in epidermal keratohyalin granules as phosphorylated cystatin α by protein kinase C. FEBS Lett 287:178–180.

Takahashi M, Tezuka T and Katunuma N (1992) Phosphorylated cystatin α is a natural substrate of epidermal transglutaminase for formation of skin cornified envelope. FEBS Lett 308:79–82.

Takahashi M, Tezuka T, Kakegawa H and Katunuma N (1994) Linkage between phosphorylated cystatin α and filaggrin by epidermal transglutaminase as a model of cornified envelope and inhibition of cathepsin L activity by cornified envelope and the conjugated cystatin α. FEBS Lett 340:173–176.

Takahashi M, Tezuka T and Katunuma N (1994) Inhibition of growth and cysteine proteinase activity of *Staphylococcus aureus* V8 by phosphorylated cystatin α in skin cornified envelop. FEBS Lett 355:275–278.

Takahashi M, Tezuka T and Katunuma N (1996) Filaggrin linker segment peptide and cystatin α are parts of a complex of the cornified envelope of epidermis. Arch Biochem Biophys 329:123–126.

Takahashi M, Tezuka T, Korant B and Katunuma N (1999) Inhibition of cysteine protease and growth of *Staphylococcus aureus* V8 and poliovirus by phosphorylated cystatin α conjugate of skin. BioFactors 10:339–345. IOS Press.

Takio K, Kominami E, Wakamatsu N, Katunuma N and Titani K (1983) Amino acid sequences of rat liver thiol proteinase inhibitors. Biochem Biophys Res Commun 115:902–908.

Takio K, Kominami E, Tada K and Katunuma N (1984). Amino acid sequence of rat epidermal thiol proteinase inhibitor. Biochem Biophys Res Commun 121:149–154.

Tsukahara T, Kominami E and Katunuma N (1987) Formation of mixed disulfide of cystatin-beta in cultured macrophages treated with various oxidants. J Biochem 101:1447–1456.

Turk V and Bode W (1991) The cystatins: Protein inhibitors of cysteine proteinases. FEBS Lett 285:213–219.

Turk V, Brzin J, Longer M et al. (1983) Protein inhibitors of cysteine proteinases. III. Amino—acid sequence of cystatin from chicken egg white. Hoppe Seylers Z Physiol Chem 364:1487–1496.

Turk V, Brzin J, Lenarcic B et al. (1985) Structure and function of lysosomal cysteine proteinases and their protein inhibitors. Prog Clin Biol Res 180:91–103.

Turk D, Podobnic M, Popovic T, Katunuma N et al. (1995) Crystal structure of cathepsin B inhibited with CA030 at 2.0A resolution: A basis for the design of specific epoxysuccinyl inhibitors. Biochemistry 34:4791–4797.

Turk V, Turk B and Turk D (2001) Lysosomal cysteine proteases: facts and opportunities. EMBO J 20:4629–4633.

Umezawa H (1972) Enzyme inhibitors of microbial origin. pp 1–117. University of Tokyo Press, Tokyo.

Umezawa H, Aoyagi T, Morishita T et al. (1970) Chymostatin, a new chymotrypsin inhibitor produced by antinomycetes. J Antibio 23:425–427.

Wakamatsu N, Kominami E, Takio K and Katunuma N (1984) Three forms of thiol proteinase inhibitor from rat liver formed depending on the oxidation-reduction state of a sulfhydryl group. J Biol Chem 259:13832–13838.

Yang M, Zhang Y, Pan J et al. (2007) Cathepsin L activity controls adipogenesis and glucose tolerance. Nat Cell Biol Lett 9:970–977.

Zhang T, Maekawa Y, Yasutomo K et al. (2000) Pepstatin A-sensitive aspartic proteases in lysosome are involved in degradation of the invariant chain and antigen-processing in antigen presenting cells of mice infected with *Leishmania major*. Biochem Biophys Res Commun 276:693–701.

Chapter 19
Metalloproteases and Proteolytic Processing

Anthony J. Turner and Natalia N. Nalivaeva*

Abbreviations

Aβ	Amyloid β-peptide
ACE	Angiotensin converting enzyme
AICD	APP intracellular domain
APP	Amyloid precursor protein
ECE	Endothelin-converting enzyme
IDE	Insulysin (insulin-degrading enzyme)
MMP	Matrix metalloproteinase
PAI	Plasminogen-activator inhibitor
RAS	Renin-angiotensin system

19.1 Introduction

Proteolysis represents one of the key processes underlying biological events from fertilization through embryonic development, cell differentiation, ageing and ultimately death via the programmed route (apoptosis) or necrosis in the case of trauma, infection or disease. The term "peptidase" (or protease) is used to refer to the complement of enzymes that cleave peptide bonds and, as such, all known proteins and peptides represent their potential substrate repertoires. According to the location of the peptide bond cleaved by the peptidases, they are conventionally sub-divided into "endopeptidases" that cleave the bond within the polypeptide chain, or exopeptidases that release amino acids from the N- or C-terminus of the substrate. However, this is an over-simplification

A.J. Turner (✉)
Institute of Molecular and Cellular Biology, University of Leeds, LS2 9JT Leeds, UK
e-mail: a.j.turner@leeds.ac.uk
and
*Dr N.N. Nalivaeva's
Institute of Evolutionary Physiology and Biochemistry, RAS, St.Petersburg, Russia

C.J. Vidal (ed.), *Post-Translational Modifications in Health and Disease*, Protein Reviews 13, DOI 10.1007/978-1-4419-6382-6_19, © Springer Science+Business Media, LLC 2011

since some peptidases, e.g. ACE, can fulfill both endo- and exopeptidase functions (Skidgel and Erdös 1985). The term "limited proteolysis" was originally introduced by Linderstrom-Lang to distinguish the restricted specificity of certain peptidases, particularly those involved in post-translational modification of proprotein precursors, from the random proteolysis accompanying protein degradation (Neurath 1999).

The number of species in which peptidases have now been identified is approaching 5,000 (http://merops.sanger.ac.uk/) distributed among five classes (metallo-, serine, cysteine, aspartic and threonine) with cysteine-, serine- and metallopeptidases representing the major ones. Among these three main classes, the known or putative metallopeptidases are more consistently represented across species varying from 105, for example, in *Trypanosoma cruzi* to 164 in mouse (and 161 in the human) (Table 19.1). The other two classes exhibit a greater diversity ranging in the number of genes for cysteine proteases from 26 in *T. cruzi* to 162 in mouse (and 160 in the human). For the serine peptidases this distribution is even more pronounced: from 14 in *T. cruzi* to 245 in mouse (210 in human). There are a smaller number of aspartic peptidases although some are of key therapeutic importance (e.g. the β- and α-secretases implicated in Alzheimer pathology) and even less variety in the number of threonine peptidases. This greater consistency of representation of metallopeptidases across species probably reflects more generic cellular functions and their distinctive catalytic mechanisms which require a water molecule activated by a divalent metal ion, commonly zinc.

The complete repertoire of peptidases expressed by a tissue or organism is now referred to as the "protein degradome," a concept originally introduced by Lopez-Otin and colleagues (Cal et al. 2003). Comparative analysis of mammalian degradomes has revealed that rodents generally have a more complex combination of peptidases compared with primates, mainly due to the expansion of peptidase families involved in reproduction and host defence (Puente et al. 2005). These authors also found several genes implicated in reproduction, nutrition and the immune system, which are functional in rat, mouse or chimpanzee, but have been inactivated by mutations in the human lineage. These findings suggested that conversion of specific peptidase genes into pseudogenes has contributed to the evolution of the human genome. A recent example of this "pseudogenisation" is found in the ACE3 gene in the human (Rella et al. 2007). Approx one-quarter of the 80 or so human inherited diseases caused by mutations in protease-coding genes (http://www.uniovi.es/degradome/) are linked to metallopeptidases (see Table 19.2). Abnormal functioning of these genes is implicated in such diverse pathologies as inflammatory and cardiovascular diseases, cancer, and neurodegeneration. Understanding the normal roles of these enzymes and their homologues is important for the design of novel and selective therapeutics. Animal venoms are a natural source of a wide range of metallopeptidases which have biological implications (Moura-da-Silva et al. 2007) and some bacterial toxins (e.g. botulinum, tetanus) exert their toxic actions as metalloproteases, in their case by blocking neurotransmitter release by proteolysis of synaptobrevin (Schiavo et al. 1992).

Table 19.1 Number of total peptidases and metallo-peptidases in genomes of some species[a]

Species	Total number of known or putative peptidases	Total non-peptidase homologues	Unique peptidases	Total known or putative inhibitors	Total non-inhibitor homologues	Unique inhibitors	Total number of known or putative metallo-peptidases	Total number of non-metallo-peptidase homologues
Homo sapiens (human)	660	387	256	179	204	34	161	66
Pan troglodytes (chimpanzee)	438	120	388	107	83	89	108	38
Rattus norvegicus (rat)	519	148	114	218	162	25	144	57
Mus musculus (mouse)	679	189	660	225	215	225	164	59
Drosophila melanogaster (fruit fly)	494	245	389	62	86	62	126	71
Caenorhabditis elegans (roundworm)	350	126	350	49	138	11	137	50
Trypanosoma cruzi (parasitic flagellate)	153	66	133	1	2	1	105 (76 of M8 family)	35 (19 of M8 family)

[a]Data collected from the MEROPS database http://merops.sanger.ac.uk/

Table 19.2 Human hereditary diseases of metallopeptidases

Metalloprotease	Class	Gene	Locus	Disease	Inheritance	Function	Animal model
Angiotensin converting enzyme	MO2	ACE	17q23	Renal tubular dysgenesis	R	Loss	KO resembles disease
Collagenase 3	M10	MMP13	11q22	Spondyloepimetaphyseal dysplasia	D	(Gain)	KO resembles disease
Gelatinase A	M10	MMP2	16q13	Multicentric osteolysis with arthritis	R	Loss	KO does not resemble disease
Enamelysin	M10	MMP20	11q22	Amelogenesis imperfecta	R	Loss	KO resembles disease
ADAMTS-10	M12	ADAMTS10	19p13	Weill-Marchesani syndrome	R	Loss	–
ADAMTS-13	M12	ADAMTS13	9q34	Thrombotic thombocytopenic purpura	R	Loss	–
Procollagen I N-endopeptidase	M12	ADAMTS2	5q23	Ehlers-Danlos syndrome type VIIC	R	Loss	KO resembles disease
Endothelin-converting enzyme 1	M13	ECE1	1p36	Hirschprung disease	D	Loss	KO partially resembles disease
Kell blood-group protein	M13	KEL	7q35	Kell blood group antigen	R	Loss	–
Neprilysin	M13	MME	3q26	Fetomaternal alloimmunisation	R	Loss	–
PHEX endopeptidase	M13	PHEX	Xp22	X-linked hypophosphatemia	D	Loss	Hyp mouse resembles disease
X-Pro dipeptidase	M24	PEPD	19q13	Prolidase deficiency	R	Loss	–
S2P protease	M50	MBTPS2	Xp22	Ichthyosis follicularis, atrichia, and photophobia syndrome (IFAP)	–	Loss	–

From a total of 80 human hereditary diseases caused by mutations in protease-coding genes, 13 are related to metallopeptidase classes discussed in this chapter. They are classified by loss or gain of function and by the type of inheritance: R (recessive) or D (dominant) (Adapted from: Quesada et al. (2009) and http://www.uniovi.es/degradome/)

19.2 Physiological Significance of Proteolysis

Organisms use proteolysis for a wide range of purposes. They include complete protein digestion of dietary proteins as a source of amino acids by the digestive proteases (trypsin, chymotrypsin, pepsin), degradation of structural components e.g. extracellular matrix by the matrix metalloproteases (MMPs), intracellular protein turnover important in post-mitotic cells to prevent accumulation of malfunctioning proteins (lysosome, ubiquitin-proteasome system) including partial digestion of external antigens in immune surveillance, degradation of cyclins at different stages of the cell cycle, conversion of precursor-proteins into their final biologically active structures (zymogens, pro-enzymes, prohormones, other cryptic peptides) and ectodomain shedding to initiate or modulate cell signalling. Highly specific proteolysis can regulate G protein receptor signalling (proteinase-activated receptors) (Bunnett 2006), and nuclear transcriptional events through regulated intramembrane proteolysis (Wolfe 2009). This chapter will predominantly focus on processing events that utilise metalloproteases, and their physiological and pathological consequences.

The process of protein biosynthesis itself involves proteolytic processing: the removal of initiator methionine, signal peptide, cell localization signal and propeptide, where present. A primary mechanism in post-translational modifications of proteins is hence the co-translational removal of the *N*-terminal methionine from a nascent protein, a process which ultimately determines the half-life of a protein by exposing a new *N*-terminal residue, as originally proposed by Varshavsky and colleagues (the *N*-end rule) (Bachmair et al. 1986). Proteins retaining the initiator methionine have shorter half-lives (Frottin et al. 2006). This has led to the definition of specific protein degradation signals or "degrons" that can allow rapid degradation through the ubiquitin-proteasome system in eukaryotes and the ClpAP protease in bacteria (Ravid and Hochstrasser 2008; Mogk et al. 2007). The initiating step, the methionine removal, is catalysed by methionyl aminopeptidases (MetAP) and, although the eukaryotic MetAP was originally thought to be cobalt-dependent, it is likely that the physiologically relevant cation for this enzyme is zinc (Bradshaw et al. 2004; Leopoldini et al. 2007). Methionine removal subsequently followed by modification of the N-terminal residue is often crucial for the function, or dysfunction, of a protein. In this context, the post-translational conversion of an N-terminal glutamate to a cyclized form (pyroglutamyl) is relatively common and is catalysed by the zinc metalloenzyme glutaminyl cyclase (QC) (Huang et al. 2005). This can modify, among other substrates, the Alzheimer's amyloid-β (Aβ) peptide rendering it more hydrophobic with a greater propensity to aggregation (Cynis et al. 2008). Small molecule inhibitors of QC may provide novel therapeutics in the treatment of AD (Schilling et al. 2008) and see below (Sect. 19.7).

19.3 Metalloproteases: Their Properties and Roles in Disease

All known metallopeptidases can be divided into 12 clans according to structural and catalytic similarities, for example in the type and number of metal ions required for their activity (Rawlings and Barrett 2004). They are also subdivided into distinct classes and in the human genome practically all classes are present. In this chapter we shall focus on zinc metallopeptidase families belonging to one of these clans, namely MA, since these are principally involved in the cell-surface or extracellular post-translational modifications of proteins and peptides. Representatives of this clan require zinc for their activity and are often referred to as "zincins." They contain histidine residues as zinc ligands in the typical zinc binding motif, His-Glu-Xaa-Xaa-His (HEXXH). They are further sub-divided according to the nature of the third zinc ligand: gluzincins use a glutamyl residue which lies 18–72 amino acids C-terminal to the HEXXH motif; metzincins use a His or Asp within an extended zinc motif HEXXHXXGXXH/D. They are termed metzincins because of a strictly conserved methionine-containing turn (the Met-turn) in their structure which is thought to provide a hydrophobic base at the zinc-binding site (Stöcker et al. 1995).

The metallopeptidase families that will be considered in particular detail are (i) the M13 family typified by neprilysin [or neutral endopeptidase (NEP)], (ii) the M2 family represented by angiotensin-converting enzyme [or peptidyl dipeptidase (ACE)], (iii) the matrix metallopeptidases (MMPs) which include collagenases and gelatinases. Finally, we shall consider the related ADAMs (A disintegrin and metalloproteinase) family involved in the shedding of surface proteins and including the ADAMS with thrombospondin motifs (ADAMTS family), which play a particular role in extracellular matrix turnover and arthritis (Jones and Riley 2005). Members of each of these families have served as potential drug targets, particularly in cardiovascular disease and cancer, but more recently have provided insight into mechanisms involved in neurodegeneration and neuroinflammation, especially from the point of view of processing of the amyloid precursor protein (APP) and its products in Alzheimer's disease.

While the metallopeptidases above utilise a single zinc ion for activity, there are examples where two zinc ions are required in a co-catalytic (binuclear) process. The first example of these to be discovered was the cytosolic enzyme, leucyl aminopeptidase (Burley et al. 1990). A key function for this enzyme is in "trimming" proteasome-produced peptides for class I antigen presentation and it is up-regulated by cytokines, e.g. γ-interferon (Beninga et al. 1998). Another example of key pathophysiological significance is the type II membrane glycoprotein, glutamate carboxypeptidase II (GCPII; NAALADase), which has two distinct roles: in the brain it inactivates the neuropeptide *N*-acetyl-aspartyl-glutamate by cleaving off the α-linked glutamate whereas in the intestine it sequentially releases γ-linked glutamates from folyl-poly-γ-glutamates (folate hydrolase activity). It is also identical with the prostate cancer marker PSMA (prostate-specific membrane antigen), which is highly expressed in aggressive prostate tumours and which might provide a target for immunological or other therapeutic approaches to the disease (Elsässer-Beile et al. 2009). The recent structural solution of a catalytically inactive mutant (E424A) of GCPII in complex with its substrate (*N*-Ac-Asp-Glu) has, for the first time, revealed the detailed reaction mechanism for this dual zinc-containing metallopeptidase (Klusák et al. 2009).

19.4 Classes of Metallopeptidases Present in the Human Genome

A brief and selected overview of the characteristics of some of the metallopeptidase families discussed herein follows before a more detailed description of individual peptidases and their roles in peptide and protein modifications in health and disease. For a full description of all classes and categories of metallopeptidases, the reader should consult (Rawlings and Barrett 2004).

19.4.1 Class M2

Family M2 contains angiotensin-converting enzyme (ACE) best known for its key role in the processing cascade of the renin-angiotensin system (RAS) and, until recently, this was the only known member of the family in humans. Arguably, ACE is better termed peptidyl-dipeptidase-A because, as a carboxydipeptidase, it has a number of important substrates and functions in addition to the processing of angiotensin, but it is generally referred to as "ACE." Relatively recently in the timescale of vertebrate evolution the ACE gene underwent an internal duplication producing the so-called somatic form of the enzyme which has two catalytic sites (the N- and C-domains) which, while similar, show some differences both in substrate specificity and inhibitor sensitivity. X-ray crystal structures for each of these domains in complex with selective inhibitors have revealed the molecular basis for this selectivity in the two peptide processing domains (Watermeyer et al. 2008). Use of an alternative initiation site produces a form of ACE with only a single catalytic site expressed exclusively in mammalian testis which is critical to male reproduction (Hagaman et al. 1998).

Vertebrates are now known to express the homologous peptidase ACE2 (Tipnis et al. 2000) whose actions seem to counterbalance those of ACE. ACE is, of course, of great importance in the regulation of blood pressure through its conversion of the decapeptide angiotensin I into the vasopressor octapeptide angiotensin II by removal of the *C*-terminal dipeptide. Potent inhibitors of ACE, such as captopril or lisinopril, are hence highly effective anti-hypertensive agents (Menard and Patchett 2001) and do not inhibit ACE2 (Tipnis et al. 2000). Figure 19.1 illustrates the similarities between ACE and ACE2 and their counter-regulatory roles in angiotensin metabolism and hence cardiovascular biology. Most notably, ACE2 comprises only a single catalytic metalloprotease extracellular domain and its transmembrane and C-terminal

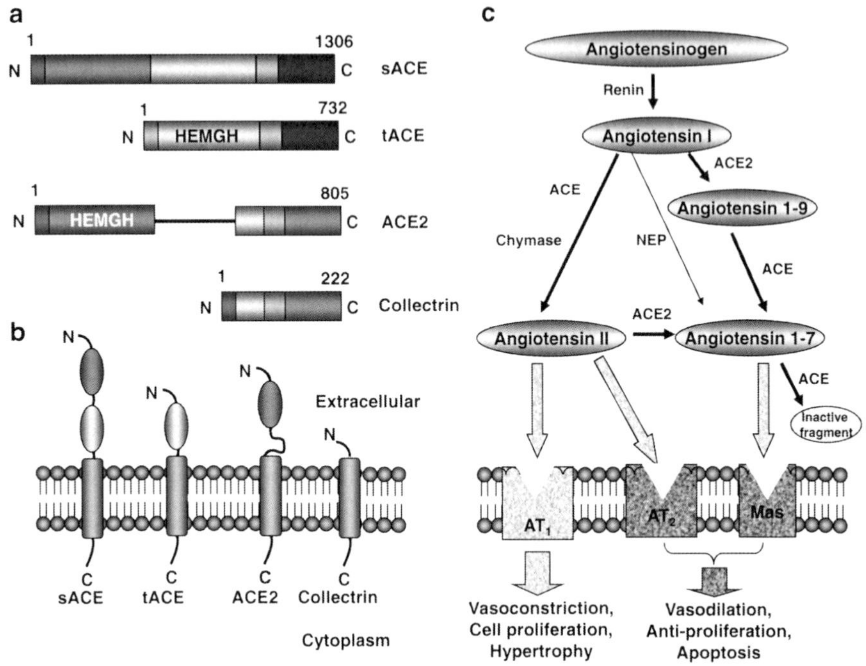

Fig. 19.1 The domain organisation (**a**) and cellular topology (**b**) of ACE isoforms and collectrin, and a schematic representation (**c**) of proteolytic processing events involved in the functioning of the renin-angiotensin system (RAS). (**a** and **b**) ACE isoforms (somatic – sACE, testicular – tACE, and ACE2) and collectrin, are type I integral membrane proteins with a signal peptide at the N-terminus and a short, C-terminal cytoplasmic domain. The numbers above the proteins represent the amino acid numbering of the sequences. tACE and ACE2 possess a single HEXXH Zn-binding motif (HEMGH in both), which forms the active site, whereas sACE possesses two of these. Collectrin lacks a catalytic site but has 47.8% sequence identity with the non-catalytic extracellular, transmembrane and cytosolic domains of ACE2 and no similarity to ACE. (**c**) shows the processing pathway in the RAS from the protein precursor angiotensinogen (452 amino acids in the human), to the biologically active angiotensin peptides (the octapeptide angiotensin II and the heptapeptide angiotensin-(1-7). ACE – angiotensin-converting enzyme, NEP – neprilysin, AT_1 – type 1 angiotensin II receptor), AT_2 – type 2 angiotensin II receptor and Mas – angiotensin (1-7) receptor. The conversion of angiotensin I to angiotensin (1-9) by ACE2 is kinetically 400 times less favourable than the conversion of angiotensin II to angiotensin (1-7) (Rice et al. 2004)

cytoplasmic domain share no similarity with ACE but with a small membrane protein, collectrin, involved in amino acid transport in the kidney (Danilczyk et al. 2006). ACE2 appears to be a dual function protein since it also is involved in amino acid transport through interactions with its collectrin-like domains but functions as a protease (a strict carboxypeptidase) through its extracellular ACE-like domain (Kowalczuk et al. 2008). ACE2 has also been subverted into acting as the cellular receptor for the severe acute respiratory syndrome (SARS) coronavirus and the associated down-regulation of surface ACE2 following infection predisposes to the acute, and potentially fatal, lung injury in the disease. Hence in a number of distinct disorders: SARS infection, cardiovascular disease, liver fibrosis and diabetes, effects all mediated through impaired angiotensin peptide metabolism, upregulation of ACE2 expression and activity can be beneficial (Turner 2008).

19.4.2 Class M10

The peptidases of family M10 represent a large family of metzincins usually occurring as secreted metalloendopeptidases, although there are some membrane-bound forms in mammals. These enzymes are principally involved in degradation of the extracellular matrix, important in tissue remodeling and repair, and hence have trivial names such as collagenases, gelatinases, etc. but are generically referred to as matrix metalloproteases, matrixins or MMPs. Excessive activation of MMPS can result in significant pathological tissue damage as seen, for example, in arthritic conditions. The matrixins are generally synthesized as inactive precursors although differences exist in the activation mechanism. The matrixin proenzymes are held in a catalytically inactive state due to the presence of a "cysteine switch" (Van Wart and Birkedal-Hansen 1990), which involves interaction between a conserved cysteine in the propeptide and the catalytic zinc ion hence preventing the binding and activation of a crucial water molecule. Because of the importance of cysteine in the latency of the proenzymes, they can commonly be activated by thiol-blocking agents such as aminophenylmercuric acetate (APMA) (see e.g. Allinson et al. 2004).

Like other metallopeptidases, the peptidases in family M10 are inhibited by chelating agents, and many potent inhibitors such as batimastat have been synthesized as possible drugs. These seemed to offer great prospects as anti-metastatic drugs in cancer chemotherapy although they have failed to live up to expectations as yet (Tu et al. 2008). The TIMP (tissue inhibitor of metalloproteases) proteins are endogenous inhibitors of these peptidases.

19.4.3 Class M12

The study of a crayfish digestive enzyme (astacin) and its mammalian homologue (meprin) (Bond and Beynon 1995) led to the subsequent discovery of one of the largest families of metallopeptidases, containing the adamalysin (ADAM) peptidases, the snake venom metallopeptidases (reprolysins) and many other related enzymes (Dumermuth et al. 1991; Seals and Courtneidge 2003). Because of their potential involvement in inflammatory and other conditions, a number of synthetic inhibitors have been described, although there is a lack of highly specific inhibitors and those available also tend to inhibit MMPs (Zhang et al. 2004; Delaet et al. 2003; Skotnicki et al. 2003). Some members of the family are inhibited by TIMP-3 (Lee et al. 2002).

The peptidases are generally membrane or secreted proteins. Two key mammalian examples are ADAM17, or tumour necrosis factor-α converting enzyme (TACE), which is important for post-translational processing of certain membrane proteins and ADAMTS13, also known as von Willebrand factor protease, the activity of which normally prevents von Willebrand disease.

19.4.4 Class M13

The prototype of this family is neprilysin, an endopeptidase (neutral endopeptidase; endopeptidase-24.11) originally discovered in renal brush border membranes by Kenny and colleagues (e.g. Kerr and Kenny 1974) and subsequently rediscovered in the brain as a key neuropeptide-degrading enzyme ("enkephalinase") (Roques et al. 1980; Matsas et al. 1983). NEP is a type II integral membrane protein composed of a short, cytoplasmic tail, a transmembrane region and a large extracellular region that folds into two domains that enclose a large, central cavity that contains the catalytic site (Oefner et al. 2000). Ten cysteine residues in the extracellular domain are conserved between all family members and are involved in intrachain disulphide bonds required for maintenance of structure and activity (Turner et al. 2001). NEP has important roles in modulating pain (through its actions on enkephalin peptides), blood pressure (through inactivation of atrial natriuretic peptide), prostate cancer (through metabolism of the mitogenic peptide, endothelin-1) and the immune system, being also identical with the common acute lymphoblastic leukaemia antigen, CALLA or CD10 (LeTarte et al. 1988). It was for some long time the only known peptidase in family M13, but additional homologues are now known, including the endothelin converting enzymes (ECE-1 and ECE-2) responsible for the post-translational processing of the protein precursors of the endothelin family of hypertensive peptides (Fig. 19.2). This processing pathway, like many protein processing events involving hormones, neuropeptides and growth factors, involves the initial removal of the N-terminal signal prepeptide to generate a physiologically inactive proprotein subsequently processed in the secretory pathway at dibasic cleavage sites to its active form by one of the seven subtilisin-like proprotein convertases, which include furin (Rholam and Fahy 2009). As intracellular, serine proteases they are not considered further in this chapter. In the endothelin pathway, it is likely that either furin or PC7 catalyses this processing event (Blais et al. 2002). However, the endothelin processing pathway is unusual in protein processing events since the product of prohormone convertase action is another physiologically inert intermediate, termed "big endothelin," which is in turn converted to the vasoactive endothelin by the metallopeptidase, ECE-1. When first identified, as the most potent, endogenous human vasoconstrictor known at the time, the endothelin system held much hope for development of novel cardiovascular therapeutics such as ECE-1 inhibitors, a promise that has yet to be fulfilled (Kirkby et al. 2008).

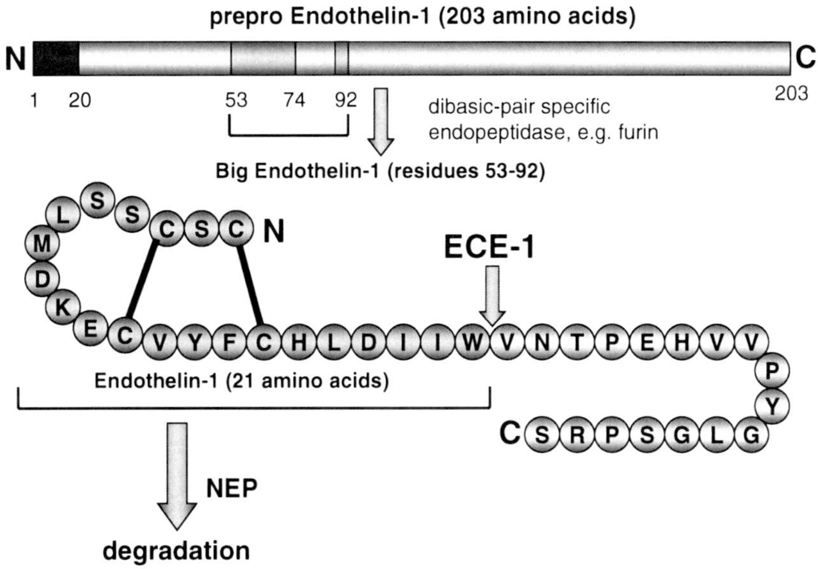

Fig. 19.2 Proteolytic processing pathway of endothelin-1. Pre-proendothelin-1 (203 amino acids) is cleaved by a dibasic endopeptidase such as furin to generate biologically inactive big endothelin-1 (39 amino acids) (Yanagisawa et al. 1988). Big ET-1 is then cleaved at a Trp-Val bond by the metalloproteinase ECE-1 to release endothelin-1 (21 amino acids). Active ET-1 contains 2 intramolecular disulphide bonds (shown as *straight lines*)

A detailed bioinformatic analysis of the phylogeny of the M13 family has recently been described (Bland et al. 2008). The human genome is now known to contain at least seven NEP-related enzymes, including ECE-1 and ECE-2, as well as a close homologue of NEP, termed NEP2 (Whyteside and Turner 2008). Several of the NEP-like enzymes are, as yet, orphan peptidases with no recognized peptide substrates. Invertebrate species contain considerably more NEP-like proteases (more than 20 genes in *Drosophila*), which are involved in events such reproduction, embryogenesis, renal function, as well as other behavioural traits (Thomas et al. 2005; Bland et al. 2009). The 3D structures of both NEP and ECE-1, each complexed with the inhibitor phosphoramidon, have been solved providing detailed mechanistic insight into the similarities and differences between these two metallopeptidases (Oefner et al. 2000; Schulz et al. 2009).

There has been much work towards the development of selective inhibitors of NEP and ECE as possible cardiovascular therapeutics and the synthetic compound thiorphan is currently the most specific and potent inhibitor of NEP (Roques et al. 1980). Endogenous physiological inhibitors of NEP have been described – sialorphin and opiorphin – which have analgesic activity of comparable potency to morphine, which is consistent with their inhibition of enkephalin inactivation by NEP (Rougeot et al. 2003; Wisner et al. 2006). In recent years, much attention has focused on a newly discovered physiological activity of both NEP and ECE, namely their ability to degrade the neurotoxic Alzheimer's amyloid peptide, Aβ (Iwata et al. 2000). Up-regulation of NEP (or ECE) may therefore provide an option for the treatment of Alzheimer's disease by stimulating clearance of Aβ – a therapeutic example of enhancing protein processing (Belyaev et al. 2009). While the homologous NEP and ECE serve complementary roles in Aβ metabolism, they play opposing roles in the progression of androgen-insensitive prostate cancer through their respective effects on endothelin maturation (ECE-1) and inactivation (NEP) (Fig. 19.3). Hence, upregulation of prostatic NEP or downregulation of ECE-1 could provide therapeutic options for this untreatable and aggressive stage of prostate cancer (Papandreou et al. 1998; Dawson et al. 2004).

19.4.5 Class M16

Peptidase family M16 (the pitrilysin family) contains metalloendopeptidases with varied specificity and physiological roles. These range from the mitochondrial processing peptidase (MPP) to the insulin-degrading enzyme (insulysin; IDE). MPP removes an N-terminal targeting signal from mitochondrial precursor proteins hence targeting nuclear-encoded proteins to their correct location within mitochondria. IDE plays a role in insulin metabolism but also may be another physiologically significant amyloid (Aβ)-degrading enzyme (Edland 2004). This family are characterized by an inverted zinc-binding domain (HXXEH rather than HEXXH) although the reaction mechanism is similar, namely nucleophilic attack of the carbonyl carbon of the peptide bond by a water molecule activated by the zinc and polarized by an active site glutamate residue. The lack of potent or specific inhibitors for peptidases of this family has significantly hindered studies of their physiological functions, particularly in the case of IDE. However, mice with targeted deletion of the IDE gene show hyperinsulinaemia and glucose intolerance as well as increased accumulation of Aβ in the brain supporting the proposed protein processing roles for the enzyme (Farris et al. 2003).

19.4.6 Class M24

Peptidase family M24 contains exopeptidases that were thought to require co-catalytic ions of cobalt or manganese. They are either aminopeptidases or dipeptidyl-peptidases. The methionyl aminopeptidases (see above) fall into this class, although it is likely that they use zinc physiologically as their cations as does the structurally related X-Pro aminopeptidase (aminopeptidase P) (Cottrell et al. 2000). Fumagillin-related angiogenesis inhibitors of type II methionyl aminopeptidase are of interest to the pharmaceutical industry for the treatment of solid tumours. Related processing enzymes include the eukaryotic X-Pro dipeptidase which plays a role in collagen recycling. Deficiency results in an increase of these proline dipeptides to toxic levels (Myara et al. 1994) (Table 19.2).

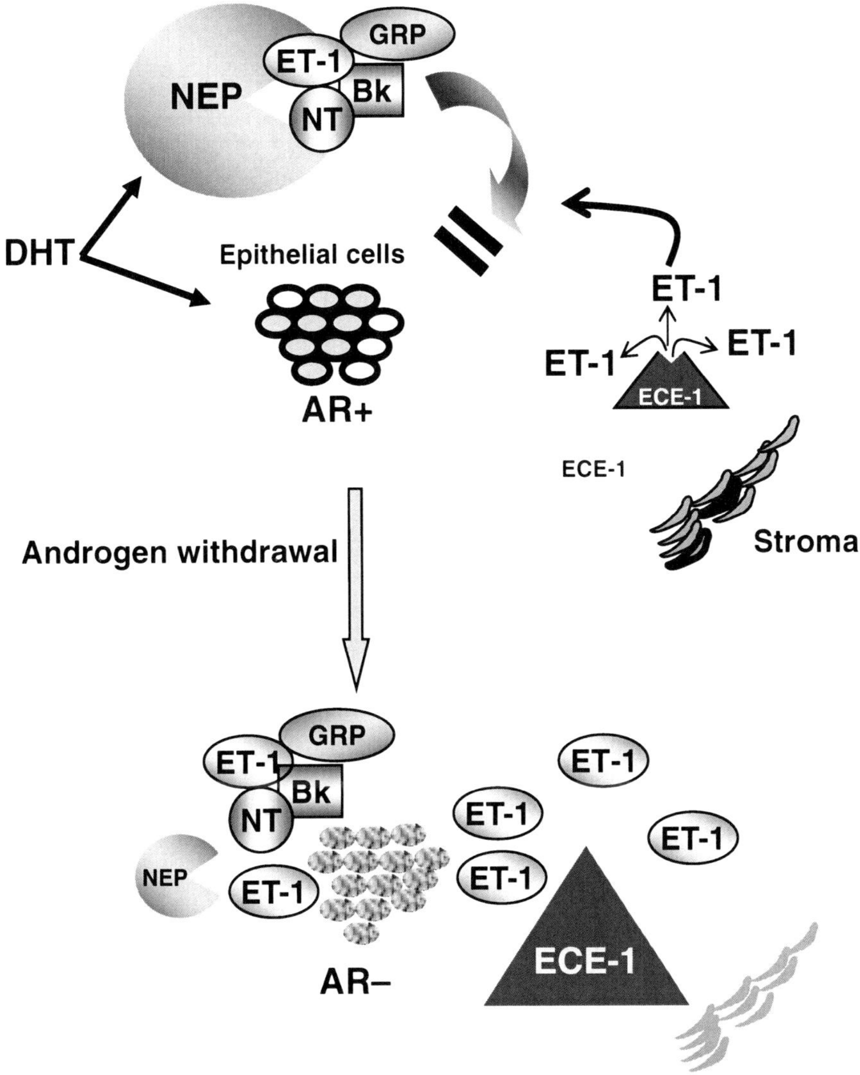

Fig. 19.3 Reciprocal roles of the metallopeptidases neprilysin (NEP) and endothelin-converting enzyme (ECE-1) in the development of androgen-independent prostate cancer. In androgen-dependent prostate cells, epithelial cells express high levels of NEP. NEP levels are up-regulated by androgens which also stimulate epithelial growth. Although epithelial cells produce the mitogenic peptide ET1 it is inactivated by NEP and does not stimulate cell growth. The stromal cells produce low levels of ECE-1 converting big ET-1 into ET-1 and this action is again opposed by the action of NEP. When androgen levels are low as in typical therapeutic surgical or drug regimes, NEP levels decrease allowing mitogenic peptides produced by epithelial cells to exhibit their growth-promoting effect. Also, under these conditions, the stromal cells express higher levels of ECE-1 producing even more ET-1 and encouraging cell invasion. ECE-1 levels then increase in epithelial cells which become metastatic and are no longer influenced by stroma. Hence, inhibition of ECE-1 activity in the stroma (or up-regulating the activity of NEP in epithelial cells) could provide a novel therapeutic strategy in prostate cancer by manipulating metallopeptidase levels. DHT, dihydrotestosterone; ET-1, NT, Bk, GRP, the mitogenic peptides endothelin-1, neurotensin, bradykinin and gastrin-releasing peptide, respectively. AR+/–, in the presence/absence of androgen receptor

19.4.7 Class M48

These enzymes are mainly metalloendopeptidases. Eukaryotic peptidases from family M48 have a requirement for substrates that are prenylated at a C-terminal motif known as a CAAX box, in which A is an aliphatic residue, and the lipid is attached to the cysteine residue. Protein prenylation enhances membrane association of proteins. Human farnesylated-protein converting enzyme 1 may be important for processing prelamin A, a farnesylated protein that is cleaved twice. In cells with a defective peptidase gene, incompletely processed prelamin A accumulates in the nuclear envelope. Farnesylated-protein converting enzyme 1 performs the second processing step for prelamin A, releasing a 15-residue peptide containing the prenylated C-terminal Cys (Pendas et al. 2002). The Oma1 endopeptidase is believed to degrade misfolded membrane proteins, thereby helping to maintain the integrity of the mitochondrial inner membrane (Kaser et al. 2003). Deficiencies in lamin A, a component of the nuclear lamina, lead to abnormalities in the nuclear architecture, and knockouts of farnesylated-protein converting enzyme 1 in mice cause growth retardation, muscular dystrophy and premature death (Bergo et al. 2002).

19.4.8 Class M50

This family is also known as the site 2 protease (S2P) family and is represented by S2P peptidase (human), sporulation factor SpoIVFB and RseP peptidase (*Escherichia coli*). They are all integral membrane proteins, and their extensive hydrophobicity, even within the sequences containing the active site residues, indicates that they cleave their substrates either within the membrane or very close to it. These peptidases are involved in the regulation of gene expression by the two-step proteolysis of transcriptional regulators. For example, following an initial cleavage in the lumen of the Golgi by the serine protease, (site-1 protease, S1P) (Bartz et al. 2008), the zinc metallopeptidase S2P releases the N-terminal transcription factor domain, sterol regulatory element binding protein (SREBP), from its membrane-bound precursor (Fig. 19.4). The released domain then enters the nucleus and activates genes that control cellular levels of cholesterol and phospholipids (Zelenski et al. 1999).

Bacterial RseP peptidase also cleaves transmembrane sequences and it now seems very likely that intramembrane proteolysis is a general activity of peptidases in family M50 related to cell regulation and signalling. The structure of a bacterial S2P-like protease has revealed elements of this unusual catalytic mechanism in which zinc and water-mediated peptide bond hydrolysis occurs in the hydrophobic environment of the lipid bilayer (Feng et al. 2007). Intramembrane proteolysis is involved in such diverse physiological and pathological events as bacterial sporulation, hepatitis C viral infection and Alzheimer's disease (Weihofen and Martoglio 2003). In the latter case, it is an aspartic protease, γ-secretase, that catalyses the intramembrane proteolytic processing event that is the final step in production of the neurotoxic amyloid Aβ peptide (see below).

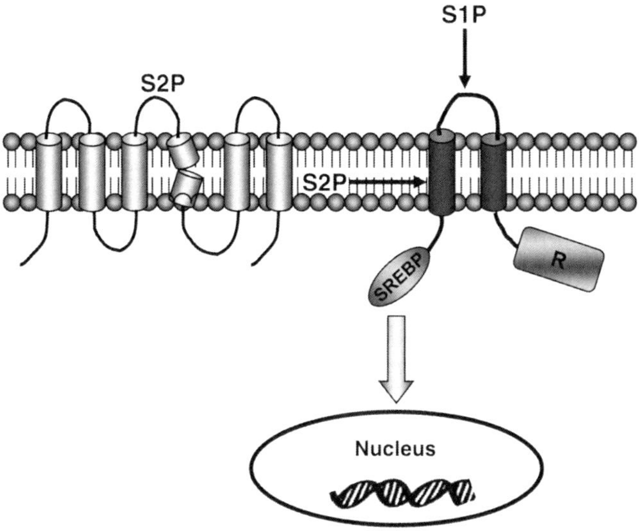

Fig. 19.4 Schematic presentation of two-step proteolytic processing of *s*terol *r*egulatory *e*lement *b*inding *p*rotein (SREBP) precursor protein – an example of intramembrane proteolysis. First, the SREBP precursor protein is cleaved within the luminal loop between two transmembrane regions by a membrane-anchored serine protease termed S1P (the site-1 protease). Then, the site-2 protease (S2P) cleaves the processed protein within the transmembrane domain but close to the cytoplasmic membrane surface, releasing the SREBP transcription factor. S2P contains the conserved HEXXH motif common for metalloproteases. The regulatory cytoplasmic domain (R) of the SREBP precursor protein interacts with a cholesterol-sensor protein. The processing events are activated only if the cholesterol level is low (adapted from Wolfe 2009)

19.5 Membrane Protein Shedding: A Primary Role for ADAMs Proteins

The limited proteolysis of precursor proteins, such as prohormones and zymogens, has been understood for decades. However, the selective proteolysis of cell-surface proteins by proteases, commonly referred to as secretases or sheddases, is a more recent discovery, which is providing insight into a number of pathological conditions (see e.g. van Goor et al. 2009 for recent review). We first observed that the metalloproteinase ACE, which itself is an ectoenzyme at the endothelial cell-surface, could be released or shed from the membrane through the action of another distinct integral membrane proteinase, inhibitable by EDTA, and hence presumably a metalloenzyme (Hooper et al. 1987). Hence, the soluble form of ACE found in plasma and used diagnostically to assess the effectiveness of ACE inhibitor therapy is derived as a result of a highly specific post-translational processing of the membrane-bound form of ACE. The physiological function of soluble ACE, if other than a down-regulation mechanism, is unknown although the soluble ectodomain could conceivably act as a ligand for another protein. ACE secretase, as we subsequently named the proteolytic activity, has turned out to be the prototype for a family of such membrane protein secretases or sheddases, which cause the release of specific membrane proteins, normally from the cell-surface into the extracellular medium (Hooper et al. 1997; Huovila et al. 2005).

A wide variety of cell-surface proteins are cleaved by membrane protein secretases, like ACE secretase, and these function in processes as diverse as cell fusion, adhesion and cell signalling. The substrates include some receptors, receptor ligands, cell adhesion molecules and ectoenzymes. The cleavage normally occurs at a single, specific site in the extracellular domain close to the membrane surface, in the case of ACE at Arg^{638}-Ser^{639} (Alfalah et al. 2001). In many cases, these shedding processes are inhibited by certain hydroxamate compounds originally developed as MMP inhibitors (e.g. batimastat). Principally through studies on one of these secretases, the tumour necrosis factor-α converting enzyme (TACE, ADAM17), which causes the post-translational release of active TNF-α into the blood from its membrane-bound precursor, pro-TNF, these secretases have mainly been identified as members of the ADAM gene family: multi-domain, membrane proteins that generally (but not always) have both metalloproteinase and adhesion activities (Seals and Courtneidge 2003).

ADAM17 is one of 23 genes encoding ADAMS proteins in humans. Ironically, given it was the first protein shedding event to be described at a molecular level, the precise identity of the ACE secretase is still unknown. The ADAMS are complex proteins featuring a pro-domain with cysteine switch like MMPs, a zinc metalloprotease domain which is inactive in some family members, a disintegrin domain involved in adhesion processes, a cysteine-rich region, an EGF-like sequence, a transmembrane region and a cytoplasmic tail. The structures of the catalytic domains of ADAM17 and ADAM33 have been solved and show many similarities, typical aspects being a central five-stranded β-sheet surrounded by five α-helices including the Met-turn, a feature of all metzincin proteases (Maskos et al. 1998; Orth et al. 2004).

A sub-group of the ADAMs comprising 19 members is the ADAMTS family in which the transmembrane domain of ADAMs proteins is replaced by one or more thrombospondin (TS) repeats and hence they act as soluble, secreted proteases. They also lack the EGF sequence of ADAMs. They were originally identified as the long sought after proteases involved in processing of pro-collagen (procollagen N-endopeptidase, ADAMTS2) and of aggrecan (ADAMTS4). Hence, members of this proteinase family are potential therapeutic targets in inflammatory and arthritic conditions (Fosang and Little 2008). Genetic deficiency of, or autoimmunity against, one member of this family (ADAMTS13) leads to the severe haematological disorder, thrombotic thrombocytopenic purpura, through its failure to process the platelet protein, von Willebrand factor (Raife and Montgomery 2000) (Table 19.2).

In addition to the shedding of the pro-inflammatory cytokine TNFα from its membrane precursor by ADAM17, many of the ligands for the EGF receptor family involved in cellular proliferation and survival are derived from membrane-bound precursors through ADAM-mediated shedding. Examples include heparin-binding EGF (HB-EGF), TGFα and neuregulin. ADAM17 seems to be important (but not exclusive) to the shedding of these ligands and itself can be activated through activation of G protein-linked receptors by, for example, thrombin, endothelin-1 and angiotensin II. This process of indirect, receptor-mediated activation of the EGF receptor through ADAM-mediated protein processing has been termed transactivation and has particular implications in vascular remodelling and in cancer. In some cases, MMPs rather than ADAMs may contribute to shedding events. For example, MMP-9 may mediate the shedding of the β2 integrin subunit (CD18) from macrophages, N-cadherin and also the soluble CD40 ligand involved in platelet activation (Vaisar et al. 2009; Dwivedi et al. 2009; Menchén et al. 2009).

Protein shedding plays a part in the processing of two key neuronal proteins involved in neurodegeneration: the prion protein and the Alzheimer's amyloid precursor protein (APP) and the latter system serves as a finale to this discussion of protein processing events combining elements of both ADAM-mediated shedding, intramembrane proteolysis and dysregulation of cell signalling. The cellular prion protein, which is essential for the development of a prion disease occurs both as a glycolipid-anchored membrane form and as a soluble, secreted form (Parkin et al. 2004). Both ADAM9 and ADAM10, but not ADAM17, contribute to prion shedding and, intriguingly, the ADAM9-mediated shedding is an indirect process acting via ADAM10, which cleaves the prion protein only three residues away from the site of addition of the lipid anchor (Gly^{228}-Arg^{229}) (Taylor et al. 2009). However, this shedding process does not appear to regulate the conversion of the prion protein to its transmissible neurotoxic form. Not surprisingly, in the light of its close involvement in the development of dementia, the complex processing of APP has been studied far more intensively and has led to a number of possible therapeutic avenues involving manipulation of protease activity or expression for treatment of the disease, some of which are currently in clinical trials.

19.6 APP Processing and Alzheimer's Disease

The sequential proteolysis of the large, transmembrane, amyloid precursor protein (APP) by the two aspartic proteases, β- and γ-secretase, releases a soluble APPβ (sAPPβ) ectodomain and generates from the C-terminal membrane fragment the 39-42 amino acid amyloid-β peptide (Aβ) together with the transcriptional regulator, APP intracellular domain (AICD) (Fig. 19.5) (Wolfe 2007). It was the Aβ peptide which was originally identified as the main constituent of the extracellular neuritic plaques which characterize AD (Glenner and Wong 1984; Masters et al. 1985) and led to the formulation of the "amyloid cascade" hypothesis of Alzheimer's disease (Hardy and Higgins 1992). Accumulation of Aβ is a characteristic feature of AD and its prevention is a primary target in therapeutic strategies. Late-onset forms of AD may primarily be due to deficiencies in Aβ clearance rather than its formation (Hama and Saido 2005) and hence understanding clearance mechanisms and their regulation could be of fundamental importance and provide novel approaches to AD treatment. The exact role of the C-terminal proteolytic product

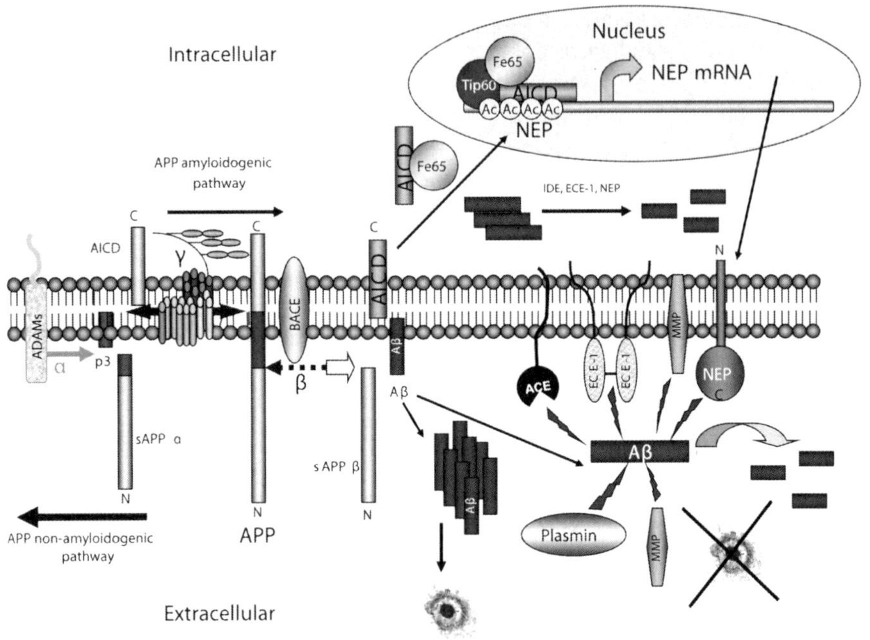

Fig. 19.5 Schematic representation of the proteolytic processing of the Alzheimer's amyloid precursor protein (APP), regulation by AICD of *NEP* gene expression and degradation of Aβ peptide by a number intracellular, membrane-bound and extracellular soluble metallopeptidases. APP is a constitutive, transmembrane protein which undergoes proteolytic cleavage by β- and γ-secretases producing a soluble ectodomain referred to as sAPPβ, the neurotoxic amyloid β peptide (Aβ) and the APP intracellular domain (AICD). Aβ has the property to form neurotoxic oligomers and then to aggregate further, ultimately forming the so-called senile plaques, which are one of the pathological hallmarks of Alzheimer's disease. This APP processing pathway is considered amyloidogenic in contrast with the cleavage of APP by the metalloproteinase α-secretase (ADAM protein) within the Aβ domain, which prevents formation of Aβ. This non-amyloidogenic processing leads to formation of the soluble ectodomain sAPPα, a short peptide p3 and AICD. AICD binds to the neprilysin (*NEP*) promoter and activates neprilysin expression (NEP) (Belyaev et al. 2009). This binding requires AICD stabilisation by Fe65 and also involves the histone-acetyl transferase, Tip60. AICD also controls expression of some other genes. If the *NEP* gene is silenced through histone deacetylation by HDACs, expression of NEP is inhibited. However, incubation of cells with VA, and hence inhibition of HDACs, allows AICD to bind to the *NEP* promoter and re-activate NEP expression (Belyaev et al. 2009). NEP, a cell-surface ectopeptidase, in turn, degrades Aβ and reverses amyloid plaque accumulation. Aβ can also be cleaved by ECE-1, ACE, matrix metallopeptidases (MMPs) and by the serine peptidase, plasmin. Cleavage of Aβ can occur within the cells, at the cell surface (both along the neurites and cell bodies) and in the extracellular space. NEP and plasmin between them can cleave monomeric and oligomeric forms of Aβ (Nalivaeva et al. 2008)

AICD remains unclear and controversial but it probably acts as a transcriptional regulator together with the APP-tail binding protein Fe65 and the histone acetyltransferase (HAT), Tip60 (Cao and Südhof 2001; Belyaev et al. 2009) (Fig. 19.5).

It is well established that APP is cleaved by selective proteases (α-, β-, and γ-secretase) into several fragments some of them possessing physiological properties

(Fig. 19.5). Thus, the action of α- or β-secretases produces the largest products of APP – the soluble ectodomain proteins $sAPP_{\alpha}$ and $sAPP_{\beta}$, which play an important role in proliferation of neuronal cells and development of the nervous system (Caille et al. 2004; Conti and Cattaneo 2005; Taylor et al. 2008). Metabolism of APP can undergo either an "amyloidogenic" pathway through the action of β- and γ-secretases and production of Aβ, or a non-amyloidogenic pathway, when APP is cleaved by α-secretase (Hooper and Turner 2002). α-Secretase cleaves APP inside of the Aβ sequence between lysine and leucine in positions 16 and 17, and prevents formation of the Aβ peptide. Hence, inhibition of the amyloidogenic processing of APP and activation of the activity of α-secretase are possible approaches for prevention of AD (Lichtenthaler and Haass 2004).

The main candidates as α-secretases are ADAM10 and ADAM17; ADAM9 may also participate but in an indirect manner via ADAM10 (Allinson et al. 2004; Hooper and Turner 2002). The constitutive process of secretion of sAPPα under the action of α-secretase occurs at the cell surface (Parvathy et al. 1999). However, the regulated α-secretase activity, stimulated by cholinergic agonists such as muscarine and carbachol (Canet-Aviles et al. 2002) or phorbol esters (Zhu et al. 2001), is reported to be localised mainly intracellularly (Jolly-Tornetta and Wolf 2000). In transgenic mice expressing human APP moderate neuronal overexpression of ADAM10 was shown to reduce the formation of Aβ and prevent its deposits in the plaques. In contrast, expression of a mutant inactive ADAM10 led to an increase in the number and size of amyloid plaques in the brain of the same transgenic mice (Postina et al. 2004). Transgenic mice lacking ADAM10 or ADAM17 genes are lethal, suggesting that α-secretase plays an important role in the viability of the organism (Killar et al. 1999), although these enzymes do have other important developmental and regulatory roles.

Formation of amyloidogenic Aβ requires the activity of the transmembrane aspartic proteases, β- and γ-secretases (Fig. 19.5). β-secretase is also known as BACE-1 and its biochemistry and cell biology are described in detail (Hunt and Turner 2009). Therapeutically, BACE1 provides an attractive drug target for AD since BACE knockout mice do not form Aβ and appear to be healthy (Luo et al. 2001). Numerous approaches have been undertaken to find a clinically effective inhibitor of human β-secretase activity, mostly in the field of peptidomimetic, noncleavable substrate analogues (Hunt and Turner 2009). However, new endogenous protein substrates are continually being discovered for BACE-1, e.g. it cleaves membrane proteins APLP-1 and -2, P-selectin glycoprotein ligand-1 (PSGL-1), membrane-bound sialyltransferase and LDL-receptor-related protein (LRP), and so caution is needed since BACE may have unexpected physiological roles. Most recently, BACE-1 is known to be required for proper myelination to occur through its processing of neuregulin (Willem et al. 2006). The cytosolic fragment of APP formed by BACE (C99) is then a substrate for γ-secretase (Fig. 19.5). The latter represents a complex of several proteins involving presenilin proteins as the catalytic core. They are intramembrane proteases like the S2P protein but representing a different catalytic class of proteinase (aspartic rather than metallo). The signal peptide peptidases also represent aspartic intramembrane proteases. There are also examples of serine-type intra-membrane proteases, e.g. rhomboid proteases (Wolfe 2009).

19.7 Amyloid-Degrading Metallopeptidases

Until recently, accumulation of Aβ in the brain and other tissues was thought to be an irreversible process, leading to development of AD. However, in the brain there are biochemical pathways for elimination of Aβ (Nalivaeva et al. 2008). In the last few years, several proteases, such as the metalloenzymes NEP, IDE, ECE-1 and ECE-2, as well as the serine proteinase plasmin have been found to be capable of degrading Aβ in vitro and in vivo (Leissring et al. 2003; Turner et al. 2004). Most of these act extracellularly either as membrane-bound (ECE-1, NEP) or as soluble extracellular proteins (IDE, plasmin). IDE is anomalous in terms of its subcellular distribution since it is largely cytosolic (Akiyama et al. 1988) but also occurs within some organelles (mitochondria, peroxisomes) as well as in a soluble, secreted form yet it lacks any secretory signal in its sequence (Zhao et al. 2009). The mechanism of its secretion remains an enigma and it has been termed an "unconventionally secreted protein" (Zhao et al. 2009). Figure 19.5 shows schematically Aβ peptide aggregation and its proteolytic degradation by these various enzymes. Pathological down-regulation of these enzymes could predispose to accumulation of Aβ and the development of AD. Thus, a possible therapeutic approach for treatment of AD might be a chronic up-regulation of one or more of these proteases, either pharmacologically or through a gene therapy approach. These proteases are localised in a variety of subcellular compartments and cell types in the brain and remove Aβ prior to its accumulation.

While intra-membrane proteolysis is now a well established mechanism of gene regulation through the action of "intracellular domains" in for example the notch signalling pathway, such a role for the ICD released from the APP tail (AICD) through the action of the presenilin/γ-secretase complex has proved controversial. This has particularly been the case in relation to the potential feedback regulation of the Aβ-degrading enzyme, NEP (Pardossi-Piquard et al. 2005; Chen and Selkoe 2007). However, chromatin immunoprecipitation studies have established that NEP expression is directly regulated through acetylation of core histones on the NEP promoter mediated via AICD and the inhibition of histone deacetylation with histone deacetylase inhibitors such as the anti-convulsant drug, valproate, or trichostatin A can potentiate NEP expression and activity (Belyaev et al. 2009). Valproate treatment in vivo in mice can reduce amyloid plaque formation (Qing et al. 2008) (Fig. 19.5). These observations add to the growing body of literature (Saha and Pahan 2006; Fischer et al. 2007) emphasizing the potential beneficial effects of HDAC inhibitors in the treatment of AD and other neurodegenerative disorders, in this case through manipulation of metalloprotease expression.

19.8 Concluding Remarks

This chapter has reviewed some of the diverse roles of metalloproteases in the cell biology and physiology of protein processing and has described aspects of their mechanisms and regulation. It is clear that metalloproteases represent important therapeutic targets not only in cardiovascular disease (ACE inhibitors being the prime example) but also with potential in inflammatory diseases, neurodegeneration and cancer. While potential targets are being explored, there are numerous "orphan" metallopeptidases whose substrates and functions are currently unknown that may themselves provide novel therapeutic opportunities. Understanding their intricate interrelationships is still a major goal of modern molecular and cell biology.

Acknowledgements We thank the Medical Research Council of Great Britain and Russian Academy of Sciences Programme "Fundamental Sciences to Medicine" for financial support.

References

Akiyama, H., Shii, K., Yokono, K., et al. (1988). Cellular localization of insulin-degrading enzyme in rat liver using monoclonal antibodies specific for this enzyme. *Biochem. Biophys. Res. Commun.* 155:914–922.

Alfalah, M., Parkin, E.T., Jacob, R., et al. (2001). A point mutation in the juxtamembrane stalk of human angiotensin I-converting enzyme invokes the action of a distinct secretase. *J. Biol. Chem.* 276: 21105–21109.

Allinson, T.M., Parkin, E.T., Condon, T.P., et al. (2004). The role of ADAM10 and ADAM17 in the ectodomain shedding of angiotensin converting enzyme and the amyloid precursor protein. *Eur. J. Biochem.* 271:2539–2547.

Bachmair, A., Finley, D., Varshavsky, A. (1986). In vivo half-life of a protein is a function of its amino-terminal residue. *Science* 234:179–186.

Bartz, R., Sun, L.P., Bisel, B., et al. (2008). Spatial separation of Golgi and ER during mitosis protects SREBP from unregulated activation. *EMBO J.* 27:948–955.

Belyaev, N.D., Nalivaeva, N.N., Makova, N.Z., et al. (2009). Neprilysin gene expression requires binding of the amyloid precursor protein intracellular domain to its promoter: implications for Alzheimer disease. *EMBO Rep.* 10:94–100.

Beninga, J., Rock, K.L., Goldberg, A.L. (1998). Interferon-gamma can stimulate post-proteasomal trimming of the N terminus of an antigenic peptide by inducing leucine aminopeptidase. *J. Biol. Chem.* 273:18734–18742.

Bergo, M.O., Gavino, B., Ross, J., et al. (2002). Zmpste24 deficiency in mice causes spontaneous bone fractures, muscle weakness, and a prelamin A processing defect. *Proc. Natl. Acad. Sci. USA.* 99:13049–13054.

Blais, V., Fugère, M., Denault, J.B., et al. (2002). Processing of proendothelin-1 by members of the subtilisin-like pro-protein convertase family. *FEBS Lett.* 524:43–48.

Bland, N.D., Pinney, J.W., Thomas, J.E., et al. (2008). Bioinformatic analysis of the neprilysin (M13) family of peptidases reveals complex evolutionary and functional relationships. *BMC Evol. Biol.* 8:16.

Bland, N.D., Robinson, P., Thomas, J.E., et al. (2009). Locomotor and geotactic behavior of Drosophila melanogaster over-expressing neprilysin 2. *Peptides* 30:571–574.

Bond, J.S., and Beynon, R.J. (1995). The astacin family of metalloendopeptidases. *Protein Sci.* 4:1247–1261.

Bradshaw, R.A., Yi, L., Walker, K.W. (2004). Methionyl aminopeptidase type 1. In: Barrett, A.J., Rawlings, N.D., Woessner, J.F. (eds) Handbook of Proteolytic Enzymes, 2 edn. Elsevier, London, pp. 911–917.

Bunnett, N.W. (2006). Protease-activated receptors: how proteases signal to cells to cause inflammation and pain. *Semin. Thromb. Hemost.* 32(Suppl 1):39–48.

Burley, S.K., David, P.R., Taylor, A., et al. (1990). Molecular structure of leucine aminopeptidase at 2.7-Å resolution. *Proc. Natl. Acad. Sci. USA.* 87:6878–6882.

Caille, I., Allinquant, B., Dupont, E., et al. (2004). Soluble form of amyloid precursor protein regulates proliferation of progenitors in the adult subventricular zone. *Development* 131:2173–2181.

Cal, S., Quesada, V., Garabaya, C., et al. (2003). Polyserase-I, a human polyprotease with the ability to generate independent serine protease domains from a single translation product. *Proc. Natl. Acad. Sci. USA.* 100:9185–9190.

Canet-Aviles, R.M., Anderton, M., Hooper, N.M., et al. (2002). Muscarine enhances soluble amyloid precursor protein secretion in human neuroblastoma SH-SY5Y by a pathway dependent on protein kinase C(α), src-tyrosine kinase and extracellular signal-regulated kinase but not phospholipase C. *Brain Res. Mol. Brain Res.* 102:62–72.

Cao, X., and Südhof, T.C. (2001). A transcriptionally active complex of APP with Fe65 and histone acetyltransferase Tip60. *Science* 293:115–120.

Chen, A.C., and Selkoe, D.J. (2007). Response to: Pardossi-Piquard et al., "Presenilin-dependent transcriptional control of the Aβ-degrading enzyme neprilysin by intracellular domains of βAPP and APLP." *Neuron* 53:479–483.

Conti, L., and Cattaneo, E. (2005). Controlling neural stem cell division within the adult subventricular zone: an APPealing job. *Trends Neurosci.* 28:57–59.

Cottrell, G.S., Hyde, R.J., Lim, J., et al. (2000). Identification of critical residues in the active site of porcine membrane-bound aminopeptidase P. *Biochemistry* 39:15129–15135.

Cynis, H., Scheel, E., Saido, T.C., et al. (2008). Amyloidogenic processing of amyloid precursor protein: evidence of a pivotal role of glutaminyl cyclase in generation of pyroglutamate-modified amyloid-β. *Biochemistry* 47:7405–7413.

Danilczyk, U., Sarao, R., Remy, C., et al. (2006). Essential role for collectrin in renal amino acid transport. *Nature* 444:1088–1091.

Delaet, N.G., Robinson, L.A., Wilson, D.M. (2003). Novel inhibitors of procollagen C-terminal proteinase. Part 1: Diamino acid hydroxamates. *Bioorg. Med. Chem. Lett.* 13:2101–2104.

Dawson, L.A., Maitland, N.J., Turner, A.J., et al. (2004). Stromal-epithelial interactions influence prostate cancer cell invasion by altering the balance of metallopeptidase expression. *Br. J. Cancer* 90:1577–1582.

Dumermuth, E., Sterchi, E.E., Jiang, W.P., et al. (1991). The astacin family of metalloendopeptidases. *J. Biol. Chem.* 266:21381–21385.

Dwivedi, A., Slater, S.C., George, S.J. (2009). MMP-9 and -12 cause N-cadherin shedding and thereby beta-catenin signalling and vascular smooth muscle cell proliferation. *Cardiovasc. Res.* 81:178–186.

Edland, S.D. (2004). Insulin-degrading enzyme, apolipoprotein ε, and Alzheimer's disease. *J. Mol. Neurosci.* 23:213–217.

Elsässer-Beile, U., Bühler, P., Wolf, P. (2009). Targeted therapies for prostate cancer against the prostate specific membrane antigen. *Curr. Drug Targets* 10:118–125.

Farris, W., Mansourian, S., Chang, Y., et al. (2003). Insulin-degrading enzyme regulates the levels of insulin, amyloid β-protein, and the β-amyloid precursor protein intracellular domain in vivo. *Proc. Natl. Acad. Sci. USA.* 100:4162–4167.

Feng, L., Yan, H., Wu, Z., et al. (2007). Structure of a site-2 protease family intramembrane metalloprotease. *Science* 318:1608–1612.

Fischer, A., Sananbenesi, F., Wang, X., et al. (2007). Recovery of learning and memory is associated with chromatin remodelling. *Nature* 447:178–182

Fosang, A.J., and Little, C.B. (2008). Drug insight: aggrecanases as therapeutic targets for osteoarthritis. *Nat. Clin. Pract. Rheumatol.* 4:420–427.

Frottin, F., Martinez, A., Peynot, P. (2006). The proteomics of N-terminal methionine cleavage. *Mol. Cell Proteomics* 5:2336–2349.

Glenner, G.G., and Wong, C.W. (1984). Alzheimer's Disease: Initial report of the purification and characterization of a novel cerebrovascular amyloid protein. *Biochem. Biophys. Res. Commun.* 120:885–890.

Hagaman, J.R., Moyer, J.S., Bachman, E.S., et al. (1998). Angiotensin-converting enzyme and male fertility. *Proc. Natl. Acad. Sci. USA.* 95:2552–2557.

Hama, E., and Saido, T.C. (2005). Etiology of sporadic Alzheimer's disease: somatostatin, neprilysin, and amyloid beta peptide. *Med. Hypotheses* 65:498–500

Hardy, J.A., and Higgins, G.A. (1992). Alzheimer's disease: the amyloid cascade hypothesis. *Science* 256: 184–185.

Hooper, N.M., and Turner, A.J. (2002). The search for α-secretase and its potential as a therapeutic approach to Alzheimer's disease. *Curr. Med. Chem.* 9:1107–1119.

Hooper, N.M., Karran, E.H., Turner, A.J. (1997). Membrane protein secretases. *Biochem J.* 321:265–279.

Hooper, N.M., Keen, J., Pappin, D.J., et al. (1987). Pig kidney angiotensin converting enzyme. Purification and characterization of amphipathic and hydrophilic forms of the enzyme establishes C-terminal anchorage to the plasma membrane. *Biochem. J.* 247:85–93.

Huang, K.F., Liu, Y.L., Wang, A.H. (2005). Cloning, expression, characterization, and crystallization of a glutaminyl cyclase from human bone marrow: a single zinc metalloenzyme. *Protein Expr. Purif.* 43:65–72.

Hunt, C.E., and Turner, A.J. (2009). Cell biology, regulation and inhibition of β-secretase (BACE-1). *FEBS J.* 276:1845–1859.

Huovila, A.P., Turner, A.J., Pelto-Huikko, M. (2005). Shedding light on ADAM metalloproteinases. *Trends Biochem. Sci.* 30:413–422.

Iwata, N., Tsubuki, S., Takaki, Y., et al. (2000). Identification of the major Aβ1-42-degrading catabolic pathway in brain parenchyma: suppression leads to biochemical and pathological deposition. *Nature Med.* 6:143–150.

Jolly-Tornetta, C., and Wolf, B.A. (2000). Protein kinase C regulation of intracellular and cell surface amyloid precursor protein (APP) cleavage in CHO695 cells. *Biochemistry* 39:15282–15290.

Jones, G.C., and Riley, G.P. (2005). ADAMTS proteinases: a multi-domain, multi-functional family with roles in extracellular matrix turnover and arthritis. *Arthritis Res. Ther.* 7:160–169.

Kaser, M., Kambacheld, M., Kisters-Woike, B., et al. (2003). Oma1, a novel membrane-bound metallopeptidase in mitochondria with activities overlapping with the m-AAA protease. *J. Biol. Chem.* 278:46414–46423.

Kerr, M., and Kenny, A.J. (1974). The purification and specificity of a neutral endopeptidase from rabbit kidney brush border. *Biochem. J.* 137:477–488.

Killar, L., White, J., Black, R. (1999). Adamalysins. A family of metzincins including TNF-alpha converting enzyme (TACE). *Ann. N Y Acad. Sci.* 878:442–452.

Kirkby, N.S., Hadoke, P.W., Bagnall, A.J., et al. (2008). The endothelin system as a therapeutic target in cardiovascular disease: great expectations or bleak house? *Br. J. Pharmacol.* 153:1105–1119.

Klusák, V., Barinka, C., Plechanovová, A., et al. (2009). Reaction mechanism of glutamate carboxypeptidase II revealed by mutagenesis, X-ray crystallography, and computational methods. *Biochemistry* 48:4126–4138.

Kowalczuk, S., Bröer, A., Tietze, N. (2008). A protein complex in the brush-border membrane explains a Hartnup disorder allele. *FASEB J.* 22:2880–2887.

Lee, M.H., Verma, V., Maskos, K., et al. (2002). The C-terminal domains of TACE weaken the inhibitory action of N-TIMP-3. *FEBS Lett.* 520:102–106.

Leissring, M.A., Lu, A., Condron, M.M., et al. (2003). Kinetics of amyloid β-protein degradation determined by novel fluorescence- and fluorescence polarization-based assays. *J. Biol. Chem.* 278:37314–37320.

Leopoldini, M., Russo, N., Toscano, M., et al. (2007). Which one among Zn(II), Co(II), Mn(II), and Fe(II) is the most efficient ion for the methionine aminopeptidase catalyzed reaction? *J. Am. Chem. Soc.* 129:7776–7784.

Letarte, M., Vera, S., Tran, R., et al. (1988). Common acute lymphocytic leukemia antigen is identical to neutral endopeptidase. *J. Exp. Med.* 168:1247–1253.

Lichtenthaler, S.F., and Haass, C. (2004). Amyloid at the cutting edge: activation of α-secretase prevents amyloidogenesis in an Alzheimer disease mouse model. *J. Clin. Invest.* 113:1384–1387.

Luo, Y., Bolon, B., Kahn, S., et al. (2001). Mice deficient in BACE1, the Alzheimer's β-secretase, have normal phenotype and abolished β–amyloid generation. *Nat. Neurosci.* 4:231–232.

Maskos, K., Fernandez-Catalan, C., Huber, R., et al. (1998). Crystal structure of the catalytic domain of human tumor necrosis factor-alpha-converting enzyme. *Proc. Natl. Acad. Sci. USA.* 95:3408–3412.

Masters, C.L., Simms, G., Weinman, N.A., et al. (1985). Amyloid plaque core protein in Alzheimer disease and Down Syndrome. *Proc. Natl. Acad. Sci. USA.* 82:4245–4249.

Matsas, R., Fulcher, I.S., Kenny, A.J., et al. (1983). Substance P and [Leu]enkephalin are hydrolyzed by an enzyme in pig caudate synaptic membranes that is identical with the endopeptidase of kidney microvilli. *Proc. Natl. Acad. Sci. USA.* 80:3111–3115.

Menard, J., and Patchett, A.A. (2001). Angiotensin-converting enzyme inhibitors. *Adv. Protein Chem.* 56:13–75.

Menchén, L., Marín-Jiménez, I., Arias-Salgado, E.G., et al. (2009). Matrix metalloproteinase 9 is involved in Crohn's disease-associated platelet hyperactivation through the release of soluble CD40 ligand. *Gut* 58:920–928.

Mogk, A., Schmidt, R., Bukau, B. (2007). The N-end rule pathway for regulated proteolysis: prokaryotic and eukaryotic strategies. *Trends Cell Biol.* 17:165–172.

Moura-da-Silva, A.M., Butera, D., Tanjoni, I. (2007). Importance of snake venom metalloproteinases in cell biology: effects on platelets, inflammatory and endothelial cells. *Curr. Pharm. Des.* 13:2893–2905.

Myara, I., Cosson, C., Moatti, N., et al. (1994). Human kidney prolidase -purification, preincubation properties and immunological reactivity. *Int. J. Biochem.* 26:207–214.

Nalivaeva, N.N., Fisk, L.R., Belyaev, N.D., et al. (2008). Amyloid-degrading enzymes as therapeutic targets in Alzheimer's disease. *Curr. Alzheimer Res.* 5:212–224.

Neurath, H. (1999). Proteolytic enzymes, past and future. *Proc. Natl. Acad. Sci. USA.* 96:10962–10963.

Oefner, C., D'Arcy, A., Hennig, M., et al. (2000). Structure of human neutral endopeptidase (Neprilysin) complexed with phosphoramidon. *J. Mol. Biol.* 296:341–349.

Orth, P., Reichert, P., Wang, W., et al. (2004). Crystal structure of the catalytic domain of human ADAM33. *J. Mol. Biol.* 335:129–137.

Papandreou, C.N., Usmani, B., Geng, Y., et al. (1998). Neutral endopeptidase 24.11 loss in metastatic human prostate cancer contributes to androgen-independent progression. *Nature Med.* 4:50–57.

Pardossi-Piquard, R., Petit, A., Kawarai, T., et al. (2005). Presenilin-dependent transcriptional control of the Aβ-degrading enzyme neprilysin by intracellular domains of βAPP and APLP. *Neuron* 46:541–554.

Parkin, E.T., Watt, N.T., Turner, A.J., et al. (2004). Dual mechanisms for shedding of the cellular prion protein. *J. Biol. Chem.* 279:11170–11178.

Parvathy, S., Hussain, I., Karran, E.H. (1999). Cleavage of Alzheimer's amyloid precursor protein by α-secretase occurs at the surface of neuronal cells. *Biochemistry* 38:9728–9734.

Pendas, A.M., Zhou, Z., Cadinanos, J., et al. (2002). Defective prelamin A processing and muscular and adipocyte alterations in Zmpste24 metalloproteinase-deficient mice. *Nat. Genet.* 31:94–99.

Postina, R., Schroeder, A., Dewachter, I. (2004). A disintegrin-metalloproteinase prevents amyloid plaque formation and hippocampal defects in an Alzheimer disease mouse model. *J. Clin. Invest.* 113:1456–1464.

Puente, X.S., Sánchez, L.M., Gutiérrez-Fernández, A., et al. (2005). A genomic view of the complexity of mammalian proteolytic systems. *Biochem. Soc. Trans.* 33:331–334.

Qing, H., He, G., Ly, P.T., et al. (2008). Valproic acid inhibits Aβ production, neuritic plaque formation, and behavioral deficits in Alzheimer's disease mouse models. *J. Exp. Med.* 205:2781–2789.

Quesada, V., Ordóñez, G.R., Sánchez, L.M., et al., (2009). The Degradome database: mammalian proteases and diseases of proteolysis. *Nucleic Acids Res.* 37 (Database issue):D239–D243.

Raife, T.J., and Montgomery, R.R. (2000). von Willebrand factor and thrombotic thrombocytopenic purpura. *Curr. Opin. Hematol.* 7:278–283.

Ravid, T., Hochstrasser, M. (2008). Diversity of degradation signals in the ubiquitin-proteasome system. *Nature Rev. Mol. Cell Biol.* 9:679–690.

Rella, M., Elliot, J.L., Revett, T.J., et al. (2007). Identification and characterisation of the angiotensin converting enzyme-3 (ACE3). gene: a novel mammalian homologue of ACE. *BMC Genomics* 8:194.

Rawlings, N.D. and Barrett, A.J. (2004). Intruduction: metallopeptidases and their clans. In: Barrett, A.J., Rawlings, N.D., Woessner, J.F. (eds) Handbook of Proteolytic Enzymes, 2 edn. Elsevier, London, pp. 231–268.

Rice, G.I., Thomas, D.A., Grant, P.J. et al. (2004). Evaluation of angiotensin-converting enzyme (ACE), its homologue ACE2 and neprilysin in angiotensin peptide metabolism. *Biochem. J.* 383:45–51.

Rholam, M., and Fahy, C. (2009). Processing of peptide and hormone precursors at the dibasic cleavage sites. *Cell Mol. Life Sci.* 66:2075–2091

Roques, B.P., Fournié-Zaluski, M.C., Soroca, E., et al. (1980). The enkephalinase inhibitor thiorphan shows antinociceptive activity in mice. *Nature* 288:286–288.

Rougeot, C., Messaoudi, M., Hermitte, V., et al. (2003). Sialorphin, a natural inhibitor of rat membrane-bound neutral endopeptidase that displays analgesic activity. *Proc. Natl. Acad. Sci. USA.* 100:8549–8554.

Saha, R.N., and Pahan, K. (2006). HATs and HDACs in neurodegeneration: a tale of disconcerted acetylation homeostasis. *Cell Death Differ.* 13:539–550.

Schiavo, G., Benfenati, F., Poulain, B., et al. (1992). Tetanus and botulinum-B neurotoxins block neurotransmitter release by proteolytic cleavage of synaptobrevin. *Nature* 359:832–835.

Schilling, S., Zeitschel, U., Hoffmann, T., et al. (2008). Glutaminyl cyclase inhibition attenuates pyroglutamate Aβ and Alzheimer's disease-like pathology. *Nature Med.* 14:1106–1111.

Schulz, H., Dale, G.E., Karimi-Nejad, Y., et al. (2009). Structure of human endothelin-converting enzyme I complexed with phosphoramidon. *J. Mol. Biol.* 385:178–187.

Seals, D.F., and Courtneidge, S.A. (2003). The ADAMs family of metalloproteases: multidomain proteins with multiple functions. *Genes Dev.* 17:7–30.

Skidgel, R.A., and Erdös, E.G. (1985). Novel activity of human angiotensin I converting enzyme: release of the NH_2- and COOH-terminal tripeptides from the luteinizing hormone-releasing hormone. *Proc. Natl. Acad. Sci. USA.* 82:1025–1029.

Skotnicki, J.S., DiGrandi, M.J., Levin, J.I. (2003). Design strategies for the identification of MMP-13 and TACE inhibitors. *Curr. Opin. Drug Discov. Devel.* 6:742–759.

Stöcker, W., Grams, F., Baumann, U., et al. (1995). The metzincins - topological and sequential relations between the astacins, adamalysins, serralysins, and matrixins (collagenases) define a superfamily of zinc-peptidases. *Protein Sci.* 4:823–840.

Taylor, D.R., Parkin, E.T., Cocklin, S.L., et al. (2009) Role of ADAMs in the ectodomain shedding and conformational conversion of the prion protein. *J. Biol. Chem.* 284(34):22590–22600.

Taylor, C.J., Ireland, D.R., Ballagh, I., et al. (2008). Endogenous secreted amyloid precursor protein-α regulates hippocampal NMDA receptor function, long-term potentiation and spatial memory. *Neurobiol. Dis.* 31:250–260.

Thomas, J.E., Rylett, C.M., Carhan, A., et al. (2005). Drosophila melanogaster NEP2 is a new soluble member of the neprilysin family of endopeptidases with implications for reproduction and renal function. *Biochem. J.* 386:357–366.

Tipnis, S.R., Hooper, N.M., Hyde, R., et al. (2000). A human homolog of angiotensin-converting enzyme. Cloning and functional expression as a captopril-insensitive carboxypeptidase. *J. Biol. Chem.* 275:33238–33243.

Tu, G., Xu, W., Huang, H., et al. (2008). Progress in the development of matrix metalloproteinase inhibitors. *Curr. Med. Chem.* 15:1388–1395.

Turner, A.J. (2008). Angiotensin-converting enzyme 2: cardioprotective player in the renin-angiotensin system? *Hypertension* 52:816–817.

Turner, A.J., Isaac, R.E., Coates, D. (2001). The neprilysin (NEP) family of zinc metalloendopeptidases: genomics and function. *Bioessays* 23:261–269.

Turner, A.J., Fisk, L., Nalivaeva, N.N. (2004). Targeting amyloid-degrading enzymes as therapeutic strategies in neurodegeneration. *Ann. N Y Acad. Sci.* 1035:1–20.

Vaisar, T., Kassim, S.Y., Gomez, I.G., et al. (2009). MMP-9 sheds the β2 integrin subunit (CD18) from macrophages. *Mol. Cell Proteomics* 8:1044–1060.

van Goor, H., Melenhorst, WB, Turner, A.J., et al. (2009). Adamalysins in biology and disease. *J. Pathol.* 219(3):277–286.

Van Wart, H.E., and Birkedal-Hansen, H. (1990). The cysteine switch: a principle of regulation of metalloproteinase activity with potential applicability to the entire matrix metalloproteinase gene family. *Proc. Natl. Acad. Sci. USA.* 87:5578–5582.

Watermeyer, J.M., Kröger, W.L., O'Neill, H.G., et al. (2008). Probing the basis of domain-dependent inhibition using novel ketone inhibitors of angiotensin-converting enzyme. *Biochemistry* 47:5942–5950.

Weihofen, A., and Martoglio, B. (2003). Intramembrane-cleaving proteases: controlled liberation of proteins and bioactive peptides. *Trends Cell Biol.* 13:71–78.

Wisner, A., Dufour, E., Messaoudi, M., et al. (2006). Human Opiorphin, a natural antinociceptive modulator of opioid-dependent pathways. *Proc. Natl. Acad. Sci. USA.* 103:17979–17984.

Whyteside, A.R., and Turner, A.J. (2008). Human neprilysin-2 (NEP2) and NEP display distinct subcellular localisations and substrate preferences. *FEBS Lett.* 582:2382–2386.

Willem, M., Garratt, A.N., Novak, B., et al. (2006). Control of peripheral nerve myelination by the beta-secretase BACE1. *Science* 314:664–666.

Wolfe, M.S. (2007). When loss is gain: reduced presenilin proteolytic function leads to increased Aβ42/Aβ40. Talking point on the role of presenilin mutations in Alzheimer disease. *EMBO Rep.* 8:136–140.

Wolfe, M.S. (2009). Intramembrane proteolysis. *Chem. Rev.* 109:1599–1612.

Yanagisawa, M., Kurihara, H., Kimura, S., et al. (1988). A novel potent vasoconstrictor peptide produced by vascular endothelial cells. *Nature* 332:411–415.

Zelenski, N.G., Rawson, R.B., Brown, M.S., et al. (1999). Membrane topology of S2P, a protein required for intramembranous cleavage of sterol regulatory element-binding proteins. *J. Biol. Chem.* 274:21973–21980.

Zhang, Y., Xu, J., Levin, J. (2004). Identification and characterization of 4-[[4-(2-butynyloxy) phenyl]sulfonyl]-N-hydroxy-2,2-dimethyl-(3S)thiomorpholinecarboxa mide (TMI-1), a novel dual tumor necrosis factor-alpha-converting enzyme/matrix metalloprotease inhibitor for the treatment of rheumatoid arthritis. *J. Pharmacol. Exp. Ther.* 309:348–355.

Zhao, J., Li, L., Leissring, M.A. (2009). Insulin-degrading enzyme is exported via an unconventional protein secretion pathway. *Mol. Neurodegener.* 4:4.

Zhu, G., Wang, D., Lin, Y.H., et al. (2001). Protein kinase Cε suppresses Aβ production and promotes activation of γ-secretase. *Biochem. Biophys. Res. Commun.* 285:997–1006.

Index